COMPUTER SOLUTION OF LARGE LINEAR SYSTEMS

STUDIES IN MATHEMATICS AND ITS APPLICATIONS

VOLUME 28

ELSEVIER
NORTH
HOLLAND
AMSTERDAM - LAUSANNE - NEW YORK - OXFORD - SHANNON - SINGAPORE - TOKYO

COMPUTER SOLUTION OF LARGE LINEAR SYSTEMS

G. MEURANT

DCSA/EC
CEA Bruyeres le Chatel, BP 12
91680 Bruyeres le Chatel
France

ELSEVIER
NORTH
HOLLAND
AMSTERDAM - LAUSANNE - NEW YORK - OXFORD - SHANNON - SINGAPORE - TOKYO

ELSEVIER B.V.
Sara Burgerhartstraat 25
P.O. Box 211, 1000 AE
Amsterdam, The Netherlands

ELSEVIER Inc.
525 B Street
Suite 1900, San Diego
CA 92101-4495, USA

ELSEVIER Ltd
The Boulevard
Langford Lane, Kidlington,
Oxford OX5 1GB, UK

ELSEVIER Ltd
84 Theobalds Road
London WC1X 8RR
UK

First edition 1999
Reprinted 2005

Library of Congress Cataloging in Publication Data
A catalog record is available from the Library of Congress.

British Library Cataloguing in Publication Data
A catalogue record is available from the British Library.

ISBN: 0-444-50169-X

∞ The paper used in this publication meets the requirements of ANSI/NISO Z39.48-1992 (Permanence of Paper).
Printed in China

This work is dedicated to Gene H. Golub

Gene has been and is always very influential about the development and growth of numerical linear algebra. He taught me most of the things I know about this field and has always been a constant source of inspiration and encouragement. Moreover, Gene is a very nice person and a very good friend.

Paris, December 1998.

"Depuis qu'on a remarqué qu'avec le temps vieilles folies deviennent sagesse, et qu'anciens petits mensonges assez mal plantés ont produit de grosses, grosses vérités, on en a de mille espèces. Et celles qu'on sait sans oser les divulguer : car toute vérité n'est pas bonne à dire; et celles qu'on vante, sans y ajouter foi : car toute vérité n'est pas bonne à croire; et les serments passionnés, les menaces des mères, les protestations des buveurs, les promesses des gens en place, le dernier mot des marchands, cela ne finit pas. Il n'y a que mon amour pour Suzon qui soit une vérité de bon aloi."

Beaumarchais (1732–1799), Le mariage de Figaro, Acte IV, Scéne I

"Chacun se dit ami; mais fol qui s'y repose :

Rien n'est plus commun que ce nom,

Rien n'est plus rare que la chose."

Jean de La Fontaine (1621–1695), Fables

Preface

This book covers both direct and iterative methods for solving non–singular sparse linear systems of equations, particularly those arising from the discretization of partial differential equations. Many problems in physics or some other scientific areas lead to solving partial differential equations or systems of partial differential equations. Most of the time these equations are non–linear. The discretization of these equations and iterative methods to deal with the non linearity (like Newton's method) lead to solving sparse linear systems. Very large sparse systems are now solved due to the progresses of numerical simulation and also of computers' speed and increases in the amounts of memory available. So, solving sparse linear systems is a problem of paramount importance for numerical computation. Many methods have been invented over the years. Unfortunately, some of the older methods are not so efficient anymore when applied to very large systems and there is still the need for a very active research in this area. During the last years, there has been a rapid development of parallel computers. Therefore, we put some emphasis on the adaptation of known methods to parallel computations and also on the development of new parallel algorithms.

This book tries to cover most of the best algorithms known so far for solving sparse linear systems. It should be useful for engineers and scientists as well as graduate students interested in numerical computations.

The first chapter recalls some mathematical results and tools that are needed in the next chapters. The direct methods we consider are different versions of Gaussian elimination and also fast solvers for solving separable partial differential equations on domains with a simple geometry. Chapter 2 is devoted to Gaussian elimination for general linear systems and Chapter 3 focuses on special techniques and variants of Gaussian elimination for sparse systems. Some fast solvers for separable partial differential equations on rectangular domains are described and analyzed in Chapter 4.

Classical iterative methods are recalled in Chapter 5. Even though there are not really efficient for solving the problems we consider, they are still of use for constructing preconditioners or for the multigrid method, so it is worthwhile to study these methods. In Chapter 6 the emphasis is put on the conjugate gradient (CG) method which is probably the most efficient method for solving sparse symmetric positive definite linear systems. We also consider some methods for symmetric indefinite problems. In Chapter 7 we consider some of the many generalizations of conjugate gradient that have appeared for non–symmetric matrices. A very important issue for using CG or CG–like methods efficiently is preconditioning. Therefore, in Chapter 8 we describe many of the attempts to derive efficient preconditioners. An introduction to the multigrid method is given in Chapter 9. This method is interesting as it can allow solving a problem with a complexity proportional to the number of unknowns. Chapter 10 describes some domain decomposition and multilevel techniques. This is a very natural framework to use when solving linear systems on parallel computers.

This book grows from lectures which have been given in Paris VI University from 1984 to 1995 and also from research and review papers written since 1982. However, we have tried to give a broad view of the field and to describe state of the art algorithms. Therefore the works of many people are described and quoted in this book, too many for thanking them individually. A large reference list gives the papers and reports we have found interesting and useful over the years.

I am very indebted to Barbara Morris who carefully tried to convert my Frenglish into good English. All the remaining mistakes are mine.

CONTENTS

1 Introductory Material

In this chapter, for the convenience of the reader, we recall some definitions and prove some well–known fundamental theorems on matrix properties that will be used in the following chapters. More details can be found in the classical books by Berman–Plemmons [142], Forsythe–Moler [572], Golub–Van Loan [676], Householder [793], Strang [1216], Stewart [1200], Varga [1300], Wilkinson [1333], and Young [1351]. We shall study some tools to be used in later chapters such as Chebyshev polynomials and Fourier analysis. We also recall some facts about computer arithmetic that will be useful in studying round off errors. More on this topic can be found in the books by Higham [781] and Chatelin and Fraysse [310]. We give a few examples of linear systems arising from discretization of partial differential equations which can be solved with the numerical methods that are considered in this book.

There are two main classes of algorithms used to compute the solution of a linear system, direct methods and iterative methods. Direct methods obtain the solution after a finite number of floating point operations by doing combinations and modifications of the given equations. Of course, as computer floating point operations can only be obtained to a certain given precision, the computed solution is generally different from the exact solution, even with a direct method. Iterative methods define a sequence of approximations that are expected to be closer and closer to the exact solution in some given norm, stopping the iterations using some predefined criterion, obtaining a vector which is only an approximation of the solution.

In chapters 2–4, we shall study some direct methods, mainly Gaussian elimination and some methods using the Fourier transform. Other chapters will be devoted to iterative methods. We shall consider classical (relaxation) methods, the Conjugate Gradient method, Krylov methods for non–symmetric systems and the Multigrid method. A chapter is devoted to the important topic of preconditioning. Finally, we shall introduce Domain Decomposition methods.

Without being explicitly stated, we shall consider matrices A of dimension $n \times n$ with real coefficients that we shall denote by $a_{i,j}$. Most of the time we shall denote matrices by capital letters, vectors by roman letters and scalars by greek letters. Elements of a vector x are denoted by x_i. Generally, elements of a vector sequence (iterates) will be denoted by x^k.

1.1. Vector and matrices norms

Let us first recall the definition of the scalar product of two vectors x and $y \in \mathbb{R}^n$

$x = (x_1, x_2, \dots, x_n)^T, \quad y = (y_1, y_2, \dots, y_n)^T.$

Definition 1.1

The Euclidean scalar product of two vectors x and y in $\mathbb{R}^n$, denoted by (x, y), is defined by

$$(x, y) = \sum_{i=1}^{n} x_i y_i$$

◇

As a matrix multiplication, this can also be written as $x^T y$. Computing a scalar product needs n multiplications and $n - 1$ additions, that is $2n - 1$ floating point operations in total. From the scalar product, we can define the Euclidean norm (sometimes called the l_2 norm) of the vector x.

Definition 1.2

The Euclidean norm of x in $\mathbb{R}^n$ is defined by

$$\|x\| = (x, x)^{\frac{1}{2}} = (\sum_{i=1}^{n} x_i^2)^{\frac{1}{2}}$$

Proposition 1.3

Every vector norm has the three following properties,

$$\begin{aligned} &\|\lambda x\| = |\lambda| \, \|x\| \ \forall \lambda \in \mathbb{R}, \\ &\|x\| \geq 0 \text{ and } \|x\| = 0 \iff x = 0, \\ &\|x + y\| \leq \|x\| + \|y\|. \end{aligned} \tag{1.1}$$

□

For the Euclidean norm, relation (1.1) can be shown using the Cauchy–Schwarz inequality.

Proposition 1.4

For the scalar product of two vectors in $\mathbb{R}^n$, we have

$$|(x, y)| \leq \|x\| \, \|y\|.$$

Equality holding if x and y are collinear.

Proof. If $y = 0$ the proof is trivial. Hence, we can assume $y \neq 0$, then we have

$$0 \leq \|x - \frac{(x,y)}{\|y\|^2} y\|^2 = \|x\|^2 - 2\frac{(x,y)^2}{\|y\|^2} + \frac{(x,y)^2}{\|y\|^4}\|y\|^2 = \frac{\|x\|^2\|y\|^2 - (x,y)^2}{\|y\|^2}.$$

So,

$$\|x\|^2\|y\|^2 - (x,y)^2 \geq 0,$$

which proves the result. □

A matrix norm can be defined in such a way that it is related to any given vector norm. Such a matrix norm is said to be induced by (or subordinate to) the vector norm.

Definition 1.5

Let $\|\cdot\|_p$ be any vector norm. The induced matrix norm is defined as

$$\|A\|_p = \max_{x \neq 0} \frac{\|Ax\|_p}{\|x\|_p}.$$

◇

Another equivalent definition of an induced matrix norm is $\|A\|_p = \max_{\|y\|_p=1} \|Ay\|_p$. Definition 1.5 implies $\|Ax\|_p \leq \|A\|_p \|x\|_p$. The following result states the properties of a matrix norm, see [676].

Proposition 1.6

A matrix norm satisfies the following properties:

$\|\lambda A\|_p = |\lambda| \, \|A\|_p, \ \forall \lambda \in \mathbb{R},$

$\|A\|_p \geq 0$ and $\|A\|_p = 0 \iff A = 0,$

$\|A + B\|_p \leq \|A\|_p + \|B\|_p.$

□

Proposition 1.7

For every induced norm,

$\|AB\|_p \leq \|A\|_p \|B\|_p.$

□

More generally, one defines every function from

$M(\mathbb{R}^n) = \{$matrices of order $n \times n$ with real coefficients$\}$

to $\mathbb{R}^+$ which satisfies proposition 1.6 as a matrix norm. We shall, however, add the property of proposition 1.7 to the definition of a norm since it is often the most useful one. Let us now turn to some examples of norms. We shall see the norm that is induced by the Euclidean norm in proposition 1.23 but this requires some additional definitions. Let us consider first other examples.

Definition 1.8

We define the maximum norm of a vector by $\|x\|_\infty = \max_i |x_i|$ *and the induced matrix norm by* $\|A\|_\infty$. ◇

Proposition 1.9

The $\|.\|_\infty$ *norm of* A *is*

$$\|A\|_\infty = \max_i \sum_{j=1}^{n} |a_{i,j}|.$$

Proof. Let $y = Ax$. Then,

$$\|y\|_\infty = \max_i \left| \sum_j a_{i,j} x_j \right| \le \max_i \sum_j |a_{i,j}| \cdot |x_j| \le \|x\|_\infty \max_i \sum_j |a_{i,j}|.$$

Hence, $\|A\|_\infty \le \max_i \sum_j |a_{i,j}|$. Now, let us suppose that $\max_i \sum_j |a_{i,j}| = \sum_j |a_{I,j}|$. Let us define x such that $x_j = \operatorname{sign} a_{I,j}$. Clearly we have $\|x\|_\infty = 1$ and $y_I = \sum_j a_{I,j} x_j = \sum_j |a_{I,j}|$. Since,

$$\|y\|_\infty = \max_j |y_i| \ge y_I = \sum_j |a_{I,j}|,$$

$$\|A\|_\infty \ge \|Ax\|_\infty = \|y\|_\infty \ge \sum_j |a_{I,j}| = \max_i \sum_j |a_{i,j}|.$$

□

The energy norm of a vector x is defined as $\|x\|_A = (Ax, x)^{\frac{1}{2}}$, where A is a given matrix. However, for this to be a norm, we need $(Ax, x) > 0, \forall x \neq 0$. This is true only for some classes of matrices (positive definite ones) which we shall consider later in this chapter. Another example of a matrix norm is given by the Frobenius norm,

Definition 1.10

The Frobenius norm of A *denoted by* $\|A\|_F$ *is defined by*

$$\|A\|_F = \left(\sum_{i=1}^{n} \sum_{j=1}^{n} a_{i,j}^2 \right)^{\frac{1}{2}} = (\operatorname{trace}[A^T A])^{\frac{1}{2}},$$

where trace$[A^T A]$ *is the sum of the diagonal elements of* $A^T A$ *and* A^T *is the transpose of* A. ◇

The Frobenius norm satisfies proposition 1.7 but is not an induced norm, since for I_n, the identity matrix of order n, we have $\|I_n\|_F = n^{\frac{1}{2}}$. For finite dimensional spaces all norms are equivalent. In particular, the Euclidean and Frobenius norms are related to each other by the following inequalities.

Proposition 1.11

The Euclidean and Frobenius norms satisfy

$$\max_{i,j} |a_{i,j}| \leq \|A\| \leq \|A\|_F \leq n\|A\|.$$

Proof. Let us prove the first inequality. For all $x, y \neq 0$ the Cauchy–Schwarz inequality (proposition 1.4) gives

$$\frac{|(x, Ay)|}{\|x\| \, \|y\|} \leq \frac{\|Ay\|}{\|y\|} \leq \|A\|.$$

Let us take $x = e^i$ and $y = e^j$ where e^i and e^j are vectors with zeros in all components except in the ith and j^{th} position where the value is 1. Then,

$$|(x, Ay)| = |a_{i,j}|,$$

and hence

$$|a_{i,j}| \leq \|A\|.$$

This inequality holds for every i and j. Therefore,

$$\max_{i,j} |a_{i,j}| \leq \|A\|.$$

We use, once again, the Cauchy–Schwarz inequality to prove the second inequality.

$$\begin{aligned}
\|Ax\|^2 &= \sum_{i=1}^{n} (\sum_{j=1}^{n} a_{i,j} x_j)^2, \\
&\leq \sum_{i=1}^{n} \left[(\sum_{j=1}^{n} a_{i,j}^2)(\sum_{j=1}^{n} x_j^2) \right], \\
&= (\sum_{i=1}^{n} \sum_{j=1}^{n} a_{i,j}^2)(\sum_{j=1}^{n} x_j^2), \\
&= \|A\|_F^2 \|x\|^2.
\end{aligned}$$

To obtain the last inequality, we bound $a_{i,j}$ by $\max |a_{i,j}|$ so,

$$\|A\|_F = (\sum_{i=1}^{n}\sum_{j=1}^{n} a_{i,j}^2)^{\frac{1}{2}} \leq n \max_{i,j} |a_{i,j}| \leq n\|A\|.$$

□

Definition 1.12

Let L be a non–singular matrix, the L–norm of A is defined by

$$\|A\|_L = \|LAL^{-1}\|.$$

◇

It is easy to prove that this definition satisfies the properties of proposition 1.6. With the scalar product of definition 1.1, we can generalize to any dimension the usual definition of orthogonality in $\mathbb{R}^3$.

Definition 1.13

Two vectors x and y are said to be orthogonal if and only if

$$(x, y) = 0.$$

◇

Every set of vectors $\{x^1, x^2, \dots, x^n\}$ can be orthogonalized in the following way.

Proposition 1.14

We can find a set of vectors $\{y^1, y^2, \dots, y^n\}$ such that $(y^i, y^j) = 0$ and $\|y^i\| = 1$, for all $i \neq j$ and $\text{span}(y^1, y^2, \dots, y^n) = \text{span}(x^1, x^2, \dots, x^n)$, *where* $\text{span}(x^1, x^2, \dots, x^n)$ *is the subspace spanned by $x^1, x^2, \dots, x^n$.*

Proof. The method to obtain y^i, is known as the Gram–Schmidt orthogonalization process. Let us consider first only two vectors, *i.e.*, $n = 2$. Let x^1

and x^2 be given. We define

$$y^1 = \frac{x^1}{\|x^1\|},$$
$$y^2 = x^2 - \frac{(x^1, x^2)}{\|x^1\|^2} x^1 = x^2 - (y^1, x^2) y^1.$$

Note that $\frac{(x^1,x^2)}{\|x^1\|^2} x^1$ is the component of x^2 in the direction x^1. Clearly, if we subtract this component from x^2 we obtain a vector y^2 which is orthogonal to x^1. The vectors y^1 and y^2 are independent, linear combinations of x^1 and x^2 and span the same subspace. This can be generalized to n vectors giving

$$y^1 = \frac{x^1}{\|x^1\|},$$
$$z^i = x^i - \frac{(y^1, x^i)}{\|y^1\|^2} y^1 - \cdots - \frac{(y^{i-1}, x^i)}{\|y^{i-1}\|^2} y^{i-1},$$
$$y^i = \frac{z^i}{\|z^i\|}, \quad i = 2, \ldots, n.$$

It is easy to check that the y^i are orthogonal and by induction that the spanned subsets are the same. □

Note that for the Gram–Schmidt algorithm we have to construct and store the previous $i-1$ vectors to compute y^i.

Definition 1.15

A matrix Q is an orthogonal matrix if and only if the columns of Q are orthogonal vectors with Euclidean norm 1. ◇

For an orthogonal matrix Q, we have $Q^T Q = I$. Note that if we normalize the vectors y^i in the Gram–Schmidt process and if we think of the vectors $\{x^1, \ldots, x^n\}$ as columns of a matrix A, this is nothing else than computing a factorization $A = QR$ where Q (whose columns are the normalized y^i) is orthogonal and R is upper triangular. In the Gram–Schmidt algorithm Q and R are computed one column at a time. Unfortunately Gram–Schmidt has poor numerical properties. Typically there is a loss of orthogonality in the computed vectors. However, there is a rearrangement of the computation, called the modified Gram–Schmidt (MGS) algorithm, that leads to much better results, see [1128], [153], [676], [781]. In MGS, as soon as one y^i is computed, all the remaining vectors are orthogonalized against it. The algorithm is the following. We introduce partially orthogonalized vectors $y^i_{(j)}$ whose initial values are

$$y^i_{(1)} = x^i, \quad i = 1, \ldots, n.$$

Then, for $i = 1, \ldots, n$,

$$y^i = \frac{y^i_{(i)}}{\|y^i_{(i)}\|},$$

and for $j = i+1, \ldots, n$,

$$y^j_{(i+1)} = y^i_{(i)} - (y^i, y^j_{(i)})y^i.$$

For stability results on MGS, see Higham [781].

1.2. Eigenvalues

In this section, we consider some matrix properties related to eigenvalues and eigenvectors that are useful even when we are not interested in computing eigenvalues.

Definition 1.16

$\lambda \in \mathbb{C}$ is an eigenvalue of A if and only if there exists at least one vector $x \neq 0$ such that

$$Ax = \lambda x.$$

The vector x is said to be an eigenvector of A. ◇

We have the following very important result:

Proposition 1.17

Suppose that A has n distinct eigenvalues $\lambda_1, \lambda_2, \ldots, \lambda_n$. Then, there exists a non–singular matrix S such that

$$S^{-1}AS = \begin{pmatrix} \lambda_1 & & & \\ & \lambda_2 & & \\ & & \ddots & \\ & & & \lambda_n \end{pmatrix},$$

i.e., A is similar to a diagonal matrix.

Proof. This can be proved using the fact that eigenvectors associated with two distinct eigenvalues are linearly independent and thus they yield an orthogonal basis for $\mathbb{R}^n$. □

Unfortunately the result of proposition 1.17 is not always true if some eigenvalues are equal.

Definition 1.18

If $\lambda_i = \lambda_{i+1} = \cdots = \lambda_{i+m-1} = \lambda$ we say that λ is of algebraic multiplicity m. The geometric multiplicity is defined as the dimension of the subspace spanned by the eigenvectors associated with λ. ◇

The two multiplicities may be different, as it is shown in the following example.

$$A = \begin{pmatrix} 0 & 1 \\ 0 & 0 \end{pmatrix}, \qquad \lambda_1 = \lambda_2 = 0.$$

The algebraic multiplicity is 2 but the geometric multiplicity is 1.

The more general result that can be proved is that A is similar to a diagonal matrix if the geometric multiplicity of each eigenvalue is the same as the algebraic multiplicity. To state a very important theorem, we must now consider complex numbers.

The scalar product of two vectors x and y with complex entries is defined by

$$(x, y) = \sum_{i=1}^{n} \overline{x_i} y_i,$$

where $\overline{x_i}$ denotes the conjugate of $x_i \in \mathbb{C}$. A^H denotes the conjugate transpose of A given by $(A^H)_{i,j} = \overline{a_{j,i}}$. A complex matrix A which is orthogonal is called unitary. A matrix A is Hermitian if and only if $A^H = A$. Then we have,

Theorem 1.19

For every matrix A there exists a unitary matrix U such that

$$U^H AU = T,$$

where T is an upper triangular matrix.

Proof. For the proof we follow the lines of Strang [1216]. Every matrix A has at least one eigenvalue, say λ_1 (which may be of algebraic multiplicity n) and at least one eigenvector x^1 that we can assume has norm 1. From the Gram–Schmidt orthogonalization process, we can find $n-1$ vectors $u^2_{(1)}, \dots, u^n_{(1)}$ such that

$$U_1 = (x^1, u^2_{(1)}, \dots, u^n_{(1)})$$

is a unitary matrix. Then,

$$AU_1 = U_1 \begin{pmatrix} \lambda_1 & * & \cdots & \cdots & * \\ 0 & & & & \\ \vdots & & A^{(2)} & & \\ 0 & & & & \end{pmatrix}.$$

We can repeat the same process on $A^{(2)}$ which has at least one eigenvalue λ_2 and an eigenvector x^2. So there exists a unitary matrix V_2 of order $n-1$ such that

$$A^{(2)} V_2 = V_2 \begin{pmatrix} \lambda_2 & * & \cdots & * \\ 0 & & & \\ \vdots & & A^{(3)} & \\ 0 & & & \end{pmatrix}.$$

Let us denote

$$U_2 = \begin{pmatrix} 1 & 0 & \cdots & 0 \\ 0 & & & \\ \vdots & & V_2 & \\ 0 & & & \end{pmatrix},$$

then,

$$\begin{pmatrix} \lambda_1 & * & \cdots & * \\ 0 & & & \\ \vdots & & A^{(2)} & \\ & & & \\ 0 & & & \end{pmatrix} U_2 = \begin{pmatrix} \lambda_1 & * & \cdots & * \\ 0 & & & \\ \vdots & & A^{(2)}V_2 & \\ & & & \\ 0 & & & \end{pmatrix}$$

$$= U_2 \begin{pmatrix} \lambda_1 & * & * & \cdots & * \\ 0 & \lambda_2 & * & \cdots & * \\ 0 & 0 & & & \\ \vdots & \vdots & & A^{(3)} & \\ 0 & 0 & & & \end{pmatrix}.$$

The result follows by induction. □

Corollary 1.20

Let A be an Hermitian matrix. Then, there exists a unitary matrix U such that

$$U^H AU = \Lambda,$$

where Λ is a diagonal matrix whose diagonal entries are the eigenvalues of A.

□

Let us now return to the real case. We have the following very useful result which is used in many proofs in numerical linear algebra.

Theorem 1.21

Let A be a real symmetric matrix. Then, there exists an orthogonal matrix Q such that

$$Q^H AQ = Q^T AQ = \Lambda,$$

where Λ is a diagonal matrix whose diagonal entries are the eigenvalues of A (which are real numbers). $Q\Lambda Q^T$ is said to be the spectral decomposition of A.

We have the following important definition.

Definition 1.22

The spectral radius $\rho(A)$ of A is defined as

$$\rho(A) = \max_{1 \le i \le n} |\lambda_i|,$$

where λ_i are the eigenvalues of A. ◇

With this definition we can give the characterization of the Euclidean norm of a symmetric matrix:

Proposition 1.23

Let A be a symmetric matrix. Then,

$$\|A\| = \rho(A).$$

Proof. To prove this result, we note that the Euclidean norm is invariant under an orthogonal transformation Q. This is because

$$\|x\|^2 = (x, x) = (Q^T Q x, x) = (Qx, Qx) = \|Qx\|^2.$$

Let us recall that

$$\|A\| = \max_{x \neq 0} \frac{\|Ax\|}{\|x\|}.$$

Theorem 1.21 shows that $\|Ax\| = \|Q\Lambda Q^T x\| = \|\Lambda Q^T x\|$, so

$$\|A\| = \max_{x \neq 0} \frac{\|\Lambda Q^T x\|}{\|Q^T x\|} \le \|\Lambda\| = \rho(A).$$

Let us order the eigenvalues as $\lambda_1 \le \lambda_2 \le \cdots \le \lambda_n$ and

$$\Lambda = \begin{pmatrix} \lambda_1 & & & \\ & \lambda_2 & & \\ & & \ddots & \\ & & & \lambda_n \end{pmatrix}.$$

Suppose λ_n is the eigenvalue of maximum modulus. If we choose x such that $Q^T x = (0, \ldots, 0, 1)^T$ then, $\|A\| = \rho(A)$. □

In the following chapters dealing with iterative methods, we shall frequently study the limit of the sequence of powers A^k for $k = 1, 2, \ldots$, of a given matrix A. The spectral radius of a matrix plays an important role in the study of the powers of a matrix. We note that the sequence of matrices $A^{(k)} = A^k$ converges to a limit A^∞ as k tends to infinity, if and only if

$$\lim_{k\to\infty} a_{i,j}^{(k)} = a_{i,j}^\infty, \quad \text{for all } i, j.$$

This is equivalent to

$$\lim_{k\to\infty} \|A^{(k)} - A^\infty\| = 0.$$

The main result here is the following.

Theorem 1.24

$$\lim_{k\to\infty} A^k = 0 \quad \textit{if and only if} \quad \rho(A) < 1.$$

Proof. For the proof we follow Householder [793] and we need several lemmas before proving the main result. Let us first remark that if $\|\cdot\|_p$ is a matrix norm and thus satisfies $\|AB\|_p \le \|A\|_p \|B\|_p$ then $\|A\|_p < 1$ implies $\lim_{k\to\infty} A^k = 0$ since $\|A^k\|_p \le \|A\|_p^k$. We then have

Lemma 1.25

For every norm $\|\cdot\|_p$,

$$\rho(A) \le \|A\|_p.$$

Proof. Let λ and x be an eigenvalue and an eigenvector of A, $Ax = \lambda x$. Then,

$$\|\lambda x\|_p = |\lambda| \cdot \|x\|_p = \|Ax\|_p \le \|A\|_p \|x\|_p.$$

Hence $|\lambda| \le \|A\|_p$ for every eigenvalue. □

Lemma 1.26

Let T be a triangular matrix (with possibly complex elements). For all $\varepsilon > 0$, there exists a norm $\|\cdot\|_p$ such that

$$\|T\|_p \leq \max_j |t_{j,j}| + \varepsilon = \rho(T) + \varepsilon.$$

Proof. Let T be upper triangular and D be a diagonal matrix,

$$D = \begin{pmatrix} 1 & & & & \\ & \delta & & & \\ & & \delta^2 & & \\ & & & \ddots & \\ & & & & \delta^{n-1} \end{pmatrix},$$

with $\delta \in \mathbb{R}$ and $\delta > 0$. Now, consider $D^{-1}TD$. It easily follows that this matrix has the same diagonal as T, the second principal diagonal is multiplied by δ, the third by δ^2 and so on... This is because

$$(D^{-1}TD)_{k,l} = t_{k,l}\frac{\delta^{l-1}}{\delta^{k-1}} \qquad \text{for} \quad l \geq k.$$

Hence, we can choose δ sufficiently small so that for fixed ε the sum along a row of the non–diagonal terms of $D^{-1}TD$ is less than ε. Denote $\|T\|_{\infty,D} = \|D^{-1}TD\|_\infty$. According to definition 1.12, this defines a norm. With the previous remarks and proposition 1.9 we have

$$\|T\|_{\infty,D} \leq \varepsilon + \max_j |t_{j,j}| = \varepsilon + \rho(T).$$

□

Now, we are able to prove theorem 1.24.

Proof. Assume that $\lim_{k\to\infty} A^k = 0$ holds. Let λ and x be an eigenvalue and an eigenvector of A. So, $A^k x = \lambda^k x$, and

$$\lim_{k\to\infty} \|\lambda^k x\| = 0.$$

As $x \neq 0$ we have $\lim_{k\to\infty} |\lambda^k| = 0$ which implies $|\lambda| < 1$. This is true for any eigenvalue of A, and therefore $\rho(A) < 1$.

Let us now assume that $\rho(A) < 1$. We first show that there exists a norm $\|\cdot\|_p$ such that

$$\|A\|_p \leq \varepsilon + \rho(A).$$

From theorem 1.19 we know that there exists a unitary matrix U such that

$$U^{-1}AU = T,$$

where T is upper triangular. The matrix T has the same eigenvalues as A which are its diagonal entries and $\rho(A) = \rho(T)$. Then, ε being fixed with $0 < \varepsilon < 1 - \rho(A)$ we define a vector norm by

$$\|x\|_\varepsilon = \|D^{-1}U^{-1}x\|_\infty,$$

where D is the matrix that arises in the proof of lemma 1.26. Let $\|\cdot\|_\varepsilon$ be the induced matrix norm

$$\begin{aligned}\|A\|_\varepsilon &= \max_x \frac{\|Ax\|_\varepsilon}{\|x\|_\varepsilon} = \max_x \frac{\|D^{-1}U^{-1}Ax\|_\infty}{\|D^{-1}U^{-1}x\|_\infty} \\ &= \max_x \frac{\|D^{-1}U^{-1}AUDD^{-1}U^{-1}x\|_\infty}{\|D^{-1}U^{-1}x\|_\infty}.\end{aligned}$$

Let $y = D^{-1}U^{-1}x$ then,

$$\begin{aligned}\|A\|_\varepsilon &= \max_y \frac{\|D^{-1}U^{-1}AUDy\|_\infty}{\|y\|_\infty}, \\ &= \max_y \frac{\|D^{-1}TDy\|_\infty}{\|y\|_\infty}, \\ &= \|T\|_{\infty,D}.\end{aligned}$$

Therefore,

$$\|A\|_\varepsilon \leq \varepsilon + \rho(T) = \varepsilon + \rho(A).$$

But we have supposed that $\varepsilon < 1 - \rho(A)$ so,

$$\|A\|_\varepsilon < 1.$$

By a preceding remark this implies $\lim_{k\to\infty} A^k = 0$. □

Now we consider series of matrices. We have the following result:

Theorem 1.27

The series $I + A + A^2 + \cdots$ converges and the limit is $(I - A)^{-1}$ if and only if $\rho(A) < 1$.

Proof. Assume that $\rho(A) < 1$ and let λ be an eigenvalue of $I - A$ then $1 - \lambda$ is an eigenvalue of A and $1 - \lambda < 1$ because $\rho(A) < 1$. Therefore $I - A$ is non–singular. Let us denote the partial sums of the series by S_k.

$$\begin{aligned} S_k &= I + A + A^2 + \cdots + A^k, \\ (I - A)S_k &= I - A^{k+1}, \\ S_k &= (I - A)^{-1}(I - A^{k+1}), \\ S_k - (I - A)^{-1} &= -(I - A)^{-1}A^{k+1}, \\ \|S_k - (I - A)^{-1}\| &\leq \|(I - A)^{-1}\| \, \|A^{k+1}\|. \end{aligned}$$

But as $\rho(A) < 1$, $\lim_{k\to\infty} \|A^{k+1}\| = 0$ and so

$$\lim_{k\to\infty} \|S_k - (I - A)^{-1}\| = 0.$$

For the reciprocal note that if $\lim_{k\to\infty} S_k$ exists, this implies $\lim_{k\to\infty} A^k = 0$ and so, $\rho(A) < 1$ by theorem 1.24. □

The series $I + A + A^2 + \cdots$ is said to be the Neumann series for $(I - A)^{-1}$ and S_k (for small k) is frequently used in numerical algorithms to approximate $(I - A)^{-1}$ when $\rho(A) < 1$.

1.3. Irreducibility and diagonal dominance

In this Section we study the properties of some classes of matrices that will be useful for studying convergence of iterative methods.

Definition 1.28

Let $\Im_n = \{1, 2, \ldots, n\}$. A matrix A is irreducible if

1) $n = 1$ and $a_{1,1} \neq 0$ or

2) $n > 1$ and for every set of integers $I_1 \in \Im_n$ and $I_2 \in \Im_n$ with $I_1 \cap I_2 = \emptyset$ and $I_1 \cup I_2 = \Im_n$ there exists $i \in I_1$, and $j \in I_2$ such that $a_{i,j} \neq 0$. ◇

This definition seems complicated but the following result helps us to understand what it means. When A is not irreducible, it is said to be reducible.

Proposition 1.29

A matrix A is reducible if and only if there exists a permutation matrix P such that

$$P^{-1}AP = \begin{pmatrix} D_1 & 0 \\ F & D_2 \end{pmatrix},$$

D_1, D_2 being square matrices.

Proof. See Young [1351]. □

Definition 1.28 is the negation of the last proposition. A being irreducible means that one cannot solve the linear system $Ax = b$ by solving two subproblems of smaller size.

Definition 1.30

A matrix A is

- *diagonally dominant if*

$$|a_{i,i}| \geq \sum_{\substack{j=1 \\ j \neq i}}^{n} |a_{i,j}|, \quad \forall i,$$

- *strictly diagonally dominant if*

$$|a_{i,i}| > \sum_{\substack{j=1 \\ j \neq i}}^{n} |a_{i,j}|, \quad \forall i,$$

- *irreducibly diagonally dominant if*

  1) A is irreducible,

  2) A is diagonally dominant,

3) there exists an i such that

$$|a_{i,i}| > \sum_{\substack{j=1\\j\neq i}}^{n} |a_{i,j}|.$$

◊

Theorem 1.31

Let A be irreducibly diagonally dominant then, A is non–singular and $a_{i,i} \neq 0$ for all i.

Proof. We prove by contradiction that $a_{i,i} \neq 0$. Assume that $n > 1$ and $a_{i,i} = 0$ for some i. Diagonal dominance implies that $a_{i,j} = 0$ for all j but, this contradicts the irreducibility of A because we can choose $I_1 = i$, $I_2 = \Im_n - \{i\}$ and have $a_{i,j} = 0$ for all $i \in I_1$ and for all $j \in I_2$. Let us now prove that A is non–singular. As $a_{i,i} \neq 0$ for all i, we can define

$$B = I - D^{-1}A,$$

where D is a diagonal matrix with $d_{i,i} = a_{i,i}$ for all i (in abbreviated form $D = \mathrm{diag}(A)$). It is obvious that

$$b_{i,i} = 0,$$
$$b_{i,j} = -\frac{a_{i,j}}{a_{i,i}}, \quad i \neq j.$$

Since A is irreducibly diagonally dominant, we have

$$\sum_{j=1}^{n} |b_{i,j}| \leq 1, \quad \text{for all} \quad i,$$

and there exists an i such that $\sum_{j=1}^{n} |b_{i,j}| < 1$. Using proposition 1.9, we obtain $\|B\|_\infty \leq 1$. So, $\rho(B) \leq 1$. Let us assume there exists an eigenvalue λ of modulus 1 and let x be the corresponding eigenvector,

$$\sum_{j=1}^{n} b_{i,j}x_j - \lambda x_i.$$

We follow the lines of Young [1351] to end the proof. Let i be such that $|x_i| = \|x\|_\infty$, then

$$\sum_{j=1}^{n} |b_{i,j}| \, \|x\|_\infty \le \|x\|_\infty \le \sum_{j=1}^{n} |b_{i,j}| \, |x_j|.$$

The second inequality follows because

$$|\lambda x_i| = |\lambda| \, |x_i| = |x_i| = \|x\|_\infty \le \sum_{j=1}^{n} |b_{i,j}| \, |x_j|.$$

Hence,

$$\sum_{j=1}^{n} |b_{i,j}| \big(|x_j| - \|x\|_\infty \big) \ge 0,$$

but

$$|x_j| - \|x\|_\infty \le 0 \quad \text{for all} \quad j.$$

This implies that $|x_j| = \|x\|_\infty$ or $b_{i,j} = 0$ but B is irreducible therefore, c.f. [1351], there exist indices $i_1, \dots, i_p$ such that

$$b_{i,i_1}, b_{i_1,i_2}, \dots, b_{i_p,j} \ne 0.$$

This shows that

$$|x_{i_1}| = |x_{i_2}| = \cdots = |x_j| = \|x\|_\infty.$$

In all cases, we have $|x_j| = \|x\|_\infty$ for all j and therefore

$$\|x\|_\infty = |x_i| \le \sum_{j=1}^{n} |b_{i,j}| \, \|x\|_\infty,$$

for all i. As $\|x\|_\infty \ne 0$, $\sum_{j=1}^{n} |b_{i,j}| \ge 1$ for all i. This contradicts the fact that there is at least one index i such that $\sum_{j=1}^{n} |b_{i,j}| < 1$. Thus we have proven that $\rho(B) < 1$ and by theorem 1.27, $I - B = D^{-1}A$ is non–singular. □

Corollary 1.32

Let A be strictly diagonally dominant then A is non–singular and $a_{i,i} \neq 0$ for all i.

□

We can characterize the properties of the eigenvalues of diagonally dominant matrices.

Theorem 1.33

Let A be irreducibly diagonally dominant and assume that $a_{i,i} > 0$ for all i. Then for every eigenvalue λ (possibly a complex number), $\mathrm{Re}\,\lambda > 0$.

Proof. From theorem 1.31, we know that A is non–singular. Let x be an eigenvector of A, then we have

$$\lambda x_i = \sum_{j=1}^{n} a_{i,j} x_j,$$

$$(\lambda - a_{i,i}) x_i = \sum_{\substack{j=1 \\ j \neq i}}^{n} a_{i,j} x_j.$$

Let us pick i such that $|x_i| = \|x\|_\infty$, then

$$|\lambda - a_{i,i}| \leq \sum_{\substack{j=1 \\ j \neq i}}^{n} |a_{i,j}|.$$

This proves that the eigenvalues belong to the union of the disks with centers $a_{i,i}$ and radius $\sum_{\substack{j=1 \\ j \neq i}}^{n} |a_{i,j}|$. These are known as the Gerschgorin disks. We note that these disks contain only complex numbers with positive real parts because if there is a $b > 0$ such that $-b$ belongs to the disk then

$$a_{i,i} < a_{i,i} + b = |-b - a_{i,i}| \leq \sum_{\substack{j=1 \\ j \neq i}}^{n} |a_{i,j}|.$$

This contradicts the diagonal dominance of A. Moreover, we cannot have $\lambda = 0$ because A is non–singular. □

The same result is true if A is strictly diagonally dominant. The definition of diagonal dominance can be generalized as follows.

Definition 1.34

- *A matrix A is generalized diagonally dominant if there exists a vector d with $d_i > 0$ for all i such that*

$$|a_{i,i}|d_i \geq \sum_{\substack{j=1 \\ j \neq i}}^{n} |a_{i,j}|d_j.$$

- *A is generalized strictly diagonally dominant if*

$$|a_{i,i}|d_i > \sum_{\substack{j=1 \\ j \neq i}}^{n} |a_{i,j}|d_j.$$

◇

Clearly, these definitions are equivalent to applying the classical definitions of diagonal dominance to the matrix AD where

$$D = \begin{pmatrix} d_1 & & & \\ & d_2 & & \\ & & \ddots & \\ & & & d_n \end{pmatrix}.$$

Hence, we have,

Proposition 1.35

Let A be generalized strictly diagonally dominant, then A is non–singular and $a_{i,i} \neq 0$ for all i.

□

Note that the definition of generalized diagonal dominance can be expressed in the following way. There exists a diagonal matrix D with strictly positive diagonal elements such that $D^{-1}AD$ is strictly diagonally dominant. To see this, write the definition in the following form

$$|a_{i,i}| \geq \sum_{\substack{j=1 \\ j\neq i}}^{n} |a_{i,j}| \frac{d_j}{d_i}.$$

This last definition is often more convenient. To end this section, we give a definition which has some formal analogy with the one of irreducibility.

Definition 1.36

The matrix A has property A if there exist two sets of indices I_1 and I_2 that partition the set $\Im_n$ and if $i \neq j$ and $a_{i,j} \neq 0$ or $a_{j,i} \neq 0$ then $i \in I_1$ and $j \in I_2$ or $j \in I_2$ and $j \in I_1$. ◇

This definition was introduced by Young [1351] who showed the following characterization for matrices having property A.

Proposition 1.37

The matrix A has property A if and only if A is diagonal or there exists a permutation matrix P such that

$$P^{-1}AP = \begin{pmatrix} D_1 & F \\ E & D_2 \end{pmatrix},$$

where D_1 and D_2 are diagonal square matrices.

□

1.4. M–Matrices and generalizations

We say that a matrix A is positive (resp. strictly positive) and we denote $A \geq 0$ (resp. $A > 0$) if $a_{i,j} \geq 0$ (resp. $a_{i,j} > 0$) for all i and j.

Definition 1.38

A matrix A is monotone if and only if A is non–singular and $A^{-1} \geq 0$. ◇

This definition is important because this is the finite dimensional version of the maximum principle: if $b \geq 0$ then the solution of $Ax = b$ is such that $x \geq 0$.

Definition 1.39

A is an L–matrix if and only if

$$a_{i,i} > 0, \quad \textit{for all } i,$$
$$a_{i,j} \leq 0, \quad i \neq j.$$

◇

From the above definition, it is obvious that every L–matrix A can be written as

$$A = \sigma I - B,$$

with $\sigma \in \mathbb{R}$, $\sigma > 0$, and $B \geq 0$. We then have the definition

Definition 1.40

Every matrix A which can be written as $A = \sigma I - B$ with $\sigma > 0, B \geq 0$ and such that $\rho(B) \leq \sigma$ is said to be an M–matrix. ◇

Note that under this definition A can be singular. If A is non–singular we have $\rho(B) < \sigma$. The following result relates the last definition to the most frequently used definition of M–matrices.

Theorem 1.41

A is a non–singular M–matrix if and only if A is non–singular with $a_{i,j} \leq 0$ for $i \neq j$ and $A^{-1} \geq 0$.

Proof. Let A be a non–singular M–matrix. Then,

$$A = \sigma I - B, \sigma \in \mathbb{R}, \quad \sigma > 0, \quad \mathrm{B} \geq 0, \quad \rho(\mathrm{B}) < \sigma.$$

All we have to show is that $A^{-1} \geq 0$. Now $\rho(\frac{B}{\sigma}) < 1$ and by theorem 1.27 $\frac{1}{\sigma}A = I - \frac{B}{\sigma}$ is non–singular. The series

$$I + \frac{B}{\sigma} + \frac{B^2}{\sigma^2} + \dots$$

converges and

$$\sigma A^{-1} = I + \frac{B}{\sigma} + \dots$$

As $B \geq 0$ and $\sigma > 0$, we have $A^{-1} \geq 0$.

Conversely, let us suppose that A is non–singular with $a_{i,j} \leq 0$ for all i and j with $i \neq j$ and $A^{-1} \geq 0$. Then we can write $A = \sigma I - B$ with $B \geq 0$ and $\sigma > 0$.

Let S_k be defined by

$$S_k = I + \frac{B}{\sigma} + \dots + \frac{B^{k+1}}{\sigma^{k+1}}.$$

Then,

$$\left(I - \frac{B}{\sigma}\right) S_k = I - \frac{B^{k+1}}{\sigma^{k+1}},$$

$$\sigma A^{-1} = \left(I - \frac{B}{\sigma}\right)^{-1} = S_k + \left(I - \frac{B}{\sigma}\right)^{-1} \frac{B^{k+1}}{\sigma^{k+1}}.$$

but $(I - \frac{B}{\sigma})^{-1} \geq 0$ and $B^{k+1} \geq 0$ hence $S_k \leq \sigma A^{-1}$. Every entry of S_k is a decreasing bounded sequence and so it converges. The series $I + \frac{B}{\sigma} + \cdots$ converges, which by theorem 1.27 implies

$$\rho\left(\frac{B}{\sigma}\right) < 1,$$

which proves that A is an M–matrix. □

One can show that an M–matrix has positive diagonal entries and so does an L–matrix. Many characterizations of non–singular M–matrices have been given; for example Berman and Plemmons [142] give 50 equivalent definitions. The following theorem gives a sufficient condition which is often useful.

Theorem 1.42

Let A be an irreducibly diagonally dominant L–matrix, then A is a non–singular M–matrix.

Proof. We know from theorem 1.31 that A is non–singular. Now let D be a diagonal matrix whose diagonal is the same as the diagonal of A, $d_{i,i} > 0$ and let $B = I - D^{-1}A$. It is clear that $B \geq 0$,

$$D^{-1}A = I - B, \quad B \geq 0.$$

In theorem 1.31 we showed that $\rho(B) < 1$. This proves that $D^{-1}A$ is an M–matrix and so is A. □

In fact, with the hypotheses of theorem 1.42, we can prove that $A^{-1} > 0$. Note that if A is a strictly diagonally dominantL–matrix, then A is a non–singular M–matrix. Along the same lines we have the next theorem.

Theorem 1.43

A is a non–singular M–matrix if and only if $a_{i,j} \leq 0$ for all $i \neq j$ and A is generalized strictly diagonally dominant.

Proof. Let A be generalized strictly diagonally dominant. Then, we know that there exists a diagonal matrix D, $d_{i,i} > 0$ such that AD is strictly diagonally dominant. It is obvious that the diagonal entries of AD are positive and the off-diagonal ones are negative, and therefore AD is a non–singular M–matrix and so is A.

Now let us suppose that A is a non–singular M–matrix. We have $a_{i,j} \leq 0$ for all $i \neq j$, $a_{i,i} > 0$ and $A^{-1} \geq 0$. Let e be such that $e^T = (1, \ldots, 1)$, and $d = A^{-1}e$, then $d > 0$ because if we suppose there exists i such that $d_i = 0$ we have, $\sum_j (A^{-1})_{i,j} = 0$ which implies $(A^{-1})_{i,j} = 0$ for all j and hence, A^{-1} is singular. Let D be a diagonal matrix with diagonal entries d_i. Of course, $De = d$ so,

$$ADe = Ad = e > 0.$$

Component-wise we have

$$a_{i,i}d_i + \sum_{\substack{j=1 \\ j \neq 1}}^{n} a_{i,j}d_j > 0.$$

But, $a_{i,j} \leq 0$ for all $i \neq j$ so $|a_{i,j}| = -a_{i,j}$. We can rewrite the previous inequality as

$$|a_{i,i}|d_i > \sum_{\substack{j=1 \\ j \neq i}}^{n} |a_{i,j}|d_j,$$

which proves that A is generalized strictly diagonally dominant. □

Corollary 1.44

The matrix A is a non–singular M–matrix if and only if A is an L–matrix and there exists a diagonal matrix D with $d_{i,i} > 0$ such that $D^{-1}AD$ is strictly diagonally dominant.

Proof. This comes from the remark following proposition 1.35. □

We have a result for M–matrices which is a generalization of theorem 1.33 for irreducibly diagonally dominant matrices.

Theorem 1.45

Let A be a matrix with $a_{i,j} \leq 0$ for all $i \neq j$. Then, A is a non–singular M–matrix if and only if $\operatorname{Re} \lambda > 0$ for all eigenvalues λ of A.

Proof. See Berman–Plemmons [142]. □

Lemma 1.46

Let A be an M–matrix written in block form as

$$A = \begin{pmatrix} B & F \\ E & C \end{pmatrix},$$

where B and C are square matrices, then the Schur complement $S = C - EB^{-1}F$ is an M–matrix.

Proof. It is obvious that the principal submatrices of an M–matrix are M–matrices. Therefore, B is an M–matrix and $B^{-1} > 0$. Since, by definition, the entries of E and F are non–positive, the entries of $EB^{-1}F$ are non–negative. Therefore, the non–diagonal entries of S are non–positive.

Now, as A is an M–matrix, we know there is a diagonal matrix D with strictly positive diagonal entries such that AD is (row) strictly diagonally dominant. Let

$$D = \begin{pmatrix} D_1 & 0 \\ 0 & D_2 \end{pmatrix},$$

AD being (row) strictly diagonally dominant means that if $e = (1 \ \dots \ 1)^T$, then $ADe > 0$. But,

$$AD = \begin{pmatrix} BD_1 & FD_2 \\ ED_1 & CD_2 \end{pmatrix},$$

and let $e = \begin{pmatrix} e_1 \\ e_2 \end{pmatrix}$. The Schur complement of AD is (row) strictly diagonally dominant (see [342]). This means that

$$0 < [CD_2 - ED_1(BD_1)^{-1}FD_2]e_2 = SD_2e_2.$$

This shows that S is (row) generalized strictly diagonally dominant. Hence, S is an M–matrix. □

We now consider a generalization of M–matrices. Let A be a matrix, define $M(A)$ as the matrix having entries $m_{i,j}$ such that

$$m_{i,i} = |a_{i,i}|, \quad m_{i,j} = -|a_{i,j}|$$

for all $i \neq j$. Clearly $M(A)$ is an L–matrix. It is obvious that many different matrices A can lead to the same $M(A)$. So, we define the set

$$\Omega(A) = \{B \,|\, |b_{i,j}| = |a_{i,j}|\}\,.$$

$\Omega(A)$ is called the equimodular set for A. Then, we have the definition of the most general class of matrices we shall consider in this book.

Definition 1.47

A is an H–matrix if and only if $M(A)$ is an M–matrix. Each definition of an M–matrix gives a corresponding definition for H–matrices. ◇

Let us recall the following result which will be useful subsequently.

Theorem 1.48

A is a non–singular H–matrix if and only if A is generalized strictly diagonally dominant.

□

1.5. Splittings

It is obvious that strictly diagonally dominant, irreducibly diagonally dominant and non–singular M–matrices are non–singular H–matrices. Many iterative methods that we shall study use a decomposition of the matrix A (which is also known as a splitting) in the form

$$A = M - N.$$

To prove some results about splittings, we have to be more specific about M and N.

Definition 1.49

$A = M - N$ is a regular splitting of A if and only if M is non–singular, $M^{-1} \geq 0$ and $N \geq 0$. ◇

Regular splittings and M–matrices are closely related as the next result shows.

Theorem 1.50

Let A with $a_{i,j} \leq 0$ for all $i \neq j$. A is a non–singular M–matrix if and only if there exists a regular splitting $A = M - N$ with $\rho(M^{-1}N) < 1$.

Proof. By definition of an M–matrix, there exists $\sigma \in \mathbb{R}, \sigma > 0$ and $B \geq 0$ such that $A = \sigma I - B$ with $\rho(B) < \sigma$. This clearly is a regular splitting of A.

Now, suppose we have a regular splitting $A = M - N$ with $\rho(M^{-1}N) < 1$,

then

$M^{-1}A = M^{-1}(M - N) = I - M^{-1}N$

is non–singular. By theorem 1.27

$(I - M^{-1}N)^{-1} = I + M^{-1}N + (M^{-1}N)^2 + \cdots$

Therefore, $(I - M^{-1}N)^{-1} \geq 0$ and $A^{-1} \geq 0$. □

It is often useful to compare two regular splittings. A useful result is given in the following theorem.

Theorem 1.51

Let A be such that $A^{-1} \geq 0$ and consider two regular splittings of A: $A = M_1 - N_1$ and $A = M_2 - N_2$ with $N_2 \leq N_1$. Then,

$\rho(M_2^{-1}N_2) \leq \rho(M_1^{-1}N_1) < 1.$

Proof. See Varga [1300]. □

1.6. Positive definite matrices

In this Section, we study the properties of another useful class of matrices.

Definition 1.52

The matrix A is positive definite if and only if $(Ax, x) > 0$ for all $x \neq 0$. ◇

These matrices are sometimes referred to as positive real. Note that if A is symmetric positive definite (SPD), we can define a vector norm $\|x\|_A$ by

$\|x\|_A^2 = (Ax, x).$

As we have seen before, this is usually called the energy norm. When A is symmetric we have the well known characterization of positive definite matrices.

Proposition 1.53

Let A be a symmetric matrix. Then A is positive definite if and only if the eigenvalues of A (which are real numbers) are positive.

□

For symmetric matrices we have a very important result, often called the Householder–John theorem.

Theorem 1.54

Let A be a non–singular symmetric M–matrix such that $A = M - N$ with M non-singular and $Q = M + M^T - A = M^T + N$ is positive definite. Then, $\rho(M^{-1}N) < 1$ if and only if A is positive definite.

Proof. Suppose that A is positive definite. Let x^0 be given. We define a sequence of vectors by

$$Mx^{k+1} = Nx^k.$$

We need to prove that x^k tends to 0 as k tends to ∞, but,

$$x^{k+1} = M^{-1}Nx^k = (I - M^{-1}A)x^k.$$

Since A is positive definite, (Ax, x) defines a norm (the energy norm) which we denote by $\|x\|_A^2$. Hence,

$$\begin{aligned}\|x^{k+1}\|_A^2 &= (Ax^{k+1}, x^{k+1}),\\ &= ((A - AM^{-1}A)x^k, (I - M^{-1}A)x^k),\\ &= (Ax^k, x^k) - (AM^{-1}Ax^k, x^k)\\ &\quad - (Ax^k, M^{-1}Ax^k) + (AM^{-1}Ax^k, M^{-1}Ax).\end{aligned}$$

Denote $y^k = M^{-1}Ax^k$, then

$$\|x^{k+1}\|_A^2 - \|x^k\|_A^2 = (Ay^k, y^k) - (My^k, y^k) - (y^k, My^k).$$

This is because $(AM^{-1}Ax^k, x^k) = (M^{-1}Ax^k, Ax^k) = (y^k, My_k)$. So,

$$\|x^{k+1}\|_A^2 - \|x^k\|_A^2 = ((A - M^T - M)y^k, y^k) = -(Qy^k, y^k) < 0.$$

If $y^k = 0$, this implies $x^k = 0$ and then $x^{k+1} = 0$. Therefore $\|x^k\|_A^2$ is a bounded decreasing sequence and it must converge to some limit. As we have

$$\|x^k\|_A^2 - \|x^{k+1}\|_A^2 = (Qy^k, y^k),$$

$(Qy^k, y^k) \to 0$ and then $y^k \to 0$ and $x^k \to 0$ which proves that $\rho(M^{-1}N) < 1$.

For the converse, let us suppose now that $\rho(M^{-1}N) < 1$ and that A is not positive definite. We have already shown that

$$(Ax^k, x^k) - (Ax^{k+1}, x^{k+1}) > 0.$$

If A is not positive definite, there exists x^0 such that $(Ax^0, x^0) < 0$. Then, by induction, the sequence x^k is such that

$$(Ax^{k+1}, x^{k+1}) < (Ax^k, x^k) < 0.$$

But this contradicts the fact that x^k tends to 0. □

We remark that the energy norms of iterates of the sequence x^k are decreasing. The last theorem can be generalized to non–symmetric matrices.

Theorem 1.55

Let A be a non–singular matrix $A = M - N$, M non–singular such that $Q = M^T A^{-T} A + N$ is positive definite. Then, $\rho(M^{-1}N) < 1$ if and only if A is positive definite.

Proof. See Ortega and Plemmons [1049]. □

We now consider the relationship between M–matrices, H–matrices and positive definite matrices for the symmetric case.

Theorem 1.56

Let A be a symmetric matrix with $a_{i,j} \leq 0$ for all $i \neq j$. Then A is a non–singular M–matrix if and only if A is positive definite.

Proof. This result follows directly from theorem 1.45. □

Theorem 1.57

Let A be a symmetric non–singular H–matrix with $a_{i,i} > 0$ for all i. Then A is positive definite.

Proof. There exists a diagonal matrix D with $d_{i,i} > 0$ such that $D^{-1}AD$ is strictly diagonally dominant and $(D^{-1}AD)_{i,i} = a_{i,i} > 0$. By theorem 1.33 we know that $D^{-1}AD$ is positive definite. Since $D^{-1}AD$ has the same eigenvalues as A, this proves the result. □

However, the converse of this theorem is not true as the following example shows.

Example 1.58

Let

$$A = \begin{pmatrix} a & e & e \\ e & a & e \\ e & e & a \end{pmatrix}.$$

The eigenvalues of A are $a + 2e, a - e, a + e$. Let $a > e > 0$. Then A is positive definite. But, as

$$M(A) = \begin{pmatrix} a & -e & -e \\ -e & a & -e \\ -e & -e & a \end{pmatrix},$$

the eigenvalues of $M(A)$ are $a + e$, $a + e$ and $a - 2e$. If we choose $0 < e < a < 2e$ then, $M(A)$ is not positive definite and by theorem 1.56 is not an M–matrix. This A is an example of a positive definite matrix which is not an H–matrix.

The following result is important for the study of direct and iterative methods.

Lemma 1.59

Let A be a symmetric positive definite matrix partitioned as

$$A = \begin{pmatrix} A_{1,1} & A_{2,1}^T \\ A_{2,1} & A_{2,2} \end{pmatrix},$$

where the blocks $A_{1,1}$ and $A_{2,2}$ are square. Then,

$$S_{2,2} = A_{2,2} - A_{2,1} A_{1,1}^{-1} A_{2,1}^T,$$

is symmetric and positive definite. $S_{2,2}$ is the Schur complement (of $A_{2,2}$ in A).

Proof. This result can be proved in many different ways. Perhaps, the simplest one is to consider A^{-1} and to compute the bottom right hand block of A^{-1}. Let

$$\begin{pmatrix} A_{1,1} & A_{2,1}^T \\ A_{2,1} & A_{2,2} \end{pmatrix} \begin{pmatrix} x_1 \\ x_2 \end{pmatrix} = \begin{pmatrix} b_1 \\ b_2 \end{pmatrix}.$$

Then,

$$A_{1,1} x_1 + A_{2,1}^T x_2 = b_1, \quad \Rightarrow x_1 = A_{1,1}^{-1}(b_1 - A_{2,1}^T x_2).$$

Therefore,

$$(A_{2,2} - A_{2,1} A_{1,1}^{-1} A_{2,1}^T) x_2 = b_2 - A_{2,1} A_{1,1}^{-1} b_1.$$

This means that the inverse of A can be written as

$$A^{-1} = \begin{pmatrix} X & Y \\ Z & S_{2,2} \end{pmatrix}.$$

With A being positive definite, the diagonal blocks of A and A^{-1} are also positive definite, so $S_{2,2}$ is positive definite. □

We also have the useful following result.

Lemma 1.60

Let A be a symmetric positive definite matrix. Then,

$$\max_{i,j} |a_{i,j}| = \max_i (a_{i,i}).$$

Proof. It is obvious that the diagonal entries of A are positive. Suppose there are indices i_0, j_0 such that $|a_{i_0,j_0}| > |a_{i,j}|, i_0 \neq j_0, \forall i, j$ different from i_0, j_0. There are two cases:

i) suppose $a_{i_0,j_0} > 0$, then let
$x = (0\dots, 0, 1, 0, \dots, 0, -1, 0, \dots, 0)^T$,
the 1 being in position i_0 and the -1 in position j_0. We have,
$(x, Ax) = a_{i_0,i_0} + a_{j_0,j_0} - 2a_{i_0,j_0} < 0$.
Therefore, A is not positive definite which is a contradiction;

ii) suppose $a_{i_0,j_0} < 0$, we choose
$x = (0\dots, 0, 1, 0, \dots, 0, 1, 0, \dots, 0)^T$,
and get
$(x, Ax) = a_{i_0,i_0} + a_{j_0,j_0} + 2a_{i_0,j_0} < 0$,
which, again, is a contradiction. □

1.7. The graph of a matrix

It is well known that a graph can be associated with every matrix. To a general non–symmetric square matrix A of order n, we associate a directed graph (or digraph). A digraph is a couple $G = (X, E)$ where X is a set of nodes (or vertices) and E is a set of directed edges. For a given matrix A of order n, there are n nodes in the graph and there is a directed edge from i to j if $a_{i,j} \neq 0$. Usually, self loops corresponding to $a_{i,i} \neq 0$ are not included.

Let

$$A = \begin{pmatrix} x & x & 0 & x \\ 0 & x & 0 & 0 \\ x & 0 & x & 0 \\ 0 & x & x & x \end{pmatrix}.$$

Then, the associated digraph is given in figure 1.1.

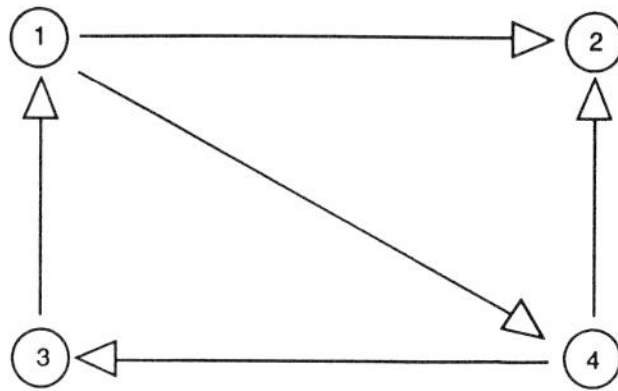

fig 1.1: A directed graph

Graphs are more commonly used in problems involving symmetric matrices. If $a_{i,j} \neq 0$, then $a_{j,i} \neq 0$. Therefore, we can consider only undirected graphs and drop the arrows on the edges. Let

$$A = \begin{pmatrix} x & x & x & 0 \\ x & x & 0 & x \\ x & 0 & x & x \\ 0 & x & x & x \end{pmatrix},$$

then, the graph of A is shown in figure 1.2.

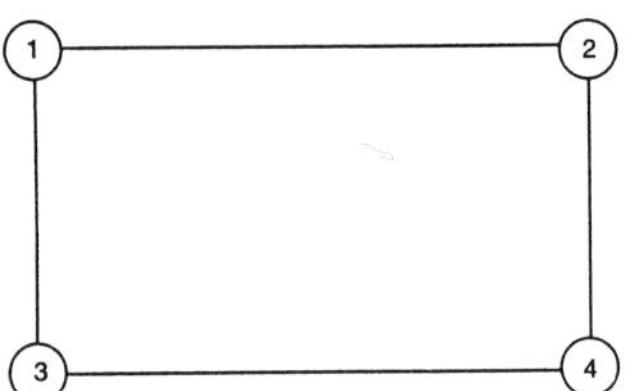

fig. 1.2: An undirected graph

Let us introduce a few definitions used for graphs. Let $G = (X, E)$ be a (undirected) graph. We denote the nodes of the graph by x_i or sometimes i.

• $G' = (X', E')$ is a subgraph of G if $X' \subset X$ and $E' \subset E$.

• Two nodes x and y of G are adjacent if $\{x, y\} \in E$. The adjacency set of a node y is defined as

$Adj(y) = \{x \in X |\ x \text{ is adjacent to } y\}.$

If $Y \subset X$, then

$Adj(Y) = \{x \in X |\ x \in Adj(y), x \notin Y, y \in Y\}.$

• The degree of a node x of G is the number of its adjacent nodes in G,

$deg(x) = |Adj(x)|.$

• Let x and $y \in X$. A path of length l from x to y is a set of nodes $\{\nu_1, \nu_2, \dots, \nu_{l+1}\}$ such that $x = \nu_1$, $y = \nu_{l+1}$ and $\{\nu_i, \nu_{i+1}\} \in E$, $1 \leq i \leq l$. A path $\{\nu_0, \nu_1, \dots, \nu_l, \nu_0\}$ is a (simple) cycle of length $l + 1$.

• A graph is connected if for every $x, y \in X$, there exists a path from x to y. This corresponds to the matrix being irreducible.

• A chord of a path is any edge joining two non–consecutive vertices in the path. A graph is chordal if every cycle of length greater than three has a chord (see [159]).

• An important kind of graph is when there are no closed paths. A particular node is labeled as the root. Then, there is a path from any node to the root. Such a (connected) graph is called a tree. If it is not connected, we have a set of trees which is called a forest.

• Let $Y \subset X$, the section graph $G(Y)$ is a subgraph $(Y, E(Y))$ with

$$E(Y) = \{\{x, y\} \in E|\ x \in Y, y \in Y\}.$$

• A set $Y \subset X$ is a separator for G (a connected graph) if $G(X/Y)$ has two or more connected components.

• The distance $d(x, y)$ between two nodes x and y of G is the length of the shortest path between x and y. The eccentricity of a node $e(x)$ is

$$e(x) = \max\{d(x, y)|y \in X\}.$$

The diameter δ of G is

$$\delta(G) = \max\{e(x)|x \in X\}.$$

A node x is peripheral if $e(x) = \delta(G)$.

• A clique is a subset of nodes which are all pairwise connected.

• A level structure of a graph G is a partition $\mathcal{L} = \{L_0, L_1, \ldots, L_l\}$ of X such that

$$Adj(L_i) \subset L_{i-1} \cup L_{i+1},\ i = 1, \ldots, l-1,$$
$$Adj(L_0) \subset L_1,$$
$$Adj(L_l) \subset L_{l-1}.$$

Note that each $L_i, i = 1, \ldots, l-1$ is a separator for G. For each node $x \in X$, a level structure $\mathcal{L}(x)$ can be defined as

$$\mathcal{L}(x) = \{L_0(x), \ldots, L_{e(x)}(x)\},$$

$L_0(x) = \{x\}$
$L_i(x) = Adj(\cup_{k=0}^{i-1} L_k(x)),\ 1 \le i \le e(x)$

where $e(x)$ is the eccentricity of x. The width of a level structure $\mathcal{L}(x)$ is

$w(x) = \max\{|L_i(x)|, 0 \le i \le e(x)\}.$

1.8. Chebyshev polynomials

We now review some facts about Chebyshev polynomials which will be important in studying some iterative methods.

Definition 1.61

Chebyshev polynomials (of the first kind) T_n are defined for integer n and $x \in \mathbb{R}$, $|x| \le 1$, by

$T_n(x) = \cos(n \arccos x).$

We can extend this definition to an interval $[a, b]$ by setting $t = \frac{a+b}{2} + \frac{b-a}{2}x$, then $t \in [a, b] \Leftrightarrow x \in [-1, +1]$. ◇

From the definition, it is not clear that T_n is a polynomial. However it follows from the properties of theorem 1.62.

Theorem 1.62

Let T_n be a Chebyshev polynomial. Then,

1) $T_0(x) = 1$, $T_1(x) = x$, $T_{n+1}(x) = 2xT_n(x) - T_{n-1}(x)$;

2) For $n \ge 1$, T_n is an n-degree polynomial whose leading coefficient is 2^{n-1};

3) $T_n(-x) = (-1)^n T_n(x)$

4) T_n has n zeros in $[-1, +1]$, namely

$$x_k = \cos\left(\frac{2k+1}{n}\frac{\pi}{2}\right), \quad k = 0, 1, \ldots, n-1,$$

and $n+1$ extremas

$$x'_k = \cos\left(\frac{k\pi}{n}\right) \quad \textit{with} \quad T_n(x'_k) = (-1)^k, \quad k = 0, 1, \ldots, n.$$

5) The Chebyshev polynomials are orthogonal with respect to the scalar product

$$((f,g)) = \int_{-1}^{+1} \frac{f(x)g(x)}{\sqrt{1-x^2}}\, dx.$$

Moreover,

$$((T_i, T_j)) = \begin{cases} 0, & i \neq j \\ \frac{\pi}{2}, & i = j \neq 0 \\ \pi, & i = j = 0 \end{cases}$$

6) For all n-degree polynomials with leading coefficient 1, $T_n/2^{n-1}$ has the smallest maximum norm, namely $1/2^{n-1}$.

Proof. See, for instance, Dahlquist and Björck [371]. □

Chebyshev polynomials are very useful especially when studying iterative methods. Their most interesting property is that they have the smallest maximum norm (item 6)). It helps in solving the following problem:

Let $\pi_n^1 = \{$ polynomials of degree n in t whose value is 1 for $t = 0$ $\}$. We shall frequently require the solution of the minimization problem

$$\min_{Q_n \in \pi_n^1} \max_{t \in [a,b]} |Q_n(t)|.$$

A solution to this problem is given by the shifted and normalized Chebyshev polynomial

$$\min_{Q_n \in \pi_n^1} \max_{t \in [a,b]} |Q_n(t)| = \max_{t \in [a,b]} \left| \frac{T_n\left(\frac{2t-(a+b)}{b-a}\right)}{T_n\left(\frac{a+b}{b-a}\right)} \right| = \frac{1}{\left|T_n\left(\frac{a+b}{b-a}\right)\right|}.$$

This solution is interesting because we know the roots of Q_n,

$$\tau_l = \frac{a+b}{2} + \frac{b-a}{2} \cos\left(\frac{2l-1}{n}\frac{\pi}{2}\right) \qquad l = 1, \ldots, n.$$

Note that if $a = -b$ then $T_n\left(\frac{a+b}{b-a}\right) = 1$ but in more general cases we need an upper bound for

$$\frac{1}{\left|T_n\left(\frac{a+b}{b-a}\right)\right|}.$$

To obtain this bound, we use the following characterization.

Lemma 1.63

$$T_n(x) = \frac{1}{2}[(x + \sqrt{x^2-1})^n + (x - \sqrt{x^2-1})^n], \quad |x| \leq 1.$$

Proof. Let $\varphi = \arccos x$ so $T_n(x) = \cos(n\varphi)$. We can write

$$\cos(n\varphi) = \frac{1}{2}(\cos(n\varphi) + i\sin(n\varphi) + \cos(n\varphi) - i\sin(n\varphi)),$$

with $i = \sqrt{-1}$. Therefore,

$$\cos(n\varphi) = \frac{1}{2}[(\cos\varphi + i\sin\varphi)^n + (\cos\varphi - i\sin\varphi)^n],$$

but,

$$x = \cos\varphi \text{ and } i\sin\varphi = \sqrt{-\sin^2\varphi} = \sqrt{x^2-1},$$

and this proves the result. $\square$

Using lemma 1.63, we obtain the bound we are looking for.

Theorem 1.64

If $0 < a < b$ *then*

$$\min_{Q_n \in \pi_n^1} \max_{t\in[a,b]} |Q_n(t)| \leq 2\left(\frac{1-\sqrt{\frac{a}{b}}}{1+\sqrt{\frac{a}{b}}}\right)^n = 2\left(\frac{\sqrt{\frac{b}{a}}-1}{\sqrt{\frac{b}{a}}+1}\right)^n.$$

Proof. We have $x = \frac{b+a}{b-a}$, so

$$\begin{aligned} x^2 - 1 &= \frac{2\sqrt{ab}}{b-a}, \\ x + \sqrt{x^2-1} &= \frac{(\sqrt{b} - \sqrt{a})^2}{b-a}, \\ &= \frac{\sqrt{b}+\sqrt{a}}{\sqrt{b}-\sqrt{a}}, \end{aligned}$$

and

$$x - \sqrt{x^2-1} = \frac{\sqrt{b}-\sqrt{a}}{\sqrt{b}+\sqrt{a}}.$$

Therefore,

$$T_n\left(\frac{b+a}{b-a}\right) \geq \frac{1}{2}\left(\frac{\sqrt{b}+\sqrt{a}}{\sqrt{b}-\sqrt{a}}\right)^n.$$

□

Chebyshev polynomials are one example of orthogonal polynomials which play a very important role in many areas of numerical analysis.

1.9. Discretization methods for partial differential equations

Although they are not the only source of systems of linear equations, problems that arise from discretization of elliptic and parabolic (systems of) partial differential equations (PDEs) are some of the most important ones. Usually, non–linear problems are handled with iterative algorithms like Newton's method (or variants) which give a linear system to solve at each non–linear iteration.

There are basically two main ways to discretize partial differential equations, namely finite difference methods and finite element methods. Finite volumes can usually be interpreted as finite elements. These two kinds of methods have many things in common and finite differences can often be seen as particular cases of finite element methods.

Let us first focus on finite differences. Most often finite differences are used on a regular (cartesian) mesh and yield matrices with a regular structure that are easy to store in the computer memory. The model problem that was widely used for testing algorithms is Poisson equation on the open unit square $\Omega = (0,1) \times (0,1)$:

$$-\Delta u = -\frac{\partial^2 u}{\partial x^2} - \frac{\partial^2 u}{\partial y^2} = f \quad \text{in} \quad \Omega,$$

$$u\big|_{\partial\Omega} = 0,$$

where $\partial\Omega$ is the boundary of Ω and f is given in an appropriate functional space (for instance $L_2(\Omega) = \{u \big| u \text{ measurable and } \int_\Omega u^2 dx < +\infty\}$). Non–homogeneous boundary conditions can be easily handled as well. To get an approximation to this infinite dimensional problem we cover Ω with a regular mesh Ω_h having m points in each direction. This gives a step size $h = \frac{1}{m+1}$ as in figure 1.3.

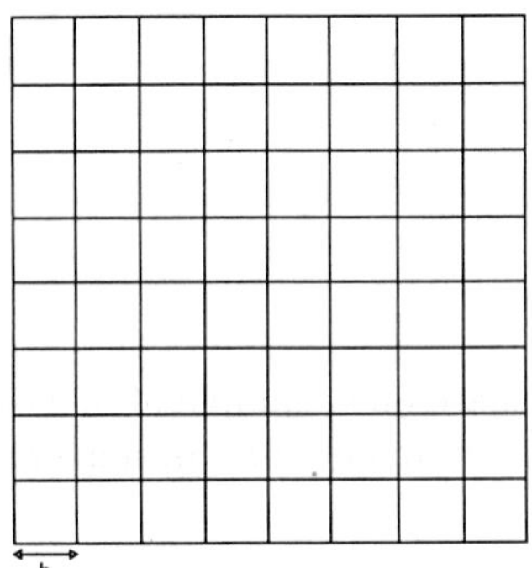

fig. 1.3: A regular mesh

We want to compute an approximate solution at each point (x_i, y_j) inside Ω and we already know u on the boundary (for i and j equal 0 or $m+1$). Denote by $u_{i,j}$ the approximation of u at (x_i, y_j). Then we approximate $\frac{\partial u}{\partial x}$ at $(i+\frac{1}{2})h, jh$ by the finite difference

$$\left(\frac{\partial u}{\partial x}\right)_{i+\frac{1}{2},j} \simeq \frac{u_{i+1,j} - u_{i,j}}{h}.$$

Hence,

$$\left(\frac{\partial^2 u}{\partial x^2}\right)_{i,j} \simeq \frac{1}{h}\left(\left(\frac{\partial u}{\partial x}\right)_{i+\frac{1}{2},j} - \left(\frac{\partial u}{\partial x}\right)_{i-\frac{1}{2},j}\right),$$

$$= \frac{u_{i-1,j} - 2u_{i,j} + u_{i+1,j}}{h^2}.$$

Doing this for both directions we get an approximation of minus the Laplacian Δ that we denote (after multiplication by h^2) by $-\Delta_5$.

$$\begin{aligned}(-\Delta_5 u)_{i,j} &= -u_{i-1,j} - u_{i,j-1} + 4u_{i,j} - u_{i,j-1} - u_{i,j+1},\\ &= h^2 f(x_i, y_j) \qquad i = 1, \ldots, m, \quad j = 1, \ldots, m\end{aligned}$$

The solution is already known for points on $\partial\Omega$, that is for $i = 0$ or $m + 1$, $j = 0$ or $m + 1$, so we have m^2 equations with m^2 unknowns and a linear system to solve. If we number the points from left to right and from bottom to top and rename the unknowns from u to x to agree with the notations that are traditionally used for linear algebra, the system can be written as

$$Ax = b$$

where

$$x = \{u_{1,1}, u_{1,2}, \ldots, u_{m,m}\}^T, \quad b = h^2\{f_{1,1}, f_{1,2}, \ldots, f_{m,m}\}^T,$$

$f_{i,j}$ being the value of f at point (i, j) and

$$A = \begin{pmatrix} T & -I & & & \\ -I & T & -I & & \\ & \ddots & \ddots & \ddots & \\ & & -I & T & -I \\ & & & -I & T \end{pmatrix},$$

where I is the $m \times m$ identity matrix and T is a tridiagonal matrix of order m,

$$T = \begin{pmatrix} 4 & -1 & & & \\ -1 & 4 & -1 & & \\ & \ddots & \ddots & \ddots & \\ & & -1 & 4 & -1 \\ & & & -1 & 4 \end{pmatrix}.$$

Of course if we use some other orderings for the mesh points, the structure of the matrix is different.

One can show that $u_{i,j} \to u(x_i, y_j)$ in some sense when $h \to 0$. In this example, the matrix coefficients are very simple, however one obtains a linear system with the same non–zero structure when solving a (diffusion) problem with variable (and possibly discontinuous) coefficients like

$$-\frac{\partial}{\partial x}\left[\lambda_1(x,y)\frac{\partial u}{\partial x}\right]-\frac{\partial}{\partial y}\left[\lambda_2(x,y)\frac{\partial u}{\partial y}\right]=f(x,y),$$
$$u\Big|_{\partial\Omega}=0.$$

For this problem we get a system like

$$Ax=b,$$

where the matrix can be written blockwise as

$$A=\begin{pmatrix} D_1 & A_2^T & & & \\ A_2 & D_2 & A_3^T & & \\ & \ddots & \ddots & \ddots & \\ & & A_{n-1} & D_{n-1} & A_n^T \\ & & & A_n & D_n \end{pmatrix}.$$

Later on, we shall consider the following examples as test problems for iterative methods:

- Test problem 1 is the Poisson model problem we have just considered.

- Test problem 2 arises from the 5–point finite difference approximation of a diffusion equation in a unit square, with Dirichlet boundary conditions.The coefficients are such that $\lambda_1(x,y)=\lambda_2(x,y)$ and they take on the value 1000 in a square $]1/4,3/4[\times]1/4,3/4[$ and to 1 otherwise.

- Test problem 3 is the same PDE but with different diffusion coefficients. The coefficient in the x direction is 100 if $x\in[1/4,3/4]$, 1 otherwise. The coefficient in the y direction is constant and equal to 1.

- Test problem 4 has a diffusion coefficient of 100 in the x direction and 1 in the y direction.

- Test problem 5 was suggested by S. Eisenstat. The PDE is

$$-\frac{\partial}{\partial x}\left(\lambda_x\frac{\partial u}{\partial x}\right)-\frac{\partial}{\partial y}\left(\lambda_y\frac{\partial u}{\partial y}\right)+\sigma u=f,$$

in the unit square with Neumann boundary conditions ($\frac{\partial u}{\partial n}|_{\partial\Omega} = 0$). The coefficients are defined as follows:

in $]0, 0.5[\times]0, 0.5[$, $\lambda_x = 1, \lambda_y = 1$,
in $]0.5, 1.[\times]0, 0.5[$, $\lambda_x = 100, \lambda_y = 1$,
in $]0, 0.5[\times]0.5, 1.[$, $\lambda_x = 1, \lambda_y = 100$,
in $]0.5, 1.[\times].50, 1.[$,$\lambda_x = 100, \lambda_y = 100$,

and $\sigma = 0.01$.

With a five point discretization stencil the equation for the point (i, j) involves only the unknowns at points (i, j), $(i-1, j)$, $(i+1, j)$, $(i, j-1)$ and $(i, j+1)$. The matrices D_i are tridiagonal and the matrices A_i are diagonal. Other types of boundary conditions give rise to matrices that are only slightly different. Let us look at the modifications given by Neumann or periodic boundary conditions on $\partial\Omega$, $\Omega = (0,1) \times (0,1)$ for the problem $-\Delta u + \sigma u = f$ where $\sigma > 0$ is a given coefficient.

With Neumann boundary conditions $\frac{\partial u}{\partial n} = 0$, n being the outward normal to the boundary, we have $m+2$ unknowns for each mesh line, *i.e.*, $(m+2)^2$ unknowns in total. Difference equations on the boundary are obtained using "shadow" unknowns as in figure 1.4.

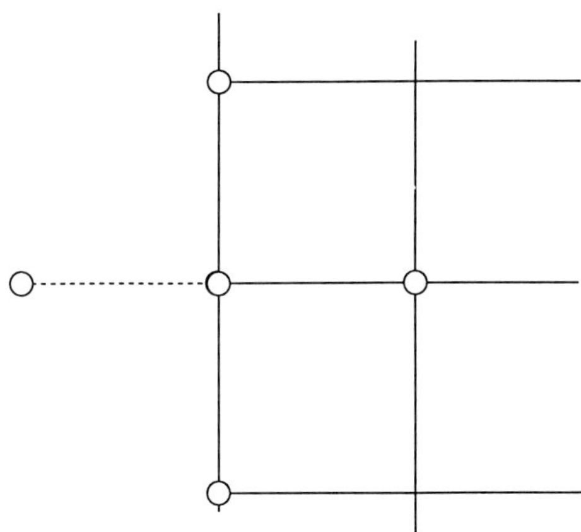

fig. 1.4: Shadow unknowns for Neumann boundary conditions

We set $u_{i,-1} = u_{i,1}$ for handling $\frac{\partial u}{\partial n} = 0$. Hence, the equation for the point $(i, 0)$ becomes

$$-u_{i-1,0} - u_{i+1,0} + (4 + \sigma h^2) u_{i,0} - 2u_{i,1} = h^2 f_{i,0}.$$

Putting together the results for the whole mesh, we get

$$A = \begin{pmatrix} T & -2I & & & \\ -I & T & -I & & \\ & \ddots & \ddots & \ddots & \\ & & -I & T & -I \\ & & & -2I & T \end{pmatrix},$$

where

$$T = \begin{pmatrix} 4+\sigma h^2 & -2 & & & \\ -1 & 4+\sigma h^2 & -1 & & \\ & \ddots & \ddots & \ddots & \\ & & -1 & 4+\sigma h^2 & -1 \\ & & & -2 & 4+\sigma h^2 \end{pmatrix}.$$

The matrix A is not symmetric but it can be symmetrized by multiplication by a suitable diagonal matrix.

For periodic boundary conditions there are $(m+1)^2$ unknowns and the matrix (for $\sigma = 0$) is

$$A = \begin{pmatrix} T & -I & & & -I \\ -I & T & -I & & \\ & \ddots & \ddots & \ddots & \\ & & -I & T & -I \\ -I & & & -I & T \end{pmatrix},$$

where

$$T = \begin{pmatrix} 4 & -1 & & & -1 \\ -1 & 4 & -1 & & \\ & \ddots & \ddots & \ddots & \\ & & -1 & 4 & -1 \\ -1 & & & -1 & 4 \end{pmatrix}.$$

Note the matrix A is singular. This is linked to the fact that the solution of the continuous problem is not unique. It is often important to make sure that $b \in \text{Range}(A)$.

Parabolic problems solved with time implicit discretization methods are usually a little easier to solve than elliptic problems. At each time step, we have to solve a linear system that looks like an elliptic problem. However, due to the discretization of the time derivative, there is an additional term (typically $h^2/\Delta t$ where Δt is the time step) added to the diagonal of the matrix giving better properties (for instance more diagonal dominance). Three dimensional problems also give matrices with the same pattern as for two dimensional ones but then, the matrices D_i are themselves block tridiagonal corresponding to a two dimensional problem.

Convection–diffusion problems are an interesting source of non–symmetric linear systems. A typical problem is

$$-\epsilon\Delta u + 2P_x\frac{\partial u}{\partial x} + 2P_y\frac{\partial u}{\partial y} = f \quad \text{in} \quad \Omega,$$
$$u\big|_{\partial\Omega} = 0.$$

First order derivatives are approximated with centered or upwind finite differences. Test problem 6 is defined by $\epsilon = 1, P_x = P_y = 50$. Test problem 7 uses $\epsilon = 1, P_x = e^{2(x^2+y^2)}, P_y = 0$. Both problems are discretized with an upwind scheme.

Dealing with sparse matrices, our aim is to be able to avoid storing the zero entries of the sparse matrix A and to avoid doing operations on these zeros. Finite differences matrices can be conveniently stored in the computer memory by diagonals (*i.e.*, a diagonal is stored in a one dimensional array). At least this is true if one wants to use an iterative method to solve the linear system where only matrix–vector products are needed.

Another way to solve elliptic and parabolic partial differential equations is to use the finite element method. Actually, there is a variety of different finite element methods. We now give the flavor of the simplest one. Let Ω be a two dimensional domain and suppose we want to solve again the Poisson model problem:

$$-\Delta u = f \quad \text{in} \quad \Omega,$$
$$u\big|_{\partial\Omega} = 0,$$

with f in $L^2(\Omega)$.

It is well known (using the Green formulae) that u is also the solution of the so called variational problem

$$a(u, v) = (f, v) \quad \text{for all} \quad v \in H_0^1(\Omega),$$

where

$$H_0^1(\Omega) = \left\{ u | u \in L^2(\Omega), \ \frac{\partial u}{\partial x_i} \in L^2, \ u\big|_{\partial\Omega} = 0 \right\},$$

and $a(u,v)$ is a given bilinear form

$$a(u,v) = \int_\Omega \nabla u \nabla v \, dx, \quad (f,v) = \int_\Omega f v \, dx.$$

An idea to approximate the solution u is to define u_h as the solution of

$$a(u_h, v_h) = (f, v_h) \quad \forall v_h \in V_h,$$

where $V_h \subset H_0^1(\Omega)$ is a finite dimensional subspace. For example, one is given a triangularization of Ω (Ω is supposed to be a polygon) as in figure 1.5.

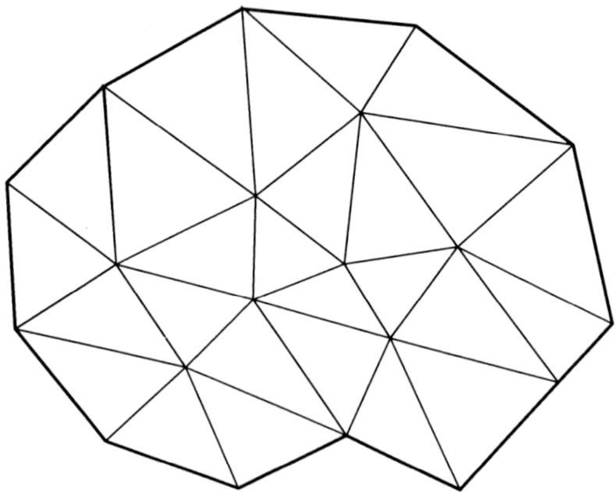

fig. 1.5: A finite element mesh in $\mathbb{R}^2$

In the simplest method V_h is the set of continuous and piecewise polynomials of degree one on each triangle. Hence, the unknowns can be taken as the values of u_h at each vertex. A basis for V_h is given by the functions $w_i \in V_h$ whose value is 1 at node i and 0 outside the triangles to which i is a vertex of. Then, the discrete problem is

$$a(u_h, w_i) = (f, w_i) \quad \text{for all basis functions } w_i.$$

But u_h can be decomposed on the basis functions as, $u_h = \sum_j u_h(i) w_j$. So, we get a linear system whose unknowns are $u_h(i)$ with a matrix A for which $a_{i,j} = a(w_i, w_j)$. The structure of the matrix depends on the numbering scheme which is used for the mesh nodes. Most often these matrices cannot be stored by diagonals. The storage mode will depend on the method that is used to solve the system (direct or iterative) and also on the type of computer that is used (scalar, vector or parallel).

The most natural way is to store only the non–zero entries $a_{i,j}$ of A, together with the row and column indices i and j. Therefore, if nz is the number of non–zeros of A, the storage needed is nz floating point numbers and 2 nz integers. In most modern computers, integers use the same number of bits as floating point numbers. In that case, the total storage is 3 nz words. However, this mode of storage (which is sometimes called the coordinate scheme) is not very convenient for direct methods as they usually require easy access to rows and/or columns of the matrix.

One common way to store a sparse matrix which is more suited to direct methods is to hold the non–zeros of each row (resp. column) as a packed sparse vector `AA`, together with the column (resp. row) index of each element in a vector `JA`. These two vectors have a length of nz words. A third vector `IA` of integers of length $n+1$ is needed for pointing to the beginning of each row in `AA` and `JA`. Let us look at this storage scheme on a small example.

Let

$$A = \begin{pmatrix} a_1 & 0 & 0 & a_2 \\ a_3 & a_4 & a_5 & 0 \\ 0 & a_6 & a_7 & 0 \\ a_8 & 0 & 0 & a_9 \end{pmatrix},$$

the stored quantities are

	1	2	3	4	5	6	7	8	9
AA	a_1	a_2	a_3	a_4	a_5	a_6	a_7	a_8	a_9
JA	1	4	1	2	3	2	3	1	4
IA	1	3	6	8	10				

Note that `IA(n+1)=nz+1`. This allows us to compute the length of row i by `IA(i+1)-IA(i)`. Equivalently, the lengths of rows can be stored in place of `IA`. Sometimes the diagonal elements are treated in a special way. As they are often all non–zero, they are stored in an additional vector of length n or the diagonal entry of each row could be stored in the first (or last) position of the row. If the matrix is symmetric only the lower or upper part is stored.

Another storage scheme which is much used is linked lists. Here, each non–zero element (together with the column index) has a pointer `IPA` to the location of the next element in the row. To be able to add or delete entries easily in the list, it is sometimes handy to have a second pointer to the location of the previous entry. Finally, we must know the beginning of the list for each row in a vector `IA`. Going back to our small example, we have the following storage when using only forward links. Note that the elements could be stored in any order,

	1	2	3	4	5	6	7	8	9
AA	a_3	a_2	a_9	a_1	a_4	a_8	a_6	a_5	a_7
JA	1	4	4	1	2	1	2	3	3
IPA	5	0	0	2	8	3	9	0	0
IA	4	1	7	6					

A zero value in `IPA(.)` indicates the end of the list for the given row.

Later on, we shall consider other schemes for storing sparse matrices that are more useful for some particular algorithms. Other storage schemes exist that are more suited to iterative methods, especially those which need only matrix vector products. Usually one tries to obtain something which is close to the diagonal storage scheme for finite difference matrices. Algorithms were devised to construct pseudo diagonals, see Melhem [963], Erhel [534]. One way is to collect the non–zero elements in vectors. The non–zero elements are flagged with 0 or 1. Initially, they are all 0. A pseudo diagonal is formed by collecting successively for each row the leftmost non–zero element whose flag is 0. When collected, its flag is set to 1. We also need to store the column index (and eventually the row index).

1.10. Eigenvalues and Fourier analysis

For the finite difference model problem (Poisson equation with Dirichlet boundary conditions) we can explicitly compute the eigenvalues of A. We start by considering the problem arising from a one dimensional periodic Poisson model problem on $[0, 1]$, namely the Poisson equation with periodic boundary conditions.

The matrix of interest is

$$A_P = \begin{pmatrix} 2 & -1 & & & -1 \\ -1 & 2 & -1 & & \\ & \ddots & \ddots & \ddots & \\ & & -1 & 2 & -1 \\ -1 & & & -1 & 2 \end{pmatrix},$$

but to compute the eigenvalues, we can restrict ourselves to

$$T_P = \begin{pmatrix} 0 & 1 & & & 1 \\ 1 & 0 & 1 & & \\ & \ddots & \ddots & \ddots & \\ & & 1 & 0 & 1 \\ 1 & & & 1 & 0 \end{pmatrix},$$

as $\lambda(A_P) = 2 - \lambda(T_P)$, T_P being of order $n+1$ (this corresponds to $h = 1/(n+1)$) and the mesh points we consider are $h, 2h, \ldots, (n+1)h$. We are interested in finding the eigenvalues λ and eigenvectors u of T_P. Looking at a generic row of T_P, we guess that

$$u_k = ae^{i(k-1)\theta} + be^{-i(k-1)\theta},$$

the kth component of the eigenvector, where $i = \sqrt{-1}$. It is easy to see, this gives directly

$$\lambda = e^{i\theta} + e^{-i\theta} = 2\cos(\theta).$$

To find θ and to satisfy the first and last equations, we look at the boundary conditions. For $j = 1$, we have

$$a(e^{in\theta} - e^{-i\theta}) + b(e^{-in\theta} - e^{i\theta}) = 0,$$

and for $j = n+1$,

$$a(1 - e^{i(n+1)\theta}) + b(1 - e^{-i(n+1)\theta}) = 0.$$

To have a non–trivial solution a, b we ask for

$$\begin{vmatrix} e^{in\theta} - e^{-i\theta} & e^{-in\theta} - e^{i\theta} \\ 1 - e^{i(n+1)\theta} & 1 - e^{-i(n+1)\theta} \end{vmatrix} = 0.$$

This gives

$$2\sin(\theta) + \sin(n\theta) - \sin((n+2)\theta) = 0.$$

It is easily verified that a solution is

$$\theta_j = \frac{2\pi j}{n+1}, \quad j = 0, \ldots, n.$$

Thus, we have the eigenvalues of T_P,

$$\lambda_j = 2\cos\left(\frac{2\pi j}{n+1}\right), \quad j = 0, \ldots, n.$$

Note $\lambda = 2$ is always a simple eigenvalue (giving the 0 eigenvalue of A). Now, suppose that n is odd ($n+1 = 2p$ even), then $\lambda = -2$ is a simple eigenvalue (corresponding to $j = p$), as well as 2. Note that if λ is an eigenvalue, then $-\lambda$

is also an eigenvalue. If p is even, 0 is a simple eigenvalue. The eigenvector corresponding to $\lambda = 2$ is clearly

$$u_0 = e^T = (1, 1, \ldots, 1)^T,$$

and the eigenvector for $\lambda = -2$ is

$$u_p = (1, -1, \ldots, 1, -1)^T.$$

For the double eigenvalues, we must select two independent vectors that span the eigenspace. We can take $a = 1$ and $b = 0$ for the first eigenvector and $a = 0$ and $b = 1$ for the second. But, we can also choose, for instance, $a = e^{\theta_j}/2, b = e^{-\theta_j}/2$ for the first eigenvector and $a = e^{\theta_j}/2\boldsymbol{i}, b = -e^{-\theta_j}/2\boldsymbol{i}$ for the second. This gives

$$\cos(k\frac{\pi j}{p}), \qquad \sin(k\frac{\pi j}{p}),$$

for the components of the eigenvectors. Note that when $h \to 0$ (or $n \to \infty$), then the components of the eigenvectors converge to values of the functions $\cos(2\pi jx)$ and $\sin(2\pi jx)$. The eigenvalues of $\frac{1}{h^2}A_P$ are

$$\frac{2}{h^2}\left(1 - \cos(2\pi jh)\right), \quad j = 0, \ldots, n.$$

For j fixed, when $h \to 0$ this eigenvalue is equal to $4\pi^2 j^2 + O(h^2)$. The values $4\pi^2 j^2$ are eigenvalues of the continuous problem. For $j = 0$, we have the constant eigenfunction.

To understand the correspondence between the periodic and the Dirichlet boundary conditions, it is important to note that the sine eigenvector has a component $u_p = 0$ corresponding to abscissa $x = 1/2$ (because $\sin(p\pi j/p) = \sin(\pi j) = 0$). Moreover $u_{n+1} = 0$, hence this eigenvector satisfies the Dirichlet boundary conditions on $]0, 1/2[$ and is also an eigenvector of the Dirichlet problem of size $p - 1$, with matrix

$$T_D = \begin{pmatrix} 0 & 1 & & & \\ 1 & 0 & 1 & & \\ & \ddots & \ddots & \ddots & \\ & & 1 & 0 & 1 \\ & & & 1 & 0 \end{pmatrix}.$$

Thus for the Dirichlet problem, we have the (simple) eigenvalues coming from the $p - 1$ double eigenvalues of the periodic problem as they are the only ones that can be associated with a sine eigenfunction. In conclusion, when n is odd, the $p-1$ double eigenvalues of the periodic problem and (the restriction) of the sine eigenvectors are eigenvalues and eigenvectors of the Dirichlet problem of size $p-1$. This can be considered as a Dirichlet problem of size $p-1$ on $[0,1]$ with a mesh size $h_d = 2h$. When $h \to 0$, we have the same conclusion as for the periodic boundary conditions, except there is no constant eigenfunction.

Now, we consider the two dimensional periodic problem for which we have the matrix:

$$T_2 = \begin{pmatrix} T_P & I & & & I \\ I & T_P & I & & \\ & \ddots & \ddots & \ddots & \\ & & I & T_P & I \\ I & & & I & T_P \end{pmatrix},$$

the matrix A_P of the problem is $A_P = 4I - T_2$, T_2 being of order $(n+1)^2$ and T_P as previously defined for the one dimensional problem.

As the matrix problem arises from a separable elliptic equation, the eigenvectors for T_2 are tensor products of the one dimensional eigenvectors. If the mesh points are labeled by integer couples (k,l) and if u and v are two eigenvectors of T_P corresponding to λ_j and λ_m, then the vector U defined, for example, as

$$U_{k,l} = u_k v_l = \sin\left(\frac{k\pi j}{p}\right) \sin\left(\frac{l\pi m}{p}\right),$$

is an eigenvector of T_2. We have to consider the tensor products of sines and cosines. U can also be written blockwise as

$$U^T = (uv_1, uv_2, \ldots, uv_{n+1}),$$

the blocks corresponding to mesh lines. Looking at one of the blocks of T_2U, we have

$$(v_{l-1} + v_{l+1})u + T_P u v_l$$

which is, as $T_P u = \lambda_j u$, equal to

$$(v_{l-1} + v_{l+1} + \lambda_j v_l)u = (\lambda_j + \lambda_m) v_l u,$$

as v is also an eigenvector of T_P with eigenvalue λ_m. Consequently, the eigenvalues of the two dimensional periodic problem are the (pairwise) sums of two of the eigenvalues of the one dimensional periodic problem.

As an example, consider $n = 7$, then the one dimensional periodic problem has the following eigenvalues with multiplicity ( ):

$$-2(1),\ -1.4142(2),\ 0(2),\ 1.4142(2),\ 2(1).$$

We combine these eigenvalues 2 by 2 to get (note that from λ and $-\lambda$ we get additional 0 eigenvalues) for the two dimensional periodic problem T_2:

$$-4(1),\ -3.4142(4),\ -2.8284(4),\ -2(4),\ -1.4142(8),\ -0.5858(4),\ 0(14),$$

$$0.5858(4),\ 1.4142(8),\ 2(4),\ 2.8284(4),\ 3.4142(4),\ 4(1).$$

The multiplicity of 0 is 14 as we combine 0 with itself (twice because of the multiplicity) which gives 4 eigenvalues, 1.4142 with -1.4142 and -1.4142 with 1.4142 which gives 8 eigenvalues. Finally, we combine 2 with -2 and -2 with 2 which gives 2 eigenvalues for a total of $4 + 8 + 2 = 14$. The multiplicity of the other eigenvalues can be computed in the same way.

The eigenvalues of A_P are:

$$0(1),\ 0.5858(4),\ 1.1716(4),\ 2(4),\ 2.5858(8),\ 3.4142(4),\ 4(14),$$

$$4.5858(4),\ 5.4142(8),\ 6(4),\ 6.8284(4),\ 7.4142(4),\ 8(1).$$

The eigenvectors are tensor products of the one dimensional eigenvectors. In the eigenspaces, there are eigenvectors of the form sine $\times$ sine, except for the 2 and -2 eigenvalues. These sine eigenvectors satisfy the Dirichlet boundary conditions on $[0, 1/2] \times [0, 1/2]$. Therefore as in the one dimensional periodic problem, we can pick only the eigenvalues with sine eigenvectors counting them only once and we combine them. In the example, we obtain for the matrix T_D of the two dimensional Dirichlet problem

$$-2.8284(1),\ -1.4142(2),\ 0(3),\ 1.4142(2),\ 2.8284(1).$$

Note that the multiplicity of the eigenvalues is smaller for Dirichlet boundary conditions than for periodic boundary conditions. We have the same conclusion as for the 1D problem: some of the eigenvalues of the periodic problem of size $(n+1)^2$ are eigenvalues of the Dirichlet problem of size $(p-1)^2$. We must exclude those eigenvectors with no sine function.

Note that the eigenspaces can also be generated by exponential functions. The invariant space for an eigenvalue of multiplicity 4 of the periodic problem can be generated by vectors whose components are

$$e^{ik\theta_j} e^{il\theta_m} \qquad e^{ik\theta_j} e^{-il\theta_m} \qquad e^{-ik\theta_j} e^{il\theta_m} \qquad e^{-ik\theta_j} e^{-il\theta_m}.$$

This space can also be generated by vectors whose components are

$$\sin(k\theta_j)\sin(l\theta_m) \quad \sin(k\theta_j)\cos(l\theta_m) \quad \cos(k\theta_j)\sin(l\theta_m) \quad \cos(k\theta_j)\cos(l\theta_m).$$

When $h \to 0$, the conclusions are the same as for the 1D problem. For the mode (j, m) the eigenvalue of $\frac{1}{h^2}A_P$ for the periodic problem is

$$\lambda = \frac{1}{h^2}\left(4 - 2\cos(2\pi jh) - 2\cos(2\pi mh)\right), \quad j, m = 0, \ldots, n.$$

When h is small, this is $4\pi^2(j^2 + m^2) + O(h^2)$, $\quad j, m = 0, \ldots$ The eigenfunctions are functions such as $\sin(2\pi jx)\cos(2\pi my)$, etc...

For the Dirichlet problem, the eigenvalues are

$$\lambda_{p,q} = 4 - 2\cos(p\pi h) - 2\cos(q\pi h), \quad h = \frac{1}{m+1}$$

The couple (p, q) refers to the mode

$$1 \le p \le m, \quad 1 \le q \le m.$$

The components of the corresponding (unnormalized) eigenvector are

$$u_{i,j}^{p,q} = \sin(ip\pi h)\sin(jq\pi h)$$

where (i, j) refers to a point in the mesh. Figure 1.6 shows the eigenvalues of the model problem with Dirichlet boundary conditions for $m = 10$.

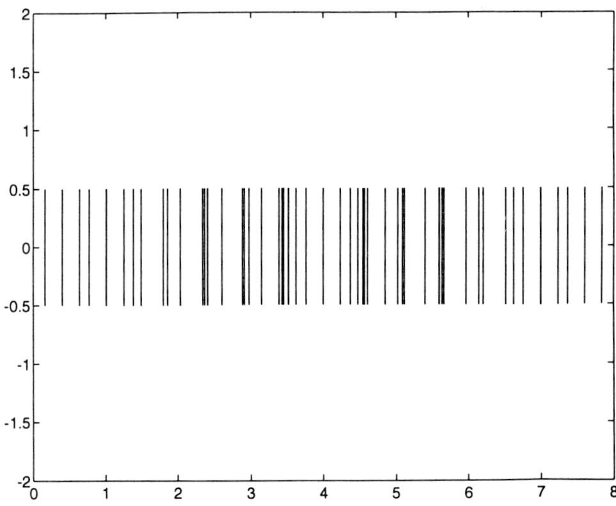

fig. 1.6: The eigenvalues of the model problem, $n = 100$

We remark that the eigenvector associated with the smallest eigenvalue as seen as the discretization of a function on the given mesh is the smoothest one. This is why it is referred as "low frequency". The eigenvector associated with the largest eigenvalue is very oscillatory. This is referred as "high frequency". See figure 1.7 which shows normalized eigenvectors.

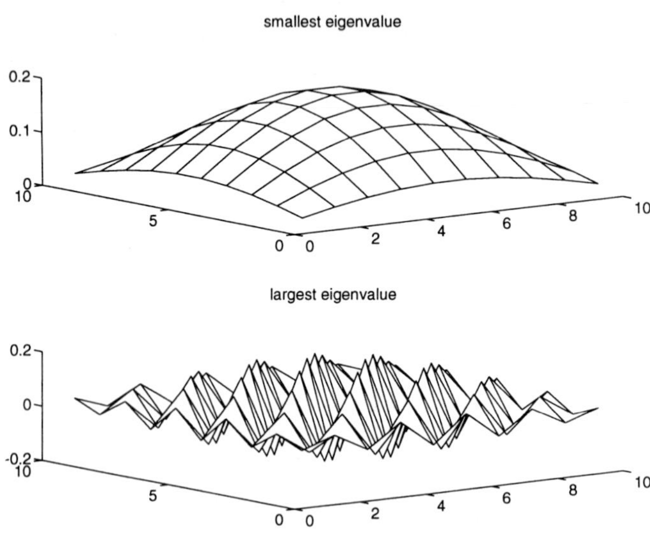

fig. 1.7: The eigenvectors of the model problem, $n = 100$ for λ_{min} and λ_{max}

Fourier analysis has been used quite extensively in the past to analyze the performance of discrete numerical methods. The main idea is to compute the effect of a discrete difference operator on the Fourier modes. If the difference operator is periodic and has constant coefficients, then its eigenfunctions are often precisely these Fourier modes (as we have just seen above for the Laplacian operator) and its symbol can be easily computed and analyzed algebraically. Examples are the classical Von Neumann stability analysis for difference schemes for time dependent PDEs, see Richtmyer and Morton [1101].

Rather surprisingly, the use of Fourier analysis for the study of iterative methods for elliptic problems has not been widespread. This is because many matrices that we have to study for non–periodic problems do not have constant coefficients, even if the original problem does. However, as is well known, the coefficients of the matrices involved in iterative methods (either splitting methods or preconditioners) often approach constant values in the interior of the domain away from the boundaries. These asymptotic values can often be easily computed by considering the corresponding periodic problem. It is thus natural to expect the Fourier analysis to produce interesting results for these problems as well.

We now illustrate the Fourier technique by applying it to several basic difference operators which we shall employ later. Consider the unit square in two dimensions and a uniform grid with m interior grid points in both the x and y directions. The Fourier modes on this grid are grid functions $u^{(s,t)}$ extended periodically to the whole space whose (j,k)th component is given by:

$$(u^{(s,t)})_{j,k} = e^{ik\theta_s} e^{ij\phi_t}$$

where $\theta_s = \frac{2\pi s}{m+1}$ and $\phi_t = \frac{2\pi t}{m+1}$.

Consider as an example the difference operator $T_1(\alpha,\beta)$ (a second order finite difference operator in one direction) defined by:

$$(T_1(\alpha,\beta)v)_{j,k} = -\beta v_{j,k-1} + \alpha v_{j,k} - \beta v_{j,k+1}.$$

The Fourier transform $F(T_1)$ (i.e. its symbol) can be computed by applying T_1 to the Fourier mode $u^{(s,t)}$ given earlier and we get:

$$T_1 u^{(s,t)} = \beta e^{ij\phi_t}(e^{i(k-1)\theta_s} + e^{i(k+1)\theta_s}) + \alpha e^{ik\theta_s} e^{ij\phi_t},$$

which after some simplifications yields:

$$T_1 u^{(s,t)} = ((\alpha - 2\beta) + 4\beta \sin^2(\theta_s/2))u^{(s,t)}.$$

Thus,

$$F(T_1(\alpha,\beta)) = (\alpha - 2\beta) + 4\beta \sin^2(\theta_s/2).$$

Similarly, the Fourier transform of the following operators (which will be useful for studying preconditioners):

$$(T_2(\alpha,\beta,\gamma)v)_{j,k} = -\beta v_{j,k-1} + \alpha v_{j,k} - \beta v_{j,k+1} - \gamma v_{j+1,k},$$

$$(T_3(\alpha,\beta,\gamma)v)_{j,k} = -\beta v_{j,k-1} + \alpha v_{j,k} - \beta v_{j,k+1} - \gamma v_{j-1,k},$$

are given by:

$$F(T_2(\alpha,\beta,\gamma)) = (\alpha - 2\beta) + 4\beta \sin^2(\theta_s/2) - \gamma e^{i\phi_t}.$$

$$F(T_3(\alpha,\beta,\gamma)) = (\alpha - 2\beta) + 4\beta \sin^2(\theta_s/2) - \gamma e^{-i\phi_t}.$$

Finally, the Fourier transform of the standard 5-point finite difference discrete Laplacian A_P in two dimensions is given by:

$$F(A_P) = 4(\sin^2(\theta_s/2) + \sin^2(\phi_t/2)).$$

We also note the Fourier transform of a product (composition) of operators is the product of the transforms of the individual factors and the transform of the inverse of an operator is the inverse of the transform.

1.11. Floating point arithmetic

The coefficients and solution of linear systems are real or complex numbers. Unfortunately, for solving these systems we use digital computers made of a finite number of parts and all real (or complex) numbers cannot be exactly represented. In particular, registers and memory words designed to store the data and the intermediate and final results have a finite length or capacity and cannot store all real numbers. Moreover, when computations are performed on a computer, each arithmetic operation $(+, -, *, /)$ is generally affected by roundoff errors as only a finite number of digits (or bits) can be retained.

The subject of roundoff error analysis is to try to understand the effects of these limitations on the result of solving a problem. Before going into these problems in subsequent chapters, we must define the floating point arithmetic model we are using. Here, we follow the expositions of Forsythe and Moler [572] and Golub and Van Loan [676]. The numbers that can be represented in the computer are a (finite) subset of the real numbers and are denoted by F. This set is characterized by four integers: the base β, the number t of base–β digits in the fractional part (also called the mantissa) and the exponent range $[e_L, e_U]$. Then, (normalized) numbers in F consists of all real numbers f of the form

$$f = \pm .d_1 d_2 \cdots d_t \; \beta^e, \quad 0 \le d_i < \beta, \; d_1 \ne 0, \; e_L \le e \le e_U,$$

where e is an integer, to which we add zero and a representation for results whose absolute value will be smaller (resp. larger) than the smallest (resp. largest) absolute value of a non–zero number in F. For a non–zero $f \in F$, we have

$$m = \beta^{e_L - 1} \le |f| \le M = \beta^{e_U}(1 - \beta^{-t}).$$

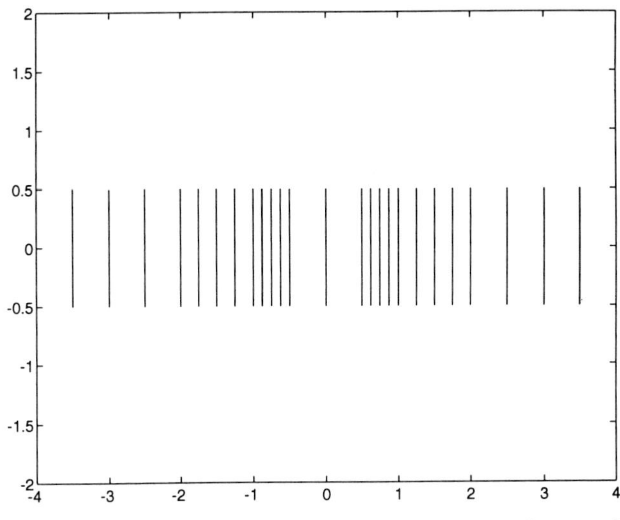

fig. 1.8: Elements of F for $\beta = 2, t = 3, e_L = 0, e_U = 2$

On the real line, the elements of F are not equally spaced (see figure 1.8 that shows the elements of F for $\beta = 2, t = 3, e_L = 0, e_U = 2$).

A real number x that we wish to represent is approximated by a number $fl(x)$ that can be defined as an operator from

$$G = \{x \in R, m \leq |x| \leq M\} \cup \{0\},$$

into F, by

$$fl(x) = \begin{cases} \text{nearest } x_{\mathrm{R}} \in F \text{ to } x \text{ if rounded arithmetic is used,} \\ \text{nearest } x_{\mathrm{C}} \in F \text{ s. t. } |x_{\mathrm{C}}| \leq |x| \text{ if chopped arithmetic is used.} \end{cases}$$

Today, most computers use $\beta = 2$, although in the past some computers used $\beta = 8$ or 16. Typical values of the other parameters for some (1997) computers are given in the following table.

Examples of floating point formats

	t	e_L	e_U
CRAY single	48	-8192	8191
IEEE single	24	-125	128
IEEE double	53	-1021	1024

For more details on floating point arithmetic particularly IEEE formats, see Golberg [652]. The following results are proved in Forsythe and Moler [572].

Theorem 1.65

If $x \in G$, then

$$x_{\mathrm{R}} = x(1 + \epsilon_{\mathrm{R}}), \ |\epsilon_{\mathrm{R}}| \le \frac{1}{2}\beta^{1-t},$$
$$x_{\mathrm{C}} = x(1 + \epsilon_{\mathrm{C}}), \ |\epsilon_{\mathrm{C}}| \le \beta^{1-t}.$$

□

Next, we have to define the operations on elements of F. Let $\bullet$ be one of the four operations $+, -, *, /$ and $\odot$ its implementation on a computer. We say that a computer arithmetic is correct if

$$x, y \in F, \ fl(x \bullet y) = x \odot y.$$

Not all arithmetics are correct. The one used on CRAYs Y–MP and C90 was not correct. The IEEE norm defines a correct arithmetic. We define the unit round off u as,

$$u = \begin{cases} u_{\mathrm{R}} = \frac{1}{2}\beta^{1-t} & \text{for rounded arithmetic,} \\ u_{\mathrm{C}} = \beta^{1-t} & \text{for chopped arithmetic.} \end{cases}$$

For IEEE single precision arithmetic (32 bits), the unit roundoff (rounded arithmetic) is $u_{\mathrm{R}} = 5.9605 \ 10^{-8}$ and $u_{\mathrm{R}} = 1.1102 \ 10^{-16}$ for double precision (64 bits). We have,

$$fl(x) = x(1 + \epsilon), \ |\epsilon| \le u.$$

We have the following result,

$$fl(x \bullet y) = (x \bullet y)(1 + \epsilon), \ |\epsilon| \le u, \tag{1.2}$$

and then

$$\frac{|fl(x \bullet y) - (x \bullet y)|}{|x \bullet y|} \le u,$$

when $|x \bullet y| \neq 0$. This shows there is a small relative error associated with each operation (provided t is large enough). Note that (1.2) was not verified by CRAY arithmetic because of the lack of guard digits. Relaxed assumptions need to be made. Fortunately, most of the following results carry over to this case and CRAY switched to IEEE arithmetic. From this last result we can construct some other error bounds. We use the following lemma from Forsythe and Moler [572].

Lemma 1.66

If $0 \leq u < 1$ *and* $n \in N$,

$$1 - nu \leq (1-u)^n.$$

If $0 \leq nu \leq 0.01$,

$$(1+u)^n \leq 1 + 1.01nu.$$

Proof. The first statement follows from Taylor's formula and the second one from

$$e^x \leq 1 + 1.01x \text{ for } 0 \leq x \leq 0.01$$

□

Note that the largest n that satisfies the inequality $0 \leq nu \leq 0.01$ is $n \approx 167772$ for IEEE single precision arithmetic and $n \approx 9\ 10^{13}$ for double precision.

Theorem 1.67

Let $w = x - yz$ *and* $e = fl(w) - w$. *If* $nu \leq 0.01$, *then*

$$|e| \leq 3.02\ u \max(|x|, |w|).$$

Proof. $fl(w)$ is defined as $fl(x - fl(yz))$. But, $fl(yz) = yz(1+\epsilon_1)$, $|\epsilon_1| \leq u$ and

$$\begin{aligned} fl(w) &= (x - yz(1+\epsilon_1))(1+\epsilon_2), \ |\epsilon_2| \leq u, \\ &= x(1+\epsilon_2) - yz(1+\epsilon_1)(1+\epsilon_2). \end{aligned}$$

This can be written with the help of lemma (1.66),

$$\begin{aligned} fl(w) &= x(1+\theta_2 u) - yz(1+2.02\,\theta_1 u), \ |\theta_i| < 1, \\ &= x - yz + u(x\theta_1 - 2.02\,yz\theta_1). \end{aligned}$$

Therefore,

$$|e| \leq u(|x| + 2.02\,|w - x|),$$

from which the result follows. □

The last result can also be written as

$$|e| \leq u(2.02|yz| + |x|).$$

We have also

$$|e| \leq u(2|yz| + |x|) + O(u^2).$$

The values of the constants involved in these inequalities such as 2.02 or 3.02 are not really important. Therefore, in most statements of the results, we shall replace them by a generic constant C or C_i. However, for the sake of completeness, we shall indicate their possible values.

Now, we shall be able to analyze the effect of roundoff errors in algorithms. There are several different ways to approach this problem. The one that is more popular today is called inverse or backward error analysis. It was introduced by W. Givens and developed, in particular for Gaussian elimination, by J. Wilkinson [1333]. In this type of analysis the roundoff errors are related to the data. The result of floating point operations is interpreted as the exact result of ordinary arithmetic on some perturbed data and some bounds are derived for the perturbations. This allows us to be able to use ordinary arithmetic.

1.12. Vector and parallel computers

Here, we make a few remarks about computers and algorithm complexity. From the first few years after World War II with the advent of the first digital computers with stored programs, there has been tremendous progress in the floating point performance of scientific computers.

In the past, people have considered the number of floating point operations involved in their algorithms as a measure of efficiency (most often the integer operations associated with address computations and loop overhead were neglected). This was supposed to give a good account of what the computer time used when running the codes. A nice feature was that this operation count was computer independent. But we cannot use the same hypothesis any more on vector and parallel computers. To understand this issue, let us briefly describe some of these machines.

At the beginning of the seventies, top scientific computers like the IBM 360–91 were running at an average of 1–2 Mflops (Millions of floating point operations per second). Roughly, one can say that a traditional computer executes basic operations in a serial way: it waits for the completion of an operation before beginning the execution of the next one. For this kind of computer, that is to say the ones we used in the fifties, sixties and the beginning of the seventies, the operation count gives a good idea of what is going on.

This was already no longer true for some computers of the seventies, for example, the CDC 7600. The 7600 was a pipelined computer. This means that a basic operation like addition or multiplication is divided into a certain number of steps, say n. When one pair of operands has moved from step 1 to step 2, step 1 is free to accept a new pair of operands. Of course, the second pair must be independent of the results of the first operation. If we (or the compiler) are able to continuously feed the pipeline with independent operations then the first result will take n cycles to be produced but after that we shall get a new result each cycle. For example, if $n = 6$ we shall produce 6 results in 12 cycles instead of 36 cycles on a serial computer. The problem for these computers was that at that time, the available compilers were not smart enough to automatically recognize that they had to do consecutive independent operations. So, there was only a slight improvement over serial computers.

This was changed at the end of the seventies with the advent of vector computers, the first one commercially available being the CRAY-1. Among other features, the CRAY-1 had 8 vector registers of 64 words each. Roughly speaking, the compiler was able to recognize that we were doing, for example, the addition of two one–dimensional arrays, *i.e.*, two vectors. It was then able by using vector instructions to feed the addition pipeline efficiently by fetching the operands from the memory into the vector registers. Moreover, this machine had independent functional units, that is to say for example one for additions and one for multiplications, that can operate simultaneously (of course on non–related data). For related operations there was a process called chaining. If x and y are vectors and we are doing, for instance, $x + \alpha y$, (α being a scalar), as soon as the first element of αy was available it was sent to the addition functional unit together with the first element of x. Doing this, the two operations were overlapped.

It is clear that with this kind of computer, the classical operation count means nothing. On typical codes, there was a ratio of 10 to 40 between scalar and vectorized codes. To run fast, an algorithm had to be expressed (if possible) in terms of vectors. Moreover, the longer the vector length, the faster the speed of execution (although eventually the peak speed is reached). The memory traffic issue is also very important in order to get good performance. Usually, the fastest vector computers are the ones with the largest memory bandwidth and the speed of programs depends also on the ratio of the number of floating point operations to the number of memory references. Despite all of their impressive features, these machines were far from reaching the speeds we need for very large scale scientific computations such as aerodynamic calculations or weather forecasting.

The next step in getting more computer power was to use multiprocessor computers. Vector computers have evolved into machines with several vector processors sharing a common large memory. Examples of these machines were the CRAY Y–MPs and C90s as well as computers from Japanese manufacturers like Fujitsu, NEC or Hitachi. Most of these machines were running (in the mid–nineties) at Gigaflops speeds on well designed programs.

Parallelism is in fact an old idea but only recently have very efficient multiprocessors been developed. At the end of the eighties and beginning of the nineties a new kind of scientific computers appeared on the market: parallel computers with a distributed memory. Almost all these machines use off–the–shelf microprocessors. They have from a few tens to a few thousand processors. Even though the memory is physically distributed, some of these machines use a global addressing system either at the hardware level or by using software devices. Teraflops speeds are reached on some applications (1997). Of course to be used efficiently, these computers need algorithms that contain enough parallelism and suitable programming models to be able to express the parallelism. It is likely that by the end of this century and the beginning of the next, parallel architectures will be clusters of shared memory parallel computers. It may be that the memory will be virtually sharable even though only a part of it is physically shared. This also raises many problems of cache coherency. By 1996 there were some studies which considered building Petaflops (10^{15} floating point operations per second) within the twenty first years of the 21st century.

Sometimes only a few modifications have to be made to classical algorithms to use them on parallel computers. But, most of the time, completely new algorithms have to be derived. Finding efficient and robust parallel algorithms is a serious challenge for numerical analysts. An important issue that has to be considered is scalability. The most powerful scientific computers are designed to solve large problems. The size of the problems that scientists and engineers want to solve is (and probably will be) always too large for the available computers. Nevertheless, as the size of the problem grows we would like, if we increase proportionally the number of processors, to keep the (elapsed) computer time constant. To obtain this, we need to have algorithms whose (serial) complexity is proportional to the number of unknowns. Unfortunately, most of the well known algorithms are not scalable.

1.13. BLAS and LAPACK

Software reusability and portability are important issues. In the seventies, Lawson *et al* [882] described a set of basic routines commonly known as the BLAS (Basic Linear Algebra Subprograms) which defined the most used kernels in linear algebra. These routines can be implemented efficiently by each manufacturer, possibly in assembly language, allowing both portability through different computers and performance.

The set of operations described in 1979 is now referred as Level 1 BLAS or BLAS1. These routines are mainly concerned with vector operations like

$$y = \alpha x + y, \quad \text{(Saxpy)}$$
$$\alpha = x^T y, \quad \text{(Sdot)}$$
$$x = \alpha x, \quad \text{(Sscal)}$$
$$x = y, \quad \text{(Scopy)}$$
$$\alpha = \|x\|_2, \quad \text{(Snrm2)}$$

where x and y are vectors and α is a scalar, the S standing for single precision. For a complete list, see Anderson *et al* [26]. The well known linear algebra package LINPACK (for linear systems solutions and least square problems) was written using BLAS1 and published in 1979 [232]. BLAS1 involves $O(n)$ floating point operations on $O(n)$ data items, n being the length of the vectors.

At the time where the BLAS1 was defined and LINPACK was completed, vector computers appeared on the market. One might think that BLAS1 would be very efficient on vector computers since it precisely defines vector operations. But unfortunately this was not the case and the BLAS1 and LINPACK performed poorly on early vector machines. The reason was that the value of the ratio of floating point operations to data loads and stores was too low to keep the processor busy all the time. The performance was lower (sometimes by a large amount) than the peak theoretical speed. Moreover, in BLAS1, there are very few possibilities of data reuse in vector registers or memory caches.

An additional set of routines called the Level 2 BLAS (BLAS2) was then designed, (Dongarra *et al* [430]) to overcome these problems and published in 1988. They are based on matrix×vector operations. Examples are

$$
\begin{aligned}
y &= \alpha A x + \beta y, \\
x &= T x, \\
A &= \alpha x y^{\mathrm{T}} + A, \\
x &= T^{-1} x,
\end{aligned}
$$

where α, β are scalars, x, y are vectors, A and T are matrices, T being triangular. Most algorithms of linear algebra operating on dense matrices can be coded using Level 2 BLAS including Gaussian elimination. Level 2 BLAS involves $O(n^2)$ floating point operations on $O(n^2)$ data items. Therefore, the ratio of Level 1 BLAS is not improved. But data locality is better than with BLAS1.

To improve further on this point and to increase the data locality, a level 3 BLAS (BLAS3) was proposed in 1990, see Dongarra *et al* [429], that defines matrix×matrix operations. Typical examples are

$$
\begin{aligned}
C &= \alpha AB + \beta C, \\
C &= \alpha AA^{\mathrm{T}} + \beta C, \\
C &= \alpha AB^{\mathrm{T}} + \alpha BA^{\mathrm{T}} + \beta C, \\
B &= \alpha TB, \\
B &= \alpha T^{-1} B,
\end{aligned}
$$

where α, β are scalars, A, B, C, T are matrices, T being triangular. Now we have $O(n^3)$ floating point operations on $O(n^2)$ data items helping to improve the data reuse. BLAS3 performs very well on vector supercomputers and computers with a memory hierarchy. Performance close to the peak speed is frequently obtained.

At the end of the eighties, at the same time as the BLAS2 and BLAS3 appeared, a new software project was developed. Its goal was to supersede both LINPACK and EISPACK (a well known package for eigenvalues computations) and also to obtain better performances. The computers targeted were parallel vector supercomputers with shared memory. Another (important) goal was to improve the quality and accuracy of the algorithms, particularly for eigenvalue computations. The first version of LAPACK (for Linear Algebra PACKage) appeared officially in 1992, see Anderson *et al* [26] and the second one in 1994. The strategy of LAPACK to obtain portable codes that are also efficient is to construct the software as much as possible using calls to the BLAS. The BLAS2 and 3 can achieve near peak performance on the target architectures. Moreover, they exploit parallelism in a transparent way.

Partitioned block forms of the LU (and other) factorizations (where point algorithms are used) in which operations have been grouped together are used in LAPACK in order to use Level 2 and 3 BLAS. Some routines also exist in LAPACK that return bounds on the componentwise backward error (see Chapter 2).

For distributed memory parallel computers, a new library known as ScaLAPACK [317] has been developed. Again, the goal is portability and performance. By 1997, LAPACK and ScaLAPACK contained the state of the art for Gaussian elimination for dense matrices. They must be used as much as possible.

The codes and literature about the BLAS, LAPACK and ScaLAPACK projects can be found in Netlib on the World Wide Web at the address http://www.netlib.org.

1.14. Bibliographical comments

Most of the results given in this chapter can be found in the classical books of Householder [793], Strang [1216], Wilkinson [1333], Varga [1300] and Young [1351].

Theorem 1.24, giving a necessary and sufficient condition for convergence of a matrix power sequence, goes back to Hensel (1926) and Oldenburger (1940). Note that most of the proofs in the literature use the Jordan normal form.

Irreducibility of a matrix was introduced by Frobenius (1912) and reintroduced by Gerringer (1949).

Diagonally dominant matrices have been studied for a very long time. Corollary 1.32 saying that a strictly diagonally dominant matrix is non–singular has been proved by many mathematicians. However, its origin seems to be in the works of L. Levy (1881) and J. Hadamard (1898). This result was then forgotten and rediscovered by Gerschgorin (1931). Olga Taussky (1949) proved that a strictly diagonally dominant matrix has eigenvalues with positive real parts.

Ostrowski (1937) introduced M and H–matrices. It is very difficult to trace where all the characterizations of M–matrices come from. The book by Berman–Plemmons [142] is a good summary of all these results.

The definition of a regular splitting comes from Varga (1960). Theorem 1.54 (Householder (1955), John (1956)) is an extension of the Ostrowski-Reich theorem (1949-1954).

Fourier analysis has been used for quite a while. But, only recently it has received attention for the analysis of iterative methods, see Chan and his co–workers [268].

The use of graphs for Gaussian elimination is due to Parter [1065] and Rose [1119].

The design of the IEEE standard for floating point arithmetic has been largely influenced by W. Kahan, see Golberg [652] and Higham [781].

The famous LINPACK package that implements direct methods for solving dense and banded linear systems appeared in 1979. Its descendant LAPACK (1992) is largely due to the efforts of Jack Dongarra.

2 Gaussian elimination for general linear systems

2.1. Introduction to Gaussian elimination

The problem we are concerned with in this chapter is obtaining by a direct method the numerical solution of a linear system

$$Ax = b, \tag{2.1}$$

where A is a square non–singular matrix (i.e. $\det(A) \neq 0$) of order n, b is a given vector, x is the solution vector. Of course, the solution x of (2.1) is given by

$$x = A^{-1}b$$

where A^{-1} denotes the inverse of A. Unfortunately, in most cases A^{-1} is not explicitly known, except for some special problems and/or for small values of n. But, it is well known that the solution can be expressed by Cramer's formulae (see [601]):

$$x_i = \frac{1}{\det(A)} \begin{vmatrix} a_{1,1} & \cdots & a_{1,i-1} & b_1 & a_{1,i+1} & \cdots & a_{1,n} \\ a_{2,1} & \cdots & a_{2,i-1} & b_2 & a_{2,i+1} & \cdots & a_{2,n} \\ \vdots & \cdots & \vdots & \vdots & \vdots & \cdots & \vdots \\ a_{n,1} & \cdots & a_{n,i-1} & b_n & a_{n,i+1} & \cdots & a_{n,n} \end{vmatrix}, \quad i = 1, \ldots, n. \tag{2.2}$$

The computation of the solution x by (2.2) requires the evaluation of $n+1$ determinants of order n. This implies that this method will require more than $(n+1)!$ operations (multiplications and additions) to compute the solution. Fortunately, there are much better methods than Cramer's rule which is almost never used.

As we said in Chapter 1, direct methods obtain the solution by making some combinations and modifications of the equations. Of course, as computer floating point operations are only performed to a certain precision (see Chapter 1), the computed solution is generally different from the exact solution. We shall return to this point later.

The most widely used direct methods for general matrices belong to a class collectively known as Gaussian elimination. There are many variations of the same basic idea and we shall describe some of them in the next sections.

2.1.1. Gaussian elimination without permutations

In this Section we describe Gaussian elimination without permutations. We give the necessary and sufficient conditions for a matrix to have an LU factorization where L (resp. U) is a lower (resp. upper) triangular matrix. Then we shall introduce permutations to handle the general case.

The first step of the algorithm is the elimination of the unknown x_1 in the equations 2 to n. This can be done through $n-1$ steps. Suppose that $a_{1,1} \neq 0$, $a_{1,1}$ is then called the first pivot. To eliminate x_1 from the second equation, we (left) multiply A by an elementary matrix,

$$E_{2,1} = \begin{pmatrix} 1 & & & & \\ -\frac{a_{2,1}}{a_{1,1}} & 1 & & & \\ 0 & 0 & 1 & & \\ \vdots & \vdots & \ddots & \ddots & \\ 0 & \dots & \dots & 0 & 1 \end{pmatrix}.$$

This corresponds to do a linear combination of the first two rows of A. More generally, to eliminate x_1 from the ith equation, we (left) multiply by

$$E_{i,1} = \begin{pmatrix} 1 & & & & & & & \\ 0 & 1 & & & & & & \\ \vdots & & \ddots & & & & & \\ 0 & & & \ddots & & & & \\ -\frac{a_{i,1}}{a_{1,1}} & 0 & \dots & 0 & 1 & & & \\ 0 & & & & & \ddots & & \\ \vdots & & & & & & \ddots & \\ 0 & \dots & & & \dots & & 0 & 1 \end{pmatrix},$$

the non–zero terms of the first column of $E_{i,1}$ being in positions $(1,1)$ and $(i,1)$. All these elementary matrices can be combined easily as it is shown in the following lemma.

Lemma 2.1

Let $j > i$, then

$$E_{i,1}E_{j,1} = \begin{pmatrix} 1 & & & & & & & & & & \\ 0 & 1 & & & & & & & & & \\ \vdots & & \ddots & & & & & & & & \\ 0 & & & \ddots & & & & & & & \\ -\frac{a_{i,1}}{a_{1,1}} & & & & \ddots & & & & & & \\ 0 & & & & & \ddots & & & & & \\ \vdots & & & & & & \ddots & & & & \\ 0 & & & & & & & \ddots & & & \\ -\frac{a_{j,1}}{a_{1,1}} & & & & & & & & \ddots & & \\ 0 & & & & & & & & & \ddots & \\ \vdots & & & & & & & & & & \ddots \\ 0 & & & & & & & & & & & 1 \end{pmatrix},$$

and

$$E_{j,1}E_{i,1} = E_{i,1}E_{j,1}.$$

Proof. The result is obtained by straightforward matrix multiplication. Note that we also have

$$E_{i,1}E_{j,1} = E_{i,1} + E_{j,1} - I.$$

Let us denote $L_1 = E_{n,1}E_{n-1,1}\cdots E_{2,1}$ and $A_2 = L_1 A$. Obviously L_1 is lower triangular. The elements of A_2 will be denoted by $a_{i,j}^{(2)}$. The matrix A_2 has the following structure,

$$A_2 = \begin{pmatrix} a_{1,1} & \text{x} & \dots & \text{x} \\ 0 & \text{x} & \dots & \text{x} \\ \vdots & \vdots & & \vdots \\ 0 & \text{x} & \dots & \text{x} \end{pmatrix},$$

the x's corresponding to (eventually) non–zero elements that are defined in the following lemma.

Lemma 2.2

$$a_{i,j}^{(2)} = a_{i,j} - \frac{a_{i,1}a_{1,j}}{a_{1,1}}, \quad 2 \le i \le n, \quad 1 \le j \le n,$$

$$a_{1,j}^{(2)} = a_{1,j}, \quad 1 \le j \le n.$$

Proof. This is simply the result of the multiplication of A by L_1. □

Let us describe the kth step of the algorithm. Let A_k be the matrix that has been obtained by zeroing the elements below the diagonal in the $k-1$ first columns,

$$A_k = \begin{pmatrix} a_{1,1}^{(k)} & \dots & \dots & \dots & \dots & a_{1,n}^{(k)} \\ & \ddots & & & & \vdots \\ & & a_{k,k}^{(k)} & \dots & \dots & a_{k,n}^{(k)} \\ & & a_{k+1,k}^{(k)} & \dots & \dots & a_{k+1,n}^{(k)} \\ & & \vdots & \vdots & \vdots & \vdots \\ & & a_{n,k}^{(k)} & \dots & \dots & a_{n,n}^{(k)} \end{pmatrix}$$

and suppose $a_{k,k}^{(k)} \neq 0$. The element $a_{k,k}^{(k)}$ is known as the kth pivot. For $i > k$, let

$$E_{i,k} = \begin{pmatrix} 1 & & & & & & & \\ & \ddots & & & & & & \\ & & 1 & & & & & \\ & & 0 & 1 & & & & \\ & & \vdots & & \ddots & & & \\ & & 0 & & & \ddots & & \\ & & -\frac{a_{i,k}^{(k)}}{a_{k,k}^{(k)}} & & & & \ddots & \\ & & 0 & & & & & \ddots \\ & & \vdots & & & & & & \ddots \\ & & 0 & & & & & & & 1 \end{pmatrix},$$

where the element $-\frac{a_{i,k}^{(k)}}{a_{k,k}^{(k)}}$ is in row i and column k. The non–diagonal elements that are not explicitly shown are 0. Let $L_k = E_{n,k}E_{n-1,k}\cdots E_{k+1,k}$ and $A_{k+1} = L_k A_k$. Then, as for the first column, we can characterize this product of matrices.

Lemma 2.3

$$L_k = \begin{pmatrix} 1 & & & & \\ & \ddots & & & \\ & & 1 & & \\ & & -\frac{a_{k+1,k}^{(k)}}{a_{k,k}^{(k)}} & 1 & \\ & & \vdots & & \ddots \\ & & -\frac{a_{n,k}^{(k)}}{a_{k,k}^{(k)}} & & & 1 \end{pmatrix},$$

and A_{k+1} has the following structure

$$A_{k+1} = \begin{pmatrix} a_{1,1}^{(k+1)} & \cdots & \cdots & \cdots & \cdots & a_{1,n}^{(k+1)} \\ & \ddots & & & & \vdots \\ & & a_{k,k}^{(k+1)} & \cdots & \cdots & a_{k,n}^{(k+1)} \\ & & 0 & a_{k+1,k+1}^{(k+1)} & \cdots & a_{k+1,n}^{(k+1)} \\ & & \vdots & \vdots & & \vdots \\ & & 0 & a_{n,k+1}^{(k+1)} & \cdots & a_{n,n}^{(k+1)} \end{pmatrix}.$$

Proof. Straightforward. □

The elements of the jth column of A_{k+1} are given by the following expressions.

Lemma 2.4

$$a_{i,j}^{(k+1)} = a_{i,j}^{(k)} - \frac{a_{i,k}^{(k)} a_{k,j}^{(k)}}{a_{k,k}^{(k)}}, \quad k+1 \le i \le n, \quad k \le j \le n,$$

$$a_{i,j}^{(k+1)} = a_{i,j}^{(k)}, \quad 1 \le i \le k, \quad 1 \le j \le n$$
$$\text{and } k+1 \le i \le n, \quad 1 \le j \le k-1,$$

Proof. This is the formula we obviously get when multiplying by L_k. □

It is useful to study some properties of the matrices L_k, characterizing their inverses.

Lemma 2.5

L_k is non–singular and

$$L_k^{-1} = \begin{pmatrix} 1 & & & & & \\ & \ddots & & & & \\ & & 1 & & & \\ & & \frac{a_{k+1,k}^{(k)}}{a_{k,k}^{(k)}} & 1 & & \\ & & \vdots & & \ddots & \\ & & \frac{a_{n,k}^{(k)}}{a_{k,k}^{(k)}} & & & 1 \end{pmatrix}$$

Proof. Let

$$l^k = \begin{pmatrix} 0 & \dots & 0 & \frac{a_{k+1,k}^{(k)}}{a_{k,k}^{(k)}} & \dots & \frac{a_{n,k}^{(k)}}{a_{k,k}^{(k)}} \end{pmatrix}^T.$$

Obviously, $L_k = (I - l^k e^{k^T})$ where $e^k = (0 \quad \dots \quad 0 \quad 1 \quad 0 \quad \dots \quad 0)^T$ with the 1 in the kth position. Then, we have

$$L_k^{-1} = I + l^k e^{k^T},$$

because,

$$\begin{aligned} L_k^{-1} L_k &= (I + l^k e^{k^T})(I - l^k e^{k^T}), \\ &= I - l^k e^{k^T} l^k e^{k^T}, \end{aligned}$$

and $e^{k^T} l^k = 0$. □

Therefore, the inverses of matrices L_k can be obtained very easily. The preceding results can be summarized in the following proposition.

Proposition 2.6

If for all k, $1 \leq k \leq n-1$, $a_{k,k}^{(k)} \neq 0$, then there exists a factorization

$$A = LU,$$

where L is lower triangular with a unit diagonal and U is upper triangular.

Proof. The elimination process is defined as follows,

$$\begin{aligned} A_1 &= A, \\ A_2 &= L_1 A_1, \\ &\vdots \\ A_n &= L_{n-1} A_{n-1}. \end{aligned}$$

The last matrix A_n is upper triangular and is therefore denoted by U. Hence,

$$L_{n-1} L_{n-2} \cdots L_1 A = U.$$

The matrices $L_i, 1 \leq i \leq n-1$ are non–singular. Then,

$$A = (L_1^{-1} \cdots L_{n-1}^{-1}) U.$$

The product of unit lower triangular matrices being unit lower triangular, we have

$$L = L_1^{-1} \cdots L_{n-1}^{-1}$$

which is a lower triangular matrix with a unit diagonal. Moreover, it is easy to show that

$$L = \begin{pmatrix} 1 & & & & & \\ \frac{a_{2,1}^{(1)}}{a_{1,1}^{(1)}} & \ddots & & & & \\ \vdots & \ddots & 1 & & & \\ \vdots & & \frac{a_{k+1,k}^{(k)}}{a_{k,k}^{(k)}} & 1 & & \\ \vdots & & \vdots & \ddots & \ddots & \\ \frac{a_{n,1}^{(1)}}{a_{1,1}^{(1)}} & \cdots & \frac{a_{n,k}^{(k)}}{a_{k,k}^{(k)}} & \cdots & \frac{a_{n,n-1}^{(n-1)}}{a_{n-1,n-1}^{(n-1)}} & 1 \end{pmatrix}.$$

As $L_i^{-1}L_{i+1}^{-1} = I + l^i e^{i^T} + l^{i+1}e^{i+1^T}$, we obtain

$$L = L_1^{-1} \cdots L_{n-1}^{-1} = I + l^1 e^{1^T} + \cdots l^{n-1}e^{n-1^T}$$

□

Theorem 2.7

If the factorization $A = LU$ exists, it is unique.

Proof. The proof is by contradiction. Suppose there exist two such factorizations, $A = L_1U_1 = L_2U_2$, then

$$L_2^{-1}L_1 = U_2U_1^{-1}.$$

The matrix on the left hand side is lower triangular with a unit diagonal and the matrix on the right hand side is upper triangular. Therefore, they are both diagonal and

$$L_2^{-1}L_1 = U_2U_1^{-1} = I$$

which shows that the decomposition is unique. □

Let us derive the conditions under which there exists an LU factorization. For this, we have to identify when all the pivots are non–zero.

Theorem 2.8

A non–singular matrix A has a unique LU factorization if and only if all the principal minors of A are non–zero. That is

$$A\begin{pmatrix} 1 & 2 & \dots & k \\ 1 & 2 & \dots & k \end{pmatrix} \neq 0, \quad k = 1, \dots, n$$

where

$$A\begin{pmatrix} i_1 & i_2 & \dots & i_p \\ k_1 & k_2 & \dots & k_p \end{pmatrix} = \begin{vmatrix} a_{i_1,k_1} & a_{i_1,k_2} & \dots & a_{i_1,k_p} \\ a_{i_2,k_1} & a_{i_2,k_2} & \dots & a_{i_2,k_p} \\ \vdots & \vdots & & \vdots \\ a_{i_p,k_1} & a_{i_p,k_2} & \dots & a_{i_p,k_p} \end{vmatrix}.$$

Proof. Suppose that there exists an LU factorization. From the proof of proposition 2.6, we have

$$A_{k+1} = L_k \dots L_1 A.$$

Therefore,

$$A = L_1^{-1} \dots L_k^{-1} A_{k+1},$$

and we have also

$$A = LU.$$

In block form, this is written as:

$$A = \begin{pmatrix} A_{1,1} & A_{1,2} \\ A_{2,1} & A_{2,2} \end{pmatrix}, \quad L = \begin{pmatrix} L_{1,1} & 0 \\ L_{2,1} & L_{2,2} \end{pmatrix}, \quad U = \begin{pmatrix} U_{1,1} & U_{1,2} \\ 0 & U_{2,2} \end{pmatrix},$$

and from proposition 2.6,

$$L_1^{-1} \dots L_k^{-1} = \begin{pmatrix} L_{1,1} & 0 \\ L_{2,1} & I \end{pmatrix}, \quad A_{k+1} = \begin{pmatrix} U_{1,1} & U_{1,2} \\ 0 & W_{2,2} \end{pmatrix},$$

where all the matrices in block position $(1,1)$ are square of order k. By equating blocks, we have

$$A_{1,1} = L_{1,1}U_{1,1}$$
$$A_{2,2} = L_{2,1}U_{1,2} + L_{2,2}U_{2,2}$$
$$A_{2,2} = L_{2,1}U_{1,2} + W_{2,2}$$

Therefore, $L_{1,1}U_{1,1}$ is the LU factorization of the leading principal submatrix of order k of A and $L_{2,2}U_{2,2}$ is the factorization of the matrix $W_{2,2}$ in the bottom right hand corner of A_{k+1} as we have $W_{2,2} = L_{2,2}U_{2,2}$.

Note that $\det(A) = \det(A_{k+1})$. We have $\det(L_{1,1}) = 1$ and $\det(A_{1,1}) = \det(U_{1,1})$. Since the matrix $U_{1,1}$ is upper triangular, its determinant is equal to the product of the diagonal elements. Therefore, for all k,

$$\det(A_{1,1}) = a_{1,1}^{(1)} \cdots a_{k,k}^{(k)}.$$

This shows that the principal minors are non–zero. The converse of the proof is easily derived by induction. □

2.1.2. Gaussian elimination with permutations (partial pivoting)

In this Section, we allow for the possibility of having zero pivots and we show that nevertheless, a factorization can be computed by exchanging rows. This is known as pivoting. If the first pivot $a_{1,1}$ is zero, we permute the first row with a row p such that $a_{p,1} \neq 0$. Finding such a p is always possible, otherwise we shall have $\det(A) = 0$. The row interchange is done by left multiplication of A by a permutation matrix P_1. P_1 is equal to the identity matrix except that rows 1 and p have been exchanged, i.e.

$$P_1 = \begin{pmatrix}
0 & 0 & \dots & 0 & 1 & 0 & \dots & 0 \\
0 & 1 & 0 & \dots & 0 & \dots & & \vdots \\
\vdots & \ddots & \ddots & \ddots & \vdots & & & \\
0 & \dots & 0 & 1 & 0 & \dots & & \\
1 & 0 & \dots & 0 & 0 & 0 & \dots & \\
0 & \dots & & \dots & 0 & 1 & \ddots & \vdots \\
\vdots & & & & \vdots & \ddots & \ddots & 0 \\
0 & \dots & & \dots & 0 & \dots & 0 & 1
\end{pmatrix}.$$

Note that $P_1^{-1} = P_1$. Then we proceed with the previous elimination algorithm on the permuted matrix. We construct L_1 such that $A_2 = L_1 P_1 A$. Let us describe the kth step. The main difference from what we have done before is that the possibility exists that the pivot is zero. If this is the case, it is possible to find a row p such that $a_{p,k}^{(k)} \neq 0$. The reason for this being that $\det(A_k) = \det(A) \neq 0$ and the determinant $\det(A_k)$ is equal to the product of the first $k-1$ (non–zero) pivots and the determinant of the matrix in the right hand bottom corner. Therefore, this matrix is non–singular. In fact, we choose a non–zero element $a_{p,k}^{(k)}, p > k$ which has the maximum modulus.

This strategy of choosing the pivot in the kth column is called partial pivoting. Then, we multiply A_k by the corresponding permutation matrix P_k and apply the elimination algorithm,

$$A_{k+1} = L_k P_k A_k.$$

Finally, we obtain

$$U = L_{n-1} P_{n-1} \cdots L_2 P_2 L_1 P_1 A.$$

It may appear that we have lost the good properties of the Gaussian algorithm as permutation matrices have appeared in between the lower triangular matrices, even if some of them are equal to the identity matrix (when there is no need for pivot interchange). However, we have the following result.

Lemma 2.9

Let P_p be a permutation matrix representing the permutation between indices p and $q > p$ then, $\forall k < p$

$$L_k P_p = P_p L'_k.$$

where L'_k is deduced from L_k by the permutation of entries in rows p and q in column k.

Proof. Note that $P_p^{-1} = P_p$ and $L_k = I - l^k {e^k}^T$,

$$L'_k = P_p L_k P_p = P_p(I - l^k {e^k}^T) P_p = I - P_p l^k {e^k}^T P_p.$$

As $p > k$, $P_p e^k = e^k$, therefore

$$L'_k = I - l^{k'} {e^k}^T$$

where $l^{k'} = P_p l^k$. The same kind of result is true for L_k^{-1} as $L_k^{-1} = I + l^k {e^k}^T$.
□

With these preliminaries, we can characterize the LU decomposition of a given matrix.

Theorem 2.10

Let A be a non–singular matrix, there exists a permutation matrix P such that

$$PA = LU,$$

where L is lower triangular with a unit diagonal and U is upper triangular.

Proof. We have seen that

$$A = P_1 L_1^{-1} P_2 \cdots P_{n-1} L_{n-1}^{-1} U.$$

From lemma 2.9,

$$A = P_1 P_2 \cdots P_{n-1} (L_1'')^{-1} \cdots (L_{n-1}'')^{-1} U,$$

where $(L_k'')^{-1} = P_{n-1} \cdots P_{k+1} L_k^{-1} P_{k+1} \cdots P_{n-1}$, corresponding to a permutation of the coefficients of column k. □

Usually, the permutation matrix P is stored as a vector of indices since row permutations are not explicitly performed during the factorization. The linear system $Ax = b$ is transformed into

$$PAx = LUx = Pb,$$

and is solved in two steps by

$$Ly = Pb,$$
$$Ux = y.$$

These two triangular solves are known as the forward and backward sweeps. For general systems, pivoting is used even when $a_{k,k}^{(k)} \neq 0$. Systematically, a permutation is done with the row p giving $\max_{p>k} |a_{p,k}^{(k)}|$.

2.1.3. Gaussian elimination with other pivoting strategies

Pivoting strategies which differ from partial pivoting may also be used. For instance, we can search for the pivot not only in the lower part of the kth column but in all the remaining columns in the submatrix. The chosen pivot is an element that realizes $\max_{i,j} |a_{i,j}^{(k)}|, \forall i, j \geq k$. This algorithm is known as complete pivoting. Then, we not only have to do row permutations but also column permutations to bring the pivot into position (k,k). This is achieved by multiplying on the right by a permutation matrix. Finally, we obtain two permutation matrices P and Q such that

$$PAQ = LU.$$

The solution of the linear system is obtained through

$$\begin{aligned} Ly &= Pb, \\ Uz &= y, \\ x &= Q^T z. \end{aligned}$$

We shall see that complete pivoting has some advantages regarding stability. However, the cost of finding the pivot is much larger than for partial pivoting. Other pivoting strategies can be used, see for instance [1076].

2.1.4. Operation counts

Despite what we said in Chapter 1, it is interesting to compute the number of floating point operations that must be done to obtain the LU factorization of A.

For computing the kth column of L, we need 1 division by the pivot and $n-k$ multiplications. To compute the updated matrix A_{k+1}, we need (after having computed the multipliers $-a_{i,k}^{(k)}/a_{k,k}^{(k)}$ which are the elements of L) $(n-k)^2$ additions and the same number of multiplications. To get the total number of operations, we sum these numbers from 1 to $n-1$

$$\sum_{k=1}^{n-1}(n-k) = n(n-1) - \frac{1}{2}n(n-1) = \frac{1}{2}n(n-1),$$

$$\begin{aligned} \sum_{k=1}^{n-1}(n-k)^2 &= n^2\sum_{k=1}^{n-1} 1 - 2n\sum_{k=1}^{n-1} k + \sum_{k=1}^{n-1} k^2, \\ &= \frac{1}{3}n(n-1)(n-\frac{1}{2}). \end{aligned}$$

Theorem 2.11

To obtain the factorization $PA = LU$ of theorem 2.10, we need

$$\frac{2n^3}{3} - \frac{n^2}{2} - \frac{n}{6}$$

floating point operations (multiplications and additions) and $n-1$ divisions. The solutions of the triangular systems to obtain the solution x give $n(n-1)$ floating point operations for L and $n(n-1)+n$ for U. Note that this is an order less than that needed for the factorization.

□

2.2. Gaussian elimination for symmetric systems

The factorization of symmetric matrices is an important special case that we shall consider in more detail. Let us specialize the previous algorithms to the symmetric case. We are considering a factorization slightly different from what we have seen:

$$A = LDL^T,$$

where L is lower triangular with a unit diagonal and D is diagonal. There are several possible ways to compute this factorization. We shall study three different algorithms that will lead to six ways of programming the factorization.

2.2.1. The outer product algorithm

The first method to construct the LDL^T factorization is the one we used for non–symmetric systems, column by column. Suppose $a_{1,1} \neq 0$ and

$$L_1 = \begin{pmatrix} 1 & 0 \\ l_1 & I \end{pmatrix}, \quad D_1 = \begin{pmatrix} a_{1,1} & 0 \\ 0 & A_2 \end{pmatrix},$$

and

$$A = \begin{pmatrix} a_{1,1} & a_1^T \\ a_1 & B_1 \end{pmatrix} = L_1 D_1 L_1^T.$$

By equating blocks, we obtain expressions for l_1 and A_2,

$$l_1 = \frac{a_1}{a_{1,1}}$$

$$A_2 = B_1 - \frac{1}{a_{1,1}} a_1 a_1^T = B_1 - a_{1,1} l_1 l_1^T$$

The matrix A_2 is a obviously symmetric. If we suppose that the $(1,1)$ element of A_2 is non–zero, we can proceed further and write

$$A_2 = \begin{pmatrix} a_{2,2}^{(2)} & a_2^T \\ a_2 & B_2 \end{pmatrix} = \begin{pmatrix} 1 & 0 \\ l_2 & I \end{pmatrix} \begin{pmatrix} a_{2,2}^{(2)} & 0 \\ 0 & A_3 \end{pmatrix} \begin{pmatrix} 1 & l_2^T \\ 0 & I \end{pmatrix}.$$

Similarly,

$$l_2 = \frac{a_2}{a_{2,2}^{(2)}},$$

$$A_3 = B_2 - \frac{1}{a_{2,2}^{(2)}} a_2 a_2^T = B_2 - a_{2,2}^{(2)} l_2 l_2^T.$$

We remark that if we denote,

$$L_2 = \begin{pmatrix} 1 & 0 & 0 \\ 0 & 1 & 0 \\ 0 & l_2 & I \end{pmatrix},$$

then,

$$D_1 = \begin{pmatrix} a_{1,1} & 0 \\ 0 & A_2 \end{pmatrix} = L_2 \begin{pmatrix} a_{1,1} & 0 & 0 \\ 0 & a_{2,2}^{(2)} & 0 \\ 0 & 0 & A_3 \end{pmatrix} L_2^T = L_2 D_2 L_2^T.$$

Therefore, after two steps, we have $A = L_1 L_2 D_2 L_2^T L_1^T$. We note that

$$L_1 L_2 = \begin{pmatrix} 1 & 0 \\ l_1 & \begin{pmatrix} 1 & 0 \\ l_2 & I \end{pmatrix} \end{pmatrix}.$$

The product of L_1 and L_2 is a lower triangular matrix. If all the pivots are non–zero, we can proceed and at the last step, we obtain

$$A = A_1 = L_1 L_2 \cdots L_{n-1} D L_{n-1}^T \cdots L_1^T = LDL^T,$$

where L is unit lower triangular and D is diagonal. There is a variant of this algorithm where a decomposition

$$A = \bar{L}\bar{D}^{-1}\bar{L}^T$$

is obtained with $\bar{L}$ being lower triangular, $\bar{D}$ diagonal and $\text{diag}(\bar{L}) = \text{diag}(\bar{D})$. We can obtain this variant from the first algorithm by writing

$$A = LDL^T = (LD)D^{-1}(DL^T),$$

and $\bar{D} = D$, $\bar{L} = LD$.

The matrix L has been constructed column by column. This method is called the outer product algorithm as, at each step, an outer product aa^T is involved.

2.2.2. The bordering algorithm

We can partition the matrix A in a different way as

$$A = \begin{pmatrix} C_n & a_n \\ a_n^T & a_{n,n} \end{pmatrix},$$

Suppose that C_n has already been factored as

$$C_n = L_{n-1} D_{n-1} L_{n-1}^T,$$

L_{n-1} being unit lower triangular and D_{n-1} diagonal. We can write,

$$A = \begin{pmatrix} L_{n-1} & 0 \\ l_n^T & 1 \end{pmatrix} \begin{pmatrix} D_{n-1} & 0 \\ 0 & d_{n,n} \end{pmatrix} \begin{pmatrix} L_{n-1}^T & l_n \\ 0 & 1 \end{pmatrix}.$$

Then equating yields

$$\begin{aligned} l_n &= D_{n-1}^{-1} L_{n-1}^{-1} a_n, \\ d_{n,n} &= a_{n,n} - l_n^T D_{n-1} l_n. \end{aligned}$$

By induction, we can start with the decomposition of the 1×1 matrix $a_{1,1}$, adding one row at a time and obtaining at each step the factorization of an enlarged matrix. The main operation we have to perform at each step is solving a triangular system. In order to proceed to the next step, we need the diagonal entries of D_n to be non–zero. For obvious reasons, this method is called the bordering algorithm.

2.2.3. The inner product algorithm

A third way of computing the factorization is explicitly writing down the formulas for the matrix product:

$$A = LDL^T.$$

Suppose $i \geq j$, we have

$$a_{i,j} = \sum_{k=1}^{j} l_{i,k} l_{j,k} d_{k,k}.$$

If we consider $i = j$ in this formula then since $l_{i,i} = 1$, we obtain

$$d_{j,j} = a_{j,j} - \sum_{k=1}^{j-1} (l_{j,k})^2 d_{k,k},$$

and for $i > j$,

$$l_{i,j} = \frac{1}{d_{j,j}} (a_{i,j} - \sum_{k=1}^{j-1} l_{i,k} l_{j,k} d_{k,k}).$$

As, locally, we have to consider the product of the transpose of a vector times a vector, this method is called the inner product algorithm or sometimes the scalar product algorithm.

The number of floating point operations required for these three variants is about $\frac{1}{2}$ of the number of operations for the general algorithm, *i.e.*, about $\frac{n^3}{6}$ multiplications and the same number of additions.

2.2.4. Coding the three factorization algorithms

First we consider the outer product algorithm. The matrix L is constructed column by column. At step k, column k is constructed by multiplying by the inverse of the pivot and then, the columns at the right of column k are modified using the values of the entries of column k. This can be done by rows or by columns and this leads to the two programs given below. We store the matrix D in a vector denoted by d and L in a separate matrix although in practice it can be stored in the lower triangular part of A (if A is not to be saved). The array `temp` is a temporary vector whose use can sometimes be avoided. We use it mainly for clarity of presentation. Note that the codings are slightly different from those given in Dongarra, Gustavson, Karp [433]. They are arranged such that in the main loop, `i` is a row index, `j` is a column index and `k` can eventually be both. The strictly lower triangular part of L is initialized to that of A.

Outer product *kij* algorithm

```
function [l,d]=kij(a)

[m,n]=size(a);

d=zeros(n,1);

temp=zeros(n,1);
```

```
d(1)=a(1,1);
l(1,1)=1;
for k=1:n-1
  dki=1./d(k);
  for i=k+1:n
    temp(i)=l(i,k)*dki;
  end
  for i=k+1:n
    for j=k+1:i
      l(i,j)=l(i,j)-temp(i)*l(j,k);
    end
  end
  for i=k+1:n
    l(i,k)=temp(i);
  end
  d(k+1)=l(k+1,k+1);
  l(k+1,k+1)=1.;
end
```

To reflect the way the three loops are nested, this algorithm is called the kij form. We can eliminate the temporary vector `temp` by using the upper part of the matrix `l`. However, we think the coding is clearer using `temp`. Modifying by rows (interchanging the loops on `i` and `j`) we get

Outer product kji algorithm

```
function [l,d]=kji(a)
[m,n]=size(a);
```

```
d=zeros(n,1);
temp=zeros(n,1);
l=init(a);
d(1)=a(1,1);
l(1,1)=1;
for k=1:n-1
  dki=1/d(k);
  for i=k+1:n
    temp(i)=l(i,k)*dki;
  end
  for j=k+1:n
    for i=j:n
      l(i,j)=l(i,j)-temp(i)*l(j,k);
    end
  end
  for i=k+1:n
    l(i,k)=temp(i);
  end
  d(k+1)=l(k+1,k+1);
  l(k+1,k+1)=1.;
end
```

Now we consider the bordering algorithm. For each row i, we have to solve a triangular system. There are two algorithms to do this. One is column oriented, the other is row oriented.

Bordering ijk algorithm

```
function [l,d]=ijk(a)
[m,n]=size(a);
d=zeros(n,1);
temp=zeros(n,1);
l=init(a);
d(1)=a(1,1);
l(1,1)=1.;
for i=2:n
  for k=1:i
    temp(k)=a(i,k);
  end
  for j=1:i
    if j ~= i
      l(i,j)=temp(j)/d(j);
    end
    for k=j+1:i
      temp(k)=temp(k)-l(k,j)*temp(j);
    end
  end
  d(i)=temp(i);
  l(i,i)=1;
end
```

There are too many divisions in the previous coding as they are in a `k` loop. These divisions can be avoided by storing the inverses of d as they are computed.

Bordering ikj algorithm

```
function [l,d]=ikj(a)
[m,n]=size(a);
d=zeros(n,1);
temp=zeros(n,1);
l=init(a);
d(1)=a(1,1);
l(1,1)=1;
for i=2:n
  for k=1:i
    temp(k)=a(i,k);
  end
  for k=1:i
    for j=1:k-1
      temp(k)=temp(k)-temp(j)*l(k,j);
    end
    if k ~= i
      l(i,k)=temp(k)/d(k);
    else
      d(i)=temp(i);
      l(i,i)=1;
```

```
    end
  end
end
```

Finally we consider the inner product algorithm.

Inner product jik algorithm

```
function [l,d]=jik(a)
[m,n]=size(a);
l=init(a);
for j=1:n
  for k=1:j-1
    l(j,k)=l(j,k)/d(k);
  end
  d(j)=a(j,j);
  for k=1:j-1
    d(j)=d(j)-l(j,k) ^ 2*d(k);
  end
  for i=j+1:n
    for k=1:j-1
      l(i,j)=l(i,j)-l(i,k)*l(j,k);
    end
  end
  l(j,j)=1.;
end
```

In the computation of $a_{i,j} - \sum_{k=1}^{j-1} l_{i,k} l_{j,k} d_{k,k}$, one can compute $a_{i,j} - l_{i,k} l_{j,k} d_{k,k}$ for a fixed value of k looping on i provided that the division by $d_{j,j}$ is done afterwards. Then we obtain the following algorithm.

Inner product jki algorithm

```
function [l,d]=jki(a)

[m,n]=size(a);

l=init(a);

for j=1:n

  for k=1:j-1

    l(j,k)=l(j,k)/d(k);

  end

  d(j)=a(j,j);

  for k=1:j-1

    d(j)=d(j)-l(j,k) ^ 2*d(k);

  end

  for k=1:j-1

    for i=j+1:n

      l(i,j)=l(i,j)-l(i,k)*l(j,k);

    end

  end

  l(j,j)=1.;

end
```

We have derived six different ways to program the LDL^T factorization of a symmetric matrix A. The same can be done for the LU factorization of a non–symmetric matrix. Of course, we are interested in finding the best implementation, that is the one requiring the smallest computing time. Unfortunately, this is dependent on the computer architecture and also on the languages used for coding and running the algorithm. It also depends on the data structure chosen for storing L since for current (1998) computers, the performance depends mainly on the way the data is accessed in the computer memory.

Suppose first that L is to be stored in a two–dimensional array or in the lower triangular part of A. In Fortran, two dimensional arrays are stored by columns, that is, consecutive elements in a column of an array have consecutive memory addresses. Therefore, it is much better to use algorithms that access data by columns. This may be different for other languages. For instance, in C two–dimensional arrays are stored by rows. Moreover, in computers with data caches, it is advantageous to perform operations on data in consecutive memory locations. This increases the cache hit ratio as data is moved into the cache by blocks of consecutive addresses.

The data access is by columns for algorithms kji and jki, by rows for ikj and jik and by rows and columns for kij and ijk. This favors algorithms kji and jki. Form kji accesses the data by columns and the involved basic operation is known as a SAXPY (for single precision a times x plus y) *i.e.*,

$$y = y + \alpha x,$$

where x and y are vectors and α is a scalar. Note that the vector y is stored after it is computed. This particular form was used in the famous LINPACK package [232]. Form jki also accesses that data by columns and the basic operation is also a SAXPY. However, the same column (j) is successively accessed many times. This is known as a generalized SAXPY or GAXPY. This can sometimes be exploited when coding in assembly language (one can keep the data in registers or in caches). These algorithms were analyzed for a vector computer in Dongarra, Gustavson and Karp [433], their notations being slightly different from ours. On vector architectures, the GAXPY jki form is generally the best one.

If L is not stored in the lower triangular part of A, it is better to store it in a one dimensional array of dimension $n(n-1)/2$. Consecutive elements can be chosen by rows or columns. If consecutive elements are chosen by rows, it is better to use algorithms ikj and jik as the data accesses will be in consecutive addresses. kji and jki forms will be chosen if the data is stored by columns.

So far, we have assumed it was not necessary to do pivoting for a symmetric system. We now describe some particular cases where it can be shown that pivoting is not needed, at least to be able to run the algorithm to completion.

2.2.5. Positive definite systems

In this part, we assume A is symmetric and positive definite. We seek an LDL^T factorization. If we examine the outer product algorithm, we see that for the first step, $A_2 = B_1 - (1/a_{1,1})a_1 a_1^T$ is a Schur complement, the matrix A being partitioned as

$$A = \begin{pmatrix} a_{1,1} & a_1^T \\ a_1 & B_1 \end{pmatrix}.$$

Therefore by using lemma 1.59, if A is positive definite, A_2 is also positive definite and the next pivot is non–zero. The process can be continued until the last step. All the square matrices involved are positive definite and so the diagonal elements are positive. All the pivots are non–zero and the algorithm can continue without any need for pivoting. This is summarized in the following result.

Theorem 2.12

A matrix A has a factorization $A = LDL^T$, where L is a unit lower triangular matrix and D is a diagonal matrix with positive diagonal elements, if and only if A is symmetric and positive definite.

Proof. Lemma 1.59 and the previous discussion have shown that if A is positive definite, it can be factored as LDL^T. Conversely, if $A = LDL^T$, then of course A is symmetric and if $x \neq 0$, then

$$x^T A x = x^T LDL^T x = y^T D y,$$

where $y = L^T x \neq 0$. Note that

$$y^T D y = \sum_{i=1}^{n} d_{i,i} y_i^2 > 0,$$

since the diagonal elements of D are positive. □

Thus we can introduce a diagonal matrix S such that $s_{i,i} = \sqrt{d_{i,i}}$, $i = 1, \ldots, n$. Therefore, $S^2 = D$ and let $\bar{L} = LS$. Then,

$$A = LDL^T = LSSL^T = \bar{L}\bar{L}^T.$$

Usually this decomposition is called the Cholesky factorization of A. However, this form of factorization is not often used today as the computation involves square roots. On modern computers, computing square roots is much slower than multiplications and additions and should be avoided whenever possible. Factorizations like LDL^T have sometimes been called the square root free Cholesky. We shall use the generic name Cholesky for any LDL^T factorization.

An interesting property of positive definite matrices is that there is no growth of the entries of the reduced matrices during the factorization.

Theorem 2.13

Let A be symmetric positive definite. Consider the matrices D_i, $i = 1, \dots, n$ of the outer product algorithm, then,

$$\max_k(\max_{i,j} |(D_k)_{i,j}|) \le \max_{i,j} |a_{i,j}| = \max_i(a_{i,i}).$$

Proof. By lemma 1.59, we know that the matrices D_k are positive definite. Therefore, by lemma 1.60, it is sufficient to consider the diagonal to find the maximum of the absolute values of the entries. We only consider the first step, since the proof is the same for the other steps. As the diagonal entries are positive, we have

$$\text{diag}(A_2) \le \text{diag}(B_1).$$

Therefore, either $\max_i(a_{i,i}) = a_{1,1}$ and then, $\max_i(D_1)_{i,i} = a_{1,1}$ or the maximum is on the diagonal of B_1 and then,

$$\max_i(D_1)_{i,i} = \max(a_{1,1}, \max_i[\text{diag}(A_2)_{i,i}]),$$

with $\max_i[\text{diag}(A_2)_{i,i}] \le \max_i[\text{diag}(B_1)_{i,i}]$. In both cases,

$$\max_i(D_1)_{i,i} \le \max_{i,j} |a_{i,j}|.$$

□

2.2.6. Indefinite systems

When factorizing an indefinite matrix (*i.e.*, one that is neither positive or negative definite), there can be some problems as shown in the following example,

$$\begin{pmatrix} \epsilon & 1 \\ 1 & 0 \end{pmatrix} = \begin{pmatrix} 1 & 0 \\ 1/\epsilon & 1 \end{pmatrix} \begin{pmatrix} \epsilon & 0 \\ 0 & -1/\epsilon \end{pmatrix} \begin{pmatrix} 1 & 1/\epsilon \\ 0 & 1 \end{pmatrix},$$

If ϵ is small, $1/\epsilon$ can be very large and the factorization can be unstable. Indeed is $\epsilon = 0$ the decomposition does not exist. One can use pivoting to avoid this. However, if symmetry is to be preserved, pivoting must be done on the diagonal but this does not always solve the problems. Moreover, zero pivots can sometimes be the only alternative. A method to solve these problems was introduced in Bunch and Parlett [235] and further developed in Bunch and Kaufman [234]. The remedy is to use diagonal pivoting with either 1×1 or 2×2 pivots. Suppose

$$P_1 A P_1^T = \begin{pmatrix} A_{1,1} & A_{1,2} \\ A_{1,2}^T & A_{2,2} \end{pmatrix},$$

where $A_{1,1}$ is of order s with $s = 1$ or 2, det $A_{1,1} \neq 0$ and P_1 is a permutation matrix. Then, this matrix can be factored as

$$P_1 A P_1^T = \begin{pmatrix} I_s & 0 \\ A_{1,2}^T A_{1,1}^{-1} & I_{n-s} \end{pmatrix} \begin{pmatrix} A_{1,1} & 0 \\ 0 & A_{2,2} - A_{1,2}^T A_{1,1}^{-1} A_{1,2} \end{pmatrix} \begin{pmatrix} I_s & A_{1,1}^{-1} A_{1,2} \\ 0 & I_{n-s} \end{pmatrix}.$$

The algorithm works provided that $A_{1,1}$ is non–singular. It can be shown that if A is non–singular, it is always possible to find a non–zero pivot ($s = 1$) or a non–singular 2×2 block ($s = 2$). A strategy was devised by Bunch and Kaufmann [234], see also [676] to find the block pivots. Another method that may be used for indefinite systems is due to Aasen [1].

2.3. Gaussian elimination for H–matrices

There are types of matrices (not necessarily symmetric) other than positive definite matrices for which there is no necessity to use pivoting (at least to obtain a factorization without permutations). Let us first consider matrices A which are diagonally dominant (see Chapter 1 for definition).

Theorem 2.14

If A is (row or column) diagonally dominant, then

$A = LU,$

where L is unit lower triangular and U is upper triangular.

Proof. Suppose A is (row) diagonally dominant. Then, $a_{1,1} \neq 0$, otherwise all the elements in the first row are 0 and A is singular. We shall prove that A_2 is also (row) diagonally dominant and then, the proof can proceed by induction. The case of the first row has already been handled. For a general row, we have

$$a_{i,j}^{(2)} = a_{i,j} - \frac{a_{i,1}a_{1,j}}{a_{1,1}}, \ 2 \leq i \leq n, \ 2 \leq j \leq n, \quad a_{i,1}^{(2)} = 0, \ 2 \leq i \leq n,$$

$$\sum_{j,j\neq i} |a_{i,j}^{(2)}| = \sum_{j,j\neq i,j\neq 1} |a_{i,j}^{(2)}| \leq \sum_{j,j\neq i,j\neq 1} |a_{i,j}| + \left|\frac{a_{i,1}}{a_{1,1}}\right| \sum_{j,j\neq i,j\neq 1} |a_{1,j}|.$$

But,

$$|a_{1,1}| \geq \sum_{j,j\neq i,j\neq 1} |a_{1,j}| + |a_{1,i}|.$$

Therefore,

$$\begin{aligned}
\sum_{j,j\neq i} |a_{i,j}^{(2)}| &\leq \sum_{j,j\neq i,j\neq 1} |a_{i,j}| + \left|\frac{a_{i,1}}{a_{1,1}}\right| (|a_{1,1}| - |a_{1,i}|), \\
&\leq \sum_{j,j\neq i} |a_{i,j}| - \frac{|a_{i,1}a_{1,i}|}{|a_{1,1}|}, \\
&\leq |a_{i,i}| - \frac{|a_{i,1}a_{1,i}|}{|a_{1,1}|}, \\
&\leq \left|a_{i,i} - \frac{a_{i,1}a_{1,i}}{a_{1,1}}\right| = |a_{i,i}^{(2)}|.
\end{aligned}$$

The reduced matrix is also diagonally dominant. This shows that all the pivots are non–zero and the computations can continue. If A is column diagonally dominant, the same proof goes through with A^T. □

We then consider M–matrices. The following result has been proved by Fiedler and Ptàk [554].

Theorem 2.15

If A is an M–matrix, then

$$A = LU,$$

where L is unit lower triangular and U is upper triangular.

Proof. See Fiedler's book [553] or Berman and Plemmons [142]. The proof uses lemma 1.46. □

We now consider H–matrices. Let B be an M–matrix, we define

$$\Omega_B = \{A | B \leq M(A)\}.$$

That is

$$|a_{i,i}| \geq b_{i,i},\ 1 \leq i \leq n,$$
$$|a_{i,j}| \leq |b_{i,j}|,\ i \neq j,\ 1 \leq i,j \leq n.$$

Note that the matrix A is at least as diagonally dominant as the matrix B.

Lemma 2.16

Let B be an M–matrix. Each matrix $A \in \Omega_B$ is (row) generalized strictly diagonally dominant.

Proof. There exists a diagonal matrix D (with $\text{diag}(D) > 0$) such that BD is strictly diagonally dominant. Let $A \in \Omega_B$. We have

$$BD \leq M(A)D = M(AD).$$

Therefore,

$$0 < BDe \leq M(AD)e,$$

which implies that AD is (row) strictly diagonally dominant. □

Theorem 2.17

Let B be an M–matrix. For each $A \in \Omega_B$,

$$A = LU,$$

where L is unit lower triangular and U is upper triangular. In particular, for every H-matrix, there exists an LU factorization.

Proof. We have seen in the proof of lemma 2.16 that AD is (row) strictly diagonally dominant. Then, by theorem 2.14, there exist $\bar{L}$ and $\bar{U}$, lower and upper triangular matrices such that

$$AD = \bar{L}\bar{U}.$$

We have,

$$A = \bar{L}\bar{U}D^{-1},$$

and the result follows. □

Let

$$\beta_D = \frac{\max_i(d_{i,i})}{\min_i(d_{i,i})}.$$

Then,

$$|l_{i,j}| \le \beta_D,$$
$$|u_{i,j}| \le 2\beta_D \max_i |a_{i,i}|.$$

We define the growth factor g_A as

$$g_A = \frac{\max_{i,j,k} |a_{i,j}^{(k)}|}{\|A\|_\infty}.$$

Sometimes, g_A is defined as

$$g_A = \frac{\max_{i,j,k} |a_{i,j}^{(k)}|}{\max_{i,j} |a_{i,j}|}.$$

For an H–matrix, we have $g_A \le 2\beta_D$ and for an M–matrix, Funderlic, Neumann and Plemmons [594] proved that $g_A \le \beta_D$.

2.4. Block methods

Block methods are obtained by partitioning the matrix A into blocks (submatrices). Consider, for instance, a 3×3 block partitioning. Then A is written as

$$A = \begin{pmatrix} A_{1,1} & A_{1,2} & A_{1,3} \\ A_{2,1} & A_{2,2} & A_{2,3} \\ A_{3,1} & A_{3,2} & A_{3,3} \end{pmatrix}.$$

The matrices $A_{i,i}$ are square of order n_i, $1 \leq n_i \leq n, \sum n_i = n$. A block LU factorization can be derived. We can do exactly the analog of the point case, provided the blocks that arise on the diagonal are non–singular. One way to proceed is the following:

$$A = \begin{pmatrix} I & & \\ L_{2,1} & I & \\ L_{3,1} & L_{3,2} & I \end{pmatrix} \begin{pmatrix} U_{1,1} & U_{1,2} & U_{1,3} \\ & U_{2,2} & U_{2,3} \\ & & U_{3,3} \end{pmatrix}.$$

The stability is investigated in Demmel, Higham and Schreiber [408]. Block LU factorization (without pivoting) is unstable in general, although it has been found to be stable for matrices which are block diagonally dominant by columns, that is

$$\|A_{j,j}^{-1}\|^{-1} \geq \sum_{i \neq j} \|A_{i,j}\|.$$

2.5. Tridiagonal and block tridiagonal systems

Tridiagonal matrices arise quite often in algorithms for solving PDEs. So, it is worth investigating their factorization. Let T be a symmetric tridiagonal matrix,

$$T = \begin{pmatrix} a_1 & -b_2 & & & \\ -b_2 & a_2 & -b_3 & & \\ & \ddots & \ddots & \ddots & \\ & & -b_{n-1} & a_{n-1} & -b_n \\ & & & -b_n & a_n \end{pmatrix}.$$

We suppose $b_i \neq 0, \forall i$. The minus sign in front of the b_is is a technical convenience. Suppose the Cholesky factorization of T exists. It is easily obtained as,

$$T = L D_L^{-1} L^T,$$

$$L = \begin{pmatrix} \delta_1 & & & & \\ -b_2 & \delta_2 & & & \\ & \ddots & \ddots & & \\ & & -b_{n-1} & \delta_{n-1} & \\ & & & -b_n & \delta_n \end{pmatrix}, \quad D_L = \begin{pmatrix} \delta_1 & & & & \\ & \delta_2 & & & \\ & & \ddots & & \\ & & & \delta_{n-1} & \\ & & & & \delta_n \end{pmatrix}.$$

By inspection, we have

$$\delta_1 = a_1, \quad \delta_i = a_i - \frac{b_i^2}{\delta_{i-1}}, \quad i = 2, \ldots, n.$$

This requires only $n-1$ additions, multiplications and divisions. Extensions are easily obtained to non–symmetric tridiagonal matrices as long as pivoting is not needed. A UL factorization is also easily obtained,

$$T = U D_U^{-1} U^T,$$

with

$$U = \begin{pmatrix} d_1 & -b_2 & & & \\ & d_2 & -b_3 & & \\ & & \ddots & \ddots & \\ & & & d_{n-1} & -b_n \\ & & & & d_n \end{pmatrix}, \quad D_U = \begin{pmatrix} d_1 & & & & \\ & d_2 & & & \\ & & \ddots & & \\ & & & d_{n-1} & \\ & & & & d_n \end{pmatrix}.$$

By inspection, we obtain

$$d_n = a_n, \quad d_i = a_i - \frac{b_{i+1}^2}{d_{i+1}}, \quad i = n-1, \ldots.$$

These factorizations of tridiagonal matrices have been used by Meurant [982] to characterize the inverse of such matrices. We have the following result.

Theorem 2.18

The inverse of T is characterized as

$$(T^{-1})_{i,j} = b_{i+1} \cdots b_j \frac{d_{j+1} \cdots d_n}{\delta_i \cdots \delta_n}, \quad \forall i, \quad \forall j > i,$$

$$(T^{-1})_{i,i} = \frac{d_{i+1} \cdots d_n}{\delta_i \cdots \delta_n}, \quad \forall i.$$

In these products, terms that have indices greater than n must be taken equal to 1.

Proof. From [113], [982], it is known that there exist two sequences $\{u_i\}, \{v_i\}, i = 1, n$ such that

$$T^{-1} = \begin{pmatrix} u_1 v_1 & u_1 v_2 & u_1 v_3 & \dots & u_1 v_n \\ u_1 v_2 & u_2 v_2 & u_2 v_3 & \dots & u_2 v_n \\ u_1 v_3 & u_2 v_3 & u_3 v_3 & \dots & u_3 v_n \\ \vdots & \vdots & \vdots & \ddots & \vdots \\ u_1 v_n & u_2 v_n & u_3 v_n & \dots & u_n v_n \end{pmatrix}.$$

Moreover, every non–singular matrix of the same form as T^{-1} (the matrices of this class have been called "matrices factorisables" in Baranger and Duc–Jacquet, [113]) is the inverse of an irreducible tridiagonal matrix. So to find all the elements of T^{-1}, it is sufficient to compute its first and last column. In fact it is enough to know $2n - 1$ quantities as u_1 can be chosen arbitrarily (note that $2n - 1$ is the number of non–zero terms determining T). The elements u_i and v_i can be computed in the following stable way. Let us first compute v. The $\{u_i\}, \{v_i\}$ are only defined up to a multiplicative constant. So, for instance, u_1 can be chosen as $u_1 = 1$. Then let $v = (v_1, \dots, v_n)^T$. Since $u_1 = 1$ the first column of T^{-1} is v, so

$$Tv = e^1,$$

where $e^1 = (1, 0, \dots, 0)^T$. Because of the special structure of the right hand side, it is natural to use a UL factorization of $T = U D_U^{-1} U^T$ for solving the linear system for v. We have

$$v_1 = \frac{1}{d_1}, \quad v_i = \frac{b_2 \cdots b_i}{d_1 \cdots d_{i-1} d_i}, \quad i = 2, \dots, n.$$

Let $u = (u_1, \dots, u_n)^T$, the last column of T^{-1} is $v_n u$ and therefore

$$v_n T u = e^n,$$

where $e^n = (0, \ldots, 0, 1)^T$. To solve this system, with the special structure of the right hand side, it is easier to use the LU factorization of $T = LD_L^{-1}L^T$,

$$u_n = \frac{1}{\delta_n v_n}, \quad u_{n-i} = \frac{b_{n-i+1} \cdots b_n}{\delta_{n-i} \cdots \delta_n v_n}, \quad i = 1, \ldots, n-1.$$

Note that

$$u_1 = \frac{b_2 \cdots b_n}{\delta_1 \cdots \delta_n v_n} = \frac{d_1 \cdots d_n}{\delta_1 \cdots \delta_n},$$

but $d_1 \cdots d_n = \delta_1 \cdots \delta_n = \det T$, so $u_1 = 1$, as the values of $\{v_i\}$ were computed with this scaling. □

This result for inverses of tridiagonal matrices has been extended to non–symmetric matrices by Naben [1001]. It gives a computationally stable and simple algorithm for computing elements of the inverse of T as it involves only Cholesky decompositions that are stable when the matrix T possesses enough properties such as, for instance, being diagonally dominant.

We are also interested in characterizing the decrease of the elements of T^{-1} along a row or column starting from the diagonal element. In Concus, Golub and Meurant [342], it is proven that if T is strictly diagonally dominant, then the sequence $\{u_i\}$ is strictly increasing and the sequence $\{v_i\}$ is strictly decreasing. From theorem 2.18, we have

$$\frac{(T^{-1})_{i,j}}{(T^{-1})_{i,j+1}} = \frac{d_{j+1}}{b_{j+1}},$$

and, therefore

$$(T^{-1})_{i,j} = \frac{d_{j+1} \cdots d_{j+l}}{b_{j+1} \cdots b_{j+l}} \, T^{-1}_{i,j+l}.$$

Theorem 2.19

If T is strictly diagonally dominant ($a_i > b_i + b_{i+1}$, $\forall i$) then the sequence d_i is such that $d_i > b_i$. Hence, the sequence $T^{-1}_{i,j}$ is a strictly decreasing function of j, for $j > i$. Similarly, we have $\delta_i > b_{i+1}$.

Proof. We prove the result by induction. We have $d_n = a_n > b_n$. Suppose $d_{i+1} > b_{i+1}$. Then

$$d_i = a_i - \frac{b_{i+1}^2}{d_{i+1}} > a_i - b_{i+1} > b_i,$$

and this proves the theorem. □

At this point, it is interesting to introduce another factorization, which in fact can be obtained from the LU and UL factorizations: the twisted factorization. This has also been called the BABE (burn at both ends) algorithm. To perform this factorization, we must first select an integer $j, 1 \leq j \leq n$. Let

$$T = (\phi + \mathcal{L})\phi^{-1}(\phi + \mathcal{L}^T),$$

where ϕ is a diagonal matrix and $\mathcal{L}$ is twisted

$$\mathcal{L} = \begin{pmatrix} 0 & & & & & & \\ -b_2 & 0 & & & & & \\ & \ddots & \ddots & & & & \\ & & -b_j & 0 & -b_{j+1} & & \\ & & & & \ddots & \ddots & \\ & & & & & 0 & -b_n \\ & & & & & & 0 \end{pmatrix},$$

where the two non–zero terms are in the jth row. By inspection, we have

$$\phi_i = \delta_i \quad i = 1, \ldots, j-1,$$

$$\phi_i = d_i \quad i = n, \ldots, j+1,$$

$$\phi_j = a_j - b_j^2 \delta_{j-1}^{-1} - b_{j+1}^2 d_{j+1}^{-1}.$$

Thus, the first $j - 1$ terms come from the LU factorization and the last $n - j$ come from the UL factorization. Therefore, every twisted factorization for any j can be computed with the knowledge of both the LU and UL factorizations. Only the jth term requires special treatment. Note that this twisted factorization leads naturally to a nice parallel method for a two processor computer. If we wish to solve $Tx = c$, we solve in sequence

$$(\phi + \mathcal{L})y = c, \quad (I + \phi^{-1}\mathcal{L}^T)x = y.$$

In parallel, we can compute

$$\phi_1 y_1 = c_1$$

$$\phi_i y_i = c_i + b_i y_{i-1}, \; i = 2, \ldots, j-1,$$

and

$$\begin{aligned} \phi_n y_n &= c_n \\ \phi_i y_i &= c_i + b_{i+1} y_{i+1}, \; i = n-1, \ldots, j+1. \end{aligned}$$

Then we have

$$\phi_j y_j = c_j + b_j y_{j-1} + b_{j+1} y_{j+1}.$$

From there, we obtain the solution as $x_j = y_j$ and in parallel we compute

$$\phi_i w_i = b_{i+1} x_{i+1}, \quad x_i = y_i - w_i, \; i = j-1, \ldots, 1,$$

$$\phi_i w_i = b_i x_{i-1}, \quad x_i = y_i - w_i, \; i = j+1, \ldots, n.$$

The twisted factorization can be used to prove the following result.

Theorem 2.20

The diagonal entries of the inverse are

$$(T)^{-1}_{j,j} = \phi_j^{-1}.$$

Proof. This comes about from solving a system with e^j as the right hand side and using the twisted factorization. □

This result can be used to easily derive the structure of the inverse of the tridiagonal matrix introduced at the beginning of this section. Let us specialize the previous results to the Toeplitz case that will be useful for separable problems. The interesting thing is that we are then able to solve analytically the recurrences arising in the Cholesky factorization. Let

$$T_a = \begin{pmatrix} a & -1 & & & \\ -1 & a & -1 & & \\ & \ddots & \ddots & \ddots & \\ & & -1 & a & -1 \\ & & & -1 & a \end{pmatrix}.$$

Lemma 2.21

Let

$$\alpha_1 = a, \quad \alpha_i = a - \frac{1}{\alpha_{i-1}}, \quad i = 2, \ldots, n.$$

Then if $a \neq 2$,

$$\alpha_i = \frac{r_+^{i+1} - r_-^{i+1}}{r_+^i - r_-^i},$$

where $r_\pm = \frac{a \pm \sqrt{a^2-4}}{2}$ *are the two solutions of the quadratic equation* $r^2 - ar + 1 = 0$. *If* $a = 2$, *then* $\alpha_i = \frac{i+1}{i}$.

Proof. We set

$$\alpha_i = \frac{\beta_i}{\beta_{i-1}}.$$

Therefore, we now have a recurrence on β_i :

$$\beta_i - a\beta_{i-1} + \beta_{i-2} = 0, \quad \beta_0 = 1, \quad \beta_1 = a.$$

The solution of this linear second order difference equation is well known:

$$\beta_i = c_0 r_+^{i+1} + c_1 r_-^{i+1}.$$

From the initial conditions we have $c_0 + c_1 = 0$. Hence the solution can be written as

$$\beta_i = c_0(r_+^{i+1} - r_-^{i+1}).$$

When $a = 2$ it is easy to see that $\beta_i = i + 1$ and the result follows. □

From lemma 2.21, the solutions of the recurrences involved in the Cholesky factorization of T_a can be deduced. When $a \neq 2$ we have,

$$d_{n-i+1} = \frac{r_+^{i+1} - r_-^{i+1}}{r_+^i - r_-^i}.$$

Solving for v the following result is obtained :

Proposition 2.22

For the sequence v_i in T_a^{-1},

$$v_i = \frac{r_+^{n-i+1} - r_-^{n-i+1}}{r_+^{n+1} - r_-^{n+1}}, \quad \forall i.$$

□

Note in particular, that

$$v_n = \frac{r_+ - r_-}{r_+^{n+1} - r_-^{n+1}}.$$

It is obvious that for the Toeplitz case, we have the relation $\delta_i = d_{n-i+1}$. Solving for u, the following result is obtained:

Proposition 2.23

For the sequence u_i in T_a^{-1},

$$u_i = \frac{r_+^i - r_-^i}{r_+ - r_-}, \quad i = 1, \ldots, n.$$

□

Now we are able to compute the elements of the inverse:

Theorem 2.24

For $j \geq i$ and when $a \neq 2$,

$$({T_a}^{-1})_{i,j} = u_i v_j = \frac{(r_+^i - r_-^i)(r_+^{n-j+1} - r_-^{n-j+1})}{(r_+ - r_-)(r_+^{n+1} - r_-^{n+1})},$$

where $r_\pm$ are the two solutions of the quadratic equation $r^2 - ar + 1 = 0$.

For $a = 2$, we have

$$(T_a)^{-1}_{i,j} = i\,\frac{n-j+1}{n+1}.$$

□

Regarding the decay of the elements of T_a^{-1}, in this simple case we can obtain useful bounds. Suppose that $a > 2$, then we have

$$\frac{u_i v_j}{u_i v_{j+1}} = \frac{r_+^{n-j+1} - r_-^{n-j+1}}{r_+^{n-j} - r_-^{n-j}} = \frac{r_+^{n-j+1}}{r_+^{n-j}} \left(\frac{1 - r^{n-j+1}}{1 - r^{n-j}} \right) > r_+ > 1,$$

and

$$u_i v_j < r_+^{i-j-1} \frac{(1-r^i)(1-r^{n-j+1})}{(1-r)(1-r^{n+1})}, \quad j \geq i+1,$$

where $r = \frac{r_-}{r_+} < 1$.

From this, the following result can be deduced:

Theorem 2.25

If $a > 2$, we have the bound

$$({T_a}^{-1})_{i,j} < (r_-)^{j-i} ({T_a}^{-1})_{i,i}, \quad \forall\, i, \quad \forall j \geq i,$$

$$({T_a}^{-1})_{i,j} < \frac{r_-^{j-i+1}}{1-r}, \quad \forall\, i, \quad \forall j \geq i+1.$$

□

The following estimate holds: let $\epsilon_1 > 0$ and $\epsilon_2 > 0$ be given. Then

$$\frac{({T_a}^{-1})_{i,j}}{({T_a}^{-1})_{i,i}} \leq \epsilon_1 \qquad \text{if} \qquad j - i \geq \frac{\log \epsilon_1^{-1}}{\log r_+},$$

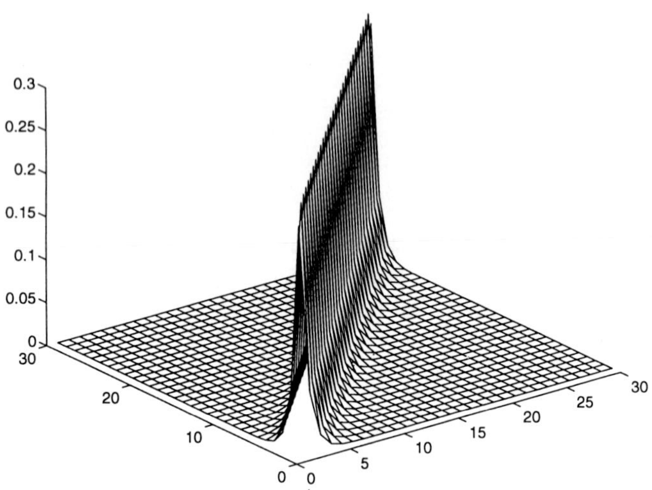

fig. 2.1: The inverse of T_4

and

$$(T_a{}^{-1})_{i,j} \le \epsilon_2 \qquad \text{if} \qquad j - i + 1 \ge \frac{\log\left[\epsilon_2(1-r)\right]^{-1}}{\log r_+}.$$

As an example, figure 2.1 shows the inverse of T_4.

The previous factorizations are easily extended to block tridiagonal symmetric matrices. Let

$$A = \begin{pmatrix} D_1 & -A_2^T & & & \\ -A_2 & D_2 & -A_3^T & & \\ & \ddots & \ddots & \ddots & \\ & & -A_{m-1} & D_{m-1} & -A_m^T \\ & & & -A_m & D_m \end{pmatrix},$$

each block being of order m. Denote by L the block lower triangular part of A then, if such a factorization exists, we have

$$A = (\Delta + L)\Delta^{-1}(\Delta + L^T),$$

where Δ is a block diagonal matrix whose diagonal blocks are denoted by Δ_i. By inspection we have

$$\Delta_1 = D_1, \quad \Delta_i = D_i - A_i(\Delta_{i-1})^{-1}A_i^T,\ i = 2, \ldots, m$$

Therefore, obtaining this block factorization only involves solving linear systems with matrices Δ_i and several right hand sides. Whatever the structure of the matrices D_i, matrices Δ_i, $i = 2, \ldots, m$ are dense matrices.

This block factorization (as well as the block twisted factorization) can also be used (see Meurant [982]) to characterize the inverse of block tridiagonal matrices. Let us write the block factorizations as

$$A = (\Delta + L)\ \Delta^{-1}\ (\Delta + L^T) = (\Sigma + L^T)\ \Sigma^{-1}\ (\Sigma + L),$$

where Δ and Σ are block diagonal matrices whose diagonal blocks are denoted by Δ_i and Σ_i. They are given by block recurrences

$$\begin{cases} \Delta_1 = D_1, \\ \Delta_i = D_i - A_i\ (\Delta_{i-1})^{-1}\ (A_i)^T \end{cases} \quad \begin{cases} \Sigma_m = D_m, \\ \Sigma_i = D_i - (A_{i+1})^T\ (\Sigma_{i+1})^{-1}\ A_{i+1}. \end{cases}$$

A block twisted factorization can be defined for each $j = 2, \ldots, m-1$ as

$$A = (\Phi + \mathcal{L})\Phi^{-1}(\Phi + \mathcal{L}^T)$$

where Φ is a block diagonal matrix and $\mathcal{L}$ has the following twisted block structure

$$\mathcal{L} = \begin{pmatrix} 0 & & & & & & \\ -A_2 & 0 & & & & & \\ & \ddots & \ddots & & & & \\ & & -A_j & 0 & -A_{j+1}^T & & \\ & & & & \ddots & \ddots & \\ & & & & & 0 & -A_m^T \\ & & & & & & 0 \end{pmatrix},$$

where the row with two non–zero terms is the jth block row. By inspection we have

$$\Phi_i = \Delta_i \quad i = 1, \ldots, j-1,$$

$$\Phi_i = \Sigma_i \quad i = m, \ldots, j+1,$$

$$\Phi_j = D_j - A_j \Delta_{j-1}^{-1} A_j^T - A_{j+1}^T \Sigma_{j+1}^{-1} A_{j+1}.$$

With the block twisted factorization at hand, the block jth column X of the inverse can be computed in a straightforward way.

Theorem 2.26

The jth block column X of A^{-1} is given by

$$X_j = \Phi_j^{-1},$$

$$X_{j-l} = \Delta_{j-l}^{-1}\, A_{j-l+1}^T\, \Delta_{j-l+1}^{-1} \cdots \Delta_{j-1}^{-1}\, A_j^T\, \Phi_j^{-1},\ l = 1, \ldots, j-1$$

$$X_{j+l} = \Sigma_{j+l}^{-1}\, A_{j+l}\, \Sigma_{j+l-1}^{-1} \cdots \Sigma_{j+1}^{-1}\, A_{j+1}\, \Phi_j^{-1},\quad l = 1, \ldots, m-j.$$

□

These expressions are valid for any symmetric block tridiagonal matrix. When matrices A_i are non–singular, A is said to be *proper*; in this case, the formulas can be simplified. Using the uniqueness of the inverse, we can prove the following.

Proposition 2.27

If A is proper, then

$$\Phi_j^{-1} = A_{j+1}^{-1}\, \Sigma_{j+1} \cdots A_n^{-1}\, \Sigma_m\, \Delta_m^{-1}\, A_m \cdots \Delta_{j+1}^{-1}\, A_{j+1}\, \Delta_j^{-1}$$
$$= A_j^{-T}\, \Delta_{j-1} \cdots A_2^{-T}\, \Delta_1\, \Sigma_1^{-1}\, A_2^T \cdots \Sigma_{j-1}^{-1}\, A_j^T\, \Sigma_j^{-1}.$$

□

From these relations, we deduce alternate formulas for the other elements of the inverse.

Theorem 2.28

If A is proper,

$$X_{j-l} = (A_{j-l}^{-T}\Delta_{j-l-1} \cdots A_2^{-T}\Delta_1)(\Sigma_1^{-1}A_2^T \cdots A_j^T\Sigma_j^{-1}),\quad l = 1, \ldots, j-1$$

$$X_{j+l} = (A_{j+l+1}^{-1}\Sigma_{j+l+1} \cdots A_m^{-1}\Sigma_m)(\Delta_m^{-1}A_m \cdots \Delta_{j+1}^{-1}A_{j+1}\Delta_j^{-1}),\quad l = 1, \ldots, m-j.$$

Proof. We use proposition 2.27 in theorem 2.26. □

As before, the elements of the inverse can be computed in a stable way using the block Cholesky factorization when the matrix is diagonally dominant or positive definite. These formulas are the block counterpart of those for tridiagonal matrices. They give a characterization of the inverse of a proper block tridiagonal matrix.

Theorem 2.29

If A is proper, there exist two (non–unique) sequences of matrices $\{U_i\}, \{V_i\}$ such that for $j \geq i$

$$(A^{-1})_{i,j} = U_i V_j^T,$$

with $U_i = A_i^{-T}\Delta_{i-1}\cdots A_2^{-T}\Delta_1$ and $V_j^T = \Sigma_1^{-1}A_2^T\cdots A_j^T\Sigma_j^{-1}$.

□

In other words, A^{-1} can be written as

$$A^{-1} = \begin{pmatrix} U_1V_1^T & U_1V_2^T & U_1V_3^T & \dots & U_1V_m^T \\ V_2U_1{}^T & U_2V_2^T & U_2V_3^T & \dots & U_2V_m^T \\ V_3U_1{}^T & V_3U_2{}^T & U_3V_3^T & \dots & U_3V_m^T \\ \vdots & \vdots & \vdots & \ddots & \vdots \\ V_mU_1{}^T & V_mU_2{}^T & V_mU_3{}^T & \dots & U_mV_m^T \end{pmatrix}.$$

The inverse of the matrix of the Poisson model problem is shown on figure 2.2.

2.6. Roundoff error analysis

Let us consider a small example to see some of the difficulties that can happen when solving a linear system using Gaussian elimination. We suppose that the arithmetic is such that $\beta = 10$, $t = 3$ (we ignore limits on the exponent range) and, for the sake of simplicity, we use chopped arithmetic. The system we solve is

$$3x_1 + 3x_2 = 6$$

$$x_1 + \delta x_2 = \delta + 1$$

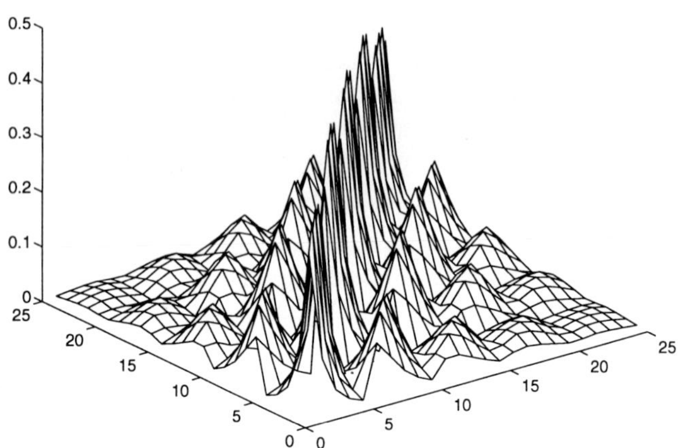

fig. 2.2: The inverse of the matrix of the Poisson problem, $m = 5$

where δ is a given parameter. The exact solution is obviously $x_1 = x_2 = 1$.

Choosing 3 as the pivot and noticing that the computed value of the multiplier $1/3$ is $0.333\ 10^0$, we obtain for the second equation

$$(\delta - 0.999\ 10^0)x_2 = \delta + 1 - 6 \times (0.333\ 10^0) = \delta - 0.990\ 10^0.$$

Therefore, if $\delta = 0.990$, then $x_2 = 0$ and $x_1 = 0.199\ 10^1$! This system is far from being singular, as the determinant is

$$\det(A) = 3 \times 0.99 - 3 = -0.03$$

Note that even though the solution is wrong, the residual $b - Ax$ is small, being $(0.03\ 0)^T$. Remark also that the pivot is not small.

Let us now consider the kth step of standard Gaussian elimination algorithm using backward error analysis. We denote the computed quantities that we consider by the same notations as in exact arithmetic as there will be no ambiguity. Also, following Wilkinson [1333], we do not pay too much attention to the values of the constants involved in the bounds. They are all independent of the order of the matrix and are of the order of 1.

Theorem 2.30

At step k we have

$$L_k A_k = A_{k+1} + E_k,$$

where

$$|(E_k)_{i,j}| \le Cu \max(|a_{i,j}^{(k+1)}|, |a_{i,j}^{(k)}|),$$

u being the unit round off.

Proof. The multipliers (e.g. the elements of L) that we denote by $l_{i,k}$ are

$$l_{i,k} = fl\left(\frac{a_{i,k}^{(k)}}{a_{k,k}^{(k)}}\right), \quad i \ge k+1,$$

$$a_{i,j}^{(k+1)} = \begin{cases} 0 & i \ge k+1,\ j = k, \\ fl(a_{i,j}^{(k)} - l_{i,k}a_{k,j}^{(k)}), & i \ge k+1,\ j \ge k+1, \\ a_{i,j}^{(k)} & \text{otherwise.} \end{cases}$$

Let us first consider $i \ge k+1$,

$$l_{i,k} = \frac{a_{i,k}^{(k)}}{a_{k,k}^{(k)}}(1+\epsilon_{i,k}), \quad |\epsilon_{i,k}| \le u.$$

This translates into

$$a_{i,k}^{(k)} - l_{i,k}a_{k,k}^{(k)} + a_{i,k}^{(k)}\epsilon_{i,k} = 0.$$

If we denote by $e_{i,k}^{(k)}$, the elements of E_k, this shows that

$$e_{i,k}^{(k)} = a_{i,k}^{(k)}\epsilon_{i,k}, \quad i \ge k+1.$$

For $i \ge k+1$ and $j \ge k+1$, we have

$$a_{i,j}^{(k)} = fl(a_{i,j}^{(k)} - fl(l_{i,k}a_{k,j}^{(k)})).$$

From Chapter 1, we have

$$|e_{i,j}^{(k)}| \le Cu \max(|a_{i,j}^{(k+1)}|, |a_{i,j}^{(k)}|).$$

This bound is also true for $e_{i,k}^{(k)}$. Therefore we have

$$|e_{i,j}^{(k)}| \le Cu \max(|a_{i,j}^{(k+1)}|, |a_{i,j}^{(k)}|), \quad i \ge k+1,\ j \ge k,$$

and the other entries of E_k are zero. □

We have the following useful technical lemma.

Lemma 2.31

If B_k is a matrix whose first k rows are zero, then

$$L_i B_k = B_k, \quad i \leq k.$$

Similarly, $(L_i)^{-1} B_k = B_k$.

Proof. Remember that $L_k = I - l^k {e^k}^T$. Then

$$L_i B_k = (I - l^i {e^i}^T) B_k = B_k - l^i {e^i}^T B_k = B_k,$$

as ${e^i}^T B_k = 0$. □

With this last result we can prove the following theorem on LU factorization.

Theorem 2.32

Let F be defined as

$$F = F_1 + \cdots + F_{n-1},$$

where

$$(F_k)_{i,j} = \begin{cases} 1 & i \geq k+1, \ j \geq k \\ 0 & \text{otherwise} \end{cases}$$

Then

$$A = LU + E,$$

with

$$|E| \leq Cu \max_{k,i,j} |a_{i,j}^{(k)}| \ F,$$

Proof. We have

$$L_k A_k = A_{k+1} + E_k \Longrightarrow A_k = L_k^{-1} A_{k+1} + E_k.$$

By lemma 2.31,

$$(L_1)^{-1}\dots(L_{k-1})^{-1}A_k = (L_1)^{-1}\dots(L_k)^{-1}A_{k+1} + E_k.$$

We sum these equalities for $k = 1$ to $n-1$. Most of the terms cancel and we obtain,

$$A = (L_1)^{-1}\dots(L_{k-1})^{-1}A_n + E_1 + \dots + E_{n-1}.$$

Therefore, $A = LU + E$ with $E = E_1 + \dots + E_{n-1}$. Finally, it is easy to see that

$$|E| \le Cu \max_{k,i,j} |a_{i,j}^{(k)}|\, F.$$

□

This result shows that matrices L and U are the exact factors of the factorization of a perturbed matrix $A - E$. The elements of F are bounded by $n-1$. Therefore $|E|$ being small depends on $\max_{i,j,k} |a_{i,j}^{(k)}|$. Using the growth factor (that we still denote by g_A for the computed quantities), this bound can be written as

$$|E| \le Cu\, g_A \|A\|_\infty F.$$

Taking norms, we have

$$\|E\|_\infty \le Cu\, g_A\, n^2 \|A\|_\infty.$$

We also have to take into account errors made in solving the triangular systems that arose from the factorization. The following result is taken from Higham [776] but it is mainly due to Wilkinson.

Theorem 2.33

$$(L+K)y = b,$$

with

$$|K| \le Cu \begin{pmatrix} |l_{1,1}| & & & & \\ |l_{2,1}| & 2|l_{2,2}| & & & \\ 2|l_{3,1}| & 2|l_{3,2}| & 2|l_{3,3}| & & \\ \vdots & \vdots & \vdots & \ddots & \\ (n-1)|l_{n,1}| & (n-1)|l_{n,2}| & (n-2)|l_{n,3}| & \dots & 2|l_{n,n}| \end{pmatrix},$$

and

$$\begin{aligned} |K| &\leq Cnu|L|, \\ \|K\|_\infty &\leq \frac{n(n+1)}{2} Cu \max_{i,j} |l_{i,j}|. \end{aligned}$$

□

With these results, we can obtain bounds for the solution of a linear system.

Theorem 2.34

The computed solution x satisfies

$$(A+H)x = b,$$

with

$$|H| \leq C_1 u \max_{k,i,j} |a_{i,j}^{(k)}| F + C_2 nu|L||U| + O(u^2).$$

Proof.

$$(L+K)(U+G)x = (LU + KU + LG + KG)x = b,$$

and

$$LU = A - E.$$

Then, if we introduce H as

$$H = KU + LG + KG - E,$$

we have $(A+H)x = b$ and

$$\begin{aligned} |H| &\leq |E| + |K||U| + |L||G| + |K||G|, \\ |H| &\leq C_1 u \max_{k,i,j} |a_{i,j}^{(k)}| F + C_2 nu|L||U| + C_3 n^2 u^2 |L||U|, \\ \|H\|_\infty &\leq (C_3 n^3 + C_4 n^2) u g_A \|A\|_\infty, \end{aligned}$$

and this proves the result. □

Pivoting techniques are not only used to avoid zero pivots, but also to reduce the growth factor. For instance, if partial pivoting is used, we have $|l_{i,j}| \leq 1$. From this and denoting $\rho_k = \max_{i,j} |a_{i,j}^{(k)}|$, it is not difficult to prove that

$$|a_{i,j}^{(k+1)}| \leq |a_{i,j}^{(k)}| + \epsilon \max(|a_{i,j}^{(k+1)}|, |a_{i,j}^{(k)}|),$$

$$\rho_{k+1} \leq 2(1+\epsilon)\rho_k.$$

By recurrence, this gives a bound for ρ_n:

$$\rho_n \leq 2^{n-1}(1+\epsilon)^{n-1}\rho_1.$$

Wilkinson [1333] gave contrived examples where such an exponential growth is observed. However, more recently, linear systems arising from practical problems for which an exponential growth is obtained have appeared in the literature, see Wright [1339] and Foster [575]. The average case stability of Gaussian elimination with partial pivoting was studied in Trefethen and Schreiber [1266]. Random matrices of order ≤ 1024 have been looked at. The average growth factor was approximately $n^{3/2}$ for partial pivoting and $n^{1/2}$ for complete pivoting. Although Gaussian elimination with partial pivoting gives accurate results in most cases, these examples show that when using this algorithm we must carefully analyze the results.

Another possible (but more costly) choice is complete pivoting where the pivot is searched in $i, j > k$. Wilkinson [1333] showed that the growth factor is bounded (in the absence of roundoff) by as

$$|a_{i,j}^{(k)}| \leq k^{\frac{1}{2}}(2 \cdot 3^{\frac{1}{2}} \cdots k^{\frac{1}{k-1}})^{\frac{1}{2}} \max_{i,j} |a_{i,j}|.$$

It was conjectured that in this case $g_A \leq n$. Cryer [352] proved that this is true for $n \leq 4$. However, the conjecture has been shown to be false for $n > 4$ if roundoff error is allowed by Gould [685]. Edelman and Ohlroch [497] modified the Gould counterexample to show that the conjecture is also false in exact arithmetic. N. Higham and D. Higham [783] have given some matrices of practical interest which have a growth factor of at least $\frac{n}{2}$ for complete pivoting.

2.7. Perturbation analysis

It is interesting to study what the effects of perturbation are on the data (the coefficients of the matrix and the right hand side). There are different ways to do this that differ mainly in the way perturbations are measured (for a thorough treatment of this topic, see Higham [781]). The oldest way is known as normwise error analysis. Let x and y be such that

$$\begin{aligned} Ax &= b, \\ (A+\Delta A)y &= b+\Delta b. \end{aligned}$$

ΔA and Δb are chosen such that

$$\|\Delta A\| \leq \alpha\omega, \quad \|\Delta b\| \leq \beta\omega,$$

ω is given, α (resp. β) will be 0 or $\|A\|$ (resp. $\|b\|$) depending on whether A, or b, or both are perturbed. ω defines the normwise relative perturbation. Then, we can bound the solution of the perturbed system.

Lemma 2.35

If $\xi = \alpha\omega\|A^{-1}\| < 1$, *then*

$$\frac{\|y\|}{\|x\|} \leq \frac{1}{1-\xi}\left(1+\frac{\xi\beta}{\alpha\|x\|}\right).$$

Proof. First, we must show that $A+\Delta A$ is non–singular. We note that

$$A+\Delta A = A(I+A^{-1}\Delta A),$$

and, with the hypothesis of the lemma,

$$\|A^{-1}\Delta A\| \leq \alpha\omega\|A^{-1}\| = \xi < 1.$$

Then, $A+\Delta A$ is non–singular by lemma 2.3.3 of Golub and Van Loan [676]. We have

$$(I+A^{-1}\Delta A)y = A^{-1}(b+\Delta b) = x + A^{-1}\Delta b.$$

Taking norms,

$$\|y\| \leq \frac{1}{1-\alpha\omega\|A^{-1}\|}(\|x\| + \beta\omega\|A^{-1}\|).$$

But,

$$\omega \leq \frac{1}{\alpha\|A^{-1}\|} \Longrightarrow \|y\| \leq \frac{1}{1-\alpha\omega\|A^{-1}\|}\left(\|x\| + \xi\frac{\beta}{\alpha}\right).$$

Therefore,

$$\frac{\|y\|}{\|x\|} \le \frac{1}{1-\alpha\omega\|A^{-1}\|}\left(1+\xi\frac{\beta}{\alpha\|x\|}\right).$$

Note that if $\alpha = \|A\|$ and $\beta = \|b\|$, then

$$\frac{\|y\|}{\|x\|} \le \frac{1+\xi}{1-\xi}.$$

□

The relative difference with the exact solution can also be bounded.

Theorem 2.36

Under the hypothesis of lemma 2.35,

$$\frac{\|x-y\|}{\|x\|} \le \omega\left(\|A^{-1}\|\frac{\alpha\|x\|+\beta}{\|x\|}\right)\left(\frac{1}{1-\xi}\right).$$

Proof. We have,

$$y - x = A^{-1}\Delta b - A^{-1}\Delta A y.$$

By taking norms, the result follows easily. □

If $\alpha = \|A\|$ and $\beta = \|b\|$, then

$$\frac{\|y-x\|}{\|x\|} \le 2\omega\|A\|\;\|A^{-1}\|\left(\frac{1}{1-\xi}\right).$$

The normwise condition number of the problem is defined as

$$K_T(A,b) = \|A^{-1}\|\frac{\alpha\|x\|+\beta}{\|x\|}.$$

Note that, when $\beta = 0$, this reduces to $K_T(A) = \|A^{-1}\|\;\|A\|$. The subscript T refers to A. Turing who first introduced this condition number. It measures the sensitivity of the solution to perturbations. Let us now introduce the normwise backward error. It measures the minimal distance to a perturbed problem which is solved exactly by the computed solution y. Let

$$\eta_T = \inf\{\omega|\;\omega \ge 0,\;\|\Delta A\| \le \omega\alpha,\;\|\Delta b\| \le \omega\beta,\;(A+\Delta A)y = b+\Delta b\}.$$

The normwise backward error has been characterized by Rigal and Gaches [1102] in the following result.

Theorem 2.37

Let $r = b - Ay$ be the residual. Then

$$\eta_T = \frac{\|r\|}{\alpha\|y\| + \beta}.$$

Proof. See Rigal and Gaches [1102]. □

We remark that the bound on the forward error, $\frac{\|y-x\|}{\|x\|}$, is approximately the product of the condition number and the backward error.

Another type of analysis has considered componentwise perturbations. This was introduced by F.L. Bauer and R. Skeel (see, Skeel [1177]). This allows to study perturbations on individual elements of A and b. It is particularly useful for sparse matrices. The same kind of results as for the normwise analysis can be proved. We consider perturbations ΔA and Δb such that

$$|\Delta A| \leq \omega E, \quad |\Delta b| \leq \omega f.$$

Theorem 2.38

If $\omega\| \, |A^{-1}|E\|_\infty < 1$,

$$\frac{\|y - x\|_\infty}{\|x\|_\infty} \leq \omega \frac{\| \, |A^{-1}|(E|x| + f)\|_\infty}{\|x\|_\infty} \frac{1}{1 - \omega\| \, |A^{-1}|E\|_\infty}.$$

□

The componentwise condition number is defined as

$$K_{BS}(A, b) = \frac{\| \, |A^{-1}|(E|x| + f)\|_\infty}{\|x\|_\infty}.$$

The subscript BS refers to Bauer and Skeel. As before, we introduce the backward error:

$$\eta_{BS} = \inf\{\omega|, \omega \geq 0, \ |\Delta A| \leq \omega E, \ |\Delta b| \leq \omega f, \ (A + \Delta A)y = b + \Delta b\}.$$

Oettli and Prager [1037] proved the following characterization of η_{BS}.

Theorem 2.39

$$\eta_{BS} = \max_i \frac{|(b - Ay)_i|}{(E|y| + f)_i}.$$

□

An algorithm is said to be backward stable when the backward error is of the order of the machine precision u. Gaussian elimination with partial pivoting is both normwise and componentwise backward unstable. There are examples where η_T or η_{BS} are large compared to machine precision. Despite this fact, Gaussian elimination can be used safely on most practical examples. Moreover, there are some remedies to this (potential) backward instability; see the section on iterative refinement.

2.8. Scaling

Scaling is a transformation of the linear system to be solved trying to give a better behaved system before using Gaussian elimination. Let D_1 and D_2 be two non–singular diagonal matrices. The system $Ax = b$ is transformed into

$$A'y = (D_1 A D_2)y = D_1 b,$$

and the solution x is recovered by $x = D_2 y$. Note that left multiplication by D_1 is a row scaling and right multiplication by D_2 is a column scaling. The element $a_{i,j}$ of A is changed into $d_i^1 d_j^2 a_{i,j}$ where d_i^l, $l = 1, 2$ are the diagonal entries of D_l.

Note that if one uses Gaussian elimination with partial pivoting to solve the scaled system, row scaling influences the choice of the pivot. Classical strategies can be found in Curtis and Reid [360]. Other proposals were described by Hager [749]. A common strategy for row scaling is to divide the entries of a row by the maximum norm of the row. Skeel [1177] showed that a good scaling matrix is made by choosing the diagonal elements of D_1 as $d_i = (|A|\,|y|)_i$ where y is the computed solution. Of course, this is impractical as the solution y depends on the scaling. However, if an approximation c of the solution is known, then A could be scaled by $(|A|\,|c|)_i$.

No scaling strategy has been shown to give consistently better results than not using any scaling although many different strategies have been proposed over the years. It has been argued that the real role of scaling is to alter the pivoting sequence. This can result in a better or worse solution. The rule of thumb given in [1077] is that if scaling can lead to a system which is more diagonally dominant, it may be useful but otherwise it should be avoided.

2.9. Iterative refinement

From the roundoff error analysis, we have shown that the computed solution satisfies

$$(A+H)y = b,$$

with

$$\|H\|_\infty \leq u\ C\|A\|_\infty,$$

if the growth factor is bounded. Let $r = b - Ay$ be the residual. Then obviously

$$\|r\|_\infty \leq \|H\|_\infty\ \|y\|_\infty \leq uC\|A\|_\infty\ \|y\|_\infty.$$

Hence, if C is not too large, Gaussian elimination usually produces a small residual. But, unfortunately, small residuals do not always imply high accuracy in the solution. Let $e = x - y$, then $Ae = b - Ay = r$. Therefore, a natural idea to improve the computed solution is to solve

$$Ae = r.$$

This will produce a computed solution $\tilde{e}$, satisfying

$$(A + \tilde{H})\tilde{e} = r$$

and we can set $\tilde{x} = y + \tilde{e}$ as the new approximation to the solution. If we like, we can iterate this process. This algorithm is known as iterative refinement (or iterative improvement). Of course, iterative refinement is just one example of a very simple iterative method. More sophisticated ones will be studied in subsequent chapters.

Here, the main question is to decide to which precision the residual r has to be computed as we are not able to get the exact answer. If we have $\tilde{r} = fl(b - Ay)$ and $(A + \tilde{H})\tilde{e} = \tilde{r}$, Skeel [1177] has shown that computing the residual vector to the same precision as the original computation is enough to make Gaussian elimination with partial pivoting backward stable. This is shown in the following theorem.

Theorem 2.40

If $u|A|\,|A^{-1}|$ *is sufficiently small, one step of iterative refinement with single precision residual computation is componentwise backward stable.*

□

However, one step of iterative refinement only gives a small backward error. It does not guarantee better accuracy. If this is desired, the residual must be computed in double precision and then rounded to single precision.

2.10. Parallel solution of general linear systems

We start by considering the parallel solution of triangular linear systems on distributed memory architectures. Although this accounts only for a small fraction of the overall computing time when solving general linear systems, it is an interesting challenge for parallel computing. At first sight, solving triangular systems might seem a very sequential task. Nevertheless, many parallel algorithms have been devised for solving triangular systems, see Heller [763]. However, several of the oldest algorithms assumed that $O(n^3)$ processors were available and are not of practical interest on present machines that have at most a few thousand processors. We mainly follow expositions by Heath and Romine [758] and Eisenstat, Heath, Henkel and Romine [514].

For solving $Lx = b$, where L is lower triangular, on a serial computer, there are basically two methods. In the first one, components of x are computed in a natural way, one after the other:

```
for i=1:n
  for j=1:i-1
    b(i)=b(i)-l(i,j)*x(j);
  end
  x(i)=b(i)/l(i,i);
end
```

This is known as the scalar product algorithm as the operation in the central loop is computing a scalar product. The other algorithm modifies the right hand side as soon as each component of x has been computed:

```
for j=1:n
  x(j)=b(j)/l(j,j);
  for i=j+1:n
    b(i)=b(i)-l(i,j)*x(j);
  end
end
```

This is known as the Saxpy algorithm by reference to the operation in the central loop. We first look at parallel implementations where the data is distributed by rows or by columns. We define a mapping `map` giving the processor number to which a row or column is mapped. The most commonly used one is known as the wrap mapping. It is defined as:

$$\begin{pmatrix} \mathtt{j}: & 1 & 2 & 3 & \dots & p & p+1 & \dots & 2p & 2p+1 & \dots & n \\ \mathtt{map(j)}: & 1 & 2 & 3 & \dots & p & 1 & \dots & p & 1 & \dots & p \end{pmatrix}$$

This mapping can be generalized by considering blocks of consecutive rows or columns instead of individual rows or columns. Doing so decreases communication time but increases load imbalance. Methods using these mappings are called panel methods in Rothberg's Ph.D. thesis [1124].

Consider first the Saxpy algorithm. It is obvious that the modifications of the components of b can be computed in parallel if the data is distributed by rows. Suppose that each processor has a set { `myrows` } containing the indices of rows (and solution components) the memory of the processor is storing. As soon as the x_j component of the solution is computed by processor number j, it must be broadcast to all other processors. This is done in a fan–out operation. `Fan-out(x,proc)` means that processor `proc` sends `x` (located in its local memory) to all other processors. The algorithm (the code running on one processor) is the following (Heath, Romine [758]):

```
for j=1:n
  if j ∈ { myrows }
```

```
        x(j)=b(j)/l(j,j);
        fan-out(x(j),map(j))
      else
        wait for x(j)
      endif
      for i ∈ { myrows }
        b(i)=b(i)-l(i,j)*x(j);
      end
    end
```

The implementation of the fan–out operation depends on the computer architecture, particularly the topology of the communication network. It provides the necessary synchronization as one processor sends data and all the others wait to receive it. One problem is that only one word (x_j) is sent at a time. Usually, sending a message of l words costs $t = t_0 + \tau l$, t_0 is the start–up time (or latency). The efficiency depends on the value of t_0 relatively to τ and l. For sending only one word, the cost is essentially the start–up time. Therefore this algorithm can be efficient only on computers with a small latency.

For the scalar product algorithm, to obtain parallelism in the inner loop, the data has to be distributed by columns. Then, each processor can compute the `l(i,j)*x(j)` term for `j` in its column index set { `mycolumns` }. These partial contributions must be added to those of the other processors. This is done in the fan–in operation: `Fan-in(x,proc)` means that processor `proc` receives the sum of all the `x`'s over all processors. Other algorithms were devised that look for parallelism in the outer loop, such as Wavefront algorithms or cyclic algorithms (see Heath and Romine [758], Li and Coleman [898], [899]).

Let us now consider the LU factorization. Many parallel algorithms have been devised over the years. They can be mainly classified by the way the matrix is stored in processor memories. If we suppose the matrix is stored by rows, then a possible algorithm is the following:

```
for k=1:n-1
  find pivot row r
  if r ∈ { myrows }
    broadcast pivot row
  else
    receive pivot row
  end
  for i>k & i ∈ { myrows }
    m(i,k)=a(i,k)/a(k,k);
    for j=k+1:n-1
      a(i,j)=a(i,j)-m(i,k)*a(k,j);
    end
  end
end
```

The drawback with this algorithm is that, as the columns are scattered amongst processors, there will be communications to find which row is the pivot row. We can also store the matrix by columns. Then, finding the pivot can be done in one processor without additional communications. But this pivot search is a sequential process. The code is the following:

```
for k=1:n-1
  if k ∈ { mycolumns }
    find pivot row r
    for i=k+1:n
      m(i,k)=a(i,k)/a(k,k)
    end
```

```
      broadcast m and pivot index
    else
      receive m and pivot index
    end
    for j> k & j ∈ { mycolumns }
      for i=k+1:n
        a(i,j)=a(i,j)-m(i,k)*a(k,j)
      end
    end
  end
```

More efficient algorithms can be obtained if the data is partitioned by blocks, see Dongarra and Walker [436]. Independent data distributions are used for rows and columns. An object m (a piece of row or column) is mapped to a couple (p, i), p being the processor number and i the location in the local memory of this processor. By using wrapping as before, we have

$$m \longrightarrow (m \bmod p, \lfloor m/p \rfloor).$$

Blocking consists of assigning contiguous entries to processors by blocks,

$$m \longrightarrow (\lfloor m/L \rfloor, m \bmod L), \quad L = \lceil m/p \rceil.$$

The block cyclic distribution is a combination of both. Blocks of consecutive data are distributed by wrapping,

$$m \longrightarrow (q, b, i),$$

where q is the processor number, b the block number in processor q and i the index in block b. If there are r data objects in a block, then

$$m \longrightarrow (\lfloor \frac{m \bmod T}{r} \rfloor, \lfloor \frac{m}{T} \rfloor, m \bmod r), \quad T = rp.$$

To distribute the matrix, independent block cyclic distributions are applied for the rows and columns. The processors are supposed to be (logically) arranged in a two dimensional mesh and referred by couples (q_1, q_2). For general data distributions, communications are required for the pivot search and the computation of the multipliers. The communications to be done are a broadcast to all processors and a broadcast to all processors in the same row (or column) in the 2D mesh of processors. Finally, the logical arrangement of processors has to be mapped to the physical layout, see Dongarra and Walker [436] and also Rothberg [1124] for details.

2.11. Bibliographical comments

Thousands of papers have been written on Gaussian elimination over the years. Elimination for solving linear systems is in fact an old idea. In the second century a.d., the Chinese were solving linear systems of very small order (see [802]). For an account of the work of Gauss on elimination, we refer to the translations of some of his papers by G.W. Stewart, although according to Stewart [1204] Lagrange was already using what we call Gaussian elimination.

Some years ago there were some questions on NA-net about Cholesky and the spelling of his name, as some authors write Choleski and some others Cholesky. It turns out that there is no choice as Cholesky was not Russian or Polish but French. André Louis Cholesky was born on October 15, 1875 in Montguyon a small village in the district of Charente Maritime in the west of France. He was the son of André Cholesky, a butler born in 1842 in the same village of French parents. Cholesky did part of his studies in Bordeaux and he attended Ecole Polytechnique in 1895. From then on he was in the military. In 1903 he was sent to the geographical service of the army. This is where he worked on his method for the solution of least squares problems. He died during the first World War on August 31, 1918. All these details were found in an interesting report written by C. Brezinski from Lille University.

When the first computers were developed after World War II, it was not obvious to everyone that Gaussian elimination could be used safely for computing solution of linear systems, see [1310]. It is mainly through the work of Wilkinson, summarized in his book [1333], that the properties of Gaussian elimination were carefully studied and it became widely known that Gaussian elimination may be successfully used. Wilkinson also developed some software to be used on the early computers available at that time.

Modern versions of Gaussian algorithms were incorporated in LINPACK (1979). Codes more suited to vector computers were developed for LAPACK

(1992). Through the work of Jack Dongarra, efficient versions of these libraries were made available on all the best scientific computers in the eighties and nineties. For a summary of the work on the stability of Gaussian elimination, see Higham [781].

The new challenge facing numerical analysts at the end of the nineties is finding efficient implementations of Gaussian elimination on parallel architectures with a large number of processors. This has been the subject of hundreds of papers during the last ten years and efficient solutions are beginning to be available.

3 Gaussian elimination for sparse linear systems

3.1. Introduction

In the Chapter 2, we addressed several properties of the matrices we considered such as symmetry or positive definiteness, but we did not examine the numerical values of the coefficients in the matrix. In this chapter we shall look at special techniques developed for handling Gaussian elimination on sparse matrices. A sparse matrix is one with many zero entries. However, there is no precise definition of what a sparse matrix is, i.e. how many zeros entries there are or the percentage of zeros. We have seen in Chapter 1 that special techniques are used to store sparse matrices. Special algorithms have been defined in order to minimize the storage and the number of operations during Gaussian elimination. A definition that has sometimes been given is that a matrix is sparse when it is beneficial (either in computer storage or in computer time) to use these special sparse techniques as opposed to the more traditional dense (or general) algorithms that we have described. Exploiting sparsity allows the solution of very large problems with millions of unknowns.

There are several good books about direct methods for sparse linear systems. We mention those of George and Liu [625] for symmetric positive definite systems and Duff, Erisman and Reid [475] for more general sparse systems.

A potential problem with most sparse techniques is that they are in general quite complex (actually much more complex than algorithms for dense matrices) and sometimes difficult to optimize. It is usually not feasible for the average user to write a sparse code from scratch. Fortunately, there are good packages available containing well tuned codes such as the Harwell Library, Duff and Reid [482], [483], [484], [485], [486], or SPARSPAK, George and Liu [625], to mention just a few. See also Gilbert et al [406] and Davis [379].

3.2. The fill–in phenomenon

At the kth step of Gaussian elimination, we compute

$$a_{i,j}^{(k+1)} = a_{i,j}^{(k)} - \frac{a_{i,k}^{(k)} a_{k,j}^{(k)}}{a_{k,k}^{(k)}}.$$

From this formula we see that, even if $a_{i,j}^{(k)} = 0$, $a_{i,j}^{(k+1)}$ can be non–zero if $a_{i,k}^{(k)} \neq 0$ and $a_{k,j}^{(k)} \neq 0$. Non–zero entries in the L and U factors in positions (i,j) for which $a_{i,j} = 0$ are known as fill–in. In general, there will be more non–zero entries in $L + U$ than in A. The storage scheme for L and U must be designed to account for this.

Consider the small example below where the matrix is symmetric and the x's stand for the non–zero entries,

$$A = \begin{pmatrix} x & x & 0 & x & 0 \\ x & x & x & 0 & 0 \\ 0 & x & x & 0 & x \\ x & 0 & 0 & x & 0 \\ 0 & 0 & x & 0 & x \end{pmatrix}.$$

Let us look at the different steps of Gaussian elimination. Fill–in is denoted by $\bullet$,

$$A_2 = \begin{pmatrix} x & x & 0 & x & 0 \\ 0 & x & x & \bullet & 0 \\ 0 & x & x & 0 & x \\ 0 & \bullet & 0 & x & 0 \\ 0 & 0 & x & 0 & x \end{pmatrix}, \quad A_3 = \begin{pmatrix} x & x & 0 & x & 0 \\ 0 & x & x & \bullet & 0 \\ 0 & 0 & x & \bullet & x \\ 0 & 0 & \bullet & x & 0 \\ 0 & 0 & x & 0 & x \end{pmatrix}.$$

Note that the fill–in in position $(4,3)$ has been created by the fill–in in position

$(4,2)$ at the previous step,

$$A_4 = \begin{pmatrix} x & x & 0 & x & 0 \\ 0 & x & x & \bullet & 0 \\ 0 & 0 & x & \bullet & x \\ 0 & 0 & 0 & x & \bullet \\ 0 & 0 & 0 & \bullet & x \end{pmatrix}, \quad A_5 = \begin{pmatrix} x & x & 0 & x & 0 \\ 0 & x & x & \bullet & 0 \\ 0 & 0 & x & \bullet & x \\ 0 & 0 & 0 & x & \bullet \\ 0 & 0 & 0 & 0 & x \end{pmatrix}.$$

Finally,

$$L = \begin{pmatrix} x & & & & \\ x & x & & & \\ 0 & x & x & & \\ x & \bullet & \bullet & x & \\ 0 & 0 & x & \bullet & x \end{pmatrix}.$$

In this example, three elements which were initially zero in the lower triangular part of A are non–zero in L. The number of fill–ins depends on the way the pivots are chosen if pivoting is allowed. As the following well known example shows, there can be large differences in the number of fill–ins with different pivoting strategies. Consider

$$A = \begin{pmatrix} x & x & x & x \\ x & x & & \\ x & & x & \\ x & & & x \end{pmatrix}.$$

Then,

$$L = \begin{pmatrix} x & & & \\ x & x & & \\ x & \bullet & x & \\ x & \bullet & \bullet & x \end{pmatrix},$$

that is, all the zero entries are filled. But, we can define a permutation matrix P such that the first element is numbered last and then

$$PAP^T = \begin{pmatrix} x & & & x \\ & x & & x \\ & & x & x \\ x & x & x & x \end{pmatrix}.$$

In this case there is no fill–in at all. This is called a perfect elimination.

The aim of sparse Gaussian elimination is to avoid doing operations on zero entries and therefore to try to minimize the number of fill–ins. This will have the effect of both minimizing the needed storage and the number of floating point operations. The way this is achieved depends on the properties of the matrix. If the matrix is symmetric and, for instance, positive definite, we do not need to pivot for numerical stability (see Chapter 2). This gives us the freedom to choose symmetric permutations only to minimize the fill–in as we did in the previous example. Moreover, the number and indices of fill–ins can be determined before doing the numerical factorization as this depends only on the structure of the matrix and not on the value of the entries. Everything can be handled within a static data structure that is constructed in a pre–processing phase called the symbolic factorization.

If the matrix is non–symmetric (and without any special properties), we have seen that we generally need to pivot to achieve an acceptable numerical accuracy. If in addition the matrix is sparse, we now have another requirement which is to minimize the fill–in. Therefore, these two (sometimes conflicting) goals have to be dealt with at the same time. This implies that the data structure for the L and U factors cannot be determined before the numerical factorization, as the pivot rows and therefore the potential fill–ins are only known when performing the numerical factorization.

It has been shown that finding an ordering that minimizes the fill–in is an NP complete problem, see Yannakakis [1349]. Therefore, all the algorithms rely on heuristics to find a "good" ordering producing a low level of fill–in.

3.3. Graphs and fill–in for symmetric matrices

The connection between sparse Gaussian elimination and graphs for symmetric matrices was first studied by Parter [1065], see also Rose [1119]. We define a sequence of graphs $G^{(i)}, i = 1, \ldots, n, G^{(1)} = G$ being the graph of A, corresponding to the steps of Gaussian elimination.

Theorem 3.1

Consider a sequence of graphs where $G^{(1)} = G$ and $G^{(i+1)}$ is obtained from $G^{(i)}$ by removing the node x_i from the graph as well as all its incident edges and adding edges such that all the remaining neighbors of x_i in $G^{(i)}$ are pairwise connected. Then, proceeding from $G^{(i)}$ to $G^{(i+1)}$ corresponds to the ith step of Gaussian elimination.

Proof. Let us prove this for the first step, eliminating the node x_1 (or the corresponding unknown in the linear system). Then,

$$a_{i,j}^{(2)} = a_{i,j} - \frac{a_{i,1}a_{1,j}}{a_{1,1}}.$$

The element $a_{i,j}^{(2)}$ is non–zero if either $a_{i,j} \neq 0$ or, $a_{i,j} = 0$ and $a_{i,1}$ and $a_{1,j}$ are non–zero. The last possibility means that x_i and x_j are neighbors of x_1 in the graph. When x_1 is eliminated, they will be connected by an edge representing the new element and $a_{i,j}^{(2)} \neq 0$. This occurs for all the neighbors of x_1. We do not consider zeros that arise by cancellation. In this way, $G^{(2)}$ is obtained corresponding to the submatrix obtained from $A^{(2)}$, by deleting the first row and the first column. The same obviously occurs for every step of Gaussian elimination. □

Starting from the graph $G(A)$ of A and adding the edges that are created in all the $G^{(i)}$s during the elimination, we obtain a graph $G_F, (F = L + L^T)$ which is called the filled graph $G_F = (X, E^F)$. A small example of an elimination graph is given below. Let

$$A = \begin{pmatrix} x & x & x & x & & & x \\ x & x & & & & & \\ x & & x & x & & & \\ x & & x & x & x & x & x \\ & & & x & x & & \\ & & & x & & x & x \\ x & & & x & & x & x \end{pmatrix}.$$

Figure 3.1 displays the graph $G(A)$. The graph $G^{(2)}$ is given in figure 3.2.

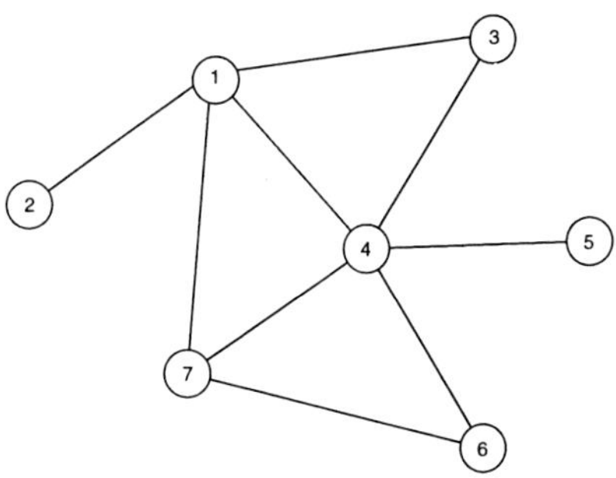

fig. 3.1: The graph $G(A)$

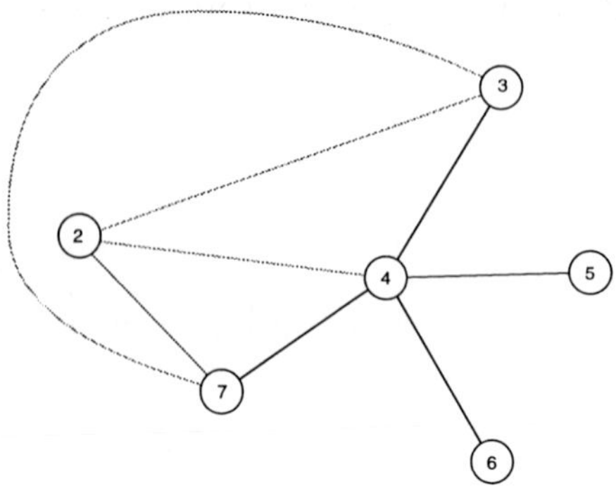

fig. 3.2: The graph G(2)

The edges corresponding to fill–ins are denoted by grey (dotted) lines. The graph $G^{(2)}$ corresponds to the matrix,

$$A^{(2)} = \begin{pmatrix} x & x & x & x & & & x \\ x & x & \bullet & \bullet & & & \bullet \\ x & \bullet & x & x & & & \bullet \\ x & \bullet & x & x & x & x & x \\ & & & x & x & & \\ & & & x & & x & x \\ x & \bullet & \bullet & x & & x & x \end{pmatrix}.$$

The filled graph G_F is given in figure 3.3.

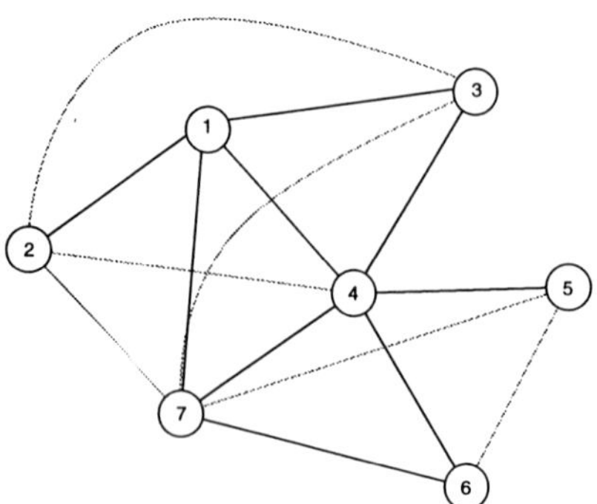

fig. 3.3: The filled graph G_F

In total there are six fill–ins in the elimination. Note that, in this example, a perfect elimination can be obtained by ordering the unknowns as

$$2, \quad 5, \quad 3, \quad 6, \quad 1, \quad 4, \quad 7$$

With this ordering, the permuted matrix is

$$A' = PAP^T = \begin{pmatrix} x & & & & x & & \\ & x & & & & x & \\ & & x & & x & x & \\ & & & x & & x & x \\ x & & x & & x & x & x \\ & x & x & x & x & x & x \\ & & & x & x & x & x \end{pmatrix}.$$

One thing that we note is that the more fill–ins we create in the early stages, the more fill-ins we shall get later on, as fill–ins usually create fill–ins. Thus, an heuristic rule is that it is likely to be beneficial to start by eliminating nodes that do not create much fill–in. These are the nodes with a small number of neighbors or nodes in cliques.

3.4. Characterization of the fill–in

Let us introduce a few definitions.

• The elimination tree of A, symmetric matrix of order n, is a graph with n nodes such that the node p is the parent of node j, if and only if

$$p = \min\{i | i > j, \ l_{i,j} \neq 0\}$$

where L is the Cholesky factor of A. The elimination tree of A will be denoted by $T(A)$ or simply T if the context makes it clear that we refer to the matrix A. Clearly, p is the index of the first non–zero element in column j of L. In the previous example of figure 3.1, $T(A')$ is given in figure 3.4 (renumbering the unknowns according to P).

This tree exhibits the fact that x'_1, x'_2, x'_3, x'_4 (corresponding to x_2, x_5, x_3, x_6 in the initial ordering) can be eliminated in any order (or even in parallel) as there are no dependencies between these variables.

• Let $S \subset X$ and $x \in X, x \notin S$, x is said to be reachable from $y \notin S$ through S if there exists a path $(y, v_1, \ldots, v_k, x)$ from y to x in G such that $v_i \in S$, $i = 1, \ldots, k$. We define

$$Reach(y, S) = \{x | x \notin S, x \text{ is reachable from } y \text{ through } S\}$$

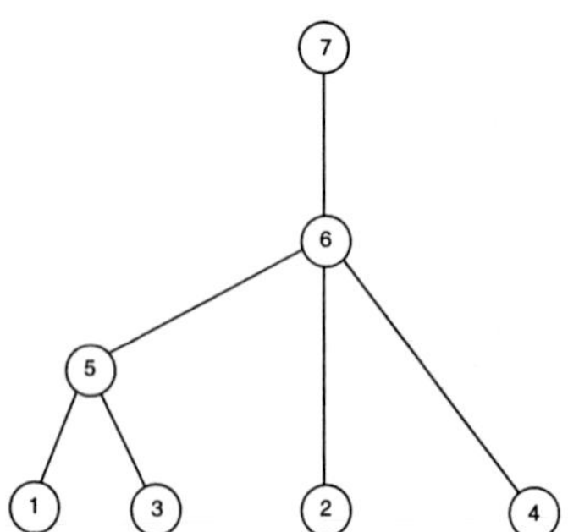

fig. 3.4: An elimination tree

There can be a fill–in between x_j and x_k only if at some step m, they are not already connected together and both neighbors of $x_m, m < j$, $m < k$. Therefore, either they were already neighbors of x_m in G or they were put in this situation by the elimination of other nodes $x_l, l < m$. Recursively, we see that at some stage, x_j was a neighbor of one of these nodes and the same for x_k with another of these nodes. This means that in G, there is at least one path between x_j and x_k and that all nodes on this path have numbers smaller than j and k. If there is no such path, there will not be a fill–in between x_j and x_k. This can be formalized in the following results. We first prove a lemma due to Parter [1065].

Lemma 3.2

$\{x_i, x_j\} \in E^F$ if and only if $\{x_i, x_j\} \in E$ or $\{x_i, x_k\} \in E^F$ and $\{x_k, x_j\} \in E^F$ for some $k < \min\{i, j\}$.

Proof. If $\{x_i, x_k\} \in E^F$ (filled graph) and $\{x_k, x_j\} \in E^F$ for some $k < \min\{i, j\}$, then the elimination of x_k will create a fill–in between x_i and x_j. Therefore, $\{x_i, x_j\} \in E^F$. Conversely, if $\{x_i, x_j\} \in E^F$ and $\{x_i, x_j\} \notin E$, then we have seen in the previous discussion that at some stage, x_i and x_j must be neighbors of a node, say x_k, that will be eliminated before x_i and x_j. Thus, $k < \min\{i, j\}$. □

The fill–in was characterized by A. George (see [625]).

Theorem 3.3

Let $k > j$, there will be a fill–in between x_j and x_k if and only if

$$x_k \in Reach(x_j, \{x_1, \ldots, x_{j-1}\}).$$

Proof. Suppose $x_k \in Reach(x_j, \{x_1, \ldots, x_{j-1}\})$. There exists a path $\{x_j, v_1, \ldots, v_l, x_k\} \in G$ with $v_i \in \{x_1, \ldots, x_{j-1}\}$, $1 \leq i \leq l$. If $l = 0$ or $l = 1$, the result follows from lemma 3.2. If $l > 1$, it is easy to show that $\{x_k, x_j\} \in E^F$ by induction.

Conversely, we assume $\{x_i, x_j\} \in E^F$, $j < k$. The proof is by induction on j. For $j = 1$, $\{x_1, x_k\} \in E^F$ implies $\{x_1, x_k\} \in E$ as there is no fill–in with the first node. Moreover, the set $\{x_1, \ldots, x_{j-1}\}$ is empty. Suppose the result is true up to $j - 1$. By lemma 3.2, there exists some $l \leq j - 1$ such that $\{x_j, x_l\} \in E^F$ and $\{x_l, x_k\} \in E^F$. By the assumption, there exists a path between x_j and x_l and another one from x_l to x_k. Clearly, this implies that there is a path from x_j to x_k whose nodes have numbers $\leq l \leq j - 1$. □

George demonstrated that reachable sets can be efficiently implemented 'y using quotient graphs, see [625]. The fill–in can also be characterized using elimination trees.

We denote by $T[x]$, the subtree of $T(A)$ rooted at node x. $y \in T[x]$ is a descendant of x and x is an ancestor of y. From the definition of $T(A)$, we have that if x_i is a proper ancestor of x_j in $T(A)$, then $i > j$.

Theorem 3.4

For $i > j$, the numerical values of columns i of L ($L_{*,i}$) depend on column j of L ($L_{*,j}$) if and only if $l_{i,j} \neq 0$.

Proof. See Liu [914]. □

3.5. Band and envelope numbering schemes for symmetric matrices

The first attempts to exploit sparsity used band or envelope storage schemes, trying to minimize the storage. Let us introduce a few definitions.

• $f_i(A) = \min\{i | a_{i,j} \neq 0\}$. $f_i(A)$ is the index of the column with the first non–zero element of row i.

• $\beta_i(A) = i - f_i(A)$ is the bandwidth of row i. The bandwidth of A is defined as

$$\beta(A) = \max_i \{\beta_i(A),\ 1 \leq i \leq n\},$$

and

$$band(A) = \{(i,j) | 0 < i - j \leq \beta(A),\ i \geq j\}.$$

• $Env(A) = \{(i,j) | 0 < i - j \leq \beta_i(A),\ i \geq j\}$ is the envelope of A. The profile of A, denoted by $P_r(A)$ is given by

$$P_r(A) = |Env(A)| = \sum_{i=1}^{n} \beta_i(A).$$

These definitions have led to ideas for storing the matrices A and L as if $\beta_i(A)$ is almost constant as a function of i, then it makes sense to store the entries corresponding to all the indices in $band(A)$. However, most of the time this is not practical as there are often a few rows with a larger bandwidth than the other ones and then, too much storage is wasted by the band scheme. Then, one can use the variable band or envelope storage scheme, see Jennings [820]. This simple storage scheme is obtained by storing for each row all the elements of the envelope in the same vector. We only need another vector of integers to point to the start of each row. The interest in this storage scheme was motivated by the following result.

Theorem 3.5

Let $Fill(A) = \{(i,j) | i > j,\ a_{i,j} = 0,\ l_{i,j} \neq 0\}$ be the index set of the fill–ins, then

$$Fill(A) \subset Env(A).$$

Proof. This is a consequence of theorem 3.3 as there cannot be any fill–in from a node x_i to a node x_j whose number is smaller than the smallest number of the neighbors of x_i. All the paths going from x_i to x_j will have a node with a number larger than x_j. □

Using these storage schemes, it was natural to try to devise orderings that minimize the bandwidth or the profile of the matrix. Unfortunately, this is an NP–complete problem. But there are some heuristics that help to obtain low profile orderings.

3.5.1. The Cuthill–McKee and reverse Cuthill–McKee orderings

The Cuthill–McKee (`CMK`) algorithm is a local minimization algorithm whose aim is to reduce the profile of A. It is clear that if at some stage of the numbering process, we would like to minimize $\beta_i(A)$, then we must immediately number all the non–numbered nodes in $Adj(x_i)$. The algorithm due to Cuthill and McKee is the following:

Algorithm CMK

1) choose a starting node,

2) for $i = 1, \ldots, n-1$ number all the (non–numbered) neighbors of x_i in $G(A)$ in increasing order of degree,

3) update the degrees of the remaining nodes.

The profile resulting from this ordering is quite sensitive to the choice of the starting node. A good choice for a starting node will be to choose a peripheral node, that is one whose eccentricity equals the diameter of the graph as this will generate a narrow level structure where the difference in number for a node and its neighbors is minimal. Peripheral nodes are not easy to find quickly. Therefore, heuristics were devised to find "pseudo–peripheral" nodes, i.e. nodes whose eccentricities are close to the diameter of the graph. Such an algorithm was proposed by Gibbs, Poole and Stockmeyer [636].

Algorithm GPS

1) choose a starting node r,

2) build the level structure $\mathcal{L}(r)$

$\mathcal{L}(r) = \{L_0(r), \ldots, L_{e(r)}(r)\}$,

3) sort the nodes $x \in L_{e(r)}(r)$ in increasing degree order,

4) for all nodes $x \in L_{e(r)}(r)$ in increasing degree order build $\mathcal{L}(x)$. If the height of $\mathcal{L}(x)$ is greater than the height of $\mathcal{L}(r)$, choose x as a starting node ($r = x$) and go to step 2).

This algorithm will eventually converge as eccentricities are bounded by the diameter of the graph. However, it can be very costly. George and Liu [620] proposed to shorten the computing time by eliminating structures with large width as soon as possible. Step 4) of the algorithm is modified as:

4′) let $w(x)$ be the width of $\mathcal{L}(x)$. For all $x \in L_{e(r)}(r)$ in order of increasing degree, build

$\mathcal{L}(x) = \{L_0(x), \ldots, L_{e(r)}(x)\}$.

At each level i if $|L_i(x)| > w(r)$, we drop the current node and we pick another x. If $w(x) \leq w(r)$ and $e(x) > e(r)$, we choose x as a starting node ($r = x$) and go to 2).

George and Liu [625] also proposed to use the following simple algorithm,

1) choose a starting node r,

2) build $\mathcal{L}(r)$,

3) choose a node x of minimum degree in $L_{e(r)}(r)$,

4) build $\mathcal{L}(x)$. If $e(x) > e(r)$, choose x as a starting node and go to 2).

Rather than using CMK, George and Liu [625] proposed to reverse the Cuthill–McKee ordering.

Algorithm Reverse Cuthill–McKee (RCM),

1) find a pseudo–peripheral starting node,

2) generate the CMK ordering,

3) reverse the numbering. Let $x_1, \ldots, x_n$ be the CMK ordering, then the RCM ordering $\{y_i\}$ is given by $y_i = x_{n+i-1}$, $i = 1, \ldots, n$.

We shall show that in terms of number of fill–ins, RCM is always as good as CMK. So, there is no reason to use CMK.

Theorem 3.6

Let A be an irreducible matrix and A_{CM} be the matrix corresponding to reordering (the graph of) A by the Cuthill–McKee scheme. Then,

$$\forall i, j,\ i \le j,\ f_i \le f_j.$$

Moreover, $f_i < i$ if $i > 1$.

Proof. Suppose that the conclusion does not hold. Then, there exist a column k and rows p, l, m, $p < l < m$ such that

$$\begin{aligned} f_p &\le k, \\ f_l &> k, \\ f_m &\le k. \end{aligned}$$

This means that

$$\begin{aligned} a_{p,k} &\neq 0 \Longrightarrow x_p \in Adj(x_k), \\ a_{m,k} &\neq 0 \Longrightarrow x_m \in Adj(x_k), \\ a_{l,k} &= 0 \Longrightarrow x_l \notin Adj(x_k). \end{aligned}$$

But this is impossible as the Cuthill–McKee algorithm has numbered successively all nodes in $Adj(x_k)$. □

Let us introduce a new definition.

- $Tenv(A) = \{(i,j) |\ j \le i,\ \exists k \ge i,\ a_{k,j} \neq 0\}$.

$Tenv(A)$ is the "transpose envelope" of A. Let us consider the example in figure 3.5. If we use the reverse Cuthill–McKee algorithm, we have to reverse the ordering. Thus, we obtain the matrix of figure 3.6. The rows of A_{RCM} are the columns of A_{CM}.

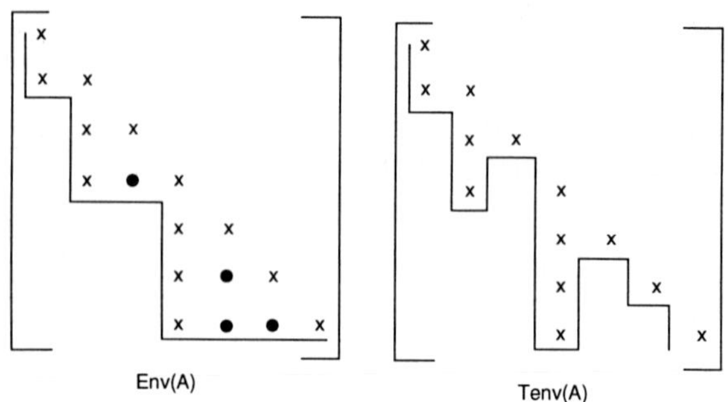

fig. 3.5: Env(A) and Tenv(A)

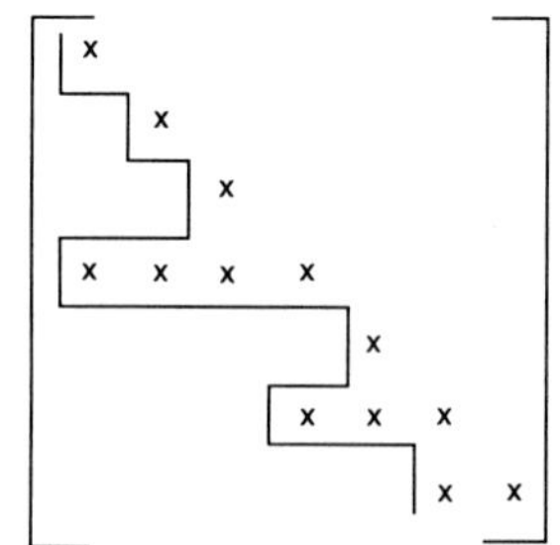

fig. 3.6: The envelope for RCM

Lemma 3.7

$|Env(A_{RCM})| = |Tenv(A_{CM})|.$

Proof. Straightforward. □

Theorem 3.8

$Tenv(A_{CM}) \subseteq Env(A_{CM}).$

Proof. Suppose we have $(i, j) \in Tenv(A_{CM})$ and $(i, j) \notin Env(A_{CM})$. Then, $\exists k \geq i$ such that $a_{k,j} \neq 0$. Either

1) $a_{i,j} \neq 0 \Longrightarrow (i,j) \in Env(A_{CM})$, or

2) $a_{i,j} = 0$. If $(i,j) \notin Env(A_{CM}) \Longrightarrow \forall l \leq j,\ a_{i,l} = 0 \Longrightarrow f_i > j$.

On the other hand, we have $f_k \leq j$. This implies $f_k < f_i$ which is impossible as $k \geq i$ by theorem 3.6. □

Obviously, we have

$$|Env(A_{RCM})| \leq |Env(A_{CM})|.$$

A. George proved the following result.

Lemma 3.9

If $\forall i > 1$, $f_i < i$, the envelope $Env(A)$ fills completely.

□

This implies

$$|Fill(A_{RCM})| \leq |Fill(A_{CM})|.$$

There are cases for which equality holds. Note that the previous results are true for every ordering such that $k \geq i \Longrightarrow f_k \geq f_i$. It has been shown that RCM can be implemented to run in $O(|E|)$ time. For a regular $N \times N$ grid and P_1 triangular finite elements, the storage for RCM varies as $O(N^3)$, ($\approx 0.7N^3$) that is $O(n^{\frac{3}{2}})$. Figure 3.7 shows the structure of the matrices of two examples. The matrix on the left arises from the Poisson model problem on a 6×6 square mesh with a lexicographic ordering. The matrix on the right is a "Wathen" matrix from the Higham's matrix toolbox. The matrix A is precisely the "consistent mass matrix" for a regular 10×10 grid of 8–node (serendipity) elements in two space dimensions. Figure 3.8 shows the structure of those matrices reordered by the Matlab RCM algorithm. For the Wathen matrix, the number of non–zeros in the Cholesky factor is 2311 for the original ordering and 2141 with RCM.

Several other algorithms have been proposed which reduce the profile of a symmetric matrix, for example the King algorithm, see George and Liu [625].

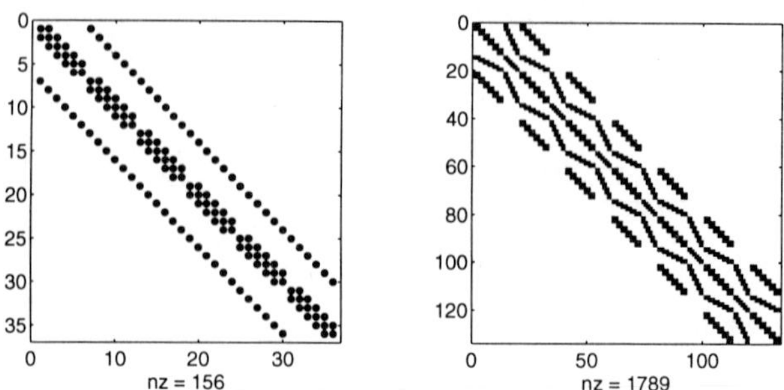

fig. 3.7: The structure of the original matrices

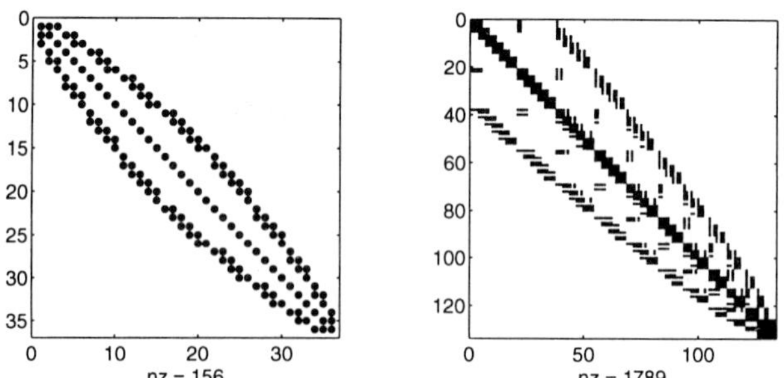

fig. 3.8: The structure of the matrices reordered by RCM

3.5.2. Sloan's algorithm

One of the drawback of the CMK algorithm is that it is a local algorithm using information about the neighbors of the last numbered nodes. Sloan [1187] has suggested an algorithm that tries to overcome this problem. The first step is the selection of pseudo–peripheral nodes.

Sloan's algorithm

- Step 1 (selection of a pseudo–diameter)

1) choose a node s with minimum degree,

2) build the level structure $\mathcal{L}(s) = \{L_0(s), \ldots, L_k(s)\}$,

3) sort the nodes of $L_k(s)$ by increasing degree, let m be the number of elements in $L_k(s)$ and Q be the $\lfloor \frac{m+2}{2} \rfloor$ first elements of the sorted set,

4) Let $w_{min} = \infty$ and $k_{max} = k$. For each node $i \in Q$ in order of ascending degree, generate $\mathcal{L}(i) = \{L_0(i), \dots, L_k(i)\}$. If $k > k_{max}$ and $w = \max_{i \le j \le k} |L_j(i)| < w_{min}$, then we set $s = i$ and go to step 3). Otherwise, if $w < w_{min}$, we set $e = i$ and $w_{min} = w$.

We exit this algorithm with a starting node s and an end node e which define a pseudo–diameter. The difference of this procedure to GPS is the shrinking strategy of step 3). This is performed as it has been observed that nodes with high degrees are not often chosen as starting nodes. The second modification is similar to what is done in the George and Liu algorithm.

- Step 2 (node labeling)

The nodes are classified in four categories according to their status: nodes which have been already assigned a label are *postactive*. Nodes which have not been assigned a number but are adjacent to postactive nodes are *active*. Nodes without a number adjacent to active nodes are *preactive*. All other nodes are *inactive*. The current degree n_i of a node i is defined as

$$n_i = m_i - c_i + k_i,$$

where m_i is the degree of i, c_i is the number of postactive or active nodes adjacent to i and $k_i = 0$ if i is active or postactive and $k_i = 1$ otherwise. The inputs of the algorithm are the two nodes s and e selected in Step 1). The algorithm maintains a list of eligible nodes each with a priority related to the current degree and the distance from the end node. Nodes with low current degree and large distance to the end have high priorities.

1) for all nodes, compute the distances $d(e, i)$ from i to e, initialize all nodes as inactive and set

$$P_i = (n_{max} - n_i) * W_1 + d(e, i) * W_2,$$

where $n_{max} = \max_i n_i$ and W_1, W_2 are integer weights. The queue of eligible nodes is initialized with s which is assigned a preactive status.

2) as long as the queue is not empty,

2.1) select the node i with highest priority in the queue (ties are broken arbitrarily),

2.2) delete i from the queue. If it is not preactive, go to 2.3). Else, consider each node j adjacent to i and set $P_j = P_j + W_1$. If j is inactive, insert j in the queue and declare it preactive,

2.3) label node i and declare it postactive,

2.4) Examine every node j adjacent to i. If j is preactive, set $P_j = P_j + W_1$, declare j as active and examine each node k adjacent to j. If k is active or preactive, set $P_k = P_k + W_1$, otherwise if k is inactive, set $P_k = P_k + W_1$, insert k in the queue and declare it as preactive.

The values of the weights W_1 and W_2 determine the balance between the local information (the current degree) and the global one (the distance to the end). Sloan [1187] recommended $W_1 = 2$ and $W_2 = 1$. He performed numerical experiments showing that on certain sets of matrices his algorithm gives lower profiles than RCM and other algorithms. Figure 3.9 gives the structures for the examples with the Sloan algorithm. For the Wathen matrix, the number of non–zeros in the Cholesky factor is 1921, slightly less than with RCM. For the finite difference example both RCM and Sloan give the same profile.

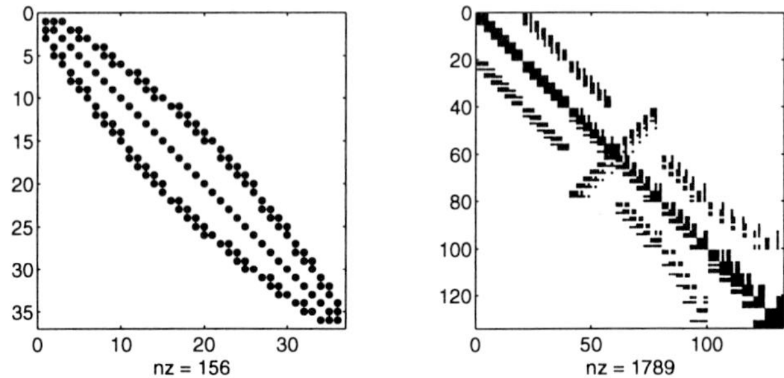

fig. 3.9: The structure of the matrices reordered by Sloan

Duff, Reid and Scott [488] have improved the Sloan algorithm by allowing it to work with weighted graphs where nodes with the same adjacency set are collapsed. Kumfert and Pothen [872] have also used the Sloan algorithm combined with other techniques. We shall come back to this in one of the following sections.

3.6. Spectral schemes

3.6.1. The basic idea

The Laplacian matrix $\mathcal{L}(G)$ of a symmetric matrix A (or the associated graph G) is defined as follows: $\mathcal{L}_{i,j} = -1$ if node j is a neighbor of node i in the graph G (or equivalently if $a_{i,j} \neq 0$). The diagonal term $\mathcal{L}_{i,i}$ is minus the sum of the other entries in row i. Clearly, $\mathcal{L}$ is a singular M–matrix. Therefore its smallest eigenvalue is zero. However, the eigenvector u_2 corresponding to the smallest positive eigenvalue λ_2 has interesting properties. This has been investigated by Fiedler [552].

The Laplacian matrix has been used both for envelope reduction and for graph partitioning. The rationale for using u_2 in envelope reducing algorithms is the following. We have seen that the profile of A is defined as

$$P_r(A) = \sum_{i=1}^{n} \beta_i(A) = \sum_{i=1}^{n} \max_{j \in row(i)} (i-j),$$

if we define $row(i) = \{j | a_{i,j} \neq 0, 1 \leq j \leq i\}$. There is a related quantity, the 1–sum σ_1:

$$\sigma_1(A) = \sum_{i=1}^{n} \sum_{j \in row(i)} (i-j).$$

The work in the Cholesky factorization is proportional to W:

$$W = \sum_{i=1}^{n} \max_{j \in row(i)} (i-j)^2.$$

This is related to the 2–sum:

$$\sigma_2^2(A) = \sum_{i=1}^{n} \sum_{j \in row(i)} (i-j)^2.$$

The following theorem is from George and Pothen [634].

Theorem 3.10

Let p be the maximum number of off diagonal non–zeros in a row of A (or the maximum vertex degree in G), then

$$W \leq \sigma_2^2(A) \leq pW,$$

□

The spectral ordering is obtained by sorting the components of u_2 in increasing order. This induces a permutation vector which gives the requested ordering. In order to justify this we consider $\sigma_2(A)$ instead of W as the quantity to minimize over all the orderings. Suppose for the sake of simplicity that n is even and let $\mathcal{P}$ be the set of vectors whose components are permutations of $\{-n/2, \ldots, -1, 0, 1, \ldots, n/2\}$. Then,

$$\min_{x \in \mathcal{P}} \sum_{i=1}^{n} \sum_{j \in row(i)} (x_i - x_j)^2 = \frac{1}{2} \min_{x \in \mathcal{P}} \sum_{a_{i,j} \neq 0} (x_i - x_j)^2.$$

Then the condition on x is relaxed to obtain an easier continuous problem. We now consider the set $\mathcal{H}$ of vectors x such that $\sum x_i = 0$ and (x, x) given. We have

$$\frac{1}{2} \min_{x \in \mathcal{H}} \sum_{a_{i,j} \neq 0} (x_i - x_j)^2 = \min_{x \in \mathcal{H}} (x, \mathcal{L}x) = \lambda_2(u_2, u_2).$$

Barnard, Pothen and Simon [116] have shown that the permutation vector induced by u_2 is the closest (in the Euclidean norm) vector in $\mathcal{P}$ to u_2. Figure 3.10 shows the result for the two examples. For the Wathen matrix the number of non–zeros in the Cholesky factor is slightly higher being 2571.

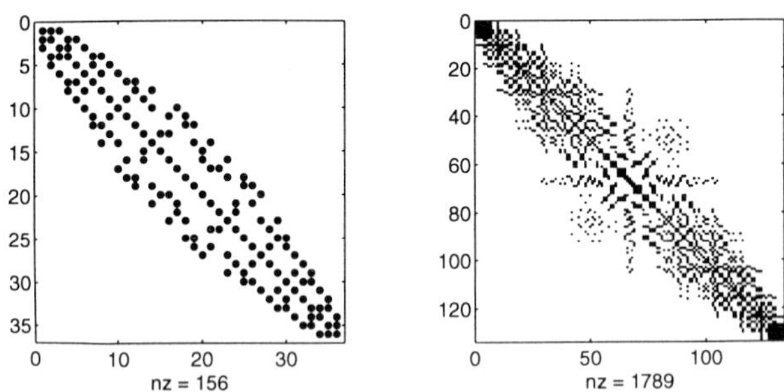

fig. 3.10: The structure of the matrices reordered by the spectral ordering

3.6.2. The multilevel spectral algorithm

Even if the eigenvalue and eigenvector do not have to be computed very accurately, the spectral algorithm as formulated above is too costly compared to other profile reduction algorithms for large examples. To overcome this problem, Barnard, Pothen and Simon proposed using the algorithm on a contracted graph with much fewer vertices than G:

1) construct a series of coarser and coarser graphs that retains the structure of the original graph,

2) compute the second eigenvector of the coarsest graph,

3) interpolate this vector to the next finer graph,

4) refine the interpolated vector (by Rayleigh Quotient Iteration, see Parlett [1062]) and go to 3) until we are back to the original graph.

There are many ways to define graph contraction. The one proposed by Barnard, Pothen and Simon was to find a maximal independent set of vertices which are to be the vertices of the contraction. The edges are found by growing domains from the selected vertices adding an edge when two domains intersect. Good results were reported in [116]. A similar algorithm has been introduced independently by Paulino, Menezes, Gattass and Mukherjee [1068].

3.6.3. The Kumfert and Pothen hybrid algorithm

Kumfert and Pothen [872] have suggested using a combination of the spectral and the Sloan algorithms. In this scheme the multilevel spectral algorithm is used to find the end nodes of a pseudo–diameter and then a modified version of the Sloan algorithm is used to number the nodes.

Kumfert and Pothen provide examples for which the spectral algorithm performs poorly. Then they introduce a variant of the Sloan algorithm for weighted graphs. They denote by $Size(i)$ the weight of a multi–vertex i. The degree of this node is the sum of the sizes of the neighboring multi–vertices. The generalization of the current degree $Cdeg(i)$ denotes the sum of the sizes of the neighbors of i for preactive or inactive vertices. Then, they define

$$Incr(i) = \begin{cases} Cdeg(i) + Size(i) & \text{if } i \text{ is preactive,} \\ Cdeg(i) & \text{if } i \text{ is active.} \end{cases}$$

Let Δ be the maximum degree in the unweighted graph, the new priority function is defined as

$$P(i) = -W_1 * \lfloor d(s,e)/\Delta \rfloor * Incr(i) + W_2 * d(i,e).$$

Considering the choices of weights, Kumfert and Pothen identify two classes of problems, one for which $W_1 = 2$ and $W_2 = 1$ gives good results, another that needs high values of W_2 to obtain small envelopes. Moreover, they implement the priority queue as a binary heap contrary to Sloan's implementation as an array. It turns out that this new implementation is much faster (by a factor of about 5) on selected sets of examples.

In the hybrid algorithm, the start and end nodes are chosen to be the first and last nodes in the spectral ordering and the priority function is given by

$$P(i) = -W_1 * \lfloor n/\Delta \rfloor * Incr(i) + W_2 * d(i,e) - W_3 * i,$$

this function being sensitive to the initial ordering through the third term. The weight W_3 is chosen by considering the eigenvector corresponding to the first non–zero eigenvalue. If the component giving the maximum of the absolute values of the components is negative, then $W_3 = -1$ and the starting and end nodes are exchanged, otherwise $W_3 = 1$. The other weights are $W_1 = W_2 = 1$. However, there are problems for which $W_1 = 1, W_2 = W_3 = 2$ give good results. Numerical results show that the hybrid algorithm gives better envelope sizes than both RCM and the Sloan algorithm.

3.6.4. The Boman–Hendrickson multilevel algorithm

In this approach [165], the original problem is approximated by a sequence of coarser ones. The vertices of the coarsest problem are then labeled and the results are interpolated back to the larger graphs with some refinement steps if needed.

The coarsening of graphs is done by edge contraction by coalescing adjacent vertices and assigning weights to the edges of the coarse graph. If two vertices are adjacent to the same neighbor then, the new edge is given a weight equal to the sum of the two old edges. In the first step, a maximum matching is found. This is a maximal set of edges for which no two edges are incident to the same vertex. The coarsest graph is numbered by the spectral ordering algorithm. The uncoarsening is done by numbering the vertices of the larger graph preserving the coarse graph ordering. A local refinement scheme is then used to improve this ordering. It is a variant of the Kernighan and Lin algorithm [856], the weighted 1–sum being the objective function. We shall explain this algorithm in more detail later on.

3.7. The minimum degree ordering

This technique was introduced in Tinney and Walker [1250]. It is one of the ordering schemes that are most often used today. Its aim is to locally minimize fill–in. The minimum degree (MD) ordering works with the elimination graphs $G^{(i)} = (X_i, E_i)$. The ith step of the algorithm is described as:

Algorithm MD

1) in $G^{(i)}$, find a node x_j such that

$$deg(x_j) = \min_{y \in X_i} \{deg(y)\},$$

and number it at the ith node.

2) form $G^{(i+1)}$ by eliminating x_j and update the degrees,

3) if $i + 1 < n$ go to 1) with $i \leftarrow i + 1$.

The biggest problem arising from the minimum degree algorithm is that, quite often, there are several nodes of minimum degree. This situation must be resolved using a tie breaking strategy. Unfortunately, the final number of fill–ins is quite sensitive to the tie breaking strategy, see George and Liu [626].

Being a local minimization algorithm, the minimum degree does not always give a globally minimum fill–in ordering. There are cases, like trees, for which it gives no fill–in at all, but there are examples for which it generates fill–in that is more than a constant time greater than the minimum fill–in, see Berman and Schnitger [143].

Over the years many improvements have been suggested to the basic algorithm, mainly to shorten the computer time needed rather than to improve the ordering. A summary of these results can be found in George and Liu [626]. The main points are the following.

∘ *Mass elimination*

When x_i is eliminated, often there are nodes in $Adj_{G^{(i)}}(x_i)$ that can be eliminated immediately. This is because, when x_i is eliminated, only the degrees of nodes in $Adj_{G^{(i)}}(x_i)$ change and some of them can be $deg(x_i) - 1$. For instance, if a node in a clique is eliminated, then the degree of all the other nodes in the clique decreases by 1. Therefore, all these nodes can be eliminated at once, before the degrees are updated thereby saving some degree updates. This leads to the concept of indistinguishable nodes.

• Two nodes u and v are indistinguishable in G if

$$Adj_G(u) \cup \{u\} = Adj_G(v) \cup \{v\}.$$

George and Liu [626] proved the following result.

Theorem 3.11

Let $z \in Adj_{G^{(i)}}(x_i)$, then $deg_{G^{(i+1)}}(z) = deg_{G^{(i)}}(x_i) - 1$ if and only if $Adj_{G^{(i)}}(z) \cup \{z\} = Adj_{G^{(i)}}(x_i) \cup \{x_i\}$.

□

By merging indistinguishable nodes, we need only to update the degrees of the representatives of these nodes.

∘ *Incomplete degree update*

Let us introduce a new definition.

• v is said to be outmatched by u in G if

$Adj_G(u) \cup \{u\} \subseteq Adj_G(v) \cup \{v\}$.

Theorem 3.12

If v is outmatched by u in $G^{(i)}$, it is also outmatched by u in $G^{(i+1)}$.

Proof. See George and Liu [626]. □

The consequence of this theorem is that if v becomes outmatched by u, it is not necessary to update the degree of v until u is eliminated.

∘ *Multiple elimination*

This slight variation of the basic scheme was proposed by Liu [908]. When x_i has been chosen, we select a node with the same degree as x_i in $G^{(i)}/(Adj_{G^{(i)}}(x_i) \cup \{x_i\})$. This process is repeated until there are no nodes of the same degree and then the degrees are updated. Therefore, at each step, an independent set of minimum degree nodes is selected. Note that the ordering that is produced is not the same as for the basic algorithm. However, it is generally as good as the genuine minimum degree ordering.

◦ *Tie breaking*

An important issue is the choice of a tie breaking strategy. Unfortunately, not much is known about how to decide which nodes to choose at a given stage. Some experiments (George and Liu, [626]) show that there can be large differences in the number of non–zeros and factorization times when several random tie breakers are chosen. Most often, the initial ordering determines the way ties are broken. It has been suggested that one could use another ordering scheme such as the Reverse Cuthill–McKee algorithm, before running the minimum degree.

◦ *Approximate minimum degree*

Amestoy, Davis and Duff [18] proposed to use some bounds on the degree of nodes instead of the real degree. This allows a faster update of the information when nodes are eliminated. The quality of the orderings that are obtained are comparable to those from the genuine minimum degree algorithm although the algorithm is much faster, see the performances in [18].

Figure 3.11 shows the result for the two examples using the Matlab minimum degree ordering. For the Wathen matrix the number of non–zeros in the Cholesky factor is 1924.

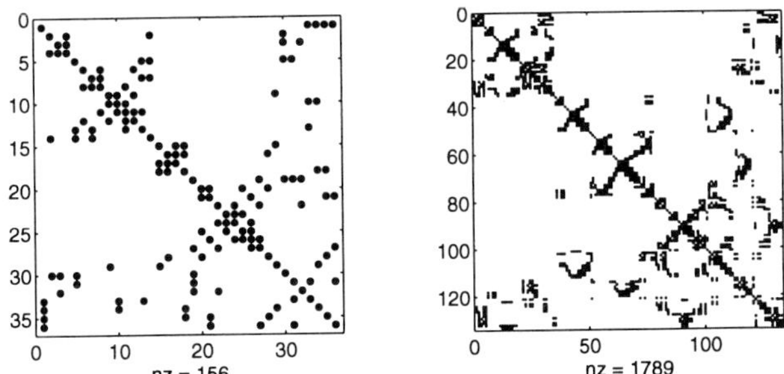

fig. 3.11: The structure of the matrices reordered by the minimum degree ordering

3.8. The nested dissection ordering

Nested dissection was introduced by Alan George [610] for finite element problems and then generalized to general sparse matrices. It is very close to an old idea used in Mechanics known as substructuring and also to what is now called domain decomposition (although there are some slight differences). This technique is based on theorem 3.3 that essentially says that there cannot be any fill–in between x_i and x_j if, on every path from x_i to x_j in G, there is a node with a number greater than x_i and x_j.

Consider the graph of figure 3.12 (arising, for instance, from a finite element matrix) and its partitioning given in the right hand side of the picture.

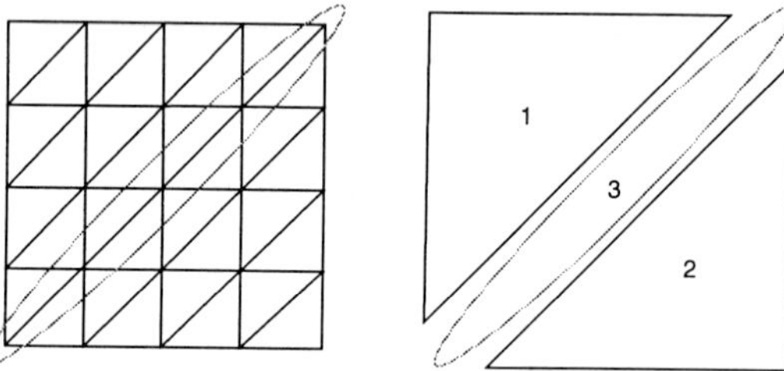

fig. 3.12: Dissection partitioning

The graph is split into three pieces. The diagonal 3 is called a separator. From theorem 3.3, it is clear that, if we first number the nodes in part 1, then the nodes in part 2 and finally the nodes of the separator 3, there cannot be any fill–in between nodes in sets 1 and 2. With this ordering and obvious notations, the matrix has the following block structure

$$A = \begin{pmatrix} A_1 & 0 & A_{3,1}^T \\ 0 & A_2 & A_{3,2}^T \\ A_{3,1} & A_{3,2} & A_3 \end{pmatrix}$$

The Cholesky factor L has the following block structure

$$L = \begin{pmatrix} L_1 & & \\ 0 & L_2 & \\ L_{3,1} & L_{3,2} & L_3 \end{pmatrix},$$

matrices L_1 and L_2 being the factors of A_1 and A_2 respectively. This means that blocks A_1 and A_2 can be factored independently. The basis of the nested dissection algorithm is to apply this idea recursively to sets 1 and 2. There are basically two ways to partition a rectangular mesh over a rectangle. The first one is to partition the graph into vertical (or horizontal) stripes. This is known as one–way dissection. The other way is to alternate between vertical and horizontal partitioning obtaining a partition into small rectangles. This is called nested dissection.

George [613] considered a mesh graph consisting of an m by l rectangular grid. It is partitioned by σ vertical grid lines. He then showed that the required storage for storing the LU factors using the one–way dissection ordering is (if $m \leq l$)

$$S(\sigma) = \frac{ml^2}{\sigma} + \frac{3\sigma m^2}{2}.$$

This (as a function of σ) is approximately minimized by

$$\sigma = l\left(\frac{2}{3m}\right)^{\frac{1}{2}},$$

giving $S_{\text{opt}} = \sqrt{6}\; m^{\frac{3}{2}} l + O(ml)$. By comparison, numbering the graph by columns would yield a storage of $m^2 l + O(ml)$. The operation count for the factorization is approximately

$$\theta = \frac{ml^3}{2\sigma^2} + \frac{7\sigma m^3}{6} + \frac{2m^2 l^2}{\sigma}.$$

This is approximately minimized by

$$\sigma = l\left(\frac{12}{7m}\right)^{\frac{1}{2}},$$

yielding $\theta_{\text{opt}} = \left(\frac{28}{3}\right)^{\frac{1}{2}} m^{\frac{5}{2}} l + O(m^2 l)$.

One–way dissection can be generalized to any sparse matrix by using level structures of the graph, see George and Liu [619].

Consider nested dissection for a square mesh. A partition function Π is defined for integers i from 0 to N $(= 2^l)$ as

$$\begin{aligned} \Pi(0) &= 1 \\ \Pi(N) &= 1 \\ \Pi(i) &= p + 1, \text{ if } i = 2^p(2q+1) \end{aligned}$$

For example, for $N = 16$, we obtain

i	0	1	2	3	4	5	6	7	8	9	10	11	12	13	14	15	16
$\Pi(i)$	1	1	2	1	3	1	2	1	4	1	2	1	3	1	2	1	1

For $k = 1, \ldots, l$ we define sets P_k of mesh nodes (i, j) as

$$P_k = \{(i, j) | \max(\Pi(i), \Pi(j)) = k\}.$$

For a 17×17 mesh we obtain the partition of figure 3.13 where the numbers refer to the set P_k to which the nodes belong and the lines separate these sets. Nodes in P_1 are numbered first, then nodes in P_2, etc...up to nodes in P_l. George [610] proved that

$$S = O(N^2 \log_2 N),$$
$$\theta = O(N^3).$$

1	1	2	1	3	1	2	1	4	1	2	1	3	1	2	1	1
1	1	2	1	3	1	2	1	4	1	2	1	3	1	2	1	1
2	2	2	2	3	2	2	2	4	2	2	2	3	2	2	2	2
1	1	2	1	3	1	2	1	4	1	2	1	3	1	2	1	1
3	3	3	3	3	3	3	3	4	3	3	3	3	3	3	3	3
1	1	2	1	3	1	2	1	4	1	2	1	3	1	2	1	1
2	2	2	2	3	2	2	2	4	2	2	2	3	2	2	2	2
1	1	2	1	3	1	2	1	4	1	2	1	3	1	2	1	1
4	4	4	4	4	4	4	4	4	4	4	4	4	4	4	4	4
1	1	2	1	3	1	2	1	4	1	2	1	3	1	2	1	1
2	2	2	2	3	2	2	2	4	2	2	2	3	2	2	2	2
1	1	2	1	3	1	2	1	4	1	2	1	3	1	2	1	1
3	3	3	3	3	3	3	3	4	3	3	3	3	3	3	3	3
1	1	2	1	3	1	2	1	4	1	2	1	3	1	2	1	1
1	1	2	1	3	1	2	1	4	1	2	1	3	1	2	1	1
2	2	2	2	3	2	2	2	4	2	2	2	3	2	2	2	2
1	1	2	1	3	1	2	1	4	1	2	1	3	1	2	1	1

fig. 3.13: Nested dissection partitioning

Duff, Erisman and Reid [474] generalized the partition function when $N \neq 2^l$.

It should be noted that the storage schemes that we have described in Chapter 1 are not well suited for nested dissection orderings. For nested dissection for mesh problems, there is a natural block structure that arises, each block corresponding to subsets of each P_k. Diagonal blocks are stored by rows in a one dimensional array together with an integer pointer that gives the position of the diagonal element. Non–diagonal blocks are stored in a one dimensional array. It is necessary to know pointers to the beginning of each block. Experiments show that for an $N \times N$ 2D grid, the storage for nested dissection is smaller than the one for RCM for $N > 37$. Figure 3. 14 shows the Poisson model problem matrix reordered with nested dissection and the corresponding Cholesky factor.

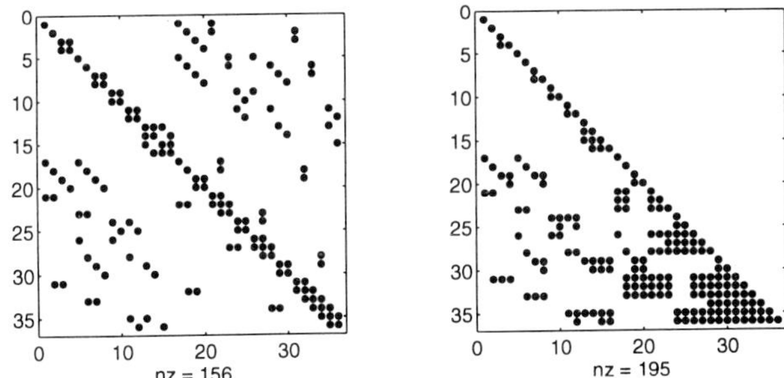

fig. 3.14: The model problem matrix with nested dissection

3.9. Generalization of dissection algorithms

For a general sparse matrix there is no underlying mesh, we have to work directly on the graph of the matrix and it is not so obvious to find a small separator that partitions the graph in two or more components of almost an equal number of nodes. There are many ways to handle this problem.

3.9.1. General dissection algorithms

General theorems have been proved using graph theory about the existence of good separators, see Lipton, Rose and Tarjan [906], Roman [1116], Charrier and Roman [308], [309]. Given a graph with weights either for the vertices or the edges, the problem is to find a vertex (or edge) small separator. Over the years many methods have been devised to solve this problem. Most of these methods proceed by recursive bisection. Geometric (or greedy) algorithms can be used that start from the nodes on the "boundary" of the graph accumulating nodes until about one half of the vertices are collected in a subset, see Ciarlet and Lamour [333], [330], [331], Farhat [546], [548]. Then, usually a refinement algorithm is used to improve the found partition.

The Kernighan and Lin [856] algorithm is one of the most used methods for partitioning graphs dating back to 1970. The goal of this heuristic algorithm is to approximately minimize the number of edges joining vertices in both subsets. Let us suppose that we have two sets of vertices V_1 and V_2. The algorithm is iterative. At every iteration, we move vertices from one set to the other as long as we cannot improve the gain which is defined as the number of edges joining V_1 to V_2. If the algorithm moves vertices from V_1 to V_2, then at the next iteration it tries to move vertices from V_2 to V_1 to improve the balance in the number of vertices in both sets. However, this process can converge to a local minima. To avoid this situation, the algorithm uses moves with negative gain for a given number of iterations. The best partition found so far is recorded and the algorithm returns to it if the negative moves do not give an improvement after a while. Fiduccia and Matheyses [551] gave an efficient implementation of this algorithm in $O(|E|)$ time where E is the set of edges. The main problem is to find an efficient way to update the gains. The results of the Kernighan and Lin algorithm depend very much on the initial partition. This is why it is now used more as a refinement step in other methods.

Recently, the spectral ordering algorithm has been much used to find a partition of a graph. Once again, the goal is to minimize the number of cutting edges. Let $\mathcal{L}$ be the Laplacian matrix of the graph and x a vector such that $x_i = 1$ if $i \in V_1$ and $x_i = -1$ if $i \in V_2$. Then,

$$\sum_{(i,j)\in E} (x_i - x_j)^2 = \sum_{\substack{i\in V_1, j\in V_2\\(i,j)\in E}} (x_i - x_j)^2 = 4|S(V_1, V_2)|,$$

where S is the edge separator between V_1 and V_2. Note that

$$(x, \mathcal{L}x) = \sum_{(i,j)\in E} (x_i - x_j)^2.$$

Therefore, the problem is to find a vector x for which

$$\min_{\substack{x_i=\pm 1\\ \sum_i x_i = 0}} (x, \mathcal{L}x).$$

This problem is too complicated, but if we relax the constraint that $x_i = \pm 1$, it is solved by the second eigenvector u_2 of the Laplacian matrix. The partition is then given by sorting the components of u_2 relative to the median component value. This algorithm is used recursively to partition the graph. Extensions have been proposed to directly generate partitions into four or eight pieces.

3.9.2. Graph bisection improvement techniques

The Kernighan and Lin algorithm is a technique to improve the partition of a graph in two domains. Therefore, it improves an edge separator. However, we are more interested in looking for improving a vertex separator. Powerful graph techniques were used and extended by Aschraft and Liu [55]. The main tool is the Dulmage–Mendelsohn decomposition which has often been used to extract a vertex separator from an edge separator (see [1080]). In order to describe the work in [55] we need a few more definitions:

- a bisector is a vertex separator S whose removal gives two components B and W, $Adj(B) \subseteq S$, $Adj(W) \subseteq S$. The partition is denoted $[S, B, W]$. A cost function γ is defined as

$$\gamma(S, B, W) = |S| \left(1 + \alpha \frac{\max\{|B|, |W|\}}{\min\{|B|, |W|\}}\right),$$

where α is a constant whose choice allows to switch between the separator size $|S|$ and the other term measuring the imbalance of the partition.

- Let Y be a subset of vertices. The interior of Y is $Int(Y) = \{y \in Y | Adj(y) \subseteq Y\}$. The boundary of Y is the set of nodes not in Y that are adjacent to Y. The border of Y is the boundary of the interior of Y.

Ashcraft and Liu tried to improve the partition by moving subsets Z that reduce the cost function. This is done by moving a subset from S to the smaller portion W. If this cannot be done, the algorithm tries to move a subset to the larger portion B. The algorithm stops when no more reduction can be obtained.

The choice of which subset of the separator to move is done by using graph matching techniques. When we move a subset Z from S to W, the separator size becomes $|S| - |Z| + |Adj(Z) \cap B|$. Therefore, if we are able to find a subset Z such that $|Z| > |Adj(Z) \cap B|$, the separator size will be improved. Liu used bipartite graph matching to choose the subset Z. In a bipartite graph the vertices can be divided into two subsets X and Y such that every edge had one endpoint in X and one in Y. A matching is a subset of edges such that no two edges in this subset have a node in common. A vertex that is incident to an edge in this subset is said to be covered, otherwise it is exposed. The number of edges in the subset is the size of the matching.

A maximum matching is of the largest possible size. A complete matching is one with a size equal to the size of the smaller of the two sets X and Y.

For the partition $[S, B, W]$, suppose that B is the largest subset and consider the bipartite graph $H = (S, border(B), E_H)$ where E_H is the set of edges between the vertices in S and those in $border(B)$. There is a result stating that there exists a subset Z of S satisfying $|Z| \leq |Adj(Z) \cap B|$ if and only if the bipartite graph H has a complete matching. Therefore, we are able to find a move that improves the size of the separator if there are exposed vertices in a maximum matching. Liu introduced the notion of an alternating path. For a matching M this is a path with no repeated vertices if the alternate edges belong to the matching. Liu proved the following result: if $x \in S$ is an exposed vertex in a maximum matching of H, let $S_x = \{s \in S | s \text{ is reachable from } x \text{ via alternating paths}\}$, then $|S_x| - |Adj(S_x \cap B)| = 1$. S_x can be found by a breadth–first search starting from x.

The Dulmage–Mendelsohn decomposition is the partition of S into three disjoint subsets: $S = S_I \cup S_R \cup S_X$ with

$S_I = \{s \in S | s$ is reachable from some exposed vertices in S via alternating paths $\}$,

$S_X = \{s \in S | s$ is reachable from some exposed vertices in B via alternating paths $\}$,

$S_R = S \setminus (S_I \cup S_X)$.

This decomposition is independent of the maximum matching used for the definition of the paths. It has been proven that S_I is the smallest subset of S with the maximum reduction of the separator size and $S_I \cup S_R$ is the largest subset with the maximum reduction. Moving S_I or $S_I \cup S_R$ give the same reduction but the balance of the sizes may be different.

When a reduction of the separator size is not possible, there is still some hope of improving the cost function by reducing the imbalance. When S_I is empty, S_R can be used to reduce the imbalance, see [55].

Ashcraft and Liu [55] extended the Dulmage–Mendelsohn decomposition to work with weighted graphs. This is useful when working with compressed graphs saving some computing time. They also relate the Dulmage–Mendelsohn decomposition to the solution of a maximum network flow problem. Solving a max flow–min cut problem can be used to improve the partition. Numerical experiments in [55] support the fact that these powerful graph techniques are in fact efficient to refine partitions.

3.9.3. The multisection algorithm

It is generally recognized that both minimum degree and nested dissection produce good orderings. However, the results are not uniformly good. We have already seen that the minimum degree algorithm can produce results far from optimal. Optimal results are given by the nested dissection ordering on grid problems. But there are more general problems for which the recursive ordering of the separators can be improved.

In the approach of Ashcraft and Liu [56], nested dissection algorithms are used to find recursive bisectors. However, the numbering is different as all the vertices in the union of the separators are numbered last. This is referred as a multisector. The constrained minimum degree algorithm is used to order the vertices in the remaining domains. To order the vertices in the multisector Ashcraft and Liu considered the elimination graph after the elimination of the vertices in the domains. The vertices of this graph are ordered by the multiple minimum degree. Even though the multisector is found by applying the dissection algorithm recursively, it is then smoothed by using the tools we have seen in the previous section. Experimental results in [56] show that this method performs uniformly well on large sets of examples, at least better than both the minimum degree and nested dissection.

3.10. The multifrontal method

The multifrontal method was introduced by Duff and Reid [483], [484] as a generalization of the frontal method developed by Irons [808] for finite element problems. The basis of the frontal method was that in finite element problems the two phases, assembly of the matrix (from integral computations) and factorization of the matrix, can be mixed together. However, a variable can be eliminated only when it has been fully assembled.

The main goal of the multifrontal method is to be able to use dense matrix technology for sparse matrices. A possible drawback of the method is that technical details are quite complex and many refinements are necessary to make the method efficient. A nice exposition of the principles of the method has been given in Liu [915]. We shall just look at a small example.

$$A = \begin{array}{c} \\ 1 \\ 2 \\ 3 \\ 4 \\ 5 \\ 6 \end{array} \begin{array}{c} \begin{array}{cccccc} 1 & 2 & 3 & 4 & 5 & 6 \end{array} \\ \begin{pmatrix} x & & & x & x & x \\ & x & & & x & \\ & & x & & x & x \\ x & & & x & & x \\ x & x & x & & x & x \\ x & & x & x & x & x \end{pmatrix} \end{array}.$$

The graph of the matrix A is given on figure 3.15.

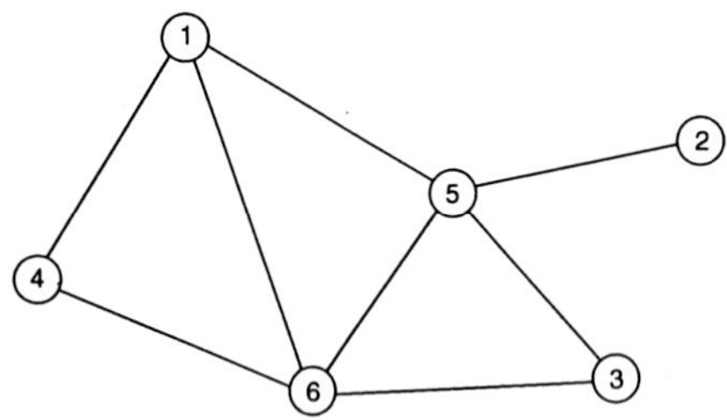

fig. 3.15: The graph of A

Gaussian elimination on A yields

$$L = \begin{array}{c} \\ 1 \\ 2 \\ 3 \\ 4 \\ 5 \\ 6 \end{array} \begin{array}{c} \begin{array}{cccccc} 1 & 2 & 3 & 4 & 5 & 6 \end{array} \\ \begin{pmatrix} x & & & & & \\ & x & & & & \\ & & x & & & \\ x & & & x & & \\ x & x & x & \bullet & x & x \\ x & & x & x & x & x \end{pmatrix} \end{array}.$$

The elimination tree $T(A)$ is shown on figure 3.16.

From the elimination tree, it is clear that we can eliminate 1, 2 and 3 independently. If we consider 1, we can restrict ourselves to the following matrix (rows and columns where there are non–zeros in the first row and first column),

$$F_1 = \begin{array}{c} \\ 1 \\ 4 \\ 5 \\ 6 \end{array} \begin{array}{c} \begin{array}{cccc} 1 & 4 & 5 & 6 \end{array} \\ \begin{pmatrix} a_{1,1} & a_{1,4} & a_{1,5} & a_{1,6} \\ a_{4,1} & & & \\ a_{5,1} & & & \\ a_{6,1} & & & \end{pmatrix} \end{array}.$$

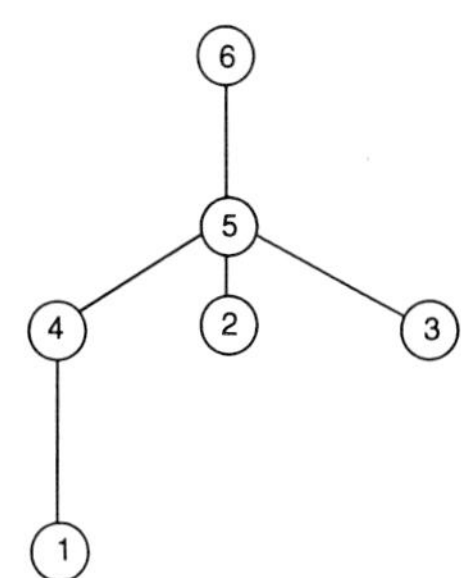

fig. 3.16: The elimination tree of A

Eliminating 1 will create contributions in a reduced matrix $\bar{U}_4$,

$$\bar{U}_4 = \begin{matrix} & \begin{matrix} 4 & 5 & 6 \end{matrix} \\ \begin{matrix} 4 \\ 5 \\ 6 \end{matrix} & \begin{pmatrix} x & \bullet & x \\ \bullet & x & x \\ x & x & x \end{pmatrix} \end{matrix},$$

where the $\bullet$ represents a fill–in. In parallel, we can eliminate 2, defining

$$F_2 = \begin{matrix} & \begin{matrix} 2 & 5 \end{matrix} \\ \begin{matrix} 2 \\ 5 \end{matrix} & \begin{pmatrix} a_{2,2} & a_{2,5} \\ a_{5,2} & \end{pmatrix} \end{matrix}.$$

Elimination of 2 will create a contribution to the $(5,5)$ term,

$$\bar{U}_5^2 = \begin{matrix} & 5 \\ 5 & (x) \end{matrix}.$$

We can also eliminate 3,

$$F_3 = \begin{matrix} & \begin{matrix} 3 & 5 & 6 \end{matrix} \\ \begin{matrix} 3 \\ 5 \\ 6 \end{matrix} & \begin{pmatrix} a_{3,3} & a_{3,5} & a_{3,6} \\ a_{5,3} & & \\ a_{6,3} & & \end{pmatrix} \end{matrix}.$$

Elimination of 3 will create contributions,

$$\bar{U}_5^3 = \begin{matrix} & \begin{matrix} 5 & 6 \end{matrix} \\ \begin{matrix} 5 \\ 6 \end{matrix} & \begin{pmatrix} x & x \\ x & x \end{pmatrix} \end{matrix}.$$

Then we eliminate 4. To do this, we have to consider the matrix resulting from the elimination of 1, that is

$$F_4 = \begin{matrix} & 4 & 5 & 6 \\ 4 & a_{4,4} & 0 & a_{4,6} \\ 5 & 0 & & \\ 6 & a_{6,4} & & \end{matrix} + \bar{U}_4.$$

Elimination of 4 creates contributions,

$$\bar{U}_5^4 = \begin{matrix} & 5 & 6 \\ 5 & x & x \\ 6 & x & x \end{matrix}.$$

Now, before eliminating node 5, we must sum the contributions from the original matrix and what we get from the eliminations of nodes 2, 3 and 4. To do this, we must extend $\bar{U}_5^2$ to the proper set of indices, *i.e.*, 5, 6. We do this as in Liu [915] by considering an operator that we denote by $\circ$. For two matrices A and B, $A \circ B$ takes as the set of indices of the result, the union of the sets of indices of A and B and whenever they coincide, the result is the sum of the entries. Let

$$\bar{U}_5 = \bar{U}_5^2 \circ \bar{U}_5^3 \circ \bar{U}_5^4,$$

then

$$F_5 = \begin{pmatrix} a_{5,5} & a_{5,6} \\ a_{6,5} & 0 \end{pmatrix} + \bar{U}_5.$$

Elimination of 5 gives a matrix of order 1 that is added to $a_{6,6}$ to give the last term of the factorization.

In this example, we have seen that all the elimination steps can be carried out by working on small dense matrices of different orders, extending and summing these matrices by looking at the elimination tree. This has been nicely formalized by Liu [915]. Software packages are available that implement the multifrontal method, for example in the Harwell Scientific Library.

3.11. Non–symmetric sparse matrices

In Chapter 2 we saw that for non–symmetric matrices we need to pivot to improve numerical stability. When dealing with sparse symmetric positive definite systems, the ordering of the unknowns can be chosen only for the purpose of maintaining sparsity as much as possible during elimination. This is not true for non–symmetric problems except in special cases.

If we choose the pivots as for dense systems (for instance, using partial pivoting), there is no room for preserving sparsity. Therefore, for sparse matrices, we have to relax the constraints for choosing the pivot. The usual strategy is to consider candidate pivots satisfying the inequality,

$$|a_{i,j}^{(k)}| \geq \omega \max_{l} |a_{l,j}^{(k)}|,$$

where ω is a user defined parameter such that $0 < \omega \leq 1$. This will limit the overall growth as

$$\max_{i} |a_{i,j}^{(k)}| \leq \left(1 + \frac{1}{\omega}\right)^{p_j} \max_{i} |a_{i,j}|,$$

where p_j is the number of off diagonal entries in column j of U, see Duff, Erisman and Reid [475]. From these candidates, one is selected that minimizes

$$(r_i^{(k)} - 1)(c_j^{(k)} - 1),$$

where $r_i^{(k)}$ is the number of non–zero entries in row i of the remaining $(n - k) \times (n - k)$ matrix in A_k. Similarly, $c_j^{(k)}$ is the number of non–zeros in column j. This is known as the Markowitz criterion, see [937].

This method modifies the smallest number of entries in the remaining submatrix. Note that if A is symmetric, this is exactly the minimum degree algorithm that was introduced historically after the Markowitz criterion. Many variations of the Markowitz criterion have been studied over the years. For a summary, see Duff, Erisman and Reid [475]. However, most of these other methods are generally not as efficient as the Markowitz criterion.

One possibility is to choose the entry (which is not too small) that introduces the least amount of fill–in at step k. Unfortunately, this is much more expensive than the Markowitz criterion. Moreover, having a local minimum fill–in does not always gives a globally optimal fill–in count. There are even some examples where the Markowitz criterion is better at globally reducing the fill–in.

As for the minimum degree algorithm, the tie–breaking strategy is quite important for the result when using the Markowitz algorithm. Details of the implementation of the Markowitz algorithm are discussed in Duff, Erisman and Reid [475]. A switch to dense matrix techniques is made when the non–zero density is sufficiently large.

The structures of the triangular factors of a non–symmetric matrix (without pivoting) can be characterized, see Gilbert and Liu [641]. This starts by considering a triangular matrix L and its digraph $G(L)$ which is acyclic (*i.e.*, there is no directed cycles). An acyclic digraph is called a dag. Let $w = (w_1, \dots, w_n)^T$, then we define

$$Struct(w) = \{i \in \{1, \dots, n\} | w_i \neq 0\}.$$

Theorem 3.13

If $Lx = b$ then $Struct(x)$ is given by the set of vertices reachable from vertices of $Struct(b)$ in the dag $G(L^T)$.

□

An economical way to represent the information contained in a dag G is to consider its transitive closure G^0. Then, in theorem 3.13, we can replace $G(L^T)$ by $G^0(L^T)$. The transitive closure G^* of a directed graph G is a graph that has an edge (u, v) whenever G has a directed path from u to v.

Let A be factored as $A = LU$ without pivoting. $G^0(L)$ and $G^0(U)$ are called the lower and upper elimination dags (edags) of A. For a symmetric matrix, $G^0(L)$ and $G^0(U)$ are both equal to the elimination tree.

If B and C are two matrices with non–zero diagonal elements then $G(B)+G(C)$ is the union of the graphs of B and C *i.e.*, the graph whose edge set is the union of those of $G(B)$ and $G(C)$. $G(B) \cdot G(C)$ is the graph with an edge (i, j) if (i, j) is an edge of $G(B)$ or (i, j) is an edge of $G(C)$ or if there is a k such that (i, k) is an edge of $G(B)$ and (k, j) is an edge of $G(C)$. Gilbert and Liu [641] proved the following result.

Theorem 3.14

If $A = LU$ and there is a path in $G(A)$ from i to j, then there exists a $k, 1 \leq k \leq n$ such that $G^0(U)$ has a path from i to k and $G^0(L)$ has a path *from k to j. That is*

$$G^*(A) \subseteq G^{0^*}(U) \cdot G^{0^*}(L).$$

□

If there is no cancellation in the factorization $A = LU$, then

$$G(L) \cdot G(U) = G(L) + G(U).$$

From these results, the row and column structures of L and U can be derived.

Theorem 3.15

If $l_{i,j} \neq 0$, then there exists a path from i to j in $G^0(L)$. Let $i > j$. Then $l_{i,j} \neq 0$ if and only if there exists $k \leq j$ such that $a_{i,k} \neq 0$ and there is a directed path in $G^0(U)$ from k to j.

$$Struct(L_{*,j}) = Struct(A_{*,j}) \cup \bigcup \{Struct(L_{*,k}) | k < j, u_{k,j} \neq 0\} - \{1, \ldots, j-1\}.$$

□

In the last statement, " $k < j, u_{k,j} \neq 0$" can be replaced by "(k, j) is an edge of $G^0(U)$". The structure of U can be characterized in the same way. From these results, an algorithm can be derived for the symbolic fill computation when there is no pivoting, see [641]. When pivoting is required for stability, edags can also be useful, this time to symbolically compute the fill at each stage of Gaussian elimination.

3.12. Numerical stability for sparse matrices

The study of componentwise error analysis for sparse systems has been considered in Arioli, Demmel and Duff [34]. They consider how to compute estimates of the backward error. The perturbation f of the right hand side is computed a posteriori and is not equal to $|b|$ to keep the perturbation on A sparse and the iterative refinement algorithm convergent (see [34]).

Let $w = |A|\ |y| + |b|$, y being the computed solution. A threshold τ_i is chosen for each w_i such that if $w_i > \tau_i$, then $f_i = |b_i|$. Otherwise if $w_i \leq \tau_i$, f_i is chosen larger. The value of τ_i suggested in [34] is $\tau_i = 1000\ n\ u(\|A_{i,*}\|_\infty \|y\|_\infty + |b_i|)$ where $A_{i,*}$ is the ith row of A.

Let $f^{(2)}$ be the components of f for which $w_i \leq \tau_i$, then $f^{(2)}$ is defined as $f^{(2)} = \|b\|_\infty e$ where e is the column vector of all ones. With this choice, we can compute an estimate of the backward error

$$\frac{|b - Ay|_i}{(|A||y| + f)_i}.$$

Recall that the condition number is

$$K_{BS}(A, b) = \frac{\|\ |A^{-1}|(E|x| + f)\|_\infty}{\|x\|_\infty}.$$

But we may use the estimate $\|\ |A^{-1}|\ |A|\ \|_\infty = |\ |A^{-1}|\ |A|e\|_\infty$. This can be estimated by an algorithm due to Hager [749] that uses multiplications by the matrix and its transpose. This is obtained by forward and backward solves using the LU factorization of A.

Numerical experiments in [34] using iterative refinement show that it is possible to guarantee solutions of sparse linear systems that are exact solutions of a nearby system with a matrix of the same structure. Estimates of the condition number and of the backward error can be easily obtained using the previous strategies giving estimates of the error.

3.13. Parallel algorithms for sparse matrices

For parallel computers, it is easier to consider sparse matrices rather than dense ones since, in the sparse case, there is more natural parallelism. Data dependencies are weaker in the sparse case as, in the factorization process, some columns are independent of each other. However, it is more difficult to obtain significant performances than for dense matrices since the granularity of independent tasks is often quite small and indirect addressing could lead to a lack of data locality.

Let us first consider symmetric matrices when there is no need to pivot for stability. Gaussian elimination proceeds in three phases: ordering, symbolic factorization and numerical factorization. The problem we are faced with is to have parallel implementations of these three phases. Therefore, for the first phase, not only do we have to find an ordering that reduces the fill–in and gives a good degree of parallelism during the solution phase, but also ideally, we need to be able to compute this ordering in parallel.

The most widely used algorithm for reducing fill–in is the minimum degree algorithm. Unfortunately, this method is rather sequential. It must be modified to run efficiently on parallel computers. Several attempts have been made in this direction. One idea is to look for multiple elimination of independent nodes of minimum degree, see Liu [908]. Another possibility is to relax the constraint of finding a node of minimum degree and to look only for nodes whose degrees are within a small additive constant from the minimum degree.

An ordering that is more promising regarding parallelism is nested dissection. Although it is a "divide and conquer" algorithm it is not easy to implement it efficiently for general sparse matrices on parallel computers.

Let us now consider the problem of finding an ordering generating some parallelism during the factorization phase. The current approach of this problem is to separate the two conflicting goals. First an ordering minimizing the fill–in is chosen. Then it is modified by restructuring the elimination tree to introduce more parallelism.

This approach has been described by Jess and Kees [826]. Their method starts by looking at PAP^T, where the permutation P was chosen to preserve sparsity. Then the natural ordering is a perfect elimination one for $F = L + L^T$. The goal is now to find a permutation matrix Q that gives also a perfect elimination but with more parallelism.

A node in G_F whose adjacency set is a clique is called simplicial. Such a node can be eliminated without causing any fill–in. Two nodes are independent if they are not adjacent in G_F. The Jess and Kees algorithm is the following:

- Until all nodes are eliminated, choose a maximum set of independent simplicial nodes, number them consecutively and eliminate these nodes.

It has been shown, see Liu [911], that the Jess and Kees method gives an ordering that has the shortest elimination tree over all orderings that yield a perfect elimination of F. The problem is to implement this algorithm in parallel. This question was not really addressed by Jess and Kees. A proposal using clique trees was described in Lewis, Peyton and Pothen [895]. Another implementation was proposed in Liu and Mirzaian [916].

Heuristically, it can be seen that larger elimination trees (having more leaf nodes) introduce more parallelism. The number of nodes being fixed, larger trees mean shorter trees. Therefore, it seems that finding an ordering that gives a shorter tree would increase the level of parallelism. Liu [911] has proposed to use tree rotations to reach this goal. The purpose of this algorithm is to find a reordering by working on the structure of PAP^T, namely the elimination tree.

A node x in a tree $T(B)$ is eligible for rotation if

$$Adj_{G(B)}(T[x]) \neq Anc(x),$$

where $Anc(x)$ is the set of ancestors of x in T and

$$Adj_{G(B)}(T[v]) = Anc(v), \; \forall v \text{ ancestor of } x.$$

A tree rotation at x is a reordering of $G(B)$ such that the nodes in $Adj_{G(B)}(T[x])$ are labeled last while keeping the relative order of the nodes.

Let $h_T(v)$ be the height of $T[v]$ and $\bar{h}_T$ be defined as -1 if every subtree of T intersects $T[v]$ or otherwise,

$$\bar{h}_T = \max\{h_T(w) | T[w] \cap T[v] = \emptyset\}.$$

The algorithm proposed by Liu [911] is the following,

- If x is an eligible node with $\bar{h}_T(x) < h_T(x)$, then apply a tree rotation at x relabeling the nodes.

Results are proven in Liu [911] that support this choice. The implementation details and experimental results are given in [911]. However, tree rotations do not always give a tree of minimum height. More than the height of the elimination tree, it will be better to minimize the parallel completion time. This was defined by Liu as:

- Let $time[v]$ be the execution time for the node v in T (for instance, a constant times the number of operations).

Then,

$$level[v] = \begin{cases} time[v] & \text{if } v \text{ is the root} \\ time[v] + level[parent\ of\ v] & \text{otherwise.} \end{cases}$$

The parallel completion time is $\max_{v \in T} level[v]$.

Liu and Mirzaian [916] proposed an implementation of the Jees and Kees algorithm. In their method, the cost of detecting simplicial nodes is $O(n\nu(F))$ where $\nu(F)$ is the number of off diagonal elements in $L + L^T$. However, the tree rotation heuristic is faster than that. The clique implementation of Lewis, Peyton and Pothen [895] is about as fast as the tree rotation heuristic.

For the numerical factorization, the first algorithms that were studied were column oriented. Traditionally, the two main operations of Gaussian elimination are denoted:

- $cmod(j, k)$: modification of column j by column k, $k < j$
- $cdiv(j)$: division of column j by a scalar (the pivot).

In the fan–out and fan–in algorithms (see Chapter 2), data distribution is such that columns are assigned to processors. As in the dense case, column k is stored on processor $p = map(k)$. Leaf nodes of the elimination tree are independent of each other and can be processed first. Let $mycols(p)$ be the set of columns owned by processor p and

$$procs(L_{*,k}) = \{map(j) | j \in Struct(L_{*,k})\}.$$

The fan–out algorithm is the following,

```
for j ∈ mycols(p)
  if j is a leaf node
    cdiv(j)
    send L_{*,j} to p' ∈ procs(L_{*,j})
    mycols(p)=mycols(p)-{ j}
  end
  while mycols(p) ≠ ∅
    receive column L_{*,k}
    for j ∈ Struct(L_{*,k})∩ mycols(p)
      cmod(j,k)
      if all cmods are done for column j
```

```
            cdiv(j)
            send L_{*,j} to p' ∈procs(L_{*,j})
            mycols(p)=mycols(p)-{ j}
         end
      end
   end
end
```

When columns are sent to other processors, it is also necessary to send their structures to be able to complete the cmods operations. There is too much communication in this algorithm. This has been improved in the fan–in algorithm Ashcraft, Eisenstat and Liu [51],

- Processing column j, processor p computes the modification $u(j,k)$ for $k \in mycols(p) \cap Struct(L_{j,*})$. If p does not own column j, it sends $u(j,k)$ to processor $map(j)$. If p owns column j, it receives and processes the aggregated modifications and then completes the $cdiv(j)$ operation.

An important issue in these column oriented algorithms is the mapping of columns to the processors. Most implementations have used a static mapping of computational tasks to processors. This can lead to load balancing problems. In the fan–out or fan–in algorithms, the assignment of columns to processors is guided by the elimination tree. The goals are good load balancing and low processor communications.

The first implementations were based on wrap mapping of the levels of the elimination tree starting from the bottom up. This gives good load balancing properties but too many communications. Another technique that has been much used is the subtree to subcube mapping. This was specifically designed for hypercube architectures but can be easily generalized to other distributed memory architectures.

This is illustrated on the following example, see figure 3.17, distributing the tree on 4 processors.

Karypis and Kumar [854] suggested a subforest to subcube mapping which seems to improve on the subtree to subcube mapping. Another algorithm that is in favor is the multifrontal algorithm. We have already seen that there is a

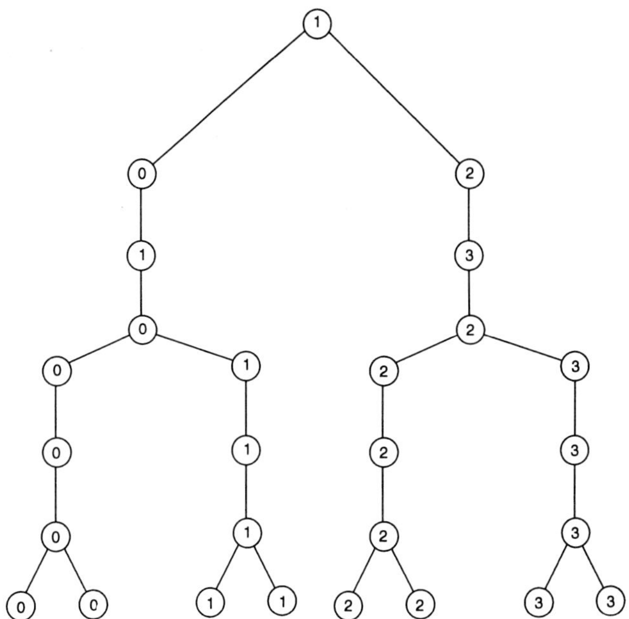

fig. 3.17: The subtree to subcube mapping

natural parallelism in the early (bottom) stages of the multifrontal method. Dense frontal matrices are assigned to one processor. The problem is that when moving towards the root of the tree, there is less and less parallelism. However, frontal matrices are getting larger and larger and dense techniques used to handle these matrices can be distributed on several processors (using BLAS3 primitives).

The multifrontal method was implemented by Kapyris and Kumar [854] using the subforest to subcube mapping and very good efficiencies were reported. In this scheme, many subtrees of the elimination tree are assigned to each subcube. They are chosen in order to balance the work. Algorithms are given in Kapyris and Kumar [854] to obtain this partitioning.

All these mappings are based on column distribution of the matrix to the processors. Rothberg and Gupta [1125] proposed using a block oriented approach of sparse Gaussian elimination. These partitionings have good communication performances. The non–symmetric version of the multifrontal method was studied by Hadfield and Davis [744] for parallel computing. As for the symmetric case, different levels of parallelism must be used. Experimental results are given in [743], [744], [745].

3.14. Bibliographical comments

The study of sparse matrices started in the sixties. Efficient sparse techniques were mainly developed through the work of Duff and Reid that gave rise to the sparse codes of the Harwell Subroutine Library for symmetric positive definite and indefinite matrices.

The main reordering techniques to minimize the bandwidth or the fill–in became popular in the seventies, see Cuthill [362], Tinney and Walker [1250]. At the same time, Alan George and his co–workers developed the dissection algorithms. These results are summarized in the book by George and Liu (625]); these methods are optimal for rectangular meshes. However, extensions for general graphs were not so obvious and less successful.

The envelope schemes and algorithms for storing sparse matrices are mainly due to A. Jennings, see [820]. J. Liu studied several improvements of the minimum degree algorithm which is probably the most used renumbering algorithm, for a summary, see [626].

Recently, there was a renewal of interest in reordering methods, mainly through the use of spectral methods and the work of H. Simon and his co–workers [116]. Combinations of dissection techniques and variants of the minimum degree seem to give good results, see Kumfert and Pothen [872], Ashcraft and Liu [56].

Another important development was the multifrontal algorithm which was devised by Duff and his team for symmetric matrices following ideas from Irons for the frontal method devised for finite element problems. The situation is not so good for non–symmetric matrices as for the symmetric case. However, progress is being made, see Davis and Duff[381].

In the last ten years, there has been much research devoted to the implementation of sparse techniques on parallel computers. Good results have been achieved for symmetric positive definite matrices.

4 Fast solvers for separable PDEs

4.1. Introduction

In Chapter 1 we introduced the finite difference discretization of the Poisson equation in a square domain with Dirichlet boundary conditions,

$$\begin{aligned} -\Delta u &= f \quad \text{in} \quad \Omega = (0,1)\times(0,1), \\ u\big|_{\partial\Omega} &= 0. \end{aligned} \tag{4.1}$$

As we have seen, this gives rise to a system of linear equations

$$Ax = b,$$

where A is a block tridiagonal matrix,

$$A = \begin{pmatrix} T & -I & & & \\ -I & T & -I & & \\ & \ddots & \ddots & \ddots & \\ & & -I & T & -I \\ & & & -I & T \end{pmatrix},$$

with I the identity matrix and T a tridiagonal matrix,

$$T = \begin{pmatrix} 4 & -1 & & & \\ -1 & 4 & -1 & & \\ & \ddots & \ddots & \ddots & \\ & & -1 & 4 & -1 \\ & & & -1 & 4 \end{pmatrix}.$$

This is the model problem for the equations we wish to solve with fast direct methods in this chapter.

The most general problem we shall consider is the following separable elliptic partial differential equation

$$a(x)\frac{\partial^2 u}{\partial x^2} + b(x)\frac{\partial u}{\partial x} + c(x)u + d(y)\frac{\partial^2 u}{\partial y^2} + e(y)\frac{\partial u}{\partial y} + f(y)u = g(x,y)$$

in a rectangle Ω with Dirichlet, Neumann (*i.e.*, $\frac{\partial u}{\partial n} = 0$) or periodic boundary conditions. These problems are said to be separable as the coefficients of the partial derivatives with respect to x (resp. y) depend only on x (resp. y). This allows us to use separation of variables. In Chapter 1, we saw the modifications introduced by Neumann or periodic boundary conditions on $\partial\Omega$, $\Omega = (0,1) \times (0,1)$ for the problem $-\Delta u + \sigma u = f$. Three dimensional separable problems can also be handled with these techniques. In this chapter, we shall study three methods to solve problems like (4.1):

- the Fourier/tridiagonal method
- the cyclic reduction method
- the FACR(l) method which is a blend of the two previous ones.

For the sake of simplicity we shall only consider the model problem most of the time. However, we shall give indications of possible extensions for more general problems.

For the Poisson model problem we explicitly know the eigenvalues of the matrix A, see Chapter 1. Since A is symmetric, the eigenvectors form an orthonormal basis (theorem 1.21) and we have the spectral decomposition

$$A = Q\Lambda Q^T,$$

where Λ is a diagonal matrix whose diagonal coefficients are the eigenvalues of A, and Q is orthogonal $Q^TQ = I$. Hence, an obvious way to solve $Ax = b$ is given by writing

$$Q\Lambda Q^T x = b.$$

Left multiplying by Q^T gives

$$\Lambda Q^T x = Q^T b,$$

and, of course,

$$x = Q\Lambda^{-1}Q^T b.$$

The algorithm is described in three steps:

1) $\hat{b} = Q^T b$,

2) $\hat{x} = \Lambda^{-1}\hat{b}$,

3) $x = Q\hat{x}$.

Looking at the values of Λ and Q for the model problem, this method is simply the Fourier analysis (sometimes called DFA, Double Fourier Analysis, as it corresponds to a Fourier decomposition in each direction). We shall see that this method, although it is the most natural, is not the most efficient. We start by introducing this method because in steps 1) and 3) we have to compute sums such as

$$x_j = \sum_{k=1}^{n} a_j \sin \frac{jk\pi}{n+1}, \quad 1 \le j \le n.$$

In the next section we study the Fast Fourier Transform (FFT) showing how to efficiently compute these sums.

4.2. Fast Fourier Transform

4.2.1. The basics of the FFT

Let us first look at a more general problem. Let $a(k)$ be a sequence of complex numbers and w be the nth root of unity *i.e.*,

$$w = e^{\frac{2i\pi}{n}}, \quad i = \sqrt{-1}.$$

We wish to compute n numbers

$$x_j = \sum_{k=0}^{n-1} a(k) w^{jk} \quad , \quad 0 \le j \le n-1. \tag{4.2}$$

Doing this computation in the obvious way, each sum requires n operations. So, n^2 operations are needed to compute the n sums, each operation being one complex multiplication and one complex addition. This gives of the order of

$8n^2$ real operations. In 1965 Cooley and Tukey [345] published an algorithm known as the Fast Fourier Transform (FFT) which allows us the computation of the n sums more efficiently. Let us suppose that we can factor the integer n as $n = n_1n_2$, n_1 and n_2 being integers greater than 1. We shall show that we can compute the sums in (4.2) with $n(n_1+n_2)$ operations. To accomplish this, we use the following decompositions of the two indices j and k

$$
\begin{aligned}
j &= j_1n_1 + j_0\\
k &= k_1n_2 + k_0.
\end{aligned}
$$

This implies that j_0 (resp. k_0) takes integer values in $[0, n_1-1]$ (resp. $[0, n_2-1]$) and j_1 (resp. k_1) in $[0, n_2-1]$ (resp. $[0, n_1-1]$). Then, we can simplify the computation of w^{jk} as

$$w^{jk} = w^{(j_1n_1+j_0)(k_1n_2+k_0)} = w^{(j_1n_1+j_0)k_0}w^{j_0k_1n_2}w^{j_1k_1n}$$

but $w^n = 1$, hence $w^{j_1k_1n} = 1$. Then,

$$x_j = \sum_{k_0,k_1} a(k)w^{jk} = \sum_{k_0} w^{(j_1n_1+j_0)k_0}a_1(j_0,k_0),$$

where a_1 is defined as

$$a_1(j_0,k_0) = \sum_{k_1} a(k_0,k_1)w^{j_0k_1n_2}.$$

We have n values of a_1 to compute, each of which is a sum over k_1 and requires n_1 operations. Knowing a_1 we have n sums over k_0 to compute. This needs nn_2 operations. Summing these results we get $n(n_1+n_2)$ operations to compute the sums (4.2). With n_1 and $n_2 \geq 2$ we always have $n_1+n_2 \leq n_1n_2$. This way of computing the sums can be applied recursively if we can further decompose n_1 or n_2. In most FFT codes, n is decomposed as

$$n = 2^p3^q4^r5^st.$$

4.2.2. The complex FFT

For the sake of simplicity we shall look at the method of computation when $n = 2^m$ (Cooley and Tukey [345]). We use the binary decomposition of j and k,

$$
\begin{aligned}
j &= j_{m-1}2^{m-1} + j_{m-2}2^{m-2} + \cdots + j_12 + j_0,\\
k &= k_{m-1}2^{m-1} + k_{m-2}2^{m-2} + \cdots + k_12 + k_0,
\end{aligned}
$$

where j_v and k_v take values 0 or 1. Then,

$$x_j = x(j_{m-1}, j_{m-2}, \ldots, j_0)$$
$$= \sum_{k_0} \sum_{k_1} \cdots \sum_{k_{m-1}} a(k_{m-1}, \ldots, k_0)\ w^{jk_{m-1}2^{m-1}+\cdots+jk_0}.$$

But, as before

$$w^{jk_{m-1}2^{m-1}} = w^{j_0 k_{m-1}2^{m-1}},$$

all other terms being powers of $w^n = 1$.

So, the sum over k_{m-1} depends only on $j_0, k_{m-2}, \ldots, k_0$. Let us denote

$$a_1(j_0, k_{m-2}, \ldots, k_0) = \sum_{k_{m-1}} a(k_{m-1}, \ldots, k_0) w^{jk_{m-1}2^{m-1}}.$$

With this notation

$$x_j = \sum_{k_0} \cdots \sum_{k_{m-2}} a_1(j_0, k_{m-2}, \cdots, k_0)\ w^{jk_{m-2}2^{m-2}+\cdots+jk_0}.$$

The sum over k_{m-2} depends only on $j_0, j_1, k_{m-3}, \cdots, k_0$. More generally, let us denote

$$a_l(j_0, j_1, \ldots, j_{l-1},\ k_{m-l-1}, \ldots, k_0)$$
$$= \sum_{k_{m-1}} a_{l-1}(j_0, \ldots, j_{l-2}, k_{m-l}, \ldots, k_0) w^{(j_{l-1}2^{l-1}+\cdots+j_0)k_{m-l}2^{m-l}}.$$

But k_{m-l} takes only values 0 and 1 and the sum reduces to

$$a_l = a_{l-1}(j_0, \ldots, j_{l-1}, 0, k_{m-l-1}, \ldots, k_0)$$
$$+ a_{l-1}(j_0, \ldots, j_{l-2}, 1, k_{m-l-1}, \ldots, k_0) w^{(j_{l-1}2^{m-1}+\cdots+j_0 2^{m-l})}$$

which we can rewrite as:

$$a_l = a_{l-1}(j_0, \ldots, j_{l-1}, 0, k_{m-l-1}, \ldots, k_0)$$
$$+ (-1)^{j_{l-1}} a_{l-1}(j_0, \ldots, j_{l-2}, 1, k_{m-l-1}, \ldots, k_0) \quad (4.3)$$
$$\times\ w^{j_{l-3}2^{m-3}+\cdots+j_0 2^{m-l}}.$$

This follows from

$$w^{j_{l-1}2^{m-1}} = \begin{cases} 1, & \text{if } j_{l-1} = 0 \\ -1, & \text{if } j_{l-1} = 1 \end{cases},$$

$$w^{j_{l-2}2^{m-2}} = \begin{cases} 1 & \text{if } j_{l-2} = 0 \\ i & \text{if } j_{l-2} = 1 \end{cases}.$$

When coding the algorithm on a computer, intermediate results are stored in an array whose elements are indexed by the binary representation of the indices. For example, a_l is stored in locations $j_0 j_1 \cdots j_{l-1} \cdot k_{m-l-1} \cdots k_0$ where the indices take all the possible values. In formula (4.3) $j_0, \ldots, j_{l-2}, k_{m-l-1}, \ldots, k_0$ being fixed, we combine two elements whose locations differ from 1 bit in the lth position and we obtain two new elements that can be stored in the same locations.

After m steps, we get

$$x(j_{m-1}, \ldots, j_0) = a_m(j_0, \ldots, j_{m-1}).$$

Note that with the storage scheme we use, the components of x are not in the natural order. We need to do a permutation which is called a bit reversal,

$$(j_0, j_1, \ldots, j_{m-1}) \rightarrow (j_{m-1}, j_{m-2}, \ldots, j_0)$$

We illustrate the algorithm graphically for $n = 2^3$ in figure 4.1. At the left of figure 4.1 we see the binary decomposition of the indices for the locations of the array used to store intermediate results. The integer in the diamond gives the power of w by which we multiply, that is $j_{l-3}2^{m-3} + \cdots + j_0 2^{m-l}$. We note that when using the FFT for solving PDEs two permutations are required and they can be omitted.

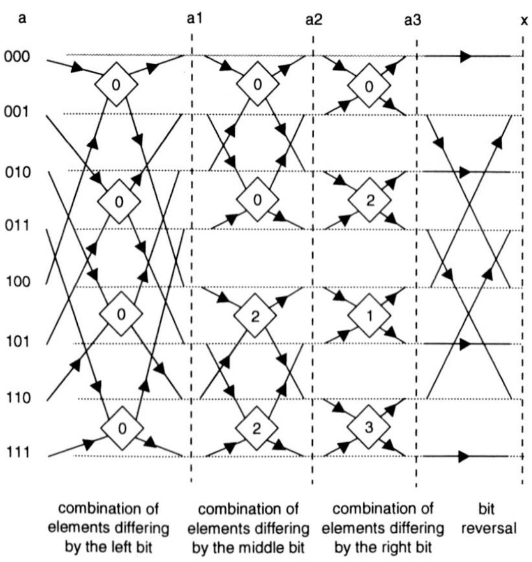

fig. 4.1: The FFT for n=8

What is the complexity of this algorithm? At each step we compute two values with one multiplication and two complex additions, *i.e.*, $\frac{n}{2}(1M+2A)$. Within the m steps we have

$$\frac{mn}{2}(1M+2A) \qquad \text{complex operations.}$$

A complex multiplication is $4M+2A$ (note that it can also be computed with 3 multiplications and 5 additions), a complex addition is 2 real additions. Finally we get $mn(2M+3A)$ or $5mn$ real operations. As $m = \log_2 n$, this is $5n\log_2 n$ real operations. This assumes that the powers of w are precomputed. At the end of this section we shall look at the use of this algorithm on vector and parallel computers. Note that many other algorithms for FFT have been derived, see for instance Temperton [1237].

4.2.3. The real transforms

We have not yet solved our problem. So, let us now look at real transforms. Most of the following is taken from Swarztrauber [1224]. Suppose that the numbers a_k are real and we would like to compute

$$z_j = \frac{1}{n}\sum_{k=0}^{n-1} a_k w^{jk}, \qquad 0 \le j \le n-1 \tag{4.4}$$

The factor $\frac{1}{n}$ is simply a normalization (regarding the inverse transforms) and is of no importance here.

Note that

$$z_{n-j} = \frac{1}{n}\sum_{k=0}^{n-1} a_k w^{(n-j)k} = \frac{1}{n}\sum_{k=0}^{n-1} a_k w^{-jk}.$$

This implies

$$\overline{z_{n-j}} = z_j,$$

where the bar denotes the complex conjugate of a complex number. For the sake of simplicity, let us suppose n is odd and let x_j and y_j denote (up to a scaling factor) the real and imaginary parts of z_j (note that z_0 is real)

$$\begin{aligned} z_0 &= \tfrac{1}{2}x_0, \\ z_j &= \tfrac{1}{2}(x_j + iy_j), \qquad j = 1, \dots, \tfrac{n-1}{2}. \end{aligned}$$

We have seen that $z_{n-j} = \frac{1}{2}(x_j - iy_j)$. By equating, we have

$$\begin{aligned} x_j &= \tfrac{2}{n}\sum_{k=0}^{n-1} a_k \cos\frac{jk2\pi}{n} \quad j = 0,\ldots,\frac{n-1}{2}, \\ y_j &= \tfrac{2}{n}\sum_{k=0}^{n-1} a_k \sin\frac{jk2\pi}{n} \quad j = 1,\ldots,\frac{n-1}{2}, \end{aligned} \tag{4.5}$$

which are the real and imaginary parts of the complex transform. To get x_j and y_j we just need to compute the complex transform of half length and then

$$x_0 = 2z_0 \quad , \quad x_j = 2Re(z_j) \quad , \quad y_j = 2Im(z_j).$$

Similar expressions exist for n even.

Let us now look at the transform we need for the Poisson equation. Suppose that n is odd and we wish to compute

$$c_j = \tfrac{2}{n}\sum_{k=1}^{n-1} a_k \sin\frac{jk\pi}{n}, \quad j = 1,\ldots,n-1. \tag{4.6}$$

It is easy to see that the inverse transform is given by

$$a_k = \sum_{j=1}^{n-1} c_j \sin\frac{jk\pi}{n}. \tag{4.7}$$

To compute c_j, let us define a sequence d_k given by

$$d_k = \tfrac{1}{2}(a_k - a_{n-k}) + \sin\frac{k\pi}{n}(a_k + a_{n-k}), \quad k = 1,\ldots,n-1. \tag{4.8}$$

Note that knowing a_k, we need $(n-1)[2A+0.5M]$ real operations to compute d_k. This is because if we change k to $l = n - k$, then

$$\sin\frac{k\pi}{n}(a_k + a_{n-k}) \rightarrow \sin\frac{l\pi}{n}(a_l + a_{n-l}). \tag{4.9}$$

Hence, we need only to compute half of these expressions. In (4.9) we replace a_k by its value (4.7),

$$\begin{aligned} d_k &= \tfrac{1}{2}\sum_{j=1}^{n-1} c_j \left(\sin\frac{jk\pi}{n} - \sin\frac{j(n-k)\pi}{n}\right), \\ &+ \sin\frac{k\pi}{n}\sum_{j=1}^{n-1} c_j \left(\sin\frac{jk\pi}{n} + \sin\frac{j(n-k)\pi}{n}\right). \end{aligned}$$

But, as $\sin(j\pi - \theta) = \sin\theta$ if j is odd, the odd terms of the first sum are equal to zero. We can rewrite it as

$$\sum_{j=1}^{\frac{n-1}{2}} c_{2j} \sin \frac{jk2\pi}{n}.$$

Similarly for j even, the terms of the second sum are zero. Therefore, we get

$$2\sum_{j=0}^{\frac{n-1}{2}} c_{2j+1} \sin \frac{k\pi}{n} \sin \frac{(2j+1)k\pi}{n}.$$

Using the trigonometric identities $\sin 2\theta = 2\sin\theta \cos\theta$ and $\cos 2\theta = 1 - 2\sin^2\theta$, we can transform this to

$$\begin{aligned}
2\sum_j c_{2j+1} &\sin\omega \left(\sin 2j\omega \cos\omega + \cos 2j\omega \sin\omega\right) \\
&= \sum_j c_{2j+1} \big(\sin 2j\omega \sin 2\omega + \cos 2j\omega(1 - \cos 2\omega)\big), \\
&= c_1 + \sum_{j=1}^{\frac{n-1}{2}} (c_{2j+1} - c_{2j-1}) \cos 2j\omega.
\end{aligned}$$

where $\omega = \frac{k\pi}{n}$. Finally we obtain

$$d_k = c_1 + \sum_{j=1}^{\frac{n-1}{2}} \left[(c_{2j+1} - c_{2j-1}) \cos 2j\frac{k\pi}{n} + c_{2j} \sin 2j\frac{k\pi}{n} \right].$$

This is a transform of the same kind as (4.4) – (4.5). We have seen how, knowing d_k, we can compute x_j and y_j satisfying

$$d_k = \tfrac{1}{2}x_0 + \sum_{j=1}^{\frac{n-1}{2}} (x_j \cos 2j\frac{k\pi}{n} + y_j \sin 2j\frac{k\pi}{n}), \quad k = 1, \ldots, n-1.$$

By equating we get

$$\begin{aligned}
c_1 &= \tfrac{1}{2}x_0, \\
c_{2j} &= y_j, \\
c_{2j+1} - c_{2j-1} &= x_j.
\end{aligned}$$

Hence the algorithm is as follows:

1) compute d_k by (4.8),

2) transform d_k (complex transform of half length),

3) compute c_k $(0.5(n-1)$ real additions).

The total number of real operations is $1.5n\log_2 n + 2.5n$ additions and $n\log_2 n + 0.5n$ multiplications if n is a power of 2.

4.2.4. FFT on vector and parallel computers

The operation counts we gave are appropriate for a serial computer. Let us consider what can be done on a vector computer. The Cooley-Tukey FFT is obviously trivially vectorizable or parallelizable because all the operations at each step are independent of each other. But, we note that the vector length (or the amount of parallelism) decreases from $\frac{n}{2}$ to 1 which is poor for the efficiency of the computation.

One can derive other FFTs where the vector length is fixed equal to $\sqrt{n}$. But it appears that the best way to vectorize or parallelize is to transform m sequences in parallel, the operations being done in scalar mode on each sequence. Having to transform m sequences is common when solving partial differential equations. For the FFT on vector computers, see the summary papers by Swarztrauber [1226], [1227]. For parallel and vector implementations, see Hockney and Jesshope [790].

4.2.5. Stability of the FFT

Errors bounds for the computation were considered in early seventies. For a clear and concise treatment, we refer to the book by Higham [781].

Theorem 4.1

Let x and $\tilde{x}$ be the exact and computed FFT *of a sequence of length $n = 2^m$. If all the weights w are such that*

$$\tilde{w} = w + \epsilon, \quad |\epsilon| \le \mu,$$

then

$$\frac{\|x-\tilde{x}\|}{\|x\|} \le n^{\frac{1}{2}} \frac{m\eta}{1-m\eta}, \quad \eta = \mu + \frac{4u}{1-4u}(1+\mu).$$

Proof. See Higham [781]. □

This theorem proves that the FFT is a stable algorithm provided the weights are computed properly.

4.2.6. Other algorithms

Many variants of the FFT algorithm have been proposed over the years. For instance, self sorting algorithms were developed. Another type of algorithm is known as the prime factor algorithm (PFA), see Good [682], [683]. In these algorithms, the length n is decomposed as the product of integers that must be mutually prime. This severely limits the set of lengths that can be used. This is unfortunate as for a given value of n, the PFA has a lower operation count than the Cooley–Tukey method. However, Temperton [1244] generalized the PFA to lengths of the form $n = 2^p 3^q 5^r$. This generalization is also self sorting, the operation count being lower than for Cooley–Tukey.

4.2.7. Double Fourier analysis

Having derived the FFT, we can solve the Poisson model problem with the DFA algorithm. For instance, $Q^T b$ reduces to computing the sums

$$\sum_{i,j} b_{i,j} \sin(ip\pi h) \sin(jq\pi h), \quad p, q \quad \text{fixed}$$

with

$$1 \leq p, q, i, j \leq m.$$

To compute $Q^T b$ we need $2m$ FFTs. The same is true for $Q\hat{x}$, so the number of operations is $6m^2 \log_2 m + 10m^2$ additions, $4m^2 \log_2 m + 2m^2$ multiplications and m^2 divisions *i.e.*, a total of order $10m^2 \log_2 m$ operations which seems very low. However, we shall see in the following sections that there are more efficient methods. But, this method is very easy to use as there are good FFT packages in most scientific computing libraries (see also in Netlib). For a fast implementation in C, see Frigo and Johnson [592].

4.3. The Fourier/tridiagonal Method

In this section we shall solve the linear system

$$Ax = b,$$

where

$$A = \begin{pmatrix} T & B & & & \\ B & T & B & & \\ & \ddots & \ddots & \ddots & \\ & & B & T & B \\ & & & B & T \end{pmatrix}.$$

Here, B and T are symmetric matrices of order m which commute *i.e.*, $BT = TB$. The following theorem will allow us to use spectral decomposition.

Theorem 4.2

Let B and T be symmetric matrices such that $BT = TB$. Then there exists an orthogonal matrix Q such that

$$T = Q\Lambda Q^T, \qquad B = Q\Omega Q^T,$$

where Λ and Ω are diagonal matrices whose diagonal elements λ_i and ω_i are the eigenvalues of T and B.

Proof. The proof of this theorem can be found in Bellman [133]. □

For the sake of simplicity, we shall take $B = -I$ in the remainder of this section.However, the method works for the more general case described above. For the model problem, we know that

$$\lambda_i = 4 - 2\cos\frac{i\pi}{m+1}, \qquad \omega_i = -1.$$

The jth component of the ith eigenvector is given by

$$\sqrt{\frac{1}{m+1}}\sin\frac{ij\pi}{m+1}.$$

To solve the linear system, we use the formulation of Meurant [965] and a block factorization of A. Let A be factorized as

$$\begin{pmatrix} \Delta_1 & & & \\ -I & \Delta_2 & & \\ & \ddots & \ddots & \\ & & -I & \Delta_n \end{pmatrix} \begin{pmatrix} \Delta_1^{-1} & & & \\ & \Delta_2^{-1} & & \\ & & \ddots & \\ & & & \Delta_n^{-1} \end{pmatrix} \begin{pmatrix} \Delta_1 & -I & & \\ & \Delta_2 & \ddots & \\ & & \ddots & -I \\ & & & \Delta_n \end{pmatrix} \tag{4.10}$$

where the matrices Δ_i, $i = 1, \ldots, m$ are square symmetric matrices of order m. By equating, we see that

$$\Delta_1 = T,$$
$$\Delta_i = T - \Delta_{i-1}^{-1}, \quad i = 2, \ldots, m.$$

Since we know the eigenvalues of T, we can compute those of Δ_i.

Lemma 4.3

For all i, Δ_i has the same eigenvectors as T. The eigenvalues σ_i^j of Δ_i are given by

$$\sigma_1^j = \lambda_j, \quad 1 \leq j \leq m,$$
$$\sigma_i^j = \lambda_j - \frac{1}{\sigma_{i-1}^j}, \quad 1 \leq j \leq m, \quad 2 \leq i \leq m.$$

Proof. Let $\sum_i$ be the diagonal matrix with diagonal elements σ_i^j. Then, since

$$\Delta_1 = T = Q\Lambda Q^T,$$

and $\sum_1 = \Lambda$ we have,

$$\begin{aligned}\Delta_i &= Q\Lambda Q^T - Q\Sigma_{i-1}^{-1}Q^T \\ &= Q(\Lambda - \Sigma_{i-1}^{-1})Q^T \\ &= Q\Sigma_i Q^T.\end{aligned}$$

This shows that $\Sigma_i = \Lambda - \Sigma_{i-1}^{-1}$. □

Let

$$b = \begin{pmatrix} b_1 \\ b_2 \\ \vdots \\ b_m \end{pmatrix}, \quad x = \begin{pmatrix} x_1 \\ x_2 \\ \vdots \\ x_m \end{pmatrix}, \quad b_i,\ x_i \in \mathbb{R}^m.$$

With the decomposition (4.10), we can solve the system by a forward and a backward block sweeps.

$$\begin{cases} \Delta_1 y_1 = b_1, \\ \Delta_i y_i = y_{i-1} + b_i, \quad i = 2, \ldots, m \quad y_i \in \mathbb{R}^m, \end{cases}$$
$$\begin{cases} x_m = y_m, \\ x_i - \Delta_i^{-1} x_{i+1} = y_i, \quad i = m-1, \ldots, 1. \end{cases}$$

Since we know the eigenvalues of Δ_i, we can give another formulation of the algorithm. Denote

$$\hat{b}_i = Q^T b_i, \qquad \hat{y}_i = Q^T y_i, \qquad \hat{x}_i = Q^T x_i.$$

Then

$$\begin{cases} \Sigma_1 \hat{y}_1 = \hat{b}_1, \\ \Sigma_i \hat{y}_i = \hat{y}_{i-1} + \hat{b}_i, \end{cases}$$

$$\begin{cases} \hat{x}_m = \hat{y}_m, \\ \hat{x}_i = \Sigma_i^{-1} \hat{x}_{i+1} + \hat{y}_i. \end{cases}$$

The algorithm is the following:

Algorithm Fourier/Tridiag

0) Precomputation of Σ_i^{-1} by $\Sigma_1 = \Lambda$, $\Sigma_i = \Lambda - \Sigma_{i-1}^{-1}$,

1) computation of $\hat{b}_i = Q^T b_i$,

2) $\hat{y}_1 = \Sigma_1^{-1}\hat{b}_1, \quad \hat{y}_i = \Sigma_i^{-1}(\hat{y}_{i-1} + \hat{b}_i)$,

3) $\hat{x}_m = \hat{y}_m, \quad \hat{x}_i = \Sigma_i^{-1}\hat{x}_{i+1} + y$,

4) computation of $x_i = Q\hat{x}_i$ for all i.

Step 0) is called precomputation since it does not depend on the right hand side b. Thus if we have several systems to solve with the same matrix A but with different right hand sides, the precomputation has to be done only once. Hence, we do not include it in the operation count.

Steps 2) and 3) require $2m^2 - m$ multiplications and $2m^2 - 2m$ additions. If we solve the model problem, we can use the FFT for steps 1) and 4). This can also be true for slightly more general problems. In that case and if m is a power of 2, for steps 1) and 4) we get $3nm^2 \log_2 m + 5nm^2$ additions and $2m^2 \log_2 m + m^2$ multiplications. Summing for all the steps, the operation count is

$$\begin{aligned} 3m^2 \log_2 m + 7m^2 - 2m & \quad \text{additions}, \\ 2m^2 \log_2 m + 3m^2 - m & \quad \text{multiplications}. \end{aligned}$$

This is clearly much better than the number of operations needed by the Double Fourier Analysis (approximately half). The method is obviously completely vectorizable, but the vector length is only m because there are two first order block recurrences. This method was used in a different formulation by Hockney in 1965, see [788]. It is equivalent to making a Fourier analysis in one direction, and to solve the tridiagonal systems we obtain by Gaussian elimination after a renumbering.

4.4. The cyclic reduction method

We first introduce the cyclic reduction method for solving a tridiagonal linear system. Let

$$\begin{pmatrix} a & -1 & & & \\ -1 & a & -1 & & \\ & \ddots & \ddots & \ddots & \\ & & -1 & a & -1 \\ & & & -1 & a \end{pmatrix} x = b, \tag{4.11}$$

the matrix being of order $n = 2^{l+1} - 1$. The idea behind this method is to use the matrix structure to first eliminate the odd unknowns. We have:

$$\begin{array}{cccccccccc} -x_{i-2} & + & ax_{i-1} & - & x_i & & & & = & b_{i-1}, \\ & - & x_{i-1} & + & ax_i & - & x_{i+1} & & = & b_i, \\ & & & - & x_i & + & ax_{i+1} & - \; x_{i+2} & = & b_{i+1}. \end{array} \tag{4.12}$$

Multiplying the middle equation by a and summing, the odd unknowns x_{i-1}, x_{i+1} disappear and we get

$$-x_{i-2} + (a^2 - 2)x_i - x_{i+2} = b_{i-1} + ab_i + b_{i+1}.$$

The new system we obtain has the form

$$\begin{pmatrix} a^2-2 & -1 & & & \\ -1 & a^2-2 & -1 & & \\ & \ddots & \ddots & \ddots & \\ & & -1 & a^2-2 & -1 \\ & & & -1 & a^2-2 \end{pmatrix} \begin{pmatrix} x_2 \\ x_4 \\ \vdots \\ \vdots \\ x_{2^{l+1}-2} \end{pmatrix} = \begin{pmatrix} b_2^{(1)} \\ b_4^{(1)} \\ \vdots \\ \vdots \\ b_{2^{l+1}-2} \end{pmatrix},$$

the matrix is now of order $2^l - 1$. So, we can apply the same reduction in a recursive way. To summarize, we define sequences of tridiagonal matrices whose diagonal elements are $a^{(r)}$, and off diagonal elements -1 and sequences of right hand sides $b^{(r)}$ given by

$$\begin{aligned} a^{(0)} &= a, \qquad b_i^{(0)} = b_i, \\ a^{(r+1)} &= (a^{(r)})^2 - 2, \quad r = 0, \ldots, l-1, \end{aligned} \tag{4.13}$$

$$b_i^{(r+1)} = b_{i-2^r}^{(r)} + a^{(r)} b_i^{(r)} + b_{i+2^r}^{(r)},$$
$$i = 2^r, 2 \times 2^r, 3 \times 2^r, \ldots, n+1-2^r.$$

At step l the tridiagonal linear system reduces to a single scalar equation

$$a^{(l)} x_{2^l} = b_{2^l}^{(l)}.$$

With this equation we compute the component x_{2^l} of the solution. We then go back and at each step we use the equations we eliminated in the forward sweep. For instance, as we know x_{i-2}, x_i, x_{i+2} we use the first equation in (4.12) to compute x_{i-1} and the third one to compute x_{i+1}. For $l = 3$ the computation is summarized in figure 4.2. In the forward sweep we circle the unknowns we keep. In the backward sweep we show the unknowns we can compute by an arrow.

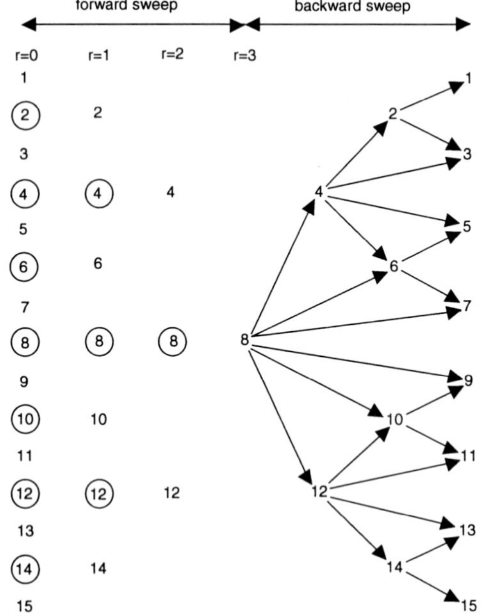

fig. 4.2: The cyclic reduction algorithm

This method, known as the cyclic reduction method, is simply the block Gaussian elimination if we number the unknowns as:

odd multiples of $2^0 = 1$,

odd multiples of 2,

odd multiples of 2^2,

and so on.

Using this numbering scheme, the system (4.11) can be written blockwise as

$$\begin{pmatrix} aI_p & L \\ L^T & aI_{p-1} \end{pmatrix} \begin{pmatrix} y_1 \\ y_2 \end{pmatrix} = \begin{pmatrix} c_1 \\ c_2 \end{pmatrix},$$

with $y_1 = (x_1, x_3, \cdots)^T$, $y_2 = (x_2, x_4, \cdots)^T$ and $p = 2^l$. Then eliminating y_1, we get

$$(aI_{p-1} - \tfrac{1}{a} L^T L) y_2 = c_2 - \tfrac{1}{a} L^T c_1. \tag{4.14}$$

One can easily verify that the product $L^T L$ has the following structure,

$$\begin{pmatrix} 2I_q & L_1 \\ L_1^T & 2I_{q-1} \end{pmatrix},$$

with $q = \frac{p}{2}$. For example, with $p = 2^3$ we have

$$\begin{pmatrix} aI_8 & L \\ L^T & aI_7 \end{pmatrix},$$

and

$$L = \begin{pmatrix} -1 & & & & & & \\ -1 & & & & -1 & & \\ & -1 & & & -1 & & \\ & -1 & & & & & -1 \\ & & -1 & & & & -1 \\ & & -1 & & & -1 & \\ & & & -1 & & -1 & \\ & & & -1 & & & \end{pmatrix}.$$

Then we can check that

$$L^T L = \begin{pmatrix} 2 & & & & 1 & & \\ & 2 & & & 1 & & 1 \\ & & 2 & & & 1 & 1 \\ & & & 2 & & 1 & \\ 1 & 1 & & & 2 & & \\ & & 1 & 1 & & 2 & \\ & 1 & 1 & & & & 2 \end{pmatrix}.$$

Hence the new system (4.14) has the same structure as the original one and we can apply the same process recursively.

Unfortunately, we cannot directly use the cyclic reduction as defined by (4.13) because for $a > 2$, $a^{(r)}$ is a very rapidly increasing sequence and the use of (4.13) leads to overflow on all computers (for instance if $a = 4$ then $a^{(10)} \sim -5 \times 10^{585}$). Of course, for a tridiagonal system we can scale the equations (as we did in fact in (4.14)) but we cannot generalize the method for a block tridiagonal system. We must use another stable way to compute $a^{(r)}$. This is provided by the result in the next lemma.

Lemma 4.4

$a^{(r)}$ defined by (4.13) is a 2^r degree polynomial in a, whose zeros are

$$2\cos\frac{2j-1}{2^{r+1}}\pi, \quad j = 1, \ldots, 2^r.$$

Proof. The proof is by induction on r. The result is clearly true for $r = 0$. Let us suppose that it is true for $r-1$, that is to say, $a^{(r-1)}$ is a 2^{r-1} degree polynomial which we denote by $p_{r-1}(a)$. Then

$$a^{(r)} = (a^{(r-1)})^2 - 2 = (p_{r-1}(a))^2 - 2.$$

Hence $a^{(r)}$ is a 2^r degree polynomial. Now we are looking for the zeros of $p_r(a)$. We begin by looking in $[-2, +2]$ and we define φ by

$$a = 2\cos\varphi.$$

Then, $a^{(1)} = a^2 - 2 = 2(2\cos^2\varphi - 1) = 2\cos 2\varphi$. By induction, let $a^{(r-1)} = 2\cos 2^{r-1}\varphi$,

$$a^{(r)} = (a^{(r-1)})^2 - 2 = 2(2\cos^2(2^{r-1}\varphi) - 1) = 2\cos 2^r\varphi.$$

This proves that

$$a^{(r)} = 2\cos(2^r \arccos\frac{a}{2}),$$

$a^{(r)}$ is zero for $\varphi = \varphi_j$ such that $2^r\varphi_j = (2j-1)\frac{\pi}{2}$ *i.e.*, $\varphi_j = (2j-1)\frac{\pi}{2^{r+1}}$, j taking the values $1, \ldots, 2^r$. We found in this way all the zeros of $a^{(r)}$, there is no zero outside $[-2, +2]$ and $a^{(r)}$ can be written as

$$a^{(r)} = \prod_{j=1}^{2^r}(a - 2\cos\frac{2j-1}{2^{r+1}}\pi),$$

which proves the result. □

We can now generalize the cyclic reduction method to block tridiagonal systems. Consider the problem

$$\begin{pmatrix} T & -I & & & \\ -I & T & -I & & \\ & \ddots & \ddots & \ddots & \\ & & -I & T & -I \\ & & & -I & T \end{pmatrix} x = b,$$

where T is a symmetric tridiagonal matrix of order m.

The use of the cyclic reduction method corresponds to defining a sequence of matrices $T^{(r)}$ and of right hand sides $b^{(r)}$ by

$$\begin{aligned} T^{(0)} &= T, \qquad b_i^{(0)} = b_i, \qquad i = 1, \ldots, m, \\ T^{(r+1)} &= (T^{(r)})^2 - 2I, \end{aligned} \tag{4.15}$$

$$\begin{aligned} b_i^{(r+1)} &= T^{(r)} b_i^{(r)} + b_{i-2^r}^{(r)} + b_{i+2^r}^{(r)}, \\ &\quad i = 2^r, 2 \times 2^r, 3 \times 2^r, \ldots, n + 1 - 2^r. \end{aligned}$$

For the model problem, T is tridiagonal, so $T^{(1)}$ is a five diagonal matrix and so on ... the matrices $T^{(r)}$ are denser and denser. However, in the same spirit as for the point case, we do not need to compute $T^{(r)}$ explicitly.

Lemma 4.5

$$T^{(r)} = \prod_{j=1}^{2r} \left(T - 2\cos\frac{2j-1}{2^{r+1}}\pi \; I\right).$$

Proof. The proof is similar to that of the point case since as T is symmetric, we can use spectral decomposition. □

$T^{(r)}$ is a product of tridiagonal matrices. To solve a linear system with $T^{(r)}$, it is enough to solve 2^r tridiagonal systems. For the right hand side in (4.15) we have to compute $T^{(r)}b_i^{(r)}$. This can be done by the following recursive algorithm. Let

$$\eta_0 = 2b_i^{(r)}, \quad \eta_1 = Tb_i^{(r)}, \quad \eta_s = B\eta_{s-1} - \eta_{s-2}.$$

It is easy to prove the following result.

Lemma 4.6

$$\eta_{2^r} = T^{(r)}b_i^{(r)}.$$

□

Unfortunately, as we suspect from the point version, this method is unstable.

Theorem 4.7

For the model problem, the cyclic reduction method is unstable.

Proof. We take the following analysis from Buzbee, Golub, Nielson [245]. Due to roundoff errors, we compute the sequence

$$\eta_0 = 2b_i^{(r)}, \quad \eta_1 = Tb_i^{(r)} + \delta_0, \quad \eta_s = B\eta_{s-1} - \eta_{s-2} + \delta_{s-1},$$

δ_s being perturbations accounting for the roundoff errors. We know that $T = Q\Lambda Q^T$, $\Lambda = (\lambda j)$. Let us denote

$$\bar{b} = Q^T b_i^{(r)}, \quad \xi_s = Q^T\eta_s, \quad \tau_s = Q^T\delta_s.$$

Then for the components $\xi_{j,s}$ of ξ_s we get

$$\begin{aligned}
\xi_{j,s+1} &= \lambda_j\xi_{j,s} - \xi_{j,s-1} + \tau_{j,s}, \\
\xi_{j,0} &= 2\bar{b}_j, \\
\xi_{j,1} &= \lambda_j\bar{b}_j + \tau_{j,0}.
\end{aligned}$$

The solutions to these recurrences are well known. Let γ_j and β_j be the two solutions of the characteristic equation

$$\alpha^2 - \lambda_j \alpha + 1 = 0.$$

Then

$$\xi_{j,s} = \frac{\beta_j^s - \gamma_j^s}{\beta_j - \gamma_j}\xi_{j,1} - \beta_j\gamma_j \frac{\beta_j^{s-1} - \gamma_j^{s-1}}{\beta_j - \gamma_j}\xi_{j,0} + \sum_{k=1}^{s-1} \frac{\beta_j^{s-k} - \gamma_j^{s-k}}{\beta_j - \gamma_j}\tau_{j,k},$$

if $\beta_j \neq \gamma_j$. For the model problem, we have $\lambda_j = 4 - 2\cos\frac{j\pi}{n+1}$. To simplify we set

$$\lambda_i = 2\cosh z_j.$$

The discriminant of the second degree equation is $4\sinh^2 z_j$. Hence

$$\beta_j = \cosh z_j + \sinh z_j = e^{z_j},$$
$$\gamma_j = \cosh z_j - \sinh z_j = e^{-z_j},$$

and

$$\beta_j\gamma_j = 1, \qquad \beta_j - \gamma_j = 2\sinh z_j.$$

The term which multiplies $\bar{b}_j$ in $\xi_{j,s}$ is

$$\frac{\beta_j^s - \gamma_j^s}{\beta_j - \gamma_j}\lambda_j - 2\beta_j\gamma_j \frac{\beta_j^{s-1} - \gamma_j^{s-1}}{\beta_j - \gamma_j},$$

which we can write as

$$\frac{2}{\sinh z_j}(\sinh sz_j \cosh z_j - \cosh(s-1)z_j) = 2\cosh sz_j.$$

Therefore

$$\xi_{j,s} = 2\bar{b}_j \cosh sz_j + \sum_{k=0}^{s-1} \frac{\sinh((s-k)z_j)}{\sinh z_j}\tau_{j,k}.$$

Consider the polynomial such that $p_s(\lambda_j) = 2\cosh sz_j$. We have

$$\cosh sz_j = 2\cosh z_j \cosh(s-1)z_j - \cosh((s-2)z_j),$$
$$p_s = \lambda_j p_{s-1} - p_{s-2},$$

and clearly $p_0 = 2$, $p_1 = \lambda_j$. ξ_s can be written as

$$\xi_s = p_s(\Lambda)\bar{b} + \sum_{k=0}^{s-1} S_{s-k}\tau_k,$$

where S_{s-k} is a diagonal matrix whose entries are $\frac{\sinh((s-k)z_j)}{\sinh z_j}$. But $\xi_s = Q^T\eta_s$ so

$$\eta_s = p_s(T)b_i^{(r)} + \sum_{k=0}^{s-1} QS_{s-k}Q^T\delta_k.$$

Because of the recurrence verified by p_s, the first term is what we would have had if there were no roundoff errors. The second one is the effect of roundoff errors. The ratio $\frac{\sinh(s-k)z}{\sinh(z)}$ is a rapidly increasing function of z. Therefore, some components of ξ_s can be large and as $\eta_s = Q\xi_s$, they can dominate in η_s. This explains the observed instability of the method. □

We have seen that this instability comes from the multiplication by T used to form the right hand side. O. Buneman [237] gave an algorithm to compute the right hand side in a stable way, the trick being to replace multiplication by solving tridiagonal systems. Note that

$$\begin{aligned} b_i^{(1)} &= Tb_i + b_{i-2^r} + b_{i+2^r}, \\ &= (T^2 - 2I)T^{-1}b_i + b_{i-2^r} + b_{i+2^r} + 2T^{-1}b_i, \\ &= T^{(1)}T^{-1}b_i + b_{i-2^r} + b_{i+2^r} + 2T^{-1}b_i. \end{aligned}$$

Denote

$$p_i^{(1)} = T^{-1}b_i, \quad q_i^{(1)} = b_{i-2^r} + b_{i+2^r} + 2p_i^{(1)}.$$

Then, $b_i^{(1)} = T^{(1)}p_i^{(1)} + q_i^{(1)}$.

More generally, we set

$$b_i^{(r)} = T^{(r)}p_i^{(r)} + q_i^{(r)}.$$

Putting this definition in (4.15), we get

$$\begin{aligned} p_i^{(r+1)} &= p_i^{(r)} + (T^{(r)})^{-1}(p_{i-2^r}^{(r)} + p_{i+2^r}^{(r)} + q_i^{(r)}), \\ q_i^{(r+1)} &= q_{i-2^r}^{(r)} + q_{i+2^r}^{(r)} + 2p_i^{(r+1)}. \end{aligned} \tag{4.16}$$

At the last step of the cyclic reduction algorithm, we have

$$T^{(l)}x_{2^l} = b_{2^l}^{(l)} = T^{(l)}p_{2^l}^{(l)} + q_{2^l}^{(l)}.$$

This means that we do not have to multiply by $T^{(l)}$ as we write

$$T^{(l)}(x_{2^l} - p_{2^l}^{(l)}) = q_{2^l}^{(l)}.$$

In the backward step we solve

$$T^{(r)}x_i = b_i^{(r)} + x_{i-2^r} + x_{i+2^r},$$

i.e.,

$$T^{(r)}(x_i - p_i^{(r)}) = q_i^{(r)} + x_{i-2^r} + x_{i+2^r}.$$

One can show that the Buneman algorithm is stable, see Buzbee, Golub and Nielson [245]. There are no more multiplications by $T^{(r)}$, but for computing the right hand side by (4.16) we have to solve tridiagonal systems. To obtain the number of operations for the algorithm we must know how many systems we have to solve for $m = 2^{l+1}-1$. In the final step we solve $2^l - 1$ systems with matrix $T^{(0)}$, at the second step 2^{l-1} systems with $T^{(1)}$ which is the product of two tridiagonal matrices. In the last step we solve one system with a matrix which the product of 2^{l-1} tridiagonal matrices. Summing these results, we get

$$\sum_{j=0}^{l-1}(2^{l-j} - 1)2^j = l2^l - 2^l + 1$$

systems to solve. Solving each system by Gaussian elimination requires $2m$ multiplications and $2m$ additions if we store some coefficients. Thus we have an operation count which is of the order of $4m^2 \log_2 m$; more precisely

$$\begin{array}{ll} 2m^2 \log_2 m + 5m^2 & \text{additions,} \\ 2m^2 \log_2 m - 5m^2 & \text{multiplications.} \end{array}$$

This method can be used easily on a vector or parallel computer because at each step we have to solve independent tridiagonal systems. The vector length or the degree of parallelism is the number of systems we solve; unfortunately it decreases to 1 during the algorithm.

For point tridiagonal systems the cyclic reduction method is often used but the problems are:

- the degree of parallelism decreases during the algorithm,
- access to the variables is with a stride of 2^l; this leads to memory bank conflicts on computers where the memory is divided into 2^k banks.

However, we are able to cure these problems. For the first one, we can apply the reduction process to all the variables. This will allow us to bypass the backward solve. At the first step the ith equation involves unknowns i, $i-2$, $i+2$, at step r the ith equation involves unknowns i, $i-2^r$, $i+2^r$. At the last step we have a diagonal matrix and we can compute all the unknowns at once. For the second problem the remedy is to store the variables in a different way.

One of the advantages of the cyclic reduction method over classical Gaussian elimination is that it can be easily extended to periodic boundary conditions without any fill–in.

Let

$$Ax = \begin{pmatrix} T & -I & & & -I \\ -I & T & -I & & \\ & \ddots & \ddots & \ddots & \\ & & -I & T & -I \\ -I & & & -I & T \end{pmatrix} x = b.$$

We only have to consider separately the first and last equations. At the first step we have

$$-x_{i-2} + (T^2 - 2I)x_i - x_{i+2} = b_{i-1} + Tb_i + b_{i+1}, \quad i = 2, \ldots, m-1$$

$$(T^2 - 2I)x_2 - x_4 - x_m = b_1 + Tb_2 + b_3,$$

$$-x_2 - x_{m-2} + (T^2 - 2I)x_m = b_{m-2} + Tb_{m-1} + b_m.$$

The matrix we get at the lth step is

$$\begin{pmatrix} T^{(l-1)} & -I & 0 & -I \\ -I & T^{(l-1)} & -I & 0 \\ 0 & -I & T^{(l-1)} & -I \\ -I & 0 & -I & T^{(l-1)} \end{pmatrix}.$$

One more step of reduction gives

$$\begin{pmatrix} T^{(l)} & -2I \\ -2I & T^{(l)} \end{pmatrix} \begin{pmatrix} x_{2^l} \\ x_{2^l+1} \end{pmatrix} = \begin{pmatrix} b^{(l)}_{2^l} \\ b^{(l)}_{2^l+1} \end{pmatrix}.$$

We eliminate $x_{2^{l+1}}$ to get

$$((T^{(l)})^2 - 4I)x_{2^l} = T^{(l)} b_{2^l}^{(l)} + 2b_{2^{l+1}}^{(l)}.$$

Thus far, we have considered the cyclic reduction method for matrices of order 2^{l-1}. The method was extended to any order by R. Sweet [1233]. One has to consider the last equation differently and there are different formulas for even and odd orders. Moreover, we have considered only a very simple problem. Swarztrauber [1224] extended the method to general separable equations which give matrices such as

$$\begin{pmatrix} B_1 & C_1 & & & \\ A_2 & B_2 & C_2 & & \\ & \ddots & \ddots & \ddots & \\ & & A_{n-1} & B_{n-1} & C_{n-1} \\ & & & A_n & B_n \end{pmatrix}.$$

Unfortunately, the formulas are much more complicated.

The cyclic reduction method and the Fourier/tridiagonal method may seem to be very different. But there is a way to see that there is some relationship between the two methods. Note that the recurrences defining the cyclic reduction methods are

$$\begin{aligned} T^{(r+1)} &= (T^{(r)})^2 - 2I \\ p_i^{(r+1)} &= p_i^{(r)} + (T^{(r)})^{-1}(p_{i-2^r}^{(r)} + p_{i+2^r}^{(r)} + q_i^{(r)}), \\ q_i^{(r+1)} &= q_{i-2^r}^{(r)} + q_{i+2^r}^{(r)} + 2p_i^{(r)}. \end{aligned}$$

Obviously, it is easy to compute the eigenvalues of $T^{(r)}$.

Lemma 4.8

The eigenvalues $\lambda^{(r)}$ of $T^{(r)}$ are given by

$$\begin{aligned} \lambda_i^{(0)} &= \lambda_i, \\ \lambda_i^{(r+1)} &= (\lambda_i^{(r)})^2 - 2, \end{aligned}$$

where the λ_is are the eigenvalues of T. The eigenvectors are the same as those of T.

Denoting $\hat{p} = Q^T p, \quad \hat{q} = Q^T q$, the cyclic reduction method can be written:

$$
\begin{array}{llcl}
1. & \hat{b}_i & = & Q^T b_i \\
2. & \hat{p}_i^{(1)} & = & \Lambda^{-1}, \\
 & \hat{q}_i^{(1)} & = & \hat{b}_{i-1} + \hat{b}_{i+1} + 2\hat{p}_i^{(1)}, \\
 & \hat{p}_i^{(r+1)} & = & p_i^{(r)} + (\Lambda^{(r)})^{-1}(\hat{p}_{i-2^r}^{(r)} + \hat{p}_{i+2^r}^{(r)} + \hat{q}_i^{(r)}), \\
 & \hat{q}_i^{(r+1)} & = & \hat{q}_{i-2^r}^{(r)} + \hat{q}_{i+2^r}^{(r)} + 2\hat{p}_i^{(r+1)}, \\
3. & \hat{x}_{2^l} & = & \hat{p}_{2^l}^{(l)} + (\Lambda^{(l)})^{-1}\hat{q}_{2^l}^{(l)}, \\
 & \hat{x}_i & = & \hat{p}_i^{(r)} + (\Lambda^{(r)})^{-1}(\hat{q}_i^{(r)} + \hat{x}_{i+2^r} + \hat{x}_{i-2^r}), \\
4. & x_i & = & Q\hat{x}_i.
\end{array}
$$

where $\Lambda = (\lambda_i)$, $\Lambda^{(r)} = (\lambda_i^{(r)})$ are diagonal matrices. The method is equivalent to doing a Fourier analysis in one direction and then solving the tridiagonal systems by the Buneman algorithm. However, as we have seen, for solving point tridiagonal linear systems we can stabilize the method by normalization of the diagonal coefficient. This means that there are ways other than the Buneman algorithm to stabilize the cyclic reduction method.

4.5. The FACR(l) method

In this section we combine the algorithms of the last two sections to obtain a more efficient one. This method uses l steps of cyclic reduction and then, gets a block tridiagonal system of smaller dimension that is solved with the Fourier/tridiagonal method. This means that we trade a reduction in the length of the FFT against solving tridiagonal systems. This method was first used in 1965 by Hockney [788] with $l = 1$. However, we shall see that $l = 1$ is usually not the optimal value. This algorithm,known as FACR, can be described as the following. For $r = 1$ to $l - 1$ compute

$$
\begin{aligned}
p_i^{(r+1)} &= p_i^{(r)} + (T^{(r)})^{-1}(p_{i-2^r}^{(r)} + p_{i+2^r}^{(r)} + q_i^{(r)}), \\
q_i^{(r+1)} &= q_{i-2^r}^{(r)} + q_{i+2^r}^{(r)} + 2p_i^{(r+1)}.
\end{aligned}
$$

Then, we get a block tridiagonal system whose generic equation is

$$-x_{i-2^l} + T^{(l)}x_i - x_{i+2^l} = T^{(l)}p_i^{(l)} + q_i^{(l)}.$$

Setting $y_i = x_i - p_i^{(l)}, \quad g_i = q_i^{(l)} + p_{i-2^l}^{(l)} + p_{i+2^l}^{(l)}$, we have

$$-y_{i-2^l} + T^{(l)}y_i - y_{i+2^l} = g_i, \quad i = 2^l, 2 \times 2^l, 3 \times 2^l, \ldots$$

Then we compute $\hat{g}_i = Q^T g_i$ and $\hat{y}_i$ by the Fourier/tridiagonal method. At the last step we get $y_i = Q\hat{y}_i$ to obtain $x_i = y_i + p_i^{(l)}$, $i = 2, 2 \times 2^l, \ldots$ The missing x_is are computed by l backward steps of cyclic reduction.

In order to determine the optimal value for l, we must compute the number of operations as a function of l. A rough estimate shows that we have to solve lm^2 tridiagonal systems. The order of the system we get after the reducing steps is $\frac{m^2}{2^l}$. The Fourier transform accounts for approximately $2\frac{m^2}{2^l}\log_2 m$ multiplications. A precise count was given by Temperton [1238]:

$$\begin{aligned} (2l + 7 + 2^{-l}(3\log_2 n))n^2 &\quad \text{additions,} \\ (2l + 2^{-l}(2\log_2 n))n^2 &\quad \text{multiplications.} \end{aligned}$$

This operation count is approximately minimized (note that l must be an integer) by choosing $l = \log_2 \log_2 n$. Then, we have

$$\begin{aligned} 2m^2 \log_2 \log_2 m + 10m^2 &\quad \text{additions,} \\ 2m^2 \log_2 \log_2 m + 2m^2 &\quad \text{multiplications.} \end{aligned}$$

For practical dimensions the number of operations is almost proportional to the total number of unknowns $n = m^2$, that is to say, close to the best possible one. As a summary, we give the number of operations for the model problem in the following table. Here m is the number of unknowns on each side of the square. Of course, it is easy to vectorize FACR(l) as we know how to vectorize the cyclic reduction algorithm and the Fourier method.

Number of operations for the model problem.

m	FFT/tridia	Buneman Cyclic Reduction	FACR(l)
64	$25m^2$ A $13m^2$ M	$17m^2$ A $7m^2$ M	$l = 2$ $14m^2$ A $6m^2$ M
4096	$43m^2$ A $25m^2$ M	$29m^2$ A $19m^2$ M	$l = 5$ $20m^2$ A $12m^2$ M

4.6. The capacitance matrix method

So far, the methods we have studied in this chapter are restricted to separable partial differential equations on rectangular domains whose boundaries are parallel to coordinate axes. We shall see that we can extend, in some way, these methods to more general domains. For the sake of simplicity, suppose we want to solve

$$\begin{aligned} -\Delta u &= f \quad \text{in} \quad \Omega, \\ u &= g \quad \text{on} \quad \partial\Omega, \end{aligned}$$

where Ω is the domain shown in figure 4.3.

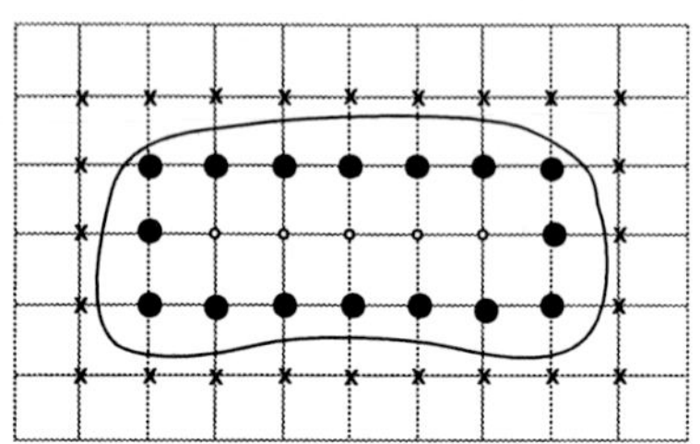

fig. 4.3: An example of embedding

To apply our previous methods, the domain Ω is embedded in a rectangle R, f is extended to 0 outside Ω and we define

- Ω_h: mesh points in Ω whose four nearest neighbors are in Ω (white dots).
- $\partial\Omega_h$: mesh points in Ω that are not in Ω_h. They are called boundary or irregular points. They are black dots in figure 4.3.
- C_h: mesh points in R but not in $\Omega_h \cup \partial\Omega_h$ (crosses).

The discrete problem is then defined by using the standard five point scheme for each point in $\Omega_h \cup C_h$ (with zero Dirichlet boundary conditions). For $\partial\Omega_h$ points, we write special equations using, for instance, the Shortley–Wellers formula (see Forsythe and Wasow [574]). These formulas use only points in $\Omega_h \cup \partial\Omega_h$. That is to say, in equations for points in $\Omega_h \cup \partial\Omega_h$, there is no connection with points in C_h. The points in C_h have relationships with the other points in $\partial\Omega_h$ but this does not matter. Solving the problem just defined in R will give us the solution of our initial problem on $\Omega_h \cup \partial\Omega_h$. We have to solve a linear system

$$Ax = b, \tag{4.17}$$

which represents the problem in R. Suppose there are p points in $\partial\Omega_h$ and let B be the matrix for the model problem on R. It is clear that B differs only from A by p lines *i.e.*, the equation for the points in $\partial\Omega_h$. For convenience, let S be the set of indices of points in $\partial\Omega_h$. For solving a linear system with matrix B, we can apply one of the fast methods of the previous sections. We are looking for a right hand side w such that

$$Bx = w,$$

where x is the solution of (4.17).

As A and B have identical ith rows for $i \notin S$, we set

$$w_i = b_i, \quad i \notin S.$$

We can look for w by writing $w = b + \sum_{i \in S} \beta_i e^i$ where $(e^i)^T = (0, \ldots, 0, 1, 0, \ldots, 0)$. We have

$$x = B^{-1}b + \sum_{i \in S} \beta_i B^{-1} e^i.$$

To find the p unknowns β_i, we need to satisfy the remaining p equations:

$$b_j = (Ax)_j = (AB^{-1}b)_j + \sum_{i \in S} \beta_i (AB^{-1}e^i)_j. \tag{4.18}$$

This is a linear system of order p. Let C be the matrix whose elements are

$$(e^j)^T AB^{-1} e^i, \qquad i, j \in S.$$

Usually C is called the capacitance matrix. Equation (4.18) is a linear system

$$C\beta = \delta.$$

For the problem we are dealing with we can show that C is non–singular.

Lemma 4.9

C is non–singular and

$$\det C = \frac{\det A}{\det B}.$$

□

The algorithm is as follows:

0) Compute

$$Bq_i = e^i, \quad i \in S,$$

by solving p problems with a fast method. Knowing the g's, we can then form C and factor it.

1) $B\hat{x} = b$,

2) $\delta_j = b_j - (A\hat{x})_j, \quad j \in S$,

3) $Bx = b + \sum_{i\in S} \beta_i e^i$.

If we store the g_is, step 3) can be replaced by $x = \hat{x} + \sum_{i\in S} \beta_i g_i$. Note that step 0) does not depend on b; it has to be done just once to solve several systems with the same matrix and different right hand sides. For this method to be efficient, p has to be relatively small. Note that if we use the FFT/tridiagonal method to solve systems in step 0) there are some simplifications because we do not need to use the FFT to compute $Q^T e^i$.

Variations of the capacitance matrix method have been studied by Proskurowski and Widlund [1084]. This method is also called the fictitious domain method in the Russian literature. Sometimes, it is also called domain embedding.

4.7. Bibliographical comments

The discovery of the Fast Fourier Transform by Cooley and Tukey in 1965 [345] was a breakthrough in the area of numerical computation. This allowed the development of many applications and particularly fast solvers for separable elliptic PDEs on rectangles. The history of the FFT is discussed by Cooley in [1006]. For a description of the different versions of the FFT, see the book by Van Loan [1306].

The Fourier tridiagonal method was described by Hockney [788], 789]. Following suggestion by Golub the cyclic reduction was developed at the same time, see Buzbee, Golub and Nielson [245]. Cyclic reduction was generalized by Swarztrauber and Sweet through a series of papers leading to the distribution of the FISHPACK package. The stable computation of the right hand side was proposed by O. Buneman [237] even though this important paper was never published in a journal. The general principle of embedding was described in Buzbee, Dorr, George and Golub [244]. For the capacitance matrix method, see the works of Proskurowski and his co–workers [1084]; [1085]. This was also considered by the Russian school, see for instance [555].

5 Classical iterative methods

5.1. Introduction

Suppose we want to solve a linear system

$$Ax = b, \tag{5.1}$$

where A is non–singular and b is given. An iterative method constructs a sequence of vectors $\{x^k\}$, $k = 0, 1, \ldots$, which is expected to converge towards x which is the solution of (5.1), x^0 being given. The method is said to be convergent if $\lim_{k\to\infty} \|x - x^k\| = 0$. Most classical iterative methods use a splitting of the matrix A, denoted as

$$A = M - N,$$

where M is a non–singular matrix. Then, the sequence x^k is defined by

$$Mx^{k+1} = Nx^k + b, \tag{5.2}$$

and x^0 is given. It is obvious that if this method is convergent, it converges towards the unique solution of (5.1). An interesting question is to determine conditions for this sequence of vectors to converge.

Let $\varepsilon^k = x - x^k$ be the error at iteration number k. As obviously $Mx = Nx + b$, we get

$$\begin{aligned} M(x - x^{k+1}) &= N(x - x^k), \\ \varepsilon^{k+1} &= M^{-1}N\varepsilon^k. \end{aligned}$$

Iterating this result, we obtain the equation for the error

$$\varepsilon^{k+1} = (M^{-1}N)^{k+1}\varepsilon^0. \tag{5.3}$$

From equation (5.3), the iterative method converges for all starting vectors if and only if $\lim_{k\to\infty}(M^{-1}N)^k = 0$. This was characterized in theorem 1.24 using the spectral radius. We then have the following fundamental result.

Theorem 5.1

The iterative method (5.2) converges for every x^0 if and only if

$\rho(M^{-1}N) < 1.$

Proof. This results from applying theorem 1.24. □

Of course, it is not easy to check if $\rho(M^{-1}N) < 1$ since for most problems the eigenvalues of $M^{-1}N$ are not explicitly known. Hence, we shall have to rely mainly on sufficient conditions for convergence. Let us now examine some very well known iterative methods.

5.2. The Jacobi method

Suppose $A = D + L + U$ where D is a diagonal matrix which has the same diagonal as A, L (resp. U) is the strictly lower (resp. upper) triangular part of A. The simplest choice for the splitting of A is to take $M = D$, $N = -(L + U)$. Then, the method (5.2) is known as the (point) `Jacobi` method. The iteration matrix of this method is usually denoted by $J(A)$,

$$J(A) = M^{-1}N = -D^{-1}(L + U). \tag{5.4}$$

It is easy to solve the linear system with matrix M in (5.2) as M is diagonal. Writing (5.2) componentwise, we get

$$x_i^{k+1} = \frac{1}{a_{i,i}}\left(b_i - \sum_{\substack{j=1\\ j\neq i}}^{n} a_{i,j}x_j^k\right). \tag{5.5}$$

We can easily obtain sufficient convergence conditions for the `Jacobi` method to converge.

Theorem 5.2

If A is strictly diagonally dominant, then the `Jacobi` *method converges.*

Proof.

$$\left(D^{-1}(L+U)\right)_{i,j} = \frac{a_{i,j}}{a_{i,i}} \forall i \neq j \quad \text{and } 0 \text{ otherwise.}$$

Hence,

$$\|D^{-1}(L+U)\|_\infty = \max_i \sum_{j\neq i} \frac{|a_{i,j}|}{|a_{i,i}|}.$$

Since A is strictly diagonally dominant, $\sum_{j\neq i} \frac{|a_{i,j}|}{|a_{i,i}|} < 1$, for all i. Then

$$\|D^{-1}(L+U)\|_\infty < 1.$$

This is a sufficient condition to have

$$\lim_{k\to\infty} J(A)^k = 0,$$

and the method converges. □

We can generalize this result for strictly diagonally dominant matrices to H–matrices.

Theorem 5.3

Let A be an H–matrix then the `Jacobi` *method converges.*

Proof. From theorem 1.48 and definition 1.34, we know that there exists a diagonal matrix E with positive diagonal elements such that $E^{-1}AE$ is strictly diagonally dominant. We can write the iteration matrix $J(A)$ as,

$$J(A) = -D^{-1}(L+U) = -D^{-1}(A-D) = I - D^{-1}A.$$

But $E^{-1}AE$ has the same diagonal as A, hence

$$J(E^{-1}AE) = I - D^{-1}(E^{-1}AE) = E^{-1}(I - D^{-1}A)E = E^{-1}J(A)E.$$

This is because two diagonal matrices commute: $E^{-1}D^{-1} = D^{-1}E^{-1}$. Hence $J(A)$ and $J(E^{-1}AE)$ are similar, therefore they have the same eigenvalues. The matrix $E^{-1}AE$ being strictly diagonally dominant, we have

$$\rho(J(A)) = \rho(J(E^{-1}AE)) < 1.$$

□

From this theorem, we can derive the following results which are useful in practical cases.

Corollary 5.4

Let A be a matrix having one of the two following properties:

1) A is irreducibly diagonally dominant,

2) A is an M–matrix.

Then, the `Jacobi` *method converges.*

Proof. Each of the two conditions implies that A is an H–matrix, so we can readily apply theorem 5.3. □

When A is an L–matrix, we have a stronger result which is also a characterization of an M–matrix.

Theorem 5.5

Let A be an L–matrix. Then, A is an M–matrix if and only if $\rho(J(A)) < 1$.

Proof. If A is an M–matrix, corollary 5.4 shows that $\rho(J(A)) < 1$. For the converse, we note that A being an L–matrix, we have

$$L \leq 0, \quad U \leq 0, \quad d_{i,i} > 0,$$

so

$$M^{-1} = D^{-1} \geq 0, \qquad N = -(L + U) \geq 0.$$

By definition 1.49, $A = M - N$ is a regular splitting with $\rho(M^{-1}N) < 1$. Hence, by theorem 1.50, A is an M–matrix. □

The (point) `Jacobi` method is one of the simplest iterative methods we can think of. There are very few operations per iteration. For the general case of a dense matrix of order n, we have n^2 multiplications and $n(n-1)$ additions per iteration provided we store the inverses of the diagonal terms. For the finite difference model problems described in Chapter 1, we have $5n$ multiplications and $4n$ additions per iteration. Moreover, this method is very well suited for vector and parallel processing. Everything can be expressed in a natural way in terms of vectors if the sparse matrix is stored by diagonals (or pseudo diagonals). The method is also inherently parallel as the computation of the components of x^{k+1} depends only on x^k. Therefore, the computation can be split into as many tasks as we need depending on how the matrix is stored and partitioned between processors.

This method looks satisfactory but unfortunately the convergence can be very slow. For the Poisson model problem we are able to compute the spectral radius $\rho(J(A))$ as

$$D = 4I, \qquad J(A) = I - \tfrac{1}{4}A.$$

The eigenvalues $\mu_{p,q}$ of $J(A)$ are

$$\mu_{p,q} = \tfrac{1}{2}(\cos p\pi h + \cos q\pi h), \quad p, q = 1, \ldots, m,$$

where $h = \frac{1}{m+1}$, $n = m^2$. It is easy to see that $\max_{p,q} |\mu_{p,q}| = |\mu_{1,1}| = |\mu_{m,m}|$ and that

$$\rho(J(A)) = \cos \pi h.$$

So, unfortunately $\lim_{h \to 0} \rho(J(A)) = 1$. As we take more discretization points, the convergence slows down and the number of iterations becomes larger and larger as

$$\rho(J(A)) = 1 - \frac{\pi^2 h^2}{2} + O(h^4).$$

An heuristic explanation for this slow convergence is obtained by considering a constant vector y (having all its components equal). Then, most components of Ay are zero. This means that the `Jacobi` method makes very small modifications to smooth modes of the error vector. The situation is the same for every matrix that arises from discretization of elliptic partial differential equations. For example, consider the problem of solving $Ax = 0$ where

$$A = \begin{pmatrix} 2 & -1 & & & \\ -1 & 2 & -1 & & \\ & \ddots & \ddots & \ddots & \\ & & -1 & 2 & -1 \\ & & & -1 & 2 \end{pmatrix},$$

is a matrix of order 100. By today standards this is a very small problem. Starting from a random vector with components in [0, 1], it takes about 28000 iterations to obtain $\max_i |x_i^k| < 10^{-6}$.

One may try to improve the performance using the `Jacobi` method in block form. Suppose A corresponds to a two dimensional PDE,

$$A = \begin{pmatrix} D_1 & A_2^T & & & \\ A_2 & D_2 & A_3^T & & \\ & \ddots & \ddots & \ddots & \\ & & A_{m-1} & D_{m-1} & A_m^T \\ & & & A_m & D_m \end{pmatrix},$$

of order m^2 where, for instance, D_i is tridiagonal and A_i diagonal. Then let

$$D = \begin{pmatrix} D_1 & & & \\ & D_2 & & \\ & & \ddots & 0 \\ & 0 & & \ddots & \\ & & & & D_m \end{pmatrix}, \qquad L = \begin{pmatrix} 0 & & & & \\ A_2 & 0 & & & \\ & A_3 & \ddots & & \\ & & \ddots & \ddots & \\ & & & A_m & 0 \end{pmatrix},$$

$$A = D + L + L^T.$$

We can use method (5.2) with $M = D$, $N = -(L + L^T)$.

This describes the block (or line) `Jacobi` method. Observe that for each iteration we have m independent tridiagonal systems to solve. If $x = (x_1, \ldots, x_m)^T$ and $b = (b_1, \ldots, b_m)^T$ blockwise,

$$\begin{aligned} D_1 x_1^{k+1} &= b_1 - A_2^T x_2^k, \\ D_i x_i^{k+1} &= b_i - A_i x_{i-1}^k - A_{i+1}^T x_{i+1}^k, \quad i = 2, \ldots, m-1, \\ D_m x_m^{k+1} &= b_m - A_m x_{m-1}^k. \end{aligned}$$

Of course, the block `Jacobi` method can also be defined for non–symmetric matrices too. This method is not in vector form because solving tridiagonal systems is not a genuine vector operation. But, fortunately there are methods like cyclic reduction (see Chapter 4) that allow us to solve tridiagonal systems in vector mode. For the line `Jacobi` method there is another way of vectorizing since we have m independent systems to solve. So, we can use the standard Gaussian elimination algorithm vectorizing on a number of systems. The vectors are then composed of all the first components of x_is, all the second components of x_is and so on. Obviously, the method contains a lot of parallelism. Regarding convergence, we can show the following.

Theorem 5.6

Let A be strictly diagonally dominant. Then the block `Jacobi` *method is convergent.*

Proof. The proof uses the same technique as the more general proof used for theorem 5.25. Hence, we refer the reader to this proof. □

As a consequence, we have

Theorem 5.7

Let A be an H–matrix. Then the block `Jacobi` *method is convergent.*

Proof. The proof is the same as for theorem 5.3. □

For the case of symmetric matrices, results can be given both for point and block `Jacobi` methods.

Theorem 5.8

Let A be a symmetric positive definite matrix. If $2D - A$ is positive definite, then the `Jacobi` *method is convergent.*

Proof. This a straightforward application of theorem 1.54. In this case

$$Q = M + M^T - A = 2D - A.$$ □

However, even the block Jacobi method is not really efficient since, for the model problem, $\rho(M^{-1}N) = 1 - \pi^2 h^2 + O(h^4)$. A common technique for improving the convergence is relaxation: *i.e.* taking an average between the last iterate and the previous one. In our case, we get

$$\begin{aligned} D\tilde{x}^{k+1} &= -(L+U)x^k + b, \\ x^{k+1} &= \omega\tilde{x}^{k+1} + (1-\omega)x^k, \end{aligned}$$

where ω is a real parameter, $\omega > 0$. When $\omega = 1$, we recover the Jacobi method. We can easily eliminate $\tilde{x}^{k+1}$ in these relations,

$$\begin{aligned} Dx^{k+1} &= -\omega(L+U)x^k + \omega b + (1-\omega)x^k, \\ \frac{1}{\omega}Dx^{k+1} &= b - (L+U)x^k + \frac{1-\omega}{\omega}Dx^k. \end{aligned} \tag{5.6}$$

Obviously, the corresponding splitting is

$$M = \frac{1}{\omega}D, \qquad N = \frac{1-\omega}{\omega}D - (L+U).$$

The iteration matrix is denoted by $J_\omega(A)$. For the relaxed Jacobi method we have similar results to the case $\omega = 1$.

Theorem 5.9

Let A be symmetric positive definite. If $\frac{2}{\omega}D - A$ is positive definite, then the relaxed Jacobi *method (5.6) converges.*

Proof. This is a straightforward application of theorem 1.54. □

The problem is knowing when $\frac{2}{\omega}D - A$ is positive definite. The answer was given by Young [1351] in his famous book. □

Proposition 5.10

Let A be symmetric with D positive definite. Then, $\frac{2}{\omega}D - A$ is positive definite if and only if $0 < \omega < \frac{2}{1-\mu_{\min}}$ where $\mu_{\min}$ is the smallest eigenvalue of $J(A)$.

Proof. One notes first that $J(A) = -D^{-1}(L + L^T)$ is similar to $-D^{-\frac{1}{2}}(L + L^T)D^{-\frac{1}{2}}$ which is a symmetric matrix and so has real eigenvalues μ_i. Let us order them as $\mu_1 \leq \mu_2 \leq \cdots \leq \mu_n$. The diagonal elements of $J(A)$ are zero. So, trace$[J(A)] = \mu_1 + \mu_2 + \cdots + \mu_n = 0$. This implies that $\mu_n \geq 0$ and $\mu_1 \leq 0$. The matrix $\frac{2}{\omega}D - A$ is positive definite if and only if $D^{-\frac{1}{2}}(\frac{2}{\omega}D - A)D^{-\frac{1}{2}}$ is positive definite.

But,

$$D^{-\frac{1}{2}}(\tfrac{2}{\omega}D - A)D^{-\frac{1}{2}} = \tfrac{2}{\omega}I - D^{-\frac{1}{2}}AD^{\frac{1}{2}} = (\tfrac{2}{\omega} - 1)I + D^{\frac{1}{2}}J(A)D^{-\frac{1}{2}},$$

because $D^{-1}A = I - J(A)$. The eigenvalues of $(\frac{2}{\omega} - 1)I + D^{\frac{1}{2}}J(A)D^{-\frac{1}{2}}$ are $\frac{2}{\omega} - 1 + \mu_i$, $i = 1, \ldots, n$. Clearly they are all positive if $\omega < \frac{2}{1 - \mu_{\min}}$. □

One may ask what is the optimal value for ω. Here, we define optimal as the value of ω for which the spectral radius is minimum. Let μ_i^ω be the eigenvalues of $J_\omega(A)$

$$\begin{aligned} J_\omega(A) &= \omega D^{-1}(\tfrac{1-\omega}{\omega}D - (L + U)), \\ &= (1-\omega)I + \omega J(A), \\ \mu_i^\omega &= (1-\omega) + \omega\mu_i = 1 - \omega(1 - \mu_i). \end{aligned}$$

We must study $|\mu_i^\omega|$ as a function of ω, see figure 5.1.

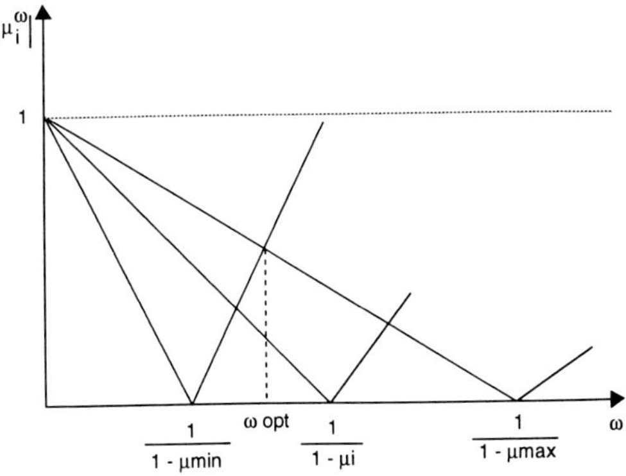

fig. 5.1: $|\mu_i^\omega|$ *as a function of* ω

Clearly, the optimal value ω_{opt} is given by the intersection of the graphs of the functions $1 - \omega(1 - \mu_n)$ and $\omega(1 - \mu_1) - 1$. This gives us

$$\omega_{opt} = \frac{2}{2 - (\mu_1 + \mu_n)}.$$

We also find again that the method converges for $0 < \omega < \frac{2}{1-\mu_{min}}$. The drawback of this method is that we are required to know the extreme eigenvalues of $J(A)$ in order to compute the optimal parameter. Usually we do not know the eigenvalues. It is interesting to note that for the model problem,

$$\mu_1 = \cos(\pi h), \qquad \mu_n = -\cos(\pi h),$$

hence $\omega_{opt} = 1$ and the `Jacobi` method is optimal! This is the case for all matrices which have property A.

5.3. The Gauss-Seidel method

Note that in the `Jacobi` method we use x^k to compute all the components of x^{k+1}. Since (at least on serial computers) we compute the elements one at a time, it is a very natural idea to use components of x^{k+1} as soon as they become available. This leads to the (point) `Gauss-Seidel` method:

$$x_i^{k+1} = \frac{1}{a_{i,i}}\left(b_i - \sum_{j=1}^{i-1} a_{i,j}x_j^{k+1} - \sum_{j=i+1}^{n} a_{i,j}x_j^k\right), \quad i = 1, \ldots, n. \tag{5.7}$$

Expressing this method in matrix form, we have

$$(D + L)x^{k+1} = b - Ux^k. \tag{5.8}$$

This formula also defines the block (or line) `Gauss-Seidel` method if D is thought of as a block diagonal matrix. The iteration matrix, traditionally denoted by $\mathcal{L}$, is

$$\mathcal{L} = -(D + L)^{-1}U.$$

As for the `Jacobi` method, we can give sufficient conditions to have the `Gauss-Seidel` method converging.

Theorem 5.11

Let A be strictly diagonally dominant. Then, the point `Gauss-Seidel` *method converges for every x^0.*

Proof. Let $\varepsilon^k = x - x^k$. We have

$$\varepsilon_i^{k+1} = -\sum_{j=1}^{i-1} \frac{a_{i,j}}{a_{i,i}} \varepsilon_j^{k+1} - \sum_{j=i+1}^{n} \frac{a_{i,j}}{a_{i,i}} \varepsilon_j^k,$$

$$|\varepsilon_i^{k+1}| \leq \sum_{j=1}^{i-1} |\frac{a_{i,j}}{a_{i,i}}| \cdot \|\varepsilon^{k+1}\|_\infty + \sum_{j=i+1}^{n} |\frac{a_{i,j}}{a_{i,i}}| \cdot \|\varepsilon^k\|_\infty.$$

Suppose $\|\varepsilon^{k+1}\|_\infty = \max_i |\varepsilon_i^{k+1}|$ is given for $i = I$, then

$$\left(1 - \sum_{j=1}^{I-1} |\frac{a_{I,j}}{a_{I,I}}|\right) \|\varepsilon^{k+1}\|_\infty \leq \sum_{j=I+1}^{n} |\frac{a_{I,j}}{a_{I,I}}| \cdot \|\varepsilon^k\|_\infty.$$

But, A being strictly diagonally dominant, we have

$$0 < \sum_{j=I+1}^{n} |\frac{a_{I,j}}{a_{I,I}}| < 1 - \sum_{j=1}^{I-1} |\frac{a_{I,j}}{a_{I,I}}|.$$

This implies that

$$\|\varepsilon^{k+1}\|_\infty < \|\varepsilon^k\|_\infty < \cdots < \|\varepsilon^0\|_\infty,$$

and ε^k is a converging sequence with limit ε. But, as A is non–singular, $A\varepsilon = 0$. Hence $\varepsilon = 0$. □

As before for the `Jacobi` method, it is easy to generalize this result for H–matrices.

Theorem 5.12

Let A be an H–matrix. Then, the (point) `Gauss-Seidel` *method converges for every x^0.*

Proof. There is a positive diagonal matrix E with positive diagonal elements such that $E^{-1}AE$ is strictly diagonally dominant. We note that

$$\mathcal{L}(E^{-1}AE) = E^{-1}\mathcal{L}(A)E.$$

This proves that the `Gauss-Seidel` method converges. □

As a corollary, we find that `Gauss-Seidel` converges if A is irreducibly diagonally dominant or if A is an M–matrix. As for the `Jacobi` method, it is easy to give sufficient conditions for the convergence of the block `Gauss-Seidel` method.

Theorem 5.13

Let A be a strictly diagonally dominant matrix. Then the block `Gauss-Seidel` *method is convergent.*

Proof. The proof uses the same technique as theorem 5.25 to which we refer the reader. □

Theorem 5.14

Let A be an H–matrix. Then the block `Gauss-Seidel` *method is convergent.*

Proof. Identical to the proof of theorem 5.12. □

In the symmetric case, we have an interesting characterization of positive definite matrices.

Theorem 5.15

Let A be symmetric with D positive definite. Then A is positive definite if and only if the `Gauss-Seidel` *method converges.*

Proof. Once again, this is a straightforward consequence of theorem 1.54. Here,

$$M = D + L \quad \text{and} \quad Q = M + M^T - A = D.$$

□

For the model problem, it is easy to compute the eigenvalues $\lambda_{p,q}$ and eigenvectors $w_{p,q}$ of $\mathcal{L}$. Recall that the eigenvalues $\mu_{p,q}$ of $J(A)$ are

$$\mu_{p,q} = \frac{1}{2}(\cos p\pi h + \cos q\pi h),$$

and the components of the corresponding (unnormalized) eigenvector $u^{p,q}$ are

$$u_{i,j}^{p,q} = \sin ip\pi h \sin jq\pi h.$$

For the sake of simplicity we drop indices p and q. Then, λ and w must satisfy

$$\lambda(D+L)w = -Uw,$$

which is written componentwise as

$$\lambda w_{i,j} = \frac{1}{4}(w_{i+1,j} + w_{i,j+1} + \lambda w_{i-1,j} + \lambda w_{i,j-1}).$$

Now let us introduce u such that $w_{i,j} = \lambda^{\frac{i+j}{2}} u_{i,j}$. Substituting, we get that $\lambda\lambda^{\frac{i+j}{2}} u_{i,j}$ is equal to

$$\frac{1}{4}\left(\lambda^{\frac{i+1+j}{2}} u_{i+1,j} + \lambda^{\frac{i+j+1}{2}} u_{i,j+1} + \lambda\lambda^{\frac{i+1+j}{2}} u_{i-1,j} + \lambda\lambda^{\frac{i+j-1}{2}} u_{i,j-1}\right).$$

Suppose $\lambda \neq 0$ then, dividing by $\lambda^{\frac{i+1+j}{2}}$,

$$\lambda^{\frac{1}{2}} u_{i,j} = \frac{1}{4}(u_{i+1,j} + u_{i,j+1} + u_{i-1,j} + u_{i,j-1}).$$

This specifies that $\lambda^{\frac{1}{2}}$ is an eigenvalue of $J(A)$. Conversely, if μ is an eigenvalue of $J(A)$, μ^2 is an eigenvalue of $\mathcal{L}$. Therefore, the spectral radius of $\mathcal{L}$ is

$$\rho(\mathcal{L}) = (\cos \pi h)^2 = 1 - \pi^2 h^2 + O(h^4).$$

This shows that, for this particular example, the `Gauss-Seidel` method converges twice as fast as the `Jacobi` method. This result is true for a larger class of matrices, namely matrices with property A (definition 1.36); see Young [1351]. For some problems, we have benefited by using `Gauss-Seidel`instead of `Jacobi` but, we also lost something since `Gauss-Seidel` seems not vectorizable or parallelizable. To compute the component x_i^{k+1} of x^{k+1} we need all the previous components x_j^{k+1}, $j < i$. We can compute one component at a time; a sequential process.

Fortunately, there are ways to partly recover the vector form. The `Gauss-Seidel` iterates clearly depend on the numbering of the unknowns. The trick for parallelization is to use a numbering scheme that decouples the unknowns. For problems arising from five point difference approximations, we can use the so-called Red Black ordering. To describe this ordering let us look at a small example on a square domain, see figure 5.2. One notes that the Red points $\circ$ (resp. Black points x) are only connected to Black points (resp. Red points). Using this ordering, the linear system can be written blockwise as

$$\begin{pmatrix} D_R & F \\ F^T & D_B \end{pmatrix} \begin{pmatrix} x_R \\ x_B \end{pmatrix} = \begin{pmatrix} b_R \\ b_B \end{pmatrix}, \tag{5.9}$$

where D_R and D_B are diagonal matrices. F represents the coupling between Red and Black points. The `Gauss-Seidel` method can then be written in block form as

$$\begin{aligned} D_R x_R^{k+1} &= b_R - F x_B^k, \\ D_B x_B^{k+1} &= b_B - F^T x_R^{k+1}. \end{aligned} \tag{5.10}$$

It is clear that we can first compute all the components of x_R^{k+1} simultaneously and then, all the components of x_B^{k+1}. Thus, this method can be used on a vector computer or even a parallel machine. The problem is knowing the convergence properties of the method defined by (5.10). The interesting fact is that for a large class of matrices, the eigenvalues of the iteration matrix for (5.10) are the same those for the natural ordering. Let us compute the eigenvalues of the `Gauss-Seidel` iteration matrix:

$$\begin{pmatrix} 0 & -D_R^{-1}F \\ 0 & D_B^{-1}F^T D_R^{-1}F \end{pmatrix}.$$

Note that 0 is a multiple eigenvalue of this matrix and that there are a set of linearly independent eigenvectors associated with the eigenvalue 0. Let

$$x = \begin{pmatrix} x_R \\ x_B \end{pmatrix},$$

be an eigenvector. Then

$$-D_R^{-1}F x_B = \lambda x_R, \qquad D_B^{-1}F^T D_R^{-1} F x_B = \lambda x_B.$$

Therefore, λ is an eigenvalue of $D_B^{-1}F^T D_R^{-1}F$. Let

$$y = \begin{pmatrix} y_R \\ y_B \end{pmatrix},$$

be an eigenvector of the Jacobi iteration matrix for (5.9) corresponding to an eigenvalue μ,

$$D_R^{-1} F y_B = \mu y_R, \qquad D_B^{-1} F^T y_R = \mu y_B.$$

Eliminating y_R gives

$$D_B^{-1} F^T D_R^{-1} F y_B = \mu^2 y_B.$$

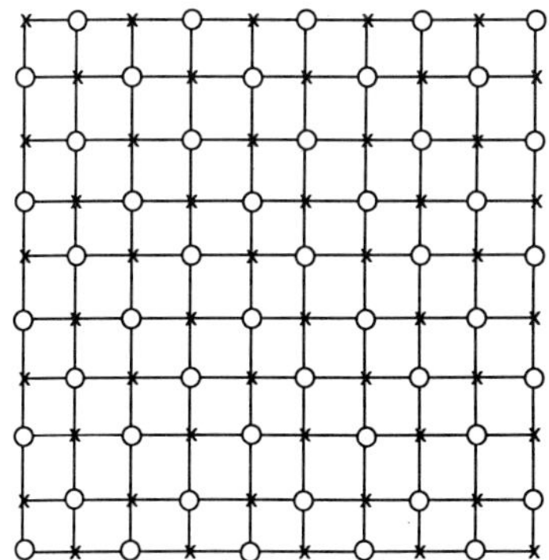

fig. 5.2: The Red–Black ordering

This shows that μ^2 is an eigenvalue of the Gauss-Seidel iteration matrix. Now, we must note that the Jacobi iterates are independent of the numbering of the unknowns. Therefore, Red-Black Jacobi has the same eigenvalues as the natural ordering Jacobi matrices. For the model problem, we have proved that the squares of the Jacobi eigenvalues are eigenvalues of $\mathcal{L}$. Hence, the Red-Black Gauss-Seidel matrix has the same eigenvalues as the matrix $\mathcal{L}$. This is more generally true for matrices having a consistent ordering in the terminology of Young [1351]. For problems with larger stencils or finite element problems, more than two colors may be used to introduce parallelism; see Adams and Jordan [4] for convergence properties.

5.4. The SOR Method

A great improvement over the Jacobi and Gauss-Seidel methods is given by the Successive Over Relaxation (SOR) method. This method gave rise to a great interest and a lot of research in the fifties; see Young [1351]. The (point) SOR method is defined by

$$x_i^{k+1} = \frac{\omega}{a_{i,i}}(b_i - \sum_{j=1}^{i-1} a_{i,j}x_j^{k+1} - \sum_{j=i+1}^{n} a_{i,j}x_j^k) + (1-\omega)x_i^k, \tag{5.11}$$

where ω is a real parameter, $\omega > 0$. Writing (5.11) in matrix form we have

$$(D + \omega L)x^{k+1} = \omega b - \omega U x^k + (1-\omega)Dx^k. \tag{5.12}$$

Relation (5.12) also defines the block SOR method if D, L and U are in block form. The iteration matrix is denoted by $\mathcal{L}_\omega$ and is given by

$$\mathcal{L}_\omega = (\tfrac{1}{\omega}D + L)^{-1}[\tfrac{1-\omega}{\omega}D - U].$$

We have the following necessary condition for convergence of the method.

Proposition 5.16

If the SOR *method converges then,* $0 < \omega < 2$.

Proof. A short manipulation shows the iteration matrix can be written as

$$\mathcal{L}_\omega = (I + \omega D^{-1}L)^{-1}[(1-\omega)I - \omega D^{-1}U].$$

Obviously, $I + \omega D^{-1}L$ is a lower triangular matrix with 1's on the diagonal, $(1-\omega)I - \omega D^{-1}U$ is an upper triangular matrix with $1-\omega$ on the diagonal. Therefore

$$\det \mathcal{L}_\omega = (1-\omega)^n.$$

As $\det \mathcal{L}_\omega$ is the product of the eigenvalues of $\mathcal{L}_\omega$, we obtain

$$|1-\omega|^n \le \rho(\mathcal{L}_\omega)^n.$$

So, $|1-\omega| \le \rho(\mathcal{L}_\omega)$. If SOR converges, $\rho(\mathcal{L}_\omega) < 1$. This implies that $|1-\omega| < 1$, which translates into

$$0 < \omega < 2.$$

□

It is easy to show that if A is strictly diagonally dominant and $0 < \omega \le 1$ then the SOR method converges. This implies that if A is an H–matrix and $0 < \omega \le 1$ the SOR method is convergent. However, this can be slightly improved by the following result.

Theorem 5.17

Let A be an H–matrix. The SOR *method converges if*

$$0 < \omega < \frac{2}{1 + \rho(|J(A)|)},$$

Proof. To prove this theorem, we first need several lemmas. □

Lemma 5.18

Let $A = M - N$ be a regular splitting. The following statements are equivalent:

1) $A^{-1} \geq 0$,

2) $A^{-1}N \geq 0$,

3) $\rho(M^{-1}N) = \frac{\rho(A^{-1}N)}{1+\rho(A^{-1}N)} < 1$.

Proof. See Varga [1300]. □

Lemma 5.19

If $|A| \leq B$ then $\rho(A) \leq \rho(B)$.

Proof. See Varga [1300]. This is a consequence of the Perron-Frobenius theorem. □

Proof of Theorem 5.17 Let $A = D + L + U$. We know that $J(A) = I - D^{-1}A$. Let $M(A) = |D| - |L| - |U|$ and $J(M(A)) = I - |D|^{-1}(|D| - |L| - |U|)$. Since matrices $|J(A)|$ and $J(M(A))$ have zero diagonal terms and the non–diagonal elements are $|\frac{a_{i,j}}{a_{i,i}}|$. Therefore $|J(A)| = J(M(A))$, but $M(A)$ is an M–matrix and by theorem 5.3, we have

$$\rho(|J(A)|) = \rho(J(M(A))) < 1.$$

Let $M_\omega = \frac{1}{\omega}D + L$, $N_\omega = \frac{1-\omega}{\omega}D - U$, then

$$\mathcal{L}_\omega = M_\omega^{-1}N_\omega.$$

Denote

$$\widetilde{M}_\omega = \tfrac{1}{\omega}|D| - |L|, \qquad \widetilde{N}_\omega = \tfrac{|1-\omega|}{\omega}|D| + |U|,$$

and

$$\widetilde{\mathcal{L}}_\omega = \widetilde{M}_\omega^{-1}\widetilde{N}_\omega, \qquad \widetilde{A}_\omega = \widetilde{M}_\omega - \widetilde{N}_\omega = \tfrac{1-|1-\omega|}{\omega}|D| - |L| - |U|.$$

It is obvious that $\widetilde{N}_\omega \geq 0$ and $|N_\omega| \leq \widetilde{N}_\omega$. Moreover

$$|M_\omega^{-1}| \leq |(\tfrac{1}{\omega}D + L)^{-1}| = |(I + \omega D^{-1}L)^{-1}\omega D^{-1}|.$$

But since $D^{-1}L$ is a strictly lower triangular matrix, we have $(D^{-1}L)^n = 0$ and

$$(I + \omega D^{-1}L)^{-1} = I - \omega D^{-1}L + \cdots + (-1)^{n-1}(\omega D^{-1}L)^{n-1},$$

so

$$\begin{aligned}|M_\omega^{-1}| &\leq \omega|I - \omega D^{-1}L + \cdots + (-1)^{n-1}(\omega D^{-1}L)^{n-1}| \cdot |D^{-1}|,\\ &\leq \omega(I + \omega|D^{-1}L| + \cdots + \omega^{n-1}|D^{-1}L|^{n-1}) \cdot |D^{-1}|.\end{aligned}$$

But, as D^{-1} is diagonal, $|D^{-1}L| = |D^{-1}| \cdot |L|$ and

$$0 \leq |M_\omega^{-1}| \leq (\tfrac{|D|}{\omega} - |L|)^{-1} = \widetilde{M}_\omega^{-1}.$$

This proves that $\widetilde{A}_\omega = \widetilde{M}_\omega - \widetilde{N}_\omega$ is a regular splitting. We want to conclude using lemma 5.18. Let us consider under which conditions we have $\widetilde{A}_\omega^{-1} \geq 0$.

Lemma 5.20

$\widetilde{A}_\omega^{-1} \geq 0$ *if* $0 < \omega < \frac{2}{1+\rho(|J(A)|)}$.

Proof.

$$\begin{aligned}\widetilde{A}_\omega &= \tfrac{1-|1-\omega|}{\omega}|D| - |L| - |U|,\\ &= |D|\left(\tfrac{1-|1-\omega|}{\omega}I - |D|^{-1}(|L| + |U|)\right),\\ &= |D|\left(\tfrac{1-|1-\omega|}{\omega}I - |J(A)|\right).\end{aligned}$$

Firstly we consider the case when $0 < \omega < 1$. Then, $\frac{1-|1-\omega|}{\omega} = 1$, so $\widetilde{A}_\omega = |D|(I - |J(A)|)$. But, as A is an H–matrix, $\rho(|J(A)|) = \rho(J(M(A))) < 1$. This, by theorem 1.27 proves that $\widetilde{A}_\omega^{-1}$ exists and

$$\widetilde{A}_\omega^{-1} = (I + |J(A)| + \cdots)|D|^{-1} \geq 0.$$

Now, suppose $1 < \omega < 2$. Then

$$\widetilde{A}_\omega = |D|(\tfrac{2-\omega}{\omega} I - |J(A)|).$$

The matrix $\frac{2-\omega}{\omega} I - |J(A)|$ is non–singular if

$$\frac{\omega}{2-\omega}\rho(|J(A)|) < 1,$$

and this implies

$$\omega < \frac{2}{1+\rho(|J(A)|)}.$$

Again, by theorem 1.27, $\widetilde{A}_\omega^{-1} \geq 0$. □

Returning to the proof of theorem 5.17, by lemma 5.18 we know that $\widetilde{A}_\omega^{-1} \geq 0$ implies $\rho(\widetilde{M}_\omega^{-1}\widetilde{N}_\omega) < 1$. But, we have also shown that

$$|\mathcal{L}_\omega| = |M_\omega^{-1} N_\omega| \leq |M_\omega^{-1}||N_\omega| \leq \widetilde{M}_\omega^{-1}\widetilde{N}_\omega = \widetilde{\mathcal{L}}_\omega.$$

By lemma 5.19, this shows that

$$\rho(\mathcal{L}_\omega) \leq \rho(\widetilde{\mathcal{L}}_\omega) < 1,$$

if $0 < \omega < \frac{2}{1+\rho(|J(A)|)}$. □

Neumann and Varga [1012] have shown that this bound is sharp. They gave examples of H–matrices for which if $\omega = \frac{2}{1+\rho(|J(A)|)}$ then $\rho(\mathcal{L}_\omega) = 1$. These results imply that SOR converges for all H–matrices if and only if $0 < \omega \leq 1$. If we now suppose that A is symmetric, we have the following result:

Theorem 5.21

Let A be symmetric with D positive definite. Then, SOR *converges for* $0 < \omega < 2$ *if and only if A is positive definite.*

Proof. This theorem is known under the name of Ostrowski and Reich. It is a straightforward consequence of the more general theorem 1.54 (Householder–John) since

$$Q = \frac{2-\omega}{\omega} D.$$

□

Regarding vector and parallel computers, SOR has just the same drawbacks as Gauss-Seidel. Therefore, we must use the same devices as for Gauss--Seidel if we want to use the method with some efficiency. An important problem is the choice of the relaxation parameter ω. Obviously, we would like to choose the ω that minimizes $\rho(\mathcal{L}_\omega)$. This problem was solved for a large class of matrices by Young (1950) in his Ph.D. thesis and is explained in great detail in his book [1351]. Young proves that, under some hypothesis, there exists an optimal value ω_b:

$$\omega_b = \frac{2}{1 + (1 - \rho(J(A))^2)^{\frac{1}{2}}}.$$

Moreover, $\rho(\mathcal{L}_{\omega_b}) = \omega_b - 1$. Young's theory relies on the fact that for matrices *with a consistent ordering* there is a relationship between the eigenvalues λ of $\mathcal{L}_\omega$ and μ of $J(A)$, namely

$$(\lambda + \omega - 1)^2 = \omega^2 \mu^2 \lambda.$$

For $\omega = 1$, we recover the relation that we proved for Gauss-Seidel for matrices having property A. It is unfortunate we need $\rho(J(A))$ to compute ω_b because in many practical problems we do not know $\rho(J(A))$. Trying to find approximate values of ω_b was the subject of many research papers. Some people devised very clever schemes to try to get an approximation of ω_b during the SOR computation. For more details, see the book by Hageman and Young [748]. For the model problem it is easy to compute ω_b. We get

$$\omega_b = \frac{2}{1 + \sin(\pi h)} \quad \text{and} \quad \rho(\mathcal{L}_\omega) = \frac{1 - \sin(\pi h)}{1 + \sin(\pi h)}.$$

Therefore, $\rho(\mathcal{L}_{\omega_b}) = 1 - 2\pi h + O(h^2)$. We recall that for $\omega = 1$ *i.e.*, the `Gauss-Seidel` method, we have $\rho(\mathcal{L}) = 1 - 2\pi^2 h^2 + O(h^4)$. So, for small h, `SOR` with ω_b is a very large improvement over `Gauss-Seidel`. This explains the great interest for this method in the fifties and sixties.

Figure 5.3 shows the $\log_{10}$ of the maximum norm of the error when solving the Poisson model problem on a 20×20 mesh for `Jacobi`, `Gauss-Seidel` and `SOR` with the optimal parameter. Figure 5.4 shows the spectral radius of the `SOR` iteration matrix as a function of ω for a 10×10 mesh. Note the spectral radius if non–differentiable for $\omega = \omega_b$.

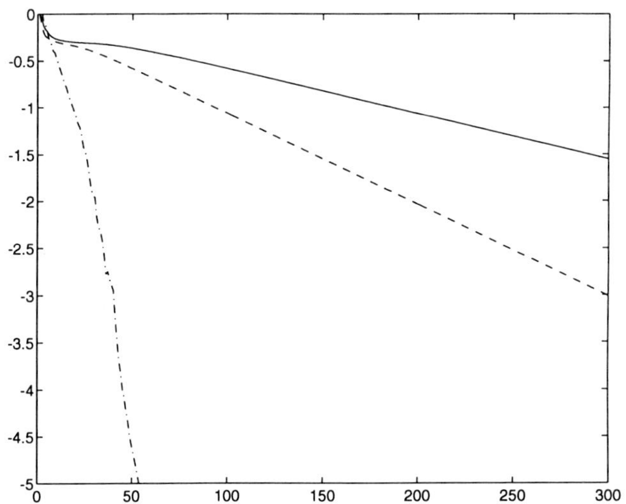

fig. 5.3: Poisson problem on a 20×20 *mesh*
solid: `Jacobi`, *dashed:* `Gauss-Seidel`, *dot–dashed:* `SOR` *with optimal* ω

5.5. The SSOR method

We have seen in previous sections that the results of `SOR` are ordering dependent. If we choose the natural ordering for systems arising for the discretization of elliptic partial differential equations (in particular diffusion equations) it may happen that this ordering is not well suited to the physical problem and it is better, for instance, to use a reverse ordering scheme. So, a natural idea is to change the ordering during the computation. That is `SOR` with numbering the unknowns from 1 to n and then from n to 1. This trick gives

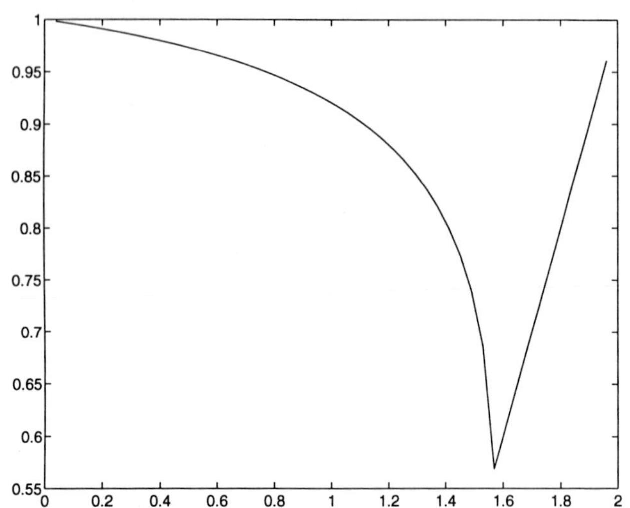

fig. 5.4: Spectral radius of the SOR *iteration matrix as a function of ω for the Poisson problem on a 10×10 mesh*

a symmetric iteration matrix. This method is denoted by SSOR (Symmetric SOR). An iteration is defined by

$$\begin{aligned}(D+\omega L)x^{k+\frac{1}{2}} &= \omega b + (1-\omega)Dx^k - \omega U x^k,\\ (D+\omega U)x^{k+1} &= \omega b + (1-\omega)Dx^{k+\frac{1}{2}} - \omega L x^{k+\frac{1}{2}}.\end{aligned} \tag{5.13}$$

To get the expression of the iteration matrix we eliminate $x^{k+\frac{1}{2}}$. We then have

$$\begin{aligned}x^{k+1} = &(\tfrac{1}{\omega}D+U)^{-1}(\tfrac{1-\omega}{\omega}D-L)(\tfrac{1}{\omega}D+L)^{-1}(\tfrac{1-\omega}{\omega}D-U)x^k\\ &+(\tfrac{1}{\omega}D+U)^{-1}(\tfrac{2-\omega}{\omega})D(\tfrac{1}{\omega}D+L)^{-1}b.\end{aligned} \tag{5.14}$$

Thus, we have a splitting of $A = M - N$ where $M = \frac{\omega}{2-\omega}(\frac{1}{\omega}D+L)D^{-1}(\frac{1}{\omega}D+U)$ and $N = \frac{\omega}{2-\omega}(\frac{1}{\omega}D+L)D^{-1}(\frac{1-\omega}{\omega}D-L)(\frac{1}{\omega}D+L)^{-1}(\frac{1-\omega}{\omega}D-U)$. The iteration matrix denoted by $\mathcal{S}_\omega$ is

$$\mathcal{S}_\omega = (\tfrac{1}{\omega}D+U)^{-1}(\tfrac{1-\omega}{\omega}D-L)(\tfrac{1}{\omega}D+L)^{-1}(\tfrac{1-\omega}{\omega}D-U).$$

As before, for the Gauss-Seidel and SOR methods, it can be shown that if A is strictly diagonally dominant and if $0 < \omega \le 1$, then the method converges. This result can be extended to H–matrices. Alefeld and Varga [14] proved a stronger result.

Theorem 5.22

Let A be an H–matrix. Then the SSOR *method converges if*

$$0 < \omega < \frac{2}{1 + \rho(|J(A)|)}.$$

Proof. The proof is similar to theorem 5.17. □

Neumaier and Varga [1009] proved the following result which gives insight on the convergence or divergence of the method. Let ν such that $0 \leq \nu < 1$, $\mathcal{H}_\nu = \{A| \ A \text{ is an H–matrix with } \rho(|J(A)|) = \nu\}$ and

$$\hat{\omega}(\nu) = \begin{cases} 2, & \text{if } 0 \leq \nu \leq \frac{1}{2}; \\ \frac{2}{1+\sqrt{2\nu - 1}}, & \frac{1}{2} < \nu < 1. \end{cases}$$

Then, for each matrix A in $\mathcal{H}_\nu$, for every ω with $0 < \omega < \hat{\omega}(\nu)$, $\rho(\mathcal{S}_\omega) < 1$. For every ω with $\omega \leq 0$ or $\omega > \hat{\omega}(\nu)$, there exists a matrix in $\mathcal{H}_\nu$ for which $\rho(\mathcal{L}_\omega) \geq 1$.

Of course one can also define block SSOR methods. We shall not comment further on these. Let us now consider the case of symmetric positive definite matrices.

Theorem 5.23

Let A be a symmetric positive definite matrix with D positive definite. Then, the SSOR *converges for $0 < \omega < 2$.*

Proof. This is a consequence of theorem 1.54, because M is symmetric and it is easy to compute that

$$Q = 2M - A = \frac{\omega}{2-\omega}(A + 2LD^{-1}L^T).$$

Since D is positive definite, $LD^{-1}L^T$ is positive semi–definite and Q is positive definite. □

We noted that the SOR iteration matrix is not symmetric even when A is and consequently may have complex eigenvalues. This will cause difficulties when we try to accelerate this method, see the following sections. One nice property of the SSOR method is that $\mathcal{S}_\omega$ has real positive eigenvalues as the next theorem shows.

Theorem 5.24

Let A be symmetric with D positive definite. Then the eigenvalues of $\mathcal{S}_\omega$ are real and non–negative.

Proof.

$$\mathcal{S}_\omega = (\tfrac{1}{\omega}D + L^T)^{-1}(\tfrac{1-\omega}{\omega}D - L)(\tfrac{1}{\omega}D + L)^{-1}(\tfrac{1-\omega}{\omega}D - L^T).$$

Let $\overline{L} = D^{-\frac{1}{2}}LD^{-\frac{1}{2}}$, then

$$\mathcal{S}_\omega = D^{-\frac{1}{2}}(\tfrac{1}{\omega}I + \overline{L}^T)^{-1}(\tfrac{1-\omega}{\omega}I - \overline{L})(\tfrac{1}{\omega}I + \overline{L})^{-1}(\tfrac{1-\omega}{\omega}I - \overline{L}^T)D^{\frac{1}{2}}.$$

We note that as $\overline{L}$ is strictly lower triangular, $(\frac{1}{\omega}I + \overline{L})^{-1}$ is a polynomial in $\overline{L}$ (since by the Cayley–Hamilton theorem $\overline{L}^n = 0$), so it commutes with $(\frac{1-\omega}{\omega}I - \overline{L})$ and we may rewrite the iteration matrix as

$$\mathcal{S}_\omega = D^{-\frac{1}{2}}(\tfrac{1}{\omega}I + \overline{L}^T)^{-1}(\tfrac{1}{\omega}I + \overline{L})^{-1}(\tfrac{1-\omega}{\omega}I - \overline{L})(\tfrac{1-\omega}{\omega}I - \overline{L}^T)D^{\frac{1}{2}}.$$

The matrix

$$\begin{aligned}\mathcal{S}'_\omega &= (\tfrac{1}{\omega}I + \overline{L}^T)D^{\frac{1}{2}}\mathcal{S}_\omega D^{-\frac{1}{2}}(\tfrac{1}{\omega}I + \overline{L}^T)^{-1},\\ &= (\tfrac{1}{\omega}I + \overline{L})^{-1}(\tfrac{1-\omega}{\omega}I - \overline{L})(\tfrac{1-\omega}{\omega}I - \overline{L}^T)(\tfrac{1}{\omega}I + \overline{L}^T),\end{aligned}$$

has the same eigenvalues as $\mathcal{S}_\omega$. If we denote $G(\omega) = (\frac{1}{\omega}I + \overline{L})^{-1}(\frac{1-\omega}{\omega}I - \overline{L})$, then

$$\mathcal{S}'_\omega = G(\omega)G(\omega)^T$$

so, $\mathcal{S}_\omega$ is similar to a symmetric positive definite matrix. □

Since A is symmetric positive definite, it can be shown (Young [1352]) that there exists an optimal parameter but, unlike for SOR, the analytical form of this optimal value is not known. However, Young [1352] showed that a "good" parameter is

$$\omega = \frac{2}{1+\sqrt{2(1-\rho(J(A)))^2}}.$$

With this parameter the convergence speed is almost twice as fast as with SOR. But, unfortunately, an SSOR iteration cost is twice that of an SOR iteration, therefore the benefits are not so obvious. However, there are two facts which are nice with SSOR. First, the number of iterations is less sensitive to the value of ω than it is with SOR and second, as we shall see later on, SSOR can be used in combination with acceleration methods. Figure 5.5 shows the spectral radius for the model problem as a function of ω for a 10×10 mesh.

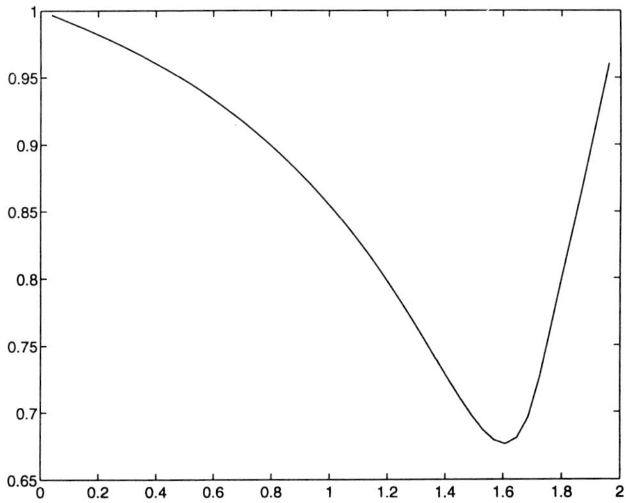

fig. 5.5: Spectral radius of the SSOR *iteration matrix for the Poisson problem on a* 10×10 *mesh*

Regarding vector or parallel computing, the problem with SSOR is clearly the same as with SOR.

5.6. Alternating direction methods

These methods originated in the fifties for solving linear problems arising from elliptic PDEs on rectangles. If one uses block methods like line `Jacobi` or line `Gauss-Seidel`, these methods emphasize one space direction over the other. So, once again, a natural idea is to switch (or alternate) directions.

Let us look again at the model problem. Let D be diagonal with $d_{i,i} = 4$

and a blockwise splitting:

$$H = \begin{pmatrix} C & 0 & & & \\ 0 & C & 0 & & \\ & 0 & \ddots & \ddots & \\ & & \ddots & \ddots & 0 \\ & & & 0 & C \end{pmatrix}, \quad \text{where} \quad C = \begin{pmatrix} 0 & -1 & & & \\ -1 & 0 & -1 & & \\ & \ddots & \ddots & \ddots & -1 \\ & & \ddots & \ddots & \ddots \\ & & & -1 & 0 \end{pmatrix},$$

and

$$V = \begin{pmatrix} 0 & -I & & & \\ -I & 0 & -I & & \\ & \ddots & \ddots & \ddots & \\ & & & & -I \\ & & & -I & 0 \end{pmatrix}.$$

Then,

$$A = D + H + V.$$

Peaceman and Rachford defined the following method which is denoted by `ADI` (Alternating Direction Implicit),

$$\begin{aligned} (\tfrac{1}{2}D + H + \rho_{k+1}I)x^{k+\frac{1}{2}} &= b - (\tfrac{1}{2}D + V - \rho_{k+1}I)x^k, \\ (\tfrac{1}{2}D + V + \rho'_{k+1}I)x^{k+1} &= b - (\tfrac{1}{2}D + H - \rho'_{k+1}I)x^{k+\frac{1}{2}}, \end{aligned} \tag{5.15}$$

where ρ_{k+1} and ρ'_{k+1} are real positive parameters. It is easy to see that at the first step we have to solve as many point tridiagonal systems as there are rows in the mesh and at the second, as many tridiagonal systems as there are columns in the mesh.

If $\rho'_{k+1} = \rho$, $\rho'_{k+1} = \rho'$, the method is called the stationary `Peaceman–Rachford` method. This latter method is a special case of a class of methods which can be written as

$$\begin{aligned} (D + H + F)x^{k+\frac{1}{2}} &= b + (F - V)x^k, \\ (D + V + G)x^{k+1} &= b + (G - H)x^{k+\frac{1}{2}}. \end{aligned} \tag{5.16}$$

The decomposition of A, $A = D + H + V$ and matrices F, G can be chosen as needed so long as $D + H + F$ and $D + V + G$ are non–singular. Different choices of F and G lead to some of the methods we have described before or to some variants:

- $F = \rho I - \frac{1}{2}D$, $G = \rho' I - \frac{1}{2}D$ gives the stationary `Peaceman-Rachford` method.

- $F = G = 0$ gives the (line) `Jacobi` alternating method.

- if $H = H_L + H_U$ where H_L (resp. H_U) is the lower (resp. upper) triangular part of H and the same for $V = V_L + V_U$. With $F = V_L$, $G = H_L$, this is the (line) `Gauss-Seidel` alternating method

- $F = -H$, $G = -V$ this is the (point) `Jacobi` method.

We shall prove that the line `Gauss-Seidel` alternating method is convergent. For a matrix B, we denote by $I_i(B) = \{j| \; b_{i,j} \neq 0\}$ the index set of columns for which there are non–zero coefficients in row i.

Theorem 5.25

Let A be strictly diagonally dominant, suppose H and V have zero diagonals and

$$I_i(V_U) \cap I_i(H + V_L) = \emptyset, \qquad I_i(H_U) \cap I_i(V + H_L) = \emptyset.$$

Then the block `Gauss-Seidel` *alternating method is convergent.*

Proof. The hypothesis on the index sets simply means that a coefficient of A is either in H or in V, but not split between the two. Let $\varepsilon^k = x - x^k$, then

$$(D + H + V_L)\varepsilon^{k+1/2} = (-V_U)\varepsilon^k,$$
$$(D + V + H_L)\varepsilon^{k+1} = (-H_U)\varepsilon^{k+1/2}.$$

Therefore,

$$\varepsilon^{k+1} = (D + V + H_L)^{-1} H_U (D + H + V_L) V_U \varepsilon^k = T\varepsilon^k.$$

Let λ, u be an eigenvalue and the related eigenvector of the iteration matrix T, $Tu = \lambda u$. Then,

$$(D + H + V_L)^{-1} V_U u = v,$$
$$(D + V + H_L)^{-1} H_U v = \lambda u.$$

Let us show that $\|v\|_\infty < \|u\|_\infty$. We have,

$$(D + H + V_L)v = V_U u.$$

Componentwise, this can be written as

$$v_i = \sum_{j\in I_i(V_U)} \frac{a_{i,j}u_j}{a_{i,i}} - \sum_{j\in I_i(H+V_L)} \frac{a_{i,j}v_j}{a_{i,i}},$$

$$|v_i| \le \|u\|_\infty \Big(\sum_{j\in I_i(V_U)} \frac{|a_{i,j}|}{|a_{i,i}|} \Big) + \|v\|_\infty \Big(\sum_{j\in I_i(H+V_L)} \frac{|a_{i,j}|}{|a_{i,i}|} \Big).$$

Suppose that $\|u\|_\infty \le \|v\|_\infty$, then

$$|v_i| \le \|v\|_\infty \Big(\sum_{j\in I_i(H+V)} \frac{|a_{i,j}|}{|a_{i,i}|} \Big) < \|v\|_\infty.$$

This is a contradiction, therefore $\|v\|_\infty < \|u\|_\infty$. In the same way we can show that

$$\|\lambda u\|_\infty < \|v\|_\infty < \|u\|_\infty.$$

Hence $|\lambda| < 1$ and the method converges. □

Remark.

- With the same technique it can be shown that the block Gauss-Seidel and Jacobi methods are convergent.
- If the diagonals of F and G are 0 and matrices $A + 2F$ and $A + 2G$ are strictly diagonally dominant, then the method (5.16) is convergent.
- The stationary Peaceman Rachford method is not covered by this theorem because of the diagonal terms.

Theorem 5.26

Let A be an H–matrix. Suppose H and V have zero diagonals and

$$I_i(V_U) \cap I_i(H + V_L) = \emptyset, \qquad I_i(H_U) \cap I_i(V + H_L) = \emptyset.$$

Then the block Gauss-Seidel *alternating method converges.*

Proof. As we did with other methods, we can prove that there exists a diagonal matrix E with a positive diagonal such that

$$T(E^{-1}AE) = E^{-1}T(A)E.$$

Again this proves that the method converges. □

Let us now consider the case of the stationary `Peaceman-Rachford method`. This method can be cast in the following way,

$$\begin{aligned} (D + D_F + H)x^{k+1/2} &= (D_F - V)x^k + b, \\ (D + D_G + V)x^{k+1} &= (D_G - H)x^{k+1/2} + b, \end{aligned} \tag{5.17}$$

where D_F and D_G are diagonal matrices with positive diagonals. To prove a convergence result, we are going to use the same techniques as in theorem 5.17.

Theorem 5.27

Let A be an H–matrix with a positive diagonal and let D_F and D_G be two diagonal matrices with positive diagonals. Suppose that the diagonals of H and V are 0 and $I_i(H) \cap I_i(V) = \emptyset$. Then the method defined by (5.17) is convergent.

Proof. Let

$$M_1 = D + D_F + H, \qquad N_1 = D_F - V.$$

Then,

$$|N_1| = |D_F - V| \le D_F + |V| = \widetilde{N}_1,$$

$$|M_1^{-1}| = |[I + (D + D_F)^{-1}H](D + D_F)^{-1}|.$$

But,

$$|(D + D_F)^{-1}H| \le (D + D_F)^{-1}|H|.$$

By lemma 5.19,

$$\rho((D + D_F)^{-1}H) \le \rho((D + D_F)^{-1}|H|).$$

But $(D + D_F,\ |H|)$ is a regular splitting of $D + D_F - |H|$ which by hypothesis is an M–matrix. Therefore $\rho((D + D_F)^{-1}|H|) < 1$. This implies that $[I + (D + D_F)^{-1}H]^{-1}$ exists and is equal to the sum of the series of matrices

$$I - (D + D_F)^{-1}H + ((D + D_F)^{-1}H)^2 - \cdots$$

Bounding the absolute value of the series we get,

$$| [I + (D + D_F)^{-1} H]^{-1} | \leq (I - (D + D_F)^{-1} |H|).$$

Hence,

$$|M_1^{-1}| \leq (D + D_F - |H|)^{-1} = \widetilde{M}_1^{-1},$$

and

$$\widetilde{M}_1 - \widetilde{N}_1 = D + D_F - |H| - D_F - |V| = D - |H| - |V|.$$

In the same way, if we denote

$$M_2 = D + D_G + V, \qquad N_2 = D_G - H.$$

Then,

$$\begin{aligned} |M_2^{-1}| &\leq (D + D_G - |V|)^{-1} = \widetilde{M}_2^{-1}, \\ |N_2| &\leq D_G + |H| = \widetilde{N}_2, \\ \widetilde{M}_2 - \widetilde{N}_2 &= D + D_G - |V| - |D_G| - |H| = D - |H| - |V|. \end{aligned}$$

By hypothesis $D - |H| - |V|$ is an M–matrix. To conclude we need the two following lemmas.

Lemma 5.28

Let B be a non–singular matrix and for all k, (M_k, N_k) be a regular splitting of B. Then the following statements are equivalent

1) $B^{-1} \geq 0$.

2) the sequence defined by $x^{k+1} = x^k - M_k^{-1}(Bx^k - c)$ *tends to a solution of* $Bx = c$.

Proof. See Moré [995]. □

Lemma 5.29

With the same hypothesis as the preceding lemma, let $\widetilde{M}_k$ (non–singular) and $\widetilde{N}_k$ such that

$$|M_k^{-1}| \leq \widetilde{M}_k^{-1}, \qquad |N_k| \leq \widetilde{N}_k,$$

and let $\widetilde{B} = \widetilde{M}_k - \widetilde{N}_k$ *for all* k. *If the sequence defined by*

$$x^{k+1} = x^k - \widetilde{M}_k^{-1}(\widetilde{B}x^k - c),$$

converges, then the sequence y^k *such that*

$$y^{k+1} = y^k - M_k^{-1}(By^k - c)$$

converges also.

Proof. Let $\widetilde{x}$ be the solution of $\widetilde{B}\widetilde{x} = c$, y be the solution of $By = c$ and $\epsilon^k = \widetilde{x} - x^k$, $\varepsilon^k = y - y^k$. Then,

$$\widetilde{M}_k \epsilon^{k+1} = \widetilde{N}_k \epsilon^k,$$
$$M_k \varepsilon^{k+1} = N_k \varepsilon^k.$$

Choose ε^0 such that $|\varepsilon^0| = \epsilon^0$, then

$$|\varepsilon^1| = |M_0^{-1} N_0 \varepsilon^0| \leq |M_0^{-1}||N_0||\varepsilon^0| \leq \widetilde{M}_0^{-1}\widetilde{N}_0 \epsilon^0 = \epsilon^1.$$

By induction, we show that

$$|\varepsilon^k| \leq \epsilon^k$$

But $\epsilon^k \to 0$ as $k \to \infty$ therefore $\varepsilon^k \to 0$. □

To conclude the proof of theorem 5.27, we first apply lemma 5.28 as $D - |H| - |V|$ is an M–matrix, its inverse is positive and therefore the sequence defined by $\widetilde{M}_k, \widetilde{N}_k$ is convergent. Then, as $|M_i^{-1}| < \widetilde{M}_i^{-1}$, $|N_i| \leq \widetilde{N}_i$ $i = 1, 2$ by lemma 5.29, the method (5.17) is convergent. □

As a consequence of the last result, we get a convergence theorem for the `Peaceman-Rachford` method.

Corollary 5.30

Let A *be a matrix satisfying the hypothesis of theorem 5.27. If* ρ *and* ρ' *are such that*

$$\rho \geq \tfrac{1}{2} \max_i a_{ii},$$

$\rho' \geq \frac{1}{2} \max_i a_{ii}$,

then the stationary `Peaceman-Rachford` *method is convergent.*

□

Remark.

- Using the same techniques, Alefeld [13] showed that the `Peaceman-Rachford` method is convergent. The problem with this method is to find the optimal parameters. This question is only solved for the model problem (see Young [1351]).

- Regarding vectorization and parallelization the situation is the same for the block `Gauss-Seidel` alternating method as for the ordinary block `Gauss-Seidel` method. It can even be worse because the data on the mesh has to be accessed by rows and by columns and that may cause conflicts on some computers. The situation is simpler for the `Peaceman-Rachford` as all the tridiagonal systems to be solved are independent of each other and can be solved in vector or parallel mode.

Let us now look at the alternating `Gauss-Seidel` method when the matrix is symmetric. To study this method, we need the following result which is interesting in itself.

Lemma 5.31

Let A be a symmetric positive definite matrix and $A = M - N = P - Q$ two splittings of A. Consider the iterative method defined by

$$\begin{aligned} Mx^{k+1/2} &= Nx^k + b, \\ Px^{k+1} &= Qx^{k+1/2} + b. \end{aligned} \tag{5.18}$$

Suppose $M^T + N$ and $P^T + Q$ are positive definite. Then the method (5.18) is convergent.

Proof. This is a direct generalization of the Householder–John theorem 1.54. It can be proved using exactly the same technique. □

Theorem 5.32

Let A be a symmetric positive definite matrix, H and V being symmetric. Suppose $D + H$ and $D + V$ are positive definite. Then the alternating block Gauss-Seidel method is convergent.

Proof. In this case, we have

$$M = D + H + V_L, \quad N = -V_U, \quad M^T + N = D + H + V_L^T - V_U = D + H,$$

$$P = D + V + H_L, \quad Q = -H_U, \quad P^T + Q = D + V + H_L^T - H_U = D + V.$$

□

Alternating directions methods are not frequently used as they are closely related to finite difference approximations on rectangular meshes. However, recently there was a renewal of interest for these methods as they can be used as preconditioners in a more general framework, see Mathew, Polyakov, Russo and Wang [946]. We shall return to this point later in the chapter on preconditioning.

5.7. Richardson methods

We now consider a method which is seldom used, but that will be useful for theoretical purposes when we look at acceleration techniques and at multigrid methods in the following chapters. This method is very simple to implement and in some cases can be shown to reduce to the `Jacobi` iteration. Let α be a strictly positive real number and consider the sequence defined by

$$x^{k+1} = x^k + \alpha(b - Ax^k). \tag{5.19}$$

This method, known as the stationary `Richardson` method, is based upon a splitting of A as,

$$\frac{1}{\alpha}x^{k+1} = \left(\frac{1}{\alpha}I - A\right)x^k + b.$$

Hence $M = \frac{1}{\alpha}I$ and $N = \frac{1}{\alpha}I - A$. For the method (5.19) to converge, we must have $\rho(I - \alpha A) < 1$.

Theorem 5.33

Let A be symmetric and positive definite and $0 < \lambda_1 \leq \lambda_2 \leq \cdots \leq \lambda_n$ be the eigenvalues of A. Then the stationary `Richardson` *method (5.19) converges if and only if $\alpha < \frac{2}{\lambda_n}$.*

Proof. It is obvious that $|1 - \alpha\lambda_i| < 1$ for all i if and only if $\alpha < \frac{2}{\lambda_n}$. □

As with every other method depending upon a parameter, the problem is to find the optimal value of α giving the smallest spectral radius.

Theorem 5.34

The optimal value of α for the stationary `Richardson` *method is $\alpha_{opt} = \frac{2}{\lambda_1+\lambda_n}$.*

Proof. The optimal value is determined by $1 - \alpha\lambda_1 = -(1 - \alpha\lambda_n)$. Hence $\alpha_{opt} = \frac{2}{\lambda_1+\lambda_n}$ and

$$\rho(I - \alpha_{opt}A) = 1 - \frac{2}{\lambda_1 + \lambda_n}\lambda_1 = \frac{\lambda_n - \lambda_1}{\lambda_1 + \lambda_n} = \frac{\kappa(A) - 1}{\kappa(A) + 1},$$

where $\kappa(A) = \frac{\lambda_n}{\lambda_1}$ is the condition number of A. The closer $\kappa(A)$ is to 1, the faster the convergence. □

Of course, the drawback of this method (as with relaxed `Jacobi` or `SOR`) is that we have to know the maximum and minimum eigenvalues of A to get the optimal parameter. A straightforward way of generalizing this method is to allow the parameter to change at every iteration,

$$x^{k+1} = x^k + \alpha_k(b - Ax^k). \tag{5.20}$$

Once again, the problem is to choose the sequence of parameters α_k's. One way is to minimize a norm of the residual at each iteration. Let μ be an integer. If A is symmetric and positive definite, then $A^{-\mu}$ has the same properties. So, let us introduce the norm

$$\|r^k\|_\mu^2 = (r^k, A^{-\mu}r^k),$$

and suppose we want to minimize $\|r^k\|_\mu^2$. Obviously,

$$r^{k+1} = r^k - \alpha_k A r^k.$$

Therefore,

$$(r^{k+1}, A^{-\mu} r^{k+1}) = (r^k, A^{-\mu} r^k) - 2\alpha_k (r^k, A^{1-\mu} r^k) + \alpha_k^2 (r^k, A^{2-\mu} r^k).$$

Hence

$$\alpha_k = \frac{(r^k, A^{1-\mu} r^k)}{(r^k, A^{2-\mu} r^k)}.$$

Of course, we need to be able to compute α_k, so usually one chooses $\mu = 1$ and in that case, we obtain,

$$\alpha_k = \frac{(r^k, r^k)}{(r^k, A r^k)}. \tag{5.21}$$

This method is called the gradient or `steepest descent` method.

Theorem 5.35

Let A be symmetric positive definite. Then, the `steepest descent` *method (5.20), (5.21) is convergent.*

Proof. We have

$$(r^{k+1}, A^{-1} r^{k+1}) = (r^k, A^{-1} r^k) - \frac{(r^k, r^k)^2}{(r^k, A r^k)},$$

$$\frac{(r^{k+1}, A^{-1} r^{k+1})}{(r^k, A^{-1} r^k)} = 1 - \frac{(r^k, r^k)^2}{(r^k, A r^k)(r^k, A^{-1} r^k)}.$$

We then apply the Kantorovitch inequality which shows that

$$\frac{(r^k, A r^k)(r^k, A^{-1} r^k)}{(r^k, r^k)^2} \le \left(\frac{\sqrt{\kappa(A)} + (\sqrt{\kappa A})^{-1}}{2} \right)^2.$$

Therefore

$$\frac{(r^{k+1}, A^{-1} r^{k+1})}{(r^k, A^{-1} r^k)} \le \left(\frac{\kappa(A) - 1}{\kappa(A) + 1} \right)^2 < 1,$$

$$0 < \|r^{k+1}\|_{-1}^2 < \|r^k\|_{-1}^2,$$

and $r^k \to 0$. □

Note that in this method we have the orthogonality relation $(r^{k+1}, r^k) = 0$. One may ask if the `steepest descent` method is optimal. The answer is negative as we shall show below.

Consider $\varepsilon^k = x - x^k$, then

$$\varepsilon^{k+1} = (I - \alpha_k A)\varepsilon^k,$$

and

$$\varepsilon^k = \prod_{i=0}^{k-1}(I - \alpha_i A)\varepsilon^0 = P_k(A)\varepsilon^0,$$

where $P_k(t)$ is a polynomial of degree k whose value is 1 for $t = 0$. Of course,

$$\|\varepsilon^k\| \le \|P_k(A)\| \, \|\varepsilon^0\|,$$

and we may try to minimize $\|P_k(A)\|$ to get the smallest value of $\|\varepsilon^k\|$ for a given ε^0. Suppose that A has l distinct eigenvalues $\lambda_1, \ldots, \lambda_l$, then the characteristic polynomial of A is

$$\prod_{i=1}^{l}(\lambda_i - \lambda).$$

By the Cayley-Hamilton theorem, we have $\prod_{i=1}^{l}(\lambda_i I - A) = 0$. So, if we choose

$$\alpha_i = \frac{1}{\lambda_{i+1}},$$

we have $P_l(A) = 0$ and $\varepsilon^l = 0$. Of course, as we said before, there is usually no way to easily obtain the eigenvalues of A. But, suppose that A is symmetric, then there exists an orthogonal matrix Q and a diagonal matrix Λ such that $A = Q^T \Lambda Q$. Then, $P_k(A) = Q^T P_k(\Lambda) Q$, and

$$\|P_k(A)\| = \|Q^T P_k(\Lambda) Q\| = \|P_k(\Lambda)\| = \max_i |P_k(\lambda_i)|.$$

An upper bound of $\|P_k(A)\|$ can be found by considering that, if for all i, $\lambda_i \in [a, b]$, then

$$\max_i |P_k(\lambda_i)| \leq \max_{\lambda \in [a,b]} |P_k(\lambda)|.$$

Therefore,

$$\|\varepsilon^k\| \leq \max_{\lambda \in [a,b]} |P_k(\lambda)| \, \|\varepsilon^0\|,$$

and the problem amounts to find the polynomial of degree k with $P_k(0) = 1$ minimizing $\max_{\lambda \in [a,b]} |P_k(\lambda)|$. The solution to this problem was given in Chapter 1,

$$P_k(\lambda) = \frac{T_k\left(\frac{a+b-2\lambda}{b-a}\right)}{T_k\left(\frac{a+b}{b-a}\right)}.$$

With this choice, we get

$$\|\varepsilon^k\| \leq 2\left(\frac{\sqrt{\kappa(A)} - 1}{\sqrt{\kappa(A)} + 1}\right)^k \|\varepsilon^0\|,$$

if

$$\frac{1}{\alpha_i} = \frac{a+b}{2} + \frac{b-a}{2} \cos\left(\frac{2i-1}{k}\frac{\pi}{2}\right), \quad 1 \leq i \leq k.$$

Note that the α_i's depend on k. Usually, one chooses an integer d and uses cyclically the parameters α_k for $k = 1, \ldots, d$. This method is clearly better than the stationary `Richardson` method regarding theoretical convergence but, unfortunately, it was noticed very early that it is very sensitive to roundoff errors. The potential instability is linked to the order we use for the parameter. Andersen and Golub [28] studied an ordering due to Lebedev and Finoguenov for which one can show that the method is stable. The α_k's are used in the order defined by the permutation

$$\chi_k = (i_1, i_2, \ldots, i_k).$$

Suppose $d = 2^p$, then $\chi_1 = 1$. If $\chi_{2^{p-1}} = (j_1, j_2, \ldots, j_{2^{p-1}})$, we take,

$$\chi_{2^p} = (j_1, 2^p + 1 - j_1, j_2, 2^p + 1 - j_2, \ldots, j_{2^{p-1}}, 2^p + 1 - j_{2^{p-1}}).$$

For example,

$$\chi_2 = (1,2), \quad \chi_4 = (1,4,2,3), \quad \chi_8 = (1,8,4,5,2,7,3,6),$$

$$\chi_{16} = (1,16,8,9,4,13,5,12,2,15,7,10,3,14,6,11).$$

The problem of finding the optimal polynomial when A is not positive (or negative) definite was solved by DeBoor and Rice [388]. Opfer and Schober [1046] studied the problem when A is non–symmetric.

This kind of method can be extended in two ways. First, one can introduce a non–singular matrix M and define

$$Mx^{k+1} = Mx^k - \alpha(b - Ax^k). \tag{5.22}$$

As we have shown that the convergence depends on the condition number, it is natural to choose M to minimize $\kappa(M^{-1}A)$. M is called the preconditioning matrix or preconditioner. Different choices for M will be studied in Chapter 8. The second possible generalization is to seek methods which use more than one previous iterate.

5.8. Acceleration techniques

Consider the usual splitting method,

$$Mx^{k+1} = Nx^k + b,$$

or

$$x^{k+1} = Bx^k + c,$$

where $B = M^{-1}N$ and $c = M^{-1}b$. We have seen that $\varepsilon^k = B^k\varepsilon^0$. To accelerate the convergence, a common technique is to average the iterates x^k to get another sequence which, hopefully, will converge faster. Let

$$y^k = \sum_{l=0}^{k} \alpha_{l,k} x^l.$$

We demand $\sum_{l=0}^{k} \alpha_{l,k} = 1$, since if $x^l = x$ for all l, we require $y^k = x$.

Let

$$\eta^k = x - y^k = \sum_{l=0}^{k} \alpha_{l,k}(x - x^l) = \sum_{l=0}^{k} \alpha_{k,l} B^l \varepsilon^0 = P_k(B)\varepsilon^0,$$

where $P_k(t)$ is a polynomial of degree k with $P_k(1) = 1$. Once again, we would like to minimize $\|P_k(B)\|$. Suppose that B has real eigenvalues belonging to $[-\rho, \rho]$ $\rho > 0$. The solution of this minimization problem is

$$P_k(t) = \frac{T_k(t/\rho)}{T_k(1/\rho)}.$$

When P_k is known, one can compute the coefficients $\alpha_{l,k}$ but it is much better to use the recurrence relations of Chebyshev polynomials to avoid storing all the x^l vectors. We know that

$$T_k(t/\rho) = T_k(1/\rho)P_k(t),$$

but,

$$T_{k+1}(x) = 2xT_k(x) - T_{k-1}(x).$$

Therefore,

$$T_{k+1}\left(\frac{1}{\rho}\right)P_{k+1}(t) = \frac{2t}{\rho}T_k\left(\frac{1}{\rho}\right)P_k(t) - T_{k-1}\left(\frac{1}{\rho}\right)P_{k-1}(t),$$

and post multiplying by ε^0,

$$T_{k+1}\left(\frac{1}{\rho}\right)\eta^{k+1} = \frac{2}{\rho}T_k\left(\frac{1}{\rho}\right)B\eta^k - T_{k-1}\left(\frac{1}{\rho}\right)\eta^{k-1},$$

or

$$y^{k+1} - x = \frac{2T_k(\frac{1}{\rho})}{\rho T_{k+1}(\frac{1}{\rho})}(By^k - x + c) - \frac{T_{k-1}(\frac{1}{\rho})}{T_{k+1}(\frac{1}{\rho})}(y^{k-1} - x).$$

Noting that

$$\left[\frac{2T_k(\frac{1}{\rho})}{\rho T_{k+1}(\frac{1}{\rho})} - \frac{T_{k-1}(\frac{1}{\rho})}{T_{k+1}(\frac{1}{\rho})}\right]x = x,$$

we obtain

$$y^{k+1} = \frac{2T_k(\frac{1}{\rho})}{\rho T_{k+1}(\frac{1}{\rho})}(By^k + c) - \frac{T_{k-1}(\frac{1}{\rho})}{T_{k+1}(\frac{1}{\rho})}y^{k-1}.$$

Therefore, the vectors y^k can be computed without any knowledge of the x^ls. Now, one further simplification can be achieved since

$$y^{k+1} = \frac{2T_k(\frac{1}{\rho})}{\rho T_{k+1}(\frac{1}{\rho})}(By^k + c - y^{k-1}) + \left(\frac{2T_k(\frac{1}{\rho})}{\rho T_{k+1}(\frac{1}{\rho})} - \frac{T_{k-1}(\frac{1}{\rho})}{T_{k+1}(\frac{1}{\rho})}\right)y^{k-1}.$$

So, this can be written as

$$y^{k+1} = \omega_{k+1}(By^k + c - y^{k-1}) + y^{k-1},$$

with

$$\omega_{k+1} = \frac{2T_k(\frac{1}{\rho})}{\rho T_{k+1}(\frac{1}{\rho})}, \quad \omega_1 = 1.$$

But,

$$\omega_{k+1} = \frac{2T_k(\frac{1}{\rho})}{2T_k(\frac{1}{\rho}) - \rho T_{k-1}(\frac{1}{\rho})} = \frac{1}{1 - \frac{\rho T_{k-1}(\frac{1}{\rho})}{2T_k(\frac{1}{\rho})}} = \frac{1}{1 - \frac{\rho^2}{4}\omega_k}.$$

This shows that we do not need to compute the Chebyshev polynomials to get the coefficients of the method. This algorithm can be rewritten as

$$y^{k+1} = \omega_{k+1}(z^k + y^k - y^{k-1}) + y^{k-1}, \tag{5.23}$$

$$Mz^k = r^k = b - Ay^k.$$

The method is known as the `Chebyshev semi-iterative` method, see Golub and Varga [677], [678]. When the eigenvalues belongs to $[a, b]$, straightforward modifications show that the method becomes

$$y^{k+1} = \frac{\omega_{k+1}}{2 - (a+b)}[2z^k + (2 - (a+b))(y^k - y^{k-1})] + y^{k-1},$$

$$Mz^k = r^k,$$

with

$$\omega_{k+1} = \frac{1}{1 - \frac{\omega_k^2}{4w^2}}, \quad \omega_2 = \frac{2w^2}{2w^2 - 1}, \quad \omega_1 = 1, \quad w = \frac{2 - (a+b)}{b - a}.$$

When the matrix A is non–symmetric the parameters can be estimated dynamically by enclosing the eigenvalues in ellipses (see Manteuffel [927],[928]), although the Chebyshev polynomials over ellipses are not necessarily optimal.

5.9. Stability of classical iterative methods

The stability of classical iterative methods has not been completely studied. For a summary of the results, one may refer to Higham [781]. There are (well conditioned) examples for which, because of roundoff errors, SOR may diverge or stagnate when it is supposed to converge. For methods derived from a splitting of A, an almost straightforward error analysis can be done, see Higham [781]. Let

$$Mx^{k+1} = Nx^k + b - \xi^k,$$

be the computed iterates. For the methods we have been examining,

$$|\xi^k| \leq cu(|M|\,|x^{k+1}| + |N|\,|x^k| + |b|) = \mu^k,$$

where c is a constant. Then the error satisfies

$$\varepsilon^{k+1} = (M^{-1}N)^{k+1}\varepsilon^0 + \sum_{i=0}^{k}(M^{-1}N)^i M^{-1}\xi^{k-i}.$$

From this equation normwise bounds can be derived, see Higham [781]. Let

$$\gamma_x = \sup_k \frac{\|x^k\|}{\|x\|},$$

and $q = \|M^{-1}N\|_\infty < 1$, then

$$\|\varepsilon^{k+1}\|_\infty \leq \|(M^{-1}N)^{k+1}\varepsilon^0\|_\infty + cu(1+\gamma_x)(\|M\|_\infty + \|N\|_\infty)\|x\|_\infty \frac{\|M^{-1}\|_\infty}{1-q}.$$

Componentwise bounds can also be obtained. Let

$$\theta_x = \sup_k \max_i \frac{((|M| + |N|)|x^k|)_i}{((|M| + |N|)|x|)_i},$$

and $c(A)$ be the smallest ϵ such that

$$\sum_{i=0}^{\infty} |(M^{-1}N)^i M^{-1}| \leq \epsilon |A^{-1}|.$$

Then

$$|\varepsilon^{k+1}| \leq |(M^{-1}N)^{k+1}\varepsilon^0| + cu(1+2\theta_x)c(A)|A^{-1}|(|M| + |N|)|x|.$$

If $M^{-1} \geq 0, N \geq 0$ then, $c(A) = 1$. This is the case if $M - N$ is a regular splitting of A. From this bound, we can deduce, for instance, that if A is an M–matrix, Jacobi and Gauss-Seidel are componentwise forward stable. For a backward error analysis, see Higham [781]. The stability of the steepest descent method was studied by Bollen [164]. The algorithm is backward stable as long as the condition number is not too large.

5.10. Bibliographical comments

The use of iterative methods for solving linear systems goes back at least to Gauss (1823). Gauss used iterations to solve small least squares problems. He was not computing the unknowns in a given a priori order. Jacobi used this method as well as his student Seidel. Seidel refers to the `Gauss-Seidel` method in 1874 but advised against its use. Block iterative methods were introduced at about the same time; Gerling (1843), Seidel (1862, 1874). The relaxation methods used by Southwell and his co–workers are in the same spirit; they corrected the component having the largest residual.

Sufficient conditions for convergence of the `Gauss-Seidel` method were given by Nekrasov in 1885. The theorem proving that `Gauss-Seidel` converges for a positive definite matrix was proved at the end of the nineteenth century and was rediscovered by Von Mises and Geiringer (1929). The converse was proved by Reich (1949). The `SOR` method was studied by Frankel (1950) and Young (1950, 1954). Since then there have been many research papers about this method that is more efficient than `Jacobi` and `Gauss-Seidel` methods. This was summarized in the seminal books by Varga [1300] and Young [1351].

Aitken (1950) introduced the `symmetric Gauss-Seidel` method (`SSOR` with $\omega = 1$). The `SSOR` method was described by Sheldon (1955). The block `SSOR` method was studied by Ehrlich (1963).

The alternating direction methods were introduced by Peaceman and Rachford (1955) and also by Douglas and Rachford (1956). For convergence results, see Birkhoff, Young and Varga (1962).

For the acceleration techniques, see Golub and Varga [677], [678] and Manteuffel [927],[928].

6 The conjugate gradient and related methods

6.1. Derivation of the method

In Chapter 5 we saw how to accelerate a basic linear iterative method. Suppose now that we start from a slight generalization of the Richardson method

$$Mx^{k+1} = Mx^k + \alpha(b - Ax^k),$$

where M is a non–singular preconditioning matrix to be chosen later on (see Chapter 8). The standard Richardson method corresponds to $M = I$. Then, we get by (5.23)

$$\begin{aligned} y^{k+1} &= \omega_{k+1}(\alpha z^k + y^k - y^{k-1}) + y^{k-1}, \\ Mz^k &= r^k, \quad r^k = b - Ax^k. \end{aligned}$$

A formal generalization of this method is to allow the parameter α to vary with the iteration number,

$$\begin{aligned} x^{k+1} &= \omega_{k+1}(\alpha_k z^k + x^k - x^{k-1}) + x^{k-1}, \\ Mz^k &= r^k. \end{aligned} \tag{6.1}$$

Let us suppose A and M to be symmetric positive definite. The problem we have to solve is to choose ω_{k+1} and α_k to get a converging sequence. We develop another method to start with and compute ω_{k+1}, α_k, $k = 0, 1, \ldots$ such that the generalized residuals z^k are mutually orthogonal for the scalar product defined by the positive definite matrix M. Independently of the convergence problem, this is of interest because of the following well known result.

Lemma 6.1

Let M be SPD and z^k, $k = 0, 1, \ldots, n$ be a sequence in $\mathbb{R}^n$ such that

$$(z^i, Mz^j) = 0, \quad i \neq j. \tag{6.2}$$

Then, $z^n = 0$.

Proof. The proof is obvious. Since the vectors z^k, $k = 0, 1, \ldots, n-1$ are orthogonal in the scalar product defined by M, they form a basis of the whole space. Therefore,

$$z^n = \sum_{j=0}^{n-1} \beta_j z^j, \quad (z^n, Mz^n) = \sum_{j=0}^{n-1} \beta_j (z^j, Mz^n) = 0.$$

Since M is positive definite, this implies that $z^n = 0$. □

Of course, this will give us a direct method because (in absence of roundoff errors) since $z^n = 0$, we have $r^n = 0$ and $x^n = x$, the exact solution of the linear system. But, we shall see in the following sections that we can efficiently use this method as an iterative one, obtaining (most of the time) a good approximation of the solution in many fewer than n iterations. Let us now show that we can construct ω_k and α_k to satisfy (6.2). We do this by induction. Suppose

$$(z^i, Mz^j) = 0, \quad i \neq j, \quad 0 \leq i, j \leq k.$$

From (6.1) it is easy to see that

$$r^{k+1} = r^{k-1} - \omega_{k+1}(\alpha_k A z^k - r^k + r^{k-1}). \tag{6.3}$$

This can be rewritten as

$$Mz^{k+1} = Mz^{k-1} - \omega_{k+1}(\alpha_k A z^k - Mz^k + Mz^{k-1}). \tag{6.4}$$

Lemma 6.2

If α_k is chosen as

$$\alpha_k = \frac{(z^k, Mz^k)}{(z^k, Az^k)},$$

then, $(z^k, Mz^{k+1}) = 0$.

Proof. Multiplying (6.4) by z^k, we have,

$$(z^k, Mz^{k+1}) = (z^k, Mz^{k-1}) - \omega^{k+1}[\alpha_k(z^k, Az^k) - (z^k, Mz^k) + (z^k, Mz^{k-1})].$$

But, by the induction hypothesis $(z^k, Mz^{k-1}) = 0$, so

$$(z^k, Mz^{k+1}) = -\omega_{k+1}[\alpha_k(z^k, Az^k) - (z^k, Mz^k)],$$

which gives the result if $\omega_{k+1} \neq 0$. □

Lemma 6.3

If ω_{k+1} is chosen as

$$\omega_{k+1} = \frac{1}{1 + \alpha_k \frac{(z^{k-1}, Az^k)}{(z^{k-1}, Mz^{k-1})}}, \tag{6.5}$$

then, $(z^{k-1}, Mz^{k+1}) = 0$.

Proof. We multiply (6.4) by z^{k-1}. Then

$$(z^{k-1}, Mz^{k+1}) = (z^{k-1}, Mz^{k-1})$$
$$- \omega_{k+1}\left[\alpha_k(z^{k-1}, Az^k) - (z^{k-1}, Mz^k) + (z^{k-1}, Mz^{k-1})\right].$$

Therefore,

$$\omega_{k+1} = \frac{(z^{k-1}, Mz^{k-1})}{\alpha_k(z^{k-1}, Az^k) + (z^{k-1}, Mz^{k-1})}.$$

□

Note that if $z^k \neq 0$, then $(z^k, Az^k) \neq 0$, so α_k can always be computed. For ω_{k+1} defined by (6.5), this is not so obvious. We shall prove this in the next section. Moreover, so far we have not used the hypothesis that A and M are symmetric to compute the coefficients. Let us now give a cheaper computational expression for ω_{k+1}. Let us split A as $A = M - N$, then

$$\begin{aligned}(z^{k-1}, Az^k) &= (z^{k-1}, (M-N)z^k),\\ &= -(z^{k-1}, Nz^k).\end{aligned}$$

By writing (6.4) at iteration k,

$$Mz^k = Mz^{k-2} - \omega_k(\alpha_{k-1}(M-N)z^{k-1} - Mz^{k-1} + Mz^{k-2}).$$

Multiplying by z^k and using the induction hypothesis, we get

$$(z^k, Mz^k) = \omega_k\alpha_{k-1}(z^k, Nz^{k-1}).$$

But, N is a symmetric matrix (note here the hypothesis comes in) so,

$$(z^{k-1}, Nz^k) = \frac{(z^k, Mz^k)}{\omega_k\alpha_{k-1}}.$$

Therefore,

$$\omega_{k+1} = \frac{1}{1 - \frac{\alpha_k(z^k, Mz^k)}{\omega_k\alpha_{k-1}(z^{k-1}, Mz^{k-1})}}. \tag{6.6}$$

Formula (6.6) is more efficient than (6.5) as we do not have to compute the additional scalar product (z^{k-1}, Az^k). As the method (6.1) involves two levels of iterations, we need x^0 and x^{-1} to start with. This is overcome by taking $\omega_1 = 1$, then

$$x^1 = \alpha_0 z^0 + x^0,$$

$$Mz^0 = r^0,$$

and we need only to define x^0. The first step is only a steepest descent iteration. Now, we must show that the induction hypothesis about orthogonality holds at level $k+1$, that is the new vector is orthogonal not only to the last two, but to all the previous ones.

Theorem 6.4

The induction hypothesis holds at iteration $k+1$:

$$(z^{k+1}, Mz^j) = 0, \qquad 0 \le j < k-1.$$

Proof. Multiplying (6.4) by z^j, $0 \le j < k-1$, we have,

$$(z^j, Mz^{k+1}) = (z^j, Mz^{k-1}) \\ - \omega_{k+1}[\alpha_k(z^j, (M-N)z^k) - (z^j, Mz^k) + (z^j, Mz^{k-1})].$$

But, since $j < k-1$,

$$(z^j, Mz^{k+1}) = \omega_{k+1}\alpha_k(z^j, Nz^k).$$

Now, we use the same technique as before and writing (6.4) at iteration $j+1$, we get

$$Mz^{j+1} = Mz^{j-1} - \omega_{j+1}(\alpha_j(M-N)z^j - Mz^j + Mz^{j-1}).$$

Multiplying by z^k and taking into account that $j+1 < k$,

$$\omega_{j+1}\alpha_j(z^k, Nz^j) = 0.$$

This is where we need the hypothesis that N is symmetric to imply $(Nz^k, z^j) = 0$. This shows that $(z^j, Mz^{k+1}) = 0$ for all j such that $j < k-1$.
□

This method is known as the preconditioned conjugate gradient (PCG) method. The computational steps in PCG are:

Let x^0 be given for $k = 0, 1, \dots$ until convergence

$$\begin{aligned} Mz^k &= r^k (= b - Ax^k), \\ \alpha_k &= \frac{(z^k, Mz^k)}{(z^k, Az^k)}, \\ \omega_{k+1} &= \frac{1}{1 - \frac{\alpha_k}{\omega_k \alpha_{k-1}} \frac{(z^k, Mz^k)}{(z^{k-1}, Mz^{k-1})}}, \qquad \omega_1 = 1, \\ x^{k+1} &= x^{k-1} + \omega_{k+1}(\alpha_k z^k + x^k - x^{k-1}). \end{aligned} \tag{6.7}$$

Note that r^{k+1} is usually computed by (6.3) and not as $b - Ax^k$. Doing this, we only need to compute two scalar products and a matrix×vector product at each iteration plus $10n$ floating point operations.

So far, we have only shown that $z^n = 0$ and thus $x^n = x$. This particular form of the method has been popularized by Concus, Golub and O'Leary [343]. We shall see in the following sections that a certain norm of the error is decreasing as a function of the iteration number and therefore the method could be used as an iterative one. Before going to that point, let us consider some extensions and other forms of the method.

6.2. Generalization and second form of PCG

In (6.7), we use the Euclidean scalar product but the method can be easily generalized to any other scalar product. Let $((.,.))$ be such a scalar product. The algorithm can be formulated as

$$\begin{aligned}
Mz^k &= r^k = b - Ax^k, \\
\alpha_k &= \frac{((r^k, r^k))}{((AM^{-1}r^k, r^k))}, \\
\omega_{k+1} &= \frac{1}{1 - \frac{\alpha_k}{\omega_k \alpha_{k-1}} \frac{((r^k, r^k))}{((r^{k-1}, r^{k-1}))}}, \qquad \omega_1 = 1, \\
x^{k+1} &= x^{k-1} + \omega_{k+1}(\alpha_k z^k + x^k - x^{k-1}).
\end{aligned} \tag{6.8}$$

We have the following result that gives conditions for (6.8) to compute the solution.

Theorem 6.5

Let A and M be non-singular matrices and suppose AM^{-1} is self adjoint e.g., $((AM^{-1}x, y)) = ((x, AM^{-1}y))$, $\forall x, y$ and positive definite e.g., $((AM^{-1}x, x)) > 0$, $\forall x \neq 0$ with respect to the given scalar product. Then, using algorithm (6.8) we have $x^n = x$.

Proof. Similar to the proof of lemmas 6.2, 6.3 and theorem 6.4. □

Note that in theorem 6.5, the matrix A is not assumed to be symmetric. Supposing M to be symmetric and using $((x, y)) = (M^{-1}x, y)$ the method

(6.7) is recovered. In the previous section, PCG was constructed as a formal generalization of acceleration techniques and like an orthogonalization method. However, PCG can also be viewed as a minimization method. For the sake of simplicity, let us take $M = I$. Then, as the matrix A is assumed to be symmetric, it is well known that finding the solution x of $Ax = b$ is equivalent to minimizing the functional

$$\varphi(x) = \frac{1}{2}(Ax, x) - (x, b).$$

It is easy to see that the gradient of φ is $\delta\varphi(x) = Ax - b$ (this is minus the residual). Let $\{p^1, p^2, ...\}$ be a sequence of vectors and suppose that we already have an approximation x^k for the exact solution x. Then, to improve the approximation, we may try to minimize φ in the direction p^k e.g., we look for γ minimizing $\varphi(x^k + \gamma p^k)$, where $\gamma \in \mathbb{R}$.

Lemma 6.6

The minimum of $\varphi(x^k + \gamma p^k)$ as a function of γ is given by

$$\gamma = \gamma_k = \frac{(p^k, r^{k-1})}{(p^k, Ap^k)}, \quad r^k = b - Ax^k.$$

Proof.

$$\begin{aligned}\varphi(x^k + \gamma p^k) &= \frac{1}{2}(x^k + \gamma p^k, Ax^k + \gamma Ap^k) - (x^k + \gamma p^k, b), \\ &= \varphi(x^k) + \frac{1}{2}\gamma^2(p^k, Ap^k) + \gamma(x^k, Ap^k) - \gamma(p^k, b), \\ &= \varphi(x^k) + \frac{1}{2}\gamma^2(p^k, Ap^k) - \gamma(r^k, p^k),\end{aligned}$$

where $r^k = b - Ax^k$. Therefore, we look at the minimum of this quadratic functional (as a function of γ) and we get

$$\gamma_k = \frac{(p^k, r^k)}{(p^k, Ap^k)}.$$

□

Let $x^{k+1} = x^k + \gamma_k p^k$. Then,

$$\varphi(x^{k+1}) = \varphi(x^k) - \frac{1}{2}\frac{(p^k, r^k)^2}{(p^k, Ap^k)}.$$

This implies that if we want to obtain a decrease in φ, we must have $(p^k, r^k) \neq 0$. So far we have just done a local minimization. The problem we need to solve is the choice of p^k to achieve the global minimization of φ. Solution to this problem can be found for instance in Golub-Van Loan [676]. If the search directions are independent and $x^{k+1} = x^0 + \text{span}(p^0, \dots, p^k)$, the functional is globally minimized. Let us write

$$x^{k+1} = x^0 + P_{k-1}y + \gamma p^k,$$

where the matrix $P_{k-1} = [p^0, \dots, p^{k-1}]$. By putting this in the functional, we have

$$\varphi(x^{k+1}) = \varphi(x^0 + P_{k-1}y) + \frac{\gamma^2}{2}(p^k, Ap^k) - \gamma(p^k, r^0),$$

if p^k is orthogonal to $\text{span}(Ap^0, \dots, Ap^{k-1})$. Then, the minimization over y and γ are decoupled. The solution of the problem for γ was given in lemma 6.6. To construct these A–orthogonal directions, see [676]. Finally, the algorithm is

Let x^0 be given, $r^0 = b - Ax^0$. For $k = 0, 1, \dots$

$$\begin{aligned}
Mz^k &= r^k, \\
\beta_k &= \frac{(z^k, Mz^k)}{(z^{k-1}, Mz^{k-1})}, \quad \beta_0 = 0, \\
p^k &= z^k + \beta_k p^{k-1}, \\
\gamma_k &= \frac{(z^k, Mz^k)}{(p^k, Ap^k)}, \\
x^{k+1} &= x^k + \gamma_k p^k, \\
r^{k+1} &= r^k - \gamma_k Ap^k.
\end{aligned} \tag{6.9}$$

We note that we need to compute two scalar products, a matrix–vector product plus $6n$ floating point operations at each iteration. We shall show that the two expressions of γ_k are equivalent by eliminating p^k in (6.9). This minimization method has very interesting properties.

Proposition 6.7

The vectors generated by the method (6.9) are such that

1) $(p^i, Ap^j) = 0, \ i \neq j,$

2) $(z^i, Mz^j) = 0, \ i \neq j,$

3) $(p^i, r^j) = 0,\ i < j.$

Proof. The proof is by induction. Suppose that the properties holds for $i, j \le k-1$,

$$r^k = r^{k-1} - \gamma_{k-1} A p^{k-1}.$$

Multiplying by p^i,

$$(r^k, p^i) = (r^{k-1}, p^i) - \gamma_{k-1}(Ap^{k-1}, p^i).$$

If $i < k-1$ then $(r^{k-1}, p^i) = 0$ and $(Ap^{k-1}, p^i) = 0$ gives $(r^k, p^i) = 0$. If $i = k-1$ then

$$\gamma_{k-1} = \frac{(r^{k-1}, p^{k-1})}{(Ap^{k-1}, p^{k-1})},$$

so $(r^k, p^{k-1}) = 0$. Therefore, $(r^k, p^i) = 0,\ i \le k-1$ proving the third property. Now, we have

$$p^i = z^i + \beta_i p^{i-1},$$

with $i \le k-1$. Multiplying by r^k gives

$$(p^i, r^k) = (z^i, r^k) + \beta_i (p^{i-1}, r^k).$$

But, $(p^i, r^k) = 0$ and $(p^{i-1}, r^k) = 0$, giving $(z^i, r^k) = 0$ for $i \le k-1$. This proves that $(z^i, Mz^k) = 0,\ i < k$. Multiplying the definition of p^i by r^i it is obvious that $(p^i, r^i) = (z^i, r^i)$. Since $p^k = z^k + \beta_k p^{k-1}$, multiplying by r^{k-1} yields

$$\beta_k = \frac{(p^k, r^{k-1})}{(z^{k-1}, r^{k-1})}.$$

Therefore,

$$(p^k, r^{k-1}) = (r^k, z^k) = (p^k, r^k) \Longrightarrow (p^k, r^{k-1} - r^k) = 0.$$

But $r^k = r^{k-1} - \gamma_k A p^{k-1}$. This shows that $(p^{k-1}, Ap^k) = 0$. □

The most important fact to note is that the method (6.9) is equivalent to the method (6.7).

Theorem 6.8

Methods (6.7) and (6.9) generate the same iterates (in exact arithmetic).

Proof. This is obvious for $k = 0$. Then, for $k > 0$ we can eliminate p^k from (6.9) as

$$p^k = \frac{1}{\gamma_k}(x^{k+1} - x^k).$$

Therefore,

$$\frac{1}{\gamma_k}(x^{k+1} - x^k) = z^k + \frac{\beta_k}{\gamma_{k-1}}(x^k - x^{k-1}).$$

A short manipulation gives

$$x^{k+1} = x^{k-1} + \gamma_k z^k + \left(1 + \frac{\gamma_k \beta_k}{\gamma_{k-1}}\right)(x^k - x^{k-1}).$$

So, we can identify both expressions for x^{k+1},

$$\omega_{k+1} = 1 + \frac{\gamma_k \beta_k}{\gamma_{k-1}},$$

$$\alpha_k = \frac{\gamma_k}{1 + \frac{\gamma_k \beta_k}{\gamma_{k-1}}}.$$

With this expression for ω_{k+1}, since $\gamma_k > 0$ and $\beta_k > 0$, it follows that $\omega_{k+1} > 1$ and it is always well defined. □

Method (6.9) is usually preferred computationally (as it is less expensive) although, as we shall see, method (6.7) is of value on parallel computers.

6.3. Optimality of PCG

In this section we shall show that PCG is optimal in a certain sense. Let us use the method in the form (6.7). Recall that

$$Mz^{k+1} = Mz^{k-1} - \omega_{k+1}(\alpha_k A z^k - Mz^k + Mz^{k-1}).$$

Since $A = M - N$, this gives

$$z^{k+1} = z^{k-1} - \omega_{k+1}(\alpha_k(I - M^{-1}N)z^k - z^k + z^{k-1}).$$

Then we have

Lemma 6.9

z^{k+1} is a polynomial in K,

$$z^{k+1} = [I - KP_k(K)]z^0,$$

where $K = M^{-1}A = I - M^{-1}N$ and P_k is a kth degree polynomial satisfying

$$P_k(\lambda) = \alpha_k \omega_{k+1} + \omega_{k+1}(1 - \alpha_k \lambda)P_{k-1}(\lambda) - (\omega_{k+1} - 1)P_{k-2}(\lambda),$$
$$P_0(\lambda) = \alpha_0,$$
$$P_1(\lambda) = \omega_2(\alpha_0 + \alpha_1 - \alpha_0 \alpha_1 \lambda).$$

Proof. The proof is straightforward by induction on k. □

Proposition 6.10

Let P_k be the polynomial defined in lemma 6.9. The iterates of PCG *satisfy*

$$x^{k+1} = x^0 + P_k(K)z^0.$$

Proof. We have

$$x^{k+1} = x^{k-1} + \omega_{k+1}(\alpha_k z^k + x^k - x^{k-1}).$$

By induction and with the help of lemma 6.9 this is written as

$$x^{k+1} = x^0 + P_{k-2}(K)z^0$$
$$+ \omega_{k+1}\left(\alpha_k[I - KP_{k-1}(K)]z^0 + P_{k-1}(K)z^0 - P_{k-2}(K)z^0\right),$$

$$x^{k+1} = x^0 + P_k(K)z^0,$$

because of the recurrence relation satisfied by P_k. □

The matrix A being symmetric positive definite, let us introduce the measure of the error in the energy norm.

$$E(x^{k+1}) = (A(x - x^{k+1}), x - x^{k+1}) = (r^{k+1}, A^{-1}r^{k+1}) = \|r^{k+1}\|^2_{A^{-1}}.$$

Then the main result is the following.

Theorem 6.11

Consider all the iterative methods that can be written as

$$\overline{x}^{k+1} = \overline{x}^0 + Q_k(K)\overline{z}^0, \quad \overline{x}^0 = x^0, \quad M\overline{z}^0 = b - A\overline{x}^0, \tag{6.10}$$

where Q_k is a kth degree polynomial. Of all these methods, PCG *is the one which minimizes $E(\overline{x}^k)$ at each iteration.*

Proof. The proof of this very important result requires a few lemmas. Denote the polynomial Q_k as

$$Q_k(\lambda) = \sum_{j=0}^{k} \beta_j^{k+1} \lambda^j.$$

First, we shall show that the coefficients β_j^{k+1} which minimize $E(\overline{x}^{k+1})$ are uniquely defined and how to compute them. By (6.10),

$$\overline{r}^{k+1} = \overline{r}^0 - A \sum_{j=0}^{k} \beta_j^{k+1} K^j \overline{z}^0.$$

Hence, if $M\overline{z}^k = \overline{r}^k$,

$$\overline{z}^{k+1} = \overline{z}^0 - K \sum_{j=0}^{k} \beta_j^{k+1} K^j \overline{z}^0.$$

If we denote $\overline{u}^j = K^j \overline{z}^0$, then

$$E(\overline{x}^{k+1}) = (M\overline{z}^0 - \sum_{j=0}^{k} \beta_j^{k+1} A\overline{u}^j, K^{-1}\overline{z}^0 - \sum_{l=0}^{k} \beta_l^{k+1} \overline{u}^l) \tag{6.11}$$

Let us assume the following induction hypothesis:

$$\begin{aligned}
&\forall j \le k, \quad \forall l < k, \; l \ne j, \qquad (M\overline{z}^j, \overline{u}^l) = 0,\\
&\forall j \le k, \quad \forall l \le k, \; l \ne j, \quad (\overline{z}^j, M\overline{z}^l) = 0,\\
&\forall j \le k, \quad \overline{r}^j \ne 0.
\end{aligned}$$

Lemma 6.12

The vectors $\overline{u}^j$, $0 \leq j \leq k$ are linearly independent.

Proof. Suppose there exist $\alpha_j \in \mathbb{R}$, $0 \leq j \leq k$ such that

$$\sum_{j=0}^{k} \alpha_j \overline{u}^j = 0.$$

Multiplying by $M\overline{z}^k$, we get $\alpha_k(M\overline{z}^k, \overline{u}^k) = 0$. But, $\overline{z}^k = \overline{z}^0 - \sum_{j=0}^{k-1} \beta_j^k \overline{u}^{j+1}$ and

$$(M\overline{z}^k, \overline{z}^k) = -\beta_{k-1}^k (M\overline{z}^k, \overline{u}^k).$$

As $(M\overline{z}^k, \overline{z}^k) \neq 0$, $\beta_{k-1}^k \neq 0$ and $(M\overline{z}^k, \overline{u}^k) \neq 0$. Therefore, $\alpha_k = 0$. To show that the other coefficients are zero, we multiply by $M\overline{z}^{k-j}$, $j = 1, \ldots, k$. □

Lemma 6.13

The coefficients β_j^{k+1} which yields the minimum of $E(\overline{x}^{k+1})$ are uniquely defined.

Proof. Expanding (6.11) gives

$$E(\overline{x}^{k+1}) = (M\overline{z}^0, K^{-1}\overline{z}^0) - 2\sum_{j=0}^{k} \beta_j^{k+1}(\overline{u}^j, M\overline{z}^0) + \sum_{j=0}^{k}\sum_{l=0}^{k} \beta_j^{k+1}\beta_l^{k+1}(A\overline{u}^j, \overline{u}^l).$$

Denote by G the matrix with entries $g_{j,l} = (A\overline{u}^j, \overline{u}^l)$ and by h the vector whose components are $h_j = (\overline{u}^j, M\overline{z}^0)$, β^{k+1} the vector with $(\beta^{k+1})_j = \beta_j^{k+1}$. Then, with these notations we can rewrite

$$E(\overline{x}^{k+1}) = (M\overline{z}^0, K^{-1}\overline{z}^0) - 2(\beta^{k+1}, h) + (G\beta^{k+1}, \beta^{k+1}).$$

We are looking for β^{k+1} minimizing $E(\overline{x}^{k+1})$. This is a quadratic functional and we have already seen that the solution is given by solving

$$G\beta^{k+1} = h.$$

Let us show that G is positive definite which will imply that the minimum is unique. We have

$$(G\gamma, \gamma) = \sum_{j=0}^{k}\sum_{l=0}^{k} \gamma_j \gamma_l (A\bar{u}^j \bar{u}^l) - \left(A\left(\sum_{j=0}^{k} \gamma_j \bar{u}^j\right), \sum_{l=0}^{k} \gamma_l \bar{u}^l\right).$$

Since A is positive definite, G is positive definite because if $\sum_{j=0}^{k} \gamma_j \bar{u}^j = 0$, $\gamma_j = 0$ by lemma 6.12 and therefore $\gamma = 0$. □

Now, we must show that the induction hypothesis holds at level $k+1$. We have,

$$\sum_{l=0}^{k} (A\bar{u}^j, \bar{u}^l)\beta_l^{k+1} = (\bar{u}^j, M\bar{z}^o),$$

because this equation is just the jth row of $G\beta^{k+1} = h$. So,

$$(M\bar{z}^o - \sum_{l=0}^{k} \beta_l^{k+1} A\bar{u}^l, \bar{u}^j) = 0.$$

That is to say, $(M\bar{z}^{k+1}, \bar{u}^j) = 0 \quad \forall j \leq k$. Moreover, $\forall j \leq k$,

$$(\bar{z}^{k+1}, M\bar{z}^j) = (M\bar{z}^{k+1}, \bar{z}^j) = (M\bar{z}^{k+1}, \bar{z}^0 - \sum_{l=0}^{j-1} \beta_l^j \bar{u}^{l+1}) = 0.$$

If $\bar{r}^{k+1} = 0$, we already get the solution. Now that we have constructed an optimal method we shall show that it is similar to PCG.

Proof of theorem 6.11

The proof is by induction. We have $\bar{x}^0 = x^0$, suppose that $\bar{x}^j = x^j$, $j \leq k$ By proposition 6.10, PCG gives

$$z^{k+1} = z^0 - \sum_{j=0}^{k} \delta_j^{k+1} K u^j, \quad u^j = K^j z^0,$$

and (6.10) gives

$$\bar{z}^{k+1} = z^0 - \sum_{j=0}^{k} \beta_j^{k+1} K u^j,$$

since $\bar{u}^j = u^j$. Then,

$$\bar{z}^{k+1} - z^{k+1} = -\sum_{j=0}^{k} (\beta_j^{k+1} - \delta_j^{k+1}) K u^j.$$

Since the z^j's are orthogonal, they are linearly independent and we have span $(u^0, \ldots, u^k) =$ span $(z^0, \ldots, z^k)$. Then, it is obvious that

$$K^{-1}(\bar{z}^{k+1} - z^{k+1}) = \sum_{j=0}^{k} c_j z^j.$$

By the properties of both methods,

$(M(\bar{z}^{k+1} - z^{k+1}), z^j) = 0, \quad \forall j \le k.$

So,

$$(M(\bar{z}^{k+1} - z^{k+1}), \sum_{j=0}^{k} c_j z^j) = 0,$$

$(M(\bar{z}^{k+1} - z^{k+1}), K^{-1}(\bar{z}^{k+1} - z^{k+1})) = 0.$

Since $Mz^k = r^k$, we have

$(\bar{r}^{k+1} - r^{k+1}, A^{-1}(\bar{r}^{k+1} - r^{k+1})) = 0.$

But A^{-1} is positive definite so $\bar{r}^{k+1} = r^{k+1}$. This proves theorem 6.11. □

The proof of theorem 6.11 could have been done using the form (6.9) of PCG. Recall that

$\varphi(x^k) = \frac{1}{2}(Ax^k, x^k) - (x^k, b).$

But,

$$E(x^k) = (b - Ax^k, A^{-1}b - x^k) = (b, A^{-1}b) - 2(b, x^k) + (x^k, Ax^k),$$
$$= (b, A^{-1}b) + 2\varphi(x^k).$$

So, if x^k minimizes φ, it does the same for E. We note that even when preconditioned, PCG minimizes the A–norm of the error.

6.4. The convergence rate of PCG

Denote the error by $\varepsilon^k = x - x^k$. Then, $A\varepsilon^k = b - Ax^k = r^k = Mz^k$, hence $K\varepsilon^k = z^k \; \forall k$, and with lemma 6.9

$\varepsilon^{k+1} = [I - KP_k(K)]\varepsilon^0,$

because,

$K[I - KP_k(K)] = [I - KP_k(K)]K.$

The matrix $K = M^{-1}A$ is not symmetric, but we have the following result.

Lemma 6.14

The eigenvalues λ_i of K are real and positive.

Proof. M being positive definite, $M^{\frac{1}{2}}$ is well defined and K is similar to $\hat{K} = M^{1/2}KM^{-1/2} = M^{-1/2}AM^{-1/2}$. □

The matrix $\hat{K}$ is symmetric positive definite by theorem 1.21. There exists an orthogonal matrix Q and a diagonal matrix Λ, $(\Lambda)_{i,i} = \lambda_i > 0$, such that

$$\hat{K} = Q\Lambda Q^T, \quad Q^TQ = I.$$

Theorem 6.15

Let

$$\bar{\varepsilon}^j = \Lambda^{1/2}Q^TM^{1/2}\varepsilon^j, \quad \forall j.$$

Then

$$E(x^{k+1}) = \sum_{i=1}^{n}(1 - \lambda_i P_k(\lambda_i))^2(\bar{\varepsilon}_i^0)^2, \tag{6.12}$$

where the λ_i's are the eigenvalues of K.

Proof.

$$\varepsilon^{k+1} = [I - KP_k(K)]\varepsilon^0,$$

and

$$P_k(K) = M^{-\frac{1}{2}}P_k(\widehat{K})M^{\frac{1}{2}}.$$

Therefore,

$$\varepsilon^{k+1} = M^{-\frac{1}{2}}[I - \widehat{K}P_k(\widehat{K})]M^{\frac{1}{2}}\varepsilon^0.$$

Since $P_k(\widehat{K}) = QP_k(\Lambda)Q^T$,

$$\varepsilon^{k+1} = M^{-\frac{1}{2}}Q\Lambda^{-\frac{1}{2}}[I - \Lambda P_k(\Lambda)]\Lambda^{\frac{1}{2}}Q^TM^{\frac{1}{2}}\varepsilon^0.$$

With the notation we introduced, this is simply
$\bar{\varepsilon}^{k+1} = [I - \Lambda P_k(\Lambda)]\bar{\varepsilon}^0$.
Let us compute $(\bar{\varepsilon}^{k+1}, \bar{\varepsilon}^{k+1})$. By definition

$$(\bar{\varepsilon}^{k+1}, \bar{\varepsilon}^{k+1}) = (\Lambda^{\frac{1}{2}} Q^T M^{\frac{1}{2}} \varepsilon^{k+1}, \Lambda^{\frac{1}{2}} Q^T M^{\frac{1}{2}} \varepsilon^{k+1}),$$
$$= (M^{\frac{1}{2}} \widehat{K} M^{\frac{1}{2}} \varepsilon^{k+1}, \varepsilon^{k+1}) = (A\varepsilon^{k+1}, \varepsilon^{k+1}).$$

Therefore,

$$([I - \Lambda P_k(\Lambda)]\bar{\varepsilon}^0, [I - \Lambda P_k(\Lambda)]\bar{\varepsilon}^0) = E(x^{k+1}).$$

□

Theorem 6.15 is important because it allows us to see PCG as an iterative method. The polynomial P_k characterizes the method. Note that for $k = 0, \ldots, n$ it defines a family of orthogonal polynomials. From this expression, we can obtain bounds on the A–norm of the error.

Theorem 6.16

$$E(x^{k+1}) \leq \max_{1 \leq i \leq n} (R_{k+1}(\lambda_i))^2 E(x^0), \tag{6.13}$$

for all polynomials R_{k+1} of degree $k+1$ such that $R_{k+1}(0) = 1$

Proof. In theorem 6.11, we show that the PCG polynomial P_k minimizes $E(x^{k+1})$. Therefore, replacing the polynomial P_k in (6.12) by any other kth degree polynomial, we shall get a greater value. This can be written as

$$E(x^{k+1}) \leq \sum_{i=1}^{n} (R_{k+1}(\lambda_i))^2 (\bar{\varepsilon}_i^0)^2,$$

for all polynomials R_{k+1} of degree $k+1$, such that $R_{k+1}(0) = 1$, equality holding only if $R_{k+1}(\lambda) = 1 - \lambda P_k(\lambda)$. Therefore,

$$E(x^{k+1}) \leq \max_{1 \leq i \leq n} (R_{k+1}(\lambda_i))^2 \sum_{i=1}^{n} (\bar{\varepsilon}_i^0)^2.$$

But, we have already noted that $(\bar{\varepsilon}^0, \bar{\varepsilon}^0) = E(x^0)$, which proves the result. □

Theorem 6.16 has many interesting consequences since we are free to choose the polynomial R_{k+1}.

Proposition 6.17

1) $E(x^n) = 0$,

2) If K *has only* p *distinct eigenvalues then,* $E(x^p) = 0$.

Proof. The first item is the finite termination property (in exact arithmetic) that was proven in lemma 6.1. To prove this again, we choose

$$R_k(\lambda) = \prod_{i=1}^{k} \left(1 - \frac{\lambda}{\lambda_i}\right).$$

Hence, $R_n(\lambda_i) = 0$, $\forall i$, $1 \le i \le n$, so $E(x^n) = 0$. To prove the other assertion, we simply take into account the distinct eigenvalues in R_k and the result follows. □

The next result is the most well known bound on the A–norm of the error. It uses the condition number of K.

Theorem 6.18

$$E(x^k) \le 4(\frac{\sqrt{\kappa} - 1}{\sqrt{\kappa} + 1})^{2k} E(x^0),$$

where $\kappa = \frac{\lambda_{\max}(K)}{\lambda_{\min}(K)}$ *is the condition number of* K, $\lambda_{\max}$ *(resp.* $\lambda_{\min}$*) being the greatest (resp. smallest) eigenvalue of* K.

Proof. $\max_{1 \le i \le n}(R_k(\lambda_i))^2$ is bounded by $\max_{\lambda_{\min} \le \lambda \le \lambda_{\max}}(R_k(\lambda))^2$. For R_k we choose the kth degree polynomial such that $R_k(0) = 1$, which minimizes $\max_{\lambda_{\min} \le \lambda \le \lambda_{\max}}(R_k(\lambda))^2$. The solution to this problem was given in Chapter 1, theorem 1.64:

$$R_k(\lambda) = \frac{T_k(\frac{\lambda_{\min} + \lambda_{\max} - 2\lambda}{\lambda_{\max} - \lambda_{\min}})}{T_k(\frac{\lambda_{\min} + \lambda_{\max}}{\lambda_{\max} - \lambda_{\min}})}.$$

where T_k are the Chebyshev polynomials. By theorem 1.64,

$$\max_{\lambda_{\min} \le \lambda \le \lambda_{\max}} |R_k(\lambda)| \le 2\left(\frac{\sqrt{\kappa} - 1}{\sqrt{\kappa} + 1}\right)^k.$$

This proves the theorem. □

Therefore, $E(x^k)$ is bounded above by a decreasing sequence that converges to 0. As k tends to infinity, $E(x^k)$ which is the square of the A–norm of the error, tends to 0. Moreover the decrease is monotone. This explains why the method can be regarded as an iterative one. In practical problems (with suitable preconditioning) one usually gets a very good approximation to the solution in much fewer than n iterations, see figure 6.1. Note that the closer κ is to one, the faster the convergence of the method. Therefore, as a first recommendation, we can try to construct the preconditioner M such that κ is as close as possible to one. However, we have seen that the distribution of the whole spectrum influences the convergence of the method. This will give us also hints for the choice of the preconditioner.

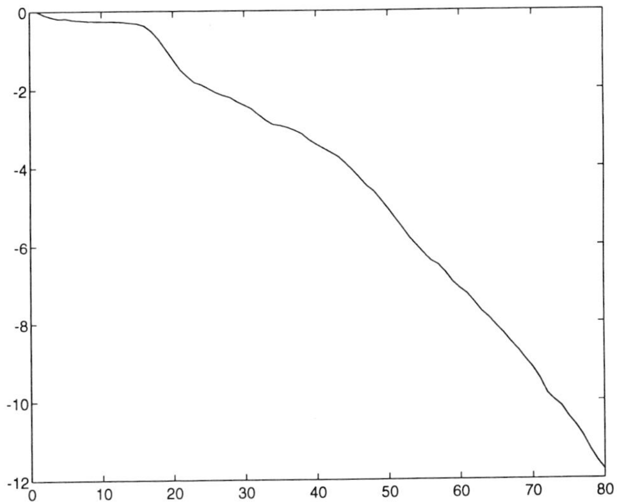

fig. 6.1: $\log_{10}$ *of the maximum norm of the error with* CG *for the Poisson problem on a* 20×20 *mesh*

That the convergence depends on the distribution of all the eigenvalues was already obvious in theorem 6.15, but now we are able to obtain bounds for some particular distributions. The following result was proven by Axelsson [60], [78].

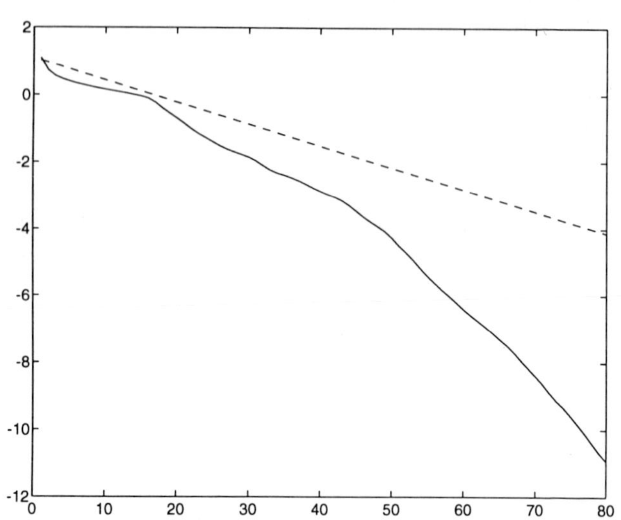

fig. 6.2: $\log_{10}$ *of the A–norm of the error with* CG
for the Poisson problem on a 20×20 *mesh and bound of theorem 6.18*

Proposition 6.19

Suppose

$$a \leq \lambda_1 \leq \lambda_2 \leq \cdots \leq \lambda_{n-m} \leq b \leq \lambda_{n-m+1} \leq \cdots \leq \lambda_n,$$

and let $k \geq m$*, then*

$$E(x^k) \leq 4\left(\frac{\sqrt{\frac{b}{a}}-1}{\sqrt{\frac{b}{a}}+1}\right)^{2(k-m)}.$$

Proof. In (6.13), we choose

$$R_k(\lambda) = \prod_{i=1}^{m}\left(1 - \frac{\lambda}{\lambda_{n-i+1}}\right)\frac{T_{k-m}\left(\frac{a+b-2\lambda}{b-a}\right)}{T_{k-m}\left(\frac{a+b}{b-a}\right)}.$$

Obviously $R_k(\lambda_i) = 0$ for all λ_i, $i = n, n-1, \ldots, n-m+1$, hence

$$\max_i (R_k(\lambda_i))^2 = \max_{i=1,\ldots,n-m} (R_k(\lambda_i))^2 \leq \max_{a \leq \lambda \leq b} (R_k(\lambda))^2.$$

Note that when $\lambda \in [a, b]$,

$$\left|1 - \frac{\lambda}{\lambda_{n-i+1}}\right| < 1, \quad \forall i = 1, \ldots m.$$

Therefore,

$$\max_{a \le \lambda \le b} (R_k(\lambda))^2 \le \max_{a \le \lambda \le b} \left| \frac{T_{k-m}\left(\frac{a+b-2\lambda}{b-a}\right)}{T_{k-m}\left(\frac{a+b}{b-a}\right)} \right|,$$

and the result follows. □

This result is useful when a few of the largest eigenvalues are well separated from the others. In this case, m is small, so we do not lose too much in the exponent and $\frac{b}{a}$ can be much less than $\kappa = \frac{\lambda_n}{\lambda_1}$. The ratio $\frac{b}{a}$ appears as an effective condition number. However, this distribution does not occur too often in problems arising from the discretization of partial differential equations. But it can be useful when some special preconditioners (the so called modified factorizations) are used.

If we have a few of the smallest eigenvalues separated from the others, we can try to obtain the same kind of bounds. Let

$$0 < \lambda_1 \le \lambda_2 \le \cdots \le \lambda_p < a \le \lambda_{p+1} \le \cdots \le \lambda_n < b.$$

Then,

$$E(x^k) < 4 \max_{a \le \lambda \le b} \left\{ \prod_{j=1}^{p} |1 - \frac{\lambda}{\lambda_j}|^2 \right\} \left(\frac{\sqrt{\frac{b}{a}} - 1}{\sqrt{\frac{b}{a}} + 1} \right)^{2(k-p)} E(x^0).$$

Unfortunately for $\lambda \in [a, b]$, $\prod_{j=1}^{p} |1 - \frac{\lambda}{\lambda_j}|$ can be large. So, this result does not give useful bounds. H. Van der Vorst and A. Van der Sluis [1275] studied the case of the smallest eigenvalues. They proved the following result. Let us suppose first that we have one isolated eigenvalue.

Theorem 6.20

Suppose $0 < \lambda_1 < a \le \lambda_2 \le \cdots \le \lambda_n \le b$, *and let* l *be an integer* $l \le k$ *which will be chosen later on. Then,*

$$E(x^k) \le 4 \left[\frac{1}{T_l(\frac{\beta}{\alpha})} \right]^2 \left(\frac{\sqrt{\frac{b}{a}} - 1}{\sqrt{\frac{b}{a}} + 1} \right)^{2(k-l)} E(x^0),$$

with $\beta = b - \alpha$,

$$\alpha = \frac{b - \lambda_1}{1 + \cos(\frac{\pi}{2l})}.$$

Proof. For this proof we are looking for a polynomial Q_l of degree l for which λ_1 is a root, with $Q_l(0) = 1$ and whose maximum over $[\lambda_1, b]$ is small. From Chapter 1, we can see that the solution is given by

$$Q_l(\lambda) = \frac{T_l(\frac{\beta-\lambda}{\alpha})}{T_l(\frac{\beta}{\alpha})},$$

with $\beta = b - \alpha$ and $\alpha = \frac{b-\lambda_1}{1+\cos(\frac{\pi}{2l})}$. Since

$$\beta = \frac{b\cos(\frac{\pi}{2l}) + \lambda_1}{1 + \cos(\frac{\pi}{2l})}, \qquad \frac{\beta - \lambda}{\alpha} = \frac{\cos(\frac{\pi}{2l})(b - \lambda) + \lambda_1 - \lambda}{b - \lambda_1},$$

we have

$$Q_l(\lambda_1) = \frac{T_l\left(\cos(\frac{\pi}{2l})\right)}{T_l(\frac{\beta}{\alpha})} = 0.$$

Let us take as a polynomial,

$$R_k(\lambda) = Q_l(\lambda)\frac{T_{k-l}\left(\frac{a+b-2\lambda}{b-a}\right)}{T_{k-l}\left(\frac{a+b}{b-a}\right)}.$$

We are interested in $\max_i(R_k(\lambda_i))^2$. Clearly $R_k(\lambda_1) = 0$. So,

$$\max_i(R_k(\lambda_i))^2 \le \max_{\lambda\in[\lambda_1,b]}(Q_l(\lambda))^2\left(\frac{\sqrt{\frac{b}{a}} - 1}{\sqrt{\frac{b}{a}} + 1}\right)^{2(k-l)}$$

But,

$$\max_{\lambda\in[\lambda_1,b]}|Q_l(\lambda)| \le \frac{1}{T_l(\frac{\beta}{\alpha})}.$$

□

To see if we have made an improvement, we need a bound for $\frac{1}{T_l(\frac{\beta}{\alpha})}$. We shall suppose that λ_1 is much less than b ($a \ll b$)

Proposition 6.21

When $\lambda_1 \ll b$, $\frac{1}{T_l(\frac{\beta}{\alpha})}$ *is of the order of* $(\frac{b}{\lambda_1})^{\frac{1}{l}}$.

Proof. Let $\rho = \frac{\beta}{\alpha} = \frac{\cos\theta + \frac{\lambda_1}{b}}{1 - \frac{\lambda_1}{b}}$ where $\theta = \frac{\pi}{2l}$. Then, with our hypothesis,

$$\rho = \frac{\lambda_1}{b} + (1 + \frac{\lambda_1}{b})\cos\theta + O((\frac{\lambda_1}{b})^2) = \cos\theta + O(\frac{\lambda_1}{b}).$$

So,

$$T_l(\rho) = T_l(\cos\theta) + (\rho - \cos\theta)T_l'(\cos\theta) + O((\frac{\lambda_1}{b})^2).$$

But,

$$\rho - \cos\theta = \frac{\lambda_1}{b}(1 + \cos\theta) + O((\frac{\lambda_1}{b})^2),$$

and $T_l(\cos\theta) = 0$. □

From the properties of the Chebyshev polynomials, it is easy to check that,

$$(1 - \cos^2\theta)T_l'(\cos\theta) = lT_{l-1}(\cos\theta),$$

so

$$T_l'(\cos\theta) = \frac{lT_{l-1}(\cos\theta)}{\sin^2\theta},$$

and

$$T_l(\rho) = \frac{\lambda_1}{b}\frac{1 + \cos\theta}{\sin^2\theta}lT_{l-1}(\cos\theta) + O((\frac{\lambda_1}{b})^2).$$

But,

$$\begin{aligned} T_{l-1}(\cos\theta) &= \cos((l-1)\arccos(\cos\theta) = \cos((l-1)\theta), \\ &= \cos\left(\tfrac{l-1}{l}\tfrac{\pi}{2}\right) = \sin\left(\frac{\pi}{2l}\right) = \sin\theta. \end{aligned}$$

Hence,

$$T_l(\rho) = \frac{\lambda_1}{b}\frac{1 + \cos\theta}{\sin\theta}l + O((\frac{\lambda_1}{b})^2).$$

Since $0 \le \frac{\sin\theta}{1+\cos\theta} \le 1$,

$$E(x^k) \le 4\left[\left(\frac{b}{\lambda_1}\right)^2 \frac{1}{l^2} + O\left(\left(\frac{\lambda_1}{b}\right)^2\right)\right]\left(\frac{\sqrt{\frac{b}{a}}-1}{\sqrt{\frac{b}{a}}+1}\right)^{2(k-l)} E(x^0).$$

This gives us an idea of what the bound looks like when the condition number $\frac{b}{\lambda_1}$ is large. It is easily seen that if we allow l to be a real number, the value of l that minimizes the upper bound is $\frac{1}{2}\sqrt{\frac{b}{a}}$ when $b \gg a$. So, this suggests we take l as the closest integer to $\frac{1}{2}\sqrt{\frac{b}{a}}$. With this value, the upper bound is almost

$$16\frac{ab}{\lambda_1^2}\left(\frac{\sqrt{\frac{b}{a}}-1}{\sqrt{\frac{b}{a}}+1}\right)^{2(k-l)} E(x^0).$$

Note that $\frac{a}{\lambda_1}$ is not necessarily very large. The important thing to note is that the convergence rate depends on the separation of the two first eigenvalues. This result can be generalized to the p smallest eigenvalues. The following theorem is due to H. Van der Vorst.

Theorem 6.22

Let

$$0 < \lambda_1 < \lambda_2 < \cdots < \lambda_p \le a \le \lambda_{p+1} \le \cdots \le \lambda_n \le b$$

and let l_j be integers such that $l = \sum_{j=1}^{p} l_j \le k$. Then,

$$E(x^k) \le 4\left(\frac{1}{\prod_{j=1}^{p} T_{l_j}(\rho_j)}\right)^2 \left(\frac{\sqrt{\frac{b}{a}}-1}{\sqrt{\frac{b}{a}}+1}\right)^{2(k-l)} E(x^0),$$

where,

$$\rho_j = \frac{b\cos\frac{\pi}{2l_j} + \lambda_j}{b - \lambda_j}.$$

□

A detailed study of the case of isolated eigenvalues has also been done by Axelsson and Lindskog [78]. These bounds are interesting to understand how PCG works but they are of little help to predict the number of iterations. We shall see later on that one can compute numerically bounds of the norm of the error.

Figure 6.3 shows the eigenvalue distributions of four matrices of order 30. The one at the top (a) is uniform between 0.1 and 100. The second one (b) has one isolated small eigenvalue, 100 has multiplicity five. The third one (c) has one isolated large eigenvalue and 0.1 has multiplicity five. The last one (d) is a combination of (b) and (c). Figure 6.4 shows the log_{10} of the A–norm of the error for this four matrices. One can see that the convergence is faster with isolated eigenvalues.

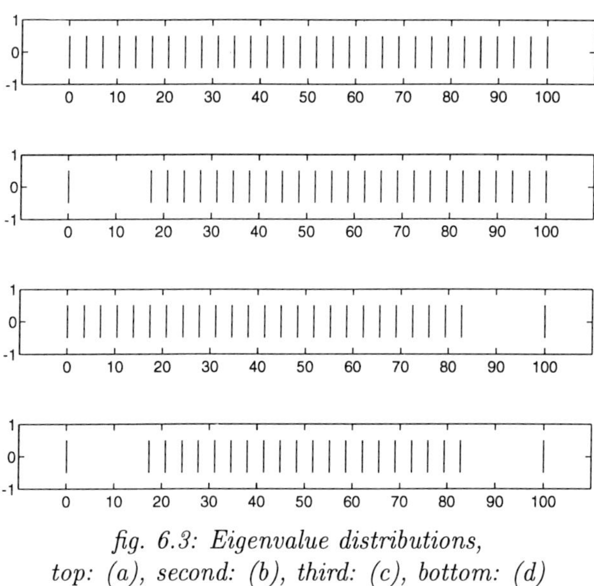

fig. 6.3: Eigenvalue distributions,
top: (a), second: (b), third: (c), bottom: (d)

It has been observed experimentally that CG exhibits what may be called a superlinear rate of convergence, that is to say, the convergence becomes faster and faster as CG proceeds, see figure 6.1. The explanation of this phenomenon is that the components of the error on the eigenvectors corresponding to the extreme eigenvalues are rapidly dampened during the iterations and then, the convergence is faster because the "effective" condition number gets smaller. To make this statement more precise, we shall show that we can derive an approximation of the extreme eigenvalues as CG proceeds. We have seen

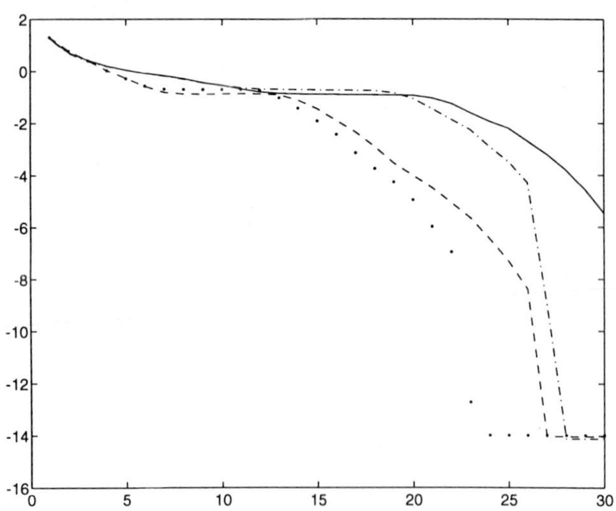

fig. 6.4: $\log_{10}$ of the A–norm of the error with CG
solid: (a), dashed: (b), dot–dashed: (c), dotted: (d)

before that,

$$z^{k+1} = z^{k-1} - \omega_{k+1}(\alpha_k K z^k + z^{k-1} - z^k).$$

This can be rewritten as

$$K z^k = c_{k-1} z^{k-1} + a_k z^k + b_{k+1} z_{k+1}, \tag{6.14}$$

with

$$c_{k-1} = \frac{1}{\omega_{k+1}\alpha_k} - \frac{1}{\alpha_k}, \quad a_k = \frac{1}{\alpha_k}, \quad b_{k+1} = -\frac{1}{\omega_{k+1}\alpha_k}.$$

Denote by Z_{k+1} the matrix whose columns are $z^0, z^1, \ldots, z^k$ and

$$T_{k+1} = \begin{pmatrix} a_0 & c_0 & & & \\ b_1 & a_1 & c_1 & & \\ & \ddots & \ddots & \ddots & \\ & & b_{k-1} & a_{k-1} & c_{k-1} \\ & & & b_k & a_k \end{pmatrix}$$

Lemma 6.23

The relation (6.14) can be rewritten for $j = 0, \dots k$ as

$$KZ_{k+1} = Z_{k+1}T_{k+1} - \frac{1}{\omega_{k+1}\alpha_k} z^{k+1}(e^{k+1})^T,$$

where $(e^{k+1})^T = (0, 0, \cdots, 1)$.

Proof. This result follows straightforwardly from (6.14). □

We have shown that CG gives $z^n = 0$, so $KZ_n = Z_n T_n$. Matrices Z_k are of full rank because vectors z^j are mutually orthogonal with respect to M, then

$$K = Z_n T_n (Z_n)^{-1}.$$

This proves that K and T_n are similar and therefore have the same eigenvalues. Let D_k be a diagonal matrix of order k, whose elements are $(z^j, Mz^j)^{\frac{1}{2}}$. Then,

$$(Z_k)^T M Z_k = (D_k)^2.$$

Denote $\tilde{Z}_k = M^{\frac{1}{2}} Z_k (D_k)^{-1}$. Then,

$$KZ_k = KM^{-\frac{1}{2}} \tilde{Z}_k D_k.$$

But,

$$KZ_k = Z_k T_k - \frac{1}{\omega_k \alpha_{k-1}} z^k (e^k)^T.$$

Multiplying this relation by $(\tilde{Z}_k)^T M^{\frac{1}{2}}$, we get

$$(\tilde{Z}_k)^T M^{\frac{1}{2}} K M^{-\frac{1}{2}} \tilde{Z}_k D_k = (\tilde{Z}_k)^T M^{\frac{1}{2}} (M^{-\frac{1}{2}} \tilde{Z}_k D_k) T_k - \frac{1}{\omega_k \alpha_{k-1}} (\tilde{Z}_k)^T M^{\frac{1}{2}} Z_k (e^k)^T.$$

But $(\tilde{Z}_k)^T M^{\frac{1}{2}} z^k = 0$ and therefore,

$$(\tilde{Z}_k)^T M^{\frac{1}{2}} K M^{-\frac{1}{2}} \tilde{Z}_k D_k = D_k T_k, \tag{6.15}$$

because $(\tilde{Z}_k)^T \tilde{Z}_k = I$. The matrix T_k is similar to $(\tilde{Z}_k)^T M^{-\frac{1}{2}} A M^{-\frac{1}{2}} \tilde{Z}_k$ which is the projection of $M^{-\frac{1}{2}} A M^{-\frac{1}{2}}$ onto the space generated by the orthonormal vectors $\tilde{z}^j$.

As we shall see in the following sections, (6.15) describes essentially the Lanczos algorithm (see Parlett [1062]). It has been proved by Paige [1056] that the eigenvalues μ_j^k of T_k approximate those of K as the computation proceeds beginning by the extreme ones. Van der Sluis and Van der Vorst [1275] showed that as soon as one of the μ_j^k converges, even moderately, to some eigenvalues of K, then it behaves as if the component of the error along the corresponding eigenvector does not exist. This explains the superconvergence property of CG: convergence is better and better as the algorithm proceeds. Moreover, what is usually most important is the distribution of the smallest eigenvalues. Figure 6.5 uses the example (b) of figure 6.3 with one isolated eigenvalue. It shows the component of the error on the eigenvector corresponding to the smallest eigenvalue (solid) and on the eigenvector corresponding to the largest eigenvalue (dashed).

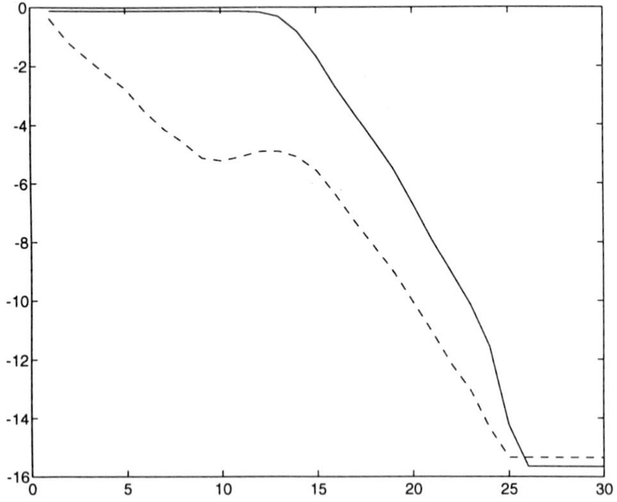

fig. 6.5: $\log_{10}$ *of the components of the error with* CG
solid: smallest eigenvalue, dashed: largest eigenvalue

On the same example, figure 6.6 shows at the top, the component of the error on the eigenvector corresponding to the smallest eigenvalue and at the bottom the approximate smallest eigenvalue that can be computed at each

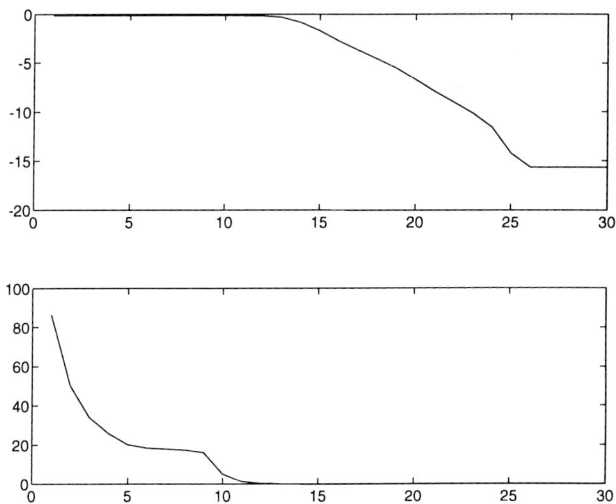

fig. 6.6: top: $\log_{10}$ *of the component of the error with* CG
bottom: convergence of the smallest eigenvalue

CG iteration. We can see that we start getting rid of this component of the error when the eigenvalue converges (to 0.1 in this case).

Bonnet and Meurant [166] proved that if we only assume exact local orthogonality, then CG still converges and as fast as steepest descent. For the sake of simplicity let us suppose $M = I$ and that we have

$$\beta_k = \frac{(r^k, r^k)}{(r^{k-1}, r^{k-1})},$$
$$(p^k, r^{k+1}) = 0,$$
$$(p^k, r^k) = (r^k, r^k),$$
$$(r^{k+1}, r^k) = 0.$$

Lemma 6.24

Supposing only the previous relations, we have

$$(Ap^k, p^k) = (Ar^k, r^k) - \beta_k^2 (Ap^{k-1}, p^{k-1})$$

and so $(Ap^k, p^k) \leq (Ar^k, r^k)$.

Proof. We have

$$Ap^k = Ar^k - \beta_k Ap^{k-1}.$$

Therefore,

$$(Ap^k, p^k) = (Ar^k, r^k) + 2\beta_k (Ar^k, p^{k-1}) + \beta_k^2 (Ap^{k-1}, p^{k-1}),$$

but

$$(Ar^k, p^{k-1}) = (r^k, Ap^{k-1}) = (p^k - \beta_k p^{k-1}, Ap^k).$$

This shows that $(Ar^k, p^{k-1}) = -\beta_k (Ap^{k-1}, p^{k-1})$. Therefore, the result holds. □

Theorem 6.25

Using local orthogonality, CG *converges and we have*

$$E(x^k) \le \left(\frac{\kappa(A) - 1}{\kappa(A) + 1} \right)^{2k} E(x^0).$$

Proof. We have

$$A^{-1} r^{k+1} = A^{-1} r^k - \gamma_k p^k.$$

Therefore

$$\begin{aligned} E(x^{k+1}) &= (A^{-1} r^{k+1}, r^{k+1}) \\ &= (A^{-1} r^k, r^{k+1}) - \gamma_k (p^k, r^{k+1}) \\ &= (A^{-1} r^k, r^{k+1}) \\ &= (A^{-1} r^k, r^k - \gamma_k A p^k) \\ &= (A^{-1} r^k, r^k) - \gamma_k (r^k, p^k). \end{aligned}$$

Hence

$$E(x^{k+1}) = E(x^k) - \frac{\|r^k\|^4}{(Ap^k, p^k)} \le E(x^k) - \frac{\|r^k\|^4}{(Ar^k, r^k)}.$$

The proof is ended by using the Kantorovitch inequality. □

Note that when using only local orthogonality the upper bound involves $\kappa(A)$ and not $\sqrt{\kappa(A)}$. This result shows that if we preserve local orthogonality, the convergence rate could be at worst that of steepest descent.

6.5. The Lanczos algorithm

Almost as the same time Hestenes and Stiefel developed the CG algorithm [772], Cornelius Lanczos introduced the method that now bears his name (see Lanczos [879]). This method is well known for computing a few of the extreme eigenvalues of sparse matrices. However, the Lanczos method can also be used to solve linear systems and, as it turns out, CG is nothing else than a particular case of the Lanczos method. This algorithm is explained in full details in Parlett's book [1062]. We shall mainly follow the exposition of Simon [1172]. In this section for the sake of simplicity, we do not consider preconditioning ($M = I$).

A being a symmetric matrix, we consider Krylov spaces

$$K_k(A, b) = \text{span}(b, Ab, \dots, A^{k-1}b).$$

Generally, $K_k(A, b)$ is of dimension k, $\{b, Ab, \dots, A^{k-1}b\}$ being a basis. Unfortunately, this basis is not well conditioned and it is much better to construct an orthonormal basis of $K_k(A, b)$. We can use the Gram–Schmidt orthogonalization method to achieve this goal. Suppose we already have orthogonal vectors $\{q^1, \dots, q^k\}$ as a basis for $K_k(A, b)$, then we have to orthogonalize $A^k b$ against $q^1, \dots, q^k$. This is the same as orthogonalizing Aq^k against $q^1, \dots, q^k$. It turns out that Aq^k is already orthogonal to $q^1, \dots, q^{k-2}$ because of the symmetry of A. We simply have to orthogonalize Aq^k against q^{k-1} and q^k. Let us denote

$$\bar{q}^k = Aq^k - \delta_k q^k - \eta_k q^{k-1},$$

with $\delta_k = (q^k, Aq^k), \eta_k = (q^{k-1}, Aq^k)$. To obtain q^{k+1} the vector $\bar{q}^k$ has to be normalized and it turns out that $\|\bar{q}^k\| = \eta_{k+1}$. Let

$$\bar{T}_k = \begin{pmatrix} \delta_1 & \eta_2 & & & \\ \eta_2 & \delta_2 & \eta_3 & & \\ & \ddots & \ddots & \ddots & \\ & & \eta_{k-1} & \delta_{k-1} & \eta_k \\ & & & \eta_k & \delta_k \end{pmatrix},$$

then, denoting $Q_k = [q^1, \cdots, q^k]$, the relation defining the vectors q^k can be written in matrix form as

$$AQ_k - Q_k \bar{T}_k = \eta_{k+1} q^{k+1} (e^k)^T.$$

Note the similarity of this relation with the equation in lemma 6.23. This can also be written as

$$AQ_k = Q_{k+1} \tilde{T}_k,$$

where

$$\tilde{T}_k = \begin{pmatrix} \bar{T}_k \\ \eta_{k+1}(e^k)^T \end{pmatrix},$$

is a $(k+1) \times k$ upper Hessenberg matrix. Of course, we have $Q_k^T A Q_k = \bar{T}_k$. $\bar{T}_k$ is the orthogonal projection of A onto $K_k(A, b)$. Therefore, it is natural to compute an approximation of the solution in $K_k(A, b)$ as $x^k = Q_k \bar{T}_k^{-1} Q_k^T b$. That is, we project the right hand side b onto $K_k(A, b)$, we solve in $K_k(A, b)$ with the projection of A and we take the solution back to the original space. Note that to compute the approximation we need all the previous basis vectors. However, we do not have to compute x^k at each iteration because we note that $Q_k^T b = \eta_1 e^1$ where e^1 is the first column of the identity matrix. Therefore $\bar{T}_k^{-1} Q_k^T b$ is η_1 times the first column of the inverse of the tridiagonal matrix $\bar{T}_k$ and the residual is

$$r^k = b - Ax^k = b - AQ_k \bar{T}_k^{-1} Q_k^T b = -\eta_{k+1} \phi_k q^{k+1},$$

where ϕ_k is the kth (last) component of $\bar{T}_k^{-1} Q_k^T b$ (that is η_1 times the last component of the first column of the inverse of $\bar{T}_k$). This implies that

$$\|r^k\| = \eta_{k+1} |\phi_k|.$$

This gives a handy way of computing the norm of the residual without even computing the vectors x^k. We remark that the vector x^k has its residual r^k orthogonal to K_k. Let us now consider the connection with CG. We write the relation $AQ_k - Q_k \bar{T}_k = \eta_{k+1} q^{k+1} (e^k)^T$ as

$$AQ_k - Q_k \bar{T}_k = G_k,$$

where the matrix G_k has only the last column being non–zero and proportional to q^{k+1}. Now, suppose that A is not only symmetric but also positive definite. Then, $\bar{T}_k = Q_k^T A Q_k$ is also positive definite and there exists a

Cholesky factorization $\bar{T}_k = L_k D_k L_k^T$ where L_k is lower bidiagonal with ones on the diagonal and D_k is diagonal. Then, with a little algebra, we have

$$AQ_k L_k^{-T} D_k^{-1} - Q_k L_k = G_k L_k^{-T} D_k^{-1}.$$

We denote $P_k = Q_k L_k^{-T}$. This can be computed as solving $P_k L_k^T = Q_k$. With this notation we write

$$AP_k D_k^{-1} - Q_k L_k = G_k L_k^{-T} D_k^{-1}.$$

The matrix L_k^{-T} is an upper triangular matrix with ones on the diagonal, therefore $G_k L_k^{-T} = G_k$. If we write the last column of the previous relation, we see that since r^k is a scalar multiple of q^{k+1}, it is a linear combination of r^{k-1} and Ap^k (which means that x^k is a linear combination of x^{k-1} and p^k). Moreover, writing $L_k P_k^T = Q_k^T$, we see that p^k is a linear combination of p^{k-1} and r^k. Therefore, up to a scaling, the Lanczos algorithm is identical to CG (without preconditioning). On a more careful examination, we can see that we have the following relationship. In CG we have

$$\begin{aligned} Ar^k &= \frac{1}{\gamma_k}(r^k - r^{k+1}) - \frac{\beta_k}{\gamma_{k-1}}(r^{k-1} - r^k) \\ &= -\frac{1}{\gamma_k} r^{k+1} + \left(\frac{1}{\gamma_k} + \frac{\beta_k}{\gamma_{k-1}}\right) r^k - \frac{\beta_k}{\gamma_{k-1}} r^{k-1}. \end{aligned}$$

As $\beta_k^2 = \|r^k\|/\|r^{k-1}\|$, we divide by the norm of r^k to get

$$Aq^{k+1} = \frac{\sqrt{\beta_{k+1}}}{\gamma_k} q^{k+2} + \left(\frac{1}{\gamma_k} + \frac{\beta_k}{\gamma_{k-1}}\right) q^{k+1} + \frac{\sqrt{\beta_k}}{\gamma_{k-1}} q^k,$$

where

$$q^{k+1} = (-1)^k \frac{r^k}{\|r^k\|}.$$

This shows that

$$\begin{aligned} \delta_k &= \frac{1}{\gamma_{k-1}} + \frac{\beta_{k-1}}{\gamma_{k-2}}, \quad \beta_0 = 0, \quad \gamma_{-1} = 1, \\ \eta_k &= \frac{\sqrt{\beta_k}}{\gamma_{k-1}}. \end{aligned}$$

This relationship of the two methods shows why (in principle) we cannot use CG for indefinite matrices. In that case, it may happen that the Cholesky factorization of $\bar{T}_k$ does not exists. We shall see how to deal with this problem in the following sections.

There is also another interpretation of the Lanczos algorithm. Let B be the matrix whose columns are $A^j b$, $j = 1, \ldots, n-1$. As Lanczos is nothing other than orthogonalizing the columns of B, we can recursively compute the columns of Q_n from those of B. Hence, we can write $Q_n = BL^{-T}\Pi^{-1}$ where L is lower triangular with ones on the diagonal and Π is a diagonal matrix whose diagonal entries are $\eta_1, \eta_1\eta_2, \ldots$ Then, we can write

$$B^T B = L\Pi^2 L^T.$$

It is easy to see that $(B^T B)_{i,j} = (b, A^{i+j} b)$. Therefore, $B^T B$ is the matrix of the moments of A and the Lanczos algorithm is nothing other than the Cholesky factorization of the matrix of moments in disguise.

6.6. A posteriori error bounds

We have seen that

$$E(x^k) = \|\varepsilon^k\|_A^2 = (A(x - x^k), x - x^k) \le 4\left(\frac{\sqrt{\kappa} - 1}{\sqrt{\kappa} + 1}\right)^{2k} \|\varepsilon^0\|_A^2,$$

where κ is the condition number. However, this is not a very sharp bound and it cannot be used for stopping the CG iterations as it is usually much too pessimistic. As we have seen $\varepsilon^k = A^{-1} r^k$, therefore

$$\|\varepsilon^k\|_A^2 = (r^k, A^{-1} r^k).$$

If we want upper and lower bounds on $\|\varepsilon^k\|_A$, we have to compute bounds for the quadratic form $(r^k, A^{-1} r^k)$. This is a special instance of a more general problem that has been considered by several authors in the past (for a bibliography, see Golub and Meurant [665]) namely: A being symmetric and positive definite, f being a smooth function and u and v being two given vectors, compute bounds for a bilinear form $(u, f(A)v)$. In our problem we have $u = v = r$ and $f(x) = 1/x$. Let us consider for a while the general problem, following the lines of Golub and Meurant [665]). Since $A = A^T$, we write A as $A = Q\Lambda Q^T$, where Q is the orthonormal matrix whose columns are the normalized eigenvectors of A and Λ is a diagonal matrix whose diagonal elements are the eigenvalues λ_i. By definition, we have $f(A) = Qf(\Lambda)Q^T$.

Therefore,

$$\begin{aligned} u^T f(A)v &= u^T Q f(\Lambda) Q^T v \\ &= \alpha^T f(\Lambda)\beta, \\ &= \sum_{i=1}^{n} f(\lambda_i)\alpha_i\beta_i. \end{aligned}$$

This last sum can be considered as a Riemann–Stieltjes integral

$$I[f] = u^T f(A)v = \int_a^b f(\lambda)\, d\alpha(\lambda),$$

where the measure α is piecewise constant and defined by

$$\alpha(\lambda) = \begin{cases} 0 & \text{if } \lambda < a = \lambda_1 \\ \sum_{j=1}^{i} \alpha_j\beta_j & \text{if } \lambda_i \leq \lambda < \lambda_{i+1} \\ \sum_{j=1}^{n} \alpha_j\beta_j & \text{if } b = \lambda_n \leq \lambda \end{cases}$$

When $u = v$, we note that α is an increasing positive function. A way to obtain bounds for the Stieltjes integral is to use Gauss, Gauss–Radau and Gauss–Lobatto quadrature formulas, see Dahlquist, Eisenstat and Golub [372], Golub [656], [657]. The general formula we shall use is

$$\int_a^b f(\lambda)\, d\alpha(\lambda) = \sum_{j=1}^{N} w_j f(t_j) + \sum_{k=1}^{M} v_k f(z_k) + R[f],$$

where the weights $[w_j]_{j=1}^N$, $[v_k]_{k=1}^M$ and the nodes $[t_j]_{j=1}^N$ are unknowns and the nodes $[z_k]_{k=1}^M$ are prescribed, see Davis and Rabinowitz [484], Gautschi [604], [605], Golub and Welsch [679]. When $u = v$, it is known (see for instance Stoer and Bulirsch [1211]) that

$$R[f] = \frac{f^{(2N+M)}(\eta)}{(2N+M)!} \int_a^b \prod_{k=1}^{M} (\lambda - z_k) \left[\prod_{j=1}^{N} (\lambda - t_j) \right]^2 d\alpha(\lambda), \quad a < \eta < b.$$

If $M = 0$, this leads to the Gauss rule with no prescribed nodes. If $M = 1$ and $z_1 = a$ or $z_1 = b$ we have the Gauss–Radau formula. If $M = 2$ and $z_1 = a, z_2 = b$, this is the Gauss–Lobatto formula. Let us briefly recall how the nodes and weights are obtained in the Gauss, Gauss–Radau and Gauss–Lobatto rules. For the measure α, it is possible to define a sequence of polynomials $p_0(\lambda), p_1(\lambda), \ldots$ that are orthonormal with respect to α:

$$\int_a^b p_i(\lambda) p_j(\lambda)\, d\alpha(\lambda) = \begin{cases} 1 & \text{if } i = j \\ 0 & \text{otherwise} \end{cases}$$

and p_k is of exact degree k. Moreover, the roots of p_k are distinct, real and lie in the interval $[a, b]$. We shall see how to compute these polynomials later on. This set of orthonormal polynomials satisfies a three term recurrence relationship:

$$\eta_{j+1}p_j(\lambda) = (\lambda - \delta_j)p_{j-1}(\lambda) - \eta_j p_{j-2}(\lambda), \quad j = 1, 2, \ldots, N$$

$$p_{-1}(\lambda) \equiv 0, \quad p_0(\lambda) \equiv 1,$$

if $\int d\alpha = 1$.

In matrix form, this can be written as

$$\lambda p(\lambda) = J_N p(\lambda) + \eta_{N+1} p_N(\lambda) e^N,$$

where

$$p(\lambda)^T = [p_0(\lambda)\ p_1(\lambda) \cdots\ p_{N-1}(\lambda)],$$

$$(e^N)^T = (0\ 0\ \cdots\ 0\ 1),$$

$$J_N = \begin{pmatrix} \delta_1 & \eta_2 & & & \\ \eta_2 & \delta_2 & \eta_3 & & \\ & \ddots & \ddots & \ddots & \\ & & \eta_{N-1} & \delta_{N-1} & \eta_N \\ & & & \eta_N & \delta_N \end{pmatrix}.$$

The eigenvalues of J_N (which are the zeros of p_N) are the nodes of the Gauss quadrature rule (i. e. $M = 0$). The weights are the squares of the first elements of the normalized eigenvectors of J_N, see [679]. We note that all the eigenvalues of J_N are real and simple. For the Gauss quadrature rule (renaming the weights and nodes w_j^G and t_j^G) we have

$$\int_a^b f(\lambda)\, d\alpha(\lambda) = \sum_{j=1}^{N} w_j^G f(t_j^G) + R_G[f],$$

with

$$R_G[f] = \frac{f^{(2N)}(\eta)}{(2N)!} \int_a^b \left[\prod_{j=1}^{N} (\lambda - t_j^G) \right]^2 d\alpha(\lambda),$$

and the next theorem follows.

Theorem 6.26

Suppose f is such that $f^{(2n)}(\xi) > 0,\ \forall n,\ \forall \xi,\ a < \xi < b$, and let

$$L_G[f] = \sum_{j=1}^{N} w_j^G f(t_j^G).$$

Then, $\forall N,\ \exists \eta \in [a,b]$ such that

$$L_G[f] \le I[f],$$

$$I[f] - L_G[f] = R_G[f].$$

Proof. See [1211]. The main idea of the proof is to use a Hermite interpolatory polynomial of degree $2N-1$ on the N nodes which allows us to express the remainder as an integral of the difference between the function and its interpolatory polynomial and to apply the mean value theorem (as the measure is positive and increasing). □

How can we construct the orthogonal polynomials related to the measure α since we do not know the distribution of the eigenvalues of A but only a lower bound a and an upper bound b?

For $u = v$, the solution of this problem is given by the Lanczos algorithm that we restate below with a slightly different notation. Let $h^{-1} = 0$ and h^0 be given such that $\|h^0\| = 1$. The Lanczos algorithm is defined by the following relations,

$$\eta_{j+1} h^j = \bar{h}^j = (A - \delta_j I)h^{j-1} - \eta_j h^{j-2}, \quad j = 1, \ldots$$

$$\delta_j = (h^{j-1}, Ah^{j-1}),$$

$$\eta_{j+1} = \|\bar{h}^j\|.$$

As we have already seen, the sequence $\{h^j\}_{j=0}^{l}$ is an orthonormal basis of the Krylov space $K_k(A, h^0)$.

Proposition 6.27

The vector h^j is given by $h_j = P_j(A)h^0$, where p_j is a polynomial of degree j defined by the three term recurrence

$$\eta_{j+1}p_j(\lambda) = (\lambda - \delta_j)p_{j-1}j(\lambda) - \eta_j p_{j-2}(\lambda), \quad p_{-1}(\lambda) \equiv 0, \quad p_0(\lambda) \equiv 1.$$

Proof. $\eta_2 h^1 = (A - \delta_1 I)h^0$ is a first order polynomial in A. Therefore, the proposition is easily obtained by induction. □

Theorem 6.28

If $h^0 = u$, we have

$$(h^k, h^l) = \int_a^b p_k(\lambda)p_l(\lambda)d\alpha(\lambda).$$

Proof. Since the h_j's are orthonormal, we have

$$\begin{aligned}(h^k, h^l) &= h^{0^T} P_k(A)^T P_l(A)h^0 \\ &= h^{0^T} QP_k(\Lambda)Q^T QP_l(\Lambda)Q^T h^0 \\ &= h^{0^T} QP_k(\Lambda)P_l(\Lambda)Q^T h^0 \\ &= \sum_{j=1}^{n} p_k(\lambda_j)p_l(\lambda_j)(\hat{h}^j)^2,\end{aligned}$$

where $\hat{h} = Q^T h^0$. Therefore, the p_j's are the orthonormal polynomials related to α that we were looking for. For generating the matrix J_N that will give us the nodes and weights, we shall just have to run some steps of the Lanczos algorithm starting with the vector u (normalized if needed).

To obtain the Gauss–Radau rule ($M = 1$), we should extend the matrix J_N in such a way that it has one prescribed eigenvalue, see Golub [656]. Assume $z_1 = a$; we wish to construct a polynomial p_{N+1} such that $p_{N+1}(a) = 0$. From the three–term recurrence relation satisfied by the polynomials, we must have

$$0 = \eta_{N+2}p_{N+1}(a) = (a - \delta_{N+1})p_N(a) - \eta_{N+1}p_{N-1}(a).$$

This gives

$$\delta_{N+1} = a - \eta_{N+1} \frac{p_{N-1}(a)}{p_N(a)}.$$

We have also

$$(J_N - aI)p(a) = -\eta_{N+1} p_N(a) e^N.$$

Let us denote $\omega(a) = [\omega_1(a), \cdots, \omega_N(a)]^T$ with

$$\omega_l(a) = -\eta_{N+1} \frac{p_{l-1}(a)}{p_N(a)} \quad l = 1, \ldots, N.$$

This gives $\delta_{N+1} = a + \omega_N(a)$ and

$$(J_N - aI)\omega(a) = \eta_{N+1}^2 e^N.$$

From these relations we obtain the solution of the problem as: 1) we generate η_{N+1} by the Lanczos process, 2) we solve the tridiagonal system for $\omega(a)$ and 3) we compute δ_{N+1}. Then the augmented tridiagonal matrix $\hat{J}_{N+1}$ defined as

$$\hat{J}_{N+1} = \begin{pmatrix} J_N & \eta_{N+1} e^N \\ \eta_{N+1} (e^N)^T & \delta_{N+1} \end{pmatrix},$$

will have a as an eigenvalue and gives the weights and the nodes of the corresponding quadrature rule. Therefore the recipe is to compute as for the Gauss quadrature rule and then to modify the last step to obtain the prescribed node. For Gauss–Radau the remainder R_{GR} is

$$R_{GR}[f] = \frac{f^{(2N+1)}(\eta)}{(2N+1)!} \int_a^b (\lambda - z_1) \left[\prod_{j=1}^N (\lambda - t_j) \right]^2 d\alpha(\lambda).$$

Again, this is proved by constructing an interpolatory polynomial for the function and its derivative on the t_js and for the function on z_1.

Theorem 6.29

Suppose f is such that $f^{(2n+1)}(\xi) < 0$, $\forall n, \forall \xi, a < \xi < b$. Let U_{GR} be defined as

$$U_{GR}[f] = \sum_{j=1}^N w_j^a f(t_j^a) + v_1^a f(a),$$

w_j^a, v_1^a, t_j^a being the weights and nodes computed with $z_1 = a$ and let L_{GR} be defined as

$$L_{GR}[f] = \sum_{j=1}^{N} w_j^b f(t_j^b) + v_1^b f(b),$$

w_j^b, v_1^b, t_j^b being the weights and nodes computed with $z_1 = b$. Then, $\forall N$ we have

$$L_{GR}[f] \le I[f] \le U_{GR}[f],$$

and

$$I[f] - U_{GR}[f] = \frac{f^{(2N+1)}(\eta)}{(2N+1)!} \int_a^b (\lambda - a) \left[\prod_{j=1}^{N} (\lambda - t_j^a) \right]^2 d\alpha(\lambda),$$

$$I[f] - L_{GR}[f] = \frac{f^{(2N+1)}(\eta)}{(2N+1)!} \int_a^b (\lambda - b) \left[\prod_{j=1}^{N} (\lambda - t_j^b) \right]^2 d\alpha(\lambda).$$

Proof. With our hypothesis the sign of the remainder is easily obtained. It is negative if we choose $z_1 = a$, positive if we choose $z_1 = b$. □

Now, consider the Gauss–Lobatto rule ($M = 2$), with $z_1 = a$ and $z_2 = b$ as prescribed nodes. Again, we should modify the matrix of the Gauss quadrature rule, see Golub [656]. Here, we would like to have

$$p_{N+1}(a) = p_{N+1}(b) = 0.$$

Using the recurrence relation for the polynomials, this leads to a linear system of order 2 for the unknowns δ_{N+1} and η_{N+1}:

$$\begin{pmatrix} p_N(a) & p_{N-1}(a) \\ p_N(b) & p_{N-1}(b) \end{pmatrix} \begin{pmatrix} \delta_{N+1} \\ \eta_{N+1} \end{pmatrix} = \begin{pmatrix} a\, p_N(a) \\ b\, p_N(b) \end{pmatrix}.$$

Let ω and μ be defined as vectors with components

$$\omega_l = -\frac{p_{l-1}(a)}{\eta_{N+1} p_N(a)}, \quad \mu_l = -\frac{p_{l-1}(b)}{\eta_{N+1} p_N(b)},$$

then

$$(J_N - aI)\omega = e^N, \quad (J_N - bI)\mu = e^N,$$

and the linear system can be written

$$\begin{pmatrix} 1 & -\omega_N \\ 1 & -\mu_N \end{pmatrix} \begin{pmatrix} \delta_{N+1} \\ \eta_{N+1}^2 \end{pmatrix} = \begin{pmatrix} a \\ b \end{pmatrix},$$

giving the unknowns that we need. The tridiagonal matrix $\hat{J}_{N+1}$ is then defined as in the Gauss–Radau rule. Having computed the nodes and weights, we have

$$\int_a^b f(\lambda) d\alpha(\lambda) = \sum_{j=1}^N w_j^{GL} f(t_j^{GL}) + v_1 f(a) + v_2 f(b) + R_{GL}[f],$$

where

$$R_{GL}[f] = \frac{f^{(2N+2)}(\eta)}{(2N+2)!} \int_a^b (\lambda - a)(\lambda - b) \left[\prod_{j=1}^N (\lambda - t_j) \right]^2 d\alpha(\lambda).$$

Then, we have the following obvious result.

Theorem 6.30

Suppose f is such that $f^{(2n)}(\eta) > 0, \ \forall n, \ \forall \eta, a < \eta < b$ and let

$$U_{GL}[f] = \sum_{j=1}^N w_j^{GL} f(t_j^{GL}) + v_1 f(a) + v_2 f(b).$$

Then, $\forall N$

$$I[f] \le U_{GL}[f],$$

$$I[f] - U_{GL}[f] = \frac{f^{(2N+2)}(\eta)}{(2N+2)!} \int_a^b (\lambda - a)(\lambda - b) \left[\prod_{j=1}^N (\lambda - t_j^{GL}) \right]^2 d\alpha(\lambda).$$

□

We remark that we need not always compute the eigenvalues and eigenvectors of the tridiagonal matrix. Let Y_N be the matrix of the eigenvectors of J_N (or $\hat{J}_N$) whose columns we denote by y_i and Λ_N be the diagonal matrix of the eigenvalues t_i which give the nodes of the Gauss quadrature rule. It is well known that the weights w_i are given by (see Wilf [1331])

$$\frac{1}{w_i} = \sum_{l=0}^{N-1} p_l^2(t_i).$$

It can be easily shown that

$$w_i = \left(\frac{(y_i)_1}{p_0(t_i)}\right)^2,$$

where $(y_i)_1$ is the first component of y_i. But, since $p_0(\lambda) \equiv 1$, we have $w_i = ((y_i)_1)^2 = ((e^1)^T y_i)^2$.

Theorem 6.31

$$\sum_{l=1}^{N} w_l f(t_l) = (e^1)^T f(J_N) e^1.$$

Proof.

$$\begin{aligned}
\sum_{l=1}^{N} w_l f(t_l) &= \sum_{l=1}^{N} (e^1)^T y_l f(t_l) y_l^T e^1 \\
&= (e^1)^T \left(\sum_{l=1}^{N} y_l f(t_l) y_l^T \right) e^1 \\
&= (e^1)^T Y_N f(\Lambda_N) Y_N^T e^1 \\
&= (e^1)^T f(J_N) e^1
\end{aligned}$$

□

The same statement is true for the Gauss–Radau and Gauss–Lobatto rules. Therefore, in those cases where $f(J_N)$ (or the equivalent) is easily computable (e.g. $f(x) = 1/x$), we do not need to compute the eigenvalues and eigenvectors of J_N. Generalizations of these results to the case $u \neq v$ and to block algorithms were introduced in Golub and Meurant [665].

Now, we have most of the necessary machinery to compute lower and upper bounds for the A–norm of the error. What remains to be seen is how to compute the $(1,1)$ element of the inverse of a symmetric tridiagonal matrix. We have

$$J_N = \begin{pmatrix} \delta_1 & \eta_2 & & & \\ \eta_2 & \delta_2 & \eta_3 & & \\ & \ddots & \ddots & \ddots & \\ & & \eta_{N-1} & \delta_{N-1} & \eta_N \\ & & & \eta_N & \delta_N \end{pmatrix}.$$

Let $x_N^T = (0 \ \ldots \ 0 \ \eta_{N+1})$, so that

$$J_{N+1} = \begin{pmatrix} J_N & x_N \\ x_N^T & \delta_{N+1} \end{pmatrix}.$$

Letting

$$\tilde{J} = J_N - \frac{x_N x_N^T}{\delta_{N+1}},$$

the upper left block of J_{N+1}^{-1} is $\tilde{J}^{-1}$. This can be obtained through the use of the Sherman–Morrison formula (see Golub and Van Loan [676]),

$$\tilde{J}^{-1} = J_N^{-1} + \frac{(J_N^{-1} x_N)(x_N^T J_N^{-1})}{\delta_{N+1} - x_N^T J_N^{-1} x_N}.$$

Let $j_N = J_N^{-1} e^N$ be the last column of the inverse of J_N. With this notation, we have

$$\tilde{J}^{-1} = J_N^{-1} + \frac{\eta_{N+1}^2 j_N j_N^T}{\delta_{N+1} - \eta_{N+1}^2 (j_N)_N}.$$

Therefore, it is clear that we only need the first and last elements of the last column of the inverse of J_N. This can be obtained using the Cholesky factorization of J_N. It is easy to check that if we define

$$d_1 = \delta_1, \quad d_i = \delta_i - \frac{\eta_i^2}{d_{i-1}}, \quad i = 2, \ldots, N$$

then

$$(j_N)_1 = (-1)^{N-1}\frac{\eta_2 \cdots \eta_N}{d_1 \cdots d_N}, \quad (j_N)_N = \frac{1}{d_N}.$$

Thus, the $(1,1)$ element of the inverse of J_N can be obtained incrementally and it can be combined with the Lanczos iterates. Putting all this together, we have the following:

Algorithm GQL$[A, u, l]$

Suppose $\|u\| = 1$, the following formulas yield a lower bound b_j of $u^T A^{-1} u$ by the Gauss quadrature rule, a lower bound $\bar{b_j}$ and an upper bound $\hat{b_j}$ through the Gauss–Radau quadrature rule and an upper bound $\breve{b_j}$ through the Gauss–Lobatto quadrature rule.

Let $h^{-1} = 0$ and $h^0 = u$, $\delta_1 = u^T A u$, $\eta_2 = \|(A - \delta_1 I)u\|$, $b_1 = \delta_1^{-1}$, $d_1 = \delta_1$, $c_1 = 1$, $\hat{d}_1 = \delta_1 - a$, $\bar{d}_1 = \delta_1 - b$, $h^1 = (A - \delta_1 I)u/\eta_2$.

Then for $j = 2, \ldots, l$ we compute

$$\delta_j = h^{j-1^T} A h^{j-1},$$

$$\tilde{h}^j = (A - \delta_j I)h^{j-1} - \eta_j h^{j-2},$$

$$\eta_{j+1} = \|\tilde{h}^j\|,$$

$$h^j = \frac{\tilde{h}^j}{\eta_{j+1}},$$

$$b_j = b_{j-1} + \frac{\eta_j^2 c_{j-1}^2}{d_{j-1}(\delta_j d_{j-1} - \eta_j^2)},$$

$$d_j = \delta_j - \frac{\eta_j^2}{d_{j-1}},$$

$$c_j = c_{j-1}\frac{\eta_j}{d_{j-1}},$$

$$\hat{d}_j = \delta_j - a - \frac{\eta_j^2}{\hat{d}_{j-1}},$$

$$\bar{d}_j = \delta_j - b - \frac{\eta_j^2}{\bar{d}_{j-1}},$$

$$\hat{\delta}_j = a + \frac{\eta_{j+1}^2}{\hat{d}_j},$$

$$\bar{\delta}_j = b + \frac{\eta_{j+1}^2}{\bar{d}_j},$$

$$\hat{b}_j = b_j + \frac{\eta_{j+1}^2 c_j^2}{d_j(\hat{\delta}_j d_j - \eta_{j+1}^2)},$$

$$\bar{b}_j = b_j + \frac{\eta_{j+1}^2 c_j^2}{d_j(\bar{\delta}_j d_j - \eta_{j+1}^2)},$$

$$\breve{\delta}_j = \frac{\hat{d}_j \bar{d}_j}{\bar{d}_j - \hat{d}_j}\left(\frac{b}{\hat{d}_j} - \frac{a}{\bar{d}_j}\right),$$

$$\breve{\eta}_{j+1}^2 = \frac{\hat{d}_j \bar{d}_j}{\bar{d}_j - \hat{d}_j}(b - a),$$

$$\breve{b}_j = b_j + \frac{\breve{\eta}_{j+1}^2 c_j^2}{d_j(\breve{\delta}_j d_j - \breve{\eta}_{j+1}^2)}.$$

Note that the bulk of the computations in Algorithm GQL is located in the matrix×vector product Ah^{j-1}. This algorithm can be used not only with iterative methods but whenever we have an approximation $\bar{x}$ of the solution of $Ax = b$. We compute the residual $r = b - A\bar{x}$, normalize it $u = r/\|r\|$ and run CGQL. Numerical experiments in Golub and Meurant [666] show that this method is quite efficient. However, this algorithm does not make much sense when the approximate solution is computed by CG. As the Lanczos algorithm and CG are equivalent, it would be strange to use the Lanczos method starting from CG residuals. Instead, it was proposed in Golub and Meurant [666] to use the following formula,

$$\begin{aligned}\|\varepsilon^k\|_A^2 = \|x - x^k\|_A^2 &= (r^0, A^{-1}r^0) - \|r^0\|^2 (J_k^{-1})_{1,1} \\ &= \|r^0\|^2((J_n^{-1})_{1,1} - (J_k^{-1})_{1,1}).\end{aligned} \tag{6.16}$$

where J_k is the tridiagonal matrix of the Lanczos coefficients. Formula (6.16) was already used in Golub and Strakos [672] for reconstructing the A–norm of the error when CG has converged but the (1, 1) element of the inverse of the tridiagonal matrix was computed by means of continued fractions. A round off error analysis given in [672] shows that below a certain value of $\|\varepsilon^k\|_A^2$, no more useful information can be obtained from this algorithm. Formula (6.16) has also been used in Fischer and Golub [559] but there, the computations of $\|\varepsilon^k\|_A$ were not done below 10^{-5} and the same difficulties arise as in [6]. It has been shown in Golub and Meurant [666] that these difficulties can be overcome and that reliable estimates of $\|\varepsilon^k\|_A$ can be computed whatever the value of the norm is.

Of course, (6.16) cannot be used directly as, at CG iteration k, we do not know $(J_n^{-1})_{1,1}$. But it is known that $(J_k^{-1})_{1,1} \to (J_n^{-1})_{1,1}$. So, we shall use the current value of $(J_k^{-1})_{1,1}$ to approximate the final value. Let b_k be the computed value of $(J_k^{-1})_{1,1}$. This is obtained in an additive way by using the Sherman–Morrison formula. Let $d_1 = \delta_1$ and

$$d_i = \delta_i - \frac{\eta_i^2}{d_{i-1}}, \quad i = 2, \dots, k$$

then, let

$$(j_k)_1 = (-1)^{k-1}\frac{\eta_2 \cdots \eta_k}{d_1 \cdots d_k}, \quad (j_k)_k = \frac{1}{d_k}.$$

Using these results, we have

$$f_k = \frac{\eta_k^2 c_{k-1}^2}{d_{k-1}(\delta_k d_{k-1} - \eta_k^2)}, \quad b_k = b_{k-1} + f_k.$$

Note that by using the definition of d_k, f_k can be computed as c_k^2/d_k. Since J_k is positive definite, this shows that $f_k > 0$. Let s_k be the estimate of $\|\varepsilon^k\|_A^2$ we are seeking and d be a positive integer (to be named the delay), at CG iteration number k, we set

$$s_{k-d} = \|r^0\|^2(b_k - b_{k-d}).$$

This will give us an estimate of the error d iterations before the current one. However, it was shown in Golub and Meurant [666] that if we compute b_k and use this formula straightforwardly, there exists a k_{max} such that if $k > k_{max}$ then, $s_k = 0$. This happens because, when k is large enough, $\eta_{k+1}/d_k < 1$ and $c_k \to 0$ and consequently $f_k \to 0$. Therefore, when $k > k_{max}$, $b_k = b_{kmax}$. But, we can compute s_{k-d} in another way as we just need to sum the last d values of f_j. The algorithm computing the iterates of CG and estimates from the Gauss (s_{k-d}), Gauss–Radau ($\underline{s}_{k-d}$ and $\bar{s}_{k-d}$) and Gauss–Lobatto ($\breve{s}_{k-d}$) rules is the following (with slight simplifications from [666]):

Algorithm CGQL

let x^0 be given, $r^0 = g - Ax^0$, $p^0 = r^0$, $\beta_0 = 0$, $\gamma_{-1} = 1$, $c_1 = 1$, for $k = 1, \dots$ until convergence

$$\gamma_{k-1} = \frac{(r^{k-1}, r^{k-1})}{(p^{k-1}, Ap^{k-1})},$$

$$\delta_k = \frac{1}{\gamma_{k-1}} + \frac{\beta_{k-1}}{\gamma_{k-2}},$$

if $k = 1$ ————————————————

$$f_1 = \frac{1}{\delta_1},$$

$$d_1 = \delta_1,$$

$$\bar{d}_1 = \delta_1 - a,$$

$$\underline{d}_1 = \delta_1 - b,$$

else ————————————————

$$c_k = c_{k-1} \frac{\eta_k}{d_{k-1}},$$

$$d_k = \delta_k - \frac{\eta_k^2}{d_{k-1}},$$

$$f_k = \frac{\eta_k^2 c_{k-1}^2}{d_{k-1}(\delta_k d_{k-1} - \eta_k^2)} = \frac{c_k^2}{d_k},$$

$$\bar{d}_k = \delta_k - a - \frac{\eta_k^2}{\bar{d}_{k-1}} = \delta_k - \bar{\delta}_{k-1},$$

$$\underline{d}_k = \delta_k - b - \frac{\eta_k^2}{\underline{d}_{k-1}} = \delta_k - \underline{\delta}_{k-1}$$

end ————————————————

$$x^k = x^{k-1} + \gamma_{k-1} p^{k-1},$$

$$r^k = r^{k-1} - \gamma_{k-1} A p^{k-1},$$

$$\beta_k = \frac{(r^k, r^k)}{(r^{k-1}, r^{k-1})},$$

$$\eta_k = \frac{\sqrt{\beta_k}}{\gamma_{k-1}},$$

$$p^k = r^k + \beta_k p^{k-1},$$

$$\bar{\delta}_k = a + \frac{\eta_k^2}{\bar{d}_k},$$

$$\underline{\delta}_k = b + \frac{\eta_k^2}{\underline{d}_k},$$

$$\breve{\delta}_k = \frac{\bar{d}_k \underline{d}_k}{\underline{d}_k - \bar{d}_k}\left(\frac{b}{\bar{d}_k} - \frac{a}{\underline{d}_k}\right),$$

$$\breve{\delta}_k^2 = \frac{\bar{d}_k \underline{d}_k}{\underline{d}_k - \bar{d}_k}(b - a),$$

$$\bar{f}_k = \frac{\eta_k^2 c_k^2}{d_k(\bar{\delta}_k d_k - \eta_k^2)},$$

$$\underline{f}_k = \frac{\eta_k^2 c_k^2}{d_k(\underline{\delta}_k d_k - \eta_k^2)},$$

$$\breve{f}_k = \frac{\breve{\eta}_k^2 c_k^2}{d_k(\breve{\delta}_k d_k - \breve{\eta}_k{}^2)},$$

if $k > d$ ——————————

$$t_k = \sum_{j=k-d+1}^{k} f_j,$$

$$s_{k-d} = \|r^0\|^2 t_k,$$

$$\bar{s}_{k-d} = \|r^0\|^2 (t_k + \bar{f}_k),$$

$$\underline{s}_{k-d} = \|r^0\|^2 (t_k + \underline{f}_k),$$

$$\breve{s}_{k-d} = \|r^0\|^2 (t_k + \breve{f}_k)$$

end ——————————

In this algorithm a and b are lower and upper bounds of the smallest and largest eigenvalues of A. Note that the value of s_k is independent of a and b, $\bar{s}_k$ depends only on a and $\underline{s}_k$ only on b. Let us now prove that this algorithm does give lower and upper bounds for the A–norm of the error.

Lemma 6.32

Let J_k, $\underline{J}_k$, $\bar{J}_k$ and $\breve{J}_k$ be the tridiagonal matrices of the Gauss, Gauss–Radau (with b and a as prescribed nodes) and the Gauss–Lobatto rules. Then, if $0 < a \le \lambda_{min}(A)$ and $b \ge \lambda_{max}(A)$, $\|r^0\|(J_k^{-1})_{1,1}$, $\|r^0\|(\underline{J}_k^{-1})_{1,1}$ are lower bounds of $\|\varepsilon^0\|^2 = (r^0, A^{-1}r^0)$, $\|r^0\|(\bar{J}_k^{-1})_{1,1}$ and $\|r^0\|(\breve{J}_k^{-1})_{1,1}$ are upper bounds of $(r^0, A^{-1}r^0)$.

Proof. See Golub and Meurant [666]. The proof is obtained easily as we know the sign of the remainder in the quadrature rules. □

Note that $\underline{J}_k$ and $\bar{J}_k$ are of order $k+1$ and $\breve{J}_k$ is of order $k+2$. We have that $\bar{f}_k > \underline{f}_k$ and therefore, $\bar{\delta}_k < \underline{\delta}_k$.

Theorem 6.33

At iteration number k of CGQL, *s_{k-d} and $\underline{s}_{k-d}$ are lower bounds of $\|\varepsilon^{k-d}\|_A^2$, $\bar{s}_{k-d}$ and $\breve{s}_{k-d}$ are upper bounds of $\|\varepsilon^{k-d}\|_A^2$.*

Proof. We have

$$\|\varepsilon^{k-d}\|_A^2 = \|r^0\|^2((J_n^{-1})_{1,1} - (J_{k-d}^{-1})_{1,1})$$

and

$$s_{k-d} = \|r^0\|^2((J_k^{-1})_{1,1} - (J_{k-d}^{-1})_{1,1}).$$

Therefore,

$$\|\varepsilon^{k-d}\|_A^2 - s_{k-d} = \|r^0\|^2((J_n^{-1})_{1,1} - (J_k^{-1})_{1,1}) > 0,$$

showing that s_{k-d} is a lower bound of $\|\varepsilon^{k-d}\|_A^2$. The same kind of proof applies for the other cases since, for instance,

$$\bar{s}_{k-d} = \|r^0\|^2((\bar{J}_k^{-1})_{1,1} - (J_{k-d}^{-1})_{1,1}).$$

□

Therefore, the quantities that we are computing in CGQL are indeed upper and lower bounds of the A–norm of the error. It turns out that the best

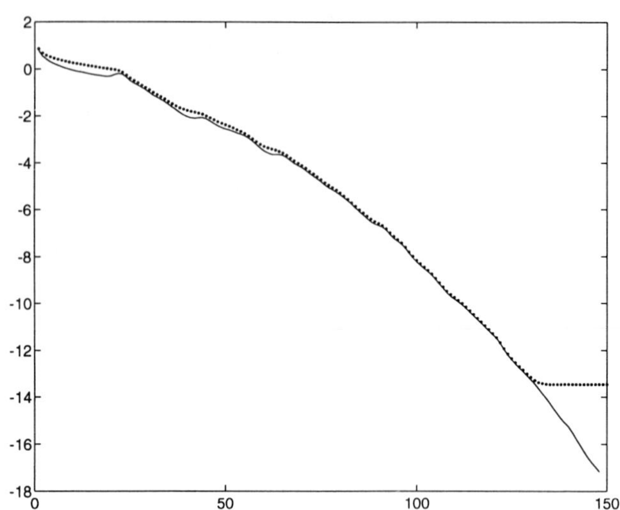

fig. 6.7: $\log_{10}$ *of the A–norm of the error (dotted line) and the estimate of the Gauss rule (solid) for* $d = 2$

bounds are the ones computed by the Gauss–Radau rule. Figure 6.7 shows the estimates computed for the Poisson model problem.

The lower bound s_k is independent of the extreme eigenvalues a and b. However, to compute the upper bound $\bar{s}_k$ we need an estimate of the smallest eigenvalue of A. We can see that $\bar{s}_k \rightarrow \infty$ if $a \rightarrow 0$. Therefore, if we take a value of a that is too small, we could obtain a large overestimate of the A–norm. Of course, there are cases where we can obtain good analytic estimates of the smallest eigenvalue, for example if the matrix A arises from finite element methods. However, as we are using CG we can obtain numerical estimates of the smallest eigenvalue when running the algorithm. The extreme eigenvalues of J_k are approximations of the extreme eigenvalues of A that get closer and closer as k increases. Therefore, we propose the following algorithm. We start the CGQL iterations with $a = a_0$ an underestimate of $\lambda_{min}(A)$. An estimate of the smallest eigenvalue can be easily obtained by inverse iteration (see Golub and Van Loan [676]) as, for computing the bounds of the norm, we already compute incrementally the Cholesky factorization of J_k. The smallest eigenvalue of J_k is obtained by repeatedly solving tridiagonal systems. We use a fixed number n_a of (inner) iterations of inverse iteration at every CG iteration, giving an estimate μ_k. When μ_k is such that

$$\frac{|\mu_k - \mu_{k-1}|}{\mu_k} \leq \varepsilon_a,$$

with a prescribed ε_a, we switch by setting $a = \mu_k$, we stop computing the eigenvalue estimate and we go on with CGQL. Numerical experiments in Golub and Meurant [666] and Meurant [984] show that these algorithms are really efficient at computing lower and upper bounds that can be used to reliably stop the CG iterations.

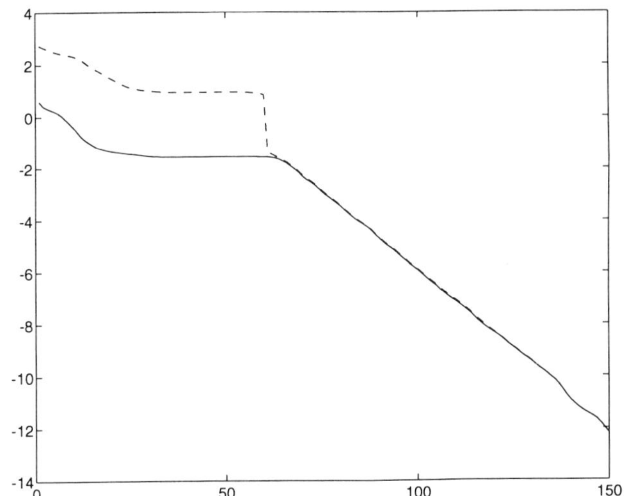

fig. 6.8: $\log_{10}$ *of the A–norm of the error (solid), example 2 dashed: adaptive algorithm,* $d = 20$

This method is easily extended to PCG by using the formula

$$\|\varepsilon^k\|_A^2 = (r^0, z^0)((J_n^{-1})_{1,1} - (J_k^{-1})_{1,1}).$$

The computation of the Euclidean norm (l_2–norm) is more complicated as there is no such formula as (6.16) for the l_2–norm. It can be computed by looking at the elements of J_N^{-2}.However, we shall use another technique introduced by Bai and Golub [97] for computing estimates of the trace of the inverse of a matrix. Consider a given iteration k of CG and for simplicity let $r = r^k$ and

$$\mu_p = \int_a^b \lambda^p \, d\alpha(\lambda) = (r, A^p r).$$

Note that we know μ_1, μ_0. The moment μ_1 can be computed during CG iterations by computing Ar^k. Then, Ap^k is computed recursively by $Ap^k =$

$Ar^k + \beta_k Ap^{k-1}$ to save a matrix multiply at the expense of storing one more vector. Moreover, we know upper and lower bounds of μ_{-1}. We are interested in computing estimates of μ_{-2}. We use a one point Gauss–Radau formula to compute an estimate $\tilde{\mu}_{-2}$. We have

$$\tilde{\mu}_p = w_0 t_0^p + w_1 t_1^p,$$

where $t_0 = a$ or b is the prescribed node. Moreover,

$$c\tilde{\mu}_p + d\tilde{\mu}_{p-1} - \tilde{\mu}_{p-2} = 0,$$

t_0 being a solution of $c\xi^2 + d\xi - 1 = 0$. In fact, we know the exact values of $\tilde{\mu}_1 = \mu_1 = (r, Ar)$ and $\tilde{\mu}_0 = \mu_0 = (r, r)$. Then,

$$\begin{pmatrix} c \\ d \end{pmatrix} = \begin{pmatrix} \mu_1 & \mu_0 \\ t_0^2 & t_0 \end{pmatrix}^{-1} \begin{pmatrix} \tilde{\mu}_{-1} \\ 1 \end{pmatrix}.$$

This gives

$$c = \frac{1}{\mu_1 t_0 - \mu_0 t_0^2}(t_0 \tilde{\mu}_{-1} - \mu_0),$$

$$d = \frac{1}{\mu_1 t_0 - \mu_0 t_0^2}(\mu_1 - t_0^2 \tilde{\mu}_{-1}).$$

Finally,

$$\tilde{\mu}_{-2} = \frac{1}{\mu_1 t_0 - \mu_0 t_0^2}[(t_0 \tilde{\mu}_{-1} - \mu_0)\mu_0 + (\mu_1 - t_0^2 \tilde{\mu}_{-1})\tilde{\mu}_{-1}].$$

If we knew the exact value μ_{-1}, setting $t_0 = a$ would give an upper bound and $t_0 = b$ a lower bound. We can compute two estimates of μ_{-2} by taking $t_0 = a$ and an upper bound of μ_{-1} from the Gauss–Radau rule for $(r, A^{-1}r)$ as well as $t_0 = b$ and a lower bound of μ_{-1}. Numerical experiments are given in Meurant [984].

6.7. The Eisenstat's trick

As we shall see in Chapter 8, there are many preconditioners that can be written as

$$M = (\Delta + L)\Delta^{-1}(\Delta + L^T),$$

where $A = D + L + L^T$, D and Δ being diagonal matrices and L being the strictly lower triangular part of A. S. Eisenstat [510] took advantage of this to derive an efficient implementation saving many floating point operations. The linear system $Ax = b$ is equivalent to

$$[(\Delta + L)^{-1}A(\Delta + L)^{-T}](\Delta + L^T)x = (\Delta + L)^{-1}b,$$

which we denote as $\hat{A}\hat{x} = \hat{b}$. Applying PCG to $Ax = b$ with M as a preconditioner is the same as using it on $\hat{A}\hat{x} = \hat{b}$ with a preconditioner $\hat{M} = \hat{D}^{-1}$. The trick is used for computing $\hat{A}\hat{p}^k$ as we have

$$\begin{aligned}\hat{A}\hat{p}^k &= (\Delta + L)^{-1}[(\Delta + L) + (\Delta + L)^T - (2\Delta - D](\Delta + L)^{-T}\hat{p}^k,\\ &= (\Delta + L)^{-1}\hat{p}^k + (\Delta + L)^{-1}[\hat{p}^k - \tilde{D}(\Delta + L)^{-T}\hat{p}^k].\end{aligned}$$

with $\tilde{D} = 2\Delta - D$. Therefore, if we define $\hat{t}^k = (\Delta + L)^{-T}\hat{p}^k$, we have

$$\hat{A}\hat{p}^k = \hat{t}^k + (\Delta + L)^{-1}(\hat{p}^k - \tilde{D}\hat{t}^k).$$

This requires $2n + nz(A)$ floating point operations, where $nz(A)$ is the number of non–zeros in A. The total cost for this PCG implementation is $8n + nz(A)$ versus $6n + 2nz(A)$ for the classical implementation. An additional n words of storage is also required. The savings depend on the structure of A. If $nz(A) = cn$, the larger is c, the larger the savings.

6.8. The Conjugate Residual method

This method (CR) is constructed by requiring the residuals to be A–orthogonal, $(r^i, Ar^j) = 0, i \neq j$. It turns out that the corresponding iterates x^k minimize the Euclidean norm (l_2–norm) of the residual

$$\|b - Ax^k\| = \min\{\|b - Ay\|, y \in x^0 + K_k(A, r^0)\}.$$

Note that here we minimize the l_2–norm of the residual when CG deals with the A–norm of the error, that is the A^{-1}–norm of the residual. For the derivation of the algorithm, see Chandra [303], Golub and Van Loan [676] or Fischer [556]. For the sake of simplicity, we do not include a preconditioner. The algorithm is the following:

Let x^0 be given, $r^0 = b - Ax^0$. For $k = 0, 1, \ldots$

$$\beta_k = \frac{(r^k, Ar^k)}{(r^{k-1}, Ar^{k-1})}, \quad \beta_0 = 0,$$

$$p^k = r^k + \beta_k p^{k-1},$$

$$Ap^k = Ar^k + \beta_k Ap^{k-1},$$

$$\gamma_k = \frac{(r^k, Ar^k)}{(Ap^k, Ap^k)},$$

$$x^{k+1} = x^k + \gamma_k p^k,$$

$$r^{k+1} = r^k - \gamma^k Ap^k.$$

This algorithm is well defined if A is positive definite. Otherwise, there can be problems computing the denominator of β_k. In the case where A is indefinite, we shall see a stable implementation later on. The conjugate residual method can be generalized by minimizing $\|r^k\|_{A^{\mu-1}}$, see Chandra [303].

6.9. SYMMLQ

We have seen that CG can be derived from the Lanczos algorithm when A is symmetric and positive definite by using the Cholesky factorization of the tridiagonal matrix $\bar{T}_k$. When A is symmetric but not positive (or negative) definite (*i.e.*,A has some negative and some positive eigenvalues), the Cholesky factorization may or may not exist and we can eventually get a zero pivot. In any case, the factorization may be unstable. One way to obtain a stable factorization of $\bar{T}_k$ would be to use eventually 2×2 pivots. This gives rise to a method known as SYMMBK. Paige and Saunders [1057], [1058]) proposed instead using an LQ factorization of $\bar{T}_k$. They named their algorithm SYMMLQ. Remember that the "CG" solution is obtained by

$$\bar{T}_k y^k = \eta_1 e^1, \quad x^k_{CG} = Q_k y^k.$$

Let us write the factorization as

$$\bar{T}_k = \bar{L}_k Z_k, \quad Z_k^T Z_k = I,$$

with $\bar{L}_k$ being lower triangular. The matrix Z_k is not constructed explicitly but as the product of matrices of plane rotations. We have

$$\bar{T}_k = \begin{pmatrix} \delta_1 & \eta_2 & & & \\ \eta_2 & \delta_2 & \eta_3 & & \\ & \ddots & \ddots & \ddots & \\ & & \eta_{k-1} & \delta_{k-1} & \eta_k \\ & & & \eta_k & \delta_k \end{pmatrix},$$

Let us look at the first steps of the reduction to triangular form. For the sake of simplicity, we consider $\bar{T}_4$,

$$\bar{T}_4 = \begin{pmatrix} \delta_1 & \eta_2 & & \\ \eta_2 & \delta_2 & \eta_3 & \\ & \eta_3 & \delta_3 & \eta_4 \\ & & \eta_4 & \delta_4 \end{pmatrix}.$$

To zero the element in position $(1,2)$, we multiply by a rotation matrix we denote by $Z_{1,2}$:

$$Z_{1,2} = \begin{pmatrix} c_1 & s_1 & & \\ s_1 & c_2 & & \\ & & 1 & \\ & & & 1 \end{pmatrix}.$$

To annihilate the $(1,2)$ element, we must have: $s_1\delta_1 = c_1\eta_2$. Let $\gamma_1 = \sqrt{\delta_1^2 + \eta_2^2}$, we take $s_1 = \eta_2/\gamma_1$, $c_1 = \delta_1/\gamma_1$ and then

$$\bar{T}_4 Z_{1,2} = \begin{pmatrix} \gamma_1 & & & \\ \omega_2 & \bar{\gamma}_2 & \eta_3 & \\ \pi_3 & \bar{\omega}_3 & \delta_3 & \\ & & \eta_4 & \delta_4 \end{pmatrix},$$

with $\bar{\gamma}_2 = (\eta_2\delta_1\delta_2)/\gamma_1$, $\pi_3 = s_1\eta_3$, $\bar{\omega}_3 = -c_1\eta_3$. Note that we have created a fill–in in position $(3,1)$. For the next step, we multiply by

$$Z_{2,3} = \begin{pmatrix} 1 & & & \\ & c_2 & s_2 & \\ & s_2 & -c_2 & \\ & & & 1 \end{pmatrix}.$$

As a result, we obtain

$$\bar{T}_4 Z_{1,2} Z_{2,3} = \begin{pmatrix} \gamma_1 & & & \\ \omega_2 & \gamma_2 & & \\ \pi_3 & \omega_3 & \bar{\gamma}_3 & \\ & \pi_4 & \bar{\omega}_4 & \delta_4 \end{pmatrix},$$

$\gamma_2 = \sqrt{\bar{\gamma}_2^2 + \eta_3^2}$, $s_2 = \eta_3/\gamma_2$, $c_2 = \bar{\gamma}_2/\gamma_2$, $\bar{\gamma}_3 = (\eta_3\bar{\omega}_3 - \bar{\gamma}_2\delta_3)/\gamma_2$, $\bar{\omega}_4 = -c_2\eta_4$, $\pi_4 = s_2\eta_4$. Now, the process is clear: $Z_{j,j+1}$ differs from the identity only in the elements $z_{j,j} = -z_{j+1,j+1} = c_j = \cos\theta_j$, $z_{j,j+1} = z_{j+1,j} = s_j = \sin\theta_j$ and

$$\bar{T}_k Z_{1,2} \cdots Z_{k-1,k} = T_k Z_k^T = \bar{L}_k = \begin{pmatrix} \gamma_1 & & & & \\ \omega_2 & \gamma_2 & & & \\ \pi_3 & \omega_3 & \gamma_3 & & \\ & \ddots & \ddots & \ddots & \\ & & \pi_k & \omega_4 & \bar{\gamma}_k \end{pmatrix}.$$

The last rotation is defined by

$$\gamma_k = \sqrt{\bar{\gamma}_k^2 + \eta_{k+1}^2}, \quad s_k = \frac{\eta_{k+1}}{\gamma_k}, \quad c_k = \frac{\bar{\gamma}_k}{\gamma_k}.$$

We have also $\bar{\omega}_{k+2} = -c_k \eta_{k+2}$, $\omega_{k+1} = \bar{\omega}_{k+1} c_k + s_k \delta_{k+1}$, $\pi_{k+2} = s_k \eta_{k+2}$. Let us define L_k as being identical to $\bar{L}_k$ except for the (k, k) element which is replaced by γ_k. Note that the principal minor of order $k-1$ of $\bar{L}_k$ is L_{k-1}. Following Paige and Saunders [1057], we denote

$$\overline{W}_k = [w^1 \cdots w^{k-1} \, \bar{w}^k] = [W_{k-1} \, \bar{w}^k] = Q_k Z_k^T,$$

$$\bar{z}^k = (\zeta_1, \ldots, \zeta_{k-1}, \bar{\zeta}_k)^T = ((z^{k-1})^T, \bar{\zeta}_k)^T = Z_k y^k.$$

With these notations,

$$\bar{L}_k \bar{z}^k = \eta_1 e^1, \quad x_{CG}^k = \bar{W}_k \bar{z}^k.$$

As we have $\bar{L}_k \bar{z}^k = \eta_1 e^1$ and $L_k z^k = \eta_1 e^1$ we get

$$\zeta_k = \bar{\zeta}_k \frac{\bar{\gamma}_k}{\gamma_k} = \bar{\zeta}_k c_k.$$

By looking at the last two columns of the matrix equality $\bar{W}_{k+1} = Q_{k+1} Z_{k+1}^T$, we see that, as $\bar{W}_{k+1} = [W_k \, \bar{w}^{k+1}]$,

$$[\bar{w}^k \, q^{k+1}] \begin{pmatrix} c_k & s_k \\ s_k & -c_k \end{pmatrix} = [w^k \, \bar{w}^{k+1}], \quad \bar{w}^1 = q^1.$$

We do not want to compute x_{CG}^k at each iteration since $\bar{L}_k$ can be singular. However, L_k is non–singular as long as $\eta_{k+1} \neq 0$, so Z^k is always well defined. Therefore Paige and Saunders chose to compute and update $x_S^k = W_k z^k$. Clearly, we have the update formula

$$x_S^k = x_S^{k-1} + \zeta_k w^k,$$

w^k is obtained by applying the plane rotation to $\bar{w}^k$ and q^{k+1}. The CG iterate can be obtained (if it exists) through

$$x_{CG}^{k+1} = x_S^k + \bar{\zeta}_{k+1} \bar{w}^{k+1}.$$

Paige and Saunders [1057] proved that the residual $r_{CG}^k = b - Ax_{CG}^k$ is given by

$$r_{CG}^k = -\frac{\eta_1 s_1 \cdots s_k}{c_k} q^{k+1}.$$

Therefore, the norm of the residual is available to stop the iteration and we can compute x_{CG}^{k+1} from x_S^k. The Paige and Saunders implementation of the algorithm is the following.

Let x^0 be given, $r^0 = b - Aw^0$, $d_1 = \|r^0\|$, $q^1 = r^0/d_1$, $\delta_1 = (q^1, Aq^1)$, $\theta = \delta_1$, $\nu = 0$, $\bar{q} = Aq^1 - \delta_1 q^1$, $w = Aq^1$, $\eta_2 = \|\bar{q}\|$, $q^2 = \bar{q}/\eta_2$, $\bar{\gamma} = \delta_1$, $\bar{\omega} = \eta_2$, for $k = 2, \dots$

$$\delta_k = (Aq^k, q^k),$$

$$\bar{q} = Aq^k - \delta_k q^k - \eta_k q^{k-1},$$

$$\eta_{k+1} = \|\bar{q}\|, \quad q^{k+1} = \bar{q}/\eta_{k+1},$$

$$\gamma = \sqrt{\bar{\gamma}^2 + \eta_k^2}, c = \bar{\gamma}/\gamma, s = \eta_k/\gamma,$$

$$\omega = c\bar{\omega} + s\delta_k, \bar{\gamma} = s\bar{\omega} - c\delta_k, \pi = s\eta_{k+1}, \bar{\omega} = -c\eta_{k+1},$$

$$\zeta = \theta/\gamma, \sigma = c\zeta, \tau = s\zeta,$$

$$x^{k+1} = x^k + (\sigma w^k + \tau q^{k+1}),$$

$$w^{k+1} = sw^k + cq^{k+1},$$

$$\theta = \nu - \omega\zeta, \quad \nu = -\pi\zeta.$$

A similar algorithm is discussed in Fischer [556] using a QR factorization instead of the LQ factorization of the Paige and Saunders method. Both methods are theoretically equivalent.

6.10. The minimum residual method

We shall use the Lanczos vectors to derive an implementation of the CR algorithm when the matrix A could be indefinite in the same way as we derive SYMMLQ from CG. One way to do this is to write $x^k = x^0 + Q_k y$ and

$$\|r^k\| = \|b - A(x^0 + Q_k y)\| = \|r^0 - AQ_k y\|$$
$$= \|Q_{k+1}(\eta_1 e^1 - \tilde{T}_k y)\| = \|\eta_1 e^1 - \tilde{T}_k y\|.$$

Then, one can use a QR factorization of the $(k+1) \times k$ matrix $\tilde{T}_k$ to obtain the solution. A generalization of this will be considered later on in Chapter 7 when A is non–symmetric. For the time being, let us consider the approach of Paige and Saunders [1057]. The minimum residual l_2–norm can also be found by solving the projected normal equations:

$$Q_k^T A^2 Q_k y^k = Q_k^T Ab, \quad x^k = Q_k y^k.$$

However, using the orthogonality of the Lanczos vectors, we have

$$Q_k^T A^2 Q_k = \bar{T}_k^2 + \eta_{k+1}^2 e^k (e^k)^T,$$
$$Q_k^T Ab = \eta_1 \bar{T}_k e^1.$$

Using the LQ factorization of $\bar{T}_k$, we have $\bar{T}_k^2 + \eta_{k+1}^2 e^k (e^k)^T = L_k L_k^T$. Then,

$$L_k L_k^T y^k = \eta_1 \bar{L}_k Z_k e^1.$$

But $\bar{L}_k$ differs only from L_k in the (k,k) element. Thus we can write $\bar{L}_k = L_k D_k$ where D_k is equal to the identity matrix of order k except for the (k,k) element which is c_k. Then,

$$L_k^T y^k = \eta_1 D_k Z_k e^1.$$

Let us denote $t^k = \eta_1 D_k Z_k e^1$. As $Z_k = Z_{k-1,k} \cdots Z_{1,2}$, it can be shown easily that

$$t_1^k = \eta_1 c_1, \quad t_j^k = \eta_1 s_1 s_2 \cdots s_{j-1} c_j, \ j = 2, \ldots, k$$

Then, the approximation is computed in the following way. Let $V_k = Q_k L_k^{-T}$. Clearly, V_k can be computed column by column starting from the first one and

$$x_M^k = Q_k y^k = Q_k L_k^{-T} L_k^T y^k = V_k t^k.$$

It is also clear that

$$x_M^k = Q_k L_k^{-T} (\eta_1 D_k Z_k e^1).$$

This implementation is known as `MINRES`. The conjugate gradient iterates (when they exist) can be computed from the minimum residual ones as $x_{CG}^k = Q_k L_k^{-T} L_k^T y^k = M_k L_k^T y^k$. Some other methods were developed for indefinite linear systems for instance, the method devised by Fridman and Fletcher, see Fischer [556] for details.

6.11. Hybrid algorithms

Hybrid methods were introduced by Brezinski and Redivo–Zaglia [204]. The principle is quite simple. Suppose we have two approximations x_1 and x_2 of the solution of $Ax = b$ (slightly changing our standard notations), we would like to combine them as

$$y = \alpha x_1 + (1-\alpha)x_2$$

to obtain a better approximation. This is done by requiring the parameter α to minimize the norm of the residual $r = b - Ay$. If r_1 and r_2 are the corresponding residuals, it is easy to see that we must choose

$$\alpha = -\frac{(r_1 - r_2, r_2)}{(r_1 - r_2, r_1 - r_2)}.$$

This choice gives

$$(r, r) \leq \min\{(r_1, r_1), (r_2, r_2)\},$$

as

$$(r, r) = (r_1, r_1) - \frac{(r_1 - r_2, r_1)^2}{(r_1 - r_2, r_1 - r_2)},$$

$$(r, r) = (r_2, r_2) - \frac{(r_1 - r_2, r_2)^2}{(r_1 - r_2, r_1 - r_2)}.$$

Many different applications were proposed in [204] (see also [202]). Suppose we have two sequences of approximations x_1^k and x_2^k, the combination being denoted by y^k then, according to [204], we can

– compute x_1^k and x_2^k by two different iterative methods. The cost is the sum of the costs of both methods or sometimes lower if both methods are related.

– compute x_1^k and take $x_2^k = x_1^{k-1}$. If a splitting method is used, this leads to semi–iterative algorithms. If `CG` is used, then

$$\|r^k\| \leq \min\{\|r_1^k\|, \|r^{k-1}\|\}.$$

– compute x_1^k and take $x_2^k = y^{k-1}$. Then, the norm of the residual is monotonely decreasing and the algorithm is called a smoothing procedure. This was introduced by Schönauer and his co–workers, see Weiss [1321].

– compute x_1^k from y^{k-1} and take $x_2^k = y^{k-1}$. In this case, if x_1^k is the `Gauss--Seidel` sequence we obtain the `SOR` iterates (with a non–optimal parameter).

– compute x_1^k by some method and x_2^k from x_1^k.

Numerical experiments in [202] show that smoothing methods are efficient in getting rid of the wriggles in the convergence histories of some methods. Moreover, there are cheap and easy to implement, so they must be considered as interesting complements to most iterative methods. However, it is useless to apply residual smoothing to algorithms that minimize the norm of the residual.

6.12. Roundoff errors of CG and Lanczos

It has been known since the discoveries of CG and Lanczos in the beginning of the fifties that these methods are prone to instabilities. There are plenty of examples (even of small sizes) for which CG takes much more than n iterations to reach an acceptable level of convergence. In fact, it is this problem that caused CG to be discarded as a direct method. In exact arithmetic the CG residuals (or the Lanczos vectors) are constructed to be orthogonal. What happens in finite precision computation is that the orthogonality of the residuals is gradually lost and moreover things get worse and worse as the algorithm proceeds. However, we have seen in theorem 6.25 that using only local orthogonality (which must be verified up to working precision) then CG converges as fast as steepest descent (whose roundoff properties have been analyzed by Bollen [164]). Figure 6.9 shows the convergence curve with CG for an example designed by Z. Strakos. The matrix is diagonal with eigenvalues given by

$$\lambda_i = a + \frac{i-1}{n-1}(b-a)\rho^{n-i}, \quad i = 1, \ldots, n$$

The computation uses $n = 48, a = 0.1, b = 100$ and $\rho = 0.875$. One can see that it takes many more than 48 iterations to get a small error.

As CG is a non–linear algorithm (because of the scalar products), it is not easy to analyze finite precision CG. So far, the most complete analysis of finite precision CG has been given by A. Greenbaum [698], see also Greenbaum and Strakos [717]. Her analysis relies on results proved by C. Paige [1056] concerning the convergence of the Lanczos algorithm. This is too technical to be reported here. To just get a rough sketch, A. Greenbaum proved that a slightly perturbed Lanczos recurrence (by the local roundoff errors at step k) can be (hypothetically) continued making small additional perturbations ($\tilde{f}^j$) up to an iteration $n + m$ where the coefficient η_k will be zero. If the norm of the perturbations is less than $\xi\|A\|$, then the eigenvalue approximations generated by a perturbed Lanczos recurrence for A are equal to those

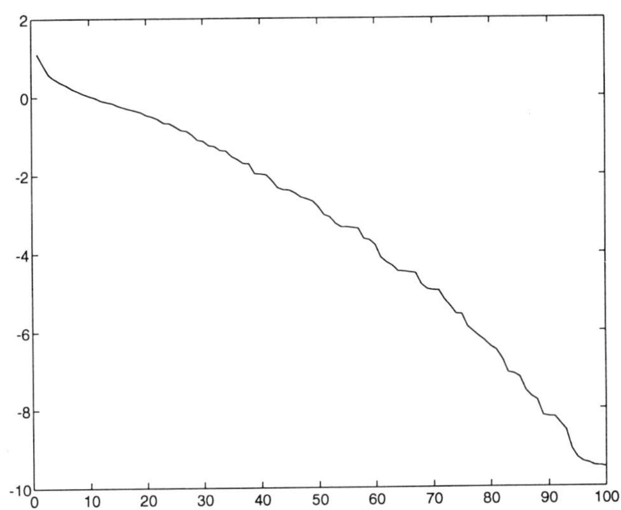

fig. 6.9: $\log_{10}$ *of the A–norm of the error with* CG *for the Strakos example:* $n = 48, a = 0.1, b = 100, \rho = 0.875$

generated by an exact Lanczos recurrence for a matrix $\tilde{A}$ whose eigenvalues lie within intervals of size $\sigma(n+m)^3 \max(u\|A\|, \xi\|A\|)$ where σ is a constant. From these results on Lanczos recurrences, the behavior of CG can be studied. The norms of the CG residuals are equal to the norms of residual vectors generated by an exact CG recurrence for the matrix $\tilde{A}$. Moreover, as long as the norms of the residuals are larger than the sum of the norms of the perturbations the computed residuals approximate well the true residuals. A. Greenbaum was also able to relate the A–norm of the true error to the $\tilde{A}$–norm of the error corresponding to the computed quantities. Both decrease at approximately the same rate. Numerical experiments in Greenbaum and Strakos [717] suggest that the behaviour of finite precision CG is close to the exact algorithm applied to any matrix $\hat{A}$ of larger size whose eigenvalues are located in tiny intervals about the eigenvalues of A. The sizes of these intervals are small multiples of machine precision. In [699] bounds are given for the relative differences between the exact and computed residuals. The same line of analysis has been followed by Notay [1021] who studied the case of isolated eigenvalues at both ends of the spectrum in presence of roundoff errors. He showed that the number of iterations necessary to get rid of the (components of the error related to) large eigenvalues is more sensitive to roundoff errors than for the other parts of the spectrum. Moreover, most of the bounds we have in exact arithmetic hold if we use the condition number of $\tilde{A}$.

H. Simon [1172], [1173], [1174] also studied the finite precision Lanczos algorithm. He made some assumptions (which are more often true) that simplify considerably the analysis. The perturbed Lanczos recurrence is written as

$$\eta_{k+1}q^{k+1} = Aq^k - \delta_k q^k - \eta_k q^{k-1} - f^k,$$

where f^k represents the local roundoff errors. Let F_k be the matrix whose columns are the vectors $f^j, j = 1, \dots, k$. It has been observed experimentally that $\|F_k\| \leq u\|A\|$, so H. Simon used that as hypothesis. Let $W_k = Q_k^T Q_k$, this matrix is equal to the identity in exact arithmetic. But this is not true in finite precision computation. If

$$|(q^i, q^j)| \leq \omega_j, \quad i = 1, \dots, k \quad j = 1, \dots, k \quad i \neq j$$

the smallest ω_j is called the level of orthogonality ω. If $\omega_j = 0$, the vectors are orthogonal, if $\omega_j = \sqrt{u}$ the vectors are said to be semiorthogonal. Simon supposed that the Lanczos vectors are exactly normalized and that local orthogonality is preserved: $|(q^{j+1}, q^j)| \leq \varepsilon_1, \forall j$, with $1 \gg \varepsilon_1 > u$. He also assumed that all η_j's do not become negligible. According to Simon (see also Grcar [691]) the loss of orthogonality comes from an amplification of the local errors because of the recurrence formula of the Lanczos algorithm. Once an error is introduced, it is propagated through the recurrence to future Lanczos vectors.

Lemma 6.34

The elements of W_k satisfy the following recurrences:

$$w_{j,j} = 1, \quad j = 1, \dots, k$$

$$w_{j,j-1} = \varepsilon_1, \quad j = 2, \dots, k$$

$$\eta_{k+1}w_{k+1,j} = \eta_{j+1}w_{k,j+1} + (\delta_j - \delta_k)w_{k,j} + \eta_j w_{k,j-1} - \eta_k w_{k-1,j} + (q^k, f^j) - (q^j, f^k), \quad 1 \leq j < k,$$

$$w_{k,j+1} = w_{j+1,k}$$

Proof. See Simon [1172]. □

This formula shows that the growth of the elements of W_k depends on the coefficients η_k's and δ_k's and is initiated by the local errors. These recurrences can also be written in matrix form. If R_k is the strictly upper triangular part of W_k and $\bar{w}^j$ its columns, we have

$$\eta_{k+1}\|w^{k+1}\| \leq 2\|A\| \max(\|\bar{w}^k\|, \|\bar{w}^{k-1}\|) + O(u\|A\|).$$

This implies that a small η_{k+1} could cause a great loss of orthogonality. The coefficients η_k and δ_k depend on the eigenvalue distribution of A and the initial vector q^1. Let $T_k S_k = S_k \Theta_k$ be the spectral decomposition of the computed T_k, Θ_k being a diagonal matrix and $y^j = Q_k s^j$, the s^j's being the columns of S_k. Let $\sigma_{k,j}$ be the bottom element of the eigenvector s^j. Then, Simon obtained that

$$(y^j, q^{k+1}) = \frac{\phi_{j,j}}{\eta_{k+1}\sigma_{k,j}},$$

where $\phi_{j,j} = (s^j, G_k s^j)$ and G_k is the strictly upper triangular part of $F_k^T Q_k - Q_k^T F_k$. The element $\phi_{j,j}$ is of the order of $u\|A\|$, so orthogonality between the Ritz vectors and the Lanczos vectors can only be lost if $\sigma_{k,j}$ is small. This occurs when a Ritz value has converged to an eigenvalue of A. Strangely enough, loss of orthogonality goes hand in hand with convergence of a Ritz value.

Theorem 6.35

If

$$\omega \leq \frac{1}{2}\frac{1}{k-2},$$

then the computed $\bar{T}_k$ is similar to a matrix $\hat{T}_k$ which is a perturbation of the orthogonal projection of A onto span(Q_k).

Proof. See Simon [1172]. □

Theorem 6.36

If ω is less than $O(\sqrt{u})$,

$$N_k^T A N_k = \bar{T}_k + V_k,$$

where the elements of V_k are $O(u\|A\|)$ and N_k is a matrix with orthonormal columns in the QR factorization of Q_k.

Proof. See Simon [1172]. □

A way to cure the loss of orthogonality among the Lanczos vectors is to reorthogonalize them. For the CG residuals, this requires computing

$$r^k = r^k - \frac{(r^k, r^j)}{(r^j, r^j)} r^j, \quad j = 1, \ldots, k-1.$$

Figure 6.10 compares CG and CG with full reorthogonalization for the Strakos example of figure 6.9

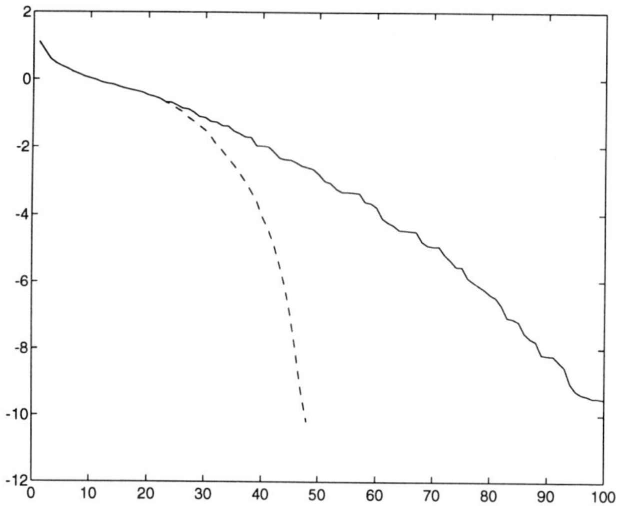

fig. 6.10: $\log_{10}$ *of the A–norm of the error with* CG
solid: CG, *dashed:* CG *with full reorthogonalization*

However, as we have seen before, full reorthogonalization (whose goal is to maintain orthogonality to working precision) is not necessary as it is enough to maintain semiorthogonality. Selective reorthogonalization was introduced by Parlett and Scott [1064] to keep semiorthogonality. Reorthogonalization works the following way for the Lanczos algorithm. Let

$$\bar{q}^k = Aq^k - \delta_k q^k - \eta_k q^{k-1} - f^k,$$

and suppose we decide to reorthogonalize. Then,

$$\eta_{k+1} q^{k+1} = \bar{q}^k - \sum_{j=1}^{k-1} \xi_j q^j - (f^k)'.$$

Ideally, we would like q^{k+1} to be orthogonal to working precision to the previous vectors. However, if we only need to maintain semiorthogonality, it is enough to have $|(q^j, q^{k+1})| \leq O(u\|A\|)$. We also need to have $\|(f^k)'\| \leq u\|A\|$. Of course, the coefficients ξ_j are given by $\xi_j = (\bar{q}^k, q^j)$. Simon [1172] introduced partial reorthogonalization to maintain semiorthogonality. Suppose we are able to compute an estimate of $w_{k+1,j}$. This will indicate against which previous Lanczos vectors orthogonality has been lost. Then, it is enough to orthogonalize against these particular vectors. Of course, it is not trivial to compute estimates of $w_{k+1,j}$ as we do not know the local roundoff errors. Simon [1172] suggested replacing the unknown quantities by random values appropriately chosen. Numerical experiments in [1172] show that this is an effective way to monitor the loss of orthogonality and to maintain semiorthogonality. Regarding solving the linear system, if semiorthogonality is maintained then, everything is fine as long as the required accuracy is not less than $\sqrt{u}$.

Sleijpen, Van der Vorst and Modersitzki [1186] analyzed the norms of the residuals and the departure of the computed residuals from the true ones for some methods that we have described before. The error in the residual Δr^k due to the floating point computation of the solution is bounded by

$$\frac{\|\Delta r^k\|}{\|r^0\|} \leq C\sqrt{k}u\kappa(A)^2 \quad \text{for MINRES}$$

where C is a constant and

$$\frac{\|\Delta r^k\|}{\|r^0\|} \leq C(4\sqrt{k} + 2\|A^{\frac{1}{2}}Q_k\|)u\kappa(A) \quad \text{for CR.}$$

Other results are given in [1186] for CG and SYMMLQ.

6.13. Solving for several right hand sides

It happens quite often that we have to solve a series of linear systems with the same matrix A and different right hand sides $b^i, i = 1, \ldots, m$. In that case if we use Gaussian elimination, we have to factor the matrix only once (with a cost proportional to n^3) and then we use the LU factors to solve all the systems (with a cost proportional to mn^2). With iterative methods, it seems difficult to exploit the fact that the matrix is the same for all systems. One thing that we can easily do is to choose the starting vectors by using solutions of the previous systems if they represent "close" physical problems.

However, it turns out that we can sometimes speed up the solution when several systems are to be solved with CG. This problem has been studied by Joly [834], Chan and Wan [299] and Erhel and Guyomarc'h [537]. We follow the exposition of Joly. This uses a general formulation for non–symmetric matrices that was introduced in Joly and Meurant [836] for matrices with complex coefficients. This general formulation allows us to derive several of the methods we have already introduced in this chapter.

Suppose we have a symmetric matrix H that defines a scalar product and another matrix K such that its symmetric part $K^* = 1/2(K + K^T)$ is either positive or negative definite and let $N = A^T HA$. The general algorithm for a single right hand side is the following:

Let x^0 be given, $r^0 = b - Ax^0$, $g^0 = A^T Hr^0$, $p^0 = Kg^0$, for $k = 0, 1, \dots$

$$
\begin{aligned}
\alpha_k &= \frac{(g^k, p^k)}{(p^k, Np^k)}, \\
x^{k+1} &= x^k + \alpha_k p^k, \\
r^{k+1} &= r^k - \alpha_k Ap^k, \\
g^{k+1} &= g^k - \alpha_k Np^k, \\
\beta_{k+1}^l &= -\frac{(Kg^{k+1}, Np^l)}{(p^l, Np^l)}, \quad 0 \le l \le k \\
p^{k+1} &= Kg^{k+1} + \sum_{l=0}^{k} \beta_{k+1}^l p^l.
\end{aligned}
$$

Different methods are recovered from appropriate choices of H and K. For instance, when A is symmetric positive definite, CG is obtained by choosing $H = A^{-1}$, $K = I$. A CR algorithm is given by $H = I$ and $K = A^{-T}$.

Theorem 6.37

The vectors defined by the previous algorithm are such that:

$$
\begin{aligned}
(p^k, Np^l) &= 0, \quad \forall k \neq l \\
(g^k, p^l) &= 0, \quad 0 \le l < k \\
(g^l, p^k) &= (g^0, p^k), \quad 0 \le l \le k \\
(g^k, p^k) &= (g^k, Kg^k), \\
(Kg^k, Np^k) &= (p^k, Np^k), \\
(g^k, Kg^l) &= 0, \quad 0 \le l < k
\end{aligned}
$$

$$span(p^0, p^1, \ldots, p^k) = span(Kg^0, Kg^1, \ldots, Kg^k) = span(p^0, KNp^0, \ldots, (KN)^k p^0).$$

Proof. See Joly and Meurant [836]. □

Moreover, we have $g^n = 0$ and many things simplify if we suppose K symmetric as then we have $\beta_{k+1}^l = 0,\ 0 \le l < k$ and

$$\beta_{k+1}^k = \frac{(g^{k+1}, Kg^{k+1})}{(g^k, Kg^k)}.$$

This method minimizes the functional $J(r) = (r, Hr)$ over the space $x^0 +$ span$(p^0, \ldots, p^k)$. In the H–norm the residuals are monotonically decreasing and if $K = L^T L$ and $M = L^T NL$ we have

$$(r^k, Hr^k) \le (r^0, Hr^0) \left(\frac{\sqrt{\kappa(M)} - 1}{\sqrt{\kappa(M)} + 1} \right)^{2k}.$$

If we have several right hand sides $b^i, i = 1, \ldots, m$ we try to minimize (r_i^k, Hr_i^k) over $x_i^0 +$ span$(p^0, \ldots, p^k)$. In other words, we want to use the same basis of the Krylov space for all systems. We set

$$x_i^{k+1} = x_i^k + \alpha_k^{(i)} p^k,$$
$$r_i^{k+1} = r_i^k - \alpha_k^{(i)} Ap^k.$$

If $g_i^k = A^T Hr_i^k$, the coefficients are given by

$$\alpha_l^{(i)} = \frac{(g_i^l, p^l)}{(p^l, Np^l)}, \quad 1 \le i \le m, \quad 0 \le l \le k.$$

The vectors p^k are computed in the same way as for a single right hand side. The first (or a chosen) linear system with right hand side b^1 is denoted as the seed system. The same kind of orthogonality properties that were given in theorem 6.37 arise in the case of multiple right hand sides.

Theorem 6.38

We have

$$(g_i^k, p^l) = 0, \quad 0 \le l < k, \quad 1 \le i \le m$$
$$(g_i^l, p^k) = (g_i^0, p^k), \quad 0 \le l \le k, \quad 1 \le i \le m$$
$$(g_i^k, p^k) = (g_i^k, Kg^k), \text{ where } g^k \text{ is associated with the seed system}$$

Proof. See Joly [834]. □

Note that only one matrix×vector product is needed per iteration whatever is the number of right hand sides. The main question is knowing what to do when the iterations for the seed system have converged (for instance at iteration k'). Then, it might be that convergence has not yet been attained for the other systems. From the remaining linear systems a new seed system has to be chosen. Joly [834] suggested choosing the system s with the largest residual norm $\|r_s^k\|$. The new search direction is computed as

$$p^{k'} = Kg_s^{k'} + \sum_{l=0}^{k'-1} \beta_{k'}^l p^l,$$

$$\beta_{k'}^l = -\frac{(Kg_s^{k'}, Np^l)}{(p^l, Np^l)}.$$

This choice preserves the orthogonality properties before and after the restart. Then, it can be proved that the restarted algorithm converges at least in n iterations (in exact arithmetic) whatever is the number of right hand sides. To summarize the algorithm in the case where K is symmetric positive definite, we have

Let $x_i^0, 1 \leq i \leq m$ be given, $r_i^0 = b^i - Ax_i^0$, $g_i^0 = A^T H r_i^0$, $s = 1$, $p^0 = Kg_s^0$, for $k = 0, 1, \ldots$

$$\alpha_k^{(i)} = \frac{(g_i^k, p^k)}{(p^k, Np^k)},$$

$$x_i^{k+1} = x_i^k + \alpha_k^{(i)} p^k,$$

$$r_i^{k+1} = r_i^k - \alpha_k^{(i)} Ap^k,$$

$$g_i^{k+1} = g_i^k - \alpha_k^{(i)} Np^k,$$

Stop the iterations for systems which have converged. If the seed system has converged, then we set $s_{old} = s$, s is such that $\|r_s^{k+1}\|$ is maximum and

$$\beta_{k+1} = \frac{(g_s^{k+1}, Kg_{s_{old}}^{k+1})}{(g_{s_{old}}^k, Kg_{s_{old}}^k)},$$

$$p^{k+1} = Kg_s^{k+1} + \beta_{k+1} p^k.$$

Numerical experiments in Joly [834] show that this method is quite effective. An analysis is also given in Chan and Wan [299] where a block seed method is proposed. These methods are particularly effective when the different right hand sides are related to each other ("close" in some sense) which is the case in many physical problems. If this is not the case the block CG method of O'Leary that we shall study in the following section may be more suitable.

6.14. Block CG and Lanczos

In this section we are considering block generalizations of the Lanczos and CG algorithms. We start with the Lanczos algorithm.

6.14.1. The block Lanczos algorithm

For the sake of simplicity let us consider the case of 2×2 blocks. Let Q_0 be an $n \times 2$ given matrix, such that $Q_0^T Q_0 = I$. Let $Q_{-1} = 0$ be also an $n \times 2$ matrix. Then, let us define

$$\Omega_j = Q_{j-1}^T A Q_{j-1},$$

$$R_j = AQ_{j-1} - Q_{j-1}\Omega_j - Q_{j-2}\Gamma_{j-1}^T,$$

$$Q_j \Gamma_j = R_j$$

The last step is the QR factorization of R_j such that Q_j is $n \times 2$ with $Q_j^T Q_j = I$ and Γ_j is 2×2. The matrix Ω_j is 2×2 and Γ_j is upper triangular. It may happen that R_j is rank deficient and in that case Γ_j is singular. The solution of this problem was given in Golub and Underwood [673]. One of the columns of Q_j can be chosen arbitrarily. To complete the algorithm, we choose this column to be orthogonal to the previous block vectors Q_k. We can, for instance, choose another vector (randomly) and orthogonalize it against the previous ones. This block algorithm generates a sequence such that $Q_j^T Q_i = \delta_{i,j} I$ where I is the 2×2 identity matrix.

Proposition 6.39

The matrices Q_i can be written as

$$Q_i = \sum_{k=0}^{i} A^k Q_0 C_k^{(i)},$$

where $C_k^{(i)}$ are 2×2 matrices.

Proof. The proof is easily obtained by induction. □

Let us look at the relation with matrix polynomials and computing estimates of $W^T f(A)W$ where W is an $n \times 2$ matrix. We define a matrix polynomial $p_i(\lambda)$, a 2×2 matrix, as

$$p_i(\lambda) = \sum_{k=0}^{i} \lambda^k C_k^{(i)}.$$

Then, we have the following result.

Theorem 6.40

$$Q_i^T Q_j = \int_a^b p_i(\lambda)^T d\alpha(\lambda) p_j(\lambda) = \delta_{i,j} I,$$

where $\alpha = W^T Q$ *is a* $2 \times n$ *matrix.*

Proof. Using the orthogonality of the Q_i's and the spectral decomposition of A, we can write

$$\begin{aligned}
\delta_{i,j} I = Q_i^T Q_j &= \left(\sum_{k=0}^{i} (C_k^{(i)})^T Q_0^T A^k \right) \left(\sum_{l=0}^{j} A^l Q_0 C_l^{(j)} \right) \\
&= \sum_{k,l} (C_k^{(i)})^T Q_0^T Q \Lambda^{k+l} Q^T Q_0 C_l^{(j)} \\
&= \sum_{k,l} (C_k^{(i)})^T \alpha \Lambda^{k+l} \alpha^T C_l^{(j)} \\
&= \sum_{k,l} (C_k^{(i)})^T \left(\sum_{m=1}^{n} \lambda_m^{k+l} \alpha_m \alpha_m^T \right) C_l^{(j)} \\
&= \sum_{m=1}^{n} \left(\sum_k \lambda_m^k (C_k^{(i)})^T \right) \alpha_m \alpha_m^T \left(\sum_l \lambda_m^l C_l^{(j)} \right).
\end{aligned}$$

□

The p_js can be considered as matrix orthogonal polynomials for the (matrix) measure α. The following recurrence relation holds.

Theorem 6.41

The matrix valued polynomials p_j satisfy

$$p_j(\lambda)\Gamma_j = \lambda p_{j-1}(\lambda) - p_{j-1}(\lambda)\Omega_j - p_{j-2}(\lambda)\Gamma_{j-1}^T,$$

$$p_{-1}(\lambda) \equiv 0, \quad p_0(\lambda) \equiv I,$$

where λ is a scalar.

Proof. From the previous definition, it is easily shown by induction that p_j can be generated by the given (matrix) recursion. □

This block recurrence can be written as

$$\lambda[p_0(\lambda), \ldots, p_{N-1}(\lambda)] = [p_0(\lambda), \ldots, p_{N-1}(\lambda)]J_N + [0, \ldots, 0, p_N(\lambda)\Gamma_N],$$

and as $P(\lambda) = [p_0(\lambda), \ldots, p_{N-1}(\lambda)]^T$,

$$J_N P(\lambda) = \lambda P(\lambda) - [0, \ldots, 0, p_N(\lambda)\Gamma_N]^T,$$

with J_N defined as

$$J_N = \begin{pmatrix} \Omega_1 & \Gamma_1^T & & & \\ \Gamma_1 & \Omega_2 & \Gamma_2^T & & \\ & \ddots & \ddots & \ddots & \\ & & \Gamma_{N-2} & \Omega_{N-1} & \Gamma_{N-1}^T \\ & & & \Gamma_{N-1} & \Omega_N \end{pmatrix},$$

is a block tridiagonal matrix of order $2N$ and a banded matrix whose half bandwidth is 2 (we have at most 5 non–zero elements in a row). The eigenvalues of J_N are approximations of those of A. One of the advantages of the block method over the (scalar) Lanczos method is that we can obtain multiple eigenvalues. It is interesting to note the following proposition.

Proposition 6.42

The eigenvalues of J_N are the zeros of $det[p_N(\lambda)]$.

Proof. Let μ be a zero of $det[p_N(\lambda)]$. As the rows of $p_N(\mu)$ are linearly dependent, there exists a vector v with two components such that

$$v^T p_N(\mu) = 0.$$

This implies that

$$\mu[v^T p_0(\mu), \ldots, v^T p_{N-1}(\mu)] = [v^T p_0(\mu), \ldots, v^T p_{N-1}(\mu)] J_N.$$

Therefore μ is an eigenvalue of J_N. $det[p_N(\lambda)]$ is a polynomial of degree $2N$ in λ. Hence, there exist $2N$ zeros of the determinant and therefore all eigenvalues are zeros of $det[p_N(\lambda)]$. □

The eigenvalues of J_N can be computed by first reducing the matrix to tridiagonal form. The block Lanczos is useful to compute eigenvalues for matrices having multiple eigenvalues. It can also be used to compute the solution of linear systems with several right hand sides. These results can be used to define block quadrature rules that allow to obtain estimates of bilinear forms in the same way it was done for a posteriori error estimates, see Golub and Meurant [665].

6.14.2. The Block CG algorithm

D.P. O'Leary [1033] proposed several block methods including a block CG algorithm. The "Hestenes and Stiefel" (two–term recurrence) version of the method for solving $AX = B$ where X and B are n by s is the following:

Let X^0 be an $n \times s$ matrix, $R^0 = B - AX^0$, $P^0 = R^0 \gamma_0$ where γ_0 is $s \times s$, for $k = 0, 1, \ldots$

$$\alpha_k = ((P^k)^T A P^k)^{-1} \gamma_k^T (R^k)^T R^k,$$

$$X^{k+1} = X^k + P^k \alpha_k,$$

$$R^{k+1} = R^k - AP^k \alpha_k,$$

$$\beta_k = \gamma_k{}^{-1}((R^k)^T R^k)^{-1}(R^{k+1})^T R^{k+1},$$

$$P^{k+1} = (R^{k+1} + P^k \beta_k)\gamma_k.$$

The block matrices satisfy

$$(R^k)^T R^j = 0, \quad j \neq k$$

$$(P^k)^T A P^j = 0, \quad j \neq k$$

The matrices γ_k can be chosen to reduce roundoff. We can also derive a three–term recurrence version of block CG and a block minimum residual algorithm. O'Leary has shown that the rate of convergence is governed by the ratio λ_n / λ_s. The block CG method can be derived from the block Lanczos algorithm by using a block LU factorization of the block tridiagonal matrix that is generated. Block CG can be used when solving systems with different right hand sides which are not related to each other.

6.15. Inner and outer iterations

When we introduced PCG we supposed that the preconditioner M is constructed in such a way that we can exactly solve the linear systems $Mz^k = r^k$ at each CG iteration. This is the case when $M = LL^T$. However, it can be the case that either the preconditioner is not explicitly computed or not in a form such that $Mz = r$ can be easily solved. Then, this system can be approximately solved by another iterative method. These iterations will be denoted as inner iterations as opposed to the CG iterations which are the outer iterations. The problem is to know if the algorithm is still converging and when the inner iterations have to be stopped.

This problem was studied for the Richardson and Chebyshev methods by Golub and Overton [669], [670] and by Munthe–Kaas for the steepest descent algorithm [1000]. Golub and Ye [680] proved results for CG. They modified CG in the following way:

Let x^0 be given, $r^0 = b - Ax^0$. For k=0,1,...

$$
\begin{aligned}
Mz^k &= r^k + y^k, \ \|y^k\|_{M^{-1}} \le \epsilon \|r^k\|_{M^{-1}} \\
\beta_k &= \frac{(z^k, r^k - r^{k-1})}{(z^{k-1}, r^{k-1})}, \quad \beta_0 = 0, \\
p^k &= z^k + \beta_k p^{k-1}, \\
\gamma_k &= \frac{(z^k, r^k)}{(p^k, Ap^k)}, \\
x^{k+1} &= x^k + \gamma_k p^k, \\
r^{k+1} &= r^k - \gamma_k A p^k.
\end{aligned}
$$

This form of PCG allows to maintain local orthogonality properties, that is $(p^{k-1}, r^{k+1}) = 0$, $(p^k, r^{k+1}) = 0$, $(z^k, r^{k+1}) = 0$, $(p^k, Ap^{k+1}) = 0$. This implies that the algorithm converges as fast as the inexactly preconditioned steepest descent method. For both theoretical and experimental reasons, Golub and Ye [680] suggested using

$$
\epsilon = \frac{1}{2\sqrt{\kappa(M^{-1}A)}},
$$

if this is computable. A CG method with variable (even non–linear) preconditioner was proposed by Axelsson and Vassilevski [88]. They show that if

$$(y, AM(y)y) \geq \delta_1(y, y), \forall y, \quad \|AM(y)\| \leq \delta_2\|y\|, \forall y,$$

the algorithm converges and

$$\|r^k\| \leq \sqrt{1 - \left(\frac{\delta_1}{\delta_2}\right)^2} \|r^{k-1}\|.$$

6.16. Constrained CG

R. Nicolaides [1018] proposed a variant of CG where the residuals verify linear constraints. This method was named "deflation" by Nicolaides, however deflation has another meaning for eigenvalue computations and we prefer to call it constrained CG. Similar methods have been developed for Krylov methods for non–symmetric systems by Wallis, Kendall and Little [1315]. The method was formulated for the three–term recurrence form of CG. Let C be a given $n \times m$ matrix ($m < n$) of rank m. We would like to modify CG to have

$$C^T r^k = 0, \; \forall k$$

Let

$$x^{k+1} = \omega_{k+1} x^k + (1 - \omega_{k+1}) x^{k-1} + \alpha_k \omega_{k+1}(r^k - Cu^k),$$

where u^k is to be defined. It follows that the residuals are defined by

$$r^{k+1} = \omega_{k+1} r^k + (1 - \omega_{k+1}) r^{k-1} + \alpha_k \omega_{k+1} A(r^k - Cu^k).$$

The vector u^k is defined by minimizing

$$(A(r^k - Cu^k), r^k - Cu^k).$$

The solution of this problem is given by solving an order m linear system

$$C^T A C u^k = C^T A r^k.$$

Denoting $\mathcal{A}_C = C^T A C$, the equation for the residuals becomes

$$r^{k+1} = \omega_{k+1} r^k + (1 - \omega_{k+1}) r^{k-1} + \alpha_k \omega_{k+1} A(I - C{\mathcal{A}_C}^{-1} C^T A) r^k.$$

Theorem 6.43

If $C^T r^0 = 0$, *then* $C^T r^k = 0, \forall k$.

Proof. This is easily proved by induction. □

The matrix $P_C = I - C\mathcal{A}_C{}^{-1}C^T A$ is the projector onto $\text{span}((AC)^\perp)$ along $\text{span}(C)$. Let $A_C = A(I - PC)$, this matrix is symmetric and singular. The coefficients of CG are determined by requiring that r^{k+1} is orthogonal to r^k and r^{k-1}. This gives

$$\alpha_k = \frac{r^k, r^k)}{(r^k, A_C r^k)}.$$

The formula for ω_{k+1} is the same as for CG. Note that at each iteration, we must solve an $m \times m$ linear system. Constrained CG can also be combined with preconditioning. The initial constraint $C^T r^0 = 0$ can be satisfied in the following way: let u be an arbitrary vector, $s = b - Au$ and

$$C^T ACt = C^T s.$$

We set $x^0 = u + Ct$. This gives $r^0 = b - A(u + Ct)$ and $C^T r^0 = C^T b - C^T Au - C^T ACt = C^T b - C^T Au - C^T(b - Au) = 0$.

The matrix of constraints C can be chosen in many different ways. Nicolaides' suggestion is related to problems arising from discretization of PDEs. It uses a coarse mesh as if the set of unknowns is partitioned into m disjoint subsets $\Omega_k, k = 1, \ldots, m$, then

$$c_{i,j} = \begin{cases} 1 & i \in \Omega_j \\ 0 & i \notin \Omega_j \end{cases}$$

Wallis, Kendall and Little [1315] used C to enforce column sums constraints.

6.17. Vector and parallel PCG

Usually the form which is computationally preferred for PCG is (6.9). Remember that for one iteration we have

1) $Mz^k = r^k$,

2) $\beta_k = \frac{(z^k, r^k)}{(z^{k-1}, r^{k-1})}$,

3) $p^k = z^k + \beta_k p^{k-1}$,

4) $r_k = \frac{(z^k, r^k)}{(p^k, Ap^k)}$,

5) $x^{k+1} = x^k + \gamma_k p^k$,

6) $r^{k+1} = r^k - \gamma_k Ap^k$.

The operations that are involved in one iteration are

i) two scalar products (sdots)

ii) three times a vector multiplied by a scalar and added to a vector (saxpys)

iii) a matrix$\times$vector product

iv) solving a linear system with matrix M.

PCG is very well suited to vector computers. It is obvious that ii) is a vector operation. Of course, fast routines for scalar products involved in i) are available on vector computers that allow us to do these operations at almost vector speed. The operation to perform is

$$(x, y) = \sum_{i=1}^{n} x_i y_i.$$

Clearly the multiplication of the components of the two vectors is a vector operation that gives us a vector w. Summing the components w_i of vector w is a reduction operation that can be done in the following way. Suppose, for the sake of simplicity, that the vector length is $n = 2^p$. Then one can sum (in vector mode) odd and even components of w, i.e., w_{2i} and w_{2i+1}, this leads to a vector of length $\frac{n}{2} = 2^{p-1}$ and we can apply the same method to the resulting vector. After $p = \log_2 n$ steps we shall get the desired scalar product. Of course, the problem is that at each step the vector length is halved.

The vectorization of the multiplication by A depends on the nature of the problem we are solving and more precisely on the storage scheme which is used for A (see Chapter 1). For problems arising from discretization of partial differential equations, finite difference methods (on regular or structured meshes) yield systems with a regular pattern that can be stored by diagonals. Then, the multiplication is done by diagonals giving an optimal vector length. Matrices arising from finite element methods (on distorted or unstructured meshes) are stored keeping only the non–zero elements and their column numbers. Then, to compute the matrix×vector product, we need indirect addressing and the vector lengths are small, typically the number of elements in a row. Hence the speeds reached with these problems are much smaller than with structured meshes. But, of course, the situation can change with progress in computers and accesses to memories. This problem can also be solved by using other storage schemes which define pseudo diagonals. Good computing speeds can be reached in this way, see Erhel [534], Melhem [961]. One may also think of finding orderings of the unknowns that give some structure to an unordered matrix.

The last problem is solving $Mz = r$ at each iteration. The vectorizability depends evidently on the choice of M, this will be studied in Chapter 8.

Regarding parallelism, the situation is not quite as satisfactory especially on today's distributed memory supercomputers because we need to split the algorithm into pieces that must be large enough because of the overhead of the synchronization procedures. Unfortunately PCG (in its usual form) is a very sequential process. Regardless of step 1 and assuming that part of the scalar product (z^k, r^k) can be computed when solving $Mz^k = r^k$, we need to have completed the (scalar) step 2 before beginning step 3. After this, we need to compute Ap^k and (p^k, Ap^k) and even if it depends on the storage used for A, it is likely that we have to wait for the completion of step 3 to do that. Then we need to finish step 4 before computing steps 5 and 6. So in an obvious way, only steps 5 and 6 can be executed in parallel. There are at least three synchronization points and more likely five, regardless of what will be needed for step 1. Moreover, the number of operations that can be done in parallel is very small.

This is very unsatisfactory. Therefore, many people have been looking for ways to modify the algorithm to increase the degree of parallelism. There are many directions for doing this (see [364], Chronopoulos and Gear [323]). The idea is to reduce the data dependencies trying to compute scalars β_k and γ_k at once.

We have

$$z^{k+1} = z^k - \gamma_k M^{-1} A p^k,$$
$$r^{k+1} = r^k - \gamma_k A p^k.$$

So,

$$\gamma_k^2 (M^{-1} A p^k, p^k) = (z^{k+1} - z^k, r^{k+1} - r^k).$$

But, we show in proposition 6.7 that $(r^i, z^j) = 0 \quad i \neq j$, therefore,

$$\gamma_k^2 (M^{-1} A p^k, p^k) = (z^{k+1}, r^{k+1}) + (r^k, z^k).$$

A natural idea would be to compute (r^k, z^k) by recurrence. However, we have to be a little careful to preserve the stability of the algorithm. One way to do this is to use a predictor–corrector method computing one more scalar product [968]. Rearranging the preconditioned algorithm we get:

$$M v^k = A p^k,$$

compute $(v^, Ap^k)$, (Ap^k, p^k), (r^k, z^k),

$$\alpha_k = \frac{(r^k, z^k)}{(Ap^k, p^k)},$$

$$s_{k+1} = \alpha_k^2 (v^k, Ap^k) - (r^k, z^k),$$

$$\beta_{k+1} = \frac{s_{k+1}}{(r^k, z^k)},$$

$$\begin{aligned} x^{k+1} &= x^k + \alpha_k p^k, \\ r^{k+1} &= r^k - \alpha_k A p^k, \\ z^{k+1} &= z^k - \alpha_k v^k, \\ p^{k+1} &= (z^k - \alpha_k v^k) + \beta_{k+1} p^k. \end{aligned}$$

We see that once the three scalar products have been computed everything can be computed in parallel. The scalar s_{k+1} is a prediction of the value of (r^{k+1}, z^{k+1}). It is corrected at the next iteration to preserve stability. The cost of the algorithm is 3 scalar products, a matrix×vector product and $8n$ floating point operations. The overhead is thus $4n$ operations, but there are much more parallelism. The gain over the standard algorithm really depends on the cost of the matrix×vector product. If it is large relative to the $8n$ operations not much can be saved. On the other hand if A is really sparse some gains may be expected.

However, it turns out that we do not need to use all these tricks. The solution to this problem is simply to use the three–term recurrences that we used to introduce CG. Remember that the algorithm is:

Let x^0 be given for $k = 0, 1, \ldots$ until convergence

$$\begin{aligned}
Mz^k &= r^k = b - Ax^k, \\
\alpha_k &= \frac{(z^k, Mz^k)}{(z^k, Az^k)}, \\
\omega_{k+1} &= \frac{1}{1 - \frac{\alpha_k}{\omega_k \alpha_{k-1}} \frac{(z^k, Mz^k)}{(z^{k-1}, Mz^{k-1})}}, \quad \omega_1 = 1, \\
x^{k+1} &= x^{k-1} + \omega_{k+1}(\alpha_k z^k + x^k - x^{k-1}), \\
r^{k+1} &= r^{k-1} - \omega_{k+1}(\alpha_k A z^k - r^k + r^{k-1}).
\end{aligned}$$

Clearly, we see that the two scalar products can be computed in parallel. Moreover, there is also natural parallelism in the computation of x^{k+1}. Therefore, this version is perfectly parallel. However, this may be counterbalanced by the fact that the number of floating point operations is larger. We have $10n$ operations plus two scalar products, a matrix×vector operation and the solve for the preconditioner. This is to be compared to $6n$ operations plus the same operations for the two–term recurrence version. Moreover it has been shown that the three–term version is less stable.

6.18. Bibliographical comments

A very interesting account of the history of CG from the beginning up to 1976 is given in the Golub and O'Leary paper [667]. The origin of CG is the Hestenes and Stiefel paper [772] which is still very interesting to read. Many of the ideas that were developed later, including preconditioning, are already described in this paper, see also Hestenes [770] and the Hestenes' book [771]. Also of interest is the Lanczos paper [879].

The authors of CG did know about the problems due to loss of orthogonality and it was also known very early that the method could be used as an iterative one, see Forsythe [569]. However, the method was not much used and was almost forgotten. We should give credit to J.K. Reid [1095] for having shown that the method could be interesting for iteratively solving "large" sparse linear systems.

What makes the success of CG in the seventies was its coupling to efficient preconditioners (see Chapter 8) which gave favorable eigenvalue distributions for fast convergence. A very influential paper in this respect was the Concus, Golub and O'Leary paper [343].

The behaviour of CG in finite precision arithmetic was mainly investigated by Greenbaum and her co–workers [695], [696], [698], [717] starting in 1980. The behaviour of the Lanczos algorithm was considered by Grcar [691] and Simon [1172], [1173], [1174].

The methods for indefinite matrices were mainly studied by Paige and Saunders [1057], [1058], see also Chandra [303] and the book by Fischer [556].

7 Krylov methods for non–symmetric systems

In Chapter 6 we studied CG and CG–like algorithms for solving linear systems for which the matrix A is symmetric either positive definite or indefinite. In this chapter we are concerned with the case where A is non–symmetric. Of course, this is such a large class of matrices that there exist many methods for solving such systems. We have already seen the "classical" methods like Jacobi, Gauss-Seidel, etc... in Chapter 5. Here we are going to concentrate on methods using Krylov spaces and we shall only study the most popular ones. As we shall see, there is a great similarity in the way these methods are derived and what we have seen for symmetric matrices in Chapter 6. Most methods construct orthogonal bases of some spaces and then, solve the projected equation directly or try to minimize some norm of the residual or the error. However, the situation for non–symmetric matrices is much less satisfactory than for the symmetric case. Unfortunately today there is no method for which we can say that it is always much better than the others. On the contrary, one can always find examples (although sometimes contrived ones) for which a given method is either the best or the worst, see for example Nachtigal, Reddy and Trefethen [1003].

We are going to start by considering the normal equations approach. This has been used for quite a long time to solve non–symmetric systems of equations. Although the rate of convergence is sometimes rather low, these methods are robust. They are also related to solving least squares problems, see

Björck [154].

7.1. The normal equations

In Chapter 6, we have seen that in its general form CG is not restricted to symmetric matrices. The algorithm has been written as

$$\begin{aligned} Mz^k &= r^k (= b - Ax^k), \\ \alpha_k &= \frac{((r^k, r^k))}{((AM^{-1}r^k, r^k))}, \\ \omega_{k+1} &= \frac{1}{1 - \frac{\alpha_k}{\omega_k \alpha_{k-1}} \frac{((r^k, r^k))}{((r^{k-1}, r^{k-1}))}}, \quad \omega_1 = 1, \\ x^{k+1} &= x^{k-1} + \omega_{k+1}(\alpha_k z^k + x^k - x^{k-1}). \end{aligned}$$

We have convergence if AM^{-1} is self adjoint for the given scalar product i.e., $((AM^{-1}x, y)) = ((x, AM^{-1}y))$, $\forall x, y$ and positive definite e.g., $((AM^{-1}x, x)) > 0$, $\forall x \neq 0$. The scalar product $((.,.))$ can be chosen at our will. Let us suppose we have two symmetric positive definite matrices S and P. Then, we use P to define the scalar product as

$$((x, y)) = (Px, y), \quad \forall x, y \tag{7.1}$$

and we choose $M = (SA^TP)^{-1}$. With these notations and choices, we have the following result.

Proposition 7.1

Let $M = (SA^TP)^{-1}$. Then AM^{-1} is self adjoint and positive definite for the scalar product $((.,.))$ defined by (7.1).

Proof. By definition $M^{-1} = SA^TP$. Thus, $AM^{-1} = ASA^TP$ and

$$((AM^{-1}x, y)) = (PASA^TPx, y) = (Px, ASA^TPy) = ((x, AM^{-1}y)).$$

In the same way,

$$((AM^{-1}x, x)) = (SA^TPx, A^TPx) > 0, \ \forall x \neq 0.$$

This shows that AM^{-1} is positive definite. □

Now, we have to choose the two matrices S and P. The only thing required is that they must be symmetric positive definite. So, there are many possibilities. To give some examples, let LU be an incomplete factorization of A ($A = LU + R$, L (resp. U) being lower (resp. upper) triangular, see Chapter 8), then we can use

i) $P = I$ and $S = (U^T L^T LU)^{-1}$.

At each iteration we have to solve $U^T L^T LU z^k = A^T r^k$.

ii) $P = (LUU^T L^T)^{-1}$ and $S = I$.

At each iteration we have to solve $z^k = A^T (LUU^T L^T)^{-1} r^k$.

iii) $P = (LL^T)^{-1}$ and $S = (U^T U)^{-1}$.

Then, $LL^T A^{-T} U^T U z^k = r^k$.

Clearly for each of these algorithms we have to solve four triangular systems and to perform a matrix×vector multiply by the transpose of A. In fact, it is easy to see that we are solving the normal equations

$$A^T Ax = A^T b \tag{7.2}$$

with PCG. Method i) corresponds to a left preconditioning, ii) to a right one and iii) to a sort of "symmetric" preconditioning. This method of using CG on the normal equations (7.2) is often denoted as CGNR. We have the following result.

Theorem 7.2

At each iteration CGNR *minimizes the Euclidean norm (l_2–norm) of the residual in $x^0 + K_k(A^T A, A^T r^0)$.*

Proof. From Chapter 6, we know that CG minimizes the norm of the error in the norm given by the matrix of the system. But,

$$\|\varepsilon^k\|^2_{A^T A} = (A^T A\varepsilon^k, \varepsilon^k) = (r^k, r^k) = \|r^k\|^2.$$

□

Figure 7.1 shows the behaviour of the norm of the error as a function of the number of floating point operations (flops) using CGNR without preconditioning for problem 7 on a 20×20 mesh.

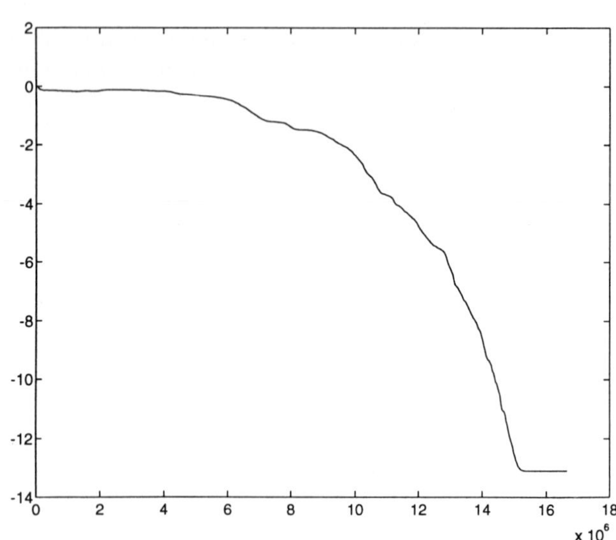

fig. 7.1: Problem 7: $\log_{10}$ *of the error with* CGNR *20×20 mesh*

If we wish we can use another variant of the normal equations, namely

$$AA^T y = b, \quad x = A^T y. \tag{7.3}$$

If we use CG on these equations, the method is denoted as CGNE.

Theorem 7.3

At each iteration CGNE *minimizes the Euclidean norm (l_2–norm) of the error in* $x^0 + A^T K_k(AA^T, r^0)$.

Proof. The iterates x^k are given by $x^k = A^T y^k$ where y^k is the CG iterate with the matrix AA^T. Then, $x - x^k = A^T(y - y^k)$ where $AA^T y = b$ and

$$\|\varepsilon^k\|^2 = (x - x^k, x - x^k) = (A^T(y - y^k), A^T(y - y^k)) = \|y - y^k\|^2_{AA^T}.$$

Moreover $y^k \in y^0 + K_k(AA^T, r^0)$. □

A drawback of the normal equations approach is that the condition number is worse than in the original problem; we have $\kappa(A^TA) = \kappa(A)^2$ where

$$\kappa(A) = \frac{\sigma_n}{\sigma_1},$$

σ_1 (resp. σ_n) being the smallest (resp. largest) singular value of A. Another possibility is to write the normal equations $A^TAx = A^Tb$ as a linear system of order $2n$:

$$\mathcal{A}\begin{pmatrix} x \\ y \end{pmatrix} = \begin{pmatrix} 0 & A^T \\ A & -I \end{pmatrix}\begin{pmatrix} x \\ y \end{pmatrix} = \begin{pmatrix} A^Tb \\ 0 \end{pmatrix}.$$

The matrix $\mathcal{A}$ is symmetric and indefinite as it has positive and negative eigenvalues, hence we can apply CG–like methods we saw in Chapter 6 (SYMMLQ or MINRES) to this system as the eigenvalues λ are such that $A^TAu = \lambda(\lambda + 1)u$. The drawback of solving the problem in this way is that the scalar products and vector updates are twice as expensive to compute as in PCG.

7.2. The Concus and Golub non–symmetric CG

There is one case where we can use CG on non–symmetric equations. This is when linear systems with the symmetric part of A are easy to solve. This algorithm was introduced by Concus and Golub [341], see also Widlund [1329]. Let $A = M - N$ with

$$M = A_S = \frac{A + A^T}{2}, \quad N = \frac{A^T - A}{2}$$

being the symmetric and antisymmetric parts of A. We apply for instance the three–term recurrence form of PCG with M as a preconditioner:

$$Mz^k = r^k,$$

$$x^{k+1} = x^{k-1} + \omega_{k+1}(\alpha_k z^k + x^k - x^{k-1}).$$

Then, simplifications occur since we have $(Nx, x) = 0 \; \forall x$. The choice $\alpha_k = 1$ gives $(z^k, Mz^{k+1}) = 0$ and to get $(z^k, Mz^{k+1}) = 0$, we have

$$\omega_{k+1} = \left(1 + \frac{(z^k, Mz^k)}{(z^{k-1}, Mz^{k-1})}\frac{1}{\omega_k}\right)^{-1}.$$

It is easy to see, [341], that this implies $(z^j, Mz^{k+1}) = 0, \ \forall j \leq k-2$. Of course, for this method to be interesting the problems with the symmetric part of A have to be easy to solve as this must be done every iteration.

This algorithm is a particular answer to a more general question that was raised by Gene Golub. The problem was to know under which conditions there exists a three–term recurrence CG method that is optimal in the same way CG is for symmetric matrices. The problem was solved by Faber and Manteuffel [542] in a more general setting. Roughly speaking, the answer is that such a method exists if the eigenvalues of the matrix are located on a straight line in the complex plane. The exact result is the following. Suppose we have a symmetric positive definite matrix H such that $H^{-1}A$ is definite, we define a Conjugate Gradient method to be an iterative method such that

$$x^k \in x^0 + K_k(A, r^0),$$

$$H\varepsilon^k = H(x - x^k) \perp K_k(A, r^0).$$

It is easy to see that this implies a monotone convergence of the error in the H–norm. If A is symmetric positive definite and $H = A$ we get CG. If we write

$$x^{k+1} = x^k + \gamma_k p^k,$$

let $CG(A, s)$ be the set of Conjugate Gradient methods such that we have $(Ap^k, Hp^j) = 0, \ k \geq j + s - 1$. This implies that the method can be written with s–term recurrences, (CG being a three–term recurrence method). Faber and Manteuffel [542] proved the following result.

Theorem 7.4

$CG(A, s)$ is not empty if and only if

1) s is larger than the degree of the minimum polynomial of A

or

2) $H^{-1}A^T H$ is a polynomial in A of degree less or equal to $s - 2$.

Proof. See Faber and Manteuffel [542]. □

If s is 3, we see that there exist α and β such that $H^{-1}A^T H = \alpha A + \beta I$. If λ_i are the eigenvalues of A, we have

$$(\alpha - 1)\lambda_i + \beta = 0, \ \forall i$$

that is, the eigenvalues are located on a straight line in the complex plane.

7.3. Construction of basis for Krylov spaces

7.3.1. The Arnoldi algorithm

The Krylov methods that we shall study all use a Krylov space $K_k(H, v)$ where

$$K_k(H, v) = \text{span}(v, Hv, \ldots, H^{k-1}v).$$

Generally $H = A$ and $v = r^0 = b - Ax^0$ as the iterates are sought in $x^0 + K_k(A, r^0)$. Therefore, there is no problem in choosing $x^0 = 0$. Note that by the Cayley–Hamilton theorem, there exists an $m \le n$ such that $x = A^{-1}b \in K_m(A, b)$. Unfortunately, the natural basis of the Krylov space $K_k(A, b)$ which is $\{b, Ab, \ldots, A^{k-1}b\}$ is ill conditioned. Moreover, the vectors $A^i b$ tend to be linearly dependent when k increases because they converge to a multiple of the eigenvector associated with the dominant eigenvalue (the one with the largest modulus). If we consider, for example, A to be the matrix of the Poisson model problem with $n = 100$ and a random vector b, the rank returned by Matlab 4.1 for the Krylov matrix of dimension 100×20 is 19. From then on, the vectors in the Krylov space are linearly dependent.

If v is a given vector, an idea is to try to construct an orthogonal basis of $K_k(A, v)$ in the same way as we did in the Lanczos algorithm for symmetric matrices. Up to roundoff errors, this basis will be perfectly conditioned. So, we look for vectors v^i such that

$$\text{span}(v^1, \ldots, v^k) = \text{span}(v, Av, \ldots, A^{k-1}v),$$

$$V_k = [v^1 \cdots v^k], \quad V_k^T V_k = I.$$

To motivate the derivation of the algorithm we can use the following presentation due to C. Vuik. Let $K_k = [v, Av, \cdots, A^{k-1}v]$ be the Krylov matrix, then it follows that

$$AK_k = K_k \begin{pmatrix} 0 & & & \\ 1 & \ddots & & \\ & \ddots & \ddots & \\ & & 1 & 0 \end{pmatrix} + A^k v (e^k)^T.$$

Now, consider the QR factorization of the $n \times k$ matrix, $K_k = QR$ with Q orthogonal and R being $k \times k$ upper triangular. Then,

$$
\begin{aligned}
Q^TAQ &= Q^TAK_kR^{-1}, \\
&= Q^T[K_k \begin{pmatrix} 0 & & & \\ 1 & \ddots & & \\ & \ddots & \ddots & \\ & & 1 & 0 \end{pmatrix} + A^k v (e^k)^T]R^{-1}, \\
&= R \begin{pmatrix} 0 & & & \vdots \\ 1 & \ddots & & R^{-1}Q^TA^kv \\ & \ddots & \ddots & \vdots \\ & & 1 & \vdots \end{pmatrix} R^{-1}.
\end{aligned}
\tag{7.4}
$$

Since R is upper triangular, this shows that Q^TAQ is an upper Hessenberg matrix. Therefore, the columns of Q span the Krylov space $K_k(A, v)$ and we can take $V_k = Q$. Computing the orthogonal basis of $K_k(A, v)$ is equivalent to computing the QR factorization of K_k. This can be done in many different ways, particularly column by column using the Gram–Schmidt orthogonalization algorithm. This is known as the `Arnoldi` method. Let us denote by $h_{i,j}$ the elements of the Hessenberg matrix and let

$$v^1 = \frac{v}{\|v\|}.$$

The algorithm for computing column j is

$$h_{i,j} = (Av^j, v^i), \quad i = 1, \dots, j$$

$$\bar{v}^j = Av^j - \sum_{i=1}^{j} h_{i,j}v^i,$$

$$h_{j+1,j} = \|\bar{v}^j\|, \quad v^{j+1} = \frac{\bar{v}^j}{h_{j+1,j}}.$$

Generally, for stability reasons one prefers using the modified Gram–Schmidt form of the algorithm. This amounts to computing $h_{i,j}$ and v^j as

$$w^j = Av^j,$$

and for $i = 1, \dots, j$

$$h_{i,j} = (w^j, v^i), \quad w^j = w^j - h_{i,j}v^i,$$

Theorem 7.5

Using the `Arnoldi` *algorithm we have* $(v^i, v^{j+1}) = 0, \quad i < j+1.$

Proof. The proof by induction is straightforward. □

If we run this algorithm for $j = 1, \ldots, k$ we construct a $k+1 \times k$ matrix $\tilde{H}_k$ such that

$$AV_k = V_{k+1}\tilde{H}_k = V_k H_k + h_{k+1,k}(e^k)^T v^{k+1},$$

where H_k denotes the upper Hessenberg square matrix constructed with the first k rows of $\tilde{H}_k$. This implies, as we have seen,

$$V_k^T A V_k = H_k.$$

We note that if A is symmetric then H_k must also be symmetric. A symmetric upper Hessenberg matrix is tridiagonal and in this case, the `Arnoldi` algorithm reduces to the Lanczos method. The symmetry allows us to orthogonalize only against the last two vectors. When the matrix is non–symmetric we need to explicitly orthogonalize against all the previous vectors. Thus, unfortunately in the `Arnoldi` method we have to keep these vectors and the storage grows linearly with k. Note that if $h_{k+1,k} = 0$, we have computed an invariant subspace for A. The orthogonal basis will be used in the following sections to construct iterative methods.

We have seen that the `Arnoldi` algorithm has some relationship with the QR factorization of the Krylov matrix. It is also very close to the QR algorithm for computing the eigenvalues of A, for a detailed account of this, see Lehoucq [890].

7.3.2. The Hessenberg algorithm

Hessenberg described a method to compute a basis of the Krylov space $K_k(A, v)$, see Wilkinson [1333]. We follow the lines of Sadok [1146]. For any $n \times k, n > k$ matrix C we define

$$C = \begin{pmatrix} C^1 \\ C^2 \end{pmatrix},$$

where C^1 is a matrix of order k. If C^1 is non–singular, C^L $(k \times n)$ is defined as $C^L = (\,(C^1)^{-1} \quad 0\,)$. Let K_k be the Krylov matrix, $K_k = [v, Av, \cdots, A^{k-1}v]$, then we can define an LU factorization $K_k = L_k U_k$ where L_k is a lower trapezoidal $n \times k$ matrix and U_k is a unit upper triangular matrix of order k. If the principal minors are non–zero the factorization exists, L_k^1 is non–singular and $K_k^1 = L_k^1 U_k$, $K_k^2 = L_k^2 U_k$.

Theorem 7.6

Let $K_k = [k^1 \dots k^k]$ be the Krylov matrix and suppose the principal minors are non–zero. Let l^j be the columns of L_k. Then

$$l^j = k^j - K_{j-1} K_{j-1}^L k^j = A l^{j-1} - L_{j-1} L_{j-1}^L A l^{j-1}, \forall j, \quad l^1 = v$$

$$span(l^1, \dots, l^k) = K_k(A, v),$$

$$A L_{k-1} = L_k \tilde{H}_{k-1},$$

where $\tilde{H}_{k-1}$ is an upper Hessenberg $k \times k-1$ matrix. The upper parts of the columns of $\tilde{H}_{k-1}$ denoted by h^j ($\in \mathbb{R}^j$) are given by $h^j = (L_j^1 \quad 0) A l^j$.

Proof. The fact that $\text{span}(l^1, \dots, l^k) = K_k(A, v)$ is obvious by construction. For the other parts, see Sadok [1146]. □

The algorithm is the following.

$l^1 = v$, for $k = 1, \dots, n$

$$u^k = A l^k,$$

$$\beta_j = \frac{(u^k)_j}{(l^j)_j}, \quad h_{j,k} = (u^k)_j, \quad (u^k)_j = 0,$$

$$(u^k)_l = (u^k)_l - \beta_j (l^j)_l, \quad l = j+1, \dots, n \quad j = 1, \dots, k$$

$$h_{k+1,k} = 1, \quad l^{k+1} = u^k.$$

Note that since we do not have to store the u^k's we can use a temporary vector u. The algorithm can break down if $(l^j)_j = 0$ or we can get problems if some components are small. As in Gaussian elimination, this can be eliminated by introducing pivoting, see Sadok [1146]. To pivot we may choose the component of u of largest modulus. This Hessenberg basis will be used later on to construct an iterative method.

7.3.3. The generalized Hessenberg process

These two previous constructions of basis for the Krylov space $K_k(A, v)$ are special cases of the general following algorithm (see Heyouni and Sadok [773]). The basis vectors are defined as

$$v^1 = v, \quad v^{j+1} = Av^j - \sum_{i=1}^{j} h_{i,j} v^i.$$

The coefficients are obtained by an orthogonality condition. Let $\{y^1, \dots, y^j\}$ be another set of linearly independent vectors. The orthogonality condition that we enforce is

$$v^{j+1} \perp Y_j = [y^1 \cdots y^j].$$

In matrix form, this is written as

$$AV_j = V_{j+1}\tilde{H}_j, \quad Y_j^T V_j = L_j,$$

where L_j is an order j lower triangular matrix and $\tilde{H}_j$ is a $(j+1) \times j$ upper Hessenberg matrix with 1's on the subdiagonal. Note that this method can be further generalized by defining the orthogonality relative to another matrix S. The elements of $\tilde{H}_j$ are computed by

$$v^1 = v, \, \eta_1 = (y^1, v^1),$$

for $j = 1, \dots, m$

$$u = Av^j,$$

for $i = 1, \dots, j$

$$h_{i,j} = \frac{(y_i, u)}{\eta_i}, \quad u = u - h_{i,j} v^i,$$

end i

$$h_{j,j+1} = 1, \quad v^{j+1} = u, \quad \eta_{j+1} = (y^{j+1}, u).$$

end j

If the vectors y^i are chosen as the columns of the identity matrix, we obtain the Hessenberg process. Breakdowns can be avoided by choosing the vectors in a different order. Note that as the vectors y^i do not have to be known in advance, they can be computed on the fly. If we choose $y^i = v^i$, we recover the Arnoldi algorithm with a slightly different scaling.

7.4. FOM and GMRES

7.4.1. Definition of FOM and GMRES

Let us suppose $x^0 = 0$. Once we have constructed an orthogonal basis of the Krylov space $K_k(A, b), k = 1, \ldots$ there are two different things we can do. The first one is defining $x^k = V_k y^k$ (this characterizes x^k as an element of $K_k(A, b)$) and solving the projected equation of order k

$$V_k^T A V_k y^k = V_k^T b,$$

to find y^k. Note that, because V_k defines an orthonormal basis and $v^1 = b$, we have $V_k^T b = \|b\| e^1$. Moreover, $V_k^T A V_k$ is an upper Hessenberg matrix H_k. This is also true for the generalized Hessenberg basis and the right hand side is also proportional to e^1. Therefore, the problem of computing y^k reduces to solving a linear system of order k

$$H_k y^k = \|b\| e^1, \tag{7.5}$$

whose matrix is upper Hessenberg. The solution is a scalar multiple of the first column of the inverse of H_k. This method is known as the Full Orthogonalization method (FOM) or the Arnoldi method. In FOM the linear system (7.5) is solved by a QR factorization that is updated at each iteration. This is quite easy as we just have to zero the lower diagonal of H_k. At each iteration a new Givens rotation is computed to annihilate the last element $h_{k+1,k}$.

The solution may also be computed by the following shooting method. Let us scale the matrix such that the subdiagonal elements $h_{j,j-1} = 1, j = 1, \ldots$. We write the solution y of $H_k y = e^1$ as $y = \alpha u, \alpha \in \mathbb{R}$. Notice that now we have $k + 1$ parameters. We fix the value of u_k say $u_k = 1$, then we are able to compute u_{k-1} using the last equation

$$u_{k-1} + h_{k,k} u_k = 0.$$

This gives $u_{k-1} = -h_{k,k}$. Recursively, we compute $u_{k-i}, i = 2, \ldots, 1$ using equations $k - 1$ to 2:

$$u_{k-i} = - \sum_{j=k-i+1}^{k} h_{k-i+1,j} u_j.$$

Of course with these values, the first equation is only satisfied if we choose appropriately the value of α, that is

$$\alpha = \frac{1}{h_{1,1}u_1 + \cdots + h_{1,k}u_k}.$$

It is well known that this algorithm is potentially dangerous as, in some cases, we can get overflows and instability. This can be handled in two ways. First of all, we can make another choice than $u_k = 1$ and secondly, we can monitor the size of the u_j's and rescale them if they grow too much but this needs additional multiplications. Therefore, for numerical computations it is better to use Givens rotations but the shooting method gives the explicit solution of the Hessenberg system as a function of the elements $h_{i,j}$ which are (Av^j, v^i) properly scaled. Moreover, when we increase the size of the problem, the solution can be updated by using the Sherman–Morrison formula.

Notice that if H_k is singular, then the algorithm breaks down as we are not able to compute y^k. In `FOM`, the residual is orthogonal to the Krylov space $K_k(A, b)$ since the definition of y^k implies $V_k^T r^k = V_k^T(b - Ax^k) = 0$. The l_2–norm of the residual can be easily computed as we have the following result.

Proposition 7.7

Let y_k^k be the last component of the solution vector y^k. Then,

$$\|r^k\| = \|b - Ax^k\| = h_{k+1,k}|y_k^k|.$$

Proof. In `FOM` the residual is given by

$$r^k = b - Ax^k = -h_{k+1,k}v^{k+1}(e^k)^T y^k,$$

which gives the result. □

The second possibility we have is to minimize the Euclidean norm (l_2–norm) of the residual by using the relation $AV_k = V_{k+1}\tilde{H}_k$ where the last matrix is a $k+1 \times k$ upper Hessenberg matrix. As in `FOM`, we set $x^k = V_k y^k$. Then,

$$\begin{aligned}
\|r^k\| &= \|b - Ax^k\| \\
&= \|b - AV_k y^k\| \\
&= \| \, \|b\| V_{k+1} e^1 - V_{k+1}\tilde{H}_k y^k \| \\
&= \| \, \|b\| e^1 - \tilde{H}_k y^k \|
\end{aligned}$$

Hence, we have a least squares problem to solve with a $k+1 \times k$ matrix. This method is known as GMRES (Generalized Minimum RESidual), see Saad and Schultz [1141]. Mathematically, the solution of this problem is written as

$$y^k = \|b\|(\tilde{H}_k)^+ e^1,$$

where $(\tilde{H}_k)^+ = (\tilde{H}_k^T \tilde{H}_k)^{-1} \tilde{H}_k^T$. In a practical way, as in FOM, we solve the least squares problem by using a QR factorization. This means that we have to use orthogonal transformation to reduce the Hessenberg matrix to upper triangular form. One way to do this is to use Givens rotations. Suppose the current matrix is

$$\begin{pmatrix} x & x & \dots & & x \\ & \ddots & & & \vdots \\ & & \ddots & & \\ & & & x & x \\ & & & 0 & r \\ & & & 0 & h \end{pmatrix}.$$

The problem we have now is to eliminate the h entry to get an upper triangular matrix. This is done by left multiplying with the Givens matrix

$$\begin{pmatrix} 1 & & & & \\ & \ddots & & & \\ & & 1 & & \\ & & & c & -s \\ & & & s & c \end{pmatrix}.$$

The coefficients s and c which are sine and cosine of the angle of rotation are given by

$$c = \frac{r}{\sqrt{r^2 + h^2}}, \qquad s = -\frac{h}{\sqrt{r^2 + h^2}}.$$

Let Q_k^T be the product of all the Givens rotation matrices, then $Q_k^T \tilde{H}_k = R_k$ and the least squares solution is obtained by solving a triangular system with matrix R_k. Note that to compute and use R_k we have to store and apply all the previous rotations to the last column of $\tilde{H}_k$ and to the right hand side. Other implementations have been proposed, for instance, using Householder transformations to generate the orthogonal basis, see Walker [1314]. We shall see that this gives a much more stable algorithm. However, Householder requires about twice as many flops as Gram–Schmidt.

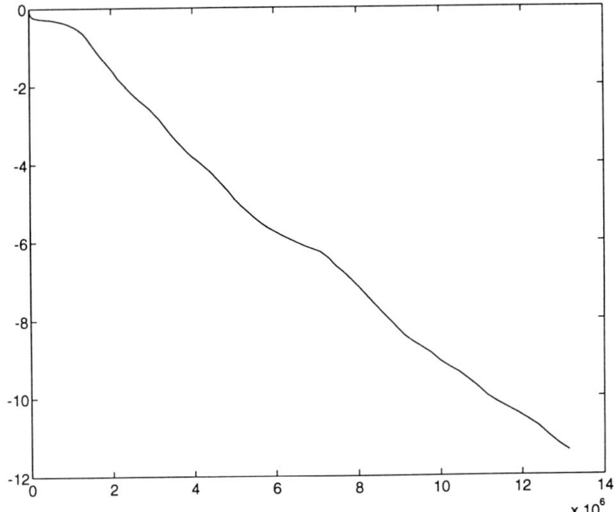

fig. 7.2: Problem 7: $\log_{10}$ *of the error with* GMRES
20×20 *mesh*

Figure 7.2 shows the behaviour of the norm of the error as a function of the number of floating point operations (flops) using GMRES without preconditioning for problem 7 on a 20×20 mesh.

The preconditioned GMRES is the following:

Let x^0 be given, $Mr^0 = b - Ax^0$,

$$v^1 = \frac{r^0}{\|r^0\|}, \quad f = \|r^0\| e^1,$$

for $k = 1, \ldots$

$$Mw = Av^k,$$

for $i = 1, \ldots, k$

$$h_{i,k} = (w, v^i), \quad w = w - h_{i,k} v^i,$$

end i

$$h_{k+1,k} = \|w\|, \quad v^{k+1} = \frac{w}{h_{k+1,k}},$$

–apply the rotations of iterations 1 to $k-1$ on $(h_{1,k} \dots h_{k+1,k})^T$. Compute the rotation $R_{k+1,k}$ to eliminate $h_{k+1,k}$, $f = R_{k+1,k} f$, solve the triangular system for y^k,

–compute the norm of the residual (which is the last component of f properly scaled), if it is small enough compute $x^k = x^0 + V_k y^k$ and stop

end k.

There are some relationships between the iterates of FOM that we denote by x_F^k and those of GMRES denoted by x_G^k.

Theorem 7.8

Let s_k and c_k represent the rotation acting on rows k and $k+1$. We have the following relations:

$$\frac{\|r_G^k\|}{\|r_G^{k-1}\|} = |s_k|,$$

$$\|r_G^k\| = |c_k| \; \|r_F^k\|.$$

Proof. See Brown [218]. □

These relations imply that

$$\|r_G^k\| = \sqrt{1 - \frac{\|r_G^k\|}{\|r_G^{k-1}\|}} \|r_F^k\|.$$

If s_k is small, GMRES reduces significantly the norm of the residual and we see that the same is true for FOM. If FOM breaks down, then GMRES stagnates.

7.4.2. Convergence results

The most well known result about convergence of GMRES was proven by Saad and Schultz [1141]. Let A_S be the symmetric part of A. We have the following result.

Theorem 7.9

Suppose A is similar to a diagonal matrix ($A = X\Lambda X^{-1}$), then the residual of the `GMRES` *algorithm is such that*

$$\|r^k\| \leq \epsilon_k \|X\| \|X^{-1}\| \|r^0\|.$$

where ϵ_k is the minimum of $\max_\lambda |p_k(\lambda)|$ over polynomials p_k of degree k that satisfy the constraint $p_k(0) = 1$ and λ is an eigenvalue of A. If the symmetric part of A is positive definite, then

$$\|r^k\| \leq \left(1 - \frac{\lambda_{min}(A_S)}{\sigma_{max}(A)}\right)^k \|r^0\|,$$

where $\sigma_{max}(A) = \lambda_{max}(A^T A)$ is the largest singular value of A.

Proof. See Saad and Schultz [1141]. We shall give another proof in one of the following theorems. □

Hochbruck and Lubich [787] proved the following upper bound on the norm of the error.

Theorem 7.10

Let ε^k be the error,

$$\|\varepsilon^k\| \leq \|A^{-1} P_k\| \min_{p_k(0)=1} \|p_k(A) b\|$$

where the minimum is taken over polynomials of degree less than or equal to k and $P_k = I - V_{k+1} \tilde{H}_k \tilde{H}_k^+ V_{k+1}^T$.

Proof. See Hochbruck and Lubich [787]. □

In particular cases the minimum can be bounded independently of b. The rate of convergence of `GMRES` was also studied by Van der Vorst and Vuik [1288].

Theorem 7.11

If the matrix A is diagonalizable by a matrix X,

$$\|r_G^{k+l}\| \leq \frac{F_k}{|c_k|}\kappa(X)\epsilon_l\|r_G^k\|,$$

where

$$F_k = \frac{|\sigma_1^{(k)}|}{|\lambda_1|}\max_{\lambda_j\neq\lambda_1}\left|\frac{\lambda_j-\lambda_1}{\lambda_j-\sigma_1^{(k)}}\right|,$$

where the λ_j are the eigenvalues of A, $\sigma_1^{(k)}$ is the smallest eigenvalue of H_k and $\epsilon_l = \min_q \max_{\lambda_i\neq\lambda_1}|q(\lambda_i)|$, q being a polynomial of degree less than l and such that $q(0) = 1$.

Proof. See Van der Vorst and Vuik [1288]. □

This helps explaining the superlinear rate of convergence that is sometimes observed with `GMRES`, see [1288]. Campbell, Ipsen, Kelley and Meyer [254] studied the convergence of `GMRES` from another perspective. They use the spectral projectors

$$X_j = \frac{1}{2\pi i}\int_{\Gamma_j}(zI-A)^{-1}\,dz, \; 1 \leq j \leq d$$

where Γ_j is a circle around λ_j with no other eigenvalue inside. Suppose there are d distinct eigenvalues. The index k_j of λ_j is the smallest integer k such that $ker(\lambda_j I - A)^k = ker(\lambda_j I - A)^{k+1}$. The minimal polynomial p_{min} is defined as

$$p_{min}(\lambda) = \prod_{j=1}^{d}(\lambda-\lambda_j)^{k_j}.$$

They studied the case where there are $d - l$ eigenvalues in a disk of center 1 and radius ρ and l eigenvalues outside (called outliers).

Theorem 7.12

Suppose the last eigenvalues are inside the disk. Let $g = \sum_{j=1}^{l} k_j$ and δ is such that $|\lambda_j - \lambda| < \delta|\lambda_j|$, $|\lambda - 1| = \rho$ for the outliers, then

$$\|r^{k+g}\| \leq C\rho^k\|r^0\|,$$

with $C = \rho\delta^g \max_{|z-1|=\rho} \|(zI - A_2)^{-1}\|$ *and* $A_2 = (I - (\sum_{j=1}^{l} X_j))A$.

Proof. See Campbell, Ipsen, Kelley and Meyer [254]. □

Note that C is independent of k. In this case, after g iterations the rate of convergence depends essentially on the size of the cluster of eigenvalues. However, C can be large if the matrix is non–normal. This result has been generalized to several clusters of eigenvalues in [254].

Sadok [1148] related the convergence of FOM and GMRES to the singular values of the matrices H_k (the so called Ritz values) and $\tilde{H}_k$. First of all, the norms of the residuals of FOM and GMRES are related to determinants of the Krylov matrices K_k. Let $C_k = AK_k$, then

Proposition 7.13

The residuals of GMRES *and* FOM *are such that*

$$\|r_G^k\|^2 = \frac{det(C_{k+1}^T C_{k+1})}{det(C_k^T C_k)},$$

$$\|r_F^k\|^2 = \frac{det(K_k^T K_k) det(K_{k+1}^T K_{k+1})}{det(K_k^T C_k)}, \quad \text{if } det(K_k^T C_k) \neq 0.$$

Proof. See Sadok [1148]. The proof uses the orthogonality relations for the residuals and Cramer's rule. □

As a consequence of this result, we have

$$\frac{\|r_G^k\|^2}{\|r_G^{k-1}\|^2} = s_k^2, \quad \|r_G^k\|^2 = c_k^2 \|r_F^k\|^2,$$

where $c_k^2 = 1 - s_k^2$, c_k^2 being the ratio of determinants,

$$c_k^2 = \frac{det(K_k^T C_k)^2}{det(C_k^T C_k) det(K_k^T K_k)}.$$

If we look at the way the Hessenberg matrix is computed in GMRES, it turns out that we have

$$h_{k+1,k}^2 = \frac{det(K_{k+1}^T K_{k+1}) det(K_{k-1}^T K_{k-1})}{det(K_k^T K_k)}.$$

This gives a proof of Brown's result [218]. Let $\alpha_j^{(k)}$ and $\beta_j^{(k)}, i = 1, \ldots, k$ be the singular values of H_k and $\tilde{H}_k$ in ascending order. Because H_k is made of the k first rows of $\tilde{H}_k$, there are some interlacing properties between the α's and β's. If A_S, the symmetric part of A, is positive definite, then $\alpha_1^{(k)} \geq \lambda_1(A_S) > 0, \forall k$.

Theorem 7.14

The norms of the residuals are related to the singular values by

$$\frac{\|r_G^k\|}{\|r_G^{k-1}\|} = \frac{\prod_{j=1}^{k-1} \beta_j^{(k-1)}}{\prod_{j=1}^{k} \beta_j^{(k)}} \sqrt{\sum_{j=1}^{k} (\beta_j^{(k)})^2 - (\alpha_j^{(k)})^2}$$

$$= \sqrt{1 - \prod_{j=1}^{k} \frac{(\alpha_j^{(k)})^2}{(\beta_j^{(k)})^2}},$$

$$\frac{\|r_F^k\|}{\|r_F^{k-1}\|} = \frac{\prod_{j=1}^{k-1} \alpha_j^{(k-1)}}{\prod_{j=1}^{k} \alpha_j^{(k)}} \sqrt{\sum_{j=1}^{k} (\beta_j^{(k)})^2 - (\alpha_j^{(k)})^2},$$

$$\frac{\|r_G^k\|}{\|r_F^k\|} = \frac{\prod_{j=1}^{k-1} \alpha_j^{(k)}}{\prod_{j=1}^{k} \beta_j^{(k)}},$$

if the denominators are different from 0.

Proof. See Sadok [1148]. The proof uses the result of proposition 7.13 and relates the determinants of $K_k^T K_k$ and $C_k^T C_k$ to those of $\tilde{H}_k^T \tilde{H}_k$. □

Moreover, we can see that

$$c_k = \frac{\prod_{j=1}^{k-1} \alpha_j^{(k)}}{\prod_{j=1}^{k} \beta_j^{(k)}}.$$

If the smallest singular value $\alpha_1^{(k)}$ is bounded below by η independently of k, we have

$$\frac{\|r_G^k\|}{\|r_G^{k-1}\|} \leq \sqrt{1 - \frac{\eta^2}{\sigma_n^2}}, \quad \frac{\|r_F^k\|}{\|r_G^k\|} \leq \frac{\sigma_n}{\eta},$$

where σ_n is the largest singular value of A.

Greenbaum, Ptàk and Strakŏs [702] proved a very interesting result about GMRES namely that given a function f such that $f(0) \geq f(1) \geq \cdots \geq f(n-1)$ it is possible to construct a matrix A and a vector r^0 with $\|r^0\| = f(0)$ such that $\|r^k\| = f(k), k = 1, \ldots, n-1$ where r^k is the residual of GMRES with an initial residual r^0. Moreover, the matrix A can be chosen to have any desired (complex) eigenvalues. They also characterize the set of these matrices. Their results show that the eigenvalues are not the relevant quantities to understand the behaviour of GMRES for non–normal matrices (that is when $A^T A \neq AA^T$). It has been argued (see Trefethen [1264]) that the pseudo spectra is the right tool to understand GMRES convergence when A is not normal. This is not contradictory to the results of Campbell, Ipsen, Kelley and Meyer [254] since (as is usual for upper bounds) the constants involved may be large when A is non–normal.

7.4.3. Truncated and restarted versions

The main problem with FOM and GMRES compared to CG and MINRES (for symmetric matrices) is that the storage grows linearly with k. When solving large problems this is not tolerable. Therefore something has to be done to keep low storage requirements. The simplest remedy is to periodically restart the algorithm. Let m be a given integer, after m GMRES (or FOM) iterations, we throw away the basis already computed and we restart the iteration with $x^0 = x^m$. This algorithm is denoted as GMRES(m). So far, we are only able to prove that GMRES(m) converges if the symmetric part of A is positive definite. Actually, there are examples for low values of m ($= 1, 2, \ldots$) where GMRES(m) does not converge or so slowly that it is not of practical interest.

Figure 7.3 shows the norm of the error as a function of the number of floating point operations (flops) using GMRES(m) with different values of m without preconditioning for problem 7 on a 20×20 mesh.

Another solution is to truncate the methods by orthogonalizing only against the last m vectors of the basis and to throw away the earlier vectors. Then, the matrix that is computed is banded Hessenberg. However, the basis of the Krylov space computed by the truncated Arnoldi method is not orthogonal anymore. However, this variant will be more appropriate for some other algorithms which are mathematically equivalent to GMRES.

7.4.4. Methods equivalent to GMRES

Over the years, many (Krylov) methods have been proposed to solve non–symmetric systems. It turns out that most of these methods belong to two

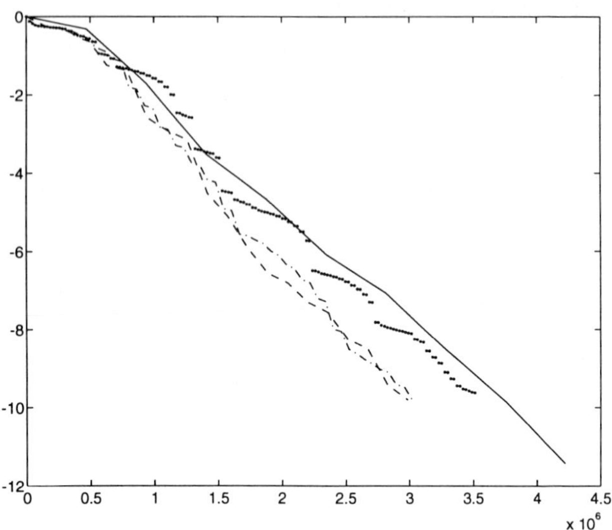

fig. 7.3: Problem 7: $\log_{10}$ *of the error with* GMRES(m)
20×20 *mesh*
solid: $m = 20$, *dashed:* $m = 10$, *dash–dotted:* $m = 5$, *dotted:* $m = 2$

classes and within each class all the methods are mathematically equivalent. This means that in exact arithmetic they compute the same approximation although the implementations could be different. Of course, this is not true in finite precision arithmetic and it can be that some implementations are more stable than others. The first class we are interested in here minimizes the l_2–norm of the residual. We have seen that GMRES belongs to this class. Other methods are the Generalized Conjugate Residual (GCR), ORTHOMIN and ORTHODIR as well as some least squares methods derived by Axelsson, see Saad and Schultz [1140]. In most methods the iterates are written as

$$x^{k+1} = x^k + \sum_{j=0}^{k} \alpha_j^{(k)} p^j,$$

where the p^j's are constructed to span the Krylov space. Then,

$$r^{k+1} = r^k - \sum_{j=0}^{k} \alpha_j^{(k)} Ap^j.$$

We minimize the l_2–norm of the residual over $x^0 + K_k$ if and only if the space spanned by the residuals is A–orthogonal to K_k. This implies that

$\alpha_j^{(k)} = 0, j \neq k$. Hence, we can rewrite

$$x^{k+1} = x^k + \alpha_k p^k.$$

The value of α_k minimizing the norm is

$$\alpha_k = \frac{(r^k, Ap^k)}{(Ap^k, Ap^k)}.$$

In GCR (see Eisenstat, Elman and Schultz [512]) the vectors p^j spanning the Krylov space are given by

$$p^{k+1} = r^{k+1} + \sum_{j=0}^{k} \beta_j^{(k+1)} p^j.$$

The coefficients are computed as

$$\beta_j^{(k+1)} = \frac{(Ar^{k+1}, Ap^j)}{(Ap^j, Ap^j)}.$$

Proposition 7.15

With the choices of GCR, *we have*

$$(Ap^k, Ap^j) = 0, \quad \forall k \neq j$$

$$(r^k, Ap^j) = 0, \quad 0 \leq j < k$$

$$(r^k, Ap^k) = (r^k, Ar^k),$$

$$(r^k, Ar^j) = 0, \quad 0 \leq j < k$$

Proof. The proof is easily obtained by induction. □

Generally, GCR is used in its restarted version that we denote by GCR(m). Every m iterations, the p^j's are thrown away and the iteration is restarted with the current value of x^{k+1}. This way the storage is kept constant. The method ORTHOMIN(m) was introduced by Vinsome [1308]. It is a truncated version of GCR where only the last m vectors p^j are kept (so, GCR=ORTHOMIN(∞)). The new direction is defined as

$$p^{k+1} = r^{k+1} + \sum_{j=k-m+1}^{k} \beta_j^{(k+1)} p^j,$$

The coefficients are computed with the same formulas as for GCR. It can be proved that this method converges if the symmetric part of A is positive definite.

Figure 7.4 shows the norm of the error as a function of the number of floating point operations (flops) using ORTHOMIN without preconditioning for problem 7 on a 20×20 mesh.

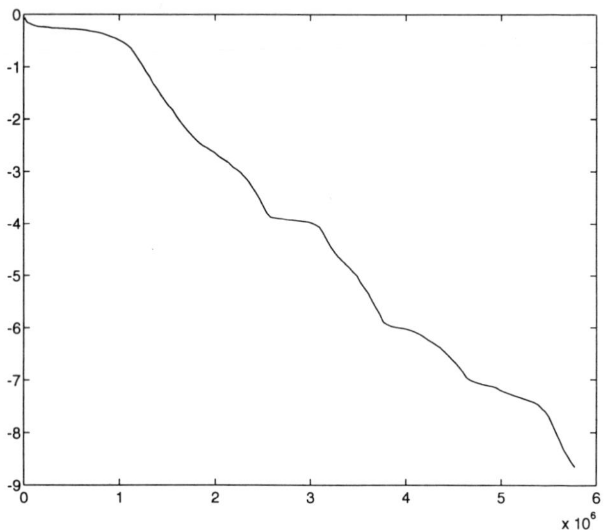

fig. 7.4: Problem 7: $\log_{10}$ *of the error with* ORTHOMIN
20×20 *mesh*

Figure 7.5 gives the norm of the error as a function of the number of floating point operations (flops) using ORTHOMIN(m) with different values of m without preconditioning for problem 7 on a 20×20 mesh.

This method can be further generalized if we introduce a definite matrix K and consider orthogonality related to $(K.,.)$. Let M be a preconditioner, then the algorithm ORTHOMIN(K, M) is

$$r^0 = b - Ax^0, \quad Mp^0 = r^0,$$

for $k = 0, 1, \ldots$

$$\psi_k = (Kp^k, p^k), \quad \alpha_k = (K\varepsilon_k, p^k)/\psi_k,$$

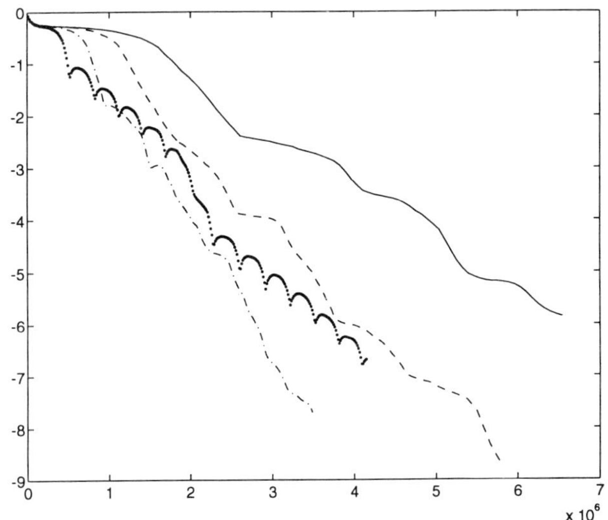

fig. 7.5: Problem 7: $\log_{10}$ *of the error with* ORTHOMIN(m)
20×20 *mesh*
solid: $m = 20$, *dashed:* $m = 10$, *dash–dotted:* $m = 5$, *dotted:* $m = 2$

$$x^{k+1} = x^k + \alpha_k p^k,$$

$$r^{k+1} = r^k - \alpha_k A p^k,$$

$$M z^{k+1} = r^{k+1},$$

$$\beta_j^{(k+1)} = -\frac{(K z^{k+1}, p^j) + \sum_{l=0}^{j-1} (K p^l, p^j)}{\psi_j}, \quad j = 0, \ldots, k$$

$$p^{k+1} = z^{k+1} + \sum_{j=0}^{k} \beta_j^{(k+1)} p^j.$$

If K is symmetric, then the formula for the $\beta_j^{(k+1)}$'s simplifies and

$$\beta_j^{(k+1)} = -\frac{(K z^{k+1}, p^j)}{\psi_j}.$$

The (unknown) error ε^k appears in these formulas. To be able to compute the coefficients we choose $K = A^T M^{-1} A$ and then $K\varepsilon^k = A^T M^{-1} A \varepsilon^k = A^T M^{-1} r^k$.

ORTHODIR has been introduced by Young and Jea (see [1354], [817]). The new direction is computed as

$$\gamma_{k+1}^{(k+1)} p^{k+1} = Ap^k + \sum_{j=0}^{k} \gamma_j^{(k+1)} p^j,$$

where the coefficients $\gamma_j^{(k+1)}, 0 \le j \le k$ are defined such that $(Ap^{k+1}, Ap^j) = 0, 0 \le j \le k$ and $\gamma_{k+1}^{(k+1)}$ is a scaling coefficient. Therefore, the same properties hold as in GCR. Note that to compute the vectors p^j's we do not need the residuals (except for $p^0 = r^0$). The truncated version of the algorithm is denoted as ORTHODIR(m). ORTHODIR can also be generalized and preconditioned. We just replace the computation of z^{k+1} by $Mz^{k+1} = Ap^k$.

Axelsson [62] defined his method by

$$p^{k+1} = r^{k+1} + \beta_{k+1} p^k,$$

with

$$\beta_{k+1} = -\frac{(Ar^{k+1}, Ap^k)}{(Ap^k, Ap^k)}.$$

Then, to minimize the norm of the residual over the Krylov space, we must modify the definition of x^k. This is done by

$$x^{k+1} = x^k + \sum_{j=0}^{k} \alpha_j^{(k)} p^j,$$

$$r^{k+1} = r^k - \sum_{j=0}^{k} \alpha_j^{(k)} Ap^j.$$

The coefficients $\alpha_j^{(k)}$ are computed by solving a linear system of order k whose matrix entry (i, j) is (Ap^i, Ap^j) and the ith component of the right hand side is (r^k, Ap^i). This algorithm can also be restarted or truncated. It is slightly more expensive than GCR per iteration.

It is unlikely that these methods which are mathematically equivalent to GMRES could be far better than GMRES. However, methods like ORTHOMIN are still popular as they can be easily truncated. Note that we can generalize the "truncation" by keeping m vectors different from the last m ones. Unfortunately, there is no theory that can help us decide which vectors must be kept to enhance the rate of convergence.

7.4.5. Methods equivalent to FOM

ORTHORES was introduced by Young and Jea [1354]. In the symmetric case, it gives the three–term form of CG (with some different scaling). The method is defined as

$$r^0 = b - Ax^0, \quad Mz^0 = r^0,$$

for $k = 0, 1, \dots$

$$\xi_k = (z^k, z^k),$$

$$\sigma_j^{(k+1)} = \frac{(Kz^k, z^j)}{\xi_j}, \quad j = 0, \dots, k$$

$$\lambda_k = (\sum_{j=0}^{k} \sigma_j^{(k+1)})^{-1},$$

$$x^{k+1} = \lambda_k (z^k + \sum_{j=0}^{k} \sigma_j^{(k+1)} x^j),$$

$$r^{k+1} = \lambda_k (\sum_{j=0}^{k} \sigma_j^{(k+1)} r^j - Az^k), \quad Mz^{k+1} = r^{k+1}.$$

In practice we choose $K = M^{-1}A$. Note that this algorithm computes a set of orthogonal generalized residuals z^k. The basis update has been replaced by a long recurrence for the iterates. This sum may also be truncated to define an ORTHORES(m) algorithm.

7.5. Roundoff error analysis of GMRES

So far there have not been many results concerning roundoff error analysis for GMRES or FOM. There are several questions that can be raised: is GMRES backward stable? Which is the best variant of GMRES regarding stability of the computation? How different are the computed residuals from the "true" residuals computed by $b - Ax^k$?

There are several ways by which roundoff errors might be introduced (and eventually propagated) in GMRES. The first one is the construction of the "orthogonal" basis using Arnoldi. One can use the (modified) Gram–Schmidt orthogonalization algorithm (eventually with iterative improvement) or Householder transformations. The outcome of that phase (as for the Lanczos algorithm we have studied in Chapter 6) is that the basis vectors might lose orthogonality. Then comes the reduction of the Hessenberg matrix to triangular form. Using Givens rotations, not many problems could arise from this phase. The last part of the algorithm is solving the triangular system. Then, again some roundoff errors may be introduced.

Z. Strakŏs [1215] has studied these questions. The first step is to look at the finite precision `Arnoldi` algorithm. This is an extension of what was done for the three–term recurrence form of the Lanczos algorithm. The computation of the orthogonal basis can be seen as

$$[r^0 \; AV_k] = V_{k+1}R_{k+1},$$

where R_{k+1} is an upper triangular matrix. Then, we are able to use known results about the QR factorization of the matrix $[r^0 \; AV_k]$. This translates into

$$AV_k = V_{k+1}\tilde{H}_k + F_k,$$

with $\|F_k\| \leq C_1 n^3 u \|A\|$, C_1 being a constant. Concerning the loss of orthogonality, for Householder, we have

$$\|I - V_k^T V_k\| \leq C_2 k^3 u.$$

For Householder and modified Gram–Schmidt, there exists an orthogonal matrix $\hat{V}_{k+1}$ such that

$$AV_k = \hat{V}_{k+1}\tilde{H}_k + \hat{F}_k,$$

with $\|\hat{F}_k\| \leq C_3 n^3 u \|A\|$. For Householder, $\|V_{k+1} - \hat{V}_{k+1}\| \leq C_4 k^3 u$. For modified Gram–Schmidt $\|V_{k+1} - \hat{V}_{k+1}\| \leq C_5 n^3 u \kappa([r^0 \; AV_k])$. Regarding residuals, Strakŏs showed that

$$\|(b - Ax^k) - V_{k+1}(\|b\| - \tilde{H}_k y^k)\| \leq 2C_1 n^3 u \|A\| \, \|y^k\| \\ + u\|b\| + (n^{\frac{3}{2}} + 3)u\|A\| \, \|x^0\|.$$

Note that we can consider that $\|x^0\| = 0$. Looking at how y^k is computed through Givens rotations,

$$\|y^k\| \leq \|b\|(1 + C_6 n^{\frac{3}{2}} u)[\sigma_1(\tilde{H}_k) - 2C_6 n^{\frac{3}{2}} u \sigma_k(\tilde{H}_k)]^{-1}.$$

Theorem 7.16

Assuming $x^0 = 0$ *and* $\sigma_1(\tilde{H}_k) > 2C_6 n^{\frac{3}{2}} u \|\tilde{H}_k\|$, *we have*

$$\frac{\|b - Ax^k\|}{\|b\|} \leq (k+1)^{\frac{1}{2}} \frac{\| \|b\|e^1 - \tilde{H}_k y^k\|}{\|b\|}$$
$$+ 2C_1 n^3 u \frac{\|A\|}{\sigma_1(\tilde{H}_k)} [1 - 2C_6 n^{\frac{3}{2}} u \kappa(\tilde{H}_k)]^{-1} + u.$$

Proof. See Strakǒs [1215]. The factor $(k+1)^{\frac{1}{2}}$ is a bound for $\|V_{k+1}\|$ but because of the local orthogonality, this is probably a crude overestimate. Then, if the second term is not too large (*i.e.*,if $\sigma_1(\tilde{H}_k) \gg 2C_6 n^{\frac{3}{2}} u \sigma_k(\tilde{H}_k)$) the true residual is a small multiple of the `Arnoldi` residual. □

This analysis can be made a little more precise for the Householder implementation if we make some hypothesis on the way orthogonality is lost, see Strakǒs [1215]. It has been shown that the Householder implementation is backward stable: the computed upper Hessenberg matrix is equal to the one generated by an exact `Arnoldi` recurrence applied to a slightly perturbed matrix.

A. Greenbaum [699] has studied a problem which may seem a little different: what is the attainable accuracy of recursively computed residuals? She considered methods that can be written as

$$x^{k+1} = x^k + \alpha_k p^k, \quad r^{k+1} = r^k - \alpha_k A p^k.$$

This cannot be readily applied to `GMRES` but to some variants, particularly those based on `GCR`. She proved that

$$\frac{\|b - Ax^k - r^k\|}{\|A\| \, \|x\|} \leq (u + O(u^2))[k + 1 + (1 + c + k(10 + 2c)) \max_{j \leq k} \frac{\|x^j\|}{\|x\|}],$$

where c is the constant involved in the matrix×vector multiply, the computing result being $Ap^k + d^k$ with

$$\|d^k\| \leq cu\|A\| \, \|p^k\|.$$

This shows that the residual computed by recurrence could be different from the “true” one if some of the intermediate iterates grow much larger than the solution x. This has been checked experimentally, for instance for CG and CGNR.

The roundoff effects in GMRES(m) have been studied experimentally by R. Karlson [853] on simple examples. Three sources of errors are identified: type 1 comes from loss of orthogonality, type 2 arises from solving the least squares problem, type 3 from updating the solution. Examples displaying the different types of roundoff errors are given. It is stated that for the chosen examples and modified Gram–Schmidt, stagnation arises from a decrease of the smallest singular value of V_k. For Householder, stagnation may also occur despite a very moderate loss of orthogonality. However, the results are usually much better using the Householder implementation.

7.6. Extensions to GMRES

In some cases, it is interesting to be able to use a different preconditioner at each iteration. This is the case, for instance, if an inner iterative method is used to solve $Mw = Av^k$.

7.6.1. Flexible GMRES

Saad has proposed a modification of GMRES denoted as Flexible GMRES (in short FGMRES), see [1138].

Let x^0 be given, $r^0 = b - Ax^0$,

$$v^1 = \frac{r^0}{\|r^0\|}, \quad f = \|r^0\| e^1,$$

for $k = 1, \ldots$

$$M_k z^k = v^k, \quad w = Az^k,$$

for $i = 1, \ldots, k$

$$h_{i,k} = (w, v^i), \quad w = w - h_{i,k} v^i,$$

end i

$$h_{k+1,k} = \|w\|, \quad v^{k+1} = \frac{w}{h_{k+1,k}},$$

Apply the rotations of iterations 1 to $k-1$ on $(h_{1,k} \ldots h_{k+1,k})^T$. Compute the rotation $R_{k+1,k}$ to eliminate $h_{k+1,k}$, $f = R_{k+1,k} f$, solve the triangular system for y^k. Compute the norm of the residual which is related to the last component of f, if it is small enough compute $x^k = x^0 + Z_k y^k$, where $Z_k = [z^1 \cdots z^k]$ and stop

end k.

The main difference with GMRES is that we have to store the vectors z^j. We have

$$AZ_k = V_{k+1}\tilde{H}_k.$$

Theorem 7.17

The approximate solution x^k of FGMRES *minimizes the Euclidean norm (l_2–norm) of the residual over $x^0 + span(z^1, \ldots, z^k)$.*

Proof. See Saad [1138]. □

It must be noted that convergence results of GMRES do not carry over to FGMRES because the space where the solution is sought is not a Krylov space.

7.6.2. GMRES*

C. Vuik and H. van der Vorst [1289] described an extension of GMRES that can be considered a recursive variant of GMRES. The idea is to precondition the linear system by an approximation of the inverse that can be derived from the first iterations. This idea comes from variants of the Broyden's method (for solving minimization problems) where approximations of the Hessian are updated at each iteration. The basic method uses iterations of GMRES itself as the preconditioner, hence the name GMRESR, the R standing for recursive. The algorithm is the following:

Let x^0 be given, $r^0 = b - Ax^0$,

for $k = 0, 1, \ldots$

$u^k_{(0)} = \mathcal{P}_{m,k} r^k, \quad c^k_{(0)} = Au^k_{(0)},$

for $i = 0, \ldots, k-1$

$$\alpha_i = (c_i, c^k_{(i)}),$$

$$c^k_{(i+1)} = c^k_{(i)} - \alpha_i c_i,$$

$$u^k_{(i+1)} = u^k_{(i)} - \alpha_i u_i,$$

end i

$$c_k = \frac{c^k_{(k)}}{\|c^k_{(k)}\|},$$

$$u_k = \frac{u^k_{(k)}}{\|c^k_{(k)}\|},$$

$$x^{k+1} = x^k + (c_k, r^k)u_k,$$

$$r^{k+1} = r^k - (c_k, r^k)c^k.$$

end k

$\mathcal{P}_{m,k}$ is the polynomial representing m steps of GMRES starting with r^k. This is the inner iteration. The previous algorithm may seem far from GMRES. However it is based on a variant described by Van der Vorst and Vuik. If the polynomial $\mathcal{P}_{m,k}$ is taken to be the identity matrix, then we obtain an algorithm which is equivalent to GCR as it can be seen by collapsing the inner i loop (which is also mathematically equivalent to GMRES). We see immediately that $\mathcal{P}_{m,k}$ can represent any other iterative method. This leads to the class of algorithms GMRES*, where the * stands for the inner iterative algorithm. Notice that the value of m may also be different in each outer iteration.

Unfortunately, the inner iteration may eventually stagnate. Then, Van der Vorst and Vuik suggested replacing the inner iteration by one step of LSQR, see Paige and Saunders [1059]. They showed that if the inner iteration is one step of GMRES(m), then the algorithm does not break down and, in exact arithmetic, the residual $r^k = 0$ for some $k < n$. This method may also be restarted (for the outer iteration) or truncated. Numerical experiments comparing GMRESR to full GMRES were provided in C. Vuik and H. van der Vorst [1289]. Further experiments and hints for choosing the number of inner iterations are given in Vuik [1311].

We remark that FGMRES is much easier to implement than GMRESR but it may break down. Note also that it can only be restarted. GMRESR can be either restarted or truncated. The behaviour of both algorithms is almost the same. In [1312], Vuik defined another method denoted as FFOM, a flexible FOM algorithm and showed that the search directions of FGMRES are constructed from the FFOM residuals. Different cheaper variants of FGMRES and GMRESR can be constructed by mixing the search directions.

De Sturler [391] proposed preserving the orthogonality relations of GCR in the inner GMRES iterations of GMRESR. Let U_k and C_k be the matrices whose columns are the vectors u^i and c^i, $i = 1, \ldots k$. Of course, we have

$C_k = AU_k$. In GCR the matrices are constructed such that $range(U_k)$ is equal to $K_k(A, r^0)$. Then, the method minimizes the l_2–norm of the residual over $range(U_k)$. In GCR the choice $u^k = r^{k-1}$ is made. The best choice for u^k will be $u^k = \varepsilon^{k-1}$ where ε^{k-1} is the error. GMRESR replaces this by a GMRES polynomial applied to r^{k-1}. At iteration $k+1$ in the inner GMRES iteration we solve

$$\min_y \|r^k - AV_m y\|.$$

However, it would have been better to solve

$$\min_y \|r^k - (I - C_k C_k^T)AV_m y\|.$$

Therefore, it seems better to use the matrix $(I - C_k C_k^T)A$ in the inner iteration. This implies that we have a globally optimal method that uses both inner and outer search vectors to compute the minimum of the norm of the residual. This is formally proved in [391]. If we solve this problem, it can be seen that

$$x^{k+1} = x^k + (I - U_k C_k^T A)V_m y,$$

and

$$\begin{aligned} c^{k+1} &= \frac{(I - C_k C_k^T)AV_m y}{\|(I - C_k C_k^T)AV_m y\|}, \\ u^{k+1} &= \frac{(I - U_k C_k^T A)V_m y}{\|(I - C_k C_k^T)AV_m y\|}. \end{aligned}$$

Since we are using a singular operator, the inner GMRES can possibly break down. However, it is shown in [391] that this is indeed a rare event and moreover some cures exist to the breakdowns. Good implementations of this algorithm are described in [391] where numerical experiments are also given. For the set of problems that were studied in [391], this algorithm (denoted as GCRO) always converges in fewer iterations than GMRESR. In fact, most of the time, the convergence was not far away from those of full GMRES. However, the computing time was not always smaller than for GMRESR. GCRO is more attractive when the matrix×vector product is expensive.

7.7. Hybrid GMRES algorithms

Some hybrid methods have been defined where the (Arnoldi) GMRES method is used, then eigenvalue estimates are extracted from the Hessenberg matrices and a Chebyshev iteration is started after an ellipse enclosing the eigenvalue estimates have been computed. Polygonal regions may also be used with optimal polynomials for different norms.

Another kind of hybrid algorithm has been defined by Nachtigal, Reichel and Trefethen [1004]. There, GMRES is run for some iterations and then, the GMRES polynomial is reapplied cyclically. The GMRES polynomial is explicitly constructed by using the relation between the orthogonal matrix V_k and the Krylov matrix. They are related by an upper triangular matrix which can be constructed column by column from the previous columns and the Hessenberg matrix. The vector of the coefficients of the GMRES polynomial is found by solving a linear system with this triangular matrix. After this first phase is completed, the polynomial is reapplied. There are several ways to do this. The simplest one is to use the Horner's scheme. However, it is well known that this method is not stable enough. Nachtigal, Reichel and Trefethen proposed using a Richardson iteration after computing the roots of the polynomial. For the first order Richardson iteration, it is important to choose a good ordering of the roots to get a stable algorithm. Nachtigal, Reichel and Trefethen used a Leja ordering, see [1004]. To make this algorithm practical we need a switching criterion. It was proposed to balance the work in the two phases. Nachtigal, Reichel and Trefethen introduced the ratio δ of the cost of one matrix$\times$vector multiplication to the cost of a vector operation. The convergence criterion being

$$\frac{\|r^k\|}{\|r^0\|} \leq \epsilon,$$

they proposed switching from phase 1 to phase 2 when

$$k + 3 + \delta \geq (1 + \delta)(\frac{\log \epsilon}{\log \tau} - 1),$$

where τ is the reduction of the residual norm relative to the norm of the initial residual in the first phase. Numerical experiments show that, in some cases, this method is better than just restarting GMRES at the switching point. Note that the two processes are not the same since the GMRES polynomial depends on the starting vector.

7.8. The non–symmetric Lanczos algorithm

We have seen that if we want to keep short (i.e. mostly three-term) recurrences, then we cannot construct an optimal method and the sequence of residuals is not orthogonal. However, we shall see that we can construct biorthogonal sequences. Note that in the generalized Hessenberg process we can use an additional sequence y^k that is constructed in the same way as v^k but using A^T instead of A. This leads to an algorithm that is known as "non–symmetric Lanczos" (or two–sided Lanczos).

7.8.1. Definition of the non–symmetric Lanczos algorithm

We use the same notations as for the Lanczos algorithm. We choose two starting vectors q^1 and $\tilde{q}^1$ such that $(q^1, \tilde{q}^1) = 1$. We set $q^{-1} = \tilde{q}^{-1} = 0$. Then for $j = 0, 1, \dots$

$$\begin{aligned} z^k &= Aq^k - \delta_k q^k - \eta_k q^{k-1}, \\ w^k &= A^T\tilde{q}^k - \delta_k \tilde{q}^k - \tilde{\eta}_k \tilde{q}^{k-1}, \end{aligned}$$

the coefficient δ_k being computed as

$$\delta_k = (\tilde{q}^k, Aq^k).$$

η_k and $\tilde{\eta}_k$ are chosen such that

$$\eta_{k+1}\tilde{\eta}_{k+1} = (z^k, w^k),$$

then

$$\begin{aligned} q^{k+1} &= \frac{z^k}{\tilde{\eta}_{k+1}}, \\ \tilde{q}^{k+1} &= \frac{w^k}{\eta_{k+1}}. \end{aligned}$$

These relations can be written in matrix form. Let

$$T_k = \begin{pmatrix} \delta_1 & \eta_2 & & & \\ \tilde{\eta}_2 & \delta_2 & \eta_3 & & \\ & \ddots & \ddots & \ddots & \\ & & \tilde{\eta}_{k-1} & \delta_{k-1} & \eta_k \\ & & & \tilde{\eta}_k & \delta_k \end{pmatrix},$$

and

$$Q_k = [q^1 \cdots q^k], \quad \tilde{Q}_k = [\tilde{q}^1 \cdots \tilde{q}^k].$$

Then, we can write

$$\begin{aligned} AQ_k - Q_k T_k &= \tilde{\eta}_{k+1} q^{k+1} (e^k)^T, \\ A^T\tilde{Q}_k - \tilde{Q}_k T_k^T &= \eta_{k+1}\tilde{q}^{k+1}(e^k)^T. \end{aligned}$$

This can also be written as

$$AQ_k = Q_{k+1}\bar{T}_k,$$
$$\tilde{Q}_k A Q_k = T_k,$$

where

$$\bar{T}_k = \begin{pmatrix} T_k \\ \tilde{\eta}_{k+1}(e^k)^T \end{pmatrix},$$

is a $k+1 \times k$ matrix.

Theorem 7.18

If the non–symmetric Lanczos algorithm does not break down, it computes biorthogonal vectors, that is such that

$$(\tilde{q}^i, q^j) = 0, i \neq j, \quad i, j = 1, \ldots, k$$

The vectors $q^1, \ldots, q^k$ span $K_k(A, q^1)$ and $\tilde{q}^1, \ldots, \tilde{q}^k$ span $K_k(A^T, \tilde{q}^1)$. The two sequences of vectors can be written as

$$q^k = P_k(A)q^1, \quad \tilde{q}^k = P_k(A^T)\tilde{q}^1,$$

where P_k is a polynomial of degree k.

Proof. Obvious from the definition. □

Notice that the non–symmetric Lanczos algorithm can be slightly generalized by using another scalar product or a different scaling, see Gutknecht [731]. The matrix T_k is an oblique projection of A onto $K_k(A, q^1)$ orthogonally to $K_k(A^T, \tilde{q}^1)$. The algorithm breaks down if at some step we have $(z^k, w^k) = 0$. Either,

a) $q^k = 0$ and/or $\tilde{q}^k = 0$. In both cases we have found an invariant subspace. If $q^k = 0$ we can compute the "exact" solution of the linear system. If only $\tilde{q}^k = 0$, almost the only way to deal with this situation is to restart the algorithm with another vector $\tilde{q}^1$. Usually using a random initial vector is enough to avoid this kind of breakdown.

b) The more dramatic situation (which is called a "serious breakdown") is when $(z^k, w^k) = 0$ with z^k and $w^k \neq 0$. Then, a way to solve this problem is to use a look–ahead strategy. We shall study this in one of the following sections, the solution being to construct also the vectors q^{k+1} and $\tilde{q}^{k+1}$ at step k maintaining bi–orthogonality in a blockwise sense. If this is not possible, we try constructing also vectors q^{k+2} and $\tilde{q}^{k+2}$ and so on. The worst case is when we reach the dimension of the system without being able to return to the normal situation. This is known as an incurable breakdown.

In finite precision arithmetic, it is unlikely that we get $(z^k, w^k) = 0$ with z^k and $w^k \neq 0$. However, it may happen that (z^k, w^k) is small. This is known as a near breakdown and it is really this problem that look–ahead strategies must deal with.

7.8.2. Variants of the non–symmetric Lanczos algorithm

Saad [1131] remarked that we have some freedom in constructing the vectors $\tilde{q}^k$. All that is needed is that they must span $K_k(A^T, \tilde{q}^1)$ as the Lanczos vectors q^k are determined by the condition that q^k must be orthogonal to $K_k(A^T, \tilde{q}^1)$. Therefore, we just need $\tilde{q}^k$ to be $t_k(A)\tilde{q}^1$ where t_k is a polynomial of exact degree k. Of course, as then bi–orthogonality does not exist anymore, we must change the formula giving the coefficients of the method. The polynomial t_k need not be given a priori. It has been suggested by Gutknecht [731] that we use a two–term recurrence such as

$$w^k = A^T\tilde{q}^k - (\tilde{q}^k, A^T\tilde{q}^k)\tilde{q}^k - (\tilde{q}^{k-1}, A^T\tilde{q}^k)\tilde{q}^{k-1},$$

or the modified Gram–Schmidt implementation of this formula. This is a kind of truncated `Arnoldi` applied to A^T. This can be further generalized by using more terms in the sum.

Another variant is constructed by using coupled two–term recurrences instead of three–term recurrences. This form of the algorithm is based on using the LU factorization of the matrix T_k. We give the algorithm in the form described by Gutknecht [731]:

Let q^1 and $\tilde{q}^1$ be given such that $\delta_1 = (q^1, \tilde{q}^1) \neq 0$ and $\delta_1' = (\tilde{q}^1, Aq^1) \neq 0$, $p^1 = q^1$, $\tilde{p}^1 = \tilde{q}^1$. For $k = 1, \dots$ choose $\gamma_k \neq 0$ and $\tilde{\gamma}_k \neq 0$ and

$$
\begin{aligned}
\phi_k &= \delta_k'/\delta_k, \\
q^{k+1} &= (Ap^k - \phi_k q^k)/\gamma_k, \\
\tilde{q}^{k+1} &= (A^T\tilde{q}^k - \phi_k \tilde{q}^k)/\tilde{\gamma}_k, \\
\delta_{k+1} &= (\tilde{q}^{k+1}, q^{k+1}), \\
\psi_k &= \tilde{\gamma}_k \delta_{k+1}/\delta_k', \\
\tilde{\psi}_k &= \gamma_k \delta_{k+1}/\delta_k', \\
p^{k+1} &= q^{k+1} - \psi_k p^k, \\
\tilde{p}^{k+1} &= \tilde{q}^{k+1} - \tilde{\psi}_k \tilde{p}^k, \\
\tilde{\delta}_{k+1} &= (\tilde{p}^{k+1}, Ap^{k+1}).
\end{aligned}
$$

It can be proved that the sequences q^k and $\tilde{q}^k$ are biorthogonal and the sequences p^k and $\tilde{p}^k$ are A–biconjugate. It turns out that this particular form of the algorithm may break down earlier than the three–term recurrence form if $\delta_k' = 0$. This is linked to the non–existence of the LU factorization of T_k.

7.8.3. Maintaining semi bi–orthogonality

In Chapter 6 we have seen how to maintain semi orthogonality in the Lanczos algorithm for symmetric matrices. D. Day [363] considered how to extend this to non–symmetric Lanczos. As the term semi bi–orthogonality is a little too long, he chose to call this maintaining semi–duality. He considered a variant of non–symmetric Lanczos where we compute Lanczos vectors of norm 1. The algorithm is the following.

$$\gamma_{k+1} q^{k+1} = Aq^k - \frac{\alpha_k}{\omega_k} q^k - \frac{\beta_k}{\omega_{k-1}} q^{k-1},$$

$$\beta_{k+1} \tilde{q}^{k+1} = A\tilde{q}^k - \frac{\alpha_k}{\omega_k} \tilde{q}^k - \frac{\gamma_k}{\omega_{k-1}} \tilde{q}^{k-1}.$$

The coefficients γ_k and β_k are computed to get $\|q^k\| = \|\tilde{q}^k\| = 1$. In matrix form this is written

$$
\begin{aligned}
AQ_k \quad - \quad Q_k \Omega_k^{-1} T_k &= \gamma_{k+1} q^{k+1} (e^k)^T, \\
A^T \tilde{Q}_k - \tilde{Q}_k \Omega_k^{-1} T_k^T &= \beta_{k+1} \tilde{q}^{k+1} (e^k)^T,
\end{aligned}
$$

T_k is tridiagonal and Ω_k is diagonal. If the algorithm produces candidate vectors which we denote as q_c^{k+1} and $\tilde{q}_c^{k+1}$ and if we decide that they are not orthogonal "enough" to the previous vectors (that is $\|\tilde{Q}_k q_c^{k+1}\|$ and $\|(\tilde{q}_c^{k+1})^T Q_k\|$ are "too large"), then we use Gram–Schmidt to correct this situation. Let q_G^{k+1} and $\tilde{q}_G^{k+1}$ be the new candidates after correction. Then, the correction step may be written as

$$|\Omega_k|^{-\frac{1}{2}} Q_k^T \tilde{q}_G^{k+1} = |\Omega_k|^{-\frac{1}{2}} (I - Q_k^T \tilde{Q}_k \Omega_k^{-1}) Q_k^T \tilde{q}_c^{k+1}$$
$$= (\text{sign}(\Omega_k) - |\Omega_k|^{-\frac{1}{2}} \text{sign}(\Omega_k) |\Omega_k|^{\frac{1}{2}} Q_k^T \tilde{Q}_k |\Omega_k|^{-\frac{1}{2}}) Q_k^T \tilde{q}_c^{k+1},$$

$$|\Omega_k|^{-\frac{1}{2}} \tilde{Q}_k q_G^{k+1} = (\text{sign}(\Omega_k) - |\Omega_k|^{-\frac{1}{2}} \tilde{Q}_k^T Q_k |\Omega_k|^{-\frac{1}{2}})$$
$$\times \text{sign}(\Omega_k) |\Omega_k|^{-\frac{1}{2}} \tilde{Q}_k^T q_c^{k+1}$$

Therefore, the interesting matrix after applying Gram–Schmidt is

$$M_k = \text{sign}(\Omega_k) - |\Omega_k|^{-\frac{1}{2}} \tilde{Q}_k^T Q_k |\Omega_k|^{-\frac{1}{2}}.$$

The measures of bi–orthogonality will be

$$\| \, |\Omega_k|^{-\frac{1}{2}} \tilde{Q}_k q^{k+1} \|_1, \quad \| \, |\Omega_k|^{-\frac{1}{2}} Q_k^T \tilde{q}^{k+1} \|_\infty.$$

From these considerations, D. Day [363] defines semi–duality as being

$$\max(\| \, \Omega_k|^{-\frac{1}{2}} Q_k^T \tilde{q}_c^{k+1} \|_\infty, \| \, \Omega_k|^{-\frac{1}{2}} \tilde{Q}_k^T q_c^{k+1} \|_1 \leq u^{\frac{1}{2}} \, |\omega_{k+1}|^{\frac{1}{4}}.$$

Local duality is said to hold at step k if

$$\max_{1 \leq i \leq k} (|(\tilde{q}^i q^{i-1}|), |(\tilde{q}^{i-1}, q^i)|) \leq 4u.$$

D. Day proved that local duality is maintained by the following algorithm.

$$r^k = A^T \tilde{q}^k - \frac{\gamma_k \omega_k}{\omega_{k-1}} \tilde{q}^{k-1}, \quad s^k = A q^k - \frac{\beta_k \omega_k}{\omega_{k-1}} q^{k-1},$$

$$\alpha_k = (\tilde{q}^k, q^k) \text{ or } (r^k, q^k),$$

$$r^k = r^k - \frac{\alpha_k}{\omega_k} \tilde{q}^k, \quad s^k = s^k - \frac{\alpha_k}{\omega_k} q^k,$$

$$\alpha_k^l = (r^k, q^k), \quad \alpha_k^r = (\tilde{q}^k, s^k),$$

$$r^k = r^k - \frac{\alpha_k^l}{\omega_k} \tilde{q}^k, \quad s^k = s^k - \frac{\alpha_k^r}{\omega_k} q^k,$$

$$\beta_{k+1} = \|r^k\|, \quad \gamma_{k+1} = \|s^k\|,$$

$$\tilde{q}^{k+1} = \frac{r^k}{\beta_{k+1}}, \quad q^{k+1} = \frac{s^k}{\gamma_{k+1}},$$

$$\omega_{k+1} = (\tilde{q}^{k+1}, q^{k+1}).$$

Let D_k^l and D_k^r be the diagonal matrices with non–zero elements α_i^l and α_i^r. At the end of a step we compute

$$x^k = Q_{k-1}^T \tilde{q}_c^{k+1}, \quad y^k = \tilde{Q}_{k-1} q_c^{k+1}.$$

Then we reorthogonalize, getting

$$\tilde{q}_G^{k+1} = \tilde{q}_c^{k+1} - \tilde{Q}_{k-1}^T \Omega_{k-1}^{-1} x^k,$$

$$q_G^{k+1} = q_c^{k+1} - Q_{k_1} \Omega_{k_1}^{-1} y^k.$$

Correction steps are applied in steps in two successive iterations to the last two pairs of Lanczos vectors. To monitor the loss of duality, recurrences are derived for $Q_k^T \tilde{q}^{k+1}$ and $\tilde{Q}_k q^{k+1}$. $\tilde{Q}_k q^{k+1}$ is estimated by h^{k+1} defined by

$$\gamma_{k+1} h^{k+1} = (T_k \Omega_k^{-1} - \frac{\alpha_k}{\omega_k} I) \begin{pmatrix} h^k \\ 0 \end{pmatrix} - \frac{\beta_k \omega_k}{\omega_{k-1}} \begin{pmatrix} h^{k-1} \\ 0 \\ 0 \end{pmatrix} \alpha_k^r e^{k-2}.$$

If a correction step is done, $u \, \mathrm{diag}(|\Omega_k^{-1} T_k|)$ is added to the right hand side. A similar recurrence is obtained for the estimate of $Q_k^T \tilde{q}^{k+1}$. Using these corrections implements a robust non–symmetric Lanczos algorithm. It can be used in all algorithms using the non–symmetric Lanczos algorithm as a first step.

7.9. The BiConjugate Gradient Algorithm

Methods for solving linear systems can be derived from the non–symmetric Lanczos algorithm in the same way `CG` was derived from the Lanczos algorithm for the symmetric case. The solution can be written as

$$x^k = x^0 + Q_k T_k^{-1} (\|b\| e^1).$$

Notice that when using the non–symmetric Lanczos algorithm, we are implicitly solving also a system with A^T, whose solution we are generally not interested in. To derive algorithms that use short recurrences we have to use the LU factorization of the non–symmetric tridiagonal matrix T_k (without permutations). Depending on the version of non–symmetric Lanczos we use, we shall get different versions of the BiConjugate Gradient (`BiCG`) algorithm, see Gutknecht [731]. `BiCG` may also be considered as a method for solving

$$\begin{pmatrix} 0 & A \\ A^T & 0 \end{pmatrix} \begin{pmatrix} y \\ x \end{pmatrix} = \begin{pmatrix} b \\ c \end{pmatrix},$$

where c is arbitrarily chosen. The classical version of `BiCG` (Fletcher [564]) is the following:

Let x^0 be given, $r^0 = b - Ax^0$, $\tilde{r}^0 = r^0$, $p^0 = r^0$, $\tilde{p}^0 = \tilde{r}^0$, for $k = 0, 1, \dots$

$$\begin{aligned}
\alpha_k &= \frac{(\tilde{r}^k, r^k)}{(\tilde{p}^k, Ap^k)}, \\
x^{k+1} &= x^k + \alpha_k p^k, \\
r^{k+1} &= r^k - \alpha_k Ap^k, \\
\tilde{r}^{k+1} &= \tilde{r}^k - \alpha_k A^T \tilde{p}^k, \\
\beta_{k+1} &= \frac{(\tilde{r}^{k+1}, r^{k+1})}{(\tilde{r}^k, r^k)}, \\
p^{k+1} &= r^{k+1} + \beta_{k+1} p^k, \\
\tilde{p}^{k+1} &= \tilde{r}^{k+1} + \beta_{k+1} \tilde{p}^k.
\end{aligned}$$

There are two reasons for which the algorithm may break down: $(\tilde{r}^k, r^k) = 0$ or $(\tilde{p}^k, Ap^k) = 0$. We may get near breakdowns if one or both of the scalar products is small. This question will be treated in one of the following sections.

Theorem 7.19

If `BiCG` *does not break down we have the following properties for* $k \neq l$*:*

$$\begin{aligned}
(\tilde{r}^k, r^l) &= 0, \\
(\tilde{p}^k, Ap^l) &= 0, \\
(\tilde{r}^k, p^l) &= 0, \\
(r^k, \tilde{p}^l) &= 0.
\end{aligned}$$

Proof. The method of proof is the same as with `CG` for symmetric systems.
□

Note that we can also choose another vector as the starting vector and residual for the auxiliary system. If the matrix A is SPD, we recover CG. Even though this algorithm has somewhat of a bad reputation, it is very simple to implement and most of the time its convergence rate is quite fast. Moreover, it is straightforward to precondition BiCG.

Figure 7.6 shows the norm of the error as a function of the number of floating point operations (flops) using BiCG without preconditioning for problem 7 on a 20×20 mesh.

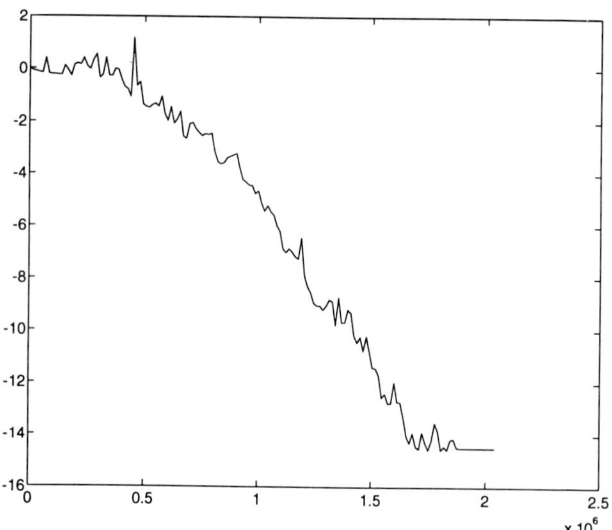

fig. 7.6: Problem 7: $\log_{10}$ *of the error with* BiCG
20×20 *mesh*

To generate other forms of the algorithm we can also consider the indefinite matrix

$$\begin{pmatrix} 0 & A \\ A^T & 0 \end{pmatrix}$$

and use one of the methods for indefinite systems we considered in Chapter 6. We can also use a biorthogonal algorithm with three–term recurrences.

7.10. Roundoff error analysis of BiCG

Not much is known about the roundoff error analysis of BiCG. Obviously, this is linked to the analysis of the non–symmetric Lanczos algorithm. Paige's results for the symmetric Lanczos algorithm were generalized to the non–symmetric case by Z. Bai [95]. C. Tong and Q. Ye [1260] have developed bounds for the finite precision computation of BiCG. Their work relies closely on the Lanczos algorithm as they eliminate the variables p. Then, if we denote $R_k = [r^0, \dots, r^k]$,

$$AR_k = R_k T_k - \frac{1}{\alpha_k} r^{k+1} (e^k)^T.$$

We suppose that the matrix×vector product is such that the computed value is $Ax + ug$ with $|g| \le n|A|\,|x| + O(u)$.

Theorem 7.20

The computed quantities in BiCG *satisfy*

$$AR_k = R_k T_k - \frac{1}{\alpha_k} r^{k+1} (e^k)^T + u\Delta_k,$$

where $\Delta_k = [\delta^0, \dots, \delta^k]$ *with*

$$|\delta^i| \le ((n+6)|A| + \frac{1}{|\alpha_i|} + \frac{|\beta_i|}{|\alpha_{i-1}|})|r^i| + (2n+7)|A|\,|p^i| + O(u).$$

Proof. See C. Tong and Q. Ye [1260]. □

Let X_k be the matrix of the residual vectors scaled by their l_2–norms. C. Tong and Q. Ye [1260] also proved some bounds for the norms of the residuals.

Theorem 7.21

If the residuals are linearly independent,

$$\|r^{k+1}\| \le (1 + C_k) \min_p \|p(A + \delta A_k) r^0\|,$$

where the minimum is taken over polynomials of degree k such that $p(0) = 1$ and $C_k = \|(AX_k - u\Delta_k)T_k^{-1}W_k\|$, the matrix W_k being the first k columns of the generalized inverse X_{k+1}^+ of X_{k+1}. The perturbation matrix δA_k is $-u\Delta_k X_k^+$, X_k being the matrix of the vectors x^k.

Proof. See C. Tong and Q. Ye [1260]. □

Notice that the right hand side is related to the GMRES method for $A + \delta A_k$. This result proves that convergence will still occur in finite precision arithmetic as long as the polynomial term balances the growth of the constant C_k. C. Tong and Q. Ye proved that

$$C_k \leq (\sqrt{n}\|A\| + u\|\Delta_k\|)\|T_k^{-1}\| \, \|X_{k+1}^+\|,$$

hence C_k must not be too large.

7.11. Handling of breakdowns

We have seen that the Lanczos biorthogonalization algorithm (as well as the biconjugate gradient method) may suffer from breakdowns and near breakdowns. Solutions to this problem have been studied by many people. The cure of the breakdown problem can be obtained by using the connection of the Lanczos algorithm to other mathematical problems. As we shall see later on, non–symmetric Lanczos is related to computing formal orthogonal polynomials (FOP). It is also related to the computation of Padé approximants and Padé tables. Solutions to the breakdown problem were already known or can be derived more easily by looking at Lanczos from these perspectives. We can also cure the breakdown problem by maintaining bi–orthogonality only in a blockwise sense. Finally, the Lanczos basis can be modified by introducing new vectors in the Krylov basis. Another goal is to find a solution of the breakdown problem which uses the least possible number of additional matrix×vector multiplies and scalar products. There are different solutions to the breakdown problem which are more or less equivalent as regards the algorithmic complexity.

7.11.1. FOP

We follow the exposition of Brezinski, Redivo Zaglia and Sadok [209]. We have seen that the residuals are written as $r^k = P_k(A)r^0$ and are orthogonal to $K_k(A^T, \tilde{r}^0)$. Therefore

$$(r^k, (A^T)^i \tilde{r}^0) = (A^i r^k, \tilde{r}^0) = (A^i P_k(A) r^0, \tilde{r}^0) = 0,\ i = 0, \ldots, k-1$$

Let c be a linear functional defined over the space of polynomials as

$$c(\xi^i) = c_i = (A^i r^0, \tilde{r}^0),\ i = 0, 1, \ldots$$

The bi–orthogonality requirements are written as

$$c(\xi^i P_k(\xi)) = 0,\ i = 0, \ldots, k-1$$

Such a family of polynomials is called formal orthogonal polynomials with respect to c. The polynomials are normalized such that $P_k(0) = 1$. It can be shown that P_k exists and is unique if and only if

$$d_k = \begin{vmatrix} c_1 & \cdots & c_k \\ \vdots & & \vdots \\ c_k & \cdots & c_{2k-1} \end{vmatrix} \neq 0.$$

A polynomial which exists is called regular. Formal orthogonal polynomials (FOP) satisfy three–term recurrences

$$P_{k+1}(\xi) = (\alpha_{k+1}\xi + \beta_{k+1}) P_k(\xi) - \gamma_{k+1} P_{k-1}(\xi).$$

A linear system for the coefficients $\alpha_{k+1}, \beta_{k+1}, \gamma_{k+1}$ is obtained by enforcing the orthogonality conditions and $P_{k+1}(0) = 1$. The polynomial P_{k+1} is computable if the 3×3 determinant for the coefficients is non–zero and if $d_{k+1} \neq 0$. If the breakdown is due to the three–term relation, we can compute the polynomial using some other relations. For instance, we can define another functional c^1 such that $c^1(\xi^i) = c_{i+1}$ and compute the polynomials P_k^1 such that $c^1(\xi^i P_k^1) = 0$, the leading coefficient of P_k^1 being 1. These polynomials lead to the ORTHORES implementation of the non–symmetric Lanczos algorithm.

We denote by n_k the degrees of regular polynomials, meaning that P_{n_k} exists and the next regular polynomial is $P_{n_{k+1}}$ of degree n_{k+1}, $n_{k+1} = n_k + m_k, m_k \geq 1$. The polynomials in between do not exist and we are interested in jumping from P_{n_k} to $P_{n_{k+1}}$. Draux [447] proved that m_k is given by

$$\begin{aligned} c^1(\xi^i P_{n_k}^1(\xi)) &= 0,\ i = 0, \ldots, n_k + m_k + 2 \\ &\neq 0,\ i = n_k + m_k + 1 \end{aligned}$$

We have an incurable breakdown if we cannot compute the next polynomial before reaching the dimension of the linear system. These $P^1_{n_k}$ polynomials can be computed by

$$P^1_{n_{k+1}}(\xi) = (\alpha_0 + \cdots + \alpha_{m_k-1}\xi^{m_k-1} + \xi^{m_k})P^1_{n_k} - \gamma_{k+1}P^1_{n_k-1}.$$

The coefficients are computed by solving a non–singular triangular system (see [209]):

$$\gamma_{k+1} = c^1(\xi^{n_k+m_k-1}P^1_{n_k})/c^1(\xi^{n_k-1}P^1_{n_k-1}),$$

$$\alpha_{m_k-1}c^1(\xi^{n_k+m_k-1}P^1_{n_k}) + c^1(\xi^{n_k+m_k}P^1_{n_k}) = \gamma_{k+1}c^1(\xi^{n_k}P^1_{n_k-1}),$$

and so on until

$$\alpha_0 c^1(\xi^{n_k+m_k-1}P^1_{n_k}) + \cdots + \alpha_{m_k-1}c^1(\xi^{n_k+2m_k-2}P^1_{n_k})$$
$$+ c^1(\xi^{n_k+2m_k-1}P^1_{n_k 1)=\gamma_{k+1}c^1(\xi^{n_k+m_k-1}P_{n_k-1}}).$$

The polynomials P_k needed in the Lanczos algorithm are recovered by

$$P_{n_{k+1}} = P_{n_k} - \xi(\beta_0 + \cdots + \beta_{m_k-1}\xi^{m_k-1})P^1_{n_k}.$$

The coefficients β_i are the solution of a non–singular triangular system:

$$\beta_{m_k-1}c^1(\xi^{n_k+m_k-1}P^1_{n_k} = c(\xi^{n_k}P_{n_k}),$$

and so on, the last equation being

$$\beta_0 c^1(\xi^{n_k+m_k-1}P^1_{n_k}) + \cdots + \beta_{m_k-1}c^1(\xi^{n_k+2m_k-2}P^1_{n_k}) = c(\xi^{n_k+m_k-1}P_{n_k}).$$

An implementation of the Lanczos algorithm using these recurrences has been defined in Brezinski, Redivo Zaglia and Sadok [207]. Of course, what is really interesting is avoiding near breakdowns. In that case we must also jump over regular polynomials that may be badly computed. Recurrences for computing the regular polynomials are given in [206]. The main problem is in fact to define when we have a near breakdown situation. One way would be to check $|c^1(\xi^i P^1_{n_k})|$ relative to a given ϵ. However, the numerical results are then too sensitive to the choice of ϵ.

7.11.2. Padé approximation

The relation of Padé approximation to the non–symmetric Lanczos algorithm was studied by Gutknecht [726]. The computation of Padé tables and the relation to Lanczos methods is considered in detail in Hochbruck [786]. Starting from algorithms for computing the Padé table, she proposed the following look–ahead Lanczos algorithm.

Let $m = 0, l = 0, v^0 = 0, w^0 = 0, \epsilon_0 = 1$ and $q^0, \tilde{q}^0$ be given. While $q^m \neq 0$ and $\tilde{q}^m \neq 0$,

$$k = 1, \quad \delta_{m,m} = (q^m, \tilde{q}^m), \quad D^l = \delta_{m,m}$$

while $\sigma_{min}(D^l) < \epsilon$ and $m + k \leq n$

$$\begin{aligned} \gamma_{k-1}^m &= -\frac{\delta_{m,m+k-1}}{\epsilon_m}, \\ q^{m+k} &= Aq^{m+k-1} + \gamma_{k-1}^m v^m, \\ \tilde{q}^{m+k} &= A^T \tilde{q}^{m+k-1} + \gamma_{k-1}^m w^m, \end{aligned}$$

$$\delta_{m+k-1,m+k-1} = (q^{m+k-1}, \tilde{q}^{m+k-1}), \quad \delta_{m+k,m+k} = (q^{m+k}, \tilde{q}^{m+k}),$$

extend the Hankel matrix D^l by one row and one column, $k = k + 1$

end while

$$\theta_{m+k} = \frac{\delta_{m,m+k-1}}{\epsilon_m},$$

Solve $D^l d^m = -z, \quad z = ((\tilde{q}^m, Aq^{m+k-1}), \ldots, (\tilde{q}^{m+k-1}, Aq^{m+k-1}))^T,$

$$\begin{aligned} q^{m+k} &= Aq^{m+k-1} + (q^m, \ldots, q^{m+k-1})d^m + \theta_{m+k} v^m, \\ \tilde{q}^{m+k} &= A^T \tilde{q}^{m+k-1} + (\tilde{q}^m, \ldots, \tilde{q}^{m+k-1})d^m + \theta_{m+k} w^m. \end{aligned}$$

if $k = 1$, $v^{m+1} = v^m, w^{m+1} = w^m, \epsilon_{m+1} = \delta_{m,m}$ else $\epsilon_{m+k} = 1$, solve $D^l \tilde{d}^m = e^{k-1}$

$$\begin{aligned} v^{m+k} &= (v^m, \ldots, v^{m+k-1})\tilde{d}^m, \\ w^{m+k} &= (w^m, \ldots, w^{m+k-1})\tilde{d}^m \end{aligned}$$

end if

$$m = m + k, \quad l = l + 1$$

In this algorithm $\sigma_{min}(D^l)$ is the smallest singular value of D^l. However, for this algorithm to be of practical interest strategies must be defined to determine the length of the look–ahead jump.

7.11.3. Block bi–orthogonality

Another perspective to solve the breakdown problem is to enforce only block bi–orthogonality. Note this is what we implicitly did in the previous subsections. We want to enforce the bi–orthogonality conditions

$$q^{n_k} \perp K_{n_k}(A^T, \tilde{q}^0), \quad \tilde{q}^{n_k} \perp K_{n_k}(A, q^0)$$

for the regular indices. For the other vectors (denoted as inner vectors) we would like to have

$$q^k \perp K_{n_k}(A^T, \tilde{q}^0), \quad \tilde{q}^k \perp K_{n_k}(A, q^0), \quad n_k \leq k < n_{k+1}$$

As Gutknecht [731] we denote

$$Q_k = (\, q^{n_k} \quad q^{n_k+1} \quad \cdots \quad q^{n_{k+1}-1}\,), \quad \tilde{Q}_k = (\, \tilde{q}^{n_k} \quad \tilde{q}^{n_k+1} \quad \cdots \quad \tilde{q}^{n_{k+1}-1}\,).$$

Let $D_k = (\tilde{Q}_k)^T Q_k$. We require $(\tilde{Q}_i)^T Q_j = 0$ if $i \neq j$. Let

$$Q_{k,l} = (\, q^{n_k} \quad q^{n_k+1} \quad \cdots \quad q^l\,), \quad \tilde{Q}_{k,l} = (\, \tilde{q}^{n_k} \quad \tilde{q}^{n_k+1} \quad \cdots \quad \tilde{q}^l\,),$$

the last block being eventually incomplete. We denote

$$\mathcal{Q}_{k+1} = (\, Q_0 \quad \cdots \quad Q_{k-1} \quad Q_{k,l}\,), \quad \tilde{\mathcal{Q}}_{k+1} = (\, \tilde{Q}_0 \quad \cdots \quad \tilde{Q}_{k-1} \quad \tilde{Q}_{k,l}\,)$$

Then, $(\tilde{\mathcal{Q}}_{k+1})^T \mathcal{Q}_{k+1} = \mathcal{D}_{k+1}$ where this last matrix is block diagonal with diagonal blocks D_i. The basis we are looking for can be defined by

$$A\mathcal{Q}_k = \mathcal{Q}_{k+1}\mathcal{T}_k, \quad A^T \tilde{\mathcal{Q}}_k = \tilde{\mathcal{Q}}_{k+1}\mathcal{T}_k^T.$$

The matrix $\mathcal{T}_k$ is shown to be block tridiagonal, the diagonal blocks being upper Hessenberg. It is easy to see how to compute the elements of $\mathcal{T}_k$ by enforcing the orthogonality conditions, all we need is to have the blocks D_i well conditioned as they must be inverted to get the solution.

As indicated, the difficulty is finding a strategy to determine whether a vector is regular or not. Monitoring the smallest singular value is generally not sufficient. Other possibilities have been considered in Freund, Gutknecht and Nachtigal [587]. It is required that the l_1 norms of (parts) of columns of $\mathcal{T}_k$ be bounded by a given prescribed tolerance for the given step of the algorithm to be considered regular. All these strategies are based on heuristics and might not work on every problem.

7.11.4. Modified Krylov spaces

Another solution to the breakdown problem has been proposed by Q. Ye [1350], see also Tong and Ye [1261]. If we decide that we have a breakdown at step k, a new start vector $\tilde{q}_{k+1}$ is chosen to be orthogonal to $K_k(A, q^0)$. The subspace $K_{k-1+2m}(A^T, \tilde{q}^0)$ is replaced by a sum of $K_{k-1+m}(A^T, \tilde{q}^0)$ and $K_m(A^T, \tilde{q}^{k+1})$ for some m. Then from iteration $k+2$ the Lanczos recurrences will involve four terms, adding a term for q^{k-2}. At the next breakdown a fifth term is added and so on. Finally, we obtain biorthogonal sequences and a banded upper Hessenberg matrix whose bandwidth is 3 plus the number of previous breakdowns. So, we get something in between Lanczos and `Arnoldi`. It has been shown that the new start vector can be chosen by enforcing only local orthogonality and we shall still have the same properties.

7.12. The Conjugate Gradient Squared algorithm

In `BiCG` it is easy to see that the residuals are polynomials applied to the initial residuals. This is formalized in the next proposition.

Proposition 7.22

Let ϕ_k and θ_k be polynomials of degree less or equal to k defined by

$$\phi_{k+1}(A) = \phi_k(A) - \alpha_k A\theta_k(A),$$
$$\theta_{k+1}(A) = \phi_{k+1}(A) + \beta_{k+1}\theta_k(A).$$

Then the vectors defined in `BiCG` *satisfy*

$$r^k = \phi_k(A)r^0, \quad p^k = \theta_k(A)r^0,$$
$$\tilde{r}^k = \phi_k(A^T)r^0, \quad \tilde{p}^k = \theta_k(A^T)r^0.$$

Proof. The proof is obvious by induction. □

Note that $(r^k, \tilde{r}^k) = (\phi_k(A)r^0, \phi_k(A^T)r^0) = (\phi_k(A)^2 r^0, r^0)$. This leads P. Sonneveld [1195] to construct an algorithm whose residual is given by $\phi_k(A)^2 r^0$ instead of $\phi_k(A)r^0$, the idea being that if $\phi_k(A)$ is a contraction for k large, then $\phi_k(A)$ will be a contraction of smaller norm. Another positive point is that the transpose matrix would not be needed anymore. To derive this method recurrences must be sought for the quantities implied in BiCG. For instance, we have

$$\phi_{k+1}^2(t) = \phi_k^2(t) - 2\alpha_k t\theta_k(t)\phi_k(t) + \alpha_k^2 t^2\theta_k^2(t).$$

The term $\phi_{k+1}\theta_k$ is used as a new unknown in the algorithm and additional recurrences can be found for the other cross term as

$$\phi_k(t)\theta_k(t) = \phi_k^2(t) + \beta_k\phi_k(t)\theta_{k-1}(t).$$

The "squared" method has been named CGS for Conjugate Gradient Squared, although it might have been better to name it BiCGS. The CGS algorithm is the following:

Let x^0 be given, $r^0 = b - Ax^0$, $q^0 = p^0 = r^0$, for $k = 0, 1, \ldots$

$$\begin{aligned}
\alpha_k &= \frac{(r^k, r^0)}{(Aq^k, r^0)}, \\
u^k &= p^k - \alpha_k Aq^k, \\
x^{k+1} &= x^k + \alpha_k(p^k + u^k), \\
r^{k+1} &= r^k - \alpha_k A(p^k + u^k), \\
\beta_{k+1} &= \frac{(r^{k+1}, r^0)}{(r^k, r^0)}, \\
p^{k+1} &= r^{k+1} + \beta_{k+1}u^k, \\
q^{k+1} &= p^{k+1} + \beta_{k+1}(u^k + \beta_{k+1}q^k).
\end{aligned}$$

Note that even though A^T is not needed anymore there are still two matrix$\times$vector products for each iteration.

Figure 7.7 shows the norm of the error as a function of the number of floating point operations (flops) using CGS without preconditioning for problem 7 on a 20×20 mesh.

When introducing a preconditioner M, we have to modify the algorithm

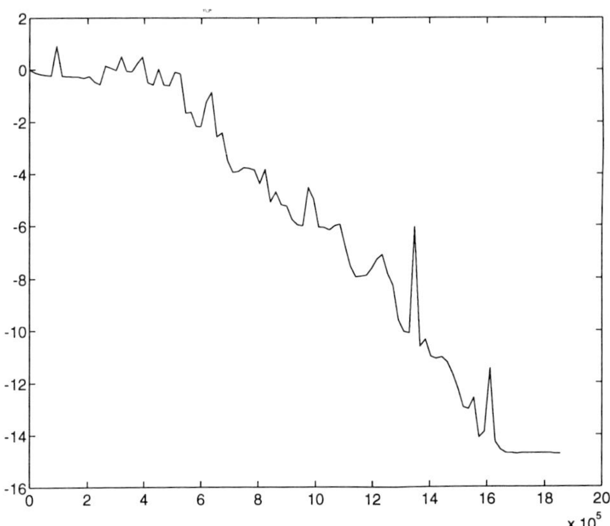

fig. 7.7: Problem 7: $\log_{10}$ *of the error with* CGS
20×20 *mesh*

such that

$$
\begin{aligned}
\hat{q}^k &= M^{-1}q^k, \\
\tilde{q}^k &= A\hat{q}^k, \\
\alpha_k &= \frac{(r^k, r^0)}{(\tilde{q}^k, r^0)}, \\
u^k &= p^k - \alpha_k \tilde{q}^k, \\
\hat{u}_k &= M^{-1}(p^k + u^k), \\
x^{k+1} &= x^k + \alpha_k \hat{u}_k, \\
r^{k+1} &= r^k - \alpha_k A\hat{u}_k, \\
\beta_{k+1} &= \frac{(r^{k+1}, r^0)}{(r^k, r^0)}, \\
p^{k+1} &= r^{k+1} + \beta_{k+1}u^k, \\
q^{k+1} &= p^{k+1} + \beta_{k+1}(u^k + \beta_{k+1}q^k).
\end{aligned}
$$

The potential problem with this method is that if there are some oscillations in BiCG, they will be amplified in CGS. Therefore, in some cases the behaviour of CGS may be more erratic than that of BiCG. This leads some researchers to generalize Sonneveld's idea to using product of different polynomials to get a smoother behaviour. Note that this may also be obtained by smoothing procedures. CGS could also have breakdown (or near breakdown) problems as with BiCG and we may have to use look–ahead techniques, see for instance Brezinski and Sadok [210] and Brezinski and Redivo Zaglia [203].

7.13. Extensions of BiCG

For the BiCG method, the vectors are computed by

$$p^k = r^{k-1} - \beta_k p^{k-1}, \quad x^{k+1} = x^k + \alpha_k p^k, \quad r^{k+1} = r^k - \alpha_k A p^k$$

with r^k and Ap^k being orthogonal to $K_k(A^T, \tilde{r}^0)$. For any polynomial ψ_k of degree k and leading coefficient θ_k, we have

$$\beta_k = \frac{\theta_{k-1}}{\theta_k} \frac{\rho_k}{\sigma_{k-1}}, \quad \alpha_k = \frac{\rho_k}{\sigma_k},$$

where

$$\rho_k = (r^k, \psi_k(A^T)\tilde{r}^0) = (y^k, \tilde{r}^0)$$

with $y^k = \psi_k(A) r^k$ and

$$\sigma_k = (Ap^k, \psi_k(A^T)\tilde{r}^0).$$

We are free to choose the polynomial ψ_k as long as $\psi_0(A^T)\tilde{r}^0, \ldots, \psi_{k-1}(A^T)\tilde{r}^0$ span the Krylov space $K_k(A^T, \tilde{r}^0)$. As the convergence of CGS could be very irregular, H. Van der Vorst [1284] suggested using another product of polynomials. He looked for residuals of the form $r^k = \psi_k(A)\phi_k(A)r^0$ where ϕ is the BiCG polynomial. For ψ he chose a product of degree one minimal residual polynomials in order to smooth the residuals. Therefore,

$$\psi_{k+1}(t) = (1 - \omega_k t) \cdots (1 - \omega_1 t),$$

and the polynomial ψ_k can be computed by the simple recurrence

$$\psi_{k+1}(t) = (1 - \omega_k t)\psi_k(t).$$

As with CGS, we can find recurrences for $\psi_k \phi_k$ and $\psi_k \theta_k$. Then, we set

$$r^k = \phi_k(A)\psi_k(A)r^0,$$
$$p^k = \psi_k(A)\theta_k(A)r^0.$$

The computation of the coefficients is a little more tricky than for CGS but it can also be done. Parameters ω_k are chosen as to minimize the Euclidean norm of the residual (i.e. a steepest descent or GMRES(1) step). This leads to the BiCGSTAB algorithm, see Van der Vorst [1284]:

Let x^0 be given, $r^0 = b - Ax^0$, $p^0 = r^0$, $\tilde{r}^0$ arbitrary for $k = 0, 1, \ldots$

$$\begin{aligned}
\alpha_k &= \frac{(r^k, \tilde{r}^0)}{(Ap^k, \tilde{r}^0)}, \\
s^k &= r^k - \alpha_k A p^k, \\
\omega_k &= \frac{(As^k, s^k)}{(As^k, As^k)}, \\
x^{k+1} &= x^k + \alpha_k p^k + \omega_k s^k, \\
r^{k+1} &= s^k - \omega_k A s^k, \\
\beta_{k+1} &= \frac{(r^{k+1}, \tilde{r}^0)}{(r^k, \tilde{r}^0)} \frac{\alpha_k}{\omega_k}, \\
p^{k+1} &= r^{k+1} + \beta_{k+1}(p^k - \omega_k A p^k),
\end{aligned}$$

Figure 7.8 shows the norm of the error as a function of the number of floating point operations (flops) using BiCGSTAB without preconditioning for problem 7 on a 20×20 mesh.

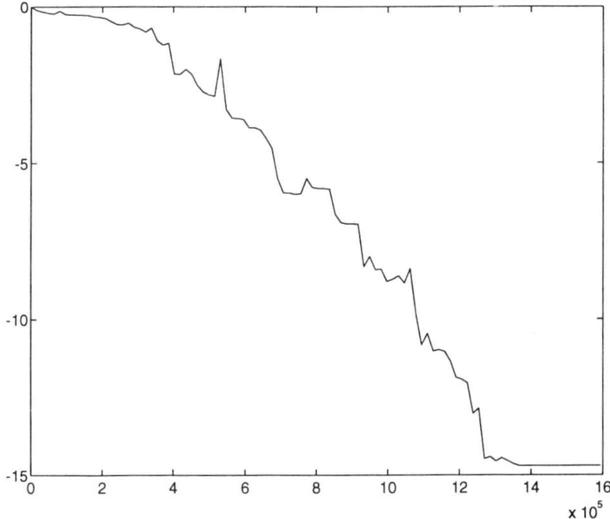

fig. 7.8: Problem 7: $\log_{10}$ *of the error with* BiCGSTAB 20×20 *mesh*

We can see on figure 7.8 that the convergence is smoother than with CGS (at least on this example). The preconditioned version is

$$
\begin{aligned}
\hat{p}^k &= M^{-1}p^k, \\
\tilde{p}^k &= A\hat{p}^k, \\
\alpha_k &= \frac{(r^k, \tilde{r}^0)}{(\tilde{p}^k, \tilde{r}^0)}, \\
s^k &= r^k - \alpha_k \tilde{p}^k, \\
\hat{s}^k &= M^{-1}s^k, \\
\tilde{s}^k &= A\hat{s}^k, \\
\omega_k &= \frac{(\tilde{s}^k, s^k)}{(\tilde{s}^k, \tilde{s}^k)}, \\
x^{k+1} &= x^k + \alpha_k \hat{p}^k + \omega_k \hat{s}^k, \\
r^{k+1} &= \hat{s}^k - \omega_k \tilde{s}^k, \\
\beta_{k+1} &= \frac{(r^{k+1}, \tilde{r}^0)}{(r^k, \tilde{r}^0)} \frac{\alpha_k}{\omega_k}, \\
p^{k+1} &= r^{k+1} + \beta_{k+1}(p^k - \omega_k \tilde{p}^k),
\end{aligned}
$$

Usually BiCGSTAB has much smoother convergence behaviour than CGS. However, the algorithm can break down if $\omega_k = 0$ or we can have troubles if ω_k is small. For recipes to avoid breakdowns, see Brezinski and Redivo Zaglia [205]. Another problem is that the minimal residual polynomial has only real roots and this can be a poor approximation for matrices with complex eigenvalues. Gutknecht [728] proposed a method denoted as BiCGSTAB2 to solve this potential problem. This was further generalized by Sleijpen and Fokkema [1180] who proposed the BiCGSTAB(l) algorithm. In this method the polynomial ψ is constructed as the product of factors of degree l, $\psi_k = \psi_{ml+l} = s_k s_{k_1} \cdots s_0$, where the polynomials s_i are of degree l, such that $s_i(0) = 1$ and s_k minimizes $\|s_k(A)\psi_{k-l}(A)r^k\|$. For $k = ml$ the operator acting on the initial residual is the product of the BiCG operator and a GMRES(l) (or more exactly a GCR(l)) operator. BiCGSTAB(l) computes the solution x^k for $k = l, 2l, \ldots$ These are the outer iterations of the algorithm. To proceed from $k = ml$ to $k = (m+1)l$ first the BiCG part computes implicitly the BiCG vectors, then the algorithm computes a minimum residual on a subspace of dimension l. Let us show the implementation of BiCGSTAB(l) given by Sleijpen and Fokkema [1180]:

Let $k = -l$, x^0 and $\tilde{r}^0$ be given, $r^0 = b - Ax^0$, $u^{-1} = 0$, $\alpha = 0$, $\omega = 1$.

Then we repeat until convergence

$$k = k + l$$
$$\hat{u}^0 = u^{k-1}, \quad \hat{r}^0 = r^k, \quad \hat{x}^0 = x^k$$
$$\rho_0 = -\omega\rho_0$$

for $j = 0, \ldots, l-1$

$$\rho_1 = (\hat{r}^j, \tilde{r}^0), \quad \beta = \beta_{k+j} = \alpha\frac{\rho_1}{\rho_0}, \quad \rho_0 = \rho_1$$
$$\hat{u}^i = \hat{r}^i - \beta\hat{u}^i, i = 0, \ldots, j$$
$$\hat{u}^{j+1} = A\hat{u}^j$$
$$\gamma = (\hat{u}^{j+1}, \tilde{r}^0), \quad \alpha = \alpha_{k+j} = \frac{\rho_0}{\gamma}$$
$$\hat{r}^i = \hat{r}^i - \alpha\hat{u}^{i+1}, i = 0, \ldots, j$$
$$\hat{r}^{j+1} = A\hat{r}^j, \quad \hat{x}^0 = \hat{x}^0 + \alpha\hat{u}^0$$

end j

for $j = 1, \ldots, l$

$$\tau_{i,j} = \frac{(\hat{r}^j, \hat{r}^i)}{\sigma_i}, \quad \hat{r}^j = \hat{r}^j - \tau_{i,j}\hat{r}^i, i = 1, \ldots, j-1$$
$$\sigma_j = (\hat{r}^j, \hat{r}^j), \quad \gamma_j' = \frac{(\hat{r}^0, \hat{r}^j)}{\sigma_j}$$

end j

$$\gamma_l = \gamma_l', \quad \omega = \gamma_l$$
$$\gamma_j = \gamma_j' - \sum_{i=j+1}^{l} \tau_{i,j}\gamma_i, j = l-1, \ldots, 1$$
$$\tilde{\gamma}_j = \gamma_{j+1} - \sum_{i=j+1}^{l-1} \tau_{i,j}\gamma_{i+1}, j = 1, \ldots, l-1$$
$$\hat{x}^0 = \hat{x}^0 + \gamma_1\hat{r}^0, \quad \hat{r}^0 = \hat{r}^0 - \gamma_1'\hat{r}^l, \quad \hat{u}^0 = \hat{u}^0 - \gamma_l\hat{u}^l$$

for $j = 1, \ldots, l-1$

$$\hat{u}^0 = \hat{u}^0 - \gamma_j\hat{u}^j$$
$$\hat{x}^0 = \hat{x}^0 + \tilde{\gamma}_j\hat{r}^j$$
$$\hat{r}^0 = \hat{r}^0 - \gamma_j'\hat{r}^j$$

end j

$$u^{k+l-1} = \hat{u}^0, \quad r^{k+l} = \hat{r}^0, \quad x^{k+l} = \hat{x}^0$$

Usually BiCGSTAB(l) is used with small values of l. For most problems $l = 2$ gives good results and improves on BiCGSTAB.

Figure 7.9 shows the norm of the error as a function of the number of floating point operations (flops) using BiCGSTAB(2) without preconditioning for problem 7 on a 20×20 mesh.

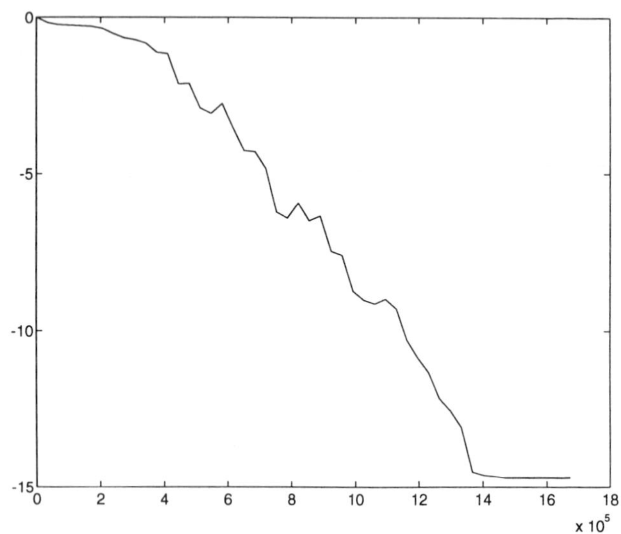

fig. 7.9: Problem 7: $\log_{10}$ *of the error with* BiCGSTAB(2) 20×20 *mesh*

A possible problem with the previous implementation is that it uses the standard (ill conditioned) basis of the CGR part. This could cause the computed residuals to differ from the "true" residuals. For large values of l it is better to use more stable (orthonormal) basis. More efficient implementations of BiCGSTAB(l) have been considered by Sleijpen and Van der Vorst [1182], [1183], [1184]. One strategy that has been proposed to improve the accuracy of computed residuals is to group the updates to the solution, that is to accumulate only the corrections for several steps:

$$x^k = x^j + q^j, \quad q^j = \alpha_j r^j + \alpha_{j+1} r^{j+1} + \cdots + \alpha_{k-1} r^{k-1}.$$

If we choose the values of $k - j$ well and/or compute the partial sums cleverly we can expect more accurate residuals. It has been suggested that we update the residuals when k is the first integer greater than j such that the norm of the residual at step k is strictly smaller than the norm of the residual at step j. Another idea is to compute the true residual from time to time, without restarting u^k, ω_k, σ_k. Sleijpen and Van der Vorst [1183] tried to avoid small values of ω_k in `BiCGSTAB` by modifying the computation as

$$\omega_k = \frac{\hat{\omega}_k}{|\hat{\omega}_k|} \max(|\hat{\omega}_k|, 0.7) \frac{\|r\|}{\|Ar\|},$$

where

$$\hat{\omega}_k = \frac{(r, Ar)}{\|r\| \, \|Ar\|}.$$

Similar solutions could be applied for `BiCGSTAB`(l), see [1183]. Moreover, the value of l can be changed during the course of the algorithm. We can increase the value of l when we get small values of ω_k. Criteria defining when to increase l are given in [1183].

Many other product type methods can be defined but, so far, there is not one method that has been proved to be really better than all the others. T. Chan and Q. Ye [300] defined an algorithm which mixes `CGS` and `BiCGSTAB`. At each iteration either a `CGS` or `BiCGSTAB` step is taken. This is done without restart which would have spoiled the convergence rate. When we apply `BiCG` we have

$$r^k = \phi_k(A) r^0, \quad p^k = \theta_k(A) r^0,$$

where

$$\phi_{k+1}(A) = \phi_k(A) - \alpha_k A \theta_k(A),$$
$$\theta_{k+1}(A) = \phi_{k+1}(A) + \beta_{k+1} \theta_k(A).$$

In the mixed method the residual is constructed as

$$r^k = \pi_k(A) \phi_k(A) r^0, \quad \pi_k(t) = \psi_l(t) \phi_{n-k}(t).$$

When constructing $r^{k+1} = \pi_{k+1}(A) \phi_{k+1}(A) r^0$ Chan and Ye [300] choose from $\psi_l(t) \phi_{k+1-l}(t)$ or $\psi_{l+1}(t) \phi_{k-l}(t)$. See the formulas in Chan and Ye [300].

7.14. The Quasi Minimal Residual algorithm

In the non–symmetric Lanczos algorithm we construct a basis of $K_k(A, r^0)$ which is (bi)orthogonal to $K_k(A^T, \tilde{r}^0)$. Therefore, we cannot do what we did with the output of the Arnoldi algorithm to derive the GMRES algorithm by minimizing the Euclidean norm (l_2–norm) of the residual. Now, we have that

$$AQ_k = Q_{k+1}\bar{T}_k.$$

The norm of the residual is

$$\begin{aligned}\|r^k\| &= \|b - Ax^k\| \\ &= \|b - AQ_k y^k\| \\ &= \|\, \|b\| Q_{k+1} e^1 - Q_{k+1}\bar{T}_k y^k \|.\end{aligned}$$

Unfortunately, the matrix Q_{k+1} is not orthogonal and cannot be taken out of the norm. It was suggested by Freund and Nachtigal [590] that we compute y^k by minimizing

$$\|\, \|b\| e^1 - \bar{T}_k y\|.$$

This method is known as QMR (Quasi Minimum Residual). In fact in the Freund and Nachtigal proposal there was also a diagonal scaling matrix involved but, so far, nobody has given a way to cleverly determine this matrix, so we do not use it. Mathematically, the solution is

$$y^k = \|b\|(\bar{T}_k)^+ e^1, \quad (\bar{T}_k)^+ = (\bar{T}_k^T \bar{T}_k)^{-1}\bar{T}_k^T.$$

Numerically the least squares problem is solved in the same way as for GMRES but everything is simpler here because of the structure of $\bar{T}_k$. Givens rotations are computed at each iteration and applied to the matrix and the right hand side.

Figure 7.10 shows the norm of the error as a function of the number of floating point operations (flops) using QMR without preconditioning and without look–ahead for problem 7 on a 20×20 mesh. It should be noted that in order to produce the plot, the least squares problem was solved at each iteration which is not necessary when the goal is just to compute the solution. Therefore, the number of operations cannot be directly compared with those for the other methods.

As there are relations between the FOM and GMRES residuals, equivalently there are relations between the BiCG and QMR residuals and iterates. The solution can be written as

$$x^k = x^{k-1} + \tau_k p^k,$$

where $\tau_k = c_k \tilde{\tau}_k$, c_k being the cosine of the Givens rotation, $\tilde{\tau}_k$ being the norm we minimize, p^k is the kth column of $Q_k R_k^{-1}$, R_k being the upper triangular matrix obtained by applying the Givens rotations. Note that p^k can be computed by short recurrences.

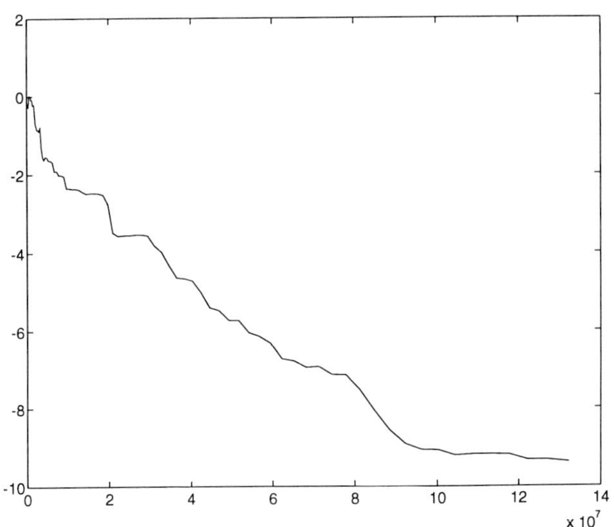

fig. 7.10: Problem 7: $\log_{10}$ *of the error with* QMR
20 × 20 mesh

Theorem 7.23

Let s_k and c_k be the sine and cosine of the Givens rotations. We have

$$\|r^k\| \leq \|r^0\|\sqrt{k+1}|s_1 s_2 \cdots s_k|,$$

$$x^k_{BiCG} = x^k_{QMR} + \frac{\tau_k |s_k|^2}{c_k^2} p^k.$$

Proof. See Freund and Nachtigal [590]. □

It has also been shown by Zhou and Walker [1364] that

$$x^k_{QMR} = x^{k-1}_{QMR} + \frac{\eta_k^2}{\rho_k^2}(x^k_{BiCG} - x^{k-1}_{QMR}),$$

with

$$\eta_k = \sqrt{\frac{1}{\sum_{j=0}^k 1/\rho_j^2}}, \quad \rho_j = \|r^j_{BiCG}\|.$$

Therefore, the BiCG iterates can be computed from the QMR ones and vice versa. Usually, QMR has a much smoother convergence curve than BiCG. QMR residuals (if Lanczos does not break down) can also be compared to GMRES residuals as

$$\|r^k_{QMR}\| \le \kappa(Q_{k+1})\|r^k_{GMRES}\|.$$

In the same spirit as what was developed for CGS, Freund [584] created a transpose free QMR method (known as TFQMR).

7.15. CMRH

GMRES can be interpreted as solving the problem

$$\min_{w\in \mathbf{R}^{k+1}, z\in K_k(A,r^0)} \|w\|, \quad Az = r^0 + V_{k+1}w.$$

Sadok [1146] introduced CMRH as solving

$$\min_{w\in \mathbf{R}^{k+1}, z\in K_k(A,r^0)} \|w\|, \quad Az = r^0 + L_{k+1}w,$$

where L_k is the matrix computed by the Hessenberg process of section 7.4. This method produces an upper Hessenberg matrix H_k. Then, the CMRH algorithm is similar to GMRES, the differences being that

$$x^k = x^0 + L_k y^k,$$

where y^k is the solution minimizing $\| \, \|r^0\|_\infty e^1 - H_k y\|$. This is very similar to GMRES. However, a CMRH iteration is cheaper than a GMRES iteration.

Theorem 7.24

If we start from the same vectors, we have that

$$\|r^k_{CMRH}\| \le \kappa(L_{k+1})\|r^k_{GMRES}\|.$$

Proof. See Sadok [1146]. □

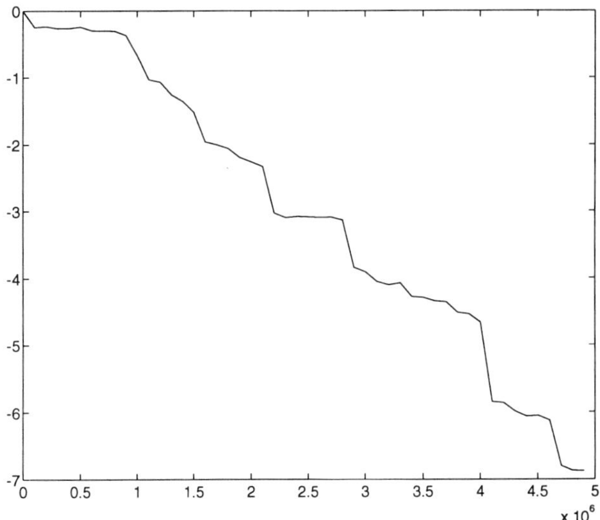

fig. 7.11: Problem 7: $\log_{10}$ *of the error with* CMRH(10)
20 × 20 *mesh*

Figure 7.11 shows the norm of the error as a function of the number of floating point operations (flops) using CMRH(10) without preconditioning and without look–ahead for problem 7 on a 20 × 20 mesh.

7.16. Which method to use?

We have seen that there are many different methods for solving non–symmetric systems. It is also likely that many new methods (or variants) will be introduced in the next few years. It would be nice for users to know which method to choose. Unfortunately, it is very difficult to give a general answer to this question. We have already said that it is always possible to find (sometimes contrived) examples for which a given method is either the best one or the worst one. However, methods which work well on average are `GMRES` and `BiCG`. The nice thing with `GMRES` is that it must converge, the disadvantage is that the storage may be too high. The nice thing with `BiCG` is that it is simple and easy to code, moreover the storage is low; the disadvantage is that sometimes the convergence is erratic. In this respect `BiCGSTAB(2)` often gives better results. An extensive set of numerical comparisons was given by C.H. Tong [1258].

7.17. Complex linear systems

Most of the time in this book we have been working with real arithmetic. However, there are interesting problems where the matrix is complex. For instance, when solving some electromagnetic problems, the matrix is complex and symmetric but not Hermitian. Special algorithms can be derived to handle these cases, see Freund [581], [583]. Other algorithms have been proposed by Joly and Meurant [836]. They used the general framework we have already introduced in Chapter 6. Let A be a complex matrix and b a complex vector. For any vector r, we introduce the functional

$$J(r) = (r, Hr)_C,$$

H being an Hermitian matrix and $(x, y)_C$ being the usual complex scalar product $\sum_{i=1}^{n} x_i \bar{y}_i$. The functional J is strictly convex therefore there exists a unique minimum which is $r^m = 0$ since $J(r) \geq 0$. We define a general minimization algorithm introducing a set of vectors $\{p^0, \ldots, p^k\}$ orthogonal in the scalar product related to the Hermitian matrix $N = A^H HA$, A^H being the Hermitian transpose of A. Let K be a definite matrix, we minimize J over subspaces $x^0 + \text{span}(Kg^0, \ldots, Kg^k)$, g^k being the gradient of J.

Let x^0 be given, $r^0 = b - Ax^0$, $g^0 = A^H Hr^0$, $p^0 = Kg^0$, for $k = 0, 1, \ldots$

$$\alpha_k = \frac{(g^k, p^k)_C}{(p^k, Np^k)_C},$$

$$x^{k+1} = x^k + \alpha_k p^k,$$

$$r^{k+1} = r^k - \alpha_k Ap^k,$$

$$g^{k+1} = g^k - \alpha_k Np^k,$$

$$\beta_{k+1}^l = -\frac{(Kg^{k+1}, Np^l)_C}{(p^l, Np^l)_C}, \quad 0 \leq l \leq k$$

$$p^{k+1} = Kg^{k+1} + \sum_{l=0}^{k} \beta_{k+1}^l p^l.$$

The directions p^k are N conjugate and the same orthogonality properties as those stated in Chapter 6 are verified. Many algorithms can be generated by an appropriate choice of the matrices H and K. For instance, we obtain a complex `GCR` by choosing $H = I$, $K = A^{-H}$. Complex `BiCG` does not fit into this framework. However, it can be obtained formally from the formulas by using:

$$H = \begin{pmatrix} 0 & A^H \\ A & 0 \end{pmatrix}^{-1}, \quad K = \begin{pmatrix} 0 & I \\ I & 0 \end{pmatrix}.$$

We can see that in this case H is not Hermitian, hence we shall not get all the properties of the general algorithm: in particular there could be breakdowns of the algorithm. The algorithm we obtain is the following.

Let x^0 be given, $r^0 = b - Ax^0$, $\tilde{r}^0 = \bar{r}^0$, $p^0 = r^0$, $\tilde{p}^0 = \tilde{r}^0$, for $k = 0, 1, \ldots$

$$\begin{aligned}
\alpha_k &= \frac{\mathrm{Re}(\tilde{r}^k, r^k)}{\mathrm{Re}(\tilde{p}^k, Ap^k)}, \\
x^{k+1} &= x^k + \alpha_k p^k, \\
r^{k+1} &= r^k - \alpha_k Ap^k, \\
\tilde{r}^{k+1} &= \tilde{r}^k - \alpha_k A^H \tilde{p}^k, \\
\beta_{k+1} &= \frac{\mathrm{Re}(\tilde{r}^{k+1}, r^{k+1})}{\mathrm{Re}(\tilde{r}^k, r^k)}, \\
p^{k+1} &= r^k + \beta_{k+1} p^k, \\
\tilde{p}^{k+1} &= \tilde{r}^k + \beta_{k+1} \tilde{p}^{k+1}.
\end{aligned}$$

In the important case where A is complex symmetric ($A = A^T$), we have $\tilde{r}^k = \bar{r}^k$, $\tilde{p}^k = \bar{p}^k$ for all k and the algorithm is as cheap as `CG` for real symmetric matrices. Notice that another complex generalization of `BiCG` was given by D. Jacobs [810]. His algorithm is only slightly different.

Let x^0 be given, $r^0 = b - Ax^0$, $\tilde{r}^0 = \bar{r}^0$, $p^0 = r^0$, $\tilde{p}^0 = \tilde{r}^0$, for $k = 0, 1, \ldots$

$$\begin{aligned}
\alpha_k &= \frac{(\tilde{r}^k, r^k)}{(\tilde{p}^k, Ap^k)}, \\
x^{k+1} &= x^k + \alpha_k p^k, \\
r^{k+1} &= r^k - \alpha_k Ap^k, \\
\tilde{r}^{k+1} &= \tilde{r}^k - \bar{\alpha}_k A^H \tilde{p}^k, \\
\beta_{k+1} &= \frac{(\tilde{r}^{k+1}, r^{k+1})}{(\tilde{r}^k, r^k)}, \\
p^{k+1} &= r^k + \beta_{k+1} p^k, \\
\tilde{p}^{k+1} &= \tilde{r}^k + \bar{\beta}_{k+1} \tilde{p}^{k+1}.
\end{aligned}$$

These extensions of `BiCG` can also be squared giving extensions of `CGS`. Numerical results and comparisons of methods were given in [836].

7.18. Krylov methods on parallel computers

Regarding vector and parallel computing, the problem is essentially the same for the Krylov algorithms as it is for `CG`. For instance, for `BiCG`, we have the choice between two-term and three–term recurrences. Although two–term recurrences are more stable, see Gutknecht and Strakŏs [732], three–term recurrences offer more parallelism since scalar products can be computed in parallel.

The parallelization of `GMRES` has been studied by several authors but an interesting variant was considered by J. Erhel [535]. The main problem in fact lies in the `Arnoldi` part of the algorithm where the basis vectors are produced sequentially by the modified Gram–Schmidt algorithm. One possible solution is to use another basis. In [535] a Newton basis was constructed (see Bai, Hu and Reichel [98]) and then orthogonalized. The basis is chosen as

$$\hat{V}_{k+1} = [\sigma_0 v, \sigma_1 (A - \mu_1 I)v, \ldots, \sigma_k \prod_{i=1}^{k} (A - \mu_i I)v],$$

where the σ_k's are scaling factors and the μ_i are the Leja points for the set S of (approximations of) eigenvalues of A (computed in a first cycle of `GMRES(m)`). This is defined as

$$\mu_1 = \max_{\lambda \in S} |\lambda|, \quad \prod_{i=1}^{j} |\mu_{j+1} - \mu_i| = \max_{\lambda \in S} \prod_{i=1}^{j} |\lambda - \mu_i|, \; j = 1, 2, \ldots$$

Then the basis is normalized and an orthonormal basis is computed by using a QR factorization of the matrix. This approach introduces much more parallelism but, unfortunately, there are examples for which convergence is considerably delayed although for many problems the convergence rate is about the same as for standard `GMRES`.

7.19. Bibliographical comments

For the use of the normal equations in least squares problems, we can refer to A. Bjorck's book [154]. The Concus and Golub method [341] has also been considered by Widlund [1329]. It is a special case of the theory by Faber and Manteuffel [542] for the existence of conjugate gradient methods for non–symmetric matrices.

The development of methods for non–symmetric matrices mimics what has been done for the symmetric case. A basis has to be computed for the Krylov space. In the symmetric case this is done with the Lanczos algorithm and the basis is orthogonal. For non–symmetric matrices, an orthogonal basis can be computed with the `Arnoldi` algorithm. This leads to the `GMRES` algorithm which is due to Saad and Schultz [1141]. Another possibility leads to `FOM` developed by Saad [1130]. Householder transformations for `GMRES` have been considered by H. Walker [1314]. The equivalence of these methods with others such as `GCR` was studied by Saad and Schultz [1140]. For`GMRES` in finite precision precision arithmetic we refer to Strakŏs [1215]. Flexible `GMRES` is due to Saad [1138]. Other extensions have been considered by Van der Vorst and his co–workers [1289], Vuik [1311].

The other possibility is based on the non–symmetric Lanczos algorithm which constructs a biorthogonal basis. This leads to the biconjugate gradient algorithm that is mainly due to Fletcher [564]. The acceleration of the methods by squaring the recurrences leading to `CGS` was considered by Sonneveld [1195]. This was further improved by Van der Vorst with `BiCGSTAB` [1284]. Other variants were developed by Gutknecht [728] and Sleijpen and Fokkema [1180].

`QMR` was inspired by `GMRES` to deal with the non–orthogonal basis. It was proposed by Freund and his co–workers, see Freund, Gutknecht and Nachtigal [587]. A transpose free `QMR` was then considered by Freund.

The breakdown problem has been handled by Freund, Gutknecht and Nachtigal [587] who proposed a look–ahead algorithm following earlier algorithms by Parlett [1061]. This has was also considered in detail by Brezinski and his co–workers [206]. They also considered extensions of these techniques to `CGS` and `BiCGSTAB`.

`CMRH` was proposed by Sadok [1146]. It is based on computing a Hessenberg basis for the Krylov space.

Linear systems with complex coefficients were considered some time ago by Jacobs [810]. More recently algorithms were proposed by Freund [585] and Joly and Meurant [836].

8 Preconditioning

8.1. Introduction

We have seen in the previous chapters that the rate of convergence of many iterative methods for solving $Ax = b$ depends on the condition number $\kappa(A)$ and/or the distribution of the eigenvalues of A. Therefore, it is a natural idea to transform the original linear system so that the new system has the same solution (or a solution from which the original one is easily recovered) and the transformed matrix has a smaller condition number and/or a better distribution of the eigenvalues. To formalize this idea, suppose that M is a non–singular matrix; we can transform the linear system $Ax = b$ into

$$M^{-1}Ax = M^{-1}b. \tag{8.1}$$

Then we can use an iterative method to solve the system (8.1) instead of the original one. This is usually called left preconditioning as we multiply matrix A from the left by M^{-1}. We could also transform the system by right preconditioning as

$$AM^{-1}y = b, \quad Mx = y. \tag{8.2}$$

If the matrix A is symmetric positive definite, we might want to keep the transformed system symmetric. In this case, we suppose M to be symmetric and positive definite and instead of (8.1) or (8.2), we use

$$M^{-\frac{1}{2}}AM^{-\frac{1}{2}}y = M^{-\frac{1}{2}}b, \tag{8.3}$$

and we recover the solution of the original system by $M^{\frac{1}{2}}x = y$. Of course, we do not want to explicitly compute M^{-1} or $M^{-\frac{1}{2}}$. But, we have already seen in previous chapters that, usually, we can eliminate this problem by a change of variables and for all methods the only thing we have to do is solving a linear system whose matrix is M at each iteration.

What are the properties that we may want the preconditioner M to have? Suppose that A is symmetric positive definite and we are using preconditioned CG to solve the linear system. Then, we might ask for

- M being symmetric and positive definite,
- if A is sparse, M being sparse, since we do not want to use much more storage for M than for A,
- M being easy to construct as we do not want to spend most of the computing time building M,
- M such that a linear system $Mz = r$ is easy to solve
- M such that we have a "good" distribution of the eigenvalues of $M^{-1}A$ whatever that means.

For most problems, it is almost impossible to construct preconditioners that fulfill all these criteria. The ones that are mandatory for sparse SPD matrices are the first two. For the third one, everything really depends on the rate of convergence that we get for the preconditioned system (8.3) and the given iterative method as there is usually a trade off between spending more time constructing a better preconditioner and having fewer iterations getting a better convergence rate. It also depends on knowing if we want to solve a single system only or if we have several systems with the same matrix and different right hand sides to solve, in which case we can spend more time constructing a good preconditioner only once. The fourth item is also important as we do not want to generate auxiliary problems that are as hard to solve as the original one. The condition about which we do not have too much control is the last one. For non–symmetric matrices we only have to consider the last four items, the last one being even more difficult to fulfill.

Of course, the optimal preconditioner giving the best rate of convergence is always $M = A$, but evidently this is not feasible as we won't get any gain over the original problem. However, this tells us that we would like to find M such that M^{-1} is, in some sense, a good sparse approximation to A^{-1}.

In the following sections, we shall describe some preconditioners. We shall be considering mostly symmetric matrices and shall indicate how things can be extended to non–symmetric matrices. We shall also consider at the end of the chapter the issue of parallelism. Again, this introduces some additional difficulties as to be efficient on a parallel computer, we need a preconditioner that gives a good improvement for the convergence rate and, at the same time, for which linear systems $Mz = r$ can be solved with a good degree of parallelism. Most of the time, these are conflicting requirements.

We shall start from the most simple ideas, going gradually towards more sophisticated methods. Sometimes, we shall consider particular applications or matrix structures as there is a rule of thumb that says that the more we know about the problem we want to solve, the easier it is to construct a good preconditioner. Nevertheless, so far, constructing preconditioners is more an art than a science and there is not much theory supporting the different choices which are mainly based on numerical experiments. We shall study preconditioners using domain decomposition ideas in Chapter 10.

Of course it is of interest to study the condition number $\kappa(M^{-1}A)$ and/or the eigenvalue distribution of $M^{-1}A$ to see by how much we have improved the convergence rate. Most of the time it is difficult to have precise results that apply to general matrices. Therefore, we shall often use model problems that can be analyzed more easily. Another possibility is to rely on Fourier analysis. However, strictly speaking this can only be used for problems with periodic boundary conditions. Right now, there is no general mathematical theory allowing us to study any preconditioner and completely explaining the numerical results.

8.2. The diagonal preconditioner

The simplest ideas (which unfortunately are by far not the most efficient ones) are based on the classical iterative methods described in Chapter 5. Besides $M = I$, the simplest choice is choosing M as a diagonal matrix whose diagonal elements are the same as the corresponding ones of A. We denote this preconditioner by `DIAG`. It corresponds to the `Jacobi` method. In some problems with varying coefficients, this scaling can improve the condition number and reduce the number of iterations. Unfortunately, there are cases for which this choice does not give any improvement at all in the convergence rate. Consider, for instance, the model problem (Poisson equation in a square). Then, the diagonal matrix is equal to 4 times the identity matrix and the condition number is unchanged with a constant diagonal scaling.

However, in many cases the diagonal preconditioner provides some improvement and, as far as we know, it is harmless, so we can recommend its use, sometimes combined with some other preconditioners.

One may ask why we choose the diagonal of A and not another diagonal matrix? This choice is supported by the following result.

Theorem 8.1

If A is symmetric and has property A, that is, it is similar to

$$\begin{pmatrix} D_1 & F \\ F^T & D_2 \end{pmatrix},$$

with D_1 and D_2 being square diagonal matrices, then the diagonal preconditioner that minimizes the condition number is D such that $d_{i,i} = a_{i,i}$,

$$\kappa(D^{-\frac{1}{2}}AD^{-\frac{1}{2}}) = \min_{\hat{D} \text{ diagonal}} \kappa(\hat{D}^{-\frac{1}{2}}A\hat{D}^{-\frac{1}{2}}).$$

Proof. See Forsythe and Straus [573]. □

If A does not have property A, we have the following result which says that D is not too far from being optimal when A is sparse.

Theorem 8.2

If A is symmetric positive definite, then

$$\kappa(D^{-\frac{1}{2}}AD^{-\frac{1}{2}}) \leq p \min_{\hat{D}} \kappa(\hat{D}^{-\frac{1}{2}}A\hat{D}^{-\frac{1}{2}}),$$

where p is the maximum number of non–zero elements in any row of A.

Proof. See Van der Sluis [1274]. □

Note that these results do not tell by how much the condition number is improved (if there is any improvement). For the Poisson model problem, the theory says that there is no diagonal matrix that can improve the condition number. A numerical check of the Forsythe and Straus result was done by Greenbaum and Rodrigue [703]. They use an optimization code to numerically compute the optimal diagonal preconditioners for several model problems. For the Poisson problem, the code converges to the diagonal of A

One of the main interests of the diagonal preconditioner is that it is perfectly parallel. It has been argued by many people that the diagonal preconditioner is the best one on parallel architectures as, even if the improvement in the convergence rate is not large, we can regain on one hand (because of the perfect parallelism) what we have lost on the other one by using a simple preconditioner. This is a misconception and other preconditioners with a smaller degree of parallelism but better conditioning properties can be constructed (polynomial preconditioners, sparse inverses or domain decomposition preconditioners) giving smaller computer times (of course, this is problem and computer dependent). For an illustration of this point, see Perlot [1071].

One can also use a block diagonal preconditioner (which we denote by `BDIAG`) if the linear systems with the chosen diagonal blocks as matrices are easy to solve. Theorem 8.2 has been generalized to block diagonal preconditioners, in which case p is the number of blocks in a block row. This result is a consequence of theorems by Demmel [398] and Eisenstat, Lewis and Schultz [515]. Their equivalence has been shown by L. Elsner [533]. The problem of the best l_2 scaling, even for rectangular matrices, has also been considered by Golub and Varah [671].

8.3. The SSOR preconditioner

We have seen in Chapter 5 that the iteration matrices of `Gauss-Seidel` and `SOR` are not symmetric. Therefore, we cannot use these matrices as preconditioners in the symmetric case. However, they can be considered as interesting candidates for the non–symmetric case. When A is symmetric, a preconditioner that has been much used is the `SSOR` preconditioner.

8.3.1. Definition of SSOR

This preconditioner has been proposed by Evans [539] and Axelsson [59], [60]. If the symmetric matrix A is written as $A = D + L + L^T$ where D is diagonal and L strictly lower triangular, then the preconditioner M is defined as

$$M = \frac{1}{\omega(2-\omega)}(D+\omega L)D^{-1}(D+\omega L^T), \quad 0 < \omega < 2. \tag{8.4}$$

The factor $\omega(2-\omega)$ is just a technical convenience (see lemma 8.3) and does not change the results (for the condition number). Note that M is straightforward to construct since it is directly obtained from the entries of A. An added bonus is that no additional storage is needed for M. Moreover, systems as $Mz = r$ are easy to solve by forward and backward sweeps computing the solutions of two triangular systems,

$$(D+\omega L)y = \omega(2-\omega)r,$$
$$(I+\omega D^{-1}L^T)z = y.$$

Therefore, most of the conditions we have set for a good preconditioner are fulfilled. However, it remains to be seen if the condition number is improved.

Note that SSOR can be straightforwardly generalized to non–symmetric matrices, $A = D + L + U$, setting

$$M = \frac{1}{\omega(2-\omega)}(D+\omega L)D^{-1}(D+\omega U).$$

8.3.2. Convergence results for SSOR

Let us first study the eigenvalues of $M^{-1}A$.

Lemma 8.3

Let A be symmetric positive definite and λ_i be the eigenvalues of $M^{-1}A$, then $\lambda_i \in]0,1], \forall i$.

Proof. Since A is positive definite, the diagonal of D is strictly positive and M positive definite. Then, $M^{-1}A$ is similar to a symmetric matrix and its eigenvalues are real. Let

$$B = \frac{2-\omega}{\omega}D, \quad C = \frac{\omega-1}{\omega}D + L.$$

With this notation, we can write M and A as

$$M = (B+C)B^{-1}(B+C^T), \quad A = B + C + C^T.$$

This implies that $M = A + CB^{-1}C^T$. From this relation, we obtain, B being positive definite,

$$(Mx, x) = (Ax, x) + (B^{-1}C^T x, C^T x) \geq (Ax, x) > 0, \ \forall x \neq 0.$$

This proves that the largest eigenvalue of $M^{-1}A$ is less or equal to 1. □

Remark. If $\omega = 1$, then 1 is an eigenvalue of $M^{-1}A$, the eigenvector being $(1, 0, \ldots, 0)^T$. If $\omega \neq 1$, then all the eigenvalues of $M^{-1}A$ are strictly less than 1.

The condition number $\kappa(M^{-1}A)$ was studied in the context of finite element methods by Axelsson [60] who proved the following algebraic result.

Theorem 8.4

Let μ and $\delta \in \mathbb{R}$ such that

$$\max_{x \neq 0} \frac{(Dx, x)}{(Ax, x)} \leq \mu, \quad \max_{x \neq 0} \frac{(LD^{-1}L^T x, x) - 1/4(Dx, x)}{(Ax, x)} \leq \delta.$$

Then there exists

$$\omega_{opt} = \frac{2}{1 + 2\sqrt{\frac{1}{\mu}(\frac{1}{2} + \delta)}},$$

such that

$$\kappa(M^{-1}A) \leq \tfrac{1}{2} + \sqrt{(\tfrac{1}{2} + \delta)\mu}.$$

For $\omega = 1$, we have

$$\kappa(M^{-1}A) \leq \tfrac{1}{2} + \frac{\mu}{4} + \delta.$$

Proof. See Axelsson [60]. □

If we apply this result to the Poisson model problem, we obtain

Proposition 8.5

For the Poisson model problem,

$$\mu = \frac{4}{\lambda_{max}(A)}, \qquad \delta = 0.$$

Proof. Since $D = 4I$, we have

$$\frac{(Dx,x)}{(Ax,x)} = 4\frac{(x,x)}{(Ax,x)} \leq \frac{4}{\lambda_{max}(A)}.$$

Moreover,

$$(LD^{-1}L^Tx,x) = (D^{-1}L^Tx, L^Tx) = \frac{1}{4}\sum_{i=1}^{n}(\alpha_i x_{i+1} + \beta_i x_{i+m})^2,$$

where α_i and β_i are 0 or -1 and m is the semi–bandwidth of A. Then it follows that

$$(LD^{-1}L^Tx,x) - \frac{1}{4}(Dx,x) = \sum_{i=1}^{n}\left(\frac{1}{4}(\alpha_i x_{i+1} + \beta_i x_{i+m})^2 - x_i^2\right),$$

$$\leq \sum_{i=1}^{n}\frac{1}{4}(\alpha_i^2 x_{i+1}^2 + \beta_i^2 x_{i+m}^2 + 2\alpha_i\beta_i x_{i+1}x_{i+m}) - x_i^2,$$

$$\leq \sum_{i=1}^{n}\frac{1}{2}(\alpha_i^2 x_{i+1}^2 + \beta_i^2 x_{i+m}^2) - x_i^2 \leq 0.$$

□

This proposition shows that for the model problem there exists an optimal $\omega(\approx 2/(1+h))$ for which we have

$$\kappa(M^{-1}A) \leq \tfrac{1}{2} + \frac{1}{2\sin\left(\frac{\pi h}{2}\right)} = O\left(\frac{1}{h}\right).$$

We should remember that $\kappa(A) = O\left(\frac{1}{h^2}\right)$, therefore we have gained an order of magnitude if we are able to compute the optimal value of ω.

8.3.3. Fourier analysis of SSOR

To obtain some idea of the eigenvalue distribution, we consider the model problem,

$$-\Delta u = f \quad \text{in } \Omega = [0,1] \times [0,1],$$

with periodic boundary conditions. This leads to a linear system with a matrix A_P

$$A_P = \begin{pmatrix} T_P & -I & & & -I \\ -I & T_P & -I & & \\ & \ddots & \ddots & \ddots & \\ & & -I & T_P & -I \\ -I & & & -I & T_P \end{pmatrix}, \quad \text{with} \quad T_P = \begin{pmatrix} 4 & -1 & & & -1 \\ -1 & 4 & -1 & & \\ & \ddots & \ddots & \ddots & \\ & & -1 & 4 & -1 \\ -1 & & & -1 & 4 \end{pmatrix}.$$

Let

$$X_P = \begin{pmatrix} 0 & & & & -1 \\ -1 & 0 & & & \\ & \ddots & \ddots & & \\ & & -1 & 0 & \\ & & & -1 & 0 \end{pmatrix}, \quad L_P = \begin{pmatrix} X_P & & & & -I \\ -I & X_P & & & \\ & \ddots & \ddots & & \\ & & -I & X_P & \\ & & & -I & X_P \end{pmatrix},$$

and D_P the diagonal of A_P. Then we define M_P as

$$M_P = (D_P + \omega L_P) D_P^{-1} (D_P + \omega L_P^T).$$

Note that we do not put the constant factor in front of the expression for M_P, but this does not change the condition number. The eigenvalues of M_P are

$$\psi_{st} = (4 - \omega + \frac{\omega^2}{2}) + 4\omega(\sin^2(\theta_s/2) + \sin^2(\phi_t/2)) + \frac{\omega^2}{2}\cos(\theta_s - \phi_t).$$

In Chan and Elman [268] it is proved that if

$$\omega = \frac{2}{1 + 2\sin(\pi h_P)},$$

then $\kappa(M_P^{-1} A_P) = O\left(\frac{1}{h}\right)$. Note that if we apply the rule $h_P = h_D/2$, this value coincides asymptotically with the one we found for Dirichlet boundary conditions. Figure 8.1 shows the distribution of the eigenvalues for the periodic problem for the optimal ω and figure 8.2 for $\omega = 1$.

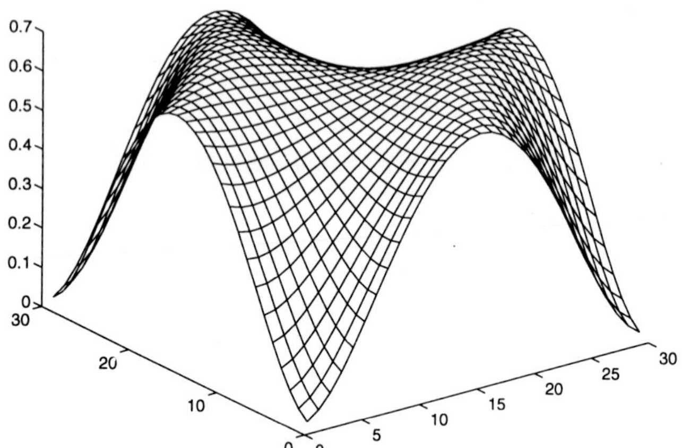

fig. 8.1: Eigenvalues for the SSOR *preconditioner*
Optimal ω*, periodic b.c.*

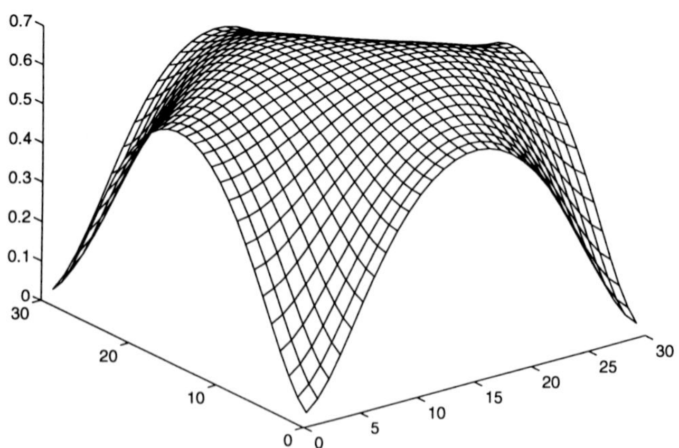

fig. 8.2: Eigenvalues for the SSOR *preconditioner*
$\omega = 1$*, periodic b.c.*

8.4. The block SSOR preconditioner

8.4.1. Definition of BSSOR

We suppose that A is symmetric positive definite and we partition the matrix in blockwise form as

$$A = \begin{pmatrix} A_{1,1} & A_{1,2} & \dots & A_{1,p} \\ A_{2,1} & A_{2,2} & \dots & A_{2,p} \\ \vdots & \vdots & & \vdots \\ A_{p,1} & A_{p,2} & \dots & A_{p,p} \end{pmatrix},$$

where $A_{i,j}$ is of order m, $A_{j,i} = A_{i,j}^T$ and $mp = n$. We denote

$$D = \begin{pmatrix} A_{1,1} & & & \\ & A_{2,2} & & \\ & & \ddots & \\ & & & A_{p,p} \end{pmatrix}, \quad L = \begin{pmatrix} 0 & & & \\ A_{2,1} & 0 & & \\ \vdots & & \ddots & \\ A_{p,1} & A_{p,2} & \dots & 0 \end{pmatrix},$$

$A = D + L + L^T$.

Then we define the block SSOR (BSSOR) preconditioner as

$$M = \frac{1}{\omega(2-\omega)}(D + \omega L)D^{-1}(D + \omega L^T), \quad 0 < \omega < 2.$$

There are some possibilities of extensions such as having a different ω for each block, but we won't pursue these ideas here. BSSOR can also be easily generalized to non–symmetric matrices.

8.4.2. Analysis of BSSOR

We have the same kind of results as for the point case concerning the location of the eigenvalues of the preconditioned matrix.

Lemma 8.6

The eigenvalues of $M^{-1}A$ are real, positive and located in $]0, 1]$.

Proof. Same as the proof of lemma 8.3. □

We note that when $\omega = 1$, 1 is an eigenvalue of multiplicity m of $M^{-1}A$. This is different from the situation of SSOR. Let us now specialize the results to block tridiagonal matrices.

Lemma 8.7

If A is block tridiagonal, then $(2-\omega)M - A$ is block diagonal.

Proof. We have

$$(2-\omega)M = A + \left(\frac{1}{\omega} - 1\right) D + \omega L D^{-1} L^T,$$

and we see that

$$(LD^{-1}L^T)_{i,j} = \begin{cases} 0 & \text{if } j \neq i \\ 0 & \text{if } i = j = 1 \\ A_{i,i-1}A_{i-1,i-1}^{-1}A_{i-1,i} & \text{if } j = i \neq 1 \end{cases}$$

therefore, the matrix $LD^{-1}L^T$ is block diagonal, the first block being 0, showing that M is block tridiagonal. □

If $A = M - R$, then $M^{-1}A$ has the same number of distinct eigenvalues as R.

Proposition 8.8

For the Poisson model problem, if $\omega = 1$, then R has only $m+1$ distinct eigenvalues.

Proof. We have

$$R = M - A = LD^{-1}L^T,$$

and we can see that the eigenvalues of R are 0 and the inverses of eigenvalues of the following tridiagonal matrix of order m

$$\begin{pmatrix} 4 & -1 & & & \\ -1 & 4 & -1 & & \\ & \ddots & \ddots & \ddots & \\ & & -1 & 4 & -1 \\ & & & -1 & 4 \end{pmatrix}.$$

□

It can be shown that for the model problem and for $\omega = 1$, the condition number can be improved from

$$\kappa(M^{-1}A) = 1 + \frac{1}{8\sin^2(\frac{\pi}{2}h)},$$

for the point SSOR preconditioner to

$$\kappa(M^{-1}A) \le 1 + \frac{1}{16(1+2\sin^2(\frac{\pi}{2}h))\sin^2(\frac{\pi}{2}h)}.$$

It can be also shown that for "quasi" optimal" values of ω using BSSOR we obtain an improvement in the condition number of $M^{-1}A$, see Chandra [303].

8.4.3. Fourier analysis of BSSOR

The following results are taken from Chan and Meurant [288]. We provide them as an example of what can be done by using Fourier analysis. As before, we consider the model problem with periodic boundary conditions. We denote

$$L_P = \begin{pmatrix} 0 & & & -I \\ -I & 0 & & \\ & \ddots & \ddots & \\ & & -I & 0 \end{pmatrix},$$

and by D_P the block diagonal of A_P. Then, we define the preconditioner BSSOR as

$$M_P = \frac{1}{\omega(2-\omega)} \, (\Delta + \omega L_P)\, \Delta^{-1} \, (\Delta + \omega L_P^T),$$

with

$$\Delta = D_P = \begin{pmatrix} T_P & & & \\ & \ddots & & \\ & & \ddots & \\ & & & T_P \end{pmatrix}.$$

In terms of the difference operators defined in Chapter 1,

$$M_P = \frac{1}{\omega(2-\omega)} T_3(4,1,\omega) T_1^{-1}(4,1) T_2(4,1,\omega).$$

We can compute the eigenvalues μ_{st} of $M_P^{-1} A_P$ as:

$$\begin{aligned}\mu_{st} &= \frac{\omega(2-\omega)F(T_1(4,1))F(A_P)}{F(T_3(4,1,\omega))F(T_2(4,1,\omega))},\\ &= \frac{4\omega(2-\omega)(\sin^2(\theta_s/2)+\sin^2(\phi_t/2))(4\sin^2(\theta_s/2)+2)}{(4\sin^2(\theta_s/2)+2-\omega e^{i\phi_t})(4\sin^2(\theta_s/2)+2-\omega e^{-i\phi_t})}\end{aligned}$$

which "simplifies" to

$$\mu_{st} = \frac{4\omega(2-\omega)(\sin^2(\theta_s/2)+\sin^2(\phi_t/2))(4\sin^2(\theta_s/2)+2)}{(4\sin^2(\theta_s/2)+2)^2+\omega^2-2\omega\cos(\phi_t)(4\sin^2(\theta_s/2)+2)},$$

where

$$\theta_s = 2\pi sh, s=1,\ldots,n; \quad \phi_t = 2\pi th, t=1,\ldots,n; \quad h=\frac{1}{n+1}.$$

Using these values, it can be seen, for instance, that for $\omega=1$, $\kappa(M_P^{-1}A_P) \geq O(h^{-2})$, but we have:

Theorem 8.9

There exists an "optimal" value $\omega^* = 2-2\sqrt{2}\pi h + O(h^2)$ of ω for which $\kappa(M_P^{-1}A_P) \leq O(h^{-1})$.

Proof. We derive upper and lower bounds for the eigenvalues of the BSSOR preconditioned matrix. This allows us to compute an "optimal" value of ω which minimizes an upper bound of the condition number. The analytic expression of the BSSOR eigenvalues was given above.

Let $x = \sin^2(\theta_s/2)$, $y = \sin^2(\phi_t/2)$. Clearly x and y take values in $[\sin^2(\pi h), 1]$, but for convenience, we shall sometimes consider $x, y \in [0,1]$, excluding the zero eigenvalue as it does not enter in the condition number.

We have

$$\mu_{st} = \mu(x,y) = \frac{4\omega(2-\omega)(x+y)(4x+2)}{(4x+2)^2+\omega^2-2\omega(4x+2)(1-2y)}.$$

Our aim is to find lower and upper bounds for $\mu(x,y)$. Before concluding the proof, we need a few technical lemmas analyzing the properties of the function μ.

Lemma 8.10

For a fixed value of x, $\mu(x,y)$ is either increasing or decreasing as a function of y, so the minimum and maximum of μ are located on the boundaries with constant y in the x,y plane.

Proof. Let $D = (4x+2)^2 + \omega^2 - 2\omega(4x+2)(1-2y)$. A little algebra shows that, for fixed x,

$$\frac{\partial \mu}{\partial y} = \frac{4\omega(2-\omega)(4x+2)}{D^2}[(4x+2)^2 + \omega^2 - 2\omega(4x+2)(1+2x)].$$

So the sign of $\frac{\partial \mu}{\partial y}$ is the same as that of $(4x+2-\omega)^2 - 4\omega x$ which is independent of y. Therefore, to find an upper bound for μ, it is sufficient to look at values for $y=0$ and $y=1$. □

Lemma 8.11

On the vertical boundaries, we have

$$\mu(x,1) < 1,$$

$$\mu(x,0) \leq 1.$$

Proof. At $y=1$, we have:

$$\mu(x,1) = \frac{4\omega(2-\omega)(1+x)(4x+2)}{(4x+2+\omega)^2}.$$

As $\omega \in [1,2[$, $\omega(2-\omega)$ is a decreasing function of ω and $4x+2+\omega$ is an increasing function of ω; and so $\mu(x,1)$ is a decreasing function of ω with x fixed. Therefore,

$$\mu(x,1) \leq \frac{4(1+x)(4x+2)}{(4x+3)^2} < 1.$$

Now, at $y=0$, we have:

$$\mu(x,0) = \frac{4\omega(2-\omega)x(4x+2)}{(4x+2-\omega)^2}$$

and

$$\frac{\partial \mu}{\partial x}\mid_{y=0} = \frac{8\omega(2-\omega)}{(4x+2-\omega)^3}[4(1-\omega)x+(2-\omega)].$$

But, $4x+2-\omega>0$, so the sign of $\frac{\partial \mu}{\partial x}\mid_{y=0}$ is the same as that of $4(1-\omega)x+(2-\omega)$. If $\omega=1$, $\frac{\partial \mu}{\partial x}\mid_{y=0}>0$ and so $\mu(x,0)$ is increasing and

$$\mu(x,0) \le \frac{24\omega(2-\omega)}{(6-\omega)^2} = \frac{24}{25} < 1.$$

If $\omega \neq 1$, there is a zero of $\frac{\partial \mu}{\partial x}\mid_{y=0}$ for $x=\bar{x}=\frac{2-\omega}{4(\omega-1)}$. Since at $x=0$, $\frac{\partial \mu}{\partial x}\mid_{y=0}>0$, the zero of the derivative corresponds to a maximum as long as $\bar{x}\in[0,1]$ i.e. for $\omega \ge \frac{6}{5}$. For $\omega \in]1,\frac{6}{5}]$, the same argument as before applies; the maximum is given for $x=1$ and $\mu \le 1$. For the remaining case, the maximum occurs for $x=\frac{2-\omega}{4(\omega-1)}$ and the minimum occurs at the boundaries.

Since $4(1-\omega)\bar{x}+2-\omega=0$, we have $4\bar{x}+2=\omega(4\bar{x}+1)$. Therefore, since $\omega>\frac{6}{5}$, we have:

$$\mu(\bar{x},0) = \frac{4\omega^2(2-\omega)\bar{x}(4\bar{x}+1)}{(4\omega\bar{x})^2} = \frac{(2-\omega)}{4\bar{x}}(4\bar{x}+1) = 1,$$

since $(4\bar{x}+1)(\omega-1)=1$. It should be noted that there are values of x and ω for which μ is equal to 1 or at least $\mu=O(1)$. □

Now, we are interested in finding lower bounds of μ for

$$x,y,\in[\sin^2(\pi h), \sin^2(\frac{\pi}{2}\frac{n}{n+1})].$$

Lemma 8.12

$\mu(x,1)$ is an increasing function of x.

Proof. A little algebra shows that:

$$\frac{\partial \mu(x,1)}{\partial x} = \frac{4\omega(2-\omega)}{(4x+2+\omega)^3}[8(\omega-1)x+6\omega-4].$$

Since $\omega \ge 1$, $\mu(x,1)$ is an increasing function. □

To summarize at this point, we already know that:

(i) for $1 \le \omega \le \frac{6}{5}$, $\mu(x,0)$ is an increasing function

(ii) for $\frac{6}{5} < \omega < 2$, $\mu(x,0)$ is increasing and then decreasing.

Since the partial derivatives are continuous functions of y, the same is true (for small enough h) for $y = \sin^2(\pi h)$ and $y = \sin^2(\frac{\pi}{2}\frac{n}{n+1}) = 1 - O(h^2)$.

In all cases, we only have three candidate points in the x, y plane to look for the minimum of μ :

$$(\sin^2(\pi h), \sin^2(\pi h)), \left(\sin^2(\frac{\pi}{2}\frac{n}{n+1}), \sin^2(\pi h)\right) \text{ and } \left(\sin^2(\pi h), \sin^2(\frac{\pi}{2}\frac{n}{n+1})\right).$$

Lemma 8.13

Let $x = \theta^2 = \sin^2(\pi h)$ *and* $\omega_0^2 = \left(\sqrt{5\theta^2+2} - \theta\right)^2 = 2 - 2\sqrt{2}\pi h + O(h^2)$. *If* $\omega < \omega_0^2$, $\mu(\theta^2, y)$ *is an increasing function of* y, *the minimum of* $\mu(\theta^2, y)$ *occurs for* $y = \sin^2(\pi h)$. *If* $\omega \ge \omega_0^2$, $\mu(\theta^2, y)$ *is a decreasing function of* y, *the minimum of* $\mu(\theta^2, y)$ *occurs for* $y = \sin^2(\frac{\pi}{2}\frac{n}{n+1})$ *and a lower bound is obtained for* $y = 1$.

Proof. As we have seen before, the sign of the derivative is the sign of

$$\left(4\theta^2 + 2 - \omega\right)^2 - 4\omega\theta^2 = (4\theta^2 + 2 - \omega - 2\theta\sqrt{\omega})(4\theta^2 + 2 - \omega + 2\theta\sqrt{\omega}).$$

So, it is also the sign of $4\theta^2 + 2 - \omega - 2\theta\sqrt{\omega}$. Let $\bar{\omega} = \sqrt{\omega}$, we look at the sign of $4\theta^2 + 2 - \bar{\omega}^2 - 2\theta\bar{\omega}$. Therefore, if $\omega < \omega_0^2$, the derivative is positive and it is negative elsewhere. □

Let us now distinguish between $x = \sin^2(\pi h)$ and $x = \sin^2(\frac{\pi}{2}\frac{n}{n+1})$.

Lemma 8.14

For sufficiently small h,

$$\mu(\sin^2(\pi h), \sin^2(\pi h)) \le \mu(\sin^2\left(\frac{\pi}{2}\frac{n}{n+1}\right), \sin^2(\pi h)).$$

Proof. A Taylor expansion of $\sin(\pi h)$ and $\sin(\frac{\pi}{2}\frac{n}{n+1})$ shows that the left hand side tends to 0 as $h \to 0$; as the right hand side is equal to

$$4\omega(2-\omega)\frac{6-4\pi^2h^2+O(h^4)}{(6-4\pi^2h^2)^2+\omega^2-2\omega(6-4\pi^2h^2+O(h^4))(1-2\pi^2h^2+O(h^4))}.$$

This is equal to

$$\frac{24\omega(2-\omega)}{(6-\omega)^2}+O(h^2).$$

□

Hence, only two points remains as candidates for the minimum: $x = \sin^2(\pi h)$ and either $y = \sin^2(\pi h)$ or $y = \sin^2(\frac{\pi}{2}\frac{n}{n+1})$.

As the maximum value is bounded by 1, our final goal is to find a value of ω which maximizes the lower bound.

Lemma 8.15

The lower bound of μ_{min} is maximized at $\omega = \omega^* \equiv \omega_0^2$.

Proof. We know that

(i) for $\omega < \omega_0^2$,

$$\mu_{min} = 4\omega(2-\omega)\frac{2\sin^2(\pi h)(4\sin^2(\pi h)+2)}{4\sin^2(\pi h)/2+4\omega\sin^2(\pi h)},$$

(ii) for $\omega \geq \omega_0^2$,

$$\mu_{min} \geq 4\omega(2-\omega)\frac{(1+\sin^2(\pi h))(4\sin^2(\pi h)+2)}{(4\sin^2(\pi h)+2+\omega)^2},$$
$$= 4\omega(2-\omega)\frac{2+O(h^2)}{(2+\omega)^2+O(h^2)}.$$

For the latter, the lower bound for μ_{min} is a decreasing function of ω; so, its maximal value is given for $\omega = \omega_0^2$. For $\omega \leq \omega_0^2$, the lower bound for μ_{min}

is also a decreasing function whose maximum value (for $\omega = 1$) tends to 0 as $h \to 0$; so, the value that maximizes the lower bound is $\omega^* = \omega_0^2$,

$$\omega^* = 2 - 2\sqrt{2}\pi h + O(h^2).$$

Then the lower bound is equal to

$$4(2 - 2\sqrt{2}\pi h)2\sqrt{2}\pi h \frac{2 + O(h^2)}{(4 - 2\sqrt{2}\pi h)^2 + O(h^2)} = \frac{32\sqrt{2}\pi h + O(h^2)}{16 - 8\sqrt{2}\pi h + O(h^2)} = O(h).$$

□

This concludes the proof of theorem 8.9. □

In Chan and Meurant [288] the Fourier results were compared to numerical results for the Dirichlet boundary conditions using $h_P = h_D/2$. Numerical results were given both for $\omega = 1$ and $\omega_{opt} = 2 - \sqrt{2}\pi h_D$. In all cases the agreements were quite good and the optimal value of ω predicted by the Fourier analysis is quite close to the actual optimum. Figure 8.3 shows the BSSOR eigenvalues for $\omega = 1$.

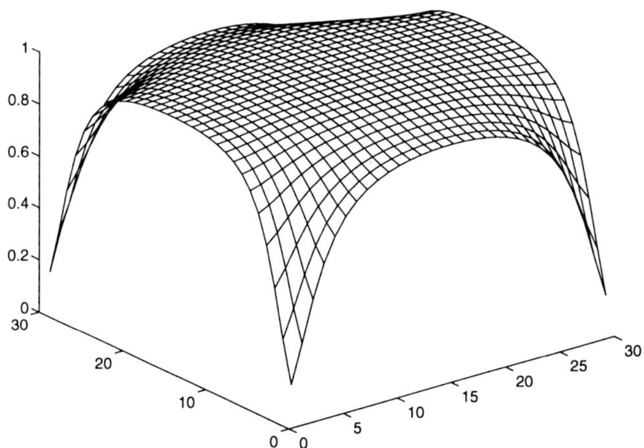

fig. 8.3: Eigenvalues for the BSSOR *preconditioner*
$\omega = 1$, periodic b.c.

8.5. The incomplete Cholesky decomposition

8.5.1. The general decomposition

We first recall the (complete) Cholesky outer product factorization that we studied in Chapter 2 using slightly different notations. Let

$$A = \bar{L}\bar{\Sigma}\bar{L}^T.$$

The first step is

$$\bar{L}_1 = \begin{pmatrix} 1 & 0 \\ \bar{l}_1 & I \end{pmatrix}, \quad \bar{\Sigma}_1 = \begin{pmatrix} a_{1,1} & 0 \\ 0 & \bar{A}_2 \end{pmatrix},$$

and

$$A = \begin{pmatrix} a_{1,1} & \bar{a}_1^T \\ \bar{a}_1 & \bar{B}_1 \end{pmatrix} = \bar{L}_1\bar{\Sigma}_1 L_1^T.$$

By equating,

$$\bar{l}_1 = \frac{\bar{a}_1}{a_{1,1}},$$

$$\bar{A}_2 = \bar{B}_1 - \frac{1}{a_{1,1}}\bar{a}_1\bar{a}_1^T.$$

Then

$$\bar{A}_2 = \begin{pmatrix} \bar{a}_{2,2}^{(2)} & \bar{a}_2^T \\ \bar{a}_2 & \bar{B}_2 \end{pmatrix} = \begin{pmatrix} 1 & 0 \\ \bar{l}_2 & I \end{pmatrix} \begin{pmatrix} \bar{a}_{2,2}^{(2)} & 0 \\ 0 & \bar{A}_3 \end{pmatrix} \begin{pmatrix} 1 & \bar{l}_2^T \\ 0 & I \end{pmatrix}.$$

We have similar formulas as for the first step, and if we denote

$$\bar{L}_2 = \begin{pmatrix} 1 & 0 & 0 \\ 0 & 1 & 0 \\ 0 & \bar{l}_2 & I \end{pmatrix},$$

then,

$$\bar{\Sigma}_1 = \begin{pmatrix} a_{1,1} & 0 \\ 0 & \bar{A}_2 \end{pmatrix} = \bar{L}_2 \begin{pmatrix} a_{1,1} & 0 & 0 \\ 0 & \bar{a}_{2,2}^{(2)} & 0 \\ 0 & 0 & \bar{A}_3 \end{pmatrix} \bar{L}_2^T = \bar{L}_2\bar{\Sigma}_2\bar{L}_2^T.$$

Therefore, after two steps, we have $A = \bar{L}_1\bar{L}_2\bar{\Sigma}_2\bar{L}_2^T\bar{L}_1^T$. Note that

$$\bar{L}_1\bar{L}_2 = \begin{pmatrix} 1 & 0 \\ \bar{l}_1 & \begin{pmatrix} 1 & 0 \\ \bar{l}_2 & I \end{pmatrix} \end{pmatrix}.$$

Finally, if all the pivots are non–zero, we obtain

$$A = \bar{L}_1 \cdots \bar{L}_{n-1} \bar{\Sigma} \bar{L}_{n-1}^T \cdots \bar{L}_1^T.$$

We saw in Chapter 2 that if A is symmetric positive definite, this algorithm can go through as all the pivots are non–zero. At each step of this algorithm some fill–in may be generated. As we know, the fill–in can only appear within the profile of the matrix A. For symmetric M–matrices or positive definite matrices, there are also some properties of decay of the elements in the L factor (see [982]). Therefore, if we want to generate an approximation of A, it is natural to neglect a part or all the fill–in during the steps of the LU factorization. There are different ways to accomplish this.

Let us suppose that we have a set of indices $G = \{(i,j),\ i > j\}$ being given. We want to construct a splitting,

$$A = L\Sigma L^T - R,$$

L being lower triangular with $l_{i,i} = 1$ and Σ diagonal. We shall construct L in such a way that $l_{i,j} = 0$ if $(i,j) \notin G$. Let us look at the first steps of the algorithm.

$$A = A_1 = \begin{pmatrix} a_{1,1} & a_1^T \\ a_1 & B_1 \end{pmatrix} = \begin{pmatrix} a_{1,1} & b_1^T \\ b_1 & B_1 \end{pmatrix} - \begin{pmatrix} 0 & r_1^T \\ r_1 & 0 \end{pmatrix} = M_1 - R_1,$$

with

$$a_1 = b_1 - r_1,$$

$$(b_1)_i = 0,\ \text{if } (i,1) \notin G \Rightarrow (r_1)_i = -(a_1)_i,$$

$$(b_1)_i = (a_1)_i,\ \text{if } (i,1) \in G \Rightarrow (r_1)_i = 0.$$

Then, we factorize M_1 (if this is feasible)

$$M_1 = \begin{pmatrix} 1 & 0 \\ l_1 & I \end{pmatrix} \begin{pmatrix} a_{1,1} & 0 \\ 0 & A_2 \end{pmatrix} \begin{pmatrix} 1 & l_1^T \\ 0 & I \end{pmatrix} = L_1 \Sigma_1 L_1^T.$$

By equating, we obtain

$$l_1 = \frac{b_1}{a_{1,1}},$$

$$A_2 = B_2 - \frac{1}{a_{1,1}} b_1 b_1^T.$$

Then, we use the same process on A_2,

$$A_2 = \begin{pmatrix} a_{2,2}^{(2)} & a_2^T \\ a_2 & B_2 \end{pmatrix} = \begin{pmatrix} a_{2,2}^{(2)} & b_2^T \\ b_2 & B_2 \end{pmatrix} - \begin{pmatrix} 0 & r_2^T \\ r_2 & 0 \end{pmatrix} = M_2 - R_2,$$

where b_2 is obtained from a_2 by zeroing the elements $(i, 2)$ whose indices do not belong to G. We set

$$L_2 = \begin{pmatrix} 1 & 0 \\ 0 & \begin{pmatrix} 1 & 0 \\ l_2 & I \end{pmatrix} \end{pmatrix}.$$

With these notations, at the end of the second step, we have

$$A = L_1 L_2 \Sigma_2 L_2^T L_1^T - L_1 \begin{pmatrix} 0 & 0 \\ 0 & R_2 \end{pmatrix} L_1^T - R_1,$$

but

$$L_1 L_2 = \begin{pmatrix} 1 & 0 \\ l_1 & \begin{pmatrix} 1 & 0 \\ l_2 & I \end{pmatrix} \end{pmatrix} \quad \text{and } L_1 \begin{pmatrix} 0 & 0 \\ 0 & R_2 \end{pmatrix} L_1^T = \begin{pmatrix} 0 & 0 \\ 0 & R_2 \end{pmatrix}.$$

Of course, it is not mandatory to have ones on the diagonal of L and we have the same variants as for the complete decomposition.

This algorithm has constructed a complete decomposition of $M = A + R$, where the matrix R is a priori unknown. The interesting question is to know the conditions under which such a decomposition is feasible. We also note that the previous algorithm (which is generically known as `IC` for Incomplete Cholesky) can be easily generalized to non–symmetric matrices (`ILU`) provided all the pivots are non–zero.

8.5.2. Incomplete decomposition of H–matrices

Let us look for conditions under which the incomplete decomposition is feasible.

Theorem 8.16

Let A be an H–matrix with a positive diagonal. Then all the pivots of the incomplete Cholesky decomposition are non–zero for any set of indices G.

Proof. We proceed in two steps. First, let us show that if we set to zero some non–diagonal elements of an H–matrix A, we still have an H–matrix.

Let $A = B + R$, R having a zero diagonal and $b_{i,j} = a_{i,j}$ or 0. We know that there exists a diagonal matrix E with a positive diagonal (whose elements are denoted by e_i) such that $E^{-1}AE$ is strictly diagonally dominant, that is to say,

$$|a_{i,i}| > \sum_{i \neq j} |a_{i,j}| \frac{e_j}{e_i}.$$

We claim that $E^{-1}BE$ is also strictly diagonally dominant since

$$|b_{i,i}| = |a_{i,i}| > \sum_{i \neq j} |a_{i,j}| \frac{e_j}{e_i} > \sum_{i \neq j} |b_{i,j}| \frac{e_j}{e_i}.$$

The second step of the proof is to show that in one stage of the complete factorization, if we start from an H–matrix, then we still obtain an H–matrix. Suppose that at step k, A_k is an H–matrix. There exists a diagonal matrix E_k whose diagonal elements are denoted by $e_i^{(k)} > 0$ and

$$a_{i,i}^{(k)} e_i^{(k)} > \sum_{\substack{j \neq i \\ j \geq k}} |a_{i,j}^{(k)}| e_j^{(k)}.$$

We have

$$a_{i,j}^{(k+1)} = a_{i,j}^{(k)} - \frac{1}{a_{k,k}^{(k)}} a_{k,i}^{(k)} a_{k,j}^{(k)}.$$

Therefore,

$$\sum_{\substack{j \neq i \\ j \geq k+1}} |a_{i,j}^{(k+1)}| e_j^{(k)} = \sum_{\substack{j \neq i \\ j \geq k+1}} |a_{i,j}^{(k)} - \frac{1}{a_{k,k}^{(k)}} a_{k,i}^{(k)} a_{k,j}^{(k)}| e_j^{(k)},$$

$$\leq \sum_{\substack{j \neq i \\ j \geq k+1}} |a_{i,j}^{(k)}| e_j^{(k)} + \frac{|a_{k,i}^{(k)}|}{a_{k,k}^{(k)}} \sum_{\substack{j \neq i \\ j \geq k+1}} |a_{k,j}^{(k)}| e_j^{(k)}.$$

But,

$$\sum_{\substack{j \neq i \\ j \geq k+1}} |a_{i,j}^{(k)}| e_j^{(k)} = \sum_{\substack{j \neq i \\ j \geq k}} |a_{i,j}^{(k)}| e_j^{(k)} - |a_{i,k}^{(k)}| e_k^{(k)} \leq a_{i,i}^{(k)} e_i^{(k)} - |a_{i,k}^{(k)}| e_k^{(k)}.$$

We also have

$$\sum_{\substack{j \neq i \\ j \geq k+1}} |a_{k,j}^{(k)}| e_j^{(k)} = \sum_{j \geq k+1} |a_{k,j}^{(k)}| e_j^{(k)} - |a_{k,i}^{(k)}| e_i^{(k)} \leq a_{k,k}^{(k)} e_k^{(k)} - |a_{k,i}^{(k)}| e_i^{(k)}.$$

With these results, we obtain

$$\sum_{\substack{j\neq i\\ j\geq k+1}} |a_{i,j}^{(k+1)}|e_j^{(k)} \leq a_{i,i}e_i^{(k)} - |a_{i,k}^{(k)}|e_k^{(k)}$$

$$+\frac{|a_{k,i}^{(k)}|}{a_{k,k}^{(k)}}(a_{k,k}^{(k)}e_k^{(k)} - |a_{k,i}^{(k)}|e_i^{(k)})$$

$$\leq (a_{i,i}^{(k)} - \frac{1}{a_{k,k}^{(k)}}|a_{k,i}^{(k)}|^2)e_i^{(k)} = a_{i,i}^{(k+1)}e_i^{(k)}.$$

This shows two things: $a_{i,i}^{(k+1)} > 0$ and the matrix at step $k+1$ of the factorization is an H–matrix. Putting together the two results that we have just established, we have that at each step of the incomplete factorization we are left with an H–matrix and the process can go on to completion whatever the set of indices G is. □

If we suppose that A is an M–matrix, we have a more precise result.

Theorem 8.17

Let A be a non–singular symmetric M–matrix. Then for all sets of indices G, the incomplete Cholesky decomposition,

$$A = L\Sigma L^T - R,$$

is a regular splitting.

Proof. With almost the same proof as for theorem 8.16, we can show that starting from an M–matrix, the matrices A_k are also M–matrices. Therefore, $a_{i,i}^{(k)} > 0$ and $a_{i,j}^{(k)} \leq 0$, $i \neq j$. This implies that $R_k \geq 0$ and $(l_k)_j \leq 0$. Then,

$$\begin{pmatrix} 1 & 0 \\ 0 & \begin{pmatrix} 1 & 0 \\ l_k & I \end{pmatrix} \end{pmatrix}^{-1} = \begin{pmatrix} 1 & 0 \\ 0 & \begin{pmatrix} 1 & 0 \\ -l_k & I \end{pmatrix} \end{pmatrix} \geq 0.$$

Hence, $R \geq 0$ and $L^{-1} \geq 0$ and this shows that we have a regular splitting. □

Corollary 8.18

Let A be a non–singular symmetric M–matrix and

$$A = L\Sigma L^T - R,$$

be the incomplete Cholesky decomposition of A. Then the iterative method

$$L\Sigma L^T x^{k+1} = Rx^k + b$$

converges for all x^0.

□

It is interesting to compare results for the incomplete decompositions of A and $M(A)$. We can see that the diagonal entries are larger and the absolute values of the non–zero off–diagonal entries are smaller for A than for $M(A)$.

Lemma 8.19

Let A be a non–singular symmetric H–matrix with a positive diagonal. Let

$$A = L\Sigma L^T - R,$$
$$M(A) = \tilde{L}\tilde{\Sigma}\tilde{L}^T - \tilde{R},$$

be the incomplete Cholesky decomposition of A and $M(A)$ for the same set of indices G. Then

$$\Sigma_{i,i} \geq \tilde{\Sigma}_{i,i} > 0, \ \forall i,$$
$$\tilde{l}_{i,j} \leq -|l_{i,j}| \leq 0, \ \forall i < j.$$

Proof. See Manteuffel [930]. The proof is obtained by induction. □

With this lemma, we can prove the following result which is a generalization of corollary 8.18.

Theorem 8.20

Let A be a non–singular symmetric H–matrix with a positive diagonal. The iterative method

$$L\Sigma L^T x^{k+1} = Rx^k + b,$$

where $A = L\Sigma L^T - R$ is an incomplete decomposition of A, converges to the solution of $Ax = b$.

Proof. We have seen that $L = L_1 L_2 \cdots L_{n-1}$ with

$$L_k = \begin{pmatrix} I & 0 \\ 0 & \begin{pmatrix} 1 & 0 \\ l_k & I \end{pmatrix} \end{pmatrix}.$$

Therefore,

$$L_k^{-1} = \begin{pmatrix} I & 0 \\ 0 & \begin{pmatrix} 1 & 0 \\ -l_k & I \end{pmatrix} \end{pmatrix}.$$

From lemma 8.19, $|l_k| \leq -\tilde{l}_k$. Hence $|L_k^{-1}| \leq \tilde{L}_k^{-1}$ and

$$|L^{-1}| \leq |L_{n-1}^{-1}| \cdots |L_1^{-1}| \leq \tilde{L}_{n-1}^{-1} \cdots \tilde{L}_1^{-1} = \tilde{L}^{-1}.$$

On the other hand, $|\Sigma^{-1}| \leq \tilde{\Sigma}^{-1}$. Altogether, this proves that

$$|(L\Sigma L^T)^{-1}| \leq (\tilde{L}\tilde{\Sigma}\tilde{L}^T)^{-1}.$$

Moreover, it is easy to prove that $|R| \leq \tilde{R}$. These results show that $\tilde{L}\tilde{\Sigma}\tilde{L}^T - \tilde{R}$ is a regular splitting of $M(A)$ which is an M–matrix. Then, we have

$$\rho((L\Sigma L^T)^{-1}R) \leq \rho((\tilde{L}\tilde{\Sigma}\tilde{L}^T)^{-1}\tilde{R}) < 1,$$

and the given iterative method converges. □

8.5.3. Incomplete decomposition of non–symmetric matrices

We have just seen that we can incompletely factorize H–matrices. However, there can be problems if we want to compute the ILU decomposition of a non–symmetric matrix which is not an H–matrix. In that case, there is the possibility of getting a zero pivot or, at least, some small pivot that can spoil the computation. Kershaw [858] proposed replacing the zero pivot by a positive quantity. Another solution is to use numerical pivoting. This was considered by Saad.

Note that the preconditioner is even more important for non–symmetric matrices since, by suitable choices, we can change the properties of the matrix. For instance, we can make the matrix symmetric or shift the spectrum to the positive half plane changing non–convergent iterative algorithms to convergent ones.

8.5.4. Different incomplete decomposition strategies

The most widely used strategy for choosing the set of indices G for the non–zero entries in L is to use a sparsity pattern of L identical to that of the lower triangular part of A. But, at least for symmetric M–matrices, the larger the set G is, the smallest is the condition number of $(L\Sigma L^T)^{-1}A$ and the fastest is the convergence of PCG. Therefore, more accurate and efficient decompositions can be obtained by retaining a part of the fill–in that is generated.

There are several ways to enlarge the set G, starting from the structure of the lower triangular part of A. One way is to associate a number called the level to any element computed in the factorization. Elements which are non–zero in the structure of A are of level 0 and all the zero elements are of level ∞. Then, in the steps of the decomposition, fill–ins generated by level 0 elements are said of to be of level 1, fill–ins generated by fill–ins are of level 2, etc... This can be formalized in the following way. Let $lev_{i,j}^{(k)}$ be the level of fill–in of $a_{i,j}^{(k)}$, then, after one step, the new level is computed as

$$lev_{i,j}^{(k+1)} = \min(lev_{i,j}^{(k)}, lev_{i,k}^{(k)} + l_{j,k}^{(k)} + 1).$$

Then, the strategy for the incomplete decomposition is to keep only the fill–in having a level which is below a given threshold p during the elimination. When the matrix is symmetric (say positive definite) we have seen that the structure of the factors can be determined in a preprocessing phase. The same is true for the incomplete factorization where the level of fill–in and the structure of the incomplete factor L can be computed before the numerical factorization takes place. The rationale behind this strategy is that if some fill–ins are small, then, the fill–ins generated by the previous fill–ins will be even smaller and so on. Therefore, neglecting the fill–ins beyond a certain level is supposed to drop only small elements.

Another strategy was introduced in the seventies by Tuff and Jennings [1267] and others. In this method, we not only look at the structure of the incomplete factors but also at the actual values of the entries. At some step, the fill–in is computed and it is discarded or not according to some dropping strategy. The most used one is to compare $a_{i,j}^{(k+1)}$ to some threshold and to drop the fill–in if it is too small. Different ways of dropping the elements have been proposed over the years. For instance, one can drop if

$$|a_{i,j}^{(k+1)}| < \epsilon[\max_{l\geq k+1} |a_{i,l}^{(k)}|, \max_{l\geq k+1} |a_{j,l}^{(k)}|].$$

Another way is to compare $|a_{i,j}^{(k+1)}|$ to the norm of the ith row of A. The main drawback of these methods is that the structure of L is not known before the elimination as when factoring non–symmetric sparse matrices with pivoting. Also, we do not know a priori how much storage is needed. However, this kind of method can be slightly modified and, for instance, the fill–ins in the kth column can be compared to a threshold, but only the p largest ones are kept provided they satisfy the given criteria. In this way the amount of storage can be controlled. This last method is one of the most efficient.

8.5.5. Finite difference matrices

Consider the pentadiagonal matrix (also to be considered as a block tridiagonal matrix) arising from the finite difference discretization of an elliptic (or parabolic) partial differential equation in a square,

$$A = \begin{pmatrix} D_1 & A_2^T & & & \\ A_2 & D_2 & A_3^T & & \\ & \ddots & \ddots & \ddots & \\ & & A_{m-1} & D_{m-1} & A_m^T \\ & & & A_m & D_m \end{pmatrix}.$$

where the matrices D_i are tridiagonal and matrices A_i are diagonal. Let m be the semi–bandwidth of A. For a generic row i of A, $a_{i,i-m}, a_{i,i-1}$ and $a_{i,i}$ are the non–zero entries in the lower triangular part of A. Let us consider the basic decomposition where no fill–in is kept outside the structure of A. We use, for convenience, a slightly different version of the incomplete Cholesky decomposition, $M = LD^{-1}L^T$, where D is diagonal and $l_{i,i} = d_{i,i}$. Then, by equating, we can see that

$$d_{i,i} = a_{i,i} - \frac{a_{i,i-1}^2}{d_{i-1,i-1}} - \frac{a_{i,i-m}^2}{d_{i-m,i-m}},$$

$$l_{i,i-1} = a_{i,i-1},$$

$$l_{i,i-m} = a_{i,i-m}.$$

In these recurrences, entries in A with negative indices (or indices larger than n) must be taken equal to 0 (and 1 for the entries of D). Clearly, for the incomplete Cholesky decomposition we just need to compute the diagonal of D which can be stored in one vector. It is easy to see that the remainder R has the structure that is shown in grey in figure 8.4.

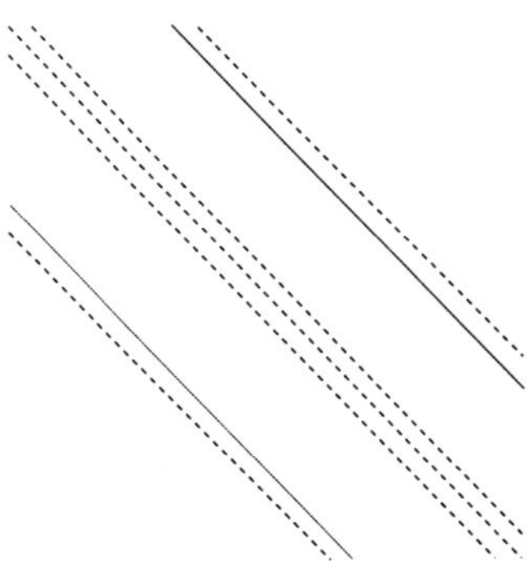

fig. 8.4: Structure of R for `IC(1,1)`

This factorization is called `IC(0)` (as we do not allow for any fill–in) or `IC(1,1)` (to remind us that the structure of L has one diagonal next to the main one and one outer diagonal). A comparison of the spectra of A and $M^{-1}A$ using `IC(1,1)` for the model problem on a 10×10 mesh is given in figure 8.5. The details of the structure of the spectrum of `IC(1,1)` is given in figure 8.6.

For a nine point operator, the non–zero entries in the lower triangular part of A are $a_{i,i-m-1}, a_{i,i-m}, a_{i,i-m+1}, a_{i,i-1}, a_{i,i}$. Then, we have

$$
\begin{aligned}
l_{i,i-m-1} &= a_{i,i-m-1},\\
l_{i,i-m} &= a_{i,i-m} - \frac{l_{i,i-m-1}l_{i-m,i-m-1}}{l_{i-m-1,i-m-1}},\\
l_{i,i-m+1} &= a_{i,i-m+1} - \frac{l_{i,i-m}l_{i-m+1,i-m}}{l_{i-m,i-m}},\\
l_{i,i-1} &= a_{i,i-1} - \frac{l_{i,i-m-1}l_{i-1,i-m-1}}{l_{i-m-1,i-m-1}} - \frac{l_{i,i-m}l_{i-1,i-m}}{l_{i-m,i-m}}\\
l_{i,i} &= a_{i,i} - \frac{l_{i,i-m-1}^2}{l_{i-m-1,i-m-1}} - \frac{l_{i,i-m}^2}{l_{i-m,i-m}}\\
&\quad - \frac{l_{i,i-m+1}^2}{l_{i-m+1,i-m+1}} - \frac{l_{i,i-1}^2}{l_{i-1,i-1}}.
\end{aligned}
$$

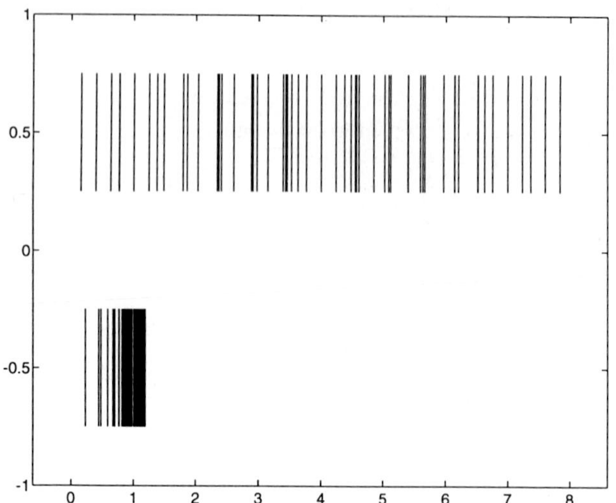

fig. 8.5: Comparison of spectra without preconditioning (top) and `IC(1,1)` *(bottom)*

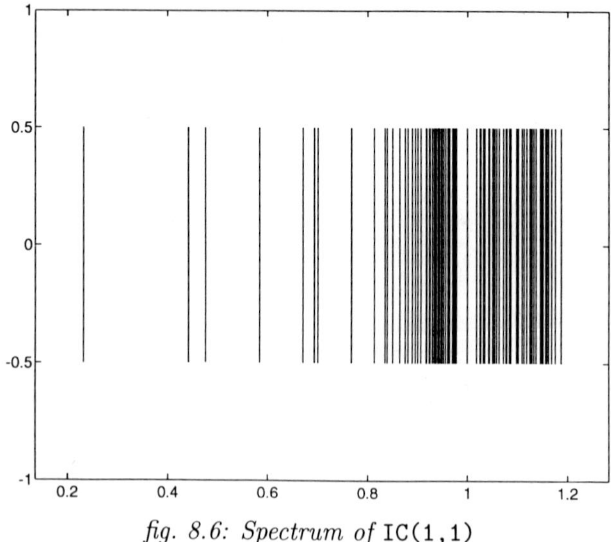

fig. 8.6: Spectrum of `IC(1,1)`

The next factorization for the five point scheme that was suggested by Meijerink and Van der Vorst [960] is obtained by keeping the fill–in in the union of the structure of L and the lower triangular part of R in `IC(1,1)`.

This is denoted by IC(1,2) and the structure of L is shown in figure 8.7. The entries are computed columnwise as

$$\begin{aligned} d_{i,i} &= a_{i,i} - \frac{l_{i,i-1}^2}{d_{i-1,i-1}} - \frac{l_{i,i-m+1}^2}{d_{i-m+1,i-m+1}} - \frac{l_{i,i-m}^2}{d_{i-m,i-m}}, \\ l_{i+1,i} &= a_{i+1,i} - \frac{a_{i+1,i-m+1} l_{i,i-m+1}}{d_{i-m+1,i-m+1}}, \\ l_{i+m-1,i} &= -\frac{a_{i+m-1,i-1} l_{i,i-1}}{d_{i-1,i-1}}, \\ l_{i+m,i} &= a_{i+m,i}. \end{aligned}$$

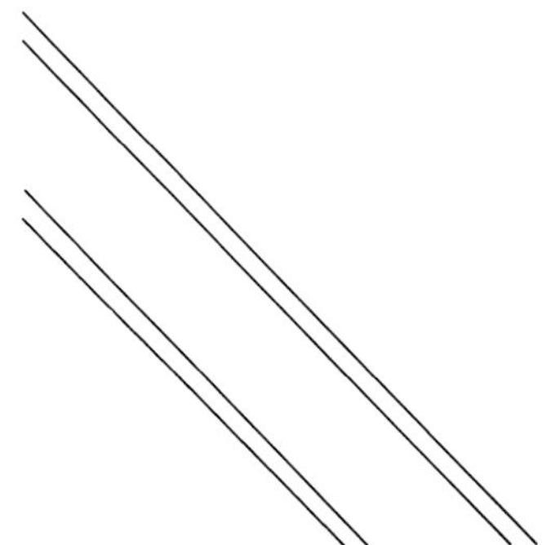

fig. 8.7: Structure of L for IC(1,2)

We need to store only three vectors for IC(1,2). The structure of the remainder is shown in grey on figure 8.8. Therefore, the next step will have one diagonal next to the main one and three outer diagonals and will be called IC(1,3). One more step gives IC(2,4). The process can be easily continued.

8.5.6. Fourier analysis of IC(1,1)

An interesting question is to determine whether using IC(1,1) (or more generally any incomplete Cholesky decomposition) improves the condition number. Chandra [303] has proven that

$$\frac{1}{17}\kappa(A) \leq \kappa(M^{-1}A) \leq 17\kappa(A),$$

for the Poisson model problem and IC(1,1). This shows that the condition number of $M^{-1}A$ is of the same order as the condition number of A. Here,

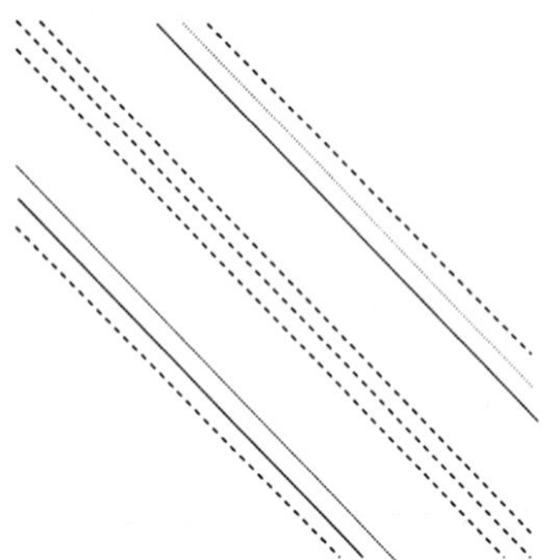

fig. 8.8: Structure of R for `IC(1,2)`

we shall consider Fourier analysis heuristics to answer this question, c.f. Chan [266]. We use the model problem with periodic boundary conditions in the unit square. Then we know that the eigenvectors denoted by $u^{(s,t)}$ are

$$(u^{(s,t)})_{j,k} = e^{ij\theta_s} \, e^{ik\phi_t}, \quad \theta_s = \frac{2\pi s}{n+1}, \quad \phi_t = \frac{2\pi t}{n+1},$$

with $i = \sqrt{-1}$. The eigenvalues of A_P are

$$\lambda^{(s,t)} = 4 \left(\sin^2 \frac{\theta_s}{2} \sin^2 \frac{\phi_t}{2} \right).$$

We cannot directly analyze the `IC(1,1)` preconditioner since the triangular factors of M do not have constant diagonals. However, it can be seen that the values of $d_{i,i}$ within a block converge quite fast to a limit value, see figure 8.9.

We consider the preconditioner where the $d_{i,i}$'s are replaced by the asymptotic value d of $d_{i,i}$.

Theorem 8.21

For the model problem and periodic boundary conditions, the (constant diagonal) preconditioner `IC(1,1)` *satisfies*

$$\kappa(M_P^{-1} A_P) = O\left(\frac{1}{h^2}\right).$$

Proof. The asymptotic value of $d_{i,i}$ is given by d satisfying

$$d = 4 - \frac{2}{d} \Rightarrow d = 2 + \sqrt{2}.$$

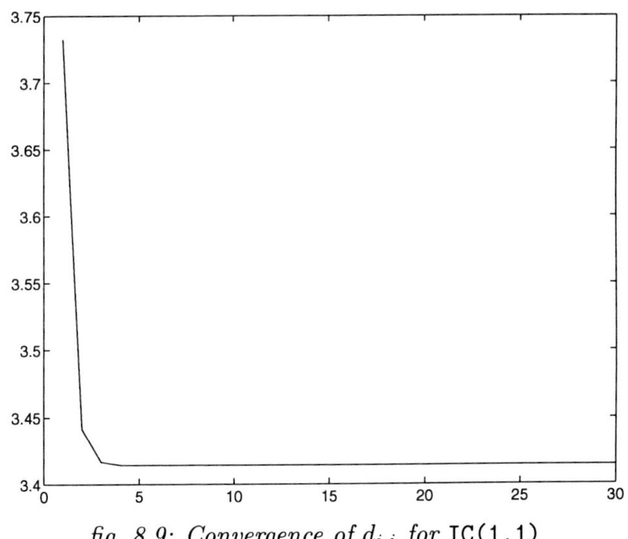

fig. 8.9: Convergence of $d_{i,i}$ for `IC(1,1)` *Poisson model problem*

The stencil of $L_P D_P^{-1} L_P^T$ is

$$\begin{pmatrix} \frac{1}{d} & -1 & 0 \\ -1 & 4 & -1 \\ 0 & -1 & \frac{1}{d} \end{pmatrix}.$$

Symbolically, we can write this as

$$M_P = L_P D_P^{-1} L_P^T = A_P + \begin{pmatrix} \frac{1}{d} & 0 & 0 \\ 0 & 0 & 0 \\ 0 & 0 & \frac{1}{d} \end{pmatrix}.$$

An eigenvector $u^{(s,t)}$ of M_P satisfies

$$M_P u^{(s,t)} = 4\left(\sin^2\frac{\theta_s}{2} + \sin^2\frac{\phi_t}{2}\right) u^{(s,t)}$$
$$+ \frac{1}{d}(e^{i(j-1)\theta_s}\, e^{i(k+1)\phi_t} + e^{i(j+1)\theta_s}\, e^{i(k-1)\phi_t})u^{(s,t)}.$$

Since

$$(e^{-i\theta_s}\, e^{i\phi_t} + e^{i\theta_s}\, e^{-i\phi_t})u^{(s,t)} = 2\cos(\theta_s - \phi_t)u^{(s,t)},$$

we have that the eigenvalues of $M_P^{-1}A_P$ are

$$\lambda^{(s,t)} = \frac{4\left(\sin^2\frac{\theta_s}{2} + \sin^2\frac{\phi_t}{2}\right)}{4\left(\sin^2\frac{\theta_s}{2} + \sin^2\frac{\phi_t}{2}\right) + \frac{2}{d}\cos(\theta_s - \phi_t)}.$$

where θ_s and ϕ_t take values in $[0, 2\pi]$. We can look at this function for continuous values of θ and ϕ. The expression for the eigenvalues is invariant when we interchange θ and ϕ. The extremas arise for $\theta = \phi$. Then,

$$\lambda_{max}(\theta) = \frac{8\sin^2\frac{\theta}{2}}{8\sin^2\frac{\theta}{2} + \frac{2}{d}}.$$

Figure 8.10 gives a plot of the eigenvalues.

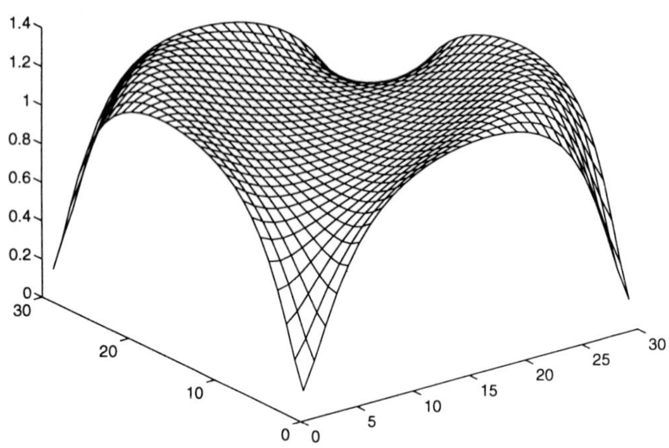

fig. 8.10: Eigenvalues for the `IC(1,1)` *preconditioner Periodic b.c.*

We easily see that $\lambda_{max} \leq 4d$ and also $\lambda^{(1,1)} = O(h^2)$. This implies that $\kappa(M^{-1}A) = O\left(\frac{1}{h^2}\right)$.

8.5.7. Comparison of periodic and Dirichlet boundary conditions

In this section, we would like to investigate the relationship between Dirichlet and periodic boundary conditions for which Fourier analysis is exact. For periodic boundary conditions we have

$$L_P D_P^{-1} L_P^T + D_P^{-1} = 4I - T_P,$$

where

$$T_P = \begin{pmatrix} 0 & 1 & & & 1 \\ 1 & 0 & 1 & & \\ & \ddots & \ddots & \ddots & \\ & & 1 & 0 & 1 \\ 1 & & & 1 & 0 \end{pmatrix}.$$

Also,

$$-D^{-1}L^T = \begin{pmatrix} -1 & 1/d & & & \\ & -1 & 1/d & & \\ & & \ddots & \ddots & \\ & & & -1 & 1/d \\ 1/d & & & & -1 \end{pmatrix} = -I + \frac{1}{d}R^T,$$

with

$$R^T = \begin{pmatrix} 0 & 1 & & & \\ & 0 & 1 & & \\ & & \ddots & \ddots & \\ & & & 0 & 1 \\ 1 & & & & 0 \end{pmatrix}.$$

Thus, the remainder $\overline{\mathcal{R}}$ (that we shall also denote when needed by $\overline{\mathcal{R}}_P$) is defined as

$$M_P = A + \overline{\mathcal{R}}, \quad \overline{\mathcal{R}} = \frac{1}{d}\mathcal{R}$$

and $\mathcal{R}$ has the structure

$$\mathcal{R} = \begin{pmatrix} 0 & R & & & R^T \\ R^T & 0 & R & & \\ & \ddots & \ddots & \ddots & \\ & & R^T & 0 & R \\ R & & & R^T & 0 \end{pmatrix}.$$

We consider what happens to the invariant spaces of A_P. Let us define the vectors u_+, u_-, v_+, v_- whose components are

$$(u_+)_k = e^{ik\theta_j}, \qquad (u_-)_k = e^{-ik\theta_j} \qquad (v_+)_k = e^{il\theta_m}, \qquad (v_-)_k = e^{-il\theta_m}.$$

The eigenspace is spanned by tensor products of these vectors. We consider the action of R and R^T on vectors $u_\pm$. It is easy to see that

$$R^T u_+ = e^{i\theta_j} u_+, \; R^T u_- = e^{-i\theta_j} u_-, \; R u_+ = e^{-i\theta_j} u_+, \; R u_- = e^{i\theta_j} u_-.$$

Therefore, $u_\pm$ are eigenvectors of R and R^T.

We must look at the tensor products of the us and vs. For one block, we have expressions like

$$R^T u v_{l-1} + R u v_{l+1}.$$

We have to compute this expression for the four combinations of $+$ and $-$.

1) $u_+ v_+$. We have (dropping the indices $+$ and $-$ for clarity)

$$R^T u v_{l-1} + R u v_{l+1} = (e^{i\theta_j} e^{-i\theta_m} + e^{-i\theta_j} e^{i\theta_m}) u v_l = a_{++} u v_l.$$

2) $u_+ v_-$.

$$R^T u v_{l-1} + R u v_{l+1} = (e^{i\theta_j} e^{i\theta_m} + e^{-i\theta_j} e^{-i\theta_m}) u v_l = a_{+-} u v_l.$$

3) $u_- v_+$.

$$R^T u v_{l-1} + R u v_{l+1} = (e^{i\theta_j} e^{i\theta_m} + e^{-i\theta_j} e^{-i\theta_m}) u v_l = a_{+-} u v_l.$$

4) $u_- v_-$.

$$R^T u v_{l-1} + R u v_{l+1} = (e^{i\theta_j} e^{-i\theta_m} + e^{-i\theta_j} e^{i\theta_m}) u v_l = a_{++} u v_l.$$

This shows that this eigenspace, which is invariant for A_P gives rise to two invariant spaces with different eigenvalues for $\mathcal{R}$. We have two double eigenvalues $a_{++} = 2\cos(\theta_j - \theta_m)$ and $a_{+-} = 2\cos(\theta_j + \theta_m)$.

This is the reason why the Fourier analysis is not exact for this preconditioner. Since the eigenspace spanned by the tensor products of u_+, u_-, v_+, v_- is not invariant by $\mathcal{R}$ with only one eigenvalue, we cannot consider eigenvectors with sine$\times$sine components as for A. There is no eigenvector whose components are zero on the boundary of the first quadrant. Therefore we cannot reduce to the Dirichlet problem as we did for A.

The only preconditioners that can be exactly analyzed are the ones which leave invariant the eigenspaces of A with only one eigenvalue. As we have seen, the eigenspaces of A are still invariant for $\mathcal{R}$ but with more than one eigenvalue.

Let us look at the form of the matrix we get if we consider the action of $\mathcal{R}$ on the sine and cosine eigenvectors of A_P. To simplify the notation, we drop the indices k of the vector components we consider and we denote by ss, sc, cs, cc the components of products of sine and cosine. Trivially, we have

$$\begin{aligned}
-4ss &= u_+v_+ - u_+v_- - u_-v_+ + u_-v_-,\\
4isc &= u_+v_+ + u_+v_- - u_-v_+ - u_-v_-,\\
4ics &= u_+v_+ - u_+v_- + u_-v_+ - u_-v_-,\\
4cc &= u_+v_+ + u_+v_- + u_-v_+ + u_-v_-.
\end{aligned}$$

Equivalently,

$$\begin{aligned}
u_+v_+ &= -ss + i(sc + cs) + cc,\\
u_+v_- &= +ss + i(sc - cs) + cc,\\
u_-v_+ &= +ss + i(-sc + cs) + cc,\\
u_-v_- &= -ss + i(-sc - cs) + cc.
\end{aligned}$$

Now, we can easily compute the action of $\mathcal{R}$ on ss, cc, sc, cs. For this eigenspace we obtain a 4×4 matrix which is

$$\begin{pmatrix} \alpha & \beta & & \\ \beta & \alpha & & \\ & & \alpha & \beta \\ & & \beta & \alpha \end{pmatrix},$$

with

$$\alpha = \frac{1}{2}(a_{++} + a_{+-}) = 2\cos(\theta_j)\cos(\theta_m), \quad \beta = \frac{1}{2}(a_{+-} - a_{++}) = -2\sin(\theta_j)\sin(\theta_m).$$

Clearly this matrix has only two distinct eigenvalues a_{++} and a_{+-} as we already know. In this basis, $\mathcal{R}$ has a block diagonal representation.

We have

$$\theta_j = \frac{2\pi j}{n+1}, \quad j = 0, \ldots, n$$

and the eigenvalues of $\mathcal{R}$ are

$$\mu_{j,m} = 2\cos(\theta_j - \theta_m), \quad j, m = 0, \dots, n$$

since the other set duplicates this one and we note that we have the right number of eigenvalues. Therefore, the eigenvalues are

$$\mu_{j,m} = 2\cos\left(\frac{2\pi}{n+1}(j-m)\right).$$

Since we have supposed that $n+1 = 2p$, the eigenvalues can be written

$$\mu_{j,m} = 2\cos\left(\frac{\pi}{p}(j-m)\right), \quad j, m = 0, \dots, n$$

or, as $j - m$ =constant gives the same value

$$\mu_{j,m} = 2\cos\left(\frac{\pi}{p}q\right), \quad q = j - m = 0, \dots, n$$

The eigenvalues 2 and -2 have multiplicity $n+1$. The remaining $(n-1)/2$ eigenvalues are of multiplicity $2(n+1)$. If p is even, 0 is an eigenvalue. Independently of the dimension of the problem, the maximum and minimum eigenvalues are 2 and -2. When $h \to 0$, the eigenvalues tend to fill the whole interval $[-2, 2]$. Since we have to divide by h^2, the maximum eigenvalue increases as $2/h^2$. The important thing to note is that there are eigenvalues of order $1/h^2$. The eigenspaces are spanned by the u_+v_+ and u_-v_- vectors or combinations of these; therefore, we can take for instance

$$\{\sin\left(\frac{2\pi}{n+1}(kj+lm)\right), \cos\left(\frac{2\pi}{n+1}(kj+lm)\right)\}.$$

When $h \to 0$, we get functions

$$\{\sin(2\pi(jx+my)), \cos(2\pi(jx+my))\}.$$

Let us consider an example with $n = 7$; the eigenvalues of $\mathcal{R}$ are

$$-2(8)\ , -1.4142(16),\ 0(16),\ 1.4142(16),\ 2(8)$$

and those of $\overline{\mathcal{R}}$ are

$$-0.5858(8)\ , -0.4142(16),\ 0(16),\ 0.4142(16),\ 0.5858(8),$$

where the multiplicity is given within the parentheses. We have the following eigenvalues for $M_P = A + \overline{\mathcal{R}}$ by combining the appropriate eigenvalues

$$0.5858(1),\ 1(4),\ 1.1716(2),\ 1.7574(2),\ 2(4),\ 2.1716(4),\ 3(8)$$

$$3.4142(8),\ 4(4),\ 4.1716(4),\ 4.5858(2),\ 5(4),\ 5.8284(4)$$

$$6(4),\ 6.8284(2),\ 7.4142(2),\ 7.8284(4),\ 8.5858(1).$$

Note that M_P is positive definite.

The eigenvalues of $M_P^{-1} A_P$ (the ratios of eigenvalues of A_P and M_P) are

$$0(1),\ 0.5858(4),\ 0.6667(2),\ 0.8619(4),\ 0.8723(2),\ 0.9210(2),$$

$$0.9289(4),\ 0.9318(1),\ 0.9471(4),\ 1(16),\ 1.0828(4),$$

$$1.0993(4),\ 1.1381(4),\ 1.1716(8),\ 1.1907(4).$$

Now, consider the Dirichlet problem of size $m = p - 1$ with $n + 1 = 2p$. In the example, $p = 4$, so $m = 3$. From what we computed before, the eigenvalues of A_D are

$$1.1716(1),\ 2.5858(2),\ 4(3),\ 5.4142(2),\ 6.8284(1).$$

Let $\mathcal{R}_D$ be defined as

$$\mathcal{R}_D = \begin{pmatrix} 0 & R_D & & & \\ R_D^T & 0 & R_D & & \\ & \ddots & \ddots & \ddots & \\ & & R_D^T & 0 & R_D \\ & & & R_D^T & 0 \end{pmatrix},$$

with

$$R_D^T = \begin{pmatrix} 0 & 1 & & & \\ & 0 & 1 & & \\ & & \ddots & \ddots & \\ & & & 0 & 1 \\ & & & & 0 \end{pmatrix}.$$

The eigenvalues of $\mathcal{R}_D$ are

$$-1.4142(1),\ -1(2),\ 0(3),\ 1(2),\ 1.4142(1).$$

Note that two of these eigenvalues also belong to the spectrum of the periodic case.

We define the Dirichlet preconditioner M_D as

$$M_D = A_D + \overline{\mathcal{R}}_D,$$

with $\overline{\mathcal{R}}_D = 1/d\, \mathcal{R}_D$. Note this is not the usual IC preconditioner for Dirichlet boundary conditions, but we shall go back to that problem later on. Then, M_D has simple eigenvalues which are for the example

$$1.4560,\ 2.4318,\ 2.7247,\ 3.6007,\ 3.7071,\ 4.1080,\ 5.2753,\ 5.5682,\ 7.1282$$

The eigenvalues of $M_D^{-1} A_D$ are

$$0.7887(1),\ 0.9228(1),\ 0.9468(1),\ 1(3),\ 1.0790(1),\ 1.0913(1),\ 1.1165(1).$$

Excluding the zero eigenvalue (which does not have to be taken into account if we use the conjugate gradient), the condition number for the periodic problem (2.032) is an upper bound for that of the Dirichlet problem (1.416).

Now, we shall explore the relationships between $\mathcal{R}$ and $\mathcal{R}_D$. We have the following result:

Theorem 8.22

Let $\mathcal{R}_D$ be of dimension q^2.

1) For $q = 2$, the eigenvalues of $\mathcal{R}_D$ are

$$-1(1),\ 0(2),\ 1(1).$$

2) Let $\Lambda(q) = \{\lambda |\ \lambda$ eigenvalue of $\mathcal{R}_D$ of dimension $q^2\}$. Then

$$\Lambda(q) = \Lambda(q-1) \cup \{2\cos\left(\frac{j\pi}{q+1}\right),\ j = 1, \ldots, q\}.$$

Proof.

1) For $q = 2$, we have

$$\mathcal{R}_D = \begin{pmatrix} 0 & 0 & 0 & 0 \\ 0 & 0 & 1 & 0 \\ 0 & 1 & 0 & 0 \\ 0 & 0 & 0 & 0 \end{pmatrix}.$$

The determinant of $\mathcal{R}_D - \lambda I$ is equal to $\lambda^2(\lambda^2 - 1)$. Consequently, the eigenvalues are 0 and ± 1. It is obvious that the eigenvectors corresponding to the 0 eigenvalue can be taken as $(1\ 0\ 0\ 0)^T$ and $(0\ 0\ 0\ 1)^T$. All the other eigenvectors have a zero first and last components. It can be computed that the eigenvector associated with eigenvalue 1 is

$$(0,\ \sin(\pi/4),\ \sin(3\pi/4),\ 0)^T$$

and that associated with -1 is

$$(0,\ \sin(\pi/4),\ -\sin(3\pi/4),\ 0)^T.$$

2) Let us prove that

$$\Lambda(q-1) \subset \Lambda(q).$$

To do this we have to analyze the relations which are verified by the components of the eigenvectors. The eigenvectors associated with the eigenvalues of the system of dimension $(q-1)^2$ can be constructed in the following way: suppose the eigenvector $x_{(q-1)}$ of size $(q-1)^2$ is divided into $q-1$ blocks of size $q-1$

$$x_{(q-1)} = (x_1, x_2, \ldots, x_{q-1})^T$$

with only one non–zero element in each block, two non–zero elements being separated by $q-2$ zero elements. Then, up to a multiplication by a constant the eigenvector $x_{(q)}$ is constructed by enlarging the size of each block of $x_{(q-1)}$ by adding a zero element at the end. In other words, if

$$x_{(q)} = (y_1, y_2, \ldots, y_q)^T,$$

and

$$x_i = (0, 0, x, 0, 0)^T,$$

then

$$y_i = (0, 0, x, 0, 0, 0)^T,$$

with the same x element.

The last extra block is filled with zeros. Since the non–zero elements are in the same relative position from the beginning of the block and since the distance between two non–zero elements has increased by one, it is obvious that we still have an eigenvector. It is also straightforward to see that if x is an eigenvector (of dimension q^2) with components x_i, then y defined as

$$y_i = x_{q^2-i+1},$$

is also an eigenvector for the same eigenvalue. This is because the matrix is persymmetric.

Now, we turn to the new eigenvalues that appear and are different from (or sometimes equal to) the eigenvalues of the smaller dimension problem. We seek eigenvectors of the form

$$x_{(q)} = (y_1, y_2, \ldots, y_q)^T,$$

with

$$y_i = (0, 0, \ldots, 0, z_i, 0, \ldots, 0)^T,$$

z_i being in position $q - i + 1$. Note this corresponds to having only non–zero components on the diagonal of the mesh points corresponding to $(1, q)$, $(q, 1)$: i.e. the equation $k + l = q + 1$.

Using the equation

$$\mathcal{R}_D x_{(q)} = \lambda x_{(q)},$$

we obtain an eigenvalue problem of dimension q for the vector z made of the z_is

$$z = (z_1, z_2, \ldots, z_q)^T.$$

This problem is

$$\begin{pmatrix} 0 & 1 & & & \\ 1 & 0 & 1 & & \\ & \ddots & \ddots & \ddots & \\ & & 1 & 0 & 1 \\ & & & 1 & 0 \end{pmatrix} z = \lambda z,$$

the matrix being of order q. Note that this is precisely the problem for Dirichlet boundary conditions in one dimension. Therefore we know the eigenvalues and eigenvectors of this problem. The eigenvalues are

$$2\cos\left(\frac{j\pi}{q+1}\right), \quad j = 1, \ldots, q$$

and the components of the eigenvectors are

$$\sin\left(k\frac{\pi j}{q+1}\right).$$

Note that these eigenvalues are simple, there are q of them and the eigenvectors also verify the relation of persymmetry we found above.

If we count the eigenvalues, we can see that if $l(q)$ is the number of eigenvalues and if $o(q)$ is the number of eigenvalues arising from dimension $(q-1)^2$, we have

$$l(q) = o(q) + q$$

$$o(q+1) = o(q) + 2q$$

with $o(2) = 2$. This shows that we have $l(q) = q^2$ as it should. Therefore we have found all the eigenvalues. Note that when q is odd, 0 is included in the "new" eigenvalues. If $z(q)$ is the multiplicity of the 0 eigenvalue, we have

$$z(2) = 2,$$

$$z(q) = z(q-1) + \begin{cases} 2, & \text{if } q \text{ is even;} \\ 1, & \text{if } q \text{ is odd.} \end{cases}$$

□

Remark.

1) Remember that for the periodic problem with matrix $\mathcal{R}$ of dimension $(n+1)^2$ with $n+1 = 2p$ the eigenvalues are

$$\mu_{j,m} = 2\cos\left(\frac{\pi}{p}k\right), \quad k = 0, \ldots, n.$$

This implies that the "new" eigenvalues of $\mathcal{R}_D$ for the Dirichlet problem of size $(p-1)^2$ are eigenvalues of the periodic problem. All eigenvalues of the periodic problem, except 2 and -2, are eigenvalues of the Dirichlet problem.

In particular, the maximum eigenvalue of $\mathcal{R}$, $2\cos(\pi/p)$ is the penultimate to the maximum eigenvalue of the periodic problem (the largest one being 2).

2) As for $\mathcal{R}$, if λ is an eigenvalue of $\mathcal{R}_D$, then $-\lambda$ is also an eigenvalue. The components of the corresponding eigenvectors differ only in sign.

3) Note that when $h \to 0$, the maximum eigenvalue of $\mathcal{R}_D$ tends to 2 which is the maximum eigenvalue of $\mathcal{R}$. The same is true for the minimum eigenvalue. The "new" eigenvalues tend to fill the interval $[-2, 2]$. Therefore, asymptotically the eigenvalues are the same in both cases.

The previous result shows that there is a simple relationship between the eigenvalues of the remainder for the periodic and the corresponding Dirichlet problem. However such simple relations for eigenvalues of $M^{-1}A$ do not hold.

Since we want to obtain information about the eigenvalues of $M_D^{-1}A_D$, we have several possible methods. We remark that as we know the eigenvalues of A_D and $\mathcal{R}_D$, we can use perturbation results to obtain bounds for the eigenvalues of $M_D^{-1}A_D$. First we show that M_D is positive definite. As we said before M_D which was constructed directly from the knowledge of M_P is not the usual IC preconditioner. We saw previously how to directly construct an incomplete preconditioner with constant diagonals. We denote by M'_D this preconditioner. It is easy to see that M_D and M'_D differ only on the diagonal. All the blocks of the difference have a zero diagonal except for the first element which is $1/d$. Only the first diagonal block is different, having two non–zero elements on the diagonal $2/d$ since the first element and $1/d$ as the last. The difference is consequently a diagonal matrix P of rank 2. Therefore, we can bound the eigenvalues of M_D in terms of those of M'_D by perturbation theorems. We have as $M_D = M'_D + P$,

$$\lambda_{min}(M_D) \geq \lambda_{min}(M'_D) + \frac{1}{d},$$

$$\lambda_{max}(M_D) \leq \lambda_{max}(M'_D) + \frac{2}{d}.$$

It is obvious that M'_D is positive definite, hence M_D is also positive definite. Note that M_D has its smallest eigenvalue bounded from below independently of the dimension of the problem, but unfortunately the Cholesky factors of M_D are not sparse.

Remembering that $M_D = A_D + \overline{\mathcal{R}}_D$, we have

$$\frac{\lambda_i(A_D)}{\lambda_{max}(M_D)} \leq \lambda_i(M_D^{-1}A_D) \leq \frac{\lambda_i(A_D)}{\lambda_{min}(M_D)}.$$

We have seen that

$$\lambda_{min}(M_D) \geq \frac{1}{d},$$

and as we also have

$$\lambda_{max}(M_D) \leq \lambda_{max}(A_D) + \lambda_{max}(\overline{\mathcal{R}}_D) \leq \lambda_{max}(A_D) + \frac{2}{2+\sqrt{2}} \leq 8 + \frac{2}{2+\sqrt{2}}.$$

Hence, we obtain

$$\frac{\lambda_i(A_D)}{6.243} \leq \lambda_i(M_D^{-1}A_D) \leq 3.4142\ \lambda_i(A_D).$$

This shows that A_D and $M_D^{-1}A_D$ are spectrally equivalent and we have

$$\frac{\kappa(A_D)}{21.3137} \leq \kappa(M_D^{-1}A_D) \leq 21.3137\ \kappa(A_D).$$

Another way to see this is the following. Let $\alpha = 1/d$, then $M_D = A_D + \alpha\mathcal{R}_D$. Hence,

$$M_D^{-1}A_D = (I + \alpha A_D^{-1}\mathcal{R}_D)^{-1},$$

and

$$\lambda(M_D^{-1}A_D) = \frac{1}{1 + \alpha\lambda(A_D^{-1}\mathcal{R}_D)}.$$

This implies that

$$\frac{8\pi^2h^2 + O(h^4)}{8\pi^2h^2 + O(h^4) + \alpha\lambda_i(\mathcal{R}_D)} \leq \lambda_i(M_D^{-1}A_D) \leq \frac{8 - 8\pi^2h^2 + O(h^4)}{8 - 8\pi^2h^2 + O(h^4) + \alpha\lambda_i(\mathcal{R}_D)}.$$

From this result we can conclude that $\kappa(M_D^{-1}A_D)$ is bounded above by a quantity which behaves as $1/([7 + 4\sqrt{2}]\pi^2h^2)$.

8.5.8. Axelsson's results

O. Axelsson [67] proved some localization results for eigenvalues of $M^{-1}A$ using some perturbation lemmas.

Theorem 8.23

Let A and M be SPD matrices and let $\mu_1, \mu_2 > 0 \in \mathbb{R}$ such that $\lambda_{max}(\mu_1 M - A) \geq 0$, $\lambda_{min}(\mu_2 M - A) \geq 0$. Then,

$$\frac{\mu_1 \lambda_i(A)}{\lambda_i(A) + \lambda_{max}(\mu_1 M - A)} \leq \lambda_i(M^{-1}A) \leq \frac{\mu_2 \lambda_i(A)}{\lambda_i(A) + \lambda_{min}(\mu_2 M - A)}.$$

□

This result can be used to derive upper bounds for $\lambda_{max}(M^{-1}A)$ and lower bounds for $\lambda_{min}(M^{-1}A)$. If $A = gI + L + L^T$ where g is a positive real, $M = (X + L)X^{-1}(X + L^T)$ with X being an M-matrix and $x = \lambda_{min}(X)$, then if $2x - g > 0$,

$$\lambda_i(M^{-1}A) \leq \begin{cases} \frac{4x\lambda_i(A)}{(2x-g+\lambda_i(A))^2} & \text{if } \lambda_i(A) \leq 2x - g \\ \frac{x}{2x-g} & \text{if } \lambda_i(A) \geq 2x - g. \end{cases}$$

For `IC(1,1)` and the model problem $x \approx 2 + \sqrt{2}$ and this gives

$$\lambda_{max}(M^{-1}A) \leq \frac{2 + \sqrt{2}}{2\sqrt{2}} \approx 1.2071\ldots$$

Moreover, we have

$$\frac{\lambda_i(A)}{\lambda_i(A) + \frac{2}{x}} \leq \lambda_i(M^{-1}A) \leq \frac{4(2 + \sqrt{2})\lambda_i(A)}{(2\sqrt{2} + \lambda_i(A))^2}.$$

This shows that $\lambda_{min(M^{-1}A)} = O(h^2)$. For the smallest eigenvalues we have

$$\lambda_i(M^{-1}A) = \frac{2 + \sqrt{2}}{2}\lambda_i(A) + O(\lambda_i(A)^2).$$

This last result essentially says that the distribution of the smallest eigenvalues of $M^{-1}A$ is almost the same as for the original problem.

8.6. The modified incomplete Cholesky decomposition

8.6.1. The DKR preconditioner

This preconditioner has its origin in a work of Dupont, Kendall and Rachford [494] following some older ideas by Buleev and Varga [1299]. It was devised for matrices arising from finite difference approximations of elliptic partial differential equations in rectangles. The idea was to try to compensate for entries left in the remainder R by summing them up to the diagonal of M. We require

$$M = LD^{-1}L^T = A + R + \delta\bar{D},$$

with $\bar{D}$ being a diagonal matrix such that $\bar{d}_{i,i} = a_{i,i}$ and R having non–zero entries in positions $(i, i-m+1), (i,i), (i, i+m-1)$ on row i. The matrix R will be such that the sums of entries of a row (this is denoted as rowsum) is zero for all rows. This implies that (generically)

$$r_{i,i} = -(r_{i,i-m+1} + r_{i,i+m-1}). \tag{8.5}$$

By equating we get,

$$d_{i,i} = (1+\delta)a_{i,i} - \frac{a_{i,i-1}}{d_{i-1,i-1}}(a_{i,i-1} + a_{i+m-1,i-1}) - \frac{a_{i,i-m}}{d_{i-m,i-m}}(a_{i,i-m} + a_{i-m+1,i-m}).$$

For Dirichlet boundary conditions a value of $\delta = ch^2$ is recommended for reasons we shall see below. This method is usually denoted as DKR. It can be extended straightforwardly to non–symmetric matrices.

8.6.2. Analysis of DKR

The following results for the model problem were proved by I. Gustafsson [720].

Lemma 8.24

For the Poisson model problem, there exists a constant c_1 such that

$$r_{i+m-1,i} \leq \frac{1}{2(1+c_1 h)}.$$

Proof. We start by proving that $d_{i,i} \geq 2(1+c_1 h)$. If $c = 0$, we have $d_{i,i} \geq 2$. This is proved by induction since $d_{1,1} = 4$ and generically, we have

$$d_{i,i} = 4 - \frac{2}{d_{i-1,i-1}} - \frac{2}{d_{i-m,i-m}}.$$

Of course, one of the two negative terms may be zero. This shows that $d_{i,i} \geq 2$. If $c \neq 0$, we have

$$d_{i,i} \geq 4(1+ch^2) - \frac{2}{1+c_1 h}.$$

If we set $c_1 = \sqrt{2c}$, we obtain $d_{i,i} \geq 2(1+c_1 h)$. Even without this hypothesis we have that

$$d_{i,i} \geq 2(1+c_1 h) + O(h^2).$$

It turns out that

$$r_{i+m-1,i} = \frac{a_{i+m-1,i-1} a_{i,i-1}}{d_{i-1,i-1}} \leq \frac{1}{d_{i-1,i-1}} \leq \frac{1}{2(1+c_1 h)}.$$

□

Lemma 8.25

For the Poisson model problem,

$$(Rx, x) \leq 0.$$

Proof.

$$(Rx,x) = -\sum_i (r_{i,i-m+1} + r_{i+m_1,i}) x_i^2 + 2 \sum_i r_{i,i+m-1} x_{i+m-1} x_i,$$

and we use the identity $2x_{i+m-1}x_i = x_{i+m-1}^2 + x_i^2 - (x_{i+m-1} - x_i)^2$ to get

$$(Rx,x) = -\sum_i (r_{i,i-m+1} + r_{i+m-1,i}) x_i^2 + \sum_i (r_{i,i+m-1} x_i^2 + r_{i,i+m-1} x_{i+m-1}^2) - \sum_i r_{i,i+m-1} (x_{i+m-1} - x_i)^2.$$

Since the sum of the first two terms is zero, we obtain the result. □

Note that this result is true for all matrices verifying (8.5). The result of lemma 8.25 shows that the eigenvalues of $M^{-1}A$ are larger than 1.

Lemma 8.26

For the Poisson model problem,

$$-(Rx, x) \le \frac{1}{1 + c_1 h}(Ax, x).$$

Proof. From lemma 8.25, we have

$$-(Rx, x) = \sum_i r_{i,i+m-1}(x_{i+m-1} - x_i)^2 \le \sum_i \frac{1}{d_{i-1,i-1}}(x_{i+m-1} - x_i)^2.$$

Using lemma 8.24,

$$-(Rx, x) \le \sum_i \frac{1}{2(1 + c_1 h)}(x_{i+m-1} - x_i)^2.$$

We have the equality $(a-b)^2 \le 2(a-e)^2 + 2(e-b)^2$ that we use with $e = x_{i-1}$,

$$-(Rx, x) \le \sum_i \frac{1}{1 + c_1 h}[(x_{i+m-1} - x_{i-1})^2 + (x_i - x_{i-1})^2].$$

With a change of variable, we have

$$-(Rx, x) \le \sum_i \frac{1}{1 + c_1 h}[(x_{i+m} - x_i)^2 + (x_{i+1} - x_i)^2] \le \frac{1}{1 + c_1 h}(Ax, x).$$

□

From these lemmas, we can bound the condition number of $M^{-1}A$.

Theorem 8.27

For the Poisson model problem,

$$\kappa(M^{-1}A) = O\left(\frac{1}{h}\right).$$

Proof.

$$\frac{(Ax, x)}{(Mx, x)} = \frac{(Ax, x)}{(Ax, x) + \delta(\bar{D}x, x) + (Rx, x)}.$$

But,

$$0 \leq \frac{\delta(\bar{D}x, x)}{(Ax, x)} = ch^2 \frac{(x, x)}{(Ax, x)} \leq \frac{c}{c_0},$$

with a constant c_0, since the smallest eigenvalue of A is written as $c_0 h^2$. Note that this is where we need $\delta = ch^2$ to cancel the h^2 term in the lower bound. Therefore,

$$\frac{1}{1 + \frac{c}{c_0}} \leq \frac{(Ax, x)}{(Mx, x)} \leq \frac{1}{1 + \frac{(Rx,x)}{(Ax,x)}} \leq \frac{1}{1 - \frac{1}{1+c_1 h}} = 1 + \frac{1}{c_1 h},$$

which proves the result. □

Note that we cannot conclude with this result for the case $c = 0$. For other boundary conditions than pure Dirichlet, the value of δ can be different, see [494]. This analysis can be extended to more general problems, see Gustafsson [720]. The DKR preconditioner is well defined for diagonally dominant matrices. This can be extended with some modifications to M–matrices.

The previous theory says that a value of the perturbation parameter $c \neq 0$ is needed to obtain a $O(1/h)$ condition number. However, it has been observed in practice for many problems that this is attained without any perturbation (in particular for the Poisson model problem with the usual row ordering). Unfortunately, there exist some orderings of the unknowns for the Poisson model problem for which the factorization with $c = 0$ fails (for example the Red–Black ordering).

8.6.3. Fourier analysis of DKR

As for IC we consider a variant of the preconditioner with a constant diagonal. With DKR, the asymptotic value d of the diagonal coefficients for the model problem is the solution of

$$d = 4 + ch^2 - \frac{4}{d} \Rightarrow d = 2 + \frac{ch^2}{2} + \tfrac{1}{2}\sqrt{8ch^2 + (ch^2)^2} = 2 + \sqrt{2}ch + O(h^2).$$

Looking at the stencils, we have

$$M_P = A_P + \frac{1}{d}\begin{pmatrix} 1 & 0 & 0 \\ 0 & 0 & 0 \\ 0 & 0 & 1 \end{pmatrix} + \left(ch^2 - \frac{2}{d}\right) I.$$

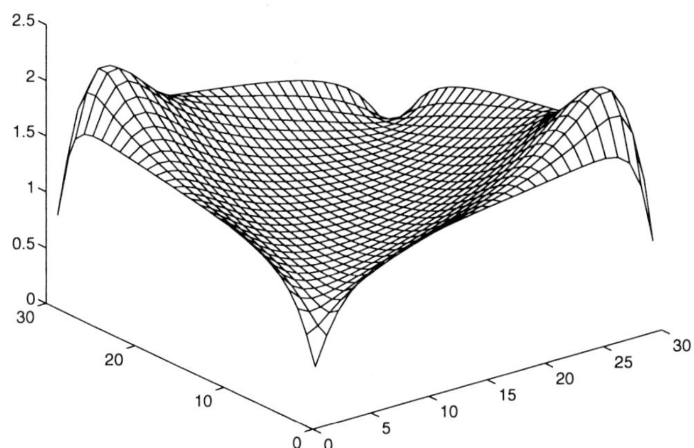

fig. 8.11: Eigenvalues for `MIC(1,1)`
$c = 8\pi^2$, *periodic b.c.*

The eigenvalues of $M_P^{-1}A_P$ are

$$\lambda^{(s,t)} = \frac{4\left(\sin^2\frac{\theta_s}{2} + \sin^2\frac{\phi_t}{2}\right)}{4\left(\sin^2\frac{\theta_s}{2} + \sin^2\frac{\phi_t}{2}\right) + \frac{2}{d}(\cos(\theta_s - \phi_t) - 1) + ch^2}.$$

The function of θ and ϕ is shown in figure 8.11 for $c = 8\pi^2$.

Theorem 8.28

For the Poisson model problem with periodic boundary conditions, if $c > 0$,

$$\kappa(M_P^{-1}A_P) = O\left(\frac{1}{h}\right).$$

Proof.

$$\lambda_{min} = \lambda(1,1) = \frac{8\sin^2(\pi h)}{8\sin^2(\pi h) + ch^2} \simeq \frac{1}{1 + \frac{c}{8\pi^2}} = O(1),$$

$$\lambda_{max} = \lambda(\theta, 2\pi - \theta) \simeq \frac{1}{\sqrt{2ch}} = O\left(\frac{1}{h}\right),$$

with $\theta \simeq \sqrt{\frac{c}{8}}h$. The optimal value of c for the periodic problem is $c_P = 8\pi^2$ and the correspondence between the values for periodic and Dirichlet boundary conditions is $c_P = 4c_D$. Therefore, $c_D = 2\pi^2$.

8.6.4. Extensions of DKR

The DKR preconditioner is easily extended to more general sparse matrices. This is denoted as MIC (MILU for non–symmetric matrices). Consider, for instance, a diagonally dominant symmetric M–matrix and the first step of the incomplete decomposition. As before for IC, we define r_1 such that $(r_1)_j = -a_{j,1}$ if $(j,1) \notin G$. Moreover, let $r_{1,1}^{(1)} = -\sum_{j=2}^{n}(r_1)_j \leq 0$ and $D_R^{(1)}$ be a diagonal matrix of order $n-1$ such that $(D_R^{(1)})_{j,j} = -(r_1)_j$. Then,

$$A = A_1 = \begin{pmatrix} a_{1,1} + r_{1,1}^{(1)} & b_1^T \\ b_1 & B_1 + D_R^{(1)} \end{pmatrix} - \begin{pmatrix} r_{1,1}^{(1)} & r_1^T \\ r_1 & D_R^{(1)} \end{pmatrix} = M_1 - R_1.$$

Clearly, we have $rowsum(R_1) = 0$ (this is $Re = 0$ where e is the vector of all ones). Moreover, since A is diagonally dominant, $a_{1,1} + r_{1,1}^{(1)} > 0$ and $B_1 + D_R^{(1)}$ is diagonally dominant. Therefore, we can proceed and we obtain that the matrices in all the steps are diagonally dominant. Adding a small ($O(h^2)$) term to the diagonal would give strict diagonal dominance. For more general M–matrices which are generalized diagonally dominant with a vector d ($Ad > 0$), we could require that $Rd = 0$.

8.7. The relaxed incomplete Cholesky decomposition

The relaxed incomplete Cholesky decomposition RIC (or RILU in the non–symmetric case) is basically the same as MIC except that the (negative for M–matrices) values that are added to the diagonal are multiplied by a relaxation parameter ω such that $0 \leq \omega \leq 1$. Generally, in this case, one uses an unperturbed method taking $c = 0$. Therefore, $\omega = 0$ gives IC, while $\omega = 1$ gives the unperturbed MIC. A Fourier analysis of the model problem (Chan [266]) has given that the optimal parameter is $\omega = 1 - 8\sin^2(\pi h_P/2)$. Then, the condition number is $O(1/h)$.

For RIC the asymptotic value of the diagonal coefficients is $d = 2 + \sqrt{2(1-\omega)}$. The eigenvalues of $M_P^{-1}A_P$ are

$$\lambda^{s,t} = \frac{4(\sin^2(\theta_s/2) + \sin^2(\phi_t/2))}{4(\sin^2(\theta_s/2) + \sin^2(\phi_t/2)) + (\frac{2}{2+\sqrt{2(1-\omega)}})(\cos(\theta_s - \phi_t) - \omega)}.$$

The condition number has a minimum for ω close to 1, but there is a sharp increase when $\omega \to 1$, so it is better practice to underestimate the optimal value of ω. In [266], it is shown that RIC is equivalent to MIC for the periodic problem if the coefficient c is allowed to vary with h.

Bounds for the eigenvalues and the condition number with Dirichlet boundary conditions are given in Axelsson [67].

8.8. More on the incomplete decompositions for the model problem

All these variants of the incomplete decompositions applied to the model problem differ only by the values of the diagonal coefficients. Therefore, it is interesting to see if we can find an optimal value for these coefficients, for instance when we use a constant diagonal whose elements are all equal to α. First, we ran some numerical experiments with Dirichlet boundary conditions on a 30×30 mesh. The numbers of PCG iterations to reach $\|r^k\| \leq 10^{-6}\|r^0\|$ are given as a function of α of figure 8.12. There exists an optimal value of α which is between 2 and 2.5 for this example. Note that for greater values of α, the number of iterations varies very slowly. The limit value for DKR is very close to the optimal value of α.

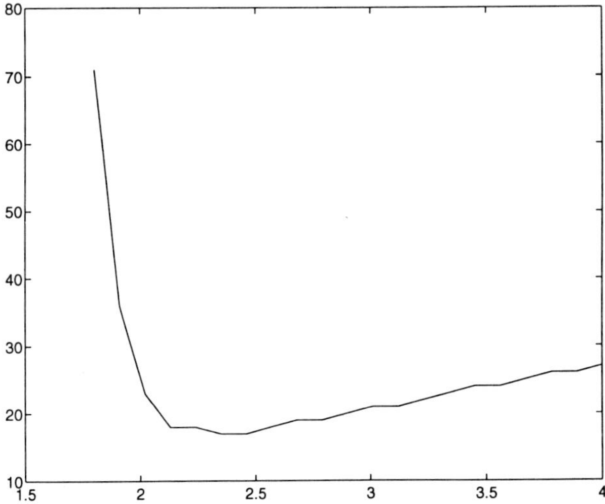

fig. 8.12: Number of PCG *iterations as a function of* α
Dirichlet b.c.

Then, for periodic boundary conditions, we analytically compute the eigenvalues of M_P. It turns out that they are given by

$$\lambda^{s,t} = 4(\sin^2(\theta_s/2) + \sin^2(\phi_t/2)) + \alpha + \frac{2}{\alpha} - 4 + \frac{2}{\alpha}\cos(\theta_s - \phi_t).$$

Now, we can look for the minimum of the condition number of $M_P^{-1}A_P$ as a function of α. The condition number of $M_P^{-1}A_P$ is given for different values of α on figure 8.13 for $m = 30$. We can see that the value of α which gives the minimum value of the condition number for the periodic problem agrees quite closely with the value giving the minimum number of iterations for Dirichlet boundary conditions.

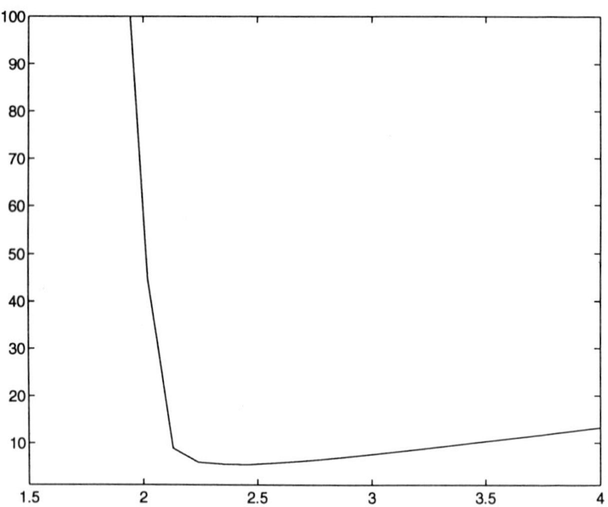

fig. 8.13: Condition number as a function of α
Periodic b.c.

It is also interesting to see if we can find the parameter α that minimizes a norm of the residual matrix R for Dirichlet boundary conditions. To do this we choose the Frobenius norm which is easily computable. Let m be the size of the blocks in A, $n = m^2$, then

$$\|R\|_F^2 = (\alpha - 4)^2 + 2(m-1)(\alpha + \frac{1}{\alpha} - 4)^2 + (m-1)^2(\alpha + \frac{2}{\alpha} - 4)^2 + 2(m-1)^2\frac{1}{\alpha^2}.$$

When m is large, only the last two terms are relevant, therefore it is sufficient to minimize

$$(\alpha + \frac{2}{\alpha} - 4)^2 + \frac{2}{\alpha^2}.$$

Using symbolic computation, it can be shown that the minimum of this expression is obtained for $\alpha = 1 + \sqrt{3 + \sqrt{10}}$ which is approximately 3.48239353... For this value of α the function value is 0.1681365259... This has to be multiplied by $(m-1)^2$ to obtain $\|R\|_F^2$. The complete function can also be minimized, the minimum being a function of m. However, the expression of α giving the minimum is quite complicated. The solution is a little larger than when neglecting lower order terms. For instance, for $m = 30$, it gives 3.5019496... (and a value of the square of the norm of 141.403...) and for $m = 100$ we obtain $m = 3.48807645$... These values are very close to what we obtain with `IC(1,1)` for which the limit value of the diagonal coefficients is $2 + \sqrt{2}$ and for $m = 30$ the square of the Frobenius norm is 142.5. This shows that minimizing the norm of the remainder does not lead to the optimal number of iterations.

Let us consider another possibility to compute an incomplete factorization. Let $M = LD^{-1}L^T$, L having the same non–zero structure as the lower triangular part of A, D being diagonal. We denote the non–zero elements of the ith row of L as c_i, b_i, d_i; d_i is also the diagonal element of D. Then, we know there are two additional diagonals in M compared to A and the elements of the ith row of $R = M - A$ are

$$c_i - a_{i,i-m}, \frac{c_i b_{i-m+1}}{d_{i-m}}, b_i - a_{i,i-1}, d_i + \frac{b_i^2}{d_{i-1}} + \frac{c_i^2}{d_{i-m}} - a_{i,i},$$

$$b_{i+1} - a_{i,i+1}, \frac{c_{i+m-1} b_i}{d_{i-1}}, c_{i+m} - a_{i,i+m}.$$

Suppose we have a set of vectors $V = (v^1, \ldots, v^k)$. We would like to compute the coefficients of the incomplete decomposition in such a way that $\|RV\|$ is small. If $v \in V$ then, the ith row of $Rv = 0$ is

$$d_i v_i + b_{i+1} v_{i+1} + c_{i+m} v_{i+m} = f_i,$$

where f_i only involves elements from the previous rows. Only the last block and the last row are different since then c_{i+m} and b_{i+1} do not exist. If we use k such vectors v with $k > 3$, we have an overdetermined system with a $k \times 3$ matrix. We compute the least squares solution of this set of equations and this gives the preconditioner. The problem which remains is knowing how to choose the set of vectors v^i. We ran some numerical experiments for the

Poisson model problem by choosing some of the eigenvectors of A or vectors of an orthonormal basis of the Krylov space $K_k(x, A)$ for some given (random) x. It turns out that in both cases the eigenvalue distributions of $M^{-1}A$ are very close to that given by IC(1,1). Figure 8.14 shows a comparison of the eigenvalues of IC and this least squares preconditioner when using a Krylov space of dimension 10 for a model problem of order 100.

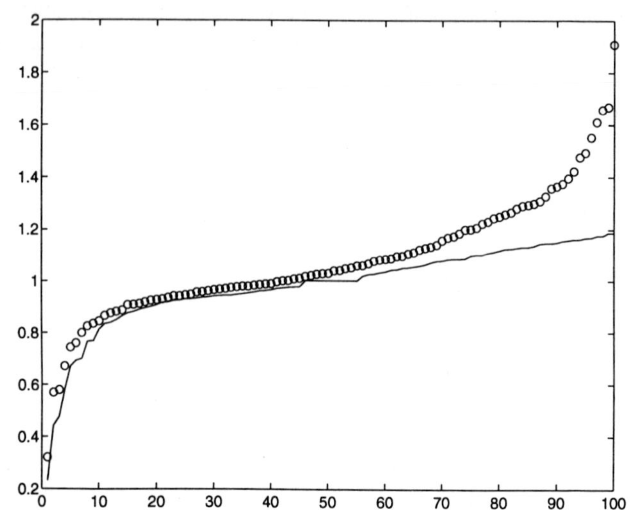

fig. 8.14: Eigenvalues of IC(1,1) *and the least squares preconditioner solid:* IC(1,1)*, circles: least squares*

Greenbaum and Rodrigue [703] have numerically computed an "optimal" preconditioner for small problems with Dirichlet boundary conditions. The structure of the incomplete triangular factors was the same as the lower and upper triangular parts of A. Their results show that neither IC or MIC are optimal with respect to minimizing the condition number.

8.9. Stability of incomplete decomposition

For incomplete decompositions it is interesting to examine the size of the pivots which govern the stability of the algorithm. In this respect, Manteufell [930] has shown that for an H–matrix the incomplete decomposition is more stable that the complete one. However, it has been noticed that there can be trouble with non–symmetric problems. Elman [520], [522] has studied the stability of the factorization and the triangular solves. He considered the following two dimensional model problem in the unit square with Dirichlet boundary conditions:

$$-\Delta u + 2P_1 \frac{\partial u}{\partial x} + 2P_2 \frac{\partial u}{\partial y} = f,$$

which is approximated with finite difference methods. For the first order terms, two schemes can be used: second–order centered or first–order upwind that depends on the signs of the coefficients P_1, P_2. Both schemes give a non–symmetric linear system whose matrix has the same structure as that of the Poisson model problem.

Elman considered numerically three problems: $(P_1 = 0, P_2 = 50)$, $(P_1 = P_2 = 50)$, $(P_1 = -50, P_2 = 50)$. It turns out that using a simple iterative method denoted as `Orthomin(1)`, `ILU` failed on the second problem and `MILU` failed on the third one with centered differences. Elman did not directly study `ILU` and `MILU` but the constant coefficients variants in the same manner as for the Fourier analysis. Then, he analyzed the triangular solves $Lv = w, Uv = w$ where L and U are the incomplete factors. This leads to studying recurrences such that

$$v_j = \frac{1}{\alpha}(w_j - \beta v_{j-1} - \gamma v_{j-m}).$$

The lower triangular solve is said to be stable if all the roots of the characteristic polynomial $(\alpha z^m + \beta z^{m-1} + \gamma)$ are less or equal to 1 in modulus.

Theorem 8.29

Necessary and sufficient conditions for the stability of the lower triangular solve are:

1) for $\beta \leq 0, \gamma \leq 0,\ \alpha + \beta + \gamma \geq 0,$

2) for $\beta \geq 0, \gamma \geq 0$ *and* m *odd,* $-\alpha + \beta + \gamma \leq 0,$

3) for $\beta \geq 0, \gamma \leq 0$ *and* m *even,* $\alpha - \beta + \gamma \geq 0,$

4) for $\beta \leq 0, \gamma \geq 0$ *and* m *even,* $\alpha - \gamma + \beta \geq 0.$

Proof. See Elman [520]. □

This translates into regions of stability in the (P_1, P_2) plane. For instance, for centered differences and ILU when $P_1 \geq 0, P_2 \geq 0$, the condition for stability is $p_1 p_2 \leq 1$ where $p_i = P_i h, i = 1, 2$. For the upwind scheme, both upper and and lower triangular solves are always stable for ILU and MILU. Numerical examples in [520] show a high correlation between these stability results and the behaviour of those preconditioners for several iterative methods. Relaxed factorizations for these problems were studied in [522]. There, stabilized incomplete factorizations were constructed that produce diagonally dominant lower and upper triangular factors. In these methods, the modifications involving the row sums are done or not adaptively based on the values of p_1 and p_2.

Bruaset, Tveito and Winther [220] studied the stability of the RIC preconditioner.

8.10. The generalized SSOR preconditioner

In this section, we look again at the problem of finding good SSOR–like preconditioner. For simplicity, we slightly change the notation. Consider the matrix A of an elliptic finite difference problem. Let L be the strictly lower triangular part of the symmetric matrix A and Δ be a diagonal matrix, we define the generalized SSOR preconditioner (GSSOR) as

$$M = (\Delta + L)\Delta^{-1}(\Delta + L^T),$$

where Δ is a diagonal matrix. The SSOR preconditioner is obtained by taking $\Delta_{i,i} = a_{i,i}$. As we have already seen before

$$M = \Delta + L + L^T + L\Delta^{-1}L^T = A + (\Delta - D) + L\Delta^{-1}L^T = A + R,$$

where D is the diagonal matrix of A. This shows that for a pentadiagonal matrix arising from finite differences approximations we have,

$$\begin{aligned} m_{i,i} &= \Delta_{i,i} + \frac{a_{i,i-1}^2}{\Delta_{i-1,i-1}} + \frac{a_{i,i-m}^2}{\Delta_{i-m,i-m}}, \\ m_{i,i-1} &= a_{i,i-1}, \\ m_{i,i-m} &= a_{i,i-m}, \\ m_{i,i-m+1} &= \frac{a_{i,i-m}a_{i-m+1,i-m}}{\Delta_{i-m,i-m}}. \end{aligned}$$

For this type of matrix, it is clear that IC(1,1) and DKR are instances of GSSOR preconditioners with different matrices Δ. Suppose that $\Delta_{i,i} \neq 0$ and let us introduce ω_i and γ_i such that

$$\omega_i = \frac{a_{i,i}}{\Delta_{i,i}}, \quad \gamma_i = \frac{a_{i,i+1} + a_{i,i+m}}{a_{i,i}}.$$

Lemma 8.30

Let A be a symmetric positive definite pentadiagonal matrix and $\delta \in \mathbb{R}$. If $\omega_i > 0$ is chosen such that

$$0 \leq \frac{1}{\omega_i} + \frac{1}{a_{i,i}}(a_{i,i-1}\omega_{i-1}\gamma_{i-1} + a_{i,i-m}\omega_{i-m}\gamma_{i-m}) \leq 1 + \delta,$$

then

$$(Mx, x) \leq (Ax, x) + \delta(Dx, x), \quad \forall x \neq 0.$$

Proof. We have

$$(L\Delta^{-1}L^T x)_i = \frac{a_{i,i-1}}{\Delta_{i-1,i-1}}(a_{i-1,i}x_i + a_{i-1,i+m-1}x_{i-m+1}) + \frac{a_{i,i-m}}{\Delta_{i-m,i-m}}(a_{i-m,i-m+1}x_{i-m+1} + a_{i-m,i}x_i).$$

With a change of variable, we get

$$(L\Delta^{-1}L^T x, x) \leq \sum_i \frac{1}{\Delta_{i,i}}(a_{i,i+1}^2 x_{i+1}^2 + a_{i,i+m}^2 x_{i+m}^2 + 2a_{i,i+1}a_{i,i+m}x_{i+1}x_{i+m}).$$

Using the identity $2x_{i+1}x_{i+m} = x_{i+1}^2 + x_{i+m}^2 - (x_{i+1} - x_{i+m})^2$, we show that $(L\Delta^{-1}L^T x, x)$ is less or equal than

$$\sum_i \frac{1}{\Delta_{i,i}}[(a_{i,i+1}^2 + a_{i,i+1}a_{i,i+m})x_{i+1}^2 + (a_{i,i+m}^2 + a_{i,i+1}a_{i,i+m})x_{i+m}^2.$$

With another change of variable,

$$(L\Delta^{-1}L^T x, x) \leq \sum_i \{\frac{1}{\Delta_{i-1,i-1}}(a_{i,i-1}^2 + a_{i,i-1}a_{i-1,i+m-1}) + \frac{1}{\Delta_{i-m,i-m}}(a_{i,i-m}^2 + a_{i,i-m}a_{i-m,i-m+1})\}x_i^2.$$

With our notations, this can written as

$$(L\Delta^{-1}L^Tx, x) \le \sum_i (\omega_{i-1}a_{i,i-1}\gamma_{i-1} + \omega_{i-m}a_{i,i-m}\gamma_{i-m})x_i^2.$$

Finally, we have

$$\begin{aligned}(Mx, x) \le (Ax, x) + \sum_i (\frac{1}{\omega_i} - 1 \\ + \frac{1}{a_{i,i}}[\omega_{i-1}a_{i,i-1}\gamma_{i-1} + \omega_{i-m}a_{i,i-m}\gamma_{i-m}])a_{i,i}x_i^2, \\ \le (Ax, x) + \delta(Dx, x).\end{aligned}$$

□

Note that we could have only required that

$$0 \le \frac{1}{\omega_i} + \frac{1}{a_{i,i}}(a_{i,i-1}\omega_{i-1}\gamma_{i-1} + a_{i,i-m}\omega_{i-m}\gamma_{i-m}) \le 1 + \delta_i$$

and then, we set $\delta = \max_i \delta_i$. In this way, some of the δ_is could be zero.

Lemma 8.31

Let A be a symmetric positive definite pentadiagonal matrix. Then there exists $\chi > 0$ such that

$$(Mx, x) \ge \chi(Ax, x).$$

Proof.

$$\begin{aligned}(L\Delta^{-1}L^Tx, x) = \sum_i (\frac{a_{i,i-1}^2}{\Delta_{i-1,i-1}} + \frac{a_{i,i-m}^2}{\Delta_{i-m,i-m}})x_i^2 \\ + \sum_i \frac{a_{i,i-1}a_{i-1,i+m-1}}{\Delta_{i-1,i-1}}(x_i^2 + x_{i+m-1}^2) \\ - \sum_i \frac{a_{i,i-1}a_{i-1,i+m-1}}{\Delta_{i-1,i-1}}(x_i - x_{i+m-1})^2.\end{aligned}$$

We can use the inequality

$$c_1c_2(a_1 - a_2)^2 \le (c_1 + c_2)[c_1(a_1 - e)^2 + c_2(a_2 - e)^2], \quad \forall e \in \mathbb{R},$$

to obtain a lower bound of the third term. Here, we take $e = \beta_{i-1}x_{i-1}$, with β_i to be chosen. With a change of variable, we obtain

$$(L\Delta^{-1}L^Tx, x) \geq \sum_i (\omega_{i-1}a_{i,i-1}\gamma_{i-1} + \omega_{i-m}a_{i,i-m}\gamma_{i-m})x_i^2$$
$$- \sum_i \omega_{i-1}\gamma_{i-1}[a_{i,i-1}(x_i - \beta_{i-1}x_{i-1})^2$$
$$+ a_{i-1,i+m-1}(\beta_{i-1}x_{i-1} - x_{i+m-1})^2].$$

By expressing the squares and using a change of variable,

$$(L\Delta^{-1}L^Tx, x) \geq -\sum_i \omega_i\gamma_i^2\beta_i^2 a_{i,i}x_i^2$$
$$+ 2\sum_i \beta_i\omega_i\gamma_i(a_{i,i+1}x_ix_{i+1} + a_{i,i+m}x_ix_{i+m}).$$

This shows that

$$(Mx, x) \geq (Ax, x) + \sum_i \left\{\frac{1}{\omega_i} - 1 - \omega_i\gamma_i^2\beta_i^2\right\} a_{i,i}x_i^2$$
$$+ 2\sum_i \beta_i\omega_i\gamma_i(a_{i,i+1}x_ix_{i+1} + a_{i,i+m}x_ix_{i+m}).$$

But

$$(Ax, x) = \sum_i a_{i,i}x_i^2 + 2\sum_i (a_{i,i+1}x_ix_{i+1} + a_{i,i+m}x_ix_{i+m}),$$

therefore since,

$$(Mx, x) \geq \sum_i (\frac{1}{\omega_i} - \omega_i\gamma_i^2\beta_i^2)a_{i,i}x_i^2$$
$$+ 2\sum_i (1 + \beta_i\omega_i\gamma_i)(a_{i,i+1}x_ix_{i+1} + a_{i,i+m}x_ix_{i+m}),$$

if we choose β_i such that

$$\beta_i\gamma_i = \frac{1}{\omega_i} - 1,$$

both factors involving β_i are equal and we have

$$(Mx, x) \geq \chi(Ax, x), \quad \chi = \min_i(2 - \omega_i).$$

□

Theorem 8.32

Let A be a symmetric positive definite pentadiagonal matrix and let ω_i satisfy the conditions of lemma 8.30. Then

$$\kappa(M^{-1}A) \leq \frac{1}{\chi}\left(1 + \frac{\delta}{\lambda_{min}(D^{-1}A)}\right), \quad \chi = \min_i(2 - \omega_i).$$

Proof. We use results of lemmas 8.30 and 8.31. □

Let us consider what we obtain for DKR and the Poisson model problem. In this case we have equality with the upper bound of lemma 8.30 and $\delta = ch^2$. The asymptotic value of ω_i is $\omega = 2 - c_0 h$. Then, $\chi = c_0 h$ and also $\lambda_{min} = O(h^2)$. This shows that

$$\kappa(M^{-1}A) \leq \frac{1}{c_0 h}(1 + c_1).$$

Unfortunately, with theorem 8.32 we cannot conclude for the non–perturbed case ($c = 0$), since we should get $\omega = 2$ and $\chi = 0$.

Note that we need only specify

$$0 \leq \frac{1}{\omega_i} + \frac{1}{a_{i,i}}(a_{i,i-1}\omega_{i-1}\gamma_{i-1} + a_{i,i-m}\omega_{i-m}\gamma_{i-m}) \leq 1 + \delta_i,$$

then the theory is the same as before with $\delta = \max_i \delta_i$. We only need to take one δ_i equal to ch^2. In practice it is not too easy to make other choices than DKR since we need also to have $\omega_i > 0$ to get a positive definite preconditioner.

8.11. Incomplete decomposition of positive definite matrices

It is easy to find examples of symmetric positive definite matrices which are not H–matrices for which the Incomplete Cholesky decomposition fails. To solve this problem, it was suggested by Manteuffel [930] that we factorize $A(\alpha) = (1 + \alpha)D + L + L^T$ instead of A, where α is a positive real that is chosen such that $A(\alpha)$ is an H–matrix.

Varga, Saff and Mehrmann [1304] have characterized the matrices that are incompletely factorizable. Let $\mathcal{F}_n$ be this set and $\mathcal{H}_n$ be the set of non–singular H–matrices. We already know that $\mathcal{H}_n \subset \mathcal{F}_n$. Let

$$\begin{aligned}
\Omega(A) &= \{B|M(B) = M(A)\},\\
\Omega^d(A) &= \{B|\ |b_{i,i}| = |a_{i,i}|,\ |b_{i,j}| \le |a_{i,j}|, i \ne j\},\\
\mathcal{F}_n^d &= \{A \in \mathcal{F}_n, \Omega^d(A) \subseteq \mathcal{F}_n\}.
\end{aligned}$$

Varga et al have shown that $\mathcal{H}_n = \mathcal{F}_n^d$. Moreover, if

$$\mathcal{F}_n^c = \{A \in \mathcal{F}_n,\ \Omega(A) \subseteq \mathcal{F}_n\},$$

then, $\mathcal{H}_n$ is strictly contained in $\mathcal{F}_n^c$ which is also strictly contained in $\mathcal{F}_n$. There exist matrices that can be incompletely factorized which are not H–matrices.

Let us now see what can be done for general symmetric positive definite matrices. This algorithm was proposed by Y. Robert [1105] following ideas from Jennings and Malik [822]. As before, we have

$$A = \begin{pmatrix} a_{1,1} & a_1^T \\ a_1 & B_1 \end{pmatrix}, \quad a_1 = b_1 - r_1.$$

However, the remainder is constructed in a different way as before. Let

$$R_1 = \begin{pmatrix} r_{1,1}^1 & r_1^T \\ r_1 & D_{R_1} \end{pmatrix},$$

with D_{R_1} diagonal such that $(D_{R_1})_{j,j} = |(r_1)_j|$ and $r_{1,1}^1 = \sum_{j=1}^{n-1} |(r_1)_j| > 0$. We split A as

$$A = A_1 = \begin{pmatrix} a_{1,1} + r_{1,1}^1 & b_1^T \\ b_1 & B_1 + D_{R_1} \end{pmatrix} - \begin{pmatrix} r_{1,1}^1 & r_1^T \\ r_1 & D_{R_1} \end{pmatrix} = M_1 - R_1.$$

Note that we add something to the (positive) diagonal element of A. Then we factorize M_1,

$$M_1 = \begin{pmatrix} 1 & 0 \\ l_1 & I \end{pmatrix} \begin{pmatrix} a_{1,1} + r_{1,1}^1 & 0 \\ 0 & A_2 \end{pmatrix} \begin{pmatrix} 1 & l_1^T \\ 0 & I \end{pmatrix}.$$

By equating,

$$l_1 = \frac{b_1}{a_{1,1} + r_{1,1}^1}, \quad A_2 = B_1 + D_{R_1} - \frac{1}{a_{1,1} + r_{1,1}^1} b_1 b_1^T.$$

We shall show that A_2 is positive definite.

Lemma 8.33

R_1 is semi positive definite.

Proof. R_1 is the sum of matrices such as

$$\begin{pmatrix} \begin{pmatrix} |r| & 0 & \dots & 0 & r \\ 0 & \ddots & & & 0 \\ \vdots & & \ddots & & \vdots \\ 0 & & & \ddots & 0 \\ r & 0 & \dots & 0 & |r| \end{pmatrix} & 0 \\ 0 & 0 \end{pmatrix},$$

with only four non–zero entries. The eigenvalues of this matrix are 0 and $2|r|$. Therefore, M_1 is semi positive definite. □

Theorem 8.34

A_2 is positive definite.

Proof. R_1 being semi positive definite, $M_1 = A + R_1$ is positive definite. Let y be a vector such that

$$y = \begin{pmatrix} \alpha \\ x \end{pmatrix},$$

$\alpha \in \mathbb{R}$. We have $(M_1 y, y) > 0$ which gives

$$\alpha^2(a_{1,1} + r^1_{1,1}) + 2\alpha(b_1, x) + ((B_1 + D_{R_1})x, x) > 0.$$

If we choose $\alpha = -\frac{(b_1,x)}{a_{1,1}+r^1_{1,1}}$, then, we have

$$-\frac{(b_1, x)^2}{a_{1,1} + r^1_{1,1}} + (x, (B_1 + D_{R_1})x) > 0,$$

which means that A_2 is positive definite. □

Note that in this decomposition R is positive definite. So the eigenvalues of $M^{-1}A$ are less than 1.

8.12. Different orderings for IC

IC has become one of the most popular preconditioners since it is relatively efficient on most problems and easy to implement. In their original work, Meijerink and Van der Vorst [959] only consider a factorization where the unknowns are numbered in a rowwise (or columnwise) fashion for finite difference approximations on rectangles. We have seen in Chapter 2 that for the direct method (Gaussian elimination), the issue of how the unknowns are ordered is of paramount importance. Some strategies as the reverse Cuthill–McKee, the minimum degree or nested dissection can reduce the work and storage for the factorization of the matrix. The effect of the ordering of the unknowns for incomplete factorizations has been studied experimentally by Duff and Meurant [480]. These experimental results have been explained theoretically to some extent by V. Eijkhout [506] and S. Doi [420], [425]. The ordering problem for incomplete factorizations is also related, as we shall see later on, to the search for parallelism. Experimental results for non–symmetric problems were given later on by Benzi, Szyld and Van Duin [138]. The conclusions are only slightly different than for symmetric problems.

8.12.1. Experimental results

Seventeen different orderings have been considered in [480]. Here, we simply report results from [480] for orderings we have already studied in Chapter 2. They are listed below together with the abbreviations used in the tables.

orderings	Abbreviation
row ordering	row
reverse Cuthill–McKee	rcm
minimum degree	mind
red black	rb
nested dissection	nest

The orderings are shown below for a 10×10 grid.

1	2	3	4	5	6	7	8	9	10
11	12	13	14	15	16	17	18	19	20
21	22	23	24	25	26	27	28	29	30
31	32	33	34	35	36	37	38	39	40
41	42	43	44	45	46	47	48	49	50
51	52	53	54	55	56	57	58	59	60
61	62	63	64	65	66	67	68	69	70
71	72	73	74	75	76	77	78	79	80
81	82	83	84	85	86	87	88	89	90
91	92	93	94	95	96	97	98	99	100

Row ordering

100	98	95	91	86	80	73	65	56	46
99	96	92	87	81	74	66	57	47	37
97	93	88	82	75	67	58	48	38	29
94	89	83	76	68	59	49	39	30	22
90	84	77	69	60	50	40	31	23	16
85	78	70	61	51	41	32	24	17	11
79	71	62	52	42	33	25	18	12	7
72	63	53	43	34	26	19	13	8	4
64	54	44	35	27	20	14	9	5	2
55	45	36	28	21	15	10	6	3	1

Reverse Cuthill–McKee ordering

4	5	52	21	20	86	19	45	7	3
6	28	53	51	77	85	59	58	27	8
75	74	44	80	60	43	84	42	62	61
22	50	81	41	82	83	40	76	49	23
18	54	95	68	39	87	69	38	91	17
93	96	37	94	88	36	92	89	35	90
16	57	72	34	99	70	33	79	67	15
46	56	32	73	31	97	78	30	65	55
9	26	64	48	100	29	71	66	25	11
2	10	63	24	14	98	13	47	12	1

Minimum degree ordering

1	51	2	52	3	53	4	54	5	55
56	6	57	7	58	8	59	9	60	10
11	61	12	62	13	63	14	64	15	65
66	16	67	17	68	18	69	19	70	20
21	71	22	72	23	73	24	74	25	75
76	26	77	27	78	28	79	29	80	30
31	81	32	82	33	83	34	84	35	85
86	36	87	37	88	38	89	39	90	40
41	91	42	92	43	93	44	94	45	95
96	46	97	47	98	48	99	49	100	50

Red–Black ordering

3	4	13	1	91	21	22	33	17	18
5	6	14	2	92	23	24	34	19	20
11	12	15	10	93	31	32	35	29	30
7	8	16	9	94	25	26	36	27	28
87	88	89	90	95	82	83	84	85	86
39	40	52	37	96	61	62	77	57	58
41	42	53	38	97	63	64	78	59	60
50	51	54	49	98	75	76	79	73	74
43	44	55	47	99	65	66	80	69	70
45	46	56	48	100	67	68	81	71	72

Nested dissection ordering

In the following tables we give the number of PCG iterations to reach $\|r^k\| \leq 10^{-6}\|r^0\|$, the number of modifications that would occur in the complete decomposition, the number of entries in the remainder R, the Frobenius norm of R and $\max_{i,j} |r_{i,j}|$. Note that the number of operations per iteration is the same for all orderings, therefore the number of iterations is a good measure of the relative merits of the orderings. All examples use a 30×30 grid, that is $n = 900$.

Results for the Poisson problem (Example 1)

ordering	# of iter.	# of modifs	# of elts in R	$\|R\|_F^2$	$\max_{i,j} \|r_{i,j}\|$
row	23	24389	841	142.5	0.293
rcm	23	16675	841	142.5	0.293
mind	39	7971	1582	467.3	0.541
rb	38	12853	1681	525.5	0.500
nest	25	15228	1012	157.1	0.293

Results for the anisotropic problem (Example 4)

ordering	# of iter.	# of modifs	# of elts in R	$\|R\|_F^2$	$\max_{i,j} \|r_{i,j}\|$
row	9	24389	841	$0.12\ 10^4$	0.87
rcm	9	16675	841	$0.12\ 10^4$	0.87
mind	48	7971	1582	$0.18\ 10^7$	49.51
rb	47	12853	1681	$0.21\ 10^7$	49.51
nest	26	15228	1012	$0.43\ 10^6$	49.51

Results for the Eist problem (Example 5)

ordering	# of iter.	# of modifs	# of elts in R	$\|R\|_F^2$	$\max_{i,j} \|r_{i,j}\|$
row	68	24389	841	$0.73\ 10^6$	38.03
rcm	69	16675	841	$0.69\ 10^6$	31.78
mind	108	7971	1582	$0.32\ 10^7$	62.34
rb	107	12853	1681	$0.35\ 10^7$	58.22
nest	83	15228	931	$0.87\ 10^6$	49.63

As one can see from these results, the number of conjugate gradient iterations is not related to the number of fill–ins we are dropping but it is almost directly related (for a given problem and a given factorization) to the norm of the remainder matrix R. The number of fill–ins is related to the structure of the matrix but the incomplete decomposition is dependent on the value of the entries. Some orderings like `mind` have very few fill–ins but a "large" matrix R and give a large number of iterations. Let us now consider the modified incomplete decomposition with no perturbation ($c = 0$).

Results for the Poisson problem (Example 1), MIC

ordering	# of iter.	# of modifs	# of elts in R	$\|R\|_F^2$	$\max_{i,j} \|r_{i,j}\|$
row	18	24389	1741	1010.	0.979
rcm	18	16675	1741	1010.	0.979
mind	>200	7971	2482	3568.	3.000
rb	>200	12853	2581	4143.	3.000
nest	38	15228	1912	1107.	1.666

We see that $\|R\|_F$ is much larger then for IC (although the number of iterations for the row ordering is smaller) and some of the orderings do not even converge if the modified factorization is used. Some other experiments were conducted keeping more fill–in. The first one uses a single level of fill–in. In the table, we add a column giving the number of entries in L.

Results for the Poisson problem (Example 1), IC, one level of fill–in

ordering	# iter.	# modifs	# elts in R	$\|R\|_F^2$	$\max_{i,j} \|r_{i,j}\|$	# elts in L
row	17	24389	1653	2.43	0.087	3481
rcm	17	16675	1653	2.43	0.087	3481
mind	23	7971	2467	38.81	2.509	4282
rb	16	12853	2016	16.47	0.090	4321
nest	19	15228	2187	35.34	0.173	3652

The second experiment uses a drop tolerance to keep or discard the fill–ins.

Results for the Poisson problem (Example 1), IC, tol=0.05

ordering	# iter.	# modifs	# elts in R	$\|R\|_F^2$	$\max_{i,j} \|r_{i,j}\|$	# elts in L
row	12	24389	1595	4.26	0.039	4293
rcm	10	16675	1540	2.846	0.041	4293
mind	10	7971	1657	2.285	0.049	5531
rb	8	12853	1484	1.683	0.042	4699
nest	12	15228	2622	3.890	0.049	5574

For the factorizations using one level of fill–in, the reduction in the number of iterations does not quite compensate for the extra work for each iteration. The reverse is true of the drop tolerance results where the greater reduction in iterations more than compensates the increased storage and work. An interesting feature is that the relative performance of the different ordering schemes has changed. For example, mind and rb do much better when some fill–in is allowed in L. A reason for this is that many of the first level fill–ins for these orderings are quite large, unlike row where the fill–ins rapidly decrease in value. For the complete factorization, this can be explained in the following way.

Theorem 8.35

In the complete Cholesky factorization, the Frobenius norm of the Cholesky factors is invariant under the choice of the ordering.

Proof. Suppose we have

$$A = LDL^T.$$

Finding another ordering is computing a permutation matrix P, a permuted matrix $A_P = PAP^T$ and a factorization $A_P = L_P D_P L_P^T$. The Frobenius norm of A is

$$\|A\|_F = \left(\sum_{i,j} a_{i,j}^2 \right)^{\frac{1}{2}}.$$

For any matrix B, we have

$$\|B\|_F^2 = trace(B^T B) = trace(BB^T).$$

Therefore,

$$\|L\sqrt{D}\|_F^2 = trace(LDL^T) = trace(A),$$

but, since P is a permutation matrix, $\|PAP^T\|_F = \|A\|_F$. Hence,

$$\|L_P\sqrt{D_P}\|_F = \|L\sqrt{D}\|_F = \sqrt{trace(A)}.$$

□

This result shows that if we have many fill–ins, most of them must be small. On the other hand, if we have a small number of fill–ins, the maximum of their absolute values is large. This was illustrated in the experiments by `rb` and `mind`. We have, in fact, a stronger result as we can consider the diagonal entries of A and A_P,

$$a_{i,i} = \sum_{j, l_{i,j} \neq 0} d_{j,j} l_{i,j}^2.$$

There is an index k for which $(a_P)_{k,k} = a_{i,i}$. Therefore

$$\sum_{j, l_{i,j} \neq 0} d_{j,j} l_{i,j}^2 = \sum_{j, (l_P)_{k,j} \neq 0} (d_P)_{j,j} (l_P)_{k,j}^2.$$

For instance, for the Poisson model problem, we have

$$\sum_{j, l_{i,j} \neq 0} d_{j,j} l_{i,j}^2 = \sum_{j, (l_P)_{k,j} \neq 0} (d_P)_{j,j} (l_P)_{k,j}^2 = 4, \forall i.$$

For the incomplete decompositions $M = LDL^T$ is the complete decomposition of $A + R$. Therefore $\|L\sqrt{D}\|_F^2 = trace(A + R)$. But for IC we have that $r_{i,i} = 0$, hence $\|L\sqrt{D}\|_F^2$ is invariant under permutations and the same conclusion holds as for complete decompositions.

8.12.2. Theory for model problems

To explain the experimental results of Duff and Meurant, Eijkhout [506] considered the model problem

$$-a\frac{\partial^2 u}{\partial^2 x} - b\frac{\partial^2 u}{\partial^2 y} = f,$$

in a square with Dirichlet boundary conditions. The five point finite difference stencil is

$$\begin{matrix} & -b & \\ -a & 2(a+b) & -a\,. \\ & -b & \end{matrix}$$

For the natural `row` ordering the diagonal coefficients satisfy

$$d_i = 2(a+b) - \frac{a^2}{d_{i-1}} - \frac{b^2}{d_{i-m}},$$

the limit value being $d = a + b + \sqrt{2ab}$. The fill–ins generated are of size ab/d_i, they converge to the value

$$\frac{a^2 b + ab^2 - ab\sqrt{2ab}}{a^2 + b^2}.$$

If one of the coefficients is much smaller than the other, the fill–in is of the order of the smallest coefficient. With the Red–Black `rb` ordering, the row sum of R is

$$\frac{a^2 + b^2 + 4ab}{2(a+b)}.$$

When one coefficient goes to infinity, so does the rowsum showing that this ordering is sensitive to anisotropy. Eijkhout [506] proved the following result.

Theorem 8.36

$\|R\|_{rb} \geq C\|R\|_{row}$,

where $C = (1 + b/a)(1 + a/b)$.

□

Therefore, the Red–Black ordering will give large fill–ins for problems with strong anisotropy. More generally, Eijkhout considered the following. Let a_i^+ (resp. a_i^-) be defined by $a_i^+ = \max\{-a_{i,i-1}, -a_{i,i+1}\}$ (resp. max replaced by min). Similarly let $b_i^+ = \max\{-a_{i,i-m}, -a_{i,i+m}\}$ with b_i^- defined accordingly.

Theorem 8.37

Eliminating a node i in the graph with at least one uneliminated neighbor in both directions generates a fill–in f of size

$$f \in \left[\frac{a_i^- b_i^-}{a_i^- + a_i^+ + b_i^- + b_i^+}, \frac{a_i^+ b_i^+}{a_i^- + b_i^-}\right].$$

Eliminating a node i with two uneliminated neighbors in the vertical direction gives

$$f \in \left[\frac{b_i^+ b_i^-}{a_i^- + a_i^+ + b_i^- + b_i^+}, \frac{b_i^- b_i^+}{b_i^- + b_i^-}\right].$$

Eliminating a node i with two uneliminated neighbors in the horizontal direction gives

$$f \in \left[\frac{a_i^- a_i^+}{a_i^- + a_i^+ + b_i^- + b_i^+}, \frac{a_i^- a_i^+}{a_i^- + a_i^+}\right].$$

□

These results lead Eijkhout to the conclusion that there is a difference between orderings that eliminate nodes between two uneliminated nodes and those that don't. This is formalized by the following definition: a node is naturally ordered if it does not have two neighbors in one direction that both have a larger number. If all nodes are naturally ordered then the ordering is said to be generalized naturally ordered. The natural ordering and the ordering by diagonals are generalized naturally ordered. From what we have seen it is clear that orderings that are generalized naturally ordered give a remainder matrix R with a smaller Frobenius norm than orderings which are not.

This problem of explaining the difference of behaviour of different orderings was also considered by Doi and Lichnewsky [425]. Let P be a permutation matrix. Changing the ordering is equivalent to considering $\tilde{A} = PAP^T$. Let M be the `IC` factorization of A and $\tilde{M}$ the one for $\tilde{A}$. We say that two orderings are equivalent if $\tilde{M} = PMP^T$. Then, the `PCG` number of iterations is the same for both orderings. For instance, it can be shown that `row` and `rcm` are equivalent explaining why we found the same number of iterations in the numerical experiments.

If A is an M–matrix and $M = A+R$, it can be shown that denoting by ρ the spectral radius of $M^{-1}R$, $\lambda_{min}(M^{-1}A) = 1-\rho$ and $1 < \lambda_{max}(M^{-1}A) < 1+\rho$. Therefore,

$$\frac{1}{1-\rho} < \kappa(M^{-1}A) < \frac{1+\rho}{1-\rho}.$$

Hence, the smaller ρ is, the better is the condition number. A smaller R gives a better preconditioner. Doi and Lichnewsky considered the model problem:

$$\frac{\partial}{\partial x}\left(k_x \frac{\partial u}{\partial x}\right) + \frac{\partial}{\partial y}\left(k_y \frac{\partial u}{\partial y}\right) + v_x \frac{\partial u}{\partial x} + v_y \frac{\partial u}{\partial y} = 0.$$

They computed the diagonal and remainder elements for a few orderings as a function of the parameters k_x, k_y, v_x, v_y and provided numerical experiments with non–symmetric problems that complemented those of Duff and Meurant [480]. Doi and Lichnewsky got conclusions similar to Eijkhout's. Then, they introduced the definition of "compatibility". An ordering is said compatible if $R \to 0$ as any of the parameters $k_x, k_y, |v_x|, |v_y|$ go to infinity. It is said partially compatible if $R \to 0$ if either $k_x, |v_x|$ or $k_y, |v_y|$ goes to infinity. The `row` ordering is compatible as well as `cm` and `rcm`. The orderings `nest`, `mind` and `rb` are incompatible. The following theorems show how to characterize compatibility or partial compatibility by looking at the digraph of L in the incomplete factorization assuming that the factorization has succeeded, that is $d_{i,i} > 0$.

Theorem 8.38

An ordering is compatible if and only if the digraph of L does not have the patterns shown in figure 8.15. This means that there is no node with four neighbors or two neighbors on the horizontal or vertical axis with larger numbers.

□

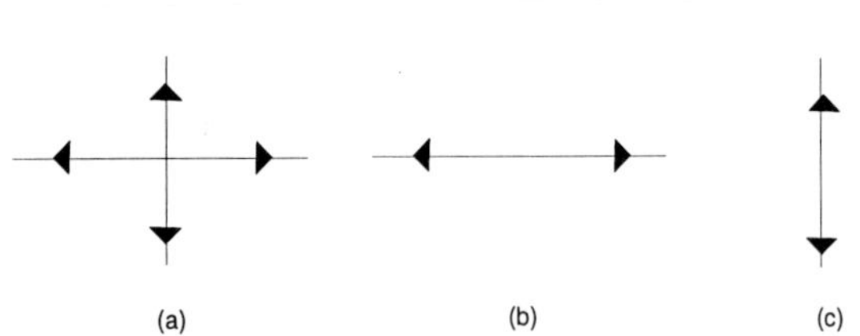

fig. 8.15: Graphs which violate compatibility

Theorem 8.39

An ordering is partially compatible if and only if the digraph of L does not have the patterns (b) and (c) shown in figure 8.15. This means that there is no node with two neighbors on the horizontal or vertical axis with larger numbers.

□

A compatible ordering is evidently partially compatible and partial compatibility is equivalent to being naturally ordered. As a rule of thumb, the less incompatible nodes there are in an ordering, the better are the results of PCG with IC.

8.12.3. Value dependent orderings

A characteristic of the orderings studied in the previous sections is that the values of the entries of A (or L) have no influence on the numbering of the nodes of the graph of A. D'Azevedo, Forsyth and W–P. Tang [367] developed an algorithm for constructing an ordering for general sparse matrices that reduces the discarded fill–in during the incomplete factorization. This is called the Minimum Discarded Fill (MDF) algorithm.

When we perform a step of IC, we would like the discarded matrix R_k to be as small as possible. We shall try to find an ordering which minimizes $\|R_k\|_F$. At each step k of the incomplete elimination for a node v_m we define

$$discard(v_m)^2 = \sum_{(v_i,v_j)\in\mathcal{F}} \left(\frac{a_{i,m}^{(k-1)} a_{m,j}^{(k-1)}}{a_{m,m}^{(k-1)}} \right)^2,$$

where

$$\mathcal{F} = \{(v_i, v_j) | (v_i, v_j) \notin E^{k-1}, (v_i, v_m) \in E^{k-1}, (v_m, v_j) \in E^{k-1}\},$$

where E^{k-1} is the set of edges of the elimination graph at step $k-1$. In the incomplete MDF factorization the condition $lev(a_{i,j}^{(k)}) > l$ is added to the definition of $\mathcal{F}$. The resulting algorithm is denoted as MDF(l). The node that is labeled next is one that has a minimum $discard(v_m)$. There are different ways for tie–breaking. For instance, one can choose the node giving the smallest number of new fill–ins.

The data structures to be used are the same as those for general sparse Gaussian elimination we saw in Chapter 2. The main drawback of MDF is that the ordering phase is quite costly, even though *discard* is not recomputed but only updated when a node is eliminated. Based on numerical experiments in [367], MDL(l) seems to be relatively efficient for small values of l, typically $l = 1$ or $l = 3$. However, we have seen that keeping the fill–ins on the basis of level is not as efficient as algorithms based on dropping by looking at the values of the coefficients and using for instance mind as the ordering, therefore it remains to be seen if MDF is competitive with these variants of IC. Attempts have been made to combine drop tolerances and MDF, see [365].

8.12.4. Multicolor orderings

In Duff and Meurant [480] the multicolor orderings that were considered were the Red–Black ordering and a four–color ordering that is also not compatible. As predicted by the theory, the numerical experiments show poor results when using the four–color ordering with IC(0) or MIC(0). Multicolor orderings are those where a node in the graph of the matrix is only connected to nodes of other colors. Multicolor orderings with a large number of colors were considered in Doi and Hoshi [423] and Fujino and Doi [593]. In earlier studies like the one by Poole and Ortega [1075] only a small number of colors was considered. The conclusion was that the convergence was slowed down. However, Doi and Hoshi considered large 3D problems (for instance using half a million mesh points). From the numerical experiments they concluded that using a large number of colors (say 75) is better than only using a small number (say 5). The explanation for this phenomenon is that, with a large number of colors, the ratio of incompatibility, which is defined as the number of incompatible nodes divided by the total number of nodes, is smaller than with a small number of colors. However, we need large problems to be able to use a large number of colors.

8.13. The repeated Red–Black decomposition

In this section, we are going to study a family of methods (first introduced by Brand [191]) where we keep more storage than in the structure of A. We shall show that for finite difference approximations of elliptic problems in rectangles this leads to very efficient preconditioners. Most of what follows is taken from the Ph.D. thesis of P. Ciarlet Jr. [326], [329].

8.13.1. Description of the methods

These methods are based on an ordering of the unknowns (or the mesh points) that is derived from the classical Red–Black ordering. In this ordering, the Red points and then the Black points are labeled sequentially. This is no longer true in the Repeated Red–Black (RRB) ordering which is based on a recursive process. At a given step k the set of nodes which have not been labeled yet is split in two halves R^k and B^k. The nodes in R^k are labeled sequentially and the process is reiterated on B^k. If the total number of nodes n is of the form $n = 2^l$, then the last set has only one element. But if this is not the case or if we choose to do so, we can stop the process after K steps and label the remaining nodes sequentially.

The splitting of the nodes occurs alternately along diagonals and parallels to the x–axis. An example is given in figures 8.16 and 8.17.

Now, we produce (eventually modified) incomplete factorizations using these orderings in the following way. It is actually easier to explain this factorization using graphs. At a given step, the nodes of R^k are removed creating some fill–in as shown in figure 8.18 (a). Of these fill–ins, we only keep those corresponding to the structure given in figure 8.18 (b).

Then, we go on to the end for the complete process or we stop after step K. In the latter case, a complete factorization is used for the last step. Clearly, the entries that are kept correspond to a (twisted) five point scheme on B^k. This preconditioner is denoted by $\mathtt{RRB}_1$. If we want a modified version, we

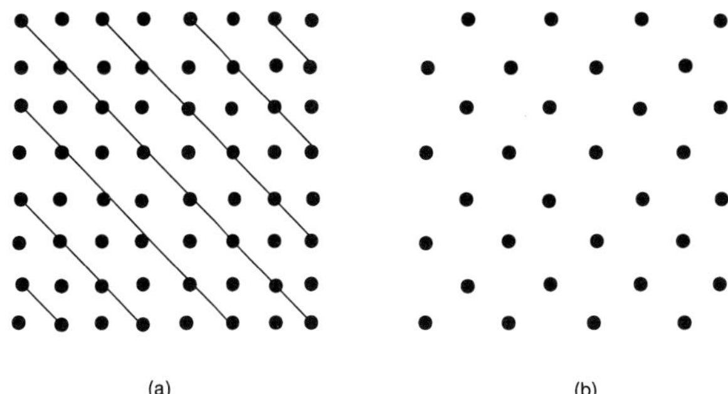

fig. 8.16: (a) extraction of R^1 nodes, (b): B^1 nodes

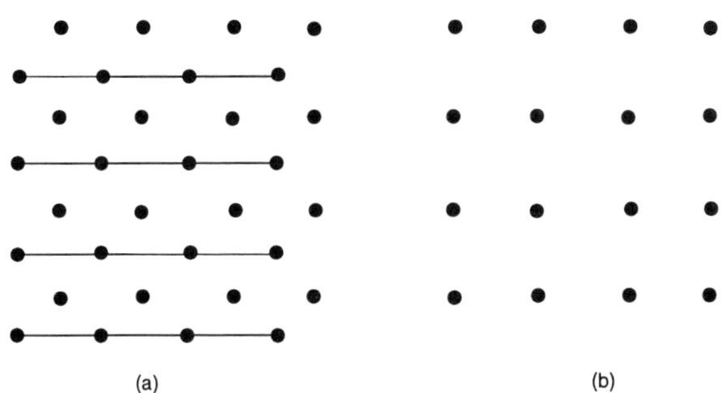

fig. 8.17: (a): extraction of R^2 nodes, (b): B^2 nodes

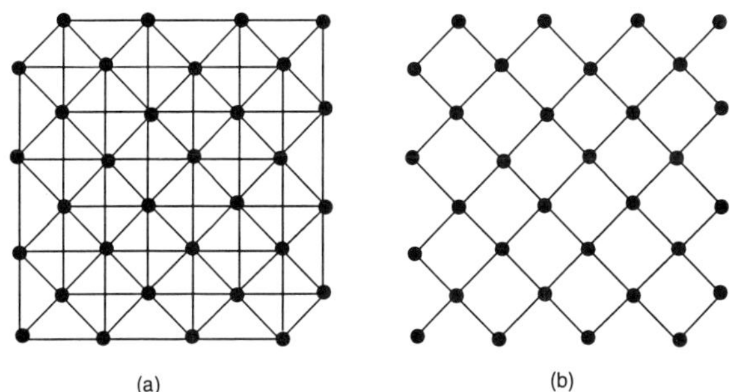

fig. 8.18: The fill–in that is (a): generated and (b): kept

just add the neglected fill–ins at each step to the diagonal giving $\mathtt{RRB}_1^m$. The second preconditioner is the one used by Brand [191]. The difference is in the structure we keep at each step which is shown in figure 8.19.

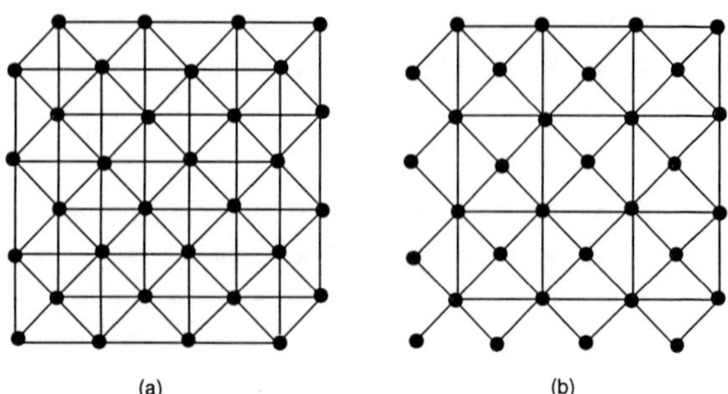

fig. 8.19: The fill–in that is (a): generated and (b): kept Brand preconditioner

Edges (fill–ins) are kept if they correspond to a five point scheme in $B^{k'}$ for $k' > k$. We denote this by $\mathtt{RRB}_2$ and $\mathtt{RRB}_2^m$ for the modified version. It turns out that these four preconditioners give matrices that have the same structure. The positions of the extra fill–ins that have been kept in $\mathtt{RRB}_2$ are reintroduced in the next steps of $\mathtt{RRB}_1$ (the value of the coefficients being of course different).

Let us consider first a fixed value of K. For example, if we choose $K = 4$, we have to store the following number of entries, see [329],

n	non–zeros	non–zeros/n
256	1182	4.62
1024	5178	5.06
4096	23154	5.65
16384	109794	6.70

The ratio of non–zeros entries divided by n is not bounded as the last matrix is factorized exactly. We could have lowered the number of non–zeros a little by using a different ordering for the nodes remaining in the last step. However, there are other ways to obtain a smaller storage. Therefore, we look for a varying K with the dimension of the problem to have a storage proportional to n and to have the number of iterations only slowly growing.

Lemma 8.40

When eliminating the nodes of R^k, the number of fill–ins generated in M is less that $5n$.

Proof. Two nodes are adjacent in the graph of M if they belong to a five point scheme in one of the steps. Since the number of nodes in $\cup_k R^k$ is less than n, we get the result. □

Lemma 8.41

During the complete factorization at step K, the number of fill–ins is of the order of $2^{-\frac{3K}{2}} n^{\frac{3}{2}}$.

Proof. The number of nodes in B^K is $2^{-K}n$. The semi–bandwith is $2^{-\frac{K}{n}}\sqrt{n}$. Since the profile fills completely, we get the result. □

Theorem 8.42

If K is $\lfloor \frac{1}{3}\log_2(n) + \frac{4}{3} \rfloor$, then the number of non–zero entries in M is at most $6n$.

Proof. We use the two preceding lemmas that show that the number of non–zero entries is less than $5.5n$. □

Note that the number of floating point operations to construct the preconditioner is $n^{\frac{4}{3}}$. With this choice of K, the storage is now the following

n	K	non–zeros	non–zeros/n
256	4	1182	4.62
1024	4	5178	5.06
4096	5	21424	5.23
16384	6	84083	5.13

8.13.2. Analysis of RRB

We would like to establish bounds on the condition number of $M^{-1}A$. For this, we use the following perturbation result.

Lemma 8.43

Let $M_i, i = 1, 2, 3$ be three symmetric positive definite matrices. Then, $\kappa(M_1^{-1}M_3) \le \kappa(M_1^{-1}M_2)\kappa(M_2^{-1}M_3)$.

Proof. Note that

$$\lambda_{min}(M_1^{-1}M_3) = \min_{x\neq 0} \frac{(M_3x, x)}{(M_1x, x)},$$
$$\lambda_{max}(M_1^{-1}M_3) = \max_{x\neq 0} \frac{(M_3x, x)}{(M_1x, x)}.$$

Since $M_1^{-1}M_3 = M_1^{-1}M_2M_2^{-1}M_3$, we obtain the result. □

If the nodes of the Poisson model problem are ordered with a Red–Black ordering, we have blockwise,

$$A = \begin{pmatrix} 4I & -L^T \\ -L & 4I \end{pmatrix},$$

where L corresponds to the stencil

$$\begin{matrix} & 1 & \\ 1 & 0 & 1 \\ & 1 & \end{matrix}$$

except near the boundary. If we remove the red nodes (in R^1), we factorize A as

$$A = \begin{pmatrix} I & \\ -\frac{1}{4}L & I \end{pmatrix} \begin{pmatrix} 4I & \\ & S \end{pmatrix} \begin{pmatrix} I & -\frac{1}{4}L^T \\ & I \end{pmatrix}.$$

Similarly, for the modified factorizations, we have

$$M_i^1 = \begin{pmatrix} I & \\ -\frac{1}{4}L & I \end{pmatrix} \begin{pmatrix} 4I & \\ & S_i^1 \end{pmatrix} \begin{pmatrix} I & -\frac{1}{4}L^T \\ & I \end{pmatrix},$$

where the upper index corresponds to the value of K. Let $B^1 = R^2 \cup B^2$. The matrix S_2^1 corresponds to a nine point operator in B^2 and a five point operator in R^2 like S and S_1^1:

$$\begin{pmatrix} & & -\frac{1}{4} & & \\ & -\frac{1}{2} & & -\frac{1}{2} & \\ -\frac{1}{4} & & 3 & & -\frac{1}{4} \\ & -\frac{1}{2} & & -\frac{1}{2} & \\ & & -\frac{1}{4} & & \end{pmatrix}, \quad \begin{pmatrix} -\frac{1}{2} & & -\frac{1}{2} \\ & 2 & \\ -\frac{1}{2} & & -\frac{1}{2} \end{pmatrix}$$

except near the boundary. Then, S, S_1^1 and S_2^1 are defined on B^1,

$$S = \begin{pmatrix} 3I - \frac{1}{4}N & -\frac{1}{2}M^T \\ -\frac{1}{2}M & 3I - \frac{1}{4}N \end{pmatrix}, \quad S_1^1 = \begin{pmatrix} 2I & -\frac{1}{2}M^T \\ -\frac{1}{2}M & 2I \end{pmatrix},$$

$$S_2^1 = \begin{pmatrix} 2I & -\frac{1}{2}M^T \\ -\frac{1}{2}M & 3I - \frac{1}{4}N \end{pmatrix},$$

where N and M correspond to

$$\begin{array}{ccccc} & & 1 & & \\ & & 0 & & \\ 1 & 0 & 0 & 0 & 1, \\ & & 0 & & \\ & & 1 & & \end{array} \qquad \begin{array}{ccc} 1 & 0 & 1 \\ 0 & 0 & 0. \\ 1 & 0 & 1 \end{array}$$

Then,

$$M_i^2 = \begin{pmatrix} I & \\ -\frac{1}{4}L & I \end{pmatrix} \begin{pmatrix} 4I & \\ & S_i^2 \end{pmatrix} \begin{pmatrix} I & -\frac{1}{4}L^T \\ & I \end{pmatrix}.$$

We have

$$S_1^1 = \begin{pmatrix} I & \\ -\frac{1}{4}M & I \end{pmatrix} \begin{pmatrix} 2I & \\ & 2I - \frac{1}{8}MM^T \end{pmatrix} \begin{pmatrix} I & -\frac{1}{4}M^T \\ & I \end{pmatrix},$$

$$\begin{aligned} S_1^2 &= \begin{pmatrix} I & \\ -\frac{1}{4}M & I \end{pmatrix} \begin{pmatrix} 2I & \\ & I - \frac{1}{4}N \end{pmatrix} \begin{pmatrix} 2I & -\frac{1}{4}M^T \\ & I \end{pmatrix} \\ &= \begin{pmatrix} 2I & -\frac{1}{2}M^T \\ -\frac{1}{2}M & I - \frac{1}{4}N + \frac{1}{8}MM^T \end{pmatrix}. \end{aligned}$$

Moreover,

$$\begin{aligned} S_2^2 &= \begin{pmatrix} I & \\ -\frac{1}{4}M & I \end{pmatrix} \begin{pmatrix} 2I & \\ & 2I - \frac{1}{2}N \end{pmatrix} \begin{pmatrix} I & -\frac{1}{4}M^T \\ & I \end{pmatrix} \\ &= \begin{pmatrix} 2I & -\frac{1}{2}M^T \\ -\frac{1}{2}M & 2I - \frac{1}{2}N + \frac{1}{8}MM^T \end{pmatrix}. \end{aligned}$$

S_2^2 corresponds to a five point operator on B^2,

$$\begin{array}{ccc} & -\frac{1}{2} & \\ -\frac{1}{2} & 2 & -\frac{1}{2} \\ & -\frac{1}{2} & \end{array}$$

except near the boundary. The idea is to study the eigenvalues of $[S_i^2]^{-1}(S - S_i^2)$ since $[S_i^2]^{-1}S = I + [S_i^2]^{-1}(S - S_i^2)$ and $S - S_i^2$ are block diagonal matrices. Let $x = (x_r, x_b)^T$ be an eigenvector, $(S - S_i^2)x = \lambda S_i^2 x$. Then by eliminating x_r, we obtain for the first preconditioner that $(2I - \frac{1}{8}MM^T)x_b$ is equal to

$$\lambda \left\{ -\frac{\lambda}{4} M([2\lambda - 1]I + \tfrac{1}{4}N)^{-1}M^T + (I - \tfrac{1}{4}N + \tfrac{1}{8}MM^T) \right\} x_b. \quad (8.6)$$

Both matrices are similar to five point operators, so it is easy to find out the eigenvectors and this leads to the following result.

Lemma 8.44

$$\kappa([M_1^2]^{-1}A) \le 4, \quad \kappa([M_1^{k+2}]^{-1}M_1^k) \le 4.$$

Proof. The proof is obtained by showing that the solutions λ of (8.6) are such that $0 < \lambda < 3$, see Ciarlet,Jr [329]. □

For the second preconditioner, we can show that $0 < \lambda < 1$. Therefore $\kappa([M_2^2]^{-1}A) \le 2$, $\kappa([M_2^{k+2}]^{-1}M_2^k) \le 2$.

Theorem 8.45

For the Poisson model problem with $h = 1/(m+1)$ *and choosing* $K = \lfloor \frac{1}{3}\log_2(n) + \frac{4}{3} \rfloor$, *we have*

$$\kappa([M_1^K]^{-1}A) \le 3m^{\frac{2}{3}}, \quad \kappa([M_2^K]^{-1}A) \le 2m^{\frac{1}{3}}.$$

Proof. We simply use lemmas 8.43 and 8.44 with the given value of K. □

The second preconditioner gives an interesting example where, keeping a small ($< 6n$) fill–in, we can reach a condition number of $O(h^{-\frac{1}{3}})$. Therefore, for practical problems dimensions, we see almost no dependence of h on the number of PCG iterations. Even, if the proof is only obtained for the model problem, numerical examples show that the same is true for more general problems, see Ciarlet,Jr [326].

8.14. The block incomplete Cholesky decomposition

8.14.1. Block tridiagonal matrices

Let us introduce a block decomposition of block tridiagonal matrices. We shall take advantage of the special structure of the matrix to derive efficient preconditioners. Of course, they are less general than the point incomplete Cholesky preconditioners which can be used for any sparse matrix. Let

$$A = \begin{pmatrix} D_1 & A_2^T & & & \\ A_2 & D_2 & A_3^T & & \\ & \ddots & \ddots & \ddots & \\ & & A_{m-1} & D_{m-1} & A_m^T \\ & & & A_m & D_m \end{pmatrix},$$

each block being of order m, $n = m^2$. Denote by L the block lower triangular part of A. In Chapter 2, we saw that the block (complete) factorization of A can be written as

$$A = (\Sigma + L)\Sigma^{-1}(\Sigma + L^T),$$

where Σ is a block diagonal matrix whose diagonal blocks are denoted by Σ_i. By inspection, we have

$$\Sigma_1 = D_1, \quad \Sigma_i = D_i - A_i(\Sigma_{i-1})^{-1}A_i^T, \; i = 2, \ldots, m$$

For instance, for finite difference approximations of elliptic (or parabolic) problems, the matrices D_i are tridiagonal and the matrices A_i are diagonal. This implies that Σ_1 is tridiagonal, but all the other matrices $\Sigma_i, i = 2, \ldots, m$ are dense. The idea to obtain an incomplete decomposition is to replace the inverses by sparse approximations. We set

$$M = (\Delta + L)\Delta^{-1}(\Delta + L^T),$$

where Δ is a block diagonal matrix whose elements are computed generically as

$$\Delta_1 = D_1,$$
$$\Delta_i = D_i - A_i \operatorname{approx}(\Delta_{i-1}^{-1})A_i^T,$$

where approx(Δ_{i-1}^{-1}) is a sparse approximation of Δ_{i-1}^{-1}. This preconditioner was first proposed in a slightly different form by R. Underwood [1272] and then generalized by Concus, Golub and Meurant [342]. There are many ways to define approx(Δ_{i-1}^{-1}), see [342]. One of the most efficient, denoted by INV, just considers tridiagonal approximations of the inverses. We denote by tridiag(B) the tridiagonal matrix whose non–zero entries are the same as the corresponding ones of B. Then, Δ is defined as

$$\Delta_i = D_i - A_i \text{tridiag}(\Delta_{i-1}^{-1}) A_i^T.$$

This choice was motivated by the case where the D_i's are tridiagonal. If A is a diagonally dominant L–matrix, then it is easy to show that the entries of the inverses of the Δ_i's decay away from the diagonal on a row. Therefore, the absolute values of the entries that we throw away are smaller than the ones we keep. This preconditioner has been used with great success in many different areas where block tridiagonal matrices arise. Note that to store the preconditioner we need only two n vectors for the Cholesky factorizations of all the Δ_i's.

8.14.2. Pointwise equivalent decomposition

When they are feasible, block decompositions for block tridiagonal matrices are usually more efficient than point incomplete decompositions using the same amount of storage. An explanation of this can be obtained by looking at the equivalent point decomposition of INV. Let

$$\Delta_i = L_i L_i^T,$$

be the Cholesky factorization of Δ_i, L_i being a lower bidiagonal matrix and $W_i = A_i L_{i-1}^{-T}$. Then, the block decomposition can be written as

$$M = \begin{pmatrix} L_1 & & & & \\ W_2 & L_2 & & & \\ & \ddots & \ddots & & \\ & & W_{m-1} & L_{m-1} & \\ & & & W_m & L_m \end{pmatrix} \begin{pmatrix} L^T & W_2^T & & & \\ & L_2^T & W_3^T & & \\ & & \ddots & \ddots & \\ & & & L_{m-1}^T & W_m^T \\ & & & & L_m^T \end{pmatrix}$$

The matrices W_i are upper triangular, the upper triangle elements being non–zero. Then, the non–zero structure of the Cholesky factor is the one shown on figure 8.20.

Of course, this equivalent point factorization is never computed, but we see that in the block factorization we implicitly keep more fill–in than in the point

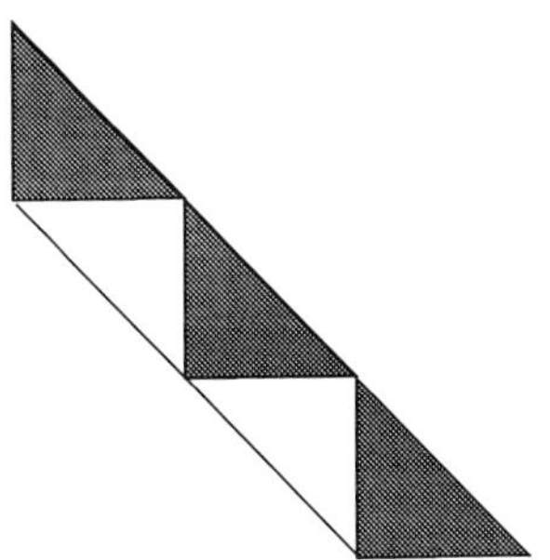

fig. 8.20: The implicit non–zero structure of INV

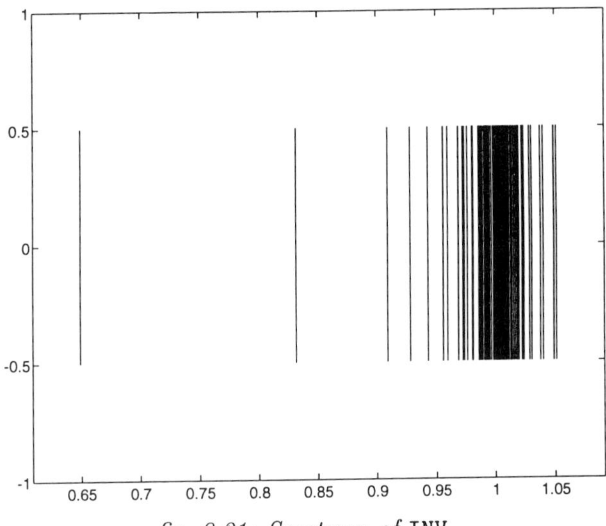

fig. 8.21: Spectrum of INV

decomposition using the same storage. Of course, we have to solve tridiagonal systems to apply the preconditioner. As an example, figure 8.21 shows the spectra for the model problem on a 10×10 mesh using INV (compare with figure 8.6).

8.14.3. The modified incomplete block decomposition

As for their point counterpart, the block preconditioners can also be modified such that, for instance, $rowsum(R) = 0$. The remainder R is easily computed,

$$R_1 = 0,$$
$$R_i = \Delta_i - D_i + A_i \Delta_{i-1}^{-1} A_i^T = A_i(\Delta_{i-1}^{-1} - \text{tridiag}(\Delta_{i-1}^{-1}))A_i^T.$$

Therefore, the rowsums of the remainder can be very easily computed and we can modify the block decomposition in the following way,

$$\Delta_1 = D_1,$$
$$\Delta_i = D_i - A_i \, \text{tridiag}(\Delta_{i-1}^{-1}) A_i^T - \mathit{rowsum}(A_i(\Delta_{i-1}^{-1} - \text{tridiag}(\Delta_{i-1}^{-1}))A_i^T).$$

This preconditioner is known as `MINV`. Note that to compute the rowsums, we only need to solve tridiagonal systems. Of course, we can also define a relaxed version where the rowsum is multiplied by a parameter ω.

8.14.4. Block incomplete decomposition for H–matrices

As for point factorizations, it is interesting to know for which classes of matrices the block decompositions are feasible. The only thing to check is that the matrices Δ_i are non–singular. It was shown in Concus, Golub and Meurant [342] that `INV` and `MINV` are well defined for M–matrices. It turns out that these results can be extended to H–matrices.

8.14.5. Generalization of the block incomplete decomposition

The idea of block incomplete Cholesky decomposition that was developed in the previous sections for block tridiagonal matrices can be straightforwardly generalized to any block structure. This was proposed by Axelsson [65]. However, there are not that many practical situations where this has been used. For general sparse matrices, it is not so easy to exhibit a natural block structure except if the ordering has been devised to do so, as in domain decomposition methods that we shall study in Chapter 10.

8.14.6. Fourier analysis of INV and MINV

In the special case we consider in this section (Poisson model problem with periodic boundary conditions), we shall use tridiagonal circulant matrices. To derive these approximations, we need a few preliminary results, see Chan and Meurant [288].

Theorem 8.46

Let $\gamma > 2$ and

$$S = \begin{pmatrix} \gamma & -1 & & & -1 \\ -1 & \gamma & -1 & & \\ & \ddots & \ddots & \ddots & \\ & & -1 & \gamma & -1 \\ -1 & & & -1 & \gamma \end{pmatrix},$$

be a symmetric tridiagonal circulant matrix. Then

a)

$$S = L_S \, D_S^{-1} \, L_S^T,$$

where

$$D_S = \begin{pmatrix} d & & & \\ & d & & \\ & & \ddots & \\ & & & d \end{pmatrix}, \quad L_S = \begin{pmatrix} d & & & -1 \\ -1 & d & & \\ & \ddots & \ddots & \\ & & -1 & d \end{pmatrix},$$

with $d = \frac{\gamma + \sqrt{\gamma^2 - 4}}{2}$.

b) S^{-1} *is a symmetric circulant matrix.*

c) The elements of the main diagonal of S^{-1} *are all the same and equal to* $\chi = \frac{d}{d^2-1}\left[\frac{d^n+1}{d^n-1}\right]$.

The second diagonal of S^{-1} *is given by* $\xi = \frac{1}{2}(\gamma\chi - 1) = \frac{1}{d^2-1}\left[1 + \gamma\frac{d}{d^n-1}\right]$.

Proof. a) Since

$$S = L_S \, D_S^{-1} \, L_S^T = \begin{pmatrix} d + \frac{1}{d} & -1 & & & -1 \\ -1 & d + \frac{1}{d} & -1 & & \\ & \ddots & \ddots & \ddots & \\ & & -1 & d + \frac{1}{d} & -1 \\ -1 & & & -1 & d + \frac{1}{d} \end{pmatrix},$$

d satisfies the equation:

$$d + \frac{1}{d} = \gamma,$$

or $d^2 - \gamma d + 1 = 0$, the positive root of which is:

$$d = \frac{\gamma + \sqrt{\gamma^2 - 4}}{2}.$$

b) It is well known that the inverse of a symmetric circulant matrix is also symmetric circulant.

c) We must determine some elements of S^{-1}. The easiest way is to first compute L_S^{-1}, where

$$L_S = \begin{pmatrix} d & & & -1 \\ -1 & d & & \\ & \ddots & \ddots & \\ & & -1 & d \end{pmatrix}.$$

Let us solve $L_S x = (1, 0, \ldots, 0)^T$ to compute the first column of L_S^{-1}. It is easily seen that $x_n = x_1/d^{n-1}$; since $dx_1 - x_n = 1$, we have

$$x_1 = \frac{d^{n-1}}{d^n - 1} \quad \text{and} \quad x_j = \frac{d^{n-j}}{d^n - 1}, \quad j = 2, \ldots, n.$$

More generally, to obtain the ith column of L_S^{-1}, we solve

$$L_S x = (0, 0, \ldots, 0, 1, 0, \ldots, 0)^T,$$

where the 1 is in the ith position.

As before, it can be seen that

$$x_{i-1} = \frac{1}{d^{i-1}} x_n \quad \text{and} \quad x_n = \frac{1}{d^{n-i}} x_i,$$

so $x_{i-1} = x_i/d^{n-1}$; as $dx_i - x_{i-1} = 1$, we have $x_{i-1} = 1/(d^n - 1)$. Therefore

$$x_i = \frac{d^{n-1}}{d^n - 1}, \quad x_{i+j} = \frac{d^{n-j-1}}{d^n - 1}, \quad x_{i-j} = \frac{d^j}{d^n - 1}.$$

Hence, we explicitly know the inverse of L_S,

$$L_S^{-1} = \frac{1}{d^n - 1} \begin{pmatrix} d^{n-1} & 1 & d & \ldots & d^{n-2} \\ d^{n-2} & d^{n-1} & 1 & \ldots & d^{n-3} \\ d^{n-3} & d^{n-2} & d^{n-1} & \ldots & d^{n-4} \\ \vdots & \vdots & \vdots & \ddots & \vdots \\ 1 & d & d^2 & \ldots & d^{n-1} \end{pmatrix}.$$

From this, we can find the elements of $L_S^{-1} D_S L_S^{-T}$ we need. For instance the $(1, 1)$ entry is

$$\begin{aligned}\chi &= \frac{1}{(d^n - 1)^2}(d^{2n-1} + d^{2n-3} + \ldots + d^3 + d) \\ &= \frac{1}{(d^n - 1)^2} \frac{d(d^{2n-1} - 1)}{d^2 - 1} \\ &= \frac{d}{d^2 - 1} \frac{d^n + 1}{d^n - 1}.\end{aligned}$$

The $(1, 2)$ term is

$$\xi = \frac{1}{(d^n - 1)^2}\left(d^n + \frac{d^2}{d^2 - 1}(d^{2n-2} - 1)\right).$$

This can also be expressed as

$$\xi = \frac{1}{2}(\gamma\chi - 1) = \frac{1}{d^2 - 1}\left[1 + \frac{\gamma d}{d^n - 1}\right].$$

□

From these results we can get the desired approximation. We define the following $Ttrid(S^{-1})$ operator as an approximation to the inverse of the circulant tridiagonal matrix S:

$$Ttrid(S^{-1}) = \begin{pmatrix} \chi & \xi & & & \xi \\ \xi & \chi & \xi & & \\ & \ddots & \ddots & \ddots & \\ & & \xi & \chi & \xi \\ \xi & & & \xi & \chi \end{pmatrix},$$

with the values of χ and ξ given in theorem 8.46. This definition can be extended by scaling to a more general tridiagonal circulant matrix.

Let

$$S(\alpha, \beta) = \begin{pmatrix} \alpha & -\beta & & & -\beta \\ -\beta & \alpha & -\beta & & \\ & \ddots & \ddots & \ddots & \\ & & -\beta & \alpha & -\beta \\ -\beta & & & -\beta & \alpha \end{pmatrix},$$

with $\alpha > 2\beta$. We define the operator $Ttrid$, which approximates the inverse of a circulant tridiagonal matrix $S(\alpha, \beta)$ by a circulant tridiagonal matrix $S(\chi, \xi)$, by:

$$Ttrid(S^{-1}(\alpha, \beta)) \equiv S(\chi, \xi),$$

where

$$\chi = \frac{d}{\beta(d^2 - 1)} \left[\frac{d^n + 1}{d^n - 1}\right],$$

$$\xi = \frac{1}{\beta(d^2 - 1)} \left[1 + \frac{\alpha}{\beta} \frac{d}{d^n - 1}\right],$$

and where

$$d = \frac{\alpha + \sqrt{\alpha^2 - 4\beta^2}}{2\beta}.$$

For periodic boundary conditions, the INV preconditioner is defined as:

$$M_P = (\Delta + L_P)\, \Delta^{-1}\, (\Delta + L_P^T),$$

with

$$\Delta = \begin{pmatrix} \Lambda & & & \\ & \ddots & & \\ & & \ddots & \\ & & & \Lambda \end{pmatrix},$$

and Λ being a tridiagonal circulant matrix satisfying the following equation:

$$\Lambda = T - Ttrid(\Lambda^{-1}).$$

With the definition of $Ttrid$, this equation can be solved for Λ.

Theorem 8.47

The INV *preconditioner defined by*

$$M_P = (\Delta + L_P)\, \Delta^{-1}\, (\Delta + L_P^T),$$

with Λ satisfying the condition $\Lambda = T - Ttrid(\Lambda^{-1})$, can be given in the following explicit form:

$$M_P \equiv T_3(\alpha, \beta, 1) T_1^{-1}(\alpha, \beta) T_2(\alpha, \beta, 1),$$

where $\Lambda = S(\alpha, \beta)$, *with*

$$\alpha = 4 - \frac{d}{\beta(d^2-1)}\left[\frac{d^n+1}{d^n-1}\right],$$

$$\beta = 1 + \frac{1}{\beta(d^2-1)}\left[1 + \frac{\alpha}{\beta}\frac{d}{d^n-1}\right],$$

and where

$$d = \frac{\alpha + \sqrt{\alpha^2 - 4\beta^2}}{2\beta}.$$

Proof. See Chan and Meurant [288]. □

Theorem 8.48

The eigenvalues of INV *are given by:*

$$\mu_{st} = \frac{4\beta(\sin^2(\theta_s/2) + \sin^2(\phi_t/2))(4\sin^2(\theta_s/2) + \zeta)}{(\beta[4\sin^2(\theta_s/2) + \zeta] - e^{i\phi_t})(\beta[4\sin^2(\theta_s/2) + \zeta] - e^{-i\phi_t})},$$

with $\zeta = \alpha/\beta - 2$, *which simplifies to:*

$$\mu_{st} = \frac{4\eta(\sin^2(\theta_s/2) + \sin^2(\phi_t/2))}{\eta^2 + 1 - 2\eta\cos(\phi_t)},$$

where $\eta = \beta(4\sin^2(\theta_s/2) + \zeta)$.

Proof. We compute the eigenvalues of the preconditioner by multiplying M_P by $u^{(s,t)}$ and use elementary algebra. □

It can be shown that $4 > \alpha > 2\beta > 2$. Thus as $n \to \infty$ (i.e. $h \to 0$), $\alpha/\beta = O(1)$. From this, we can deduce:

Theorem 8.49

For the INV *preconditioner,*

$$\kappa(M_P^{-1}A_P) \geq O\left(\frac{1}{h^2}\right).$$

Proof. First consider μ_{st} for $\theta_s = \phi_t = \pi$. Then $\eta = \beta(4+\zeta) = O(1)$;

$$\mu_{st} = \frac{8\eta}{\eta^2 + 2\eta + 1} = O(1).$$

Hence, $\lambda_{max} \geq O(1)$. Now, consider $\theta_s = \phi_t = 2\pi h$. Then $\eta = O(1)$, $\cos(\phi_t) = O(1)$ and $\mu_{st} = O(h^2)$. Therefore $\lambda_{min} \leq O(h^2)$. This implies $\kappa(M_P^{-1} A_P) \geq O(h^{-2})$. □

This result was first observed in numerical results for Dirichlet boundary conditions in Concus, Golub and Meurant [342].

It is useful to consider the asymptotic case, when $n \to \infty$, since this reflects what happens inside the domain, far enough from the boundaries. Then, the values for α and β we obtain for the constant coefficient periodic problem are the same as for the asymptotic interior values of the Dirichlet problem. The limiting values of α and β are solution of the non–linear equations

$$\alpha = 4 - \frac{d}{\beta(d^2-1)},$$

$$\beta = 1 + \frac{1}{\beta(d^2-1)}.$$

By computation: $\alpha = 3.6539...$ and $\beta = 1.1138....$

The modified preconditioner `MINV` is expressed in the same way as for `INV`; the only difference being that the values of diagonal elements of M_P are modified such that the sum of the elements of each row is equal to the sum of the elements of the corresponding row of A_P (plus possibly an $O(h^2)$ term). Hence, to be able to define the `MINV` preconditioner for periodic boundary conditions, we must first compute the row sums of $S^{-1}(\alpha,\beta)$ where $S(\alpha,\beta)$ has been defined previously.

Lemma 8.50

Let $r(S^{-1})$ be the vector of row sums of S^{-1} and $e = (1,1,\dots,1)^T$. Then

$$r(S^{-1}(\alpha,\beta)) = \frac{1}{\alpha - 2\beta}\; e.$$

Proof. It is obvious that $r(S^{-1}) = S^{-1}\, e$; but $Se = (\alpha - 2\beta)\, e$, and hence

$$S^{-1}e = \frac{1}{\alpha - 2\beta}\; e,$$

giving the value of the row sums. □

It is easy to see that $L_P\Delta^{-1}L_P^T$ is a block diagonal matrix with all the blocks equal to Λ^{-1}; hence R_P is also block diagonal with blocks R whose value is

$$R = \Lambda - T + \Lambda^{-1}.$$

Let $\Lambda = S(\alpha, \beta)$ and $Mtrid(\Lambda^{-1})$ be the circulant tridiagonal approximation we shall use for Λ^{-1}. Analogously to `INV`, Λ is defined by

$$\Lambda = T - Mtrid(\Lambda^{-1}),$$

and therefore we have

$$R = \Lambda^{-1} - Mtrid(\Lambda^{-1}).$$

The row sum condition we require is $r(R) = ch^2$, where c is a constant independent of h. This implies that:

$$r(Mtrid(\Lambda^{-1})) = r(\Lambda^{-1}) - ch^2 = \frac{1}{\alpha - 2\beta} - ch^2.$$

Therefore, using the approximation $Ttrid$ of Λ^{-1} and modifying its diagonal to satisfy the row sum condition above, we arrive at the following approximation of the inverse of a circulant tridiagonal matrix. We define the operator $Mtrid$, which approximates the inverse of the constant coefficient circulant tridiagonal matrix as $S(\alpha, \beta)$ by a circulant tridiagonal matrix $S(\chi, \xi)$, by:

$$Mtrid(S^{-1}(\alpha, \beta)) \equiv S(\chi, \xi),$$

with

$$\chi = \frac{1}{\alpha - 2\beta} - \frac{2}{\beta(d^2 - 1)}\left[1 + \frac{\alpha}{\beta}\frac{d}{d^n - 1}\right] - ch^2,$$

$$\xi = \frac{1}{\beta(d^2 - 1)}\left[1 + \frac{\alpha}{\beta}\frac{d}{d^n - 1}\right],$$

and

$$d = \frac{\alpha + \sqrt{\alpha^2 - 4\beta^2}}{2\beta}.$$

The condition $\Lambda = T - Mtrid(\Lambda^{-1})$ now gives a set of non–linear equations for α and β.

Theorem 8.51

The MINV *preconditioner defined by*

$$M_P = (\Delta + L_P)\Delta^{-1}(\Delta + L_P^T),$$

with Λ *satisfying the condition* $\Lambda = T - Mtrid(\Lambda^{-1})$ *can be given in the following explicit form:*

$$M_P \equiv T_3(\alpha, \beta, 1)T_1^{-1}(\alpha, \beta)T_2(\alpha, \beta, 1),$$

where $\Lambda = S(\alpha, \beta)$ *and* α *and* β *are the solutions of the following equations:*

$$\alpha = 4 + \frac{2}{\beta(d^2 - 1)}\left[1 + \frac{\alpha}{\beta}\frac{d}{d^n - 1}\right] - \frac{1}{\alpha - 2\beta} + ch^2,$$

$$\beta = 1 + \frac{1}{\beta(d^2 - 1)}\left[1 + \frac{\alpha}{\beta}\frac{d}{d^n - 1}\right],$$

where

$$d = \frac{\alpha + \sqrt{\alpha^2 - 4\beta^2}}{2\beta}.$$

Proof. See Chan and Meurant [288]. □

As before the limiting values of α and β when $n \to \infty$ can be found numerically; for example, $\alpha = 3.3431...$ and $\beta = 1.1715...$ when $c = 0$. However, even without knowledge of the actual values of α and β, we can deduce the following:

Theorem 8.52

For any n, *the* α *and* β *computed by the* MINV *preconditioner satisfy the "*MINV *condition":*

$$(\alpha - 2\beta - 1)^2 = (\alpha - 2\beta)ch^2.$$

Proof. From theorem 8.51, we have

$$\alpha - 2\beta = 2 - \frac{1}{\alpha - 2\beta} + ch^2.$$

Define $\nu = \alpha - 2\beta$. Then $\nu^2 = (2 + ch^2)\nu - 1$ and therefore

$$(\nu - 1)^2 = \nu^2 - 2\nu + 1 = (2 + ch^2)\nu - 2\nu = \nu ch^2.$$

□

It should be emphasized that this "MINV condition" arises independently of the approximation we choose for Λ^{-1}, as long as the row sum criterion is satisfied. The expression for the eigenvalues μ_{st} of $M_P^{-1} A_P$ is exactly the same as for INV, the only difference being the actual values of α and β. So we have:

$$\mu_{st} = \frac{4\eta(\sin^2(\theta_s/2) + \sin^2(\phi_t/2))}{\eta^2 + 1 - 2\eta\cos(\phi_t)},$$

$$\eta = \beta(4\sin^2(\theta_s/2) + \zeta), \quad \zeta = \frac{\alpha}{\beta} - 2.$$

The main result is:

Theorem 8.53

For the MINV *preconditioner, with α, β being any values that satisfy the "*MINV *condition"*

$$(\alpha - 2\beta - 1)^2 = (\alpha - 2\beta)ch^2,$$

then

$$\kappa(M_P^{-1} A_P) \leq O\left(\frac{1}{h}\right).$$

Proof. We recall that the eigenvalues are

$$\mu_{st} = \frac{4\eta(\sin^2(\theta_s/2) + \sin^2(\phi_t/2))}{\eta^2 + 1 - 2\eta\cos(\phi_t)},$$

$$\eta = \beta(4\sin^2(\theta_s/2) + \zeta), \quad \zeta = \frac{\alpha}{\beta} - 2$$

and that α and β satisfy the "MINV condition": $(\alpha - 2\beta - 1)^2 = (\alpha - 2\beta)ch^2$.

Let $x = \sin^2(\theta_s/2)$, $y = \sin^2(\phi_t/2)$. Clearly, $C_0 h^2 \leq x \leq 1$, and $C_0 h^2 \leq y \leq 1$. In this proof, C_i denotes a generic constant independent of h. Now,

$$\cos(\phi_t) = 1 - 2\sin^2(\phi_t/2) = 1 - 2y \quad \text{and} \quad \eta = \beta(4x + \zeta) = 4\beta x + C_1;$$

so,

$$\mu(x,y) = \frac{4\beta(x+y)(4x+\zeta)}{\eta^2 + 1 - 2\eta(1-2y)} = \frac{4(x+y)(4\beta x + C_1)}{(\eta-1)^2 + 4\eta y}.$$

But, because of the "`MINV` condition"

$$\eta - 1 = 4\beta x + \alpha - 2\beta - 1 = 4\beta x + C_2 h.$$

Therefore

$$\mu(x,y) = \frac{4(x+y)(4\beta x + C_1)}{(4\beta x + C_2 h)^2 + 4y(4\beta x + C_1)}.$$

Our aim is to find upper and lower bounds for $\mu(x,y)$ in the range $C_0 h^2 \leq x, y \leq 1$.

For the lower bound, we have:

$$\mu \geq \frac{4C_1(x+y)}{(4\beta x + C_2 h)^2 + 4y(4\beta + C_1)}.$$

If $4\beta x \geq C_2 h$, then since $x^2 < x$, we have

$$\mu \geq \frac{4C_1(x+y)}{(8\beta x)^2 + 4y(4\beta + C_1)} \geq \frac{4C_1(x+y)}{C_3 x + C_4 y} \geq \frac{4C_1}{\max(C_3, C_4)} = C.$$

And if $4\beta x < C_2 h$, then

$$\mu \geq \frac{4C_1(x+y)}{C_5 h^2 + C_6 y} \geq \frac{4C_1 y}{C_5 h^2 + C_6 y}.$$

The function on the right hand side is an increasing function of y. Since $y = C_0 h^2$, this function is bounded from below by a constant and we have

$$\mu \geq C.$$

That is to say, the eigenvalues are bounded away from zero when $h \to 0$.

Next we find an upper bound for μ. We first have:

$$\mu \leq \frac{C_7(x+y)}{C_8 x^2 + C_9 y} \leq C_{10} \frac{x+y}{x^2+y}.$$

Let $g(x,y) = \frac{x+y}{x^2+y}$,

$$\frac{\partial g}{\partial y} = \frac{x^2+y-x-y}{(x^2+y)^2} = \frac{x^2-x}{(x^2+y)^2} \leq 0 \quad \text{as} \quad x \leq 1.$$

Hence, in the range $C_0 h^2 \leq x, y \leq 1$, we have

$$g(x,y) \leq \frac{x + C_0 h^2}{x^2 + C_0 h^2} \leq \frac{2x}{x^2 + C_0 h^2}.$$

Let $m(x) = \frac{2x}{x^2+C_0 h^2}$. The function m has a maximum for $x = C_0 h$; hence

$$g(x,y) \leq \frac{2 C_0 h}{C_0^2 h^2 + C_0 h^2} = \frac{2}{C_0^2 + C_0} \frac{1}{h}.$$

Altogether these results imply that

$$\kappa(M_P^{-1} A_P) \leq O(h^{-1}).$$

□

The bound on the condition number holds for any value of $c \geq 0$, which is different from the analogous situation with `MIC` whose behaviour depends on whether $c = 0$ or $c > 0$. The important condition is the "`MINV`" condition, not the value of c, nor the values of α and β. Comparing with numerical results for Dirichlet boundary conditions, Chan and Meurant [288] show that the agreement for the condition number and the minimum and maximum eigenvalues is excellent for `INV`. For `MINV` the agreement is quite good for $c = 0$, but less good for $c \neq 0$. However, the optimal value of c given by Fourier analysis is close to the optimal value for Dirichlet boundary conditions.

8.14.7. Axelsson's results

The perturbation theorems that have been used for point incomplete factorizations can also give results for block methods, see [67]. In particular, upper bounds for the largest eigenvalue can be obtained. Under certain hypotheses, it can be shown that $\lambda_{max}(M^{-1}A) \leq 2m$ where m is the number of blocks. Bounds for the condition number were studied by Axelsson and Lu.

8.14.8. Block–size reduction

Chan and Vassilevski [298] described an extension of the block incomplete factorization algorithm. They introduce matrices R_i that are restriction operators transforming vectors of the same dimension as D_i to a lower dimensional space. For PDE problems this corresponds to a restriction to a coarser mesh. We use the same notation as for the block incomplete factorizations. Let $Z_1 = D_1, \tilde{Z}_1 = R_1 Z_1 R_1^T$, and

$$\begin{aligned} V_{i-1} &= approx(\tilde{Z}_{i-1}^{-1}), \\ Z_i &= D_i - A_i R_{i-1}^T V_{i-1} R_{i-1} A_i^T, \\ \tilde{Z}_i &= R_i Z_i R_i^T. \end{aligned}$$

This is not precisely the preconditioner of Chan and Vassilevski since they introduced another level of approximation to compute an approximate inverse of Δ_i for parallel computing purposes. The main idea is that since the correction if of low rank, expressions of the inverses are provided by the Sherman–Morrison–Woodbury formula. In [298], it is proved that if A is an M–matrix and $0 \leq V_{i-1} \leq \tilde{Z}_{i-1}^{-1}$, then $\tilde{Z}_i$ is an M–matrix and therefore the algorithm can go through to completion. Numerical experiments are given using

$$R_i = \begin{pmatrix} e^T & & & \\ & e^T & & \\ & & \ddots & \\ & & & e^T \end{pmatrix},$$

where e is the vector of all ones of dimension m/p, p being the dimension of the coarse space. Numerical results show that this gives some improvement over the classical block preconditioner. For piecewise constant restrictions and the model problem, the condition number is $O(H^2/h^2)$ where H is the mesh size of the coarse mesh.

8.15. The block Cholesky decomposition for 3D problems

It is relatively easy to derive sparse approximations of the inverses involved in block factorizations when the matrices D_i are tridiagonal and matrices A_i are diagonal. This corresponds generally to the approximation of a two dimensional PDE problem. The situation is much more complicated for systems arising from three dimensional problems. Consider for instance a finite difference method in a cube using a seven point stencil. The matrix is again block tridiagonal of order $n = pqm$ and the diagonal blocks are

$$D_i = \begin{pmatrix} D_i^1 & (B_i^2)^T & & & \\ B_i^2 & D_i^2 & (B_i^3)^T & & \\ & \ddots & \ddots & \ddots & \\ & & B_i^{p-1} & D_i^{p-1} & (B_i^p)^T \\ & & & B_i^p & D_i^p \end{pmatrix}, \quad 1 \le i \le m$$

of order $r = pq$, D_i^j being of order q. With the seven point stencil, the matrices A_i and B_i^j are diagonal. The matrices D_i^j are tridiagonal. Each matrix D_i corresponds to connections in a plane and each D_i^j corresponds to connections on a mesh line.

The matrix A can be considered in three different ways:

– as a point seven diagonal matrix,

– as a block pentadiagonal matrix (using the "small" blocks corresponding to mesh lines),

– as a block tridiagonal matrix (using the "large" blocks corresponding to planes).

This leads to three ways of constructing incomplete Cholesky preconditioners. For the first, we use the point incomplete Cholesky decomposition. We shall consider only `IC(1,1,1)`. Such a method can also be modified in order for the remainder R to have zero row sums (plus eventually an $O(h^2)$ term). In the second, we regard A as a block five diagonal matrix with $q \times q$ blocks. Then, we can use the generalization of the block incomplete decomposition proposed by Axelsson. Finally, we can consider A as a block tridiagonal matrix with $pq \times pq$ large blocks corresponding to two dimensional problems in planes and try using `INV`–like preconditioners.

8.15.1. Point preconditioners

Let us consider A as a point matrix with seven diagonals. We denote the non–zero elements of the ith line of the lower triangular part of A by e_i, c_i, b_i, a_i. We set $M = LD^{-1}L^T$ with L being a point lower triangular matrix with the same structure as the lower triangular part of A. We denote the non–zero elements of the ith row of L by $\tilde{e}_i$, $\tilde{c}_i$, $\tilde{b}_i$, $\tilde{a}_i$ and the elements of D by d_i with $d_i = \tilde{a}_i$. By equating, we find that

$$\tilde{e}_i = e_i, \quad \tilde{c}_i = c_i, \quad \tilde{b}_i = b_i,$$

$$d_i = a_i - \frac{b_i^2}{d_{i-1}} - \frac{c_i^2}{d_{i-q}} - \frac{e_i^2}{d_{i-r}}.$$

For obvious reasons, this method is denoted as `IC(1,1,1)`. It can also be modified as was done for `MIC` for two dimensional problems, denoting

$$M = A + R.$$

It is easy to compute the remainder R. For a generic row of R, there are six non–zero elements in positions $i-r+1$, $i-r+q$, $i-q+1$, $i+q-1$, $i+r-q$, $i+r-1$. We compute these elements and subtract their sum from the diagonal. Then, the formulas for `MIC(1,1,1)` are

$$\tilde{e}_i = e_i, \quad \tilde{c}_i = c_i, \quad \tilde{b}_i = b_i,$$

$$d_i = a_i - b_i \frac{b_i + c_{i+q-1} + e_{i+r-1}}{d_{i-1}} - c_i \frac{c_i + b_{i-q+1} + e_{i+r-q}}{d_{i-q}} - e_i \frac{e_i + c_{i-r+q} + b_{i-r+1}}{d_{i-r}}.$$

8.15.2. 1D block preconditioners

We consider the matrix A as a block five diagonal matrix with "small" $q \times q$ blocks that correspond to lines in the mesh. For the sake of simplicity we slightly change the notation and we denote the blocks in the lower part of the ith block row of A as G_i, F_i and E_i. We set

$$M = (\mathcal{D} + \mathcal{L})\mathcal{D}^{-1}(\mathcal{D} + \mathcal{L}^T),$$

where

$$\mathcal{L} = \begin{pmatrix} 0 & & & & \\ \tilde{F}_2 & 0 & & & \\ & \ddots & \ddots & & \\ \tilde{G}_{p+1} & & \ddots & \ddots & \\ & \ddots & & \ddots & \ddots & \\ & & \tilde{G}_s & & \tilde{F}_s & 0 \end{pmatrix}, \quad \mathcal{D} = \begin{pmatrix} \tilde{E}_1 & & & \\ & \tilde{E}_2 & & \\ & & \ddots & \\ & & & \tilde{E}_s \end{pmatrix}.$$

By equating,

$$\tilde{F}_i = F_i, \quad i = 2, \ldots, s$$

$$\tilde{G}_i = G_i, \quad i = p+1, \ldots, s$$

To compute $\mathcal{D}$ we have to deal with inverses of tridiagonal matrices. As for two dimensional problems, we use tridiagonal approximations of these inverses. Then,

$$\tilde{E}_i = E_i - F_i \text{tridiag}(\tilde{E}_{i-1}^{-1}) F_i^T - G_i \text{tridiag}(\tilde{E}_{i-p}^{-1}) G_i^T, \quad i = 1, \ldots, s$$

As for the two dimensional problem, we denote this preconditioner by INV. These formulas can also be modified to yield a remainder R with zero row sums. The block structure of M is the block analog of what we get for point two dimensional matrices. Comparing to the block structure of A, we have an additional block diagonal next to the outer one. On the ith block row, the non–zero blocks in the lower triangular part are G_i, K_i, F_i, H_i. It is easy to see that

$$H_i = \tilde{E}_i + F_i \tilde{E}_{i-1}^{-1} F_i^T + G_i \tilde{E}_{i-p}^{-1} G_i^T,$$

$$K_i = G_i \tilde{E}_{i-p}^{-1} F_{i-p+1}^T.$$

Hence, the remainder R has three block diagonals. To compute the row sums of R, we have to compute

$$rowsum(G_i \tilde{E}_{i-p}^{-1} F_{i-p+1}^T + F_i \tilde{E}_{i-1}^{-1} G_{i+p-1}^T$$
$$+ F_i(\tilde{E}_{i-1}^{-1} - \text{tridiag}(\tilde{E}_{i-1}^{-1})) F_i^T + G_i(\tilde{E}_{i-p}^{-1} - \text{tridiag}(\tilde{E}_{i-p}^{-1}) G_i^T)).$$

Since we explicitly know $\text{tridiag}(\tilde{E}_j^{-1})$, it is easy to compute

$$rowsum(F_i \text{tridiag}(\tilde{E}_{i-1}^{-1}) F_i + G_i \text{tridiag}(\tilde{E}_{i-p}^{-1}) G_i^T).$$

To compute a term like $rowsum(G_i \tilde{E}_{i-p}^{-1} F_{i-p+1}^T)$, we have to compute $G_i \tilde{E}_{i-p}^{-1} F_{i-p+1}^T e$ where e is the vector of all ones. Then, let $w = F_{i-p+1}^T e$. We solve the tridiagonal system $\tilde{E}_{i-p} z = w$ and we obtain

$$rowsum(G_i \tilde{E}_{i-p}^{-1} F_{i-p+1}^T) = G_i z.$$

The other terms in the row sum are computed in the same way by solving tridiagonal systems. As for the two dimensional problem, we denote this preconditioner by MINV.

8.15.3. 2D point preconditioners

We consider the block incomplete Cholesky decomposition with "large" blocks corresponding to two dimensional problems. So, we set

$$M = (\Delta + L)\Delta^{-1}(\Delta + L^T).$$

Then, similar to the two dimensional case,

$$\Delta_1 = D_1,$$
$$\Delta_i = D_i - A_i \text{approx}(\Delta_{i-1}^{-1}) A_i^T, \quad i = 2, \ldots, m$$

The whole problem reduces to finding a sparse approximation to the inverse of the matrix representing a two dimensional problem. Some of the solutions that we shall consider in the following section could do the job. But first, we shall look at some other solutions.

There are ways of computing the exact inverse of the matrix of a two dimensional finite difference problem, see Meurant [982] but, as we only need a sparse approximation, let us first try to use incomplete factorizations of Δ_i to obtain the approximation of the inverse. Suppose Δ_i is a point five diagonal matrix, let us denote the elements of the jth row of the lower triangular part of Δ_i by c_i^j, b_i^j, a_i^j. Then, we use the IC(1,1) decomposition of Δ_i that we denote by

$$(\tilde{D}_i + \tilde{L}_i)\tilde{D}_i^{-1}(\tilde{D}_i + \tilde{L}_i^T),$$

where $\tilde{L}_i$ is the strictly lower triangular part of Δ_i. $\tilde{D}_i$ is a diagonal matrix with diagonal elements $\tilde{d}_i^j$. We know that

$$\tilde{d}_i^j = a_i^j - \frac{(b_i^j)^2}{\tilde{d}_i^{j-1}} - \frac{(c_i^j)^2}{\tilde{d}_i^{j-q}}.$$

The exact inverse of $(\tilde{D}_i + \tilde{L}_i)\tilde{D}_i^{-1}(\tilde{D}_i + \tilde{L}_i^T)$ is $(\tilde{D}_i + \tilde{L}_i^T)^{-1}\tilde{D}_i(\tilde{D}_i + \tilde{L}_i)^{-1}$. But we only need an approximation of this matrix with the same sparsity as $\Delta - i$. Thus, we approximate $(I + \tilde{D}_i\tilde{L}_i)^{-1}$ by $I - \tilde{D}_i^{-1}\tilde{L}_i$, so the approximation we are looking for is

$$\tilde{D}_i^{-1} - \tilde{D}_i^{-1}\tilde{L}_i\tilde{D}_i^{-1} - \tilde{D}_i^{-1}\tilde{L}_i^T\tilde{D}_i^{-1} + \tilde{D}_i^{-1}\tilde{L}_i^T\tilde{D}_i^{-1}\tilde{L}_i\tilde{D}_i^{-1}.$$

The first three terms have the structure that we are looking for; therefore we simply add the diagonal of the fourth term. Finally, the approximation we choose is

$$\tilde{D}_i^{-1} - \tilde{D}_i^{-1}\tilde{L}_i\tilde{D}_i^{-1} - \tilde{D}_i^{-1}\tilde{L}_i^T\tilde{D}_i^{-1} + \text{diag}(\tilde{D}_i^{-1}\tilde{L}_i^T\tilde{D}_i^{-1}\tilde{L}_i\tilde{D}_i^{-1}).$$

The lower triangular elements of this matrix are easy to compute. The elements of the outer diagonal are

$$-\frac{c_i^j}{\tilde{d}_i^{j-q}\tilde{d}_i^j}.$$

The elements of the inner diagonal are

$$-\frac{b_i^j}{\tilde{d}_i^{j-1}\tilde{d}_i^j},$$

and the diagonal elements are

$$\frac{1}{\tilde{d}_i^j}+\frac{1}{\tilde{d}_i^{j+1}}\left(\frac{b_i^{j+1}}{\tilde{d}_i^j\tilde{d}_i^{j+1}}\right)^2+\frac{1}{\tilde{a}_i^{j+q}}\left(\frac{c_i^{j+q}}{\tilde{d}_i^j\tilde{d}_i^{j+q}}\right)^2.$$

We denote this preconditioner by BKIC. This method can also be easily modified (MBKIC).

8.15.4. 2D block preconditioners

Here, we take the approximation as being the INV incomplete factorization and we approximate its inverse. Let us denote the (block) matrix for which we want to approximate the inverse as $\bar{\Delta}_i$,

$$\bar{\Delta}_i=\begin{pmatrix} O_i^1 & P_i^2 & & & \\ P_i^2 & O_i^2 & P_i^3 & & \\ & \ddots & \ddots & \ddots & \\ & & P_i^{p-1} & O_i^{p-1} & P_i^p \\ & & & P_i^p & O_i^p \end{pmatrix},$$

where P_i^j is diagonal and O_i^j is tridiagonal. Let N_i be the block strictly lower triangular part of $\bar{\Delta}_i$ and Σ_i a block diagonal matrix whose diagonal blocks are denoted by Σ_i^j. Let Δ_i be the INV incomplete block factorization of $\bar{\Delta}_i$,

$$\Delta_i=(\Sigma_i+N_i)\Sigma_i^{-1}(\Sigma_i+N_i^T).$$

Matrices Σ_i^j are tridiagonal. We want to compute an approximation Λ_i of

$$\Delta_i^{-1}=(\Sigma_i+N_i^T)^{-1}\Sigma_i(\Sigma_i+N_i)^{-1}.$$

We proceed in the same way as for the point case. However, the approximations that we shall derive involve inverses of tridiagonal matrices which are dense matrices. At the end of our derivation, we shall replace these dense inverses by sparse approximations to avoid having too much storage. We set

$$(\Sigma_i+N_i)^{-1}\simeq\begin{pmatrix} I & & & & \\ -(\Sigma_i^2)^{-1}P_i^2 & I & & & \\ & \ddots & \ddots & \ddots & \\ & & -(\Sigma_i^{p-1})^{-1}P_i^{p-1} & I & \\ & & & -(\Sigma_i^p)^{-1}P_i^p & I \end{pmatrix}\Sigma_i^{-1},$$

and then

$$(\Sigma_i + N_i^T)^{-1}\Sigma_i(\Sigma_i + N_i)^{-1} \simeq \begin{pmatrix} Y_i^1 & Z_i^1 & & \\ W_i^2 & Y_i^2 & Z_i^2 & \\ & & \ddots & \\ & & W_i^p & Y_i^p \end{pmatrix}$$

with

$$\begin{aligned} Y_i^j &= \Sigma_i^j)^{-1} + (\Sigma_i^j)^{-1}P_i^{j+1}(\Sigma_i^{j+1})^{-1}P_i^{j+1}(\Sigma_i^j)^{-1} \\ Z_i^j &= -(\Sigma_i^j)^{-1}P_i^{j+1}(\Sigma_i^{j+1})^{-1} \\ W_i^j &= -(\Sigma_i^j)^{-1}P_i^j(\Sigma_i^{j-1})^{-1} \\ Y_i^p &= (\Sigma_i^p)^{-1} \end{aligned}$$

We have to compute approximations of

$$(\Sigma_i^j)^{-1} + (\Sigma_i^j)^{-1}P_i^{j+1}(\Sigma_i^{j+1})^{-1}P_i^{j+1}(\Sigma_i^j)^{-1}$$

and $-(\Sigma_i^j)^{-1}P_i^j(\Sigma_i^{j-1})^{-1}$. It is natural to replace $(\Sigma_i^j)^{-1}$ by a tridiagonal approximation tridiag$((\Sigma_i^j)^{-1})$ and to keep only the non–zero elements corresponding to the structure of $\bar{\Delta}_i$.

We recall that P_i^j is diagonal and let b_l, a_l, b_{l+1} be the non–zero elements of the lth row of tridiag$((\Sigma_i^j)^{-1})$, f_l, e_l, f_{l+1} those of the matrix tridiag$(P_i^{j+1}(\Sigma_i^{j+1})^{-1})P_i^{j+1})$. If we denote by h_l, g_l, h_{l+1} the non–zero elements of the approximation of $(\Sigma_i^j)^{-1} + (\Sigma_i^j)^{-1}P_i^{j+1}(\Sigma_i^{j+1})^{-1}P_i^{j+1}(\Sigma_i^j)^{-1}$, we have

$$g_l = a_l + f_l(b_l a_{l-1} + a_l b_l) + 2a_l b_{l+1} f_{l+1} + e_l b_l^2 + a_l^2 e_l + b_{l+1}^2 e_{l+1},$$

$$h_l = b_l + f_{l-1} b_l b_{l-1} + f_l(b_l^2 + a_l a_{l-1}) + f_{l+1} b_l b_{l+1} + b_l a_{l-1} e_{l-1} + a_l b_l e_l.$$

For $-(\Sigma_i^j)^{-1}P_i^j(\Sigma_i^{j-1})^{-1}$, we simply compute a diagonal approximation using a similar formula. We denote this preconditioner by BKINV. It can also be modified to have zero row sums (MBKINV).

Other approximations of the inverse of the two dimensional problems are feasible. We can, for instance, use BSSOR instead of INV to generate the approximation of the inverse. Another possibility is to use polynomial approximation of the inverse. Analytic expressions for the inverse of block tridiagonal matrices are known, see Meurant [982]. From these results, other sparse approximations of the inverse can be derived. Let

$$A = \begin{pmatrix} D_1 & A_2^T & & & \\ A_2 & D_2 & A_3^T & & \\ & \ddots & \ddots & \ddots & \\ & & A_{m-1} & D_{m-1} & A_m^T \\ & & & A_m & D_m \end{pmatrix},$$

and

$$\Sigma_1 = D_1, \quad \Sigma_i = D_i - A_i \Sigma_{i-1}^{-1} A_i^T, \quad i = 2, \dots, m$$

$$\bar{\Sigma}_m = D_m, \quad \bar{\Sigma}_i = D_i - A_{i+1}^T \bar{\Sigma}_{i+1}^{-1} A_{i+1}, \quad i = m-1, \dots, 1.$$

We define

$$\Phi_i = D_i - A_i \Sigma_{i-1}^{-1} A_i^T - A_{i+1}^T \bar{\Sigma}_{i+1}^{-1} A_{i+1}.$$

Then,

$$\{A^{-1}\}_{i,i} = \Phi_i^{-1}, \quad \{A^{-1}\}_{i,i-1} = \Sigma_{i-1}^{-1} A_i^T \Phi_i^{-1}, \quad \{A^{-1}\}_{i,i+1} = \bar{\Sigma}_{i+1}^{-1} A_{i+1} \Phi^{-1}.$$

From these formulas, approximations can be computed. However, it is difficult to obtain positive definite matrices in this way. A preconditioner that seems efficient is the following: let $(\Delta + L)\Delta^{-1}(\Delta + L^T)$ be the INV incomplete factorization of A. We set $M^{-1} = \mathcal{M}^{-T} \Delta \mathcal{M}^{-1}$ with $\mathcal{M}$ lower block bidiagonal,

$$\{\mathcal{M}^{-1}\}_{i,i} = \Delta_i^{-1}, \quad \{\mathcal{M}^{-1}\}_{i,i-1} = \Delta_i^{-1} A_i^T \Delta_{i-1}^{-1}.$$

Of course, to apply this preconditioner or to use it for 3D problems, we have to solve tridiagonal systems. Note that M^{-1} is clearly positive definite. For numerical experiments, see Perlot [1071].

8.16. Nested factorization

This preconditioner was derived at the beginning of the eighties by Appleyard, Cheshire and Pollard (see [33], [32]) for block tridiagonal matrices arising from finite difference approximations of three dimensional oil reservoir problems. They used the modified form of this preconditioner, but let us first describe the unmodified one which we denote by ACP.

8.16.1. The ACP preconditioner for 3D problems

In this preconditioner we consider A with large blocks corresponding to two dimensional problems. Let L be the block strictly lower triangular part of A,

$$L = \begin{pmatrix} 0 & & & & \\ A_2 & 0 & & & \\ & \ddots & \ddots & & \\ & & A_{n-1} & 0 & \\ & & & A_n & 0 \end{pmatrix}$$

and Δ be a block diagonal matrix with diagonal blocks Δ_i. The `ACP` preconditioner M is chosen as

$$M = (\Delta + L)\Delta^{-1}(\Delta + L^T).$$

Each Δ_i is in turn chosen as

$$\Delta_i = (\theta_i + M_i)\theta_i^{-1}(\theta_i + M_i^T),$$

where

$$M_i = \begin{pmatrix} 0 & & & & \\ B_i^2 & 0 & & & \\ & \ddots & \ddots & & \\ & & B_i^{p-1} & 0 & \\ & & & B_i^p & 0 \end{pmatrix}$$

and θ_i is a block diagonal matrix with diagonal blocks θ_i^j. In the unmodified method we choose $\theta_i^j = D_i^j, i = 1, \dots, n, \quad j = 1, \dots, p$. We only need to compute the Cholesky factorization of each D_i^j to use this preconditioner. We remark that if we define the same kind of preconditioner for two dimensional problems, it reduces to `BSSOR` with $\omega = 1$ (i.e. block symmetric Gauss–Seidel).

8.16.2. The preconditioner of Appleyard, Cheshire and Pollard

They use the modified form of `ACP` that we denote by `MACP`. To define this preconditioner we must compute the row sums of R. As before, we have

$$M = \Delta + L + L^T + L\Delta^{-1}L^T.$$

We know that $L\Delta^{-1}L^T$ is block diagonal,

$$L\Delta^{-1}L^T = \begin{pmatrix} 0 & & & & \\ & A_2\Delta_1^{-1}A_2^T & & & \\ & & \ddots & & \\ & & & A_{n-1}\Delta_{n-2}^{-1}A_{n-1}^T & \\ & & & & A_n\Delta_{n-1}^{-1}A_n^T \end{pmatrix},$$

hence R is also a block diagonal matrix. Let us denote its block diagonal elements as $R_{i,i}$,

$$R_{i,i} = \Delta_i - D_i + A_i\Delta_{i-1}^{-1}A_i^T.$$

Since $\Delta_i = \theta_i + M_i + M_i^T + M_i\theta_i^{-1}M_i^T$ and $\theta_i + M_i + M_i^T = D_i$, we have

$$\Delta_i - D_i = M_i\theta_i^{-1}M_i^T,$$

and

$$R_{i,i} = M_i\theta_i^{-1}M_i^T + A_i\Delta_{i-1}^{-1}A_i^T.$$

It is easy to see that the matrix $M_i\theta_i^{-1}M_i^T$ is

$$\begin{pmatrix} 0 & & & & \\ & B_i^2(\theta_i^1)^{-1}(B_i^2)^T & & & \\ & & \ddots & & \\ & & & B_i^{p-1}(\theta_i^{p-2})^{-1}(B_i^{p-1})^T & \\ & & & & B_i^p(\theta_i^{p-1})^{-1}(B_i^p)^T \end{pmatrix}.$$

To compute the row sums of R, after we have computed Δ_{i-1}, we must solve $\Delta_{i-1}w_i = A_i^Te$ where $e = (1,\ldots,1)^T$. Then, $rowsum(A_i\Delta_{i-1}^{-1}A_i^T) = s_i = A_iw_i$, where $s_i = (s_i^1,\ldots,s_i^p)^T$. During the solve for each plane, we compute $\theta_i^{j-1}z_i^j = (B_i^j)^Te$ and $rowsum(B_i^j(\theta_i^{j-1})^{-1})(B_i^j)^T) = t_i^j = B_i^jz_i^j$. When we compute the next block diagonal element, we have to subtract $s_i + t_i^j$ from the diagonal of θ_i^j.

8.16.3. Improvements of ACP

In ACP we set

$$M = (\Delta + L)\Delta^{-1}(\Delta + L^T)$$

and then we choose a preconditioner for each plane problem (*i.e.*, the diagonal block Δ_i). The choice of nested factorization is a block symmetric Gauss–Seidel (BSSOR with $\omega = 1$). Of course, one may choose other preconditioners for the two dimensional problems that are usually better (although less general) than BSSOR. In the following, we choose INV. The resulting preconditioner will be denoted by ACPINV. We have

$$\Delta_i = (\theta_i + M_i)\theta_i^{-1}(\theta_i + M_i^T), \quad i = 1, \dots, n$$

and then

$$\theta_i^j = D_i^j - B_i^j \text{tridiag}((\theta_i^{j-1})^{-1})(B_i^j)^T, \quad j = 1, \dots, p.$$

This method can also be modified by computing the row sums of the remainder. Obviously, these preconditioners are just examples among many other possibilities. We can use, for instance, more refined two dimensional preconditioners using more storage or use variants specially designed for vector or parallel computers. However, it must be understood that preconditioners like ACP and ACPINV neglect the physical coupling that occurs between planes (except for the terms in L). This means that for problems for which there is a strong coupling between planes, ACP and the derived methods will probably give poor results. It also means that ACP is very sensitive to the ordering of the unknowns. It is much better (if possible) to order the unknowns in such a way that the strongest coupling is within planes (*i.e.*,the largest non–diagonal elements are in the B_i^j matrices).

8.17. Sparse approximate inverses

We have seen that we would like $M^{-1}A$ to behave like the identity matrix. Therefore it is natural to try computing $C = M^{-1}$ such that some norm of $AC - I$ or $CA - I$ is minimized. If we are able to do this, then applying the preconditioner simply amounts to a matrix multiply. Huckle and Grote [798] considered this type of preconditioner for non–symmetric matrices and the Frobenius norm. This was also done earlier by Grote and Simon [714].

The Frobenius norm is specially interesting to choose because we have

$$\|AC - I\|_F^2 = \sum_{k=1}^{n} \|(AC - I)e^k\|^2,$$

where e^k is the kth column of the identity matrix. Therefore, we just have to consider minimizing the l_2–norms $\|Ac_k - e^k\|$, $k = 1, \dots, n$ where the c_ks are the columns of C. Note that the n least squares problems are independent of each other. Normally since A^{-1} is full, c_k will be a full vector but as we would like to compute a sparse C, we shall choose a given sparsity for the vectors c_k. Hence, we can just consider a vector $\hat{c}_k$ which contains only the non–zero components of c_k. Accordingly let $\hat{A}_k$ be the matrix whose columns are the columns of A corresponding to the set of indices $G_k = \{j|(c_k)_j \neq 0\}$ and whose rows i are such that there exists $a_{i,j} \neq 0, j \in G_k$. Then, everything reduces to a set of (small) least squares problems,

$$\min_{\hat{c}_k} \|\hat{A}_k \hat{c}_k - e^k\|, \quad k = 1, \ldots, n$$

Since $\hat{A}$ is of full rank, we solve the least squares problems with the QR algorithm, see for instance Golub and Van Loan [676]. The main question is to know how to choose the set of indices G. This is where different approaches can be used.

8.17.1. The sparse inverses of Huckle and Grote

Huckle and Grote [798] proposed an incremental method for choosing the sparsity patterns. They start from set of indices G_k^0 (usually corresponding to a diagonal C or to the structure of the matrix A), they solve the least squares problems and then, iteratively, they enlarge the sets of indices and solve again the least squares problems until some criteria are satisfied. Suppose we are at iteration p, to extend the set of indices, we consider the residual $r = Ac_k^p - e_k$, the goal being to reduce $\|r\|$.

Let $\mathcal{L} = \{j|(r)_j \neq 0\}$ and $\forall l \in \mathcal{L}$ let $\mathcal{N}_l = \{j|a_{l,j} \neq 0, j \notin G_k^p\}$. Then the candidates for indices of new non–zero elements in the solution vector are chosen in $\cup_{l \in \mathcal{L}} \mathcal{N}_l$. For j in this set of indices, we consider solving the problem

$$\min_{\mu_j \in \mathbf{R}} \|r + \mu_j A e_j\| \quad \Longrightarrow \quad \mu_j = -\frac{(r, Ae_j)}{\|Ae_j\|^2}.$$

It can be shown that the norm of the new residual is

$$\|r\|^2 - \frac{(r, Ae_j)^2}{\|Ae_j\|^2}.$$

There exist indices such that $(r, Ae_j) \neq 0$. From these indices, we choose those which give the smallest residuals. This is repeated until the norm of the residual is smaller than a prescribed criterion or until we have reached the maximum storage that we have allowed for that column. Note that putting new indices in the solution vector will add some columns and some rows to the matrices of the least squares problems and we have to update the QR factorization and there are some efficient techniques for doing this, see [676].

Huckle and Grote gave some bounds for a few norms of $AC-I$ and also sufficient conditions for C being non–singular (although they are difficult to use in practice) and provided some numerical experiments. Their implementation is denoted as SPAI (Sparse Approximate Inverse). Further details concerning the implementation (for instance the variants of the QR algorithm) are discussed in Huckle [797]. The implementation of SPAI on a distributed memory parallel computer is considered in Deshpande, Grote, Messmer and Sawyer [414].

8.17.2. The sparse inverses of Gould and Scott

Gould and Scott [686] described some improvements to the preceding algorithm. Let $\hat{A}$ be the matrix with the compressed set of column indices. Suppose we add a new column c, then the new least squares problem becomes

$$\sigma_{+c} = \min_{z,\xi} \|\hat{A}z + \xi c - d\|^2.$$

Huckle and Grote based the choice of the new column on the solution of the minimization of $\|\xi c - r\|^2$ where r is the residual $\hat{A}z - d$. But the solution σ_{+c} is known as

$$\sigma_{+c} = \sigma - \frac{(c, P_{\hat{A}}d)^2}{\|P_{\hat{A}}c\|^2} = \sigma - \frac{(c, r)^2}{\|P_{\hat{A}}c\|^2},$$

where σ is the previous solution and $P_{\hat{A}} = I - \hat{A}(\hat{A}^T\hat{A})^{-1}\hat{A}^T$. This can be computed from the QR factorization of $\hat{A}$. It turns out that the value of $\|P_{\hat{A}}c\|^2$ can be easily updated. Then, the choice of the updating column is based on σ_{+c}. Numerical results in [686] indicate that it is not more costly to use the exact improvement formula to choose the new column and the quality of the approximate inverse preconditioner is sometimes better on the basis of the computing time for solving a set of non–symmetric problems. Numerical comparisons with ILU(0) are given in [686].

8.17.3. The sparse inverses of Chow and Saad

In [318] Chow and Saad proposed a few methods for computing sparse inverses. One of these methods is to solve approximately $Ac_j = e_j$ by using an iterative method. Of course, this problem is as hard to solve as the problem we would like to precondition, therefore only a crude solution is sought.

Some of the elements of this crude solution are dropped to preserve a given sparsity structure of the approximate inverse. Chow and Saad also proposed to precondition the iterations with the columns already computed although this has a negative impact on parallelism. They also suggested to recursively compute preconditioners M_k that gives

$$\min_{M_{i+1}} \|I - M_1 M_2 \cdots M_i M_{i+1}\|_F^2.$$

8.17.4. Sparse approximate inverses for symmetric matrices

If one of the preceding methods for generating sparse approximate inverses is used for a symmetric matrix, the sparse inverse it produces is generally not symmetric. Therefore something special has to be done for symmetric matrices.

One way to solve the problem is to compute the columns of the sparse inverse recursively. There is no change for the first column. When computing the second column, the entry in position $(1, 2)$ (if not zero) is known from the first column and is not anymore an unknown of the problem. Then, for each column to be computed, the upper part is known from previous solves. This effectively produces a symmetric sparse inverse. However, it is tricky to implement this algorithm on a parallel computer. Moreover, if we start from an SPD matrix, there is no guarantee that the result will be positive definite. This can be fixed if, for instance, dropping some of the smallest elements (in absolute value) gives rise to a diagonally dominant sparse inverse.

Another way to deal with this problem is to look for the sparse inverse in the form KK^T, solving

$$\min \|AKK^T - I\|_F^2.$$

If K has no particular structure, this is a non–linear least squares problem which is difficult to solve. If K is lower triangular, solving the columns one after another and setting some constraints on the diagonal terms, the problem is not more difficult to solve than the non–symmetric one. If K is non–singular, then the sparse inverse is positive definite. Another possibility is to minimize $\|I - LK\|_F^2$ where L is the Cholesky factor of A. It turns out that for doing this we do not need the explicit knowledge of the entries of L. However, there are some other ways to compute the sparse inverse in factored form as we shall see in the next section.

8.17.5. The sparse inverses of Benzi, Meyer and Tuma

The preconditioner of Benzi, Meyer and Tuma [137] for SPD matrices arises from the remark that if $Z = [z_1, z_2, \dots, z_n]$ is a set of conjugate directions for A, we have $Z^T A Z = D$ a diagonal matrix with diagonal elements $d_i = (z_i, A z_i)$. This gives an expression for the inverse $A^{-1} = Z D^{-1} Z^T$. A set of conjugate directions can be constructed by a Gram–Schmidt orthogonalization like algorithm applied to a set of linearly independent vectors $v_1, v_2, \dots, v_n$. If $V = [v_1, v_2, \dots, v_n] = I$, then Z is upper triangular. The algorithm from [137] is the following:

1) $z_i^{(0)} = e_i, \quad i = 1, \dots, n$

2) for $i = 1, \dots, n \quad d_j^{(i-1)} = (a_i, z_j^{(i-1)}), \quad j = i, \dots, n$ where a_i is the ith column (or row) of A

$$\text{if } j \neq n,\ z_j^{(i)} = z_j^{(i-1)} - \left(\frac{d_j^{(i-1)}}{d_i^{(i-1)}} \right) z_i^{(i-1)}, \quad j = i+1, \dots, n$$

3) $z_i = z_i^{(i-1)}, d_i = d_i^{(i-1)}, \quad i = 1, \dots, n$

This method is an application of the more general principle of biconjugation, see Chu, Funderlic and Golub [324]. The idea to obtain a sparse approximate inverse is, as usual, to preserve sparsity by dropping some fill–in (in Z) outside some prescribed positions or the fill–ins whose absolute values are below a preset drop tolerance and by controlling the amount of storage used. This method is denoted by `AINV`.

Benzi, Meyer and Tuma proved that `AINV` is feasible for H–matrices. The incomplete algorithm is the same as the complete one applied to a matrix where some elements have been set to zero. We have seen that the H–matrix class is invariant under this operation. Then, it is proved that the pivots d_i produced by the incomplete algorithm for an M–matrix are larger than the pivots for the complete decomposition. The last step is, for an H–matrix, to compare the factorizations for A and $M(A)$ which is an M-matrix. It turns out that the pivots for A are larger than the pivots for $M(A)$, see [137]. Numerical examples are given in [137] comparing `AINV`, dropping fill–ins by values, to `IC` using the same strategy. Experiments show that for using approximately the same amount of storage the number of iterations of both algorithms are not much different.

This method of constructing an approximate inverse was generalized to non–symmetric matrices by Benzi and Tuma [139]. Here, two sets of vectors $Z = [z_1, \dots, z_n]$ and $W = [w_1, \dots, w_n]$ are constructed, such that $W^T A Z = D$ diagonal. W and Z are constructed by a biconjugation process applied to $W^{(0)} = Z^{(0)} = I$. The formulas are quite similar to the ones for the symmetric case. Numerical experiments comparing `SPAI` and `AINV` are given in Benzi and Tuma [140]. Figure 8.22 shows the spectrum of `AINV` for the model problem on a 10×10 mesh. The threshold that is used ($\varepsilon = 6\ 10^2$) gives approximately the same fill–in as in `IC(1,1)`. Figure 8.23 compares $\varepsilon = 6\ 10^2$ (top) and $\varepsilon = 2\ 10^2$ (bottom) which gives a better spectrum but uses more storage.

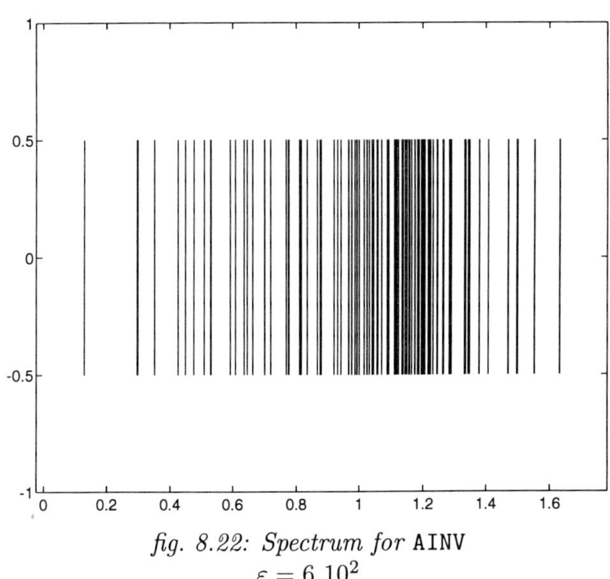

fig. 8.22: Spectrum for `AINV`
$\varepsilon = 6\ 10^2$

8.18. Polynomial preconditioners

In this type of preconditioners, we look for the inverse of M to be a polynomial in A of a given degree k,

$$M^{-1} = P_k(A) = \sum_{j=0}^{k} \alpha_j A^j.$$

This idea is both natural and "bizarre". It is natural as, by the Cayley–Hamilton theorem, we know there exists a polynomial Q of degree l, $l \leq n$ with coefficients β_j, such that $Q(A) = 0$. Q is known as the characteristic polynomial. Its roots are the distinct eigenvalues of A. By multiplying

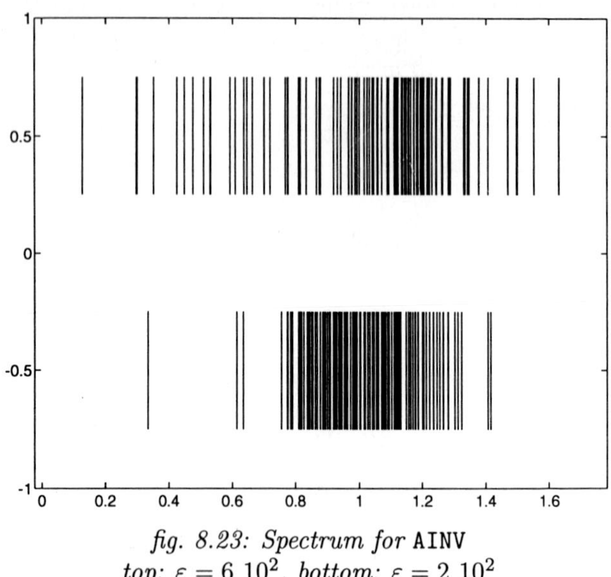

fig. 8.23: Spectrum for AINV
top: $\varepsilon = 6\ 10^2$, *bottom:* $\varepsilon = 2\ 10^2$

$Q(A) = 0$ by A^{-1} we obtain that

$$A^{-1} = -\frac{1}{\beta_0}(\beta_1 I + \cdots + \beta_l A^{l-1}).$$

Therefore A^{-1} (which gives the best preconditioner!) is a polynomial in A. However, we usually do not know the coefficients β_j and the degree of the polynomial can be quite large since it is the number of distinct eigenvalues.

On the other hand, this idea of using a polynomial as a preconditioner is strange since (for SPD matrices) it is going to be used with CG and we know that the polynomial generated by CG is, in some sense, optimal. Therefore, applying m iterations of CG to $P_k(A)A$ will generate a polynomial of degree $k+m$ that will be less efficient than $k+m$ iterations of CG for reducing the A–norm of the error. But, in this approach it could be that there will be fewer scalar products that are a bottleneck on certain computer architectures.

We note that for a polynomial preconditioner, since the polynomial commutes with A, applying the preconditioner from the left or from the right give the same results.

8.18.1. Truncated Neumann series

Let A be a symmetric positive definite matrix. We write $A = D - L -$

L^T where D is a diagonal matrix, the minus signs being just a technical convenience. Then, we symmetrically scale A by its diagonal,

$$A = D^{\frac{1}{2}}(I - D^{-\frac{1}{2}}(L + L^T)D^{-\frac{1}{2}})D^{\frac{1}{2}},$$

$$A^{-1} = D^{-\frac{1}{2}}(I - D^{-\frac{1}{2}}(L + L^T)D^{-\frac{1}{2}})^{-1}D^{-\frac{1}{2}}.$$

We note that

$$D^{-\frac{1}{2}}(L + L^T)D^{-\frac{1}{2}} = I - D^{-\frac{1}{2}}AD^{-\frac{1}{2}},$$

and

$$\rho(I - D^{-\frac{1}{2}}AD^{-\frac{1}{2}}) = \rho(I - D^{-1}A).$$

If $\rho(I - D^{-1}A) < 1$, then we can expand $I - D^{-\frac{1}{2}}AD^{-\frac{1}{2}}$ in Neumann series. The simplest idea to generate a polynomial preconditioner is to use a few terms of the Neumann series. Generally, only one or three terms are used. For instance, the first order Neumann polynomial `NEUM1` is

$$M^{-1} = D^{-1} + D^{-1}(L + L^T)D^{-1} = 2D^{-1} - D^{-1}AD^{-1}.$$

Generally, polynomials of even degree are not used as it was shown in Dubois, Greenbaum and Rodrigue [459] that a Neumann polynomial of odd degree k is more efficient than the Neumann polynomial of degree $k+1$. In practice, only degrees 1 and 3 (`NEUM3`) are used. Figure 8.24 plots the three first Neumann polynomials ($xp_k(x)$) on $[0, 8]$ corresponding to the interval of eigenvalues for the Poisson model problem. We can see that the eigenvalues are mapped to a larger interval for the even degree.

8.18.2. The minmax polynomial

Since we have $M^{-1}A = P_k(A)A$, it is natural to consider the polynomial q_k such that $q_{k+1}(\lambda) = \lambda p_k(\lambda)$. Note that we have $q_k(0) = 0$ as a constraint. The ideal situation would be to have $q_k(\lambda) \equiv 1, \forall\lambda$. Unfortunately this is not possible because of the constraint at the origin. The eigenvalues of the preconditioned system are the images of the eigenvalues of A by q_{k+1}. However, we do not know the minimum and maximum of the transformed eigenvalues (even if it is likely that the smallest eigenvalue will be the image of $\lambda_{min}(A)$). Johnson, Michelli and Paul [830]) defined a generalized condition number in the following way. Let a and b such that $\lambda_i(A) \in [a, b], \forall i$, we are interested in considering the “condition number”:

$$cond(q) = \frac{\max_{\lambda\in[a,b]} q(\lambda)}{\min_{\lambda\in[a,b]} q(\lambda)}.$$

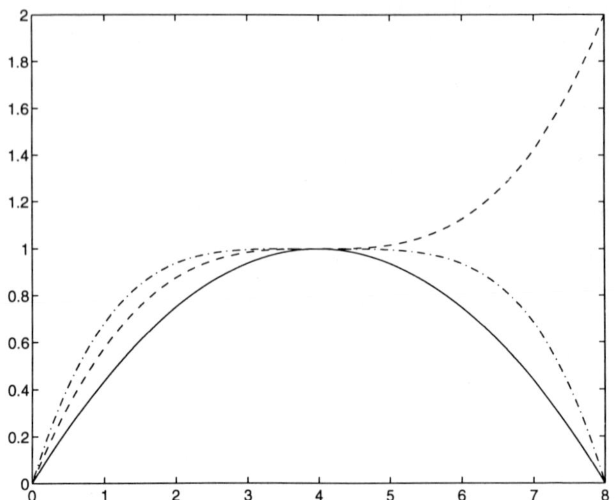

fig. 8.24: Three first Neumann polynomials
solid: degree 1, dashed: degree 2, dash–dotted: degree 3

Let $\mathcal{Q}_k = \{\text{polynomials } q_k | \forall\lambda \in [a,b], q_k(\lambda) > 0, q_k(0) = 0\}$. The first constraint will give us a positive definite preconditioner. Then we look for the solution of:

$$\text{Find } q_k \in \mathcal{Q}_k \text{ such that } \forall q \in \mathcal{Q}_k, cond(q_k) \leq cond(q).$$

The solution of this problem was given by Johnson, Michelli and Paul [830], see also Ashby [42]. First of all, we map $[a,b]$ to $[-1,1]$ by defining

$$\mu(\lambda) = \frac{2\lambda - b - a}{b - a}.$$

Theorem 8.54

Let

$$q_k(\lambda) = 1 - \frac{T_k(\mu(\lambda))}{T_k(\mu(0))},$$

where T_k is the Chebyshev polynomial of order k. Then, q_k is the solution of the previous minimization problem.

Proof. Let $\theta = |T_k(\mu(0))|$. Then

$$cond(q_k) = \frac{\theta + 1}{\theta - 1},$$

since we have

$$cond(q_k) = \frac{\max_{\lambda \in [a,b]} \{T_k(\mu(0)) - T_k(\mu(\lambda))\}}{\min_{\lambda \in [a,b]} \{T_k(\mu(0)) - T_k(\mu(\lambda))\}} = \frac{T_k(\mu(0)) + 1}{T_k(\mu(0) - 1},$$

because $\min T_k = -1$ and $\max T_k = 1$. Moreover, $|T_k(\mu(\lambda))| \leq 1$ on $[a, b]$ and $T_k(\mu(0)) > 1$. Let $q \in \mathcal{Q}_k$ and $v = \min_{\lambda \in [a,b]} q(\lambda)$, $V = \max_{\lambda \in [a,b]} q(\lambda)$. We would like to show that

$$\frac{\theta + 1}{\theta - 1} \leq \frac{V}{v},$$

or equivalently

$$\frac{V + v}{V - v} \leq \theta.$$

Note that outside $[-1, 1]$ the Chebyshev polynomials are the ones with the fastest increase. Let u_k be a polynomial of degree k. Then, for $\mu(\lambda) \geq 1$,

$$\frac{|u_k(\mu(\lambda))|}{\max_{|t| \leq 1} |u(t)|} \leq |T_k(\mu(\lambda))|.$$

We use this result with $u_k(\mu(\lambda)) = 1 - \frac{2q(\lambda)}{V+v}$. Clearly,

$$\max_{|t| \leq 1} |u_k(\lambda)| = \frac{V - v}{V + v}.$$

Therefore,

$$\frac{V + v}{V - v} \leq |T_k(\mu(0))|.$$

□

We have also the following result that relates $cond(q_k)$ to the condition number of Λ.

Theorem 8.55

Let $\kappa = b/a$ and $\nu = \frac{\sqrt{\kappa}-1}{\sqrt{\kappa}+1}$. Then, we have $cond(q_k) = \left(\frac{1+\nu^k}{1-\nu^k}\right)^2$.

Proof. See Perlot [1071]. □

We denote this preconditioner by `MINMAX`. Figure 8.25 plots the minmax polynomial ($xp(x)$) using $[a, 8]$ as the eigenvalue interval for different values of a. For the last three values a is greater than the smallest eigenvalue of A. Clearly, λ_{min} is not the optimal value for a.

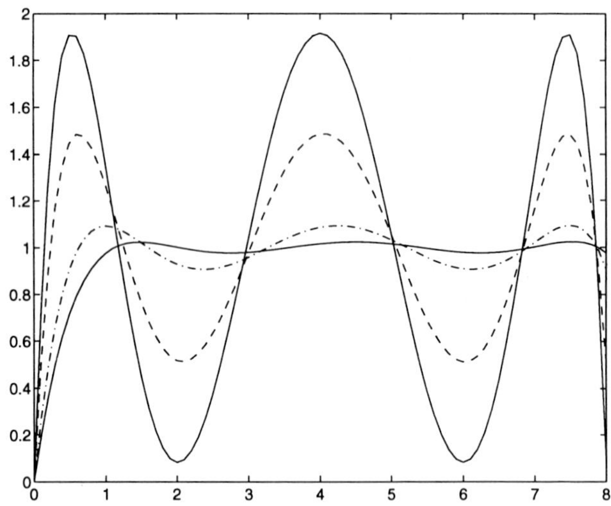

fig. 8.25: The minmax polynomial over $[a, 8]$
solid: $a = 0.01$, *dashed:* $a = 0.1$, *dash–dotted:* $a = 0.5$, *solid:* $a = 1$

Figure 8.26 shows the minmax polynomial over $[0.1, 8]$ for different degrees.

8.18.3. Least squares polynomials

We wish to have the polynomial $\lambda p_k(\lambda)$ as close as possible to 1 in some sense. One way of achieving this is to look for the polynomial p_k of degree k that minimizes

$$\int_a^b (1 - \lambda q(\lambda))^2 w(\lambda)\, d\lambda, \quad q \in \mathcal{Q}_k,$$

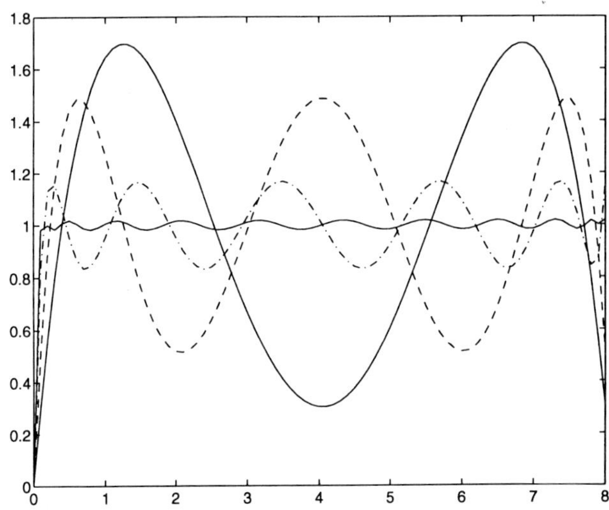

fig. 8.26: The minmax polynomial over $[0.1, 8]$
degree= solid: 3, *dashed:* 5, *dash–dotted:* 10, *solid:* 20

where $w(\lambda)$ is a positive weight. Usually, one chooses the Jacobi weights,

$$w(\lambda) = (b - \lambda)^{\alpha}(\lambda - a)^{\beta}, \quad \alpha \geq \beta \geq -\tfrac{1}{2},$$

because we know the orthogonal polynomials associated with these weights. Let $s_i(\lambda)$ be this normalized orthogonal polynomial and

$$J_k(\sigma, \lambda) = \sum_{j=0}^{k} s_j(\sigma) s_j(\lambda).$$

To derive the solution of this problem, let us prove a few technical results.

Lemma 8.56

Let $r(\lambda)$ *be a polynomial of degree* $\leq k$. *Then*

$$1 - r(\sigma) = \int_a^b J_k(\sigma, \lambda)(1 - r(\lambda)) w(\lambda)\, d\lambda.$$

Proof. We can write $1 - r(\lambda) = \sum_l \alpha_l s_l(\lambda)$, then

$$\begin{aligned}\int_a^b J_k(\sigma,\lambda)(1 - r(\lambda))w(\lambda)\,d\lambda \\ &= \sum_j s_j(\sigma) \sum_l \int_a^b s_j(\lambda)s_l(\lambda)\alpha_l w(\lambda)\,d\lambda, \\ &= \sum_j s_j(\sigma) \sum_l \alpha_l \int_a^b s_j(\lambda)s_l(\lambda)w(\lambda)\,d\lambda, \\ &= \sum_j \alpha_j s_j(\sigma) = 1 - r(\sigma).\end{aligned}$$

□

This result of lemma 8.56 explains why $J(\sigma, \lambda)$ is known as a reproducing kernel. The next result gives a lower bound for $J_k(0,0)$.

Lemma 8.57

Let $r(\lambda)$ be a polynomial of degree $\leq k$ such that $r(0) = 0$. Then

$$1 \leq J_k(0,0) \int_a^b (1 - r(\lambda))^2 w(\lambda)\,d\lambda.$$

Proof. We apply the Cauchy–Schwarz inequality to the integral,

$$\begin{aligned}1 = (1 - r(0))^2 &= \int_a^b J_k(0,\lambda)(1 - r(\lambda))w(\lambda)\,d\lambda, \\ &\leq \int_a^b J_k(0,\lambda)^2 w(\lambda)\,d\lambda \int_a^b (1 - r(\lambda))^2 w(\lambda)\,d\lambda.\end{aligned}$$

But,

$$\int_a^b J_k(0,\lambda)^2 w(\lambda)\,d\lambda = \sum_j s_j(0)^2 = J_k(0,0).$$

□

Theorem 8.58

The solution of the least squares minimization problem is

$$\lambda q(\lambda) = r_k(\lambda) = 1 - \frac{J_k(0,\lambda)}{J_k(0,0)}.$$

Proof. $1 - r_k(\lambda)$ is collinear to $J_k(0,\lambda)$ and equal to 1 in $\lambda = 0$, therefore we have equality in the Cauchy–Schwarz inequality,

$$\frac{1}{J_k(0,0)} = \int_a^b (1 - r_k(\lambda))^2 w(\lambda)\, d\lambda,$$

which is the minimum values as lemma 8.57 shows. □

The solution can be rewritten as

$$q_k(\lambda) = \sum_{j=0}^{k+1} b_j t_j(\lambda),$$

where

$$b_j = \frac{s_j(0)}{\sum_{l=0}^{k+1} s_l(0)^2}, \quad t_j(\lambda) = \frac{s_j(0) - s_j(\lambda)}{\lambda}.$$

In the following, we shall consider two special cases of the Jacobi weights: the Chebyshev weight $\alpha = \beta = -\frac{1}{2}$ (LSQUARE) and the Legendre weight $\alpha = \beta = 0$ (LEG). For these weights, we can derive bounds for the generalized condition number that was used in MINMAX. We denote by $\mathcal{P}_k$ the space of real polynomials of degree less or equal to k.

Lemma 8.59

For all $p \in \mathcal{P}_k$, for all $\lambda \in [a,b]$,

$$|p(\lambda)|^2 \le C_k \int_a^b |p(t)|^2\, dt,$$

with $C_k = (k+1)^2/(b-a)$.

Proof. See [830] and [1071]. □

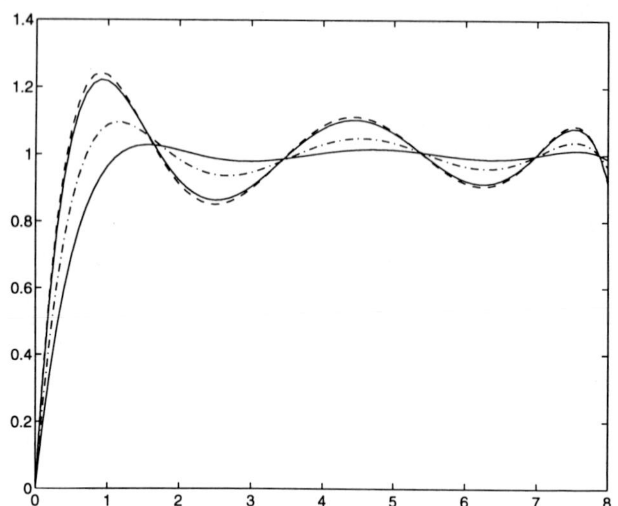

fig. 8.27: The least squares polynomial over $[a, 8]$
solid: $a = 0.01$, *dashed:* $a = 0.1$, *dash–dotted:* $a = 0.5$, *solid:* $a = 1$

Theorem 8.60

The polynomial preconditioner LEG *satisfies*

$$\frac{\sqrt{cond(q_k)} - 1}{\sqrt{cond(q_k)} + 1} \leq \sqrt{2}(k+1)\nu^k,$$

where ν *is defined as* $\nu = \frac{\sqrt{\kappa}-1}{\sqrt{\kappa}+1}$, $\kappa = b/a$.

Proof. See [1071]. □

Now, we extend these results to LSQUARE.

Lemma 8.61

For all $p \in \mathcal{P}_k$, *for all* $\lambda \in [a, b]$ *and the Jacobi weights* ω,

$$|p(\lambda)|^2 \leq \bar{C}_k \int_a^b |p(t)|^2 \, \omega(t) dt,$$

with $\bar{C}_k = (2k+1)/\pi$.

Proof. See [1071]. □

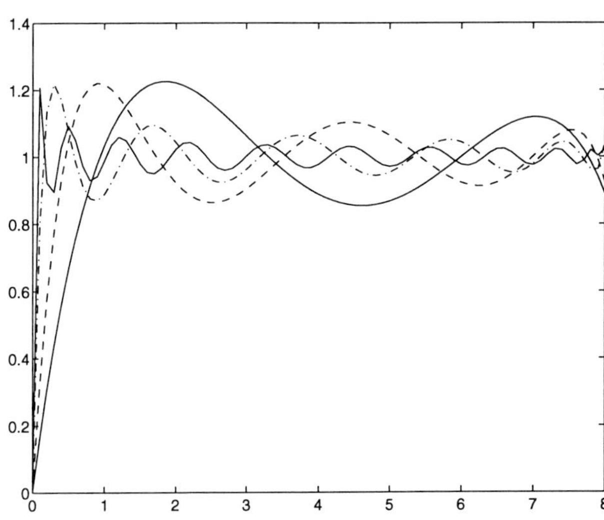

fig. 8.28: The least squares polynomial over [0.1, 8]
degree= solid: 3, *dashed:* 5, *dash–dotted:* 10, *solid:* 20

Theorem 8.62

The polynomial preconditioner `LSQUARE` *satisfies*

$$\frac{\sqrt{cond(q_k)} - 1}{\sqrt{cond(q_k)} + 1} \le \sqrt{4k + 2}\nu^k,$$

where ν *is defined as in theorem 8.60.*

Proof. See [1071]. □

These results give estimates of what we can expect when using these polynomial preconditioners with `PCG`. Figure 8.27 plots the least squares polynomial over $[a, 8]$ for different values of a.

Figure 8.28 shows the least squares polynomial over $[0.1, 8]$ for different degrees. Figure 8.29 compares polynomials of degree 5: `MINMAX` over $[0.1, 8]$

and **LSQUARE** over $[0, 8]$. With the least squares polynomial, the eigenvalue are more clustered around 1 than with `MINMAX`. However, the smallest eigenvalues are less well separated.

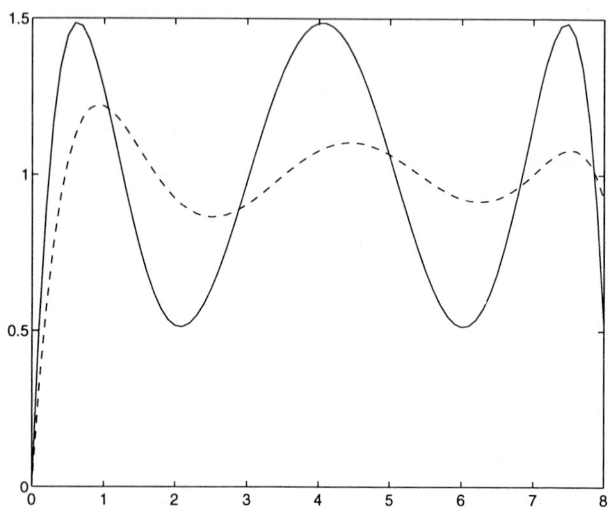

fig. 8.29: solid: minmax polynomial over $[0.1, 8]$*,*
dashed: least squares over $[0, 8]$

8.18.4. Stable evaluation of polynomials

Usually, for a polynomial $p_k(\lambda) = \sum_i \alpha_i \lambda^i$, the Horner scheme is used to evaluate $z = P_k(A)r$:

$$b_k = \alpha_k r,$$

$$b_i = \alpha_i r + A b_{i+1}, \quad i = k-1, \ldots, 0$$

then, $z = b_0$. However, when the degree of the polynomial is large and when the computation is done in single (IEEE) precision, there can be some stability problems in these evaluations. Although the Horner scheme can be improved, it is of interest to have alternate ways of evaluating the polynomials. This can be done by finding the recurrence relations for the various polynomials we have defined. Let us start with `MINMAX`. It satisfies a three-term recurrence relation.

Lemma 8.63

The MINMAX *polynomial* p_k *satisfies*

$$p_k(\lambda) = \frac{4}{a-b}\frac{c_k}{c_{k+1}} + 2\mu(\lambda)\frac{c_k}{c_{k+1}}p_{k-1}(\lambda) - \frac{c_{k-1}}{c_{k+1}}p_{k-2}(\lambda),$$

with

$$p_0(\lambda) = \frac{2}{a+b}, \quad p_1(\lambda) = \frac{8(a+b-\lambda)}{a^2+b^2+6ab},$$

and $c_k = T_k(\mu(0))$.

Proof. The Chebyshev polynomials satisfy

$$T_{k+1}\mu(\lambda)) = 2\mu(\lambda)T_k(\mu\lambda)) - T_{k-1}(\mu(\lambda)).$$

This translates also into

$$c_{k+1} = 2\mu(0)c_k - c_{k-1}.$$

The MINMAX polynomial is given by

$$p_k(\lambda) = \frac{1}{\lambda}\left(1 - \frac{T_{k+1}(\mu(\lambda))}{T_{k+1}(\mu(0))}\right).$$

Conversely,

$$\frac{T_k(\mu(\lambda))}{\lambda} = c_k\left(\frac{1}{\lambda} - p_{k-1}(\lambda)\right).$$

Then,

$$p_k(\lambda) = \frac{1}{\lambda} - \frac{1}{c_{k+1}}\left(2\mu(\lambda)\frac{T_k(\mu(\lambda))}{\lambda} - \frac{T_{k-1}(\mu(\lambda))}{\lambda}\right).$$

By substituting the values for Chebyshev polynomials, we get

$$p_k(\lambda) = \frac{1}{\lambda}\left(1 - 2\mu(\lambda)\frac{c_k}{c_{k+1}} + \frac{c_{k-1}}{c_{k+1}}\right) + 2\mu(\lambda)\frac{c_k}{c_{k+1}}p_{k-1}(\lambda) - \frac{c_{k-1}}{c_{k+1}}p_{k-2}(\lambda).$$

Finally,

$$\frac{1}{\lambda}\left(1 - 2\mu(\lambda)\frac{c_k}{c_{k+1}} + \frac{c_{k-1}}{c_{k+1}}\right) = \frac{2}{\lambda}\frac{c_k}{c_{k+1}}(\mu(0) - \mu(\lambda)) = \frac{4}{a-b}\frac{c_k}{c_{k+1}}.$$

It is also easy to compute the initial conditions. □

Theorem 8.64

The scheme for evaluating the MINMAX *polynomial* $z = P_k(A)r$ *is*

$$z_o = \frac{2}{a+b}r,$$
$$z_1 = \frac{8}{a^2+b^2+6ab}((a+b)I - A)r,$$
$$z_j = \frac{4}{a-b}\frac{c_j}{c_{j+1}}r + \frac{2}{b-a}\frac{c_j}{c_{j+1}}(2A-(a+b)I)z_{j-1} - \frac{c_{j-1}}{c_{j+1}}z_{j-2},$$
$$j = 2, \dots, k$$

and $z = z_k$.

□

For the least squares polynomials, the situation is a little more complex and we have to use a generalization of Clenshaw's formula.

Lemma 8.65

$$t_{j+1}(\lambda) = \delta_{j+1} + \alpha_j \nu(\lambda) t_j(\lambda) - \gamma_{j-1} t_{j-1}(\lambda), \ \forall j \geq 1,$$

with $\delta_{j+1} = K\alpha_j s_j(0),\ K = \frac{\nu(0)-\nu(\lambda)}{\lambda}$.

Proof. We have

$$\begin{aligned} t_{j+1}(\lambda) &= \frac{s_{j+1}(0) - s_{j+1}(\lambda)}{\lambda}, \\ &= \frac{1}{\lambda}(\alpha_j \nu(0) s_j(0) - \gamma_{j-1} s_{j-1}(0) \\ &\quad - \alpha_j \nu(\lambda) s_j(\lambda) + \gamma_{j-1} s_{j-1}(\lambda)), \\ &= \alpha_j \frac{\nu(0) - \nu(\lambda)}{\lambda} + \alpha_j \nu(\lambda) \frac{s_j(0) - s_j(\lambda)}{\lambda} \\ &\quad - \gamma_{j-1} \frac{s_{j-1}(0) - s_{j-1}(\lambda)}{\lambda}, \\ &= K\alpha_j s_j(0) + \alpha_j \nu(\lambda) t_j(\lambda) - \gamma_{j-1} t_{j-1}(\lambda). \end{aligned}$$

□

Theorem 8.66

For LSQUARE *and* LEG, *we have*

$$p_k(\lambda) = t_1(\lambda)\eta_1(\lambda) + \omega_2(\lambda),$$

where η_j *are defined by*

$$\eta_j(\lambda) = b_j + \alpha_j \nu(\lambda)\eta_{j+1}(\lambda) - \gamma_j \eta_{j+2}(\lambda), \quad j = k+1, \ldots, 1$$

and $\eta_{k+3} \equiv \eta_{k+2} \equiv 0$. ω_j *is defined by*

$$\omega_j(\lambda) = \delta_j \eta_j(\lambda) + \omega_{j+1}(\lambda), \quad j = k+1, \ldots, 2, \quad \omega_{k+2} \equiv 0.$$

Proof. Let $A_k(\lambda) = \sum_{j=0}^{k} b_j t_j(\lambda)$ and let H_l denote the hypothesis that

$$p_k(\lambda) = A_{k-l}(\lambda) + \omega_{k-l+2}(\lambda) + \eta_{k-l+1}(\lambda)t_{k-l+1}(\lambda) - \gamma_{k-l}\eta_{k-l+2}(\lambda)t_{k-l}(\lambda).$$

We know that $p_k(\lambda) = A_{k+1}(\lambda)$, therefore H_{-1} holds. Let us suppose that H_l is true, then we show that H_{l+1} holds.

$$\begin{aligned}
p_k(\lambda) &= A_{k-l-1}(\lambda) + b_{k-l}t_{k-l}(\lambda) + \omega_{k-l+2}(\lambda) \\
&\quad + \eta_{k-l+1}(\lambda)t_{k-l+1}(\lambda) - \gamma_{k-l}\eta_{k-l+2}(\lambda)t_{k-l}(\lambda), \\
&= A_{k-l-1}(\lambda) + b_{k-l}t_{k-l}(\lambda) + \omega_{k-l+2}(\lambda) - \gamma_{k-l}\eta_{k-l+2}(\lambda) \\
&\quad + \eta_{k-l+1}(\lambda)(\delta_{k-l+1}\alpha_{k-l}\nu(\lambda)t_{k-l}(\lambda) - \gamma_{k-l-1}t_{k-l-1}(\lambda)), \\
&= A_{k-l-1}(\lambda) + (b_{k-l} + \alpha_{k-l}\nu(\lambda)\eta_{k-l+1}(\lambda) \\
&\quad - \gamma_{k-l}\eta_{k-l+2}(\lambda))t_{k-l}(\lambda) + \omega_{k-l+2}(\lambda) + \eta_{k-l+1}(\lambda)\delta_{k-l+1} \\
&\quad - \gamma_{k-l-1}\eta_{k-l+1}(\lambda)t_{k-l-1}(\lambda) \\
&\quad A_{k-l-1}(\lambda) + \eta_{k-l}(\lambda)t_{k-l}(\lambda) + \omega_{k-l+1}(\lambda) \\
&\quad - \gamma_{k-l-1}\eta_{k-l+1}(\lambda)t_{k-l-1}(\lambda).
\end{aligned}$$

Finally, using $l = k$, we have

$$p_k(\lambda) = A_0(\lambda) + \omega_2(\lambda) + \eta_1(\lambda)t_1(\lambda) - \gamma_0\eta_2(\lambda)t_0(\lambda) = \omega_2(\lambda) + \eta_1(\lambda)t_1(\lambda).$$

□

Let us now apply this result to LSQUARE.

Theorem 8.67

For LSQUARE, $z = P_k(A)r$ *can be computed in the following way: let*

$$s_0(0) = \frac{1}{\sqrt{\pi}}, \quad s_1(0) = \sqrt{\frac{2}{\pi}}\frac{a+b}{a-b}, \quad s_2(0) = \sqrt{\frac{2}{\pi}}\left[2\left(\frac{a+b}{a-b}\right)^2 - 1\right],$$

and

$$s_j(0) = 2\mu(0)s_{j-1}(0) - s_{j-2}(0), \quad j = 3, \ldots, k+1$$

$$b_j = \frac{s_j(0)}{\sum_{i=0}^{k+1} s_i^2(0)}, \quad j = 1, \ldots, k+1.$$

Then,

$$z_{k+1} = b_{k+1}r, \quad z_k = b_k r + \frac{2}{b-a}(2A - (a+b)I)z_{k+1},$$

$$z_j = b_j r + \frac{2}{b-a}(2A - (a+b)I)z_{j+1} - z_{j+2}, \quad j = k-1, \ldots, 1$$

and

$$u_{k+1} = \frac{4}{a-b}s_k(0)z_{k+1},$$

$$u_{j+1} = \frac{4}{a-b}s_j(0)z_{j+1} + u_{j+2}, \quad j = k-1, \ldots, 1$$

Finally

$$z = \sqrt{\frac{2}{\pi}}\frac{2}{a-b}z_1 + u_2.$$

□

For the LEG polynomial, the recurrence relation is

$$s_j(\lambda) = \frac{(4j^2-1)^{\frac{1}{2}}}{j}\mu(\lambda)s_{j-1}(\lambda) - \frac{j-1}{j}\left(\frac{2j+1}{2j-3}\right)^{\frac{1}{2}} s_{j-2}(\lambda).$$

Theorem 8.68

For LEG, $z = P_k(A)r$ *can be computed in the following way:*

let

$$\phi_j = \frac{(4j^2-1)^{\frac{1}{2}}}{j}, \quad \psi_j = \frac{j-1}{j}\left(\frac{2j+1}{2j-3}\right), \quad j = 2, \ldots, k+1$$

$$s_0(0) = \sqrt{\frac{1}{b-a}}, \quad s_1(0) = \sqrt{\frac{3}{b-a}}\frac{a+b}{a-b},$$

$$s_j(0) = \phi_j \mu(0) s_{j-1}(0) - \psi_j s_{j-2}(0), \quad j = 2, \ldots, k+1$$

$$b_j = \frac{s_j(0)}{\sum_{i=0}^{k+1} s_i^2(0)}, \quad j = 1, \ldots, k+1.$$

Then,

$$z_{k+1} = b_{k+1} r, \quad z_k = b_k r + \frac{1}{b-a}\phi_{k+1}(2A - (a+b)I) z_{k+1},$$

$$z_j = b_j r + \frac{1}{b-a}\phi_{j+1}(2A - (a+b)I) z_{j+1} - \psi_{j+2} z_{j+2}, \quad j = k-1, \ldots, 1$$

and

$$u_{k+1} = \frac{2}{a-b}\phi_{k+1} s_k(0) z_{k+1},$$

$$u_{j+1} = \frac{2}{a-b}\phi_{j+1} s_j(0) z_{j+1} + u_{j+2}, \quad j = k-1, \ldots, 1$$

Finally

$$z = \frac{2}{a-b}\sqrt{\frac{3}{b-a}} z_1 + u_2.$$

□

8.18.5. A polynomial independent of eigenvalue estimates

The drawback of most of these polynomial preconditioners is that they need estimates of the smallest and largest eigenvalues of A. For our PDEs examples a scaling by the diagonal yields $b = 2$. However, we still need an estimate a of the smallest eigenvalue. We shall see in the next section that there are some adaptive methods to improve the eigenvalue estimate. Nevertheless it is desirable to look for polynomials whose coefficients are independent of the extreme eigenvalues.

For MINMAX, it turns out that the exact smallest eigenvalue is not always the optimal value regarding the number of iterations of PCG. Moreover, we cannot use $a = 0$ as then the number of iterations blows up. For a given degree, the MINMAX polynomial oscillates more and more around 1 when $a \to 0$. When $a = 0$, $P_k(A)A$ is only positive semi definite. On the contrary, LSQUARE is rather insensitive to the choice of a and $a = 0$ can also be chosen as the polynomial stays positive definite. When $a = 0$, LSQUARE can be computed in a simpler way:

let $d_0 = 1, d_1 = \frac{3}{2}$ and

$$d_{j+1} = \frac{2j+1}{j+1} d_j - \frac{(j+1/2)(j-1/2)}{j(j+1)} d_{j-1}, \quad j = 1, \ldots, k.$$

Then the solution of $z = M^{-1}r$ is computed as

$$z_0 = \frac{2}{3}r, \quad z_1 = -\frac{4}{5}Ar + 2r,$$

$$z_j = \frac{2j+1}{j+1} \frac{d_j}{d_{j+1}} (r + (I - A)z_{j-1}) - \frac{(j^2 - 1/4)}{j(j+1)} \frac{d_{j-1}}{d_{j+1}} z_{j-2}, \quad j = 2, \ldots, k$$

finally, $z = z_k$. This implementation is faster than the more general implementation of LSQUARE, we denote it by NORM.

8.18.6. Adaptive algorithms for SPD matrices

This kind of algorithm tries to improve the estimates of the smallest and largest eigenvalues during the PCG iterations. The algorithm is the following:

every prescribed number of PCG iterations with an interval $S = [a, b]$,

1) compute eigenvalues estimates of $P_k(A)A$ by computing eigenvalues of a tridiagonal matrix,

2) extract the eigenvalue estimates for A and update S,

3) compute the new polynomial,

4) resume (or restart) the iterations.

When an estimate μ for $P_k(A)A$ is computed, we must obtain an eigenvalue estimate for A by computing the inverse image(s) of μ. We must choose the one in $[\lambda_{min}, \lambda_{max}]$. This can always be done for SPD matrices if $k+1$ is odd and if we use the MINMAX polynomial, see Ashby [41]. Let $\|q\|_S = \max_{\lambda \in S} |q(\lambda)|$. The image of S under $q_{k+1}(\lambda) = p_k(\lambda)\lambda$ is $J_\varepsilon = [1-\varepsilon, 1+\varepsilon]$ where $\varepsilon = \|1 - q_{k+1}\|_S$. If $\mu \in J_\varepsilon$ then it must be discarded since the reciprocal of μ lies in S. If $\mu < 1-\varepsilon$, there is a unique $\lambda_1 < a$ solution and S can be extended. Similarly, if $\mu > 1+\varepsilon$ there is a unique solution $\lambda_2 > b$.

The problem of finding adaptive algorithms for non–positive definite matrices is much more difficult, see Ashby [41].

8.18.7. Polynomials for symmetric indefinite problems

The problem of computing optimal polynomials for symmetric indefinite matrices is more difficult than for SPD matrices. Let us suppose that the spectrum is contained in $[a, b] \cup [c, d]$, $b < 0$, $c > 0$. There are cases where the optimal polynomial is explicitly known, see Fischer [556]. For the general case, DeBoor and Rice [388] formulated an algorithm for the numerical computation of the optimal polynomial. For this reason, we shall refer to this polynomial as the DR polynomial. This method is a Remez type algorithm. The classical Remez algorithm is an iterative technique for computing the minimax polynomial approximation to a real function f on a compact set S. It can be also modified to solve a constrained minimization problem. A detailed description is given in Ashby [41]. Roughly speaking, it looks like the following.

Suppose we want to satisfy a constraint $p(0) = 0$. Let $n = k+1$ where k is the degree of the approximation polynomial and let $X^0 = \{x_1^0, \ldots, x_n^0\} \subset S, x_1^0 < \cdots < x_n^0$. The discrete minimax approximation of f on X^0 is denoted by p^0. Then, either $\|f - p^0\|_{X^0} = \|f - p^0\|_S = e^0$ or we are looking for a new set of points $X^1 = \{x_1^1, \ldots, x_n^1\}, x_1^1 < \cdots x_n^1$ satisfying $|(f - p^0)(x_l)| = e^0$ for some $x_l \in X^1$ and such that $f - p^0$ alternates on X^1. Such an X^1 can always be found and a new discrete minimax problem is solved on X^1. Let us now look at how this problem is solved on $X = \{x_1, \ldots, x_n\}$. There is a characterization theorem stating that the polynomial p satisfies

$$e(x_i) = f(x_i) - p(x_i) = (-1)^{i-l_i}\varepsilon, \quad i = 1, \ldots, n$$

where l_i is the number of constraints between x_1 and x_i (in our case 0 or 1) and $\varepsilon = \|e\|_X$. We are looking for ε and the coefficients of p. If we take $p = xg$ where $g(x) = \sum_{i=0}^{k-1} \gamma_i x^i$, the problem reduces to the solution of a linear system:

$$\begin{pmatrix} x_1 & x_1^2 & \dots & x_1^{n-1} & -1 \\ x_2 & x_2^2 & \dots & x_2^{n-1} & (-1)^{2-l_2} \\ \vdots & \vdots & & \vdots & \vdots \\ x_n & x_n^2 & \dots & x_n^{n-1} & (-1)^{n-l_n} \end{pmatrix} \begin{pmatrix} \gamma_0 \\ \gamma_1 \\ \vdots \\ \gamma_{k-1} \\ \varepsilon \end{pmatrix} = \begin{pmatrix} f(x_1) \\ \vdots \\ f(x_n) \end{pmatrix}.$$

It remains to be seen how the set of points X is updated. In the simplest method, an extreme point of $f-p$ is found and exchanged for a point currently in X. To improve on this, several points can be simultaneously exchanged. Zeros of Chebyshev polynomials are chosen to form the initial set X^0. Moreover, we usually choose a Chebyshev basis to expand g since it is more stable, see Ashby [41] for details. There exist public domain implementations of this algorithm. Other polynomials can be considered. For instance, an Hermite constraint $p'(0) = 0$ can be added.

8.18.8. Polynomials for non–symmetric problems

Using polynomial preconditioners for non–symmetric matrices is much more difficult than for the symmetric case as, now, the eigenvalues of A are complex numbers and we have first to locate these eigenvalues and to find a region S of the complex plane that contains the eigenvalues. This problem was considered by Manteuffel [927], [928] who suggested computing an ellipse (not containing 0) that contains the convex hull of the spectrum. Then, the minimax polynomial can (sometimes) be expressed in terms of Chebyshev polynomials. An algorithm for locating the convex hull of the spectrum and for computing the best ellipse was proposed by Manteuffel.

Good eigenvalue estimates can be also obtained by enclosing the eigenvalues in polygonal regions. Then, the approximation problem can be solved by a Remez algorithm. Least squares polynomials may also be considered, see Smolarski and Saylor [1194] and Saad [1136].

8.19. Double preconditioners

So far in this chapter, we have described many different preconditioners. They are more or less efficient depending on the problem we are solving (and on our definition of efficiency). It might seem interesting to try to combine two of these preconditioners to achieve a better efficiency. In fact, there is a preconditioner that can be combined with all the other ones as the matrix A

can always be diagonally scaled before constructing a preconditioner. That is one can combine `DIAG` with other preconditioners.

Polynomial preconditioners are particularly easy to combine with other ones. First of all we can combine a polynomial P_k with a polynomial Q_l by solving

$$P_k(Q_l(A)A)Q_l(A)Ax = P_k(Q_l(A)A)Q_l(A)b.$$

This can be interesting for generating high degree polynomials for which we have seen that the Horner scheme is unstable. Constructing the polynomial by products allows us to keep the degree of the polynomial small and then the Horner scheme can be used safely. This may lead to some computer time savings, see Perlot [1071] for numerical experiments. For example, combining the polynomial preconditioner `MINMAX` with itself gives some improvement.

As another example, we can combine `IC` and a polynomial preconditioner. We write the `IC` preconditioner as $M = LL^T$ and we solve

$$P_k(L^{-1}AL^{-T})L^{-1}AL^{-T}y = P_k(L^{-1}AL^{-T})L^{-1}b, \quad L^T x = y.$$

Perlot [1071] combined `IC` and `NORM`. The numerical experiments show that low degree polynomials give an improvement in the computer times over both `IC` and `NORM`, the number of `PCG` iterations being much smaller. This can help on computers where the scalar products are costly. Combining with block incomplete preconditioners will be even more efficient.

8.20. Other ideas

We have already described many preconditioners. However, many more ideas have been published since this topic of preconditioning appeared in the seventies. Below, we briefly mention some other ideas that can be useful in some particular context. We shall see other classes of preconditioners in Chapter 10.

8.20.1. ADI preconditioner

Using the same notation as when we studied the Alternating Direction method, we have $A = D + H + V$. Then, Chandra [303] proposed using the following `ADI` preconditioner:

$$M = \frac{1}{\rho + \rho'}(\frac{D}{2} + H + \rho I)(\frac{D}{2} + V + \rho' I).$$

Note that M is symmetric only if H and V commute. However, this preconditioner can also be used for non–symmetric problems.

8.20.2. ADDKR preconditioner

Since many preconditioners have been derived from classical direct or iterative methods (for instance, Gaussian elimination or SSOR iterations), it is interesting to ask if we can obtain preconditioners by using a sequence of different ordering algorithms. This has been considered by Chan, Jackson and Zhu [280]. Using these methods, we are concerned with matrix problems arising from finite differences approximations in rectangles (for 2D problems). Then, we can look at approximate factorizations corresponding to ordering the mesh nodes first in the `row` ordering; the rows are ordered from bottom to top and on each row the mesh nodes are ordered left to right. The other ordering considered by Chan et al. is to have the mesh nodes ordered from right to left on each row. Considering DKR preconditioners for both orderings, we obtain incomplete factorizations (resp. remainders) L_1U_1 and L_2U_2 (resp. R_1 and R_2). Then, the ADDKR preconditioner is defined as

$$M = L_1U_1(A + R_1 + R_2)^{-1}L_2U_2.$$

Generally M is non–symmetric. If a symmetric preconditioner is sought, we can consider

$$\tilde{M}^{-1} = \frac{1}{2}[M^{-1} + M^{-T}].$$

Chan et al. proved that if the modification term that is added to the diagonal elements of the DKR factorizations is ch^p with $p = \frac{4}{3}$ (versus ch^2 for DKR), we have $\kappa(M^{-1}A) = O(h^{-\frac{2}{3}})$.

8.20.3. Element by element preconditioner

This kind of preconditioner arises from the finite element community. Quite often, in finite element methods, the matrix of the problem A is not fully assembled and is only known at the element level. We can write this as

$$A = \sum_e A^e,$$

where A_e is a very sparse matrix representing what happens on the element e. Of course, only the non–zeros of A^e are stored. Therefore, there is no global representation of A. This does not matter if we use CG to solve the problem since in this algorithm we do not need A but only the action of A on given vectors. This can be computed at the element level without the need to assemble A. However, the troubles arise when looking for a preconditioner. Most of the different methods that we have seen require knowledge of A, excepting the diagonal and the polynomial preconditioners. Therefore some researchers have tried to construct preconditioners by only using information at the element level. There are several variants of these ideas. The most well known element by element preconditioner was defined by Hughes, Levit and Winget [800].

Let W be the diagonal of A and W^e the diagonal of A^e. We define

$$\bar{A}^e = I + W^{-\frac{1}{2}}(A^e - W^e)W^{-\frac{1}{2}}.$$

This is called the Winget reguralization. Then, we introduce a Cholesky factorization of $\bar{A}^e = \mathcal{L}^e\mathcal{D}^e(\mathcal{L}^e)^T$. This is easily computed at the element level. The element by element preconditioner EBE is defined as

$$M = W^{-\frac{1}{2}}[\prod_e \mathcal{L}^e][\prod_e \mathcal{D}^e][\prod_e (\mathcal{L}^e)^T]W^{-\frac{1}{2}},$$

where the products run on the number of elements. This preconditioner has been theoretically studied by Wathen (see Lee and Wathen [889]). They show that the element by element preconditioner M is spectrally equivalent to the diagonal preconditioner W. For a one dimensional problem it is shown that asymptotically

$$\kappa(W^{-\frac{1}{2}}AW^{-\frac{1}{2}}) \leq 9\kappa(M^{-1}A).$$

Numerical results show that the same is true for 2D problems using distorted elements. The condition number in experiments with discontinuous coefficients is strongly dependent on the discontinuities.

8.20.4. Fast solvers

Fast solvers can only be used efficiently for separable problems with constant coefficients. When we have more general problems to solve, fast solvers using the Fast Fourier Transform are not directly usable. Concus and Golub [340] proposed using fast solvers as preconditioners. For instance, average constant coefficients can be derived from the continuous problems and then, FFTs can be used to rapidly solve systems with the preconditioners.

8.20.5. Wavelets

Chan, W.P. Tang and Wan [295] considered using wavelets to help define sparse inverse preconditioners. Let W be the orthogonal matrix transforming vectors from the standard basis to the wavelet basis. Then, $\bar{A} = WAW^T$ is the representation of the matrix A in the wavelet basis. The transform of the inverse is the inverse of the transform. Chan, Tang and Wan defined the following algorithm:

1) transform A to the wavelet basis $\bar{A} = WAW^T$,

2) apply an approximate inverse (for instance SPAI or AINV) to $\bar{A}$ to obtain $\bar{M}$,

3) use $\bar{M}$ as a preconditioner to solve $\bar{A}\bar{x} = Wb$,

4) recover $x = W^T\bar{x}$.

Numerical results in [295] show that on some test problems, using this algorithm gives better results than directly using the sparse inverses of A.

8.21. Vector and parallel computing

Vector computers came to the scientific computing market at about the same time as the interest in preconditioning grew at the end of the seventies. Therefore, it was a natural question for numerical analysts to ask which preconditioners were efficient on vector computers. Later on, with the advent of parallel computers, particularly distributed memory parallel computers, the question arose of which are the best parallel preconditioners.

Regarding vector or parallel computing, there are two issues to be considered. The first one is the computation of the preconditioner itself. If this computation is not vectorizable or parallelizable, this step can take much too long relative to the other steps of the computation. However, if the preconditioner is going to be reused for solving many linear systems, then it is less important to have an optimal coding of the preconditioner computation. The second issue is the use of preconditioner at each iteration of the iterative method. In all cases, it is important to vectorize or parallelize this part. Moreover, for parallel computing, we need to know if the preconditioner solve is scalable or not when the problem size is increased. It is quite difficult to devise good scalable preconditioners.

The preconditioners we have considered in this chapter can be classified into two categories: implicit or explicit. The implicit preconditioners are those that need the solution of linear systems (generally triangular) when applying the preconditioner, the generic example being IC. The explicit preconditioners are those where applying the preconditioner amounts to a (series of) matrix multiply. Examples of explicit methods are the diagonal and polynomial preconditioners as well as the sparse inverses. This second class of preconditioners is naturally vectorizable or parallelizable as a matrix multiply can be easily parallelized. The only problem is how to distribute the data.

To be vectorized or parallelized, the implicit preconditioners have to be modified. Usually this gives a new preconditioner different from the original one and we have to study its rate of convergence. Generally, the modifications to introduce more parallelism lead to an increase in the number of iterations and we have to check if the gains given by parallelization are not canceled by the slower rate of convergence.

It should be mentioned that the domain decomposition methods we shall study in Chapter 10 give a natural framework for the development of efficient parallel preconditioners.

8.21.1. Vectorization of IC(1,1)

Let us consider the Incomplete Cholesky factorization `IC(1,1)` for 2D finite differences matrices:

$$M = LD^{-1}L^T,$$

where the non–zero elements of the diagonal matrix D are computed by

$$d_{i,i} = a_{i,i} - \frac{a_{i,i-1}^2}{d_{i-1,i-1}} - \frac{a_{i,i-m}^2}{d_{i-m,i-m}}.$$

Of course, the computation of the preconditioner is mostly sequential but it represents a small part of the computation. What is more interesting is to consider the two triangular solves that we have to perform for every `PCG` iteration. This was handled by H. van der Vorst [1277] whose goal was to modify the `IC` preconditioner to achieve efficient implementations on vector computers. First, we scale the two triangular matrices by writing $Mz = r$ as

$$(D^{-\frac{1}{2}}LD^{-\frac{1}{2}})(D^{-\frac{1}{2}}L^T D^{-\frac{1}{2}})D^{\frac{1}{2}}z = D^{-\frac{1}{2}}r.$$

The matrix involved in the first triangular solve can be written blockwise as

$$D^{-\frac{1}{2}}LD^{-\frac{1}{2}} = \begin{pmatrix} E_1 & & & & \\ B_2 & E_2 & & & \\ & \ddots & \ddots & & \\ & & B_{m-1} & E_{m-1} & \\ & & & B_m & E_m \end{pmatrix},$$

where the E_i are bidiagonal matrices with unit diagonals. Blockwise the triangular solve is

$$E_1 y_1 = D_1^{-\frac{1}{2}},$$

$$E_i y_i = -B_i y_{i-1} + D_i^{-\frac{1}{2}} r_i, \quad i = 2, \ldots, m.$$

If we solve these smaller systems exactly in sequence with matrices E_i we get the IC preconditioner. Van der Vorst's idea was to replace this by an approximate solve to obtain a better efficiency on vector computers. So, we write $E_i = I + F_i$ with F_i being a lower triangular matrix with zero elements everywhere except on the diagonal next to the main one (that is $i - j = 1$. The algorithm that we shall denote as ICVDV replaces E_i^{-1} by three terms of the Neumann expansion

$$E_i^{-1} \approx I - F_i + F_i^2 - F_i^3 = (I - F_i)(I + F_i^2).$$

Therefore, we only need F_i and F_i^2. The latter matrix has also only one non–zero diagonal ($i - j = 2$) and is computed once for all in the initialization phase. Note that we still have a block recursion. Of course, using this trick we have modified the preconditioner and we won't get the same convergence rate. Usually it is a little worse than for IC. However, the difference depends on the degree of diagonal dominance of the matrices E_i. The larger the diagonal dominance, the faster is the decrease of the elements in the true inverse and the better is the approximation. Of course, the vectorization problem is exactly the same for MIC and the other variants of the modified incomplete factorization. However, the remedy may not be so efficient since the diagonal dominance is smaller for modified preconditioners.

8.21.2. Parallel orderings

As we have already seen, one way to use (point) incomplete factorizations on vector and parallel computers is to use orderings for the nodes of the graph of the matrix that decouple the unknowns. An example is multicolor orderings in which the graph nodes are labeled with colors in such a way that any node of a given color is only connected to nodes of other colors. Then, the nodes of one color are numbered sequentially. Using multicolor orderings for incomplete factorizations have been considered, for instance, by Poole and Ortega [1075].

When we considered the results of Duff and Meurant [480] on IC we saw that a four color ordering gave a larger number of iterations for finite difference model problems. However, using a larger number of colors the convergence rate is better. For instance, it is shown in Jones and Plassmann

[841] that many color orderings can be used efficiently on some finite element problems.

8.21.3. Vectorization of INV

This problem was considered by Meurant [966]. The preconditioner is defined in the block form

$$M = (\Delta + L)\Delta^{-1}(\Delta + L^T),$$

Δ being a block diagonal matrix with tridiagonal blocks Δ_i. When we apply the preconditioner, we have to solve in sequence tridiagonal systems like $\Delta_i y_i = c_i$. If we use standard Gaussian elimination, this is a sequential process. However, there exist parallel algorithms to solve tridiagonal systems, but, as we are only looking for a preconditioner we do not need to solve those systems exactly. Therefore again the idea is to use a sparse inverse and to replace Δ_i^{-1} by an approximation. Since elements of inverses of tridiagonal matrices are easy to compute (see [982]) we have chosen to use the matrix $\Omega(j, \Delta_i)$ which is a banded matrix constructed by using the $2j+1$ main diagonals of Δ_i^{-1}. Then, we replace the tridiagonal solves by matrix multiplies,

$$y_i = \Omega(j, \Delta_i)c_i.$$

This preconditioner is denoted by `INVV(j)`. Of course, we still have to choose the value of j. For most problems $j = 3$ is fine. Computational speeds similar to those for the diagonal preconditioner were obtain by Meurant and the number of iterations is only slightly larger than for `INV`. This type of algorithms was also considered by Axelsson and his coworkers [82], [72].

8.21.4. Twisted incomplete block factorizations

Twisted factorizations can be used to introduce more parallelism in block preconditioners. Let us introduce this kind of incomplete factorization. Suppose that we have a computer with only two processors and that A is block tridiagonal with the number of block rows m being even. We set $M = \Pi\Delta^{-1}\Pi^T$, with

$$\Pi = \begin{pmatrix} \Delta_1 & 0 & & & & & & & \\ A_2 & \Delta_2 & 0 & & & & & & \\ & \ddots & \ddots & \ddots & & & & & \\ & & A_{\frac{m}{2}-1} & \Delta_{\frac{m}{2}-1} & 0 & & & & \\ & & & A_{\frac{m}{2}} & \Delta_{\frac{m}{2}} & A^T_{\frac{m}{2}+1} & & & \\ & & & & 0 & \Delta_{\frac{m}{2}+1} & A^T_{\frac{m}{2}+2} & & \\ & & & & & \ddots & \ddots & \ddots & \\ & & & & & & 0 & \Delta_{m-1} & A^T_m \\ & & & & & & & 0 & \Delta_m \end{pmatrix}.$$

Note that Π is block lower bidiagonal for the upper part and block upper bidiagonal for the lower part. By equating, we can compute

$$\begin{aligned} \Delta_1 &= D_1, \\ \Delta_i &= D_i - A_i trid(\Delta_{i-1}^{-1})A_i^T, \quad i = 2, \ldots, m/2 \\ \Delta_m &= D_m, \\ \Delta_i &= D_i - A_{i+1}^T trid(\Delta_{i+1}^{-1})A_{i+1}, \quad i = n, \ldots, m/2 - 1 \end{aligned}$$

Note that the term in the mth block row should have been computed as

$$\Delta_{\frac{m}{2}} = D_{\frac{m}{2}} - A_{\frac{m}{2}} trid(\Delta_{\frac{m}{2}-1})A^T_{\frac{m}{2}} - A^T_{\frac{m}{2}+1} trid(\Delta^{-1}_{\frac{m}{2}+1})A_{\frac{m}{2}+1}.$$

However, this does not make too much difference if m is large. We denote this preconditioner by `INV2P`. From the structure of Π, it is clear that when solving $\Pi y = c$, we can start in parallel from $i = 1$ and $i = m$ (this kind of algorithm has been used for point tridiagonal matrices and has been termed as "burn at both ends"). Therefore both "triangular" solves are parallel.

This method can be easily generalized. Suppose, for instance, that we want to split into four parts and m is a multiple of 4. Then, we choose $M = \Theta \Delta^{-1} \Theta^T$, with

$$\Theta = \begin{pmatrix} L_1 & N_2^T & 0 & 0 \\ 0 & L_2^T & 0 & 0 \\ 0 & N_3 & L_3 & N_4^T \\ 0 & 0 & 0 & L_4^T \end{pmatrix}, \quad L_1 = \begin{pmatrix} \Delta_1 & & & \\ A_2 & \Delta_2 & & \\ & \ddots & \ddots & \\ & & A_{\frac{m}{4}} & \Delta_{\frac{m}{4}} \end{pmatrix},$$

$$L_2^T = \begin{pmatrix} \Delta_{\frac{m}{4}+1} & A^T_{\frac{m}{4}+2} & & \\ & \ddots & \ddots & \\ & & \Delta_{\frac{m}{2}-1} & A^T_{\frac{m}{2}} \\ & & & \Delta_{\frac{m}{2}} \end{pmatrix},$$

$$L_3 = \begin{pmatrix} \Delta_{\frac{m}{2}+1} & & & \\ A_{\frac{m}{2}+2} & \Delta_{\frac{m}{2}+2} & & \\ & \ddots & \ddots & \\ & & A_{\frac{3m}{4}} & \Delta_{\frac{3m}{4}} \end{pmatrix},$$

$$L_4^T = \begin{pmatrix} \Delta_{\frac{3m}{4}+1} & A^T_{\frac{3m}{4}+2} & & \\ & \ddots & \ddots & \\ & & \Delta_{m-1} & A^T_m \\ & & & \Delta_m \end{pmatrix}, \quad N_2^T = \begin{pmatrix} 0 & 0 & \dots & 0 \\ \vdots & \vdots & & \vdots \\ 0 & 0 & \dots & 0 \\ A^T_{\frac{m}{4}+1} & 0 & \dots & 0 \end{pmatrix},$$

$$N_3 = \begin{pmatrix} 0 & 0 & \dots & A_{\frac{m}{2}+1} \\ 0 & 0 & \dots & 0 \\ \vdots & \vdots & & \vdots \\ 0 & 0 & \dots & 0 \end{pmatrix}, \quad N_4^T = \begin{pmatrix} A_{\frac{m}{4}+1} & 0 & \dots & 0 \\ 0 & 0 & \dots & 0 \\ \vdots & \vdots & & \vdots \\ 0 & 0 & \dots & 0 \end{pmatrix}.$$

The formulas for computing the diagonal blocks are the following:

$$\begin{aligned}
\Delta_1 &= D_1, \\
\Delta_i &= D_i - A_i trid(\Delta_{i-1}^{-1})A_i^T, \quad i = 2, \dots, m/4 \\
\Delta_{\frac{m}{2}} &= D_{\frac{m}{2}}, \\
\Delta_i &= D_i - A_{i+1}^T trid(\Delta_{i+1}^{-1})A_{i+1}, \quad i = m/2-1, \dots, m/4+1 \\
\Delta_i &= D_i - A_i trid(\Delta_{i-1}^{-1})A_i^T, \quad i = m/2+1, \dots, 3m/4 \\
\Delta_m &= D_m, \\
\Delta_i &= D_i - A_{i+1}^T trid(\Delta_{i+1}^{-1})A_{i+1}, \quad i = n, \dots, m/2-1.
\end{aligned}$$

During the forward and backward solves, the four subsets can be computed in parallel. Using the same technique, this preconditioner can be generalized to k processors. We denote it by INVkP; it is very similar to domain decomposition methods that we shall describe in Chapter 10. The main drawback of these techniques is that, of course, the larger k, the larger is the number of iterations as some couplings between the blocks are discarded. Numerical experiments were described in Meurant [968], [970], [976].

8.21.5. Incomplete block cyclic reduction

Almost any direct method for solving linear systems can be turned into a preconditioner. Therefore, it does not come as a surprise that block cyclic reduction was considered in this respect. Rodrigue and Wolitzer [1113] described an incomplete block cyclic reduction where inverses that arise in this method are replaced by tridiagonal approximation. Numerical experiments in [1113] show that the number of iterations is about the same as for IC(1,1) but cyclic reduction has a better degree of parallelism.

8.21.6. A massively parallel preconditioner

As an example of preconditioners that can be derived for parallel computers (having a large number of processors) we consider the algorithm proposed by Alouges and Loreaux [80]. This starts by considering a linear system $(I - E)x = y$, where E is strictly lower triangular and has at most only one non–zero element per row, whose column index is denoted by $j(i)$. Note that E can be stored in a one dimensional array and also that the squares of E can be easily computed. Then, the linear system is first transformed to

$$(I - E^2)x = (I + E)y.$$

Then, we multiply by $I + E^2$ and so on until $k = \lceil \log_2 n \rceil$. Remarking that $E^n = 0$, we obtain

$$x = (I + E^{2^k})(I + E^{2^{k-1}}) \cdots (I + E^2)(I + E)y.$$

Note that from the Euler's formula:

$$(I + E^{2^k})(I + E^{2^{k-1}}) \cdots (I + E^2)(I + E) = I + E + E^2 + \cdots + E^{n-1}.$$

Thus, the solve can be computed in parallel. If this method is applied to a tridiagonal matrix, this is simply cyclic reduction (applied to all the unknowns). For a symmetric matrix A, Alouges and Loreaux [80] considered preconditioners in the form

$$M = (I - E_1) \cdots (I - E_k) D (I - E_k^T) \cdots (I - E_1^T),$$

where D is diagonal and the matrices E_i have at most only one non–zero element per row. It remains to define the matrices E_i. The matrix A is decomposed as

$$A = D_0 + D_1 + \cdots + D_p + D_1^T + \cdots + D_p^T,$$

where D_i is generalized subdiagonal as having only one non–zero element per row. D_0 is the diagonal of A. Let B be the lower triangular part of A, D_1 is a strictly lower triangular matrix whose non–zero element of row i is defined as $\forall i, i > j, j = \max_l((B - D_0)_{i,l} \neq 0$. Recursively, the column index of the non–zero element of D_m is $j = \max_l(B - \sum_{q=0}^{m-1} D_q)_{i,l} \neq 0$. We also define operators M_i such that $D_i = M_i(A)$. This extracts from the matrix A the generalized diagonals. To define the preconditioner, we start with

$$A_1 = A = (I - E_1) A_2 (I - E_1^T).$$

We require $M_1(A_2) = 0$. Applying M_0 and M_1, we obtain a system of matrix equations. However, there are some terms which cannot be easily computed. Thus we use the following approximate equations:

$$\begin{aligned} M_0(A_1) &= M_0(A_2) + M_1(A_1)M_0(A_2)^{-1}M_1(A_1)^T, \\ E_1 &= -M(A_1)M_0(A_2)^{-1}. \end{aligned}$$

The first equation gives the diagonal of A_2. The other diagonals are (approximately) computed by $M_i(A_2) = M_i(A_1)$. Then, we repeat the process on A_2. Parallel algorithms for solving the matrix equations are given in [80].

The second algorithm proposed by Alouges and Loreaux attempts to directly compute a sparse approximate inverse in the form

$$A^{-1} \approx (I - E_1^T) \cdots (I - E_k^T)D^{-1}(I - E_k) \cdots (I - E_1).$$

As before, the computation is defined inductively,

$$A_2 = (I - E_1)A_1(I - E_1^T).$$

The matrix equations we consider are the same as before:

$$\begin{aligned} M_0(A_2) &= M_0(A_1) - M_1(A_1)M_0(A_2)^{-1}M_1(A_1)^T, \\ E_1 &= M_1(A_1)M_0(A_2)^{-1}. \end{aligned}$$

The last matrix is approximated by its diagonal.

Numerical experiments show that both methods give results that can compare favorably with polynomial preconditioners. However, so far there are no theoretical results characterizing the class of matrices for which these methods are feasible and efficient. As some generalized diagonals are kept in the incomplete factorizations of A or its inverse, it is likely that these methods are efficient for matrices for which there is a strong enough decrease in the elements of the Cholesky factors and/or the inverse of A. Note that these preconditioners can be generalized in block form.

8.22. Bibliographical comments

There have been a huge number of research papers on preconditioning since the seventies. The idea of preconditioning goes back at least to Gauss and Jacobi (after all, formulating the normal equations is using some form of preconditioning). See also L. Cesari (1937). One may also consider that preconditioning is already included in the 1952 paper of Hestenes and Stiefel. Early contributions to incomplete decompositions are Varga (1960), Buleev (1960), Oliphant (1962) and Stone (1968). SSOR preconditioners were proposed by Evans (1967) and Axelsson (1974).

All these ideas lead to the development of the Incomplete Cholesky factorization of Meijerink and Van der Vorst (1977) which was the starting point of the popularity of this type of preconditioners. Block incomplete factorizations were first proposed by R. Underwood (1976) and then developed by Concus, Golub and Meurant (1982–85). The modified variants of incomplete factorizations have their origin in the work of Dupont, Kendall and Rachford (1968). They were further developed by Gustafsson (1978). The stability of the incomplete factorizations were studied by H. Elman (1986). The effect of orderings on the performance of incomplete factorizations was demonstrated numerically by Duff and Meurant (1989) and explained theoretically by Eijkhout (1991) and Doi (1990).

The most interesting contributions to sparse inverses are Huckle and Grote (1994) and Benzi, Meyer and Tuma (1996).

Simple polynomial preconditioners were introduced by Dubois, Greenbaum and Rodrigue (1979). Important contributions with more sophisticated preconditioners are Johnson, Michelli and Paul (1983), Saad (1985) and Ashby (1987).

9 Multigrid methods

9.1. Introduction

In this chapter we shall solve algebraic problems arising from the discretization of elliptic and parabolic partial differential equations. We have already specifically addressed these problems in Chapter 4 but there, we developed methods only for a restricted class of (separable) equations and (rectangular) domains. The multigrid method can handle much more general problems and is (at least asymptotically) faster than these specialized methods. However, for the sake of simplicity and for expository purposes, we shall only look into details for simple problems.

Let n be the order of the linear system. To reach the level of truncation error the classical iterative methods we studied in Chapter 5 need approximately the following number of operations (the Greek letters being constants independent of n):

`Jacobi`: λn^3

`Gauss-Seidel`: βn^3

`SOR` with optimal ω: γn^2

Some variants of the multigrid method will give us an operation count which is $O(n)$, *i.e.*,optimal (remember that the methods of Chapter 4 give at best $O(n \log \log n)$ although the constants are small).

Although some methods using several grids had been developed previously, the multigrid method was first studied in the Soviet Union in the sixties (Fedorenko [549]) but the method received very little attention and applications until A. Brandt popularized it in the seventies by solving many different and difficult problems. Since that time the method has received considerable attention and has emerged almost as a new branch of numerical mathematics. This chapter is only an introduction to the method. An excellent reference for both theory and application of the method is the book edited by W. Hackbush and U. Trottenberg [740] and the books by Hackbush [736], [739] and Briggs [212]. We start by considering a method using two grids which is not of practical interest but which allows us to introduce all the ideas that are used in the multigrid method.

9.2. The two–grid method

Suppose we are solving a second order linear elliptic partial differential equation

$$Lu = f$$

in a domain Ω, say the unit square, with Dirichlet boundary conditions. Ω is discretized with a regular mesh of stepsize $h = \frac{1}{m+1}$, m being odd. Only the values of u at interior points are unknowns, so there are m^2 such points. Suppose also that the equation is discretized using the classical five point finite difference stencil. An example of such a problem is given by the Poisson model problem we studied in the preceding chapters.

Let

$$A_h u_h = b_h$$

be the resulting linear system. Each time where there is no ambiguity we shall drop the h index, writing

$$Au = b.$$

This change of notation from x to u means we are solving a system arising from a continuous problem and is motivated by using notation which is almost standard in the multigrid literature. The two–grid method we are going to describe will use a coarse grid whose stepsize is $H = 2h$. Basically the two–grid method can be introduced from two different viewpoints:

1) One can study why classical iterative methods like `Jacobi` or `Gauss-Seidel` have a poor convergence rate and try to remedy this flaw by using the coarse grid.

2) Define a two–grid method, study why it does not work as well as we would like and correct this using for instance `Jacobi` or `Gauss-Seidel` "smoothing".

We shall use the second viewpoint since it leads to further generalizations. So, suppose that we have an approximation u^k of the solution u of $Au = b$. Let $\varepsilon^k = u - u^k$ be the error, then we have already seen that

$$A\varepsilon^k = Au - Au^k = b - Au^k = r^k.$$

Knowing u^k we are able to compute r^k, but solving $A\varepsilon^k = r^k$ is as difficult as solving the original problem. However, if we know an approximation v^k to ε^k, then $v^k + u^k$ will probably be a better approximation for u (this is the essence of iterative refinement).

The idea behind multigrid is to compute an approximation w^k of ε^k on a coarser grid Ω_H consisting (for instance) of every other point in each direction. The fine grid Ω_h has m^2 points, the coarse grid Ω_h has p^2 points, where $p = \frac{m-1}{2}$.

The problem with the error being defined on Ω_h is that we need a method to go from Ω_h to Ω_H so we define a linear restriction operator R

$$R : \Omega_h \to \Omega_H, \quad (\mathbb{R}^{m^2} \to \mathbb{R}^{p^2})$$

The representation of R is a $p^2 \times m^2$ rectangular matrix. When we have solved the problem on Ω_H, we need to go back to Ω_h, so a linear prolongation (interpolation) operator P is defined.

$$P : \Omega_H \to \Omega_h, \quad (\mathbb{R}^{p^2} \to \mathbb{R}^{m^2}).$$

Now, we must define the problem we are going to solve on Ω_H to compute w^k. There are two basic ways of finding a non–singular coarse grid matrix A_H.

1) Use the same approximation as for A_h but on Ω_H. In our example this leads to a five point approximation.

2) $A_H = RA_hP$. With usual choices for R and P, this will give a nine point approximation scheme.

One step of the two–grid algorithm is now defined by:

1) $r^k = b - Au^k$,

2) $r_H^k = Rr^k$,

3) Solve exactly $A_H \varepsilon_H^k = r_H^k$,

4) $v^k = P\varepsilon_H^k$,

5) $u^{k+1} = v^k + u^k$. If no convergence, go to step 1.

It is easy to exhibit the iteration matrix of this iterative method as

$$v^k = P\varepsilon_H^k = PA_H^{-1} r_H^k = PA_H^{-1} R(b - Au^k).$$

Hence,

$$u^{k+1} = (I - PA_H^{-1}RA)u^k + PA_H^{-1}Rb.$$

The iteration matrix is $K = I - PA_H^{-1}RA$. As we know, a necessary and sufficient condition for convergence is $\rho(I - PA_H^{-1}R) \leq 1$. Unfortunately, this is usually not true.

Lemma 9.1

Let A be symmetric positive definite and suppose $A_H = RAP$ and $P = R^T$, then the eigenvalues of $I - PA_H^{-1}RA$ are 0 and 1.

Proof. $I - PA_H^{-1}RA$ is similar to $I - A^{1/2}PA_H^{-1}RA^{1/2}$ which is symmetric. Let $z \neq 0$ and λ be an eigenvector and an eigenvalue of this matrix, *i.e.*,

$$(I - A^{1/2}PA_H^{-1}RA^{1/2})z = \lambda z.$$

Multiplying by $RA^{1/2}$, we get

$$(RA^{1/2} - RAPA_H^{-1}RA^{1/2})z = \lambda RA^{1/2}z,$$

but the left hand side matrix is zero so $\lambda RA^{1/2}z = 0$. Therefore, $\lambda = 0$, or $RA^{1/2}z = 0$. The question which arises is: does there exist $z \neq 0$ such that $RA^{1/2}z = 0$? The answer is yes. Denoting by $Ran(B)$ the range of B, we have

$$\dim \ker(RA^{1/2}) + \dim Ran(RA^{1/2}) = m^2,$$

so

$$\dim \ker(RA^{1/2}) \geq m^2 - p^2 > 1 \quad \text{when} \quad m > 1.$$

Hence, there are $z \neq 0$ in $\ker(RA^{1/2})$. But, when $z \neq 0$ and $RA^{1/2}z = 0$, we have $z = \lambda z$ so $\lambda = 1$. As a conclusion $\rho(I - PA_H^{-1}RA) = 1$. □

There is also a heuristic explanation for the non–convergence of the two–grid algorithm. Since A is symmetric, the eigenvectors span a basis of $\mathbb{R}^n$. This basis corresponds to (or converges towards when $h \to 0$) the eigenfunctions of L. Some of these eigenfunctions vary rapidly (high frequencies), some others are more smooth (low frequencies). It is intuitive that components of any vector on the high frequencies eigenvectors cannot be well approximated on the coarse grid. Hence, to make the method work, we need smooth residuals such that r_H^k is a good approximation of r^k

Generally (for elliptic problems) the operator L and the matrix A are regularizing operators since $r^k = A\varepsilon^k$, a smooth error will give an even smoother residual. The natural idea now is to add some features to the method to get smooth residuals. It is well known that classical iterative methods such as relaxed `Jacobi` or `Gauss-Seidel` (which in the context of this chapter will be called relaxation methods) give smoother and smoother errors as the iteration proceeds. A few iterations of these methods can be used as a smoother for residuals. Conversely, if one considers the `Gauss-Seidel` method, we see that the components of the error on high frequencies eigenvectors are very rapidly dampened and that the slow convergence is accounted for the smooth eigenvectors. But, when the error is smooth, it is likely that the problem can be solved on a coarser grid. We shall denote by S the iteration matrix of the chosen relaxation method.

The two–grid algorithm is now the following:

1) Do ν_1 iterations of the iterative method whose iteration matrix is S. Let $\bar{u}^k$ be the resulting vector.

2) $\bar{r}^k = b - A\bar{u}^k$,

3) $\bar{r}_H^k = R\bar{r}^k$,

4) solve exactly $A_H \varepsilon_H^k = \bar{r}_H^k$,

5) $v^k = P\varepsilon_H^k$,

6) starting from $\bar{u}^k + v^k$, do ν_2 iterations of the method whose matrix is S.

Let u^{k+1} be the result.

7) if no convergence go to 1. □

The iteration matrix of this method is

$$M = S^{\nu_2}(I - PA_H^{-1}RA)S^{\nu_1} = S^{\nu_2}KS^{\nu_1}.$$

Suppose S is non–singular. Then M is similar to

$$S^{-\nu_2}MS^{\nu_2} = KS^{\nu_1+\nu_2}$$

or $S^{\nu_1+\nu_2}K$. Clearly convergence depends only on $\nu_1 + \nu_2$.

What we have just defined is only a general framework. To get a practical method, several choices have to be made. By combining the different possibilities, many variants can be generated. We have to choose

- the relaxation method S,
- the integers ν_1 and ν_2,
- how to construct the coarse grid,
- the restriction operator R,
- the prolongation operator P,
- how to define A_H.

We shall return to some of the possible choices. Before doing this, however, we shall look at an example choosing the components of the method for the sake of the analysis. More efficient choices will be discussed when we have understood how the method works.

9.3. A one dimensional example

We shall completely study a simple one dimensional example. Of course, this is not of any practical use but will give us insight in how this method works. A similar analysis can be done for the two dimensional model problem (see Stuben and Trottenberg [1220]), but the technical details are more involved and it does not lead to more knowledge for general problems. The continuous problem is

$$-\frac{d^2u}{dx^2} = f \quad \text{in} \quad \Omega = (0,1),$$
$$u(0) = u(1) = 0.$$

Ω is divided into $n+1$ equal intervals, so the mesh Ω_h has stepsize $h = \frac{1}{n+1}$. We suppose n odd and $n > 3$ *e.g.*, $h \le 1/4$ in order to be able to define a coarse grid with more than just one mesh point. The coarse grid Ω_H is defined with $H = 2h$. If we denote by u_i the approximation of u at point number i, then the finite difference approximation of the continuous problem is

$$\frac{-u_{i-1} + 2u_i - u_{i+1}}{h^2} = f_i,$$
$$u_0 = u_{n+1} = 0.$$

The linear system of order n is

$$Au = \frac{1}{h^2}\begin{pmatrix} 2 & -1 & & & \\ -1 & 2 & -1 & & \\ & \ddots & \ddots & \ddots & \\ & & -1 & 2 & -1 \\ & & & -1 & 2 \end{pmatrix} u = b.$$

We already know that the eigenvalues of A are $\lambda_k = \frac{1}{h^2}(2 - 2\cos(k\pi h))$. The related eigenvector ϕ_k is such that

$$(\phi_k)_i = \sin(ik\pi h).$$

We must now define the components of the two–grid algorithm: S, ν_1,ν_2, R, P, A_H.

9.3.1. The choice of the smoothing

We choose $\nu_1 + \nu_2 = 1$ and as a smoother the relaxed `Jacobi` method, but since the diagonal of A is constant it is exactly the same as the Richardson method. Let

$$L = \frac{1}{h^2}\begin{pmatrix} 0 & & & & \\ -1 & 0 & & & \\ & -1 & 0 & & \\ & & \ddots & \ddots & \\ & & & -1 & 0 \end{pmatrix}.$$

Then

$$A = \frac{2}{h^2} I + L + L^T = D + L + L^T.$$

From Chapter 5 we know that the iteration matrix is

$$S = I - \alpha D^{-1} A = I - \frac{\alpha h^2}{2} A.$$

The eigenvalues of S are $\mu_k = 1 - \alpha(1 - \cos(k\pi h))$. The eigenvectors are the same as those of A. Note that for the Richardson method to converge we must have $\alpha < \frac{2}{1+\cos(\pi h)}$ and $\rho(S) = 1 - \alpha(1 - \cos(\pi h)) = 1 - O(h^2)$. The spectral radius $\rho(S)$ is given by μ_1 and corresponds to the most "regular" eigenvector (in the sense that it approximates the smoothest eigenfunction of L). We have already seen that for convergence the optimal value of α is 1. But, now we are not interested in the convergence of the Richardson method but by smoothing properties and more precisely at how the Richardson method smoothes the "high frequencies eigenvectors", *i.e.*,those which cannot be approximated on the coarse grid. Thus, we define as

low frequencies: vectors φ_k with $k < \frac{n}{2}$,

high frequencies: vectors φ_k with $\frac{n}{2} \le k \le n$.

Note that whatever h is, the values of $\varphi_{\frac{n+1}{2}}$ restricted to Ω_H are always 0. Then, following Stuben and Trottenberg [1220], we define a quantity which is analogous to the spectral radius but restricted to high frequencies.

$$\mu(h, \alpha) = \max_k \{|\lambda_k|, \tfrac{n}{2} \le k \le n\}.$$

It is also interesting to know what we get when h becomes small. So, let

$$\mu^*(\alpha) = \sup_h \{\mu(h, \alpha), h \le \tfrac{1}{4}\}.$$

Obviously, we have

Lemma 9.2

For our model example,

$$\begin{aligned}
\mu(h, \alpha) &= \max\{|1 - \alpha + \alpha \cos(\tfrac{n}{2}\pi h)|, |1 - \alpha - \alpha\cos(\pi h)|\} \\
\mu^*(\alpha) &= \max\{|1 - \alpha|, |1 - 2\alpha|\}.
\end{aligned}$$

The value of α which minimizes $\mu^*(\alpha)$ is $\alpha = \frac{2}{3}$ and $\mu^*(\frac{2}{3}) = \frac{1}{3}$.

Proof. We use the same techniques as in Chapter 5. □

So, if μ is a good measurement of the smoothing properties, the best smoothing value of α is different from the best value for the rate of convergence. The interesting feature is that the limiting value when $h \to 0$ is $\frac{1}{3}$ strictly less than one. It means that the smoothing will be efficient regardless of how h is chosen. The asymptotic convergence of the Richardson method is only driven by the smooth eigenvectors which are not of interest for our purposes.

9.3.2. The choice of the restriction

The simplest choice would be to restrict a vector by taking the value on the coarse grid to be the same as the corresponding value on the fine grid but we shall make a different choice here. We choose a weighted average and denote

$$w_H = Rw.$$

For the sake of simplicity, points on the coarse grid (*i.e.*,components of the vectors) keep the same labels as they have on the fine grid. So, a vector v defined on Ω_H will be $(v_2, v_4, v_6, \ldots)$. The restricted vector w_H is defined as the average

$$(w_H)_{2i} = \frac{1}{4}(w_{2i-1} + 2w_{2i} + w_{2i+1}).$$

9.3.3. The choice of prolongation

We choose a linear interpolation to extend vectors from the coarse grid to the fine grid. Hence

$$w = Pw_H$$

with

$$\begin{aligned} (w)_{2i} &= (w_H)_{2i}, \\ (w)_{2i+1} &= \frac{1}{2}((w_H)_{2i} + (w_H)_{2i+2}). \end{aligned}$$

Lemma 9.3

For the previous choices of restriction and prolongation, we have $P = 2R^T$.

Proof. To prove this result, we just write P and R as rectangular matrices. □

9.3.4. The choice of the coarse grid matrix

For the approximation on Ω_H we choose to use the same stencil as on Ω_h. Hence, at point $2i$ we have

$$(A_H v)_{2i} = \frac{-v_{2i-2} + 2v_{2i} - v_{2i+2}}{H^2}.$$

Let Φ_k be the eigenvectors of A_H. It is obvious that

$$(\Phi_k)_{2i} = \sin(ik\pi H) = \sin(2ik\pi h) = (\varphi_k)_{2i}.$$

Lemma 9.4

Let k be such that $k < \frac{n}{2}$, then

$$(\varphi_{n+1-k})_{2i} = -(\varphi_k)_{2i},$$
$$(\varphi_{n+1-k})_{2i+1} = (\varphi_k)_{2i+1},$$
$$(\varphi_{\frac{n+1}{2}})_{2i} = 0.$$

Proof. Since $h = \frac{1}{n+1}$,

$$\begin{aligned}
(\varphi_{n+1-k})_{2i} &= \sin((n+1-k)2i\pi h) \\
&= \sin(2i\pi - 2ik\pi h) \\
&= -\sin(2ik\pi h) \\
&= -(\varphi_k)_{2i}, \\
(\varphi_{n+1-k})_{2i+1} &= \sin((n+1-k)(2i+1)\pi h), \\
&= \sin((2i+1)\pi - (2i+1)k\pi h) \\
&= \sin((2i+1)k\pi h) \\
&= (\varphi_k)_{2i+1} \\
(\varphi_{\frac{n+1}{2}})_{2i} &= \sin(i(n+1)\pi h) \\
&= \sin(i\pi) = 0,
\end{aligned}$$

and this proves the result. □

We are now going to look at what happens to the eigenvectors of A when the operator K is applied. We do this in four stages.

1) $\varphi_k \to A\varphi_k$.

Since φ_k is an eigenvector $\varphi_k \to \lambda\varphi_k$. As we noted before $\lambda_k = \frac{2}{h^2}((1 - \cos(k\pi h)) = \frac{4}{h^2}\sin^2(\frac{k\pi h}{2})$. For $k < \frac{n}{2}$,

$$\lambda_{n+1-k} = \frac{2}{h^2}(1 + \cos(k\pi h)) = \frac{4}{h^2}\cos^2(\frac{k\pi h}{2}).$$

2) $A\varphi_k \to RA\varphi_k$.

Lemma 9.5

Let $k < \frac{n}{2}$,

$$(R\varphi_k)_{2i} = \cos^2(\frac{k\pi h}{2})(\Phi_k)_{2i},$$

$$(R\varphi_{n+1-k})_{2i} = \sin^2(\frac{k\pi h}{2})(\varphi_{n+1-k})_{2i} = -\sin^2(\frac{k\pi h}{2})(\Phi_k)_{2i},$$

$$(R\varphi_{\frac{n+1}{2}})_{2i} = 0.$$

Proof.

$$\begin{aligned}(R\varphi)_{2i} &= \frac{1}{4}\sin((2i-1)k\pi h) + 2\sin(2ik\pi h) + \sin((2i+1)k\pi h),\\ &= \frac{1}{4}(2 + 2\cos(k\pi h))\sin(2ik\pi h),\\ &= \cos^2(\frac{k\pi h}{2})\sin(2ik\pi h).\end{aligned}$$

The other formulas are also obtained by straightforward trigonometric identities. □

3) $RA\varphi_k \to A_H^{-1}RA\varphi_k$.

The eigenvalues of A_H are

$$\begin{aligned}\frac{2}{H^2}(1 - \cos(k\pi H)) &= \frac{1}{2h^2}(1 - \cos(2k\pi h)),\\ &= \frac{1}{h^2}\sin^2(k\pi h).\end{aligned}$$

Lemma 9.6

$$(A_H^{-1}RA)\varphi_k = \mu\Phi_k$$

with

$$\mu = \begin{cases} 1 & \text{if } k < \frac{n+1}{2} \\ -1 & \text{if } k > \frac{n+1}{2} \\ 0 & \text{if } k = \frac{n+1}{2} \end{cases}$$

Proof. Let $k < \frac{n+1}{2}$. Starting with φ_k. Applying A multiplies by $\frac{4}{h^2}\sin^2(\frac{k\pi h}{2})$, applying R we get Φ_k and multiplies by $\cos^2(\frac{k\pi h}{2})$ and then we divide by the eigenvalue of A_H, that is $\frac{1}{h^2}\sin^2(k\pi h)$. Therefore

$$(A_H^{-1}RA)\varphi_k = \frac{\frac{4}{h^2}\sin^2(\frac{k\pi h}{2})\cos^2(\frac{k\pi h}{2}))}{\frac{1}{h^2}\sin^2(k\pi h)}\Phi_k.$$

Clearly, the multiplicative factor is 1. The proof is similar for $k > \frac{n+1}{2}$. □

4) $\Phi_k \rightarrow P\Phi_k$.

Lemma 9.7

$$\begin{aligned}(P\Phi_k)_{2i} &= (\varphi_k)_{2i}, \\ (P\Phi_k)_{2i+1} &= \cos(k\pi h)(\varphi_k)_{2i+1}.\end{aligned}$$

Proof. The first equality is obvious and for the other one

$$(P\Phi_k)_{2i+1} = \frac{\sin(2ik\pi h) + \sin((2i+2)k\pi h)}{2} = \cos(k\pi h)\sin((2i+1)k\pi h).$$

Putting all these four steps together, we get

Proposition 9.8

$$(PA_H^{-1}RA)\varphi_k = \psi_k.$$

Let $k < \frac{n+1}{2}$. *Then*

$$\begin{aligned} (\psi_k)_{2i} &= (\varphi_k)_{2i}, \\ (\psi_k)_{2i+1} &= \cos(k\pi h)(\varphi_k)_{2i+1}, \\ \psi_{n+1-k} &= -\psi_k, \\ \psi_{\frac{n+1}{2}} &= 0. \end{aligned}$$

Proof. This proposition summarizes the previous lemmas if we note that

$$\begin{aligned} \cos(k\pi h)(\varphi_k)_{2i+1} &= \cos(k\pi h)(\varphi_{n+1-k})_{2i+1}, \\ &= -\cos((n+1-k)\pi h)(\varphi_{n+1-k})_{2i+1}. \end{aligned}$$

This shows that $\psi_{n+1-k} = -\psi_k$. □

The problem that we face is that the subspace spanned by φ_k is not invariant under $PA_H^{-1}RA$ but fortunately ψ_k can be expressed in a simple way from φ_k and φ_{n+1-k}.

Lemma 9.9

ψ_k *is uniquely written as*

$$\psi_k = \frac{1+c_k}{2}\varphi_k + \frac{c_k - 1}{2}\varphi_{n+1-k},$$

where $c_k = \cos(k\pi h)$.

Proof. Suppose

$$\psi_k = \beta\varphi_k + \gamma\varphi_{n+1-k},$$

but

$$(\psi_k)_{2i} = \beta(\varphi_k)_{2i} - \gamma(\varphi_k)_{2i} = (\varphi_k)_{2i}.$$

Hence $\beta - \gamma = 1$ and

$(\psi_k)_{2i+1} = (\beta + \gamma)(\varphi_k)_{2i+1} = c_k(\varphi_k)_{2i+1}$.

Hence $\beta + \gamma = c_k$, therefore $\beta = \frac{1+c_k}{2}$ and $\gamma = \frac{c_k - 1}{2}$. □

This decomposition is unique because φ_k and φ_{n+1-k} are independent as being eigenvectors of A. As a consequence of this result we have

Proposition 9.10

Let $k < \frac{n+1}{2}$. Then the subspace spanned by φ_k and φ_{n+1-k} is invariant under K.

Proof.

$$\begin{aligned} K\varphi_k &= \varphi_k - \psi_k = \left(1 - \tfrac{1+c_k}{2}\right)\varphi_k - \tfrac{c_k-1}{2}\varphi_{n+1-k}, \\ K\varphi_{n+1-k} &= \varphi_{n+1-k} - \psi_k = -\tfrac{1+c_k}{2}\varphi_k + \left(1 + \tfrac{c_k-1}{2}\right)\varphi_{n+1-k}, \\ K\varphi_{\frac{n+1}{2}} &= \varphi_{\frac{n+1}{2}}. \end{aligned}$$

□

The φ_k's are an orthonormal basis of the whole space so K can be written in that basis. Since $\text{span}(\varphi_k, \varphi_{n+1-k})$ is invariant it is convenient to take the vector basis in the order $\varphi_1, \varphi_n, \varphi_2, \varphi_{n-1}, \varphi_3, \ldots, \varphi_{\frac{n+1}{2}}$. Then, the matrix representing K is block diagonal, all blocks being 2 by 2 except the last block which is one by one. The kth block is

$$\frac{1}{2}\begin{pmatrix} 1 - c_k & 1 + c_k \\ 1 - c_k & 1 + c_k \end{pmatrix}.$$

The last block is equal to 1. In this form it is very easy to find the eigenvalues of K.

Proposition 9.11

The eigenvalues of K are 0 *and* 1.

Proof. By direct computation in this simple example we find the result that we have shown for more general problems in lemma 9.1. □

To conclude for the two–grid algorithm, we must include the operator S and study $M = SK$. The φ_k's being eigenvectors of S, we obtain

$$\begin{aligned} S\varphi_k &= (1-\alpha(1-c_k))\varphi_k, \\ S\varphi_{n+1-k} &= (1-\alpha(1+c_k))\varphi_{n+1-k}, \\ S\varphi_{\frac{n+1}{2}} &= (1-\alpha)\varphi_{\frac{n+1}{2}}. \end{aligned}$$

Proposition 9.12

M can be represented by a block diagonal matrix whose blocks are

$$\begin{aligned} M_k &= \frac{1}{2}\begin{pmatrix} (1-c_k)[1-\alpha(1-c_k)] & (1+c_k)[1-\alpha(1-c_k)] \\ (1-c_k)[1-\alpha(1+c_k)] & (1+c_k)[1-\alpha(1+c_k)] \end{pmatrix}, \\ M_{\frac{n+1}{2}} &= 1-\alpha. \end{aligned}$$

The eigenvalues of M_k are 0 and $1-\alpha(1+c_k^2)$.

Proof. Because $\det K = 0$, it is obvious that $\det M_k = 0$ and $\mathrm{trace}(M_k) = 1-\alpha(1+c_k^2)$. □

This gives us conditions of convergence for the two–grid method.

Theorem 9.13

For our 1D example, the two–grid method converges if $\alpha < \frac{2}{1+c_1^2}$. The optimal value of α is $\alpha_{\mathrm{opt}} = \frac{2}{2+c_1^2}$ and $\rho(M) = \frac{c_1^2}{2+c_1^2} \le \frac{1}{3}$.

Proof. The eigenvalues of M are 0, $1-\alpha$, $1-\alpha(1+c_k^2)$ for $k = 1, \ldots, \frac{n-1}{2}$. □

We remark that as $h \to 0$, $\alpha_{\mathrm{opt}} \to \frac{2}{3}$ and $\rho(M) \to \frac{1}{3}$. These two values were already found for the smoothing operator. It means that, in this case, a study of the smoothing operator alone gives a good insight into the final result. This fact is more generally true for the multigrid method.

If one takes $\nu = 2$, then $\alpha_{\mathrm{opt}} \approx \frac{2}{3}$ and $\rho(M) \approx \frac{1}{9}$. In these results the important thing to note is that $\rho(M)$ is bounded independently of h. The efficiency will be the same whatever the value of h. The spectral radius gives insight into the asymptotic convergence but as it is interesting to get bounds for the error reduction, we must study some norms of M. Let us consider, for instance, the spectral norm

$$\|M\| = \rho(MM^T)^{1/2}$$

and $\nu_1 = 1$, $\nu_2 = 0$, so that $M = KS$. In the eigenvector basis M can be written, as before, as a block diagonal matrix. A straightforward computation gives

Lemma 9.14

The eigenvalues of $M_k M_k^T$ are 0 and $\frac{1}{2}[(1-c_k)^2(1-\alpha(1-c_k))^2 + (1+c_k^2)(1-\alpha(1+c_k))^2]$.

□

If we choose the almost optimal value $\alpha = \frac{2}{3}$ the eigenvalues are 0, $\frac{1}{9}$ and $\frac{1-3c_k^2+4c_k^4}{9}$ for $k = 1, \ldots, \frac{n-1}{2}$. By lemma 9.14, $\sup_k \|M_k\| = \frac{\sqrt{2}}{3}$. This is of course greater than $\rho(M)$ whose limit is $\frac{1}{3}$.

The same kind of analysis can be done for the 2D model problem (see Stuben and Trottenberg [1220]) if we choose

1) the relaxed `Jacobi` method,

2) a weighted average as restriction

$$(Ru)_{i,j} = \frac{1}{16}[4u_{i,j} + 2(u_{i-1,j} + u_{i+1,j} + u_{i,j-1} + u_{i,j+1}) + u_{i-1,j-1} + u_{i-1,j+1} + u_{i+1,j+1} + u_{i+1,j-1}],$$

3) a bilinear interpolation,

4) for A_H the same 5 point stencil as on the fine grid. Note that, in this case, this is different from choosing PAR

For the smoothing operator Stuben and Trottenberg get the following results:

α	1	$\frac{4}{5}$	$\frac{1}{2}$
$\mu(h,\alpha)$	c_1	$\frac{1+2c_1}{5}$	$\frac{2+c_1}{4}$
$\mu^*(\alpha)$	1	$\frac{3}{5}$	$\frac{3}{4}$

$\alpha = \frac{4}{5}$ is optimal, giving the smallest value of μ^*. Setting $\rho^* = \sup_h \rho(M)$ we have

ν	μ^*	ρ^*
1	0.6	0.6
2	0.36	0.36
3	0.216	0.216
4	0.130	0.137
5	0.078	0.113

For small values of ν the study of the smoothing operator (μ^*) gives a good idea of the behaviour of $M(\rho^*)$. It should be stressed again that the different choices we made were mainly for the sake of simplicity in the theoretical analysis. Practical problems usually require different and specific choices.

9.4. The choices of components

Let us look at some of the choices that can be made for all the components of the method.

9.4.1. The smoothing

The relaxed `Jacobi` method was only chosen for the preceding analysis because, for the model problem, the iteration matrix has the same eigenvectors as A. But, in some practical computations, even for the model problem, the `Gauss-Seidel` method is a better choice. Unfortunately, not only does the iteration matrix has different eigenvectors than those of A, but they don't even give a complete basis. To study the smoothing effect of `Gauss-Seidel`, Brandt [192] introduced a heuristic technique that we have already used for preconditioners: the local Fourier analysis. The analysis is local in the sense that the coefficients of the elliptic operator are held fixed at the mesh point we consider. The boundary conditions are neglected *i.e.*,we study the problem in an infinite domain or with periodic boundary conditions. Furthermore, we only study the Fourier modes corresponding to "high frequencies".

Let us study the `Gauss-Seidel` smoothing for the 2D model problem. Consider one iteration of `Gauss-Seidel` and let u be the initial value and $\bar{u}$ the final value. In this section i will denote the complex number, square root of -1, hence the generic grid point will be denoted by (k, j). The iteration formula is

$$4\bar{u}_{k,j} - \bar{u}_{k-1,j} - \bar{u}_{k,j-1} = h^2 b_{k,j} + u_{k+1,j} + u_{k,j+1}. \tag{9.1}$$

Let ε be the error. Then equation (9.1) gives

$$4\bar{\varepsilon}_{k,j} - \bar{\varepsilon}_{k-1,j} - \bar{\varepsilon}_{k,j-1} = \varepsilon_{k+1,j} + \varepsilon_{k,j+1}. \tag{9.2}$$

Consider at point (k, j), the Fourier mode (l, m)

$$\varepsilon_{k,j}^{l,m} = \exp[i(lkh + mjh)\pi].$$

By substituting in (9.2), we get

Lemma 9.15

The error can be written as

$$\bar{\varepsilon}_{k,j}^{l,m} = \rho^{l,m} \varepsilon_{k,j}^{l,m}$$

with

$$\rho^{l,m} = \frac{\exp(ilh\pi) + \exp(imh\pi)}{4 - \exp(-ilh\pi) - \exp(-imh\pi)}.$$

□

We are interested in the values of $|\rho^{l,m}|$ for those modes which cannot be well approximated on the coarse grid with stepsize $H = 2h$. The indices l and m take values in a finite set but to simplify denote $\theta = lh\pi$ and $\varphi = mh\pi$, θ and φ taking values in $[0, \pi]$. Low frequencies are functions which are smooth in both directions *e.g.*,

$$G_{low} = \{(\theta, \varphi) | 0 \leq \theta < \frac{\pi}{2} \quad \text{and} \quad 0 \leq \varphi < \frac{\pi}{2}\}.$$

Then, $G_{high} = [0, \pi]^2 \setminus G_{low}$ and we want to compute the values θ, φ that give

$$\sigma = \max_{(\theta,\varphi) \in G_{high}} |\rho(\theta, \varphi)|,$$

$$\rho(\theta, \varphi) = \frac{\exp(i\theta) + \exp(i\varphi)}{4 - \exp(-i\theta) - \exp(-i\varphi)}.$$

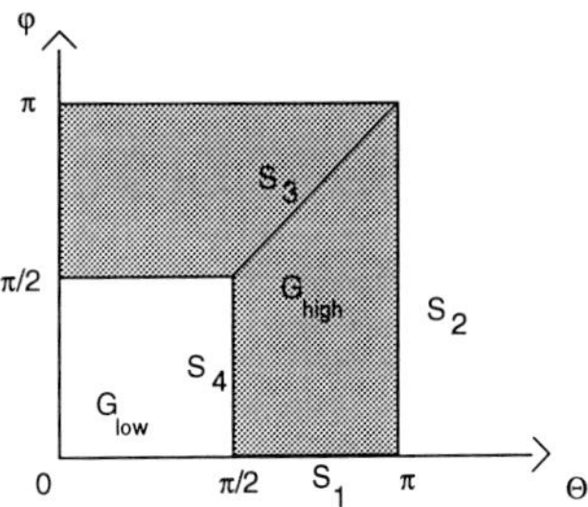

fig. 9.1: Definition of G_{high} in the (θ, ϕ) plane

Theorem 9.16

The maximum of $|\rho(\theta,\varphi)|$ is obtained for $\theta = \frac{\pi}{2}$, $\cos\varphi = \frac{4}{5}$, $\sin\varphi = \frac{3}{5}$ and then $\sigma = \frac{1}{2}$.

Proof. Let $z = \frac{1}{2}(\exp(i\theta) + \exp(i\varphi))$. Hence, $\sigma = \max|\frac{z}{2-\bar{z}}|$ where $\bar{z}$ denotes the complex conjugate of z. Since

$$|\frac{z}{2-\bar{z}}| = \frac{|z|}{|2-z|},$$

the only extrema of this function are for $z = 0$ and $z = 2$. On a compact domain σ is reached on the boundary. Therefore, we have to compute the function on segments S_1, S_2, S_3, S_4 defined in figure 9.1.

On segment S_1, $\rho^2(\theta,\varphi) = \frac{1+\cos\theta}{5-3\cos\theta}$ and the maximum $\frac{1}{5}$ is given for $\cos\theta = 0$.

On segment S_2, $\rho^2(\theta,\varphi) = \frac{1-\cos\varphi}{13-5\cos\varphi}$ and the maximum $\frac{1}{9}$ is given for $\cos\varphi = -1$.

On segment S_3, $\theta = \varphi$ and $\rho^2(\theta,\theta) = \frac{1}{5-4\cos\theta}$ and the maximum is $\frac{1}{5}$ for $\cos\theta = 0$.

On segment S_4, $\rho^2(\theta,\varphi) = \frac{1+\sin\varphi}{9-4\cos\varphi+\sin\varphi}$ and the maximum is $\frac{1}{4}$ for $\cos\varphi = \frac{4}{5}$, $\sin\varphi = \frac{3}{5}$. □

The conclusion of this heuristic analysis is that, for the model problem, `Gauss-Seidel` is a better smoother that relaxed `Jacobi` because we get a reduction factor of 0.5 instead of 0.6 without any need for a relaxation parameter. Other basic iteration schemes can be used as smoothers. Some examples are:

1) the `Gauss-Seidel` iteration with other orderings

   - the red–black ordering (this gives $\mu^* = 0.074$ if $\nu_1 = \nu_2 = 1$),
   - many color orderings, etc...

2) block relaxations

   - by lines,
   - by columns,
   - line zebra (odd–even ordering on the lines),
   - column zebra,
   - alternating directions (lines–columns or zebra lines–zebra columns),
   - Douglas, Malhotra and Schultz [446] suggested using one iteration of `Gauss-Seidel` to solve the tridiagonal systems of the ADI smoothers.

3) `ILU` or block `ILU`

4) Conjugate gradient

C. Douglas remarked that some of these choices must not be called "smoothers" as one iteration does not damp all the components of the error. Some of them can in fact increase, in which case the method should be called a "rougher". Moreover, a careful analysis has shown that the optimal number of smoothing steps is usually small (2–3) although this is problem dependent.

9.4.2. The coarsening

For finite difference problems in two dimensions, the standard coarsening is taking every other point in each direction as shown in figure 9.2. The coarse mesh has step sizes $(2h_x, 2h_y)$.

For some cases (particularly anisotropic problems) it can be useful to coarsen only in one direction (see figure 9.3) giving coarse meshes $(2h_x, h_y)$ or $(h_x, 2h_y)$. Another possibility is to use a Red–Black coarsening, see figure 9.4.

For finite element meshes, the fine mesh is usually derived from a coarse

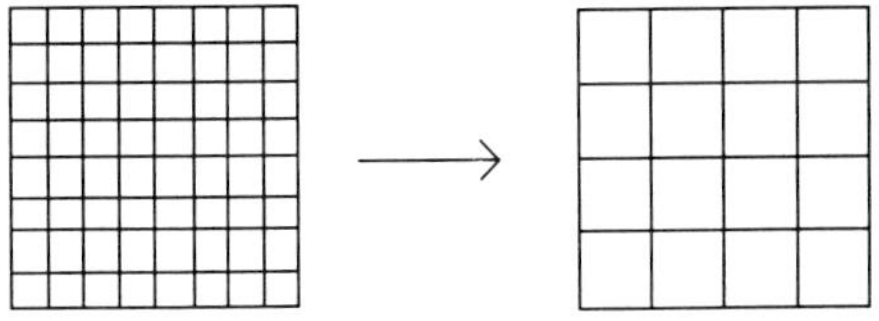

fig. 9.2: The standard coarsening

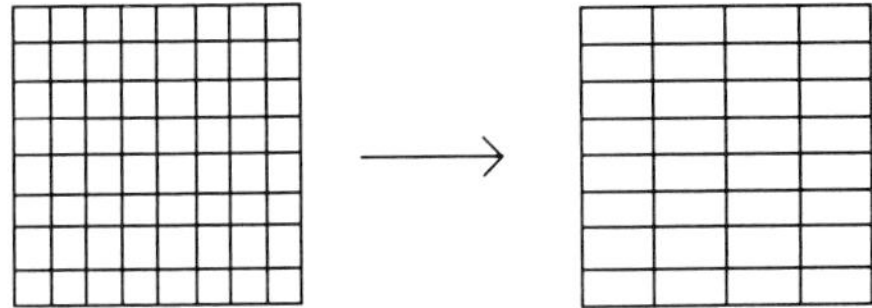

fig. 9.3: Semi coarsening

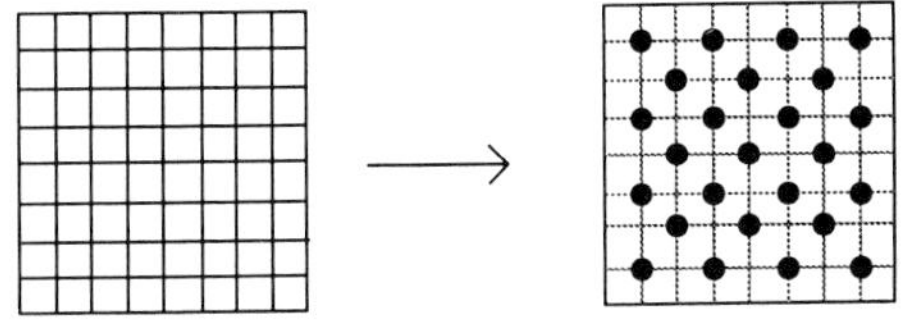

fig. 9.4: The RB coarsening

mesh by dividing each triangle into four triangles, see figure 9.5, if using linear elements. The same can be done with quadratic (having also the mid–points as unknowns) and cubic elements. However, there exist algorithms for automatically coarsening a mesh (in fact a planar graph).

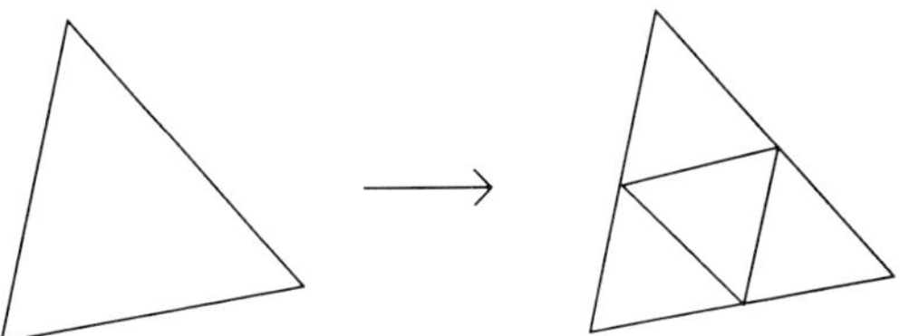

fig. 9.5: Refinement of triangles

9.4.3. Grid transfers

We have to choose the prolongation P ($\Omega_H \to \Omega_h$) and the restriction R

$(\Omega_h \rightarrow \Omega_H)$. For the prolongation (interpolation) we have the following choices:

1) bilinear interpolation. This is symbolically written as the stencil

$$\frac{1}{4}\begin{bmatrix} 1 & 2 & 1 \\ 2 & 4 & 2 \\ 1 & 2 & 1 \end{bmatrix}.$$

2) a seven point interpolation

$$\frac{1}{2}\begin{bmatrix} 1 & 1 & 0 \\ 1 & 2 & 1 \\ 0 & 1 & 1 \end{bmatrix}.$$

3) quadratic or cubic interpolation.

4) interpolation using the equations. For example if we are using 9 point finite differences on a regular mesh, the value in the middle of a coarse cell is given by

$$\frac{1}{a_{0,0}}(-\sum_{\substack{i,j=-1 \\ (i,j)\neq 0}}^{1} a_{i,j}v(x+ih, y+jh)),$$

where the $a_{i,j}$ are the coefficients of the matrix corresponding to the equation for that unknown. Remark that if we use the interpolation with the equations for the Laplacian, then we get the same result as with bilinear interpolation. This type of interpolation is generally used for PDE problems with discontinuous coefficients and also for singularly perturbed problems.

For the restriction operator R, we have the following choices:

1) it can be deduced from P by $R = cP^T$, where c is a constant ($c = 1/4$ for finite differences). If P is the bilinear interpolation, the corresponding R is known as full weighting (FW).

2) half weighting (HW). This is defined by

$$\frac{1}{8}\begin{bmatrix} 0 & 1 & 0 \\ 1 & 4 & 1 \\ 0 & 1 & 0 \end{bmatrix}.$$

3) the trivial restriction which is simply taking the value we have on the fine grid. Generally the trivial restriction is not considered to be robust and is not used.

The prolongation and restriction must satisfy a "compatibility" condition. If p and r are the orders of P and R and $2l$ is the order of the PDE, we must have $p + r \geq 2l - 1$. An analysis by C. Douglas shows that high order interpolation methods are not cost effective.

9.4.4. The coarse grid operator

For this choice there are also several possibilities:

1. using the same discretization as for the fine grid,

2. using a "Galerkin" approximation: $A_H = RA_hP$. This is quite natural in the finite element framework.

9.5. The multigrid method

So far we have solved the problem $Au = b$ using only two grids. In step 4 of the two–grid algorithm we made the assumption that we can exactly solve $A_H\varepsilon_H^k = \bar{r}_H^k$. This is certainly not realistic since with a 2D problem and standard coarsening, Ω_H has only roughly one fourth of the number of points in Ω_h. Therefore, the "coarse" problem can still be quite large. So, a natural idea is to approximately solve $A_H\varepsilon_H^k = \bar{r}_H^k$ by using the same two–grid algorithm defining a coarser grid of stepsize $2H = 4h$. To be able to use this idea recursively down to some grids containing 1 or 4 points, it is necessary that the finest grid has the correct number of points. For example, for the 2D model problem, we need in each direction $m = 2^p - 1$ for some integer p. To describe these ideas in a precise way, let us introduce an index l to label the grids we use. So, we have a sequence Ω_l of grids whose stepsizes are h_l, $l = 0$ corresponding to the coarsest grid and $l = L$ to the finest one.

Let us denote

- A_l the approximation of A on Ω_l,
- R_l the restriction operator: $\Omega_l \to \Omega_{l-1}$,
- P_{l-1} the interpolation operator: $\Omega_{l-1} \to \Omega_l$,
- S_l the iteration matrix of the smoothing operator on Ω_l.

Let $\bar{w}_l = smooth^{\nu}(w_l, A_l, b_l)$ be the result of ν smoothing iterations for the problem $A_l u_l = b_l$ starting from w_l.

The multigrid algorithm for $l+1$ grids is the following:

If $l = 1$ apply the two–grid algorithm.

If $l > 1$

1) $\bar{u}_l^k = smooth^{\nu_1}(u_l^k, A_l, b_l)$,

2) $\bar{r}_l^k = b_l - A_l \bar{u}_l^k$,

3) $\bar{r}_{l-1}^k = R_l \bar{r}_l^k$,

4) compute $\bar{v}_{l-1}^k$ as the approximate solution of

$$A_{l-1} v_{l-1}^k = \bar{r}_{l-1}^k$$

on Ω_{l-1} by doing γ iterations of the l-grid algorithm $(\Omega_{l-1}, \ldots, \Omega_0)$ starting from 0.

5) $\bar{v}_l^k = P_{l-1} \bar{v}_{l-1}^k$,

6) $u_l^{k+1} = smooth^{\nu_2}(\bar{u}_l^k + \bar{v}_l^k, A_l, b_l)$.

This method is given in recursive mode as step 4 refers to the same method with one grid less and so on, until $l = 1$ and we can apply the two–grid algorithm. There is also a new parameter γ which indicates how precisely we solve the problem on grid Ω_{l-1}.

To clarify this algorithm, we shall graphically show its behaviour graphically. We denote

$\searrow$ the restriction,

$\swarrow$ the interpolation,

- • the smoothing iterations,
- ∘ the exact solution of a problem.

When $L = 1$ this is the two–grid algorithm which graphically looks like figure 9.6.

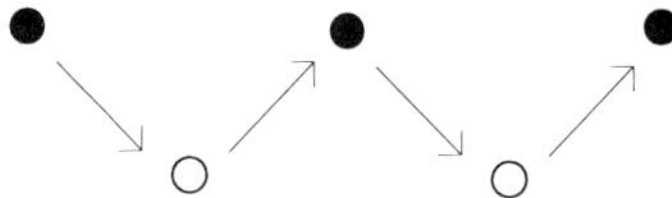

fig. 9.6: Two grids

Since we exactly solve when $l = 0$, γ has no meaning in this algorithm. With three grids we may have what is shown in figure 9.7.

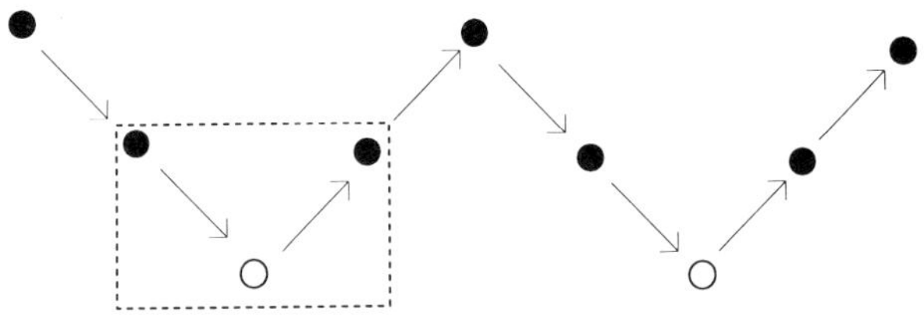

fig. 9.7: Three grids, $\gamma = 1$

Inside the dotted box is one ($\gamma = 1$) iteration of the two–grid method. For obvious graphical reasons, this is usually called a V-cycle. Using $\gamma = 2$, we get figure 9.8.

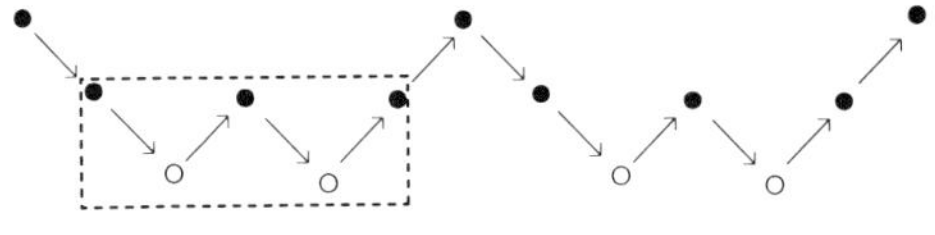

fig. 9.8: Three grids, $\gamma = 2$

This is called a W-cycle. With four grids we may have figure 9.9.

Usually the number of grids is not very large, as for instance $l = 8$ corresponds to a 512×512 points fine grid. Let us compute the iteration matrix for the multigrid method.

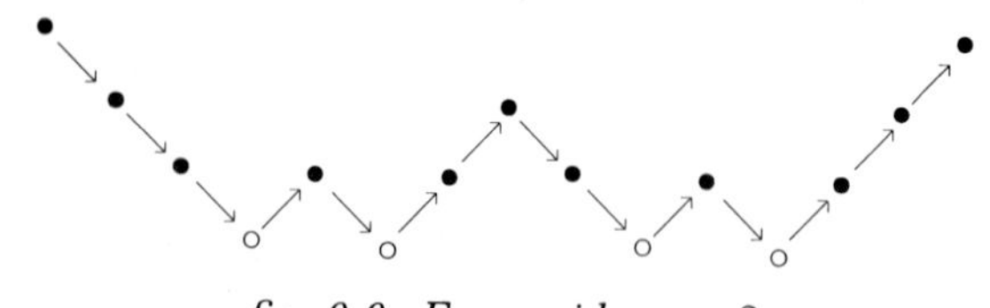

fig. 9.9: Four grids, $\gamma = 2$

Lemma 9.17

If one solves $Ax = b$ with the iterative method defined by

$$x^{k+1} = Sx^k + c \tag{9.3}$$

with $x^0 = 0$, then

$$x^{k+1} = (I - S^{k+1})A^{-1}b.$$

Proof. As (9.3) must be consistent we have $x = Sx + c$, so

$$x^{k+1} - x = S(x^k - x) = S^{k+1}(x^0 - x) = -S^{k+1}x.$$

Therefore

$$x^{k+1} = (I - S^{k+1})x.$$

□

Denote by M_l^{l-1} the iteration matrix for the two–grid method with grids Ω_l and Ω_{l-1}. We have shown previously that

$$M_l^{l-1} = S_l^{\nu_2}(I - P_{l-1}A_{l-1}^{-1}R_lA_l)S_l^{\nu_1}.$$

Going from the two–grid method to the multigrid method is replacing A_{l-1}^{-1} by γ iterations of the l–grid method starting from 0, *e.g.*,

$$A_{l-1}^{-1} \to (I - M_{l-1}^{\gamma})A_{l-1}^{-1},$$

M_{l-1} denoting the iteration matrix of the l–grid method. Therefore

$$M_l = S_l^{\nu_2}(I - P_{l-1}(I - M_{l-1}^{\gamma})A_{l-1}^{-1}R_lA_l)S_l^{\nu_1}$$

for $l > 2$ and

$$M_1 = S_1^{\nu_2}(I - P_0 A_0^{-1} R_1 A_1) S_1^{\nu_1}.$$

By straightforward algebra

$$M_l = M_l^{l-1} + (S_l^{\nu_2} P_{l-1}) M_{l-1}^{\gamma} (A_{l-1}^{-1} R_l A_l S_l^{\nu_1}).$$

To end this section we remark that the multigrid method we have defined is not the most general one we could have derived. For instance, the post–smoothing operator can be different from the pre–smoothing one, the prolongation and restriction operators can depend on the level, the coarse grid operators A_l may be defined using another operator Q_l instead of R_l, etc. . .

9.6. Convergence theory

We are going to show that if the two–grid algorithms as well as the interpolation and restriction operators satisfy certain criteria, then the multigrid method converges. There are quite a few different ways to prove multigrid convergence depending on the assumptions that are made. Those assumptions are more or less difficult to verify in practical situations.

Lemma 9.18

Let σ and C be two positive real numbers such that

$$\|M_l^{l-1}\| \le \sigma$$

and

$$\|S_l^{\nu_2} P_{l-1}\| \cdot \|A_{l-1}^{-1} R_l A_l S_l^{\nu_1}\| \le C$$

then

$$\|M_l\| \le \eta_l$$

where $\eta_1 = \sigma$ *and* $\eta_{l+1} = \sigma + C\eta_l^{\gamma}, \quad l \ge 1.$

Proof. The proof is straightforward by induction. □

To prove convergence of the multigrid method we must study the behavior of the sequence η_l.

Theorem 9.19

Suppose $\gamma = 2$ and $4C\sigma \le 1$. Then

$$\|M_l\| \le \frac{1-\sqrt{1-4C\sigma}}{2C} \le 2\sigma.$$

Proof. η_l is an increasing bounded sequence if $1 - 4C\sigma \ge 0$ whose limit is $\frac{1-\sqrt{1-4C\sigma}}{2C}$. □

This theorem shows that if technical conditions given by the constant C are satisfied, the multigrid method converges if σ is small enough, *e.g.*, if the two–grid method converges fast enough. W. Hackbush developed a theory based on different hypotheses. The theory is based on two properties. The first one is known as the smoothing property:

there are functions $\eta(\nu)$ and $\bar{\nu}(h)$ independent of l such that

$$\|A_l S_l^{\nu}\| \le \eta(\nu)\|A_l\| \text{ for all } 0 \le \nu \le \bar{\nu}(h_l).$$

These functions are such that $\lim_{\nu\to\infty} \eta(\nu) = 0$ and $\lim_{h\to 0} \bar{\nu}(h) = \infty$ or $\bar{\nu}(h) = \infty$. The smoothing property can be proved for the Richardson method, the relaxed `Jacobi` iteration, the Red–Black `Gauss-Seidel` method and also for the symmetric `Gauss-Seidel`, the `SSOR` and the `ILU` iterations, see Hackbusch [739]. The second property is the approximation property:

$$\|A_l^{-1} - PA_{l-1}^{-1}R\| \le \frac{C}{\|A_l\|}, \forall l \ge 1.$$

Generally, proofs of the approximation property are using properties of the PDE we are solving. Using these two properties, Hackbush proved the following result:

Theorem 9.20

If we assume the smoothing and approximation properties and $\|S_l^{\nu}\| \le C$, $C_1\|x_{l-1}\| \le \|Px_{l-1}\| \le C_2\|x_{l-1}\|$, $\gamma \ge 2$, then

$$\|M_l\| < 1, \text{ for } \nu \le \min \bar{\nu}(h_l),$$

provided h is small enough. If $\bar{\nu}(h) = \infty$ the choice of the grid size is not restricted.

Proof. See Hackbusch [739] □

Convergence can be proved for all ν if h is sufficiently small. This theory can be refined in some special cases, mainly if the smoothing iteration is symmetric.

Proofs of the approximation properties use the regularity of the PDE problem. Therefore, several researchers have looked for a convergence theory without regularity assumptions. Such a theory has been provided by Bramble, Pasciak, Wang and Xu [187]. The analysis is set in the finite element framework with a bilinear form $a(u, v)$. A nested sequence of spaces

$$\mathcal{M}_0 \subset \mathcal{M}_1 \subset \cdots \subset \mathcal{M}_l$$

is defined and operators A_k given by

$$(A_k u, v) = a_k(u, v) = a(u, v) = (Au, v), \quad \forall v \in \mathcal{M}_k.$$

The hypothesis (with simplifying assumptions) is that:

$$A((I - M_k A_k)u, (I - M_k A_k)u) \leq \delta_k A(u, u), \quad \forall u \in \mathcal{M}_k.$$

Linear operators Q_k: $\mathcal{M}_l \to \mathcal{M}_k$ are assumed to verify

$$\|(Q_k - Q_{k-1})u\|^2 \leq C_1 \frac{(Au, u)}{\lambda_k}, \quad \forall k = 1, \ldots, l$$

where λ_k is the largest eigenvalue of A_k,

$$A(Q_k u, Q_k u) \leq C_2 A(u, u), \quad \forall k = 0, \ldots, l - 1$$

$$\frac{\|u\|^2}{\lambda_k} \leq C_3 (S_k u, u), \quad \forall u \in \text{the range of } S_k.$$

Under these hypotheses, Bramble, Pasciak, Wang and Xu proved that

$$\delta_k = 1 - \frac{1}{Ck},$$

where $C = [1 + C_2^{1/2} + (C_3 C_1)^{1/2}]^2$. Note that these results do not depend on the number of smoothing iterations. However, they can be applied using standard hypothesis on finite element approximations. Another application is given in [187] to finite elements with local refinement. Multigrid convergence can also be studied by specializing general theories about multilevel methods, see J. Xu's Ph.D. thesis [1340].

C. Douglas (see [444]) has defined a variant of the multigrid algorithm that can be analyzed with algebraic hypothesis. Again, a set of approximation spaces $\{\mathcal{M}\}_{k=1}^{l}$ is given but not necessarily nested. In addition to the restrictions R_k and prolongation P_k, a mapping Q_k is defined such that $A_{k-1} = Q_k A_k P_{k-1}$. An extra level $l+1$ is introduced such that

$$\mathcal{M}_{l+1} = \mathcal{M}_l, \quad P_l = R_{l+1} = Q_{l+1} = I, \quad A_{l+1} = A_l,$$

and the initial residual on level $l+1$, $z_{l+1} = A_{l+1}x_{l+1}^{(-1)} - b$. The algorithm NSMG$(k, z_{k+1}, x_k^{(-1)})$ is the following:

1) compute $R_{k+1}z_{k+1} \in \mathcal{M}_k$,

2) smoothing: $x_k^{(0)} = M_k^{(1)} x_k^{(-1)}$ such that

$$A_k x_k^{(0)} + R_{k+1}z_{k+1} = z_k^{(0)},$$

where $\|z_k^{(0)}\| \le \rho_k^{(1)} \|z_{k+1}\|$.

3) let $\hat{x}_k^{(1)} = x_k^{(0)}, \hat{z}_k^{(1)}$ and $\gamma_1^{(1)} = 0$,

4) for $i = 1, \ldots, \mu_k$

4a) if $i > 1$

$$A_k x_k^{(i-1)} + R_{k+1}z_{k+1} = \hat{\theta}_k^{(i)},$$

smoothing: $\hat{x}_k^{(i)} = M_k^{(i)} x_k^{(i-1)}$ such that

$$A_k x_k^{(i)} + R_{k+1}z_{k+1} = \hat{z}_k^{(i)}, \quad \|\hat{z}_k^{(i)}\| \rho_k^{(i)} \|\hat{\theta}_k^{(i)}\|,$$

4b) if $k > 1$

correction: $\gamma_k^{(i)} =_{k-1} \bar{x}_{k-1}^{(i)}$ where $\bar{x}_{k-1}^{(i)} =$ NSMG$(k - 1, \hat{z}_k^{(i)}, 0)$,

4c) $A_k(\hat{x}_k^{(i)} + \gamma_k^{(i)}) + R_{k+1}z_{k+1} = \theta_k^{(i)}$,

4d) smoothing: $x_k^{(i)} = N_k^{(i)}(\hat{x}_k^{(i)} + \gamma_k^{(i)})$ such that

$$A_k x_k^{(i)} + R_{k+1}z_{k+1} = z_k^{(i)}, \quad \|z_k^{(i)}\| \le \epsilon_k^{(i)} \|\theta_k^{(i)},$$

5) return $x_k^{(\mu_k)}$.

Note that, in this formulation, the number of smoothing steps can vary at each iteration if $\rho_k^{(i)}$ and $\epsilon_k^{(i)}$ are fixed. Two types of results were given in C. Douglas and J. Douglas [444]. The first one only assumes the sets $\mathcal{M}_k$ are nested and that

$$\|(I - Q_k^{-1} R_k)u\| \leq \delta_k \|u\|, \forall u \in \mathcal{M}_k.$$

Remark that usually Q_k cannot be inverted. Therefore, we must give a meaning to Q_k^{-1}. For some finite element problems, we can take Q_k^{-1} as being the injection of $\mathcal{M}_{k-1}$ into $\mathcal{M}_k$. Otherwise, it is chosen as a pseudo–inverse. Then, if P_k is the imbedding into $\mathcal{M}_{k+1}$, we have

$$\|Q_l^{-1} z_l^{(\mu_l)}\| \leq C_l^{(\mu_l)} \|z_{l+1}\|,$$

where the constants $C_k^{(\mu_k)}$ are computed as

$$C_1^{(1)} = \epsilon_1^{(1)} \rho_1^{(1)}, \quad C_k^{(\mu_k)} = \prod_{i=1}^{\mu_k} (\epsilon_k^{(i)} \rho_k^{(i)} [\delta_k + C_{k-1}^{(\mu_{k-1})}]), \ k > 1.$$

On some simple examples like the Poisson model problem, these bounds are far from being sharp. To get better bounds Douglas and Douglas [444] introduced a decomposition $\mathcal{M}_k = \mathcal{S}_k \oplus \mathcal{T}_k$ where $\mathcal{T}_k = \mathcal{M}_k$ and $\mathcal{M}_{k-1}^{\perp} \cap \mathcal{M}_k$. The set $\mathcal{S}_k$ contains the high frequency components and $\mathcal{T}_k$ the low frequencies. Then, refined bounds can be obtained by considering what happens to the high and low frequencies before and after smoothing, see [444].

9.7. Complexity of multigrid

In this section we would like to estimate the number of floating point operations for the standard multigrid method. Note that what is really interesting in this kind of elliptic problem solvers is knowing the amount of work required to make the error between the exact solution u of the PDE and the solution u_h^k given by the discrete algorithm smaller than a given threshold. This error is the sum of the difference between u and u_h the exact solution of the discrete problem $\|u - u_h\|$ and the difference $\|u_h - u_h^k\|$. If the cost per step of the iterative method is Cn and if the factor of reduction of the norm of the error is independent of h ($i.e.$,n), then to have $\|u - u_h^k\| \leq \varepsilon$, the cost is $O(n \log n)$.

Let us denote by n_l the number of grid points in Ω_l, W_l the number of operations for one multigrid iteration, W_{l+1}^l the number of operations for one two–grid iteration *without* solving on the coarse grid. Then it is obvious that

$$W_{l+1} = W_{l+1}^l + \gamma W_l,$$
$$W_1 = W_1^0 + W_0,$$

W_0 being the number of operations to solve exactly on Ω_0. The solution of the recurrence relation for W_l is

$$W_l = \sum_{k=1}^{l} \gamma^{l-k} W_k^{k-1} + \gamma^{l-1} W_0.$$

Let us examine the 2D model problem. Then, $n_l = 4n_{l-1}$ so $n_l = 4^l n_0$ and we shall see later on that $W_l^{l-1} = Cn_l$, C being a small constant. Therefore, we have the following result:

Theorem 9.21

For the 2D model problem and with $\gamma \leq 3$ the number of floating point operations for the multigrid method is of the order of n the number of points on the finest grid.

Proof.

$$\frac{W_l}{C} = (\gamma^l \sum_{k=1}^{l} (\frac{4}{\gamma}^k) + \gamma^{l-1}) n_0,$$

but $\sum_{k=1}^{l} (\frac{4}{\gamma})^k = (\frac{4^{l+1}}{\gamma^l} - 4)/(4-\gamma)$ so,

$$\frac{W_l}{C} = \frac{4^{l+1} - \gamma^l}{4-\gamma} n_0.$$

$$\frac{W_l}{C} = \begin{cases} \frac{4^{l+1}}{3} n_0 \leq \frac{4}{3} n_l & \text{for } \gamma = 1, \\ \frac{4^{l+1}-2^l}{2} n_0 \leq 2n_l & \text{for } \gamma = 2, \\ (4^{l+1} - 3^l) n_0 \leq 4n_l & \text{for } \gamma = 3. \end{cases}$$

But note that if $\gamma = 4$, then

$$\frac{W_l}{C} = (\gamma^l \sum_{k=1}^{l} 1 + \gamma^{l-1}) n_0 = l n_l + \frac{n_l}{4}.$$

Since l is almost $\log_2 n_l$, $\frac{W_l}{C} = O(n_l \log_2 n_l)$. Therefore, if $\gamma = 1, 2, 3$, $W_l = O(n_l)$. □

Of course this result is true not only for the model problem but more generally for every problem on a square. Let us now try to estimate C for this problem. We also study the components defined at the beginning, that is relaxed `Jacobi`, bilinear interpolation and weighted average restriction for a five point stencil. The `Jacobi` iteration can be conveniently expressed in terms of the residual. We need to restrict the residual to the coarse grid. Therefore, the first thing to do is to compute the residual. This requires 5 multiplications and 5 additions for each point. Knowing the residual, the relaxed `Jacobi` method needs only 1 multiplication and 1 addition per point if the reciprocals of the diagonal entries of the matrix are stored (remember that division is usually much more costly than a multiply). Then, we have to restrict the residual but only for points of the coarse grid, this amounts for 1 multiplication and 1 addition for each point of the fine grid. Bilinear interpolation requires 1/2 multiplication and 5/4 additions per point. The total is approximately 8 mults and 8 adds per grid point of the fine grid, therefore C is 16. One can see that for usual dimensions the number of operations is roughly the same as, for instance, for the `FACR`(l) method of Chapter 4, but multigrid methods can handle much more general problems.

The storage needed for the multigrid method is a little larger than for the more classical iterations but not by much. For instance, for the 2D model problem, the storage is almost $\frac{4}{3}n_l$. Generally for 2D problems the extra storage (for storing the vectors on the grid hierarchy) is about 30%.

9.8. The full multigrid method

In the multigrid method one starts iterating from an initial vector defined on the finest grid and doing relaxation steps and going down to coarser grids gives corrections to this starting vector. A natural idea is to construct the starting vector for multigrid using the coarse grids. This is known as the full multigrid method (`FMG`) or nested iteration (`NI`). To make this idea clear, let us denote by $M^r(w, A_l, b_l)$ the operator corresponding to r iterations of the multigrid method to solve $A_l u = b_l$ on $\Omega_l, \ldots, \Omega_0$ starting from w. Then, the `FMG` method is defined as

1) let $\widetilde{u}_0$ be the exact solution of $A_0 \widetilde{u}_0 = b_0$

2) for $k = 1, \ldots, l$

$$u_k^0 = \Pi_{k-1}\widetilde{u}_{k-1},$$
$$\widetilde{u}_k = M^r(u_k^0, A_k, b_k),$$

where Π_{k-1} is an interpolation operator, $\Omega_{k-1} \rightarrow \Omega_k$ which may be distinct from P_{k-1}. $\widetilde{u}_l$ is said to be an approximation of the true solution.

Clearly, the full multigrid method starts from the coarsest grid and at each step interpolates to the next finest grid doing r iterations of the multigrid method before going upwards. As examples and with the same notation as before and denoting interpolation by Π_{k-1} as $\nearrow$, the method is depicted on figure 9.10 with three grids and $\gamma = 1$, $r = 1$.

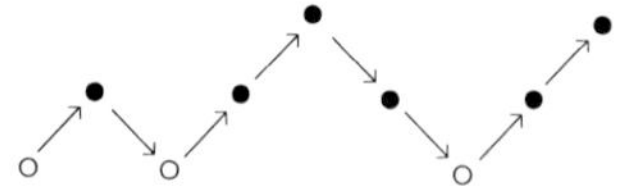

fig. 9.10: Full multigrid, three grids, $\gamma = 1, r = 1$

With four grids and $\gamma = 2$, $r = 1$, we have figure 9.11.

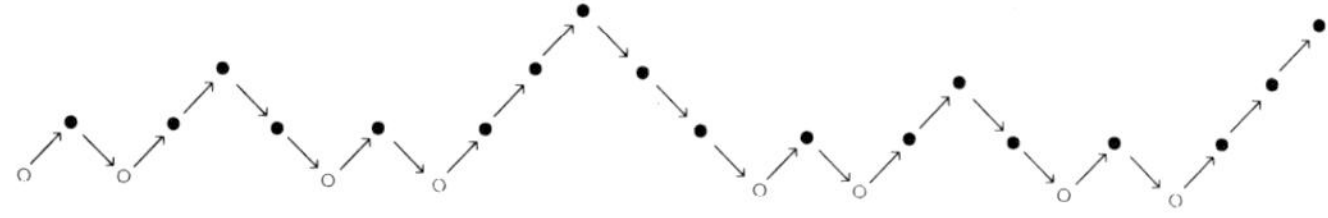

fig. 9.11: Full multigrid, four grids, $\gamma = 2, r = 1$

The question is to know how far $\widetilde{u}_l$ is from the solution of the problem. Let u_l be the exact solution of $A_l u_l = b_l$, we shall see that the difference between $\widetilde{u}_l$ and u_l is of the same order as the truncation error, *i.e.*,the difference between u_l and u the solution of the underlying continuous problem. Therefore, it is usually enough to stop with $\widetilde{u}_l$. To prove this result we need some technical hypotheses:

1) $\|M_l\| \leq \eta < 1$, this means that the multigrid method we use at each step is convergent. We assume that $h_{l-1} = 2h_l$.

2) $\|\Pi_{l-1}\| \leq C$, C being a constant independent of h_l.

3) $\|u - u_l\| \leq K_1 h_1^k$, *i.e.*,the truncation error is of order $h_l^{k_1}$.

4) $\|w - \Pi_{l-1}w\| \leq K_2 h_l^{k_2}$.

One can note that u is a function and u_l a vector so it seems that $\|u - u_l\|$ has no meaning. If we use the l_2–norm, this is a shorthand notation for

$$\sum_{(i,j)\in\Omega_l} (u(x_i, y_j) - (u_l)_{i,j})^2)^{1/2}.$$

Theorem 9.22

If $k_2 > k_1$ and $\eta^r C 2^{k_1} < 1$, then

$$\|\widetilde{u}_l - u_l\| \leq \delta h_l^{k_1}, \quad \textit{where} \quad \delta = \frac{\eta^r K_1(1 + C_1)}{1 - \eta^r C_1} + O(1),$$

with $C_1 = C 2^{k_1}$.

Proof. By definition

$$\widetilde{u}_l - u_l = M_l^r(u_l^0 - u_l),$$

and

$$u_l^0 = \Pi_{l-1}\widetilde{u}_{l-1},$$

then

$$u_l^0 - u_l = \Pi_{l-1}(\widetilde{u}_{l-1} - u_{l-1}) + \Pi_{l-1}(u_{l-1} - u) + (\Pi_{l-1}u - u) + u - u_l.$$

Let $\delta_l = \frac{\|\widetilde{u}_l - u_l\|}{h_l^{k_1}}$, then

$$\delta_l \leq \eta^r \left(C\frac{\|\widetilde{u}_{l-1} - u_{l-1}\|}{h_l^{k_1}} + C\frac{\|u - u_{l-1}\|}{h_l^{k_1}} + K_2 h_l^{k_2-k_1} + K_1 \right).$$

As $h_{l-1} = 2h_l$, we have

$$\frac{\|\widetilde{u}_{l-1} - u_{l-1}\|}{h_l^{k_1}} = 2^{k_1}\frac{\|\widetilde{u}_{l-1} - u_{l-1}\|}{h_{l-1}^{k_1}} = 2^{k_1}\delta_{l-1}$$

and

$$\frac{\|u - u_{l-1}\|}{h_l^{k_1}} = 2^{k_1}\frac{\|u - u_{l-1}\|}{h_{l-1}^{k_1}} \leq 2^{k_1}K_1.$$

Therefore

$$\delta_l \leq \eta^r (C 2^{k_1} \delta_{l-1} + C K_1 2^{k_1} + K_1 + K_2 h_l^{k_2 - k_1}).$$

Hence,

$$\delta_l \leq \eta^r (C \delta_{l-1} + K_1 (1 + C_1) + K_2 h_l^{k_2 - k_1}).$$

If $k_2 > k_1$, we have

$$\delta_l \leq \eta^r (C_1 \delta_{l-1} + K_1 (1 + C_1) + O(h_l)).$$

The result follows from this inequality when $l \to \infty$. □

When we use FMG, then the cost to reach the order of the discretization error (which is all we can expect) is proportional to n_l. The number of multigrid iterations that are done at each step is usually quite small. For instance, for the model problem in 2D, it is enough to do only one iteration.

9.9. Vector and parallel multigrid

Regarding vectorization, the problem is similar to classical iterative methods. The chosen smoothing method must be vectorizable, for instance Red-Black Gauss-Seidel. When we are going to the coarsest grids, we take every other point which means there could be a stride different from 1 in the vectors we use and this can give memory bank conflicts. This problem can be avoided by using special storage schemes. Another problem which is more important is the following: as we go down to coarsest grids the vector length becomes shorter and shorter, the computational speed decreases and below some levels it is more efficient to do scalar computation.

The work about parallelizing the multigrid method has been well summarized by C. Douglas [443] and also R. Tuminaro [1268]. The most basic way to parallelize multigrid is to use the standard method and the techniques that are usually invoked to introduce parallelism in classical iterative methods. Domain decomposition methods that we shall study in Chapter 10 are good candidates to parallelize the smoothers. Of course, when we reach the coarsest grids, it may be that the number of unknowns assigned to each processor becomes very small and some processors may even be idle. This could or could not be a problem, the main goal being to compute the solution faster than on a serial computer at a reasonable cost. Asynchronous iterations can also be used as smoothers although this is not too much in favor, mainly because only a small number of smoothing iterations are usually used.

There are two main kinds of multigrid variants for parallel computers:

- Concurrent iterations,
- Convergence acceleration.

The first approach tries to process all levels concurrently. The problem is distributed over all grids, keeping all the processors busy all the time. Relaxation sweeps are done also on all levels and, then the solutions are combined. Such an algorithm was proposed by Gannon and Van Rosendale [600]. It assumes that $n \log_2 n$ processors are available for 2D problems. All operations are performed simultaneously on all unknowns on all levels. The solution is obtained as

$$\hat{u} = \sum_{k=coarse}^{fine} I_k^l u_k,$$

where I_k^l are interpolation operators from grid level k to level l. The residual is redistributed over all grids and relaxations are performed on all grids. The convergence rate is slower than standard multigrid but, as all processors are kept busy, the solution can be obtained much faster on parallel computers. For an analysis of this method, see Douglas [443] and Tuminaro [1268].

To introduce the second approach, consider that if we are solving a 2D problem with one grid point per processor, on the next level 3/4 of the processors are idle. Therefore, a natural idea is to try solving 3 additional coarse problems and combine the result to accelerate convergence. This leads to having the same number of unknowns on each level as this idea is applied recursively. There are several methods using this idea. One method, the Parallel Superconvergent Multigrid, was developed by Frederickson and Mc Bryan [580]. They used the same interpolation and projection as in the standard method. An optimized Richardson method is used as the smoothing. Experiments show that the convergence rate is better than the standard one.

Chan and Tuminaro [296] proposed a method where the residual r is split as

$$r_2 = Zr, \quad r_1 = r - r_2.$$

$Ax_1 = r_1$ is projected on coarse grids and $Ax_2 = r_2$ is solved by relaxation sweeps. Z is a filter chosen to split the low and high frequencies.

Another kind of methods use approximate A–orthogonal spaces. A matrix P_i is A–orthogonal to P_j if $P_iAP_j = 0$. Suppose that $R_iAP_j = 0, 1 \geq i, j \geq q, i \neq j$, let

$$A_i\tilde{x}_i = R_ib, \quad A_i = R_iAP_i.$$

Then the approximate solution is

$$\hat{x} = \sum_{i=1}^{q} P_i\tilde{x}_i.$$

The residual $\hat{r} = b - A\hat{x}$ is orthogonal to the subspace spanned by the columns of R_i. Note that if R_i span the entire space, $\hat{x}$ is the exact solution. However, this not really practical, therefore it is interesting to study methods using approximate A–orthogonal spaces. As was done by Tuminaro [1268] let us consider an algorithm using four coarse grids. The iteration matrix for the two–grid version is

$$M = (A^{-1} - \sum_{i=1}^{4} P_iA_i^{-1}R_i)AS^{\nu}.$$

Let

$$X = \begin{pmatrix} R_1 \\ R_2 \\ R_3 \\ R_4 \end{pmatrix}.$$

Each R_i satisfies $R_iX^{-1} = E_i$ where

$$E_i = (\,\Delta_{i,1} \quad \Delta_{i,2} \quad \Delta_{i,3} \quad \Delta_{i,4}\,),$$

$\Delta_{i,j}$ being I if $i = j$ and 0 otherwise. T is similar to

$$(I - \sum_{i=1}^{4} \begin{pmatrix} R_1AP_i(R_iAP_i)^{-1} \\ R_2AP_i(R_iAP_i)^{-1} \\ R_3AP_i(R_iAP_i)^{-1} \\ R_4AP_i(R_iAP_i)^{-1} \end{pmatrix} E_i)\bar{S}^{\nu},$$

with $\bar{S} = XASA^{-1}X^{-1}$. This can be written as $W\bar{S}^{\nu}$ with

$$W = -\begin{pmatrix} 0 & \Phi_{1,2} & \Phi_{1,3} & \Phi_{1,4} \\ \Phi_{2,1} & 0 & \Phi_{2,3} & \Phi_{2,4} \\ \Phi_{3,1} & \Phi_{3,2} & 0 & \Phi_{3,4} \\ \Phi_{4,1} & \Phi_{4,2} & \Phi_{4,3} & 0 \end{pmatrix},$$

where $\Phi_{i,j} = (R_i A P_j)(R_j A P_j)^{-1}$. The matrix W measures the deviation from A–orthogonality. If we construct two operators such that $R_2 P_1 = 0$, then R_2 and P_1 are also A–orthogonal if $Ax \in span(P_1)$ for all $x \in span(P_1)$. Methods by Douglas and Miranker, see Douglas [440] satisfy this requirement. For some problems, the operators are exactly A–orthogonal and this leads to a direct method.

The methods introduced by Ta'asan and also by Hackbush [737], [738]) fit into this framework. These methods use interleaved grids which for a small problem in a square look like

$$
\begin{matrix}
3 & 4 & 3 & 4 & 3 & 4 & 3 & 4 \\
1 & 2 & 1 & 2 & 1 & 2 & 1 & 2 \\
3 & 4 & 3 & 4 & 3 & 4 & 3 & 4 \\
1 & 2 & 1 & 2 & 1 & 2 & 1 & 2 \\
3 & 4 & 3 & 4 & 3 & 4 & 3 & 4 \\
1 & 2 & 1 & 2 & 1 & 2 & 1 & 2
\end{matrix}
$$

In Hackbush's method the four coarse spaces correspond to low–low frequencies, high–high in x and y, low(x)–high(y), high(x)–low(y). The restriction operators are defined as

$$
R_1 = \frac{1}{8}\begin{pmatrix} 1 & 2 & 1 \\ 2 & 4 & 2 \\ 1 & 2 & 1 \end{pmatrix}, \quad R_2 = \frac{1}{8}\begin{pmatrix} -1 & 2 & -1 \\ -2 & 4 & -2 \\ -1 & 2 & -1 \end{pmatrix},
$$

$$
R_3 = \frac{1}{8}\begin{pmatrix} -1 & -2 & -1 \\ 2 & 4 & 2 \\ -1 & -2 & -1 \end{pmatrix}, \quad R_4 = \frac{1}{8}\begin{pmatrix} 1 & -2 & 1 \\ -2 & 4 & -2 \\ 1 & -2 & 1 \end{pmatrix},
$$

where R_1 is applied at the even points, R_2 at the even points in y, odd in x, R_3 is applied at the odd points in y, even in x and R_4 is applied at the odd points. The prolongation operator is $P_i = R_i^T$ and $A_i = R_i A P_i$. A Fourier analysis explaining how this method works is given in the Ph.D. thesis of R. Tuminaro [1268] and subsequent papers [1269],[1270]. It turns out that this method is more robust than standard multigrid and can be used for discontinuous coefficients or anisotropic problems.

9.10. Algebraic multigrid

The standard multigrid method that we have explained in the previous sections cannot be used as a black box solver for any PDE problem. For some problems like singularly perturbed equations, anisotropic problems or even problems with strong variations in the coefficients, the standard multigrid method is not robust. There are some (problem dependent) solutions to fix these problems. For instance, semi coarsening (i.e. coarsening in only one dimension) can be used and/or interpolation defined by using the fine grid matrix. Another solution is to use more robust smoothers like line `Gauss-Seidel` or `ILU`. As an example, see the method defined by Dendy [409]. Nevertheless, in all these variants, there is still a geometric type of coarsening related to the mesh of the discretization of the problem. A major problem is that the user needs some good knowledge of multigrid methods to be able to devise a robust method for his/her own problem.

Another type of methods is known as algebraic multigrid (`AMG`). Then, only the matrix A is used regardless of the underlying PDE problem if any. This was introduced by Ruge and Stuben [1127]. It was hoped that this method could be used for much more general problems than the standard multigrid method. However, it is not so easy to obtain a $O(n)$ algorithm.

To describe the algorithm, we use the notation of Grauschopf, Griebel and Regler [690]. At each level we have to partition the set of unknowns (denoted also as points) into coarse grid points and fine grid points. Let N be the set of unknowns. The algorithm is defined for a symmetric M–matrix. The idea is to identify points which are strongly coupled. Let I be a subset of N and

$$d(i,I) = \frac{\sum_{j\in I} -a_{i,j}}{\max_{j\neq i}(-a_{i,j})}.$$

Let α be a given parameter and

$$S_i = \{j \in N | d(i,\{j\}) \geq \alpha\}, \quad U_i = \{j \in N | i \in S_j\}.$$

The selection of the set of coarse grid points C and fine grid points F is done in two stages:

[I]:

1) Set $C = \emptyset, F = \emptyset$

2) While $C \cup F \neq N$

Pick $i \in N \backslash (C \cup F)$ with maximal numbers of elements: $|U_i| + |U_i \cap F|$

if $|U_i| + |U_i \cap F| = 0$, set $F = N \backslash C$ else set $C = C \cup \{i\}$ and $F = F \cup (U_i \backslash C)$.

[II]:

1) Set $T = \emptyset$

2) While $T \subset F$

Pick $i \in F \backslash T$ and set $T = T \cup \{i\}$

Set $\tilde{C} = \emptyset$, $C_i = S_i \cap C$ and $F_i = S_i \cap F$

While $F_i \neq \emptyset$

Pick $j \in F_i$ and set $F_i = F_i \backslash \{j\}$

if $d(j, C_i)/d(i, \{j\}) \leq \beta$

if $|\tilde{C}| = 0$, set $\tilde{C} = \{j\}$ and $C_i = C_i \cup \{j\}$

else set $C = C \cup \{i\}$, $F = F \backslash \{i\}$ and go to 2.

$C = C \cup \tilde{C}$, $F = F \backslash \tilde{C}$.

At level l when the set of points N_l is divided into sets C_l and F_l we can define the interpolation operator P^l_{l-1}. For a given vector x we have $(P^l_{l-1}x)_i = x_i$ if $i \in C_l$ and

$$(P^l_{l-1}x)_i = -\frac{\sum_{j \in C^l_i} (a^l_{i,j} + c_{i,j}) x_j}{a^l_{i,i} + c_{i,i}}, \quad i \in F_l,$$

where

$$c_{i,j} = \sum_{\substack{k \notin C^l_i \\ k \neq i}} \frac{a^l_{i,k} a^l_{k,j}}{a^l_{k,i} + \sum_{p \in C^l_i} a^l_{k,p}}.$$

The next matrix A^{l-1} is given by a Galerkin identity $A^{l-1} = (P^l_{l-1})^T A^l P^l_{l-1}$. It remains to choose the parameters α and β. In [690], it is recommended to set $\alpha = 0.25$ and $\beta = 0.35$. Numerical experiments are given in [690].

Many other algebraic methods have been defined over the years. Let us describe the multilevel method of Shapira [1166]. This method uses three parameters $\alpha, 0 \leq \alpha < 1$, β and γ ($\beta = \gamma = 0$ for symmetric problems). Shapira defines a matrix $A^{(\alpha,\beta,\gamma)}$. Then $a^{(\alpha,\beta,\gamma)}_{i,j} = 1$ if

$$j \neq i, |a_{i,j} - a_{j,i}| \geq -\gamma + \beta \min_{l \in \{i,j\}} \max_{l \leq k \leq n} |a_{l,k} - a_{k,l}|$$

and either

$$|a_{i,j}| > \alpha \max_{1\le k\le n, k\neq i} |a_{i,k}| \text{ or } |a_{j,i}| > \alpha \max_{1\le k\le n, k\neq j} |a_{j,k}|,$$

and 0 otherwise. N being an index set, one defines a recursive procedure to obtain the coarse (c) and (f) fine points.

coarsen(S, A, c):

1) Set $f = F = \emptyset$,

2) Pick $i \in S$ and set $c = C = \{i\}$,

3) Until $c \cup f = S$:

3.1 $F = \cup_{i\in C}\{j \in S | a_{i,j}^{(\alpha,\beta,\gamma)} \neq 0\}\backslash c\backslash f$,

3.2) $f = f \cup F$,

3.3) coarsen$(\cup_{i\in F}\{j \in S | a_{i,j}^{(\alpha,\beta,\gamma)} \neq 0\}\backslash c\backslash f, A, C)$,

3.4) if $C = \emptyset$, pick $i \in S\backslash c\backslash f$, preferably $i \in \cup_{k\in F}\{j \in S | a_{k,j}^{(\alpha/2,\beta/2,2\gamma)} \neq 0\}$ and set $C = \{i\}$,

3.5) $c = c \cup C$.

After applying this procedure the sets c and f are further refined by using a small parameter δ: if for some $i \in f$ and

$$\max\{a_{i,i} - \sum_{j\neq i, j\in f \text{ or } a_{i,j}^{(\alpha,\beta\gamma)}=0} |a_{i,j}|, \sum_{j\in c, a_{i,j}^{(\alpha,\beta,\gamma)}\neq 0} |a_{i,j}|\} \le \delta |a_{i,i},$$

then $c = c \cup \{i\}, f = f\backslash\{i\}$. For any set $g \in N$, J_g denotes the injection operator. Let $rs(B)$ be a diagonal matrix with diagonal elements equal to the row sums of B and $\hat{A}$ be a matrix whose elements are defined as $a_{i,j}$ if $a_{i,j}^{(\alpha,\beta,\gamma)} = 0$ and either $i \in f$ and $j \in c$ or $i \in c$ and $j \in f$ and 0 otherwise. For any matrix B we denote

$$B = \begin{pmatrix} B_{f,f} & B_{f,c} \\ B_{c,f} & B_{c,c} \end{pmatrix}.$$

Let $d(B)$ be a diagonal matrix whose diagonal elements are those of B and

$$\tilde{A} = \begin{pmatrix} d(A_{f,f}) - rs(|A_{f,f} - d(A_{f,f})|) - rs(|\hat{A}_{f,c}|) & A_{f,c} - \hat{A}_{f,c} \\ A_{c,f} - \hat{A}_{c,f} & A_{c,c} \end{pmatrix}.$$

$\tilde{A}$ is subsequently modified to obtain a non–singular $\tilde{A}_{f,f}$. If $i \in f$ and $(\tilde{A}_{f,f})_{i,i} \leq \delta|a_{i,i}|$, $(\tilde{A}_{f,f})_{i,i} = rs(|A_{f,c} - \hat{A}_{f,c}|)_{i,i}$. The prolongation and restriction matrices are defined by

$$P^{-1} = \begin{pmatrix} \tilde{A}_{f,f} & \tilde{A}_{f,c} \\ 0 & I \end{pmatrix}, \quad R^{-1} = \begin{pmatrix} \tilde{A}_{f,f} & 0 \\ \tilde{A}_{c,f} & I \end{pmatrix}.$$

The coarse grid matrix A_C is defines as $A_C = J_c RAPJ_c^T$. Let

$$W = R_{f,f} d(A_{f,f}) P_{f,f}, \quad F = \begin{pmatrix} W & 0 \\ 0 & Q \end{pmatrix}.$$

Shapira [1166] defines a two level iteration by

$$x^{k+1} = x^k + PF^{-1}R(b - Ax^k).$$

Relaxations may also be added before and after as in multigrid. The multilevel method is defined as the following. Let A_i be a block diagonal matrix whose diagonal blocks are $W_1, W_2, \ldots, W_i, Q_i$. Then,

$$x^{k+1} = x^k + P_{L,1} A_L^{-1} R_{L,1}(b - Ax^k),$$

where for $i > k, P_{k,i} = R_{k,i} = I, P_{i,i} = P_i, R_{i,i} = R_i$ and for $i < k, P_{k,i} = P_i P_{i+1} \cdots P_k, R_{k,i} = R_k R_{k-1} \cdots R_i$. Relaxations steps may be added in a multigrid spirit. For symmetric positive definite matrices, Shapira [1166] shows a bound proportional to the number of levels for the condition number of the iteration matrix or to its square depending on the problem. This multilevel method may be accelerated by any iterative method of Chapter 7.

Many other methods have been proposed over the years, see for instance Chan, Go and Zikatanov [274], Braess [175], Kickinger [863], Mandel and al [924], [1291], [1292], Francescatto [578], [579], Notay [1025].

9.11. Bibliographical comments

The origins of the multigrid method is in the Russian literature in the sixties (Fedorenko [549]). Then, it was developed and popularized by A. Brandt [192], [193], [194]. A great deal of work was also done by the German school, see [740]. Since the seventies, many papers have been written about multigrid, see the bibliography in MGNET maintained by C. Douglas. For a good account of the method, see the book by Hackbusch [736]. With the latest research there is less difference between multilevel and domain decomposition methods, see Chapter 10 about this point.

10 | Domain decomposition and multilevel methods

10.1. Introduction to domain decomposition

Domain decomposition (DD) is a very natural framework in which to develop solution methods for parallel computers. Although the idea is quite old and it has been used for many years, mainly in structural mechanics, the interest in domain decomposition was renewed from the end of the eighties. This interest was motivated by the advent of parallel computers with physically distributed memories. There are a lot of DD methods and it is not possible to describe all of them in a book like this. However, we shall look at some of the main trends in this area.

Although this is not completely the case, DD has generally been used for linear systems arising from PDEs discretization. Since most of the methods are closely related to partitioning the domain on which the PDE is to be solved, it is not always possible to study all DD methods from a purely algebraic point of view. Therefore, from time to time, we shall have to go back to the continuous problem and the discretization of the PDE problem. We shall restrict ourselves to problems which give rise to symmetric positive definite matrices although some DD methods have been developed for indefinite and non–symmetric problems.

For constructing algorithms for parallel computers, a good principle is to divide the problem into smaller pieces, solve the subproblems in parallel and then to paste the local results together. This strategy has often been called "divide and conquer". DD methods proceed in a similar way. First of all, the domain Ω (or preferably the problem) is split into subdomains (or subproblems), a problem is defined and solved on each subdomain in parallel and then the partial solutions are glued together to get the global solution. Originally DD algorithms were devised to use existing software for solving subproblems where separation of variables can be used (like FFT) and also for computers with small memories, as the subproblems can then fit into the memories of these machines. Today, memories are much larger and cheaper and DD methods are used to introduce parallelism in very strongly coupled problems. The modern perspective on DD is to use these techniques to construct preconditioners that will be used in some Krylov methods (CG for SPD matrices). We can divide DD methods into two main categories: with and without overlapping. We shall first study some methods with overlapping. They are generically known as Schwarz methods as Hermann Schwarz in 1869 was probably the first to use a domain decomposition with overlapping. His goal was to prove existence of the solution of a PDE problem on a domain which was the union of a disk and a rectangle. He used an iterative method (known today as the Schwarz alternating method) and proved that it converges to the solution of the PDE.

We shall then turn to methods without overlapping. In general, they reduce the problem to another one of smaller dimension for the unknowns on the interfaces. They are called Schur complement or substructuring methods. There are many variants of these methods. Another important distinction between methods is the algorithm which is used for solving the subproblems. One can use either a direct method (like Gaussian elimination) or iterative methods or simply a preconditioner for the matrix of the subproblem. Combining all these possibilities gives rise to a very large number of algorithms. A good source of information on DD methods are the proceedings of the annual DD conferences that started in 1987 in Paris [649], [272], [273], [650], [861], etc..., see also the book by Smith, Bjørstad and Gropp [1192].

The ultimate goal being to develop a method whose complexity is proportional to the number of unknowns n, if we use CG, we need a preconditioner M such that the condition number $\kappa(M^{-1}A)$ is independent of n. Moreover, it is desirable that the condition number does not depend on the number of subdomains or jumps in the coefficients. During the development of DD methods, it was realized that to reach this goal some sort of global transport of information is needed. This gives rise to multilevel methods which can be also be seen as extensions of the multigrid method.

The domain decomposition framework has also been used to match different physical modelings. For instance, in studying flows around airplanes, one would like to use the potential equations in some regions, the Euler equations in others or even the Navier–Stokes equations. Another example for computations of flows around space shuttles is the matching of the Boltzmann equation and the Navier–Stokes equations. Domain decomposition offers a very natural framework to match different models. This has been particularly developed by Glowinski, Périaux and their co–workers, see, for instance, Glowinski, Périaux and Terrasson [651]. It can also be used in some combustion problems to handle boundary layers, see Garbey [602].

10.2. Schwarz methods

10.2.1. The classical Schwarz alternating method

Suppose we are solving a second order elliptic PDE in a bounded two–dimensional domain Ω. For simplicity, we consider the domain Ω split into two overlapping subdomains Ω_1 and Ω_2. Let Γ_i, $i = 1, 2$, be the part of the boundary of Ω_i enclosed in Ω (see figure 10.1).

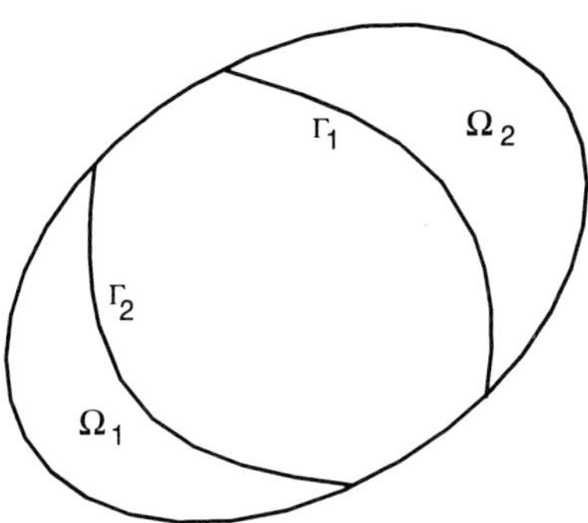

fig. 10.1: Overlapping subdomains

Roughly speaking, the Schwarz alternating method is the following: we guess a value for the unknowns on the inner boundary Γ_1, solve the problem exactly in Ω_1, use the computed values on the inner boundary Γ_2 to solve exactly in Ω_2 and repeat the process until convergence. This very simple method "almost always" converges. A general theory has been given for PDE problems by P.L. Lions [903],[904], [905] in terms of projections in Hilbert spaces. Of course, the rate of convergence of the method depends on the extent of overlapping. The larger the overlapping, the faster the convergence. This is expected since, when $\Omega_1 = \Omega_2 = \Omega$, the method converges in one iteration, but note that when the overlapping is larger, the cost of solving the subproblems is higher. Therefore a trade–off has to be found between the number of iterations and the cost of solving the subproblems. The following exposition is taken from Lions [903]. We consider the Poisson model problem but the results are more generally true for bilinear forms. Let

$$-\Delta u = f \text{ in } \Omega, \quad u|_{\partial\Omega} = 0.$$

The basic Schwarz alternating algorithm can be formulated at the PDE level when u^1 is given as

$$-\Delta u^{2k} = f \text{ in } \Omega_1, \quad u^{2k}|_{\Gamma_1} = u^{2k-1}|_{\Gamma_1},$$
$$-\Delta u^{2k+1} = f \text{ in } \Omega_2, \quad u^{2k+1}|_{\Gamma_2} = u^{2k}|_{\Gamma_2},$$

and the given boundary conditions on the other parts of the boundary. The bilinear form a of the problem is defined as

$$a(u,v) = \int_\Omega \nabla u \cdot \nabla v \, dx.$$

We saw in Chapter 1 that the model problem can be written in variational form as

$$a(u,v) = (f,v), \quad \forall v \in H_0^1(\Omega).$$

Let $V_1 = H_0^1(\Omega_1)$ and $V_2 = H_0^1(\Omega_2)$ and the projectors P_1 and P_2 defined by

$$a(P_i v, w) = a(v,w), \ \forall w \in V_i, i = 1,2.$$

The functions defined only on subdomains are extended by 0 to $H_0^1(\Omega)$. Then we have

$$a(u^{2k} - u, v_1) = 0, \quad \forall v_1 \in V_1, \quad u^{2k} - u^{2k-1} \in V_1,$$
$$a(u^{2k+1} - u, v_2) = 0, \quad \forall v_2 \in V_2, \quad u^{2k+1} - u^{2k} \in V_2.$$

It is easy to see that

$$u - u^{2k} = (I - P_1)(u - u^{2k-1}),$$
$$u - u^{2k+1} = (I - P_2)(u - u^{2k}).$$

Therefore,

$$u - u^{2k+1} = (I - P_2)(I - P_1)(u - u^{2k-1}).$$

This equation explains why this type of algorithm is more generally known as a multiplicative Schwarz method. The mathematical formulation of the problem for studying convergence is

$$v^0 \in V, \quad v^{2k} = (I - P_1)v^{2k-1}, \quad v^{2k+1} = (I - P_2)v^{2k}.$$

We have to examine convergence of iterated projections. The following result shows that the method converges.

Theorem 10.1

If $V = \overline{V_1 + V_2}$, *where the overbar denotes the closure of the set, then* $v^k \longrightarrow 0$.

Proof. See P.L. Lions [903]. □

Moreover, if $V = V_1 + V_2$ then

$$\|(I - P_2)(I - P_1)\| \le c, \quad c < 1.$$

10.2.2. The matrix form of the Schwarz alternating method

The previous framework is very general. Let us now specialize to solving a second order elliptic equation in a rectangle using a five point finite difference scheme with the natural (rowwise) ordering. We solve $Ax = b$ and we saw in Chapter 1 that the matrix is written blockwise as

$$A = \begin{pmatrix} D_1 & -B_2^T & & & \\ -B_2 & D_2 & -B_3^T & & \\ & \ddots & \ddots & \ddots & \\ & & -B_{m-1} & D_{m-1} & -B_m^T \\ & & & -B_m & D_m \end{pmatrix}.$$

Suppose the mesh is partitioned as in figure 10.2

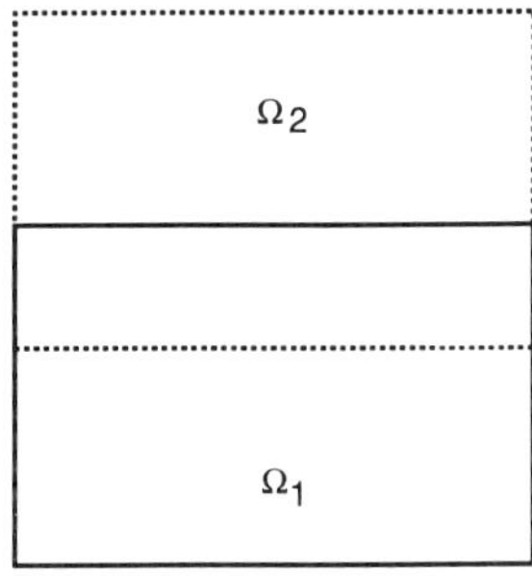

fig. 10.2: Partitioning of the rectangle with overlapping

Then, the matrix $A^{(1)}$ corresponding to Ω_1 is

$$A^{(1)} = \begin{pmatrix} D_1 & -B_2^T & & & & \\ -B_2 & D_2 & -B_3^T & & & \\ & \ddots & \ddots & \ddots & & \\ & & -B_{p-2} & D_{p-2} & -B_{p-1}^T \\ & & & -B_{p-1} & D_{p-1} \end{pmatrix},$$

and the matrix $A^{(2)}$ corresponding to Ω_2 is

$$A^{(2)} = \begin{pmatrix} D_{l+1} & -B_{l+2}^T & & & & \\ -B_{l+2} & D_{l+2} & -B_{l+3}^T & & & \\ & \ddots & \ddots & \ddots & & \\ & & -B_{m-1} & D_{m-1} & -B_m^T \\ & & & -B_m & D_m \end{pmatrix}.$$

Let us denote the matrix A in block form as

$$A = \begin{pmatrix} A^{(1)} & A^{(1,2)} \\ X & X \end{pmatrix} \text{ and } A = \begin{pmatrix} Y & Y \\ A^{(2,1)} & A^{(2)} \end{pmatrix},$$

and let b_1 and b_2 be the restrictions of the right hand side b to Ω_1 and Ω_2. Note that $A^{(1,2)}$ has only one non–zero block in the left lower corner and $A^{(2,1)}$ is zero except for the upper right block. We denote by x_1 and x_2 the unknowns in Ω_1 and Ω_2 keeping the natural block numbering that is

$$x_1 = ((x_1)_1 \quad \cdots \quad (x_1)_{p-1}), \quad x_2 = ((x_2)_{l+1} \quad \cdots \quad (x_2)_m).$$

We extend the vectors x_1 and x_2 to Ω by completing with the components of the previous iterate and we define x^{2k} by x_1^{2k} for the first $p-1$ (block) components and the blocks p to m of x_2^{2k-1} (which we denote by $x_{1,2}^{2k-1}$) for the remaining ones. Similarly x^{2k+1} is defined by the l first components of x_1^{2k} (which we denote by $x_{2,1}^{2k}$) and then the components of x_2^{2k+1}. With this notation, we can write the Schwarz alternating method as

$$A^{(1)} x_1^{2k} = b_1 + \begin{pmatrix} 0 \\ \vdots \\ 0 \\ B_p^T (x_2^{2k-1})_p \end{pmatrix}, \quad A^{(2)} x_2^{2k+1} = b_2 + \begin{pmatrix} B_{l+1} (x_1^{2k})_l \\ 0 \\ \vdots \\ 0 \end{pmatrix}.$$

By adding and subtracting suitable quantities, this is also

$$\begin{aligned} x_1^{2k} &= x_1^{2k-1} + (A^{(1)})^{-1} (b_1 - A^{(1)} x_1^{2k-1} - A^{(1,2)} x_{1,2}^{2k-1}), \\ x_2^{2k+1} &= x_2^{2k} + (A^{(2)})^{-1} (b_2 - A^{(2)} x_2^{2k} - A^{(2,1)} x_{2,1}^{2k}). \end{aligned}$$

We note that the expression within parentheses in the first equation is simply the restriction to Ω_1 of the residual. In the same way the parenthesis in the second equation is the restriction of the residual to Ω_2. Then, we can write globally:

$$x^{2k} = x^{2k-1} + \begin{pmatrix} (A^{(1)})^{-1} & 0 \\ 0 & 0 \end{pmatrix} (b - Ax^{2k-1}),$$

$$x^{2k+1} = x^{2k} + \begin{pmatrix} 0 & 0 \\ 0 & (A^{(2)})^{-1} \end{pmatrix} (b - Ax^{2k}).$$

By eliminating x^{2k} we obtain

$$x^{2k+1} = x^{2k-1} + [\begin{pmatrix} (A^{(1)})^{-1} & 0 \\ 0 & 0 \end{pmatrix} + \begin{pmatrix} 0 & 0 \\ 0 & (A^{(2)})^{-1} \end{pmatrix}$$
$$- \begin{pmatrix} 0 & 0 \\ 0 & (A^{(2)})^{-1} \end{pmatrix} A \begin{pmatrix} (A^{(1)})^{-1} & 0 \\ 0 & 0 \end{pmatrix}] r^{2k-1},$$

$$r^{2k-1} = b - Ax^{2k-1}.$$

This shows that the Schwarz alternating method is nothing other than a preconditioned Richardson iteration (see Chapters 5 and 8). This method can also be written with another notation that will be useful later on. We introduce restriction operators R_1 and R_2 such that

$$x_1^k = R_1 x^k, \quad x_2^k = R_2 x^k.$$

R_1 is simply $(\, I_{p-1} \quad 0 \,)$ and $R_2 = (\, 0 \quad I_{m-l+1} \,)$. The transposes of R_1 and R_2 are extension operators. Then, we see easily that

$$A^{(1)} = R_1 A R_1^T, \quad A^{(2)} = R_2 A R_2^T.$$

In the first half step of the iteration, we have to restrict the residual by R_1, apply the inverse of $R_1 A R_1^T$ and extend the result by R_1^T, so this is written as

$$x^{2k} = x^{2k-1} + R_1^T (R_1 A R_1^T)^{-1} R_1 (b - Ax^{2k-1}).$$

Similarly, the second half step is

$$x^{2k+1} = x^{2k-1} + R_2^T (R_2 A R_2^T)^{-1} R_2 (b - Ax^{2k}).$$

Proposition 10.2

The matrix $P_i = R_i^T(R_iAR_i^T)^{-1}R_iA, i = 1,2$ *is an orthogonal projection in the scalar product defined by* A.

Proof. We have

$$P_iP_i = R_i^T(R_iAR_i^T)^{-1}R_iAR_i^T(R_iAR_i^T)^{-1}R_iA = P_i.$$

Moreover

$$AP_i = AR_i^T(R_iAR_i^T)^{-1}R_iA = (AP_i)^T.$$

□

If ε^k is the error, we have

$$\varepsilon^{2k} = (I - P_1)\varepsilon^{2k-1}, \quad \varepsilon^{2k+1} = (I - P_2)\varepsilon^{2k}.$$

Therefore, P_i is the discrete version of the projection operator we introduced earlier. We can easily generalize this method to more than two subdomains. We remark that if we choose the restrictions R_i properly the point and block `Gauss-Seidel` iterations fit into this general framework. Hence, the Schwarz alternating method is simply a generalization of the block `Gauss-Seidel` algorithm where the restriction operators allow for overlapping of the blocks.

The Schwarz alternating method can also be viewed as directly applying the block `Gauss-Seidel` algorithm to a larger enhanced problem, see Rodrigue and Simon [1112] and W.P. Tang [1235].

10.2.3. The rate of convergence

We can study the rate of convergence of the Schwarz alternating method for general problems; however, for the sake of simplicity let us consider a one dimensional Poisson model problem. We use the analog of the previous notation. The matrices which are involved are

$$A^{(1)} = \begin{pmatrix} 2 & -1 & & & \\ -1 & 2 & -1 & & \\ & \ddots & \ddots & \ddots & \\ & & -1 & 2 & -1 \\ & & & -1 & 2 \end{pmatrix}$$

of order $p-1$ and $A^{(2)}$ which is the same matrix but of order $n-l$. We saw in Chapter 2 that we explicitly know the inverses of these matrices.

Proposition 10.3

We have

$$\varepsilon_i^{2k} = \frac{i}{p}\varepsilon_p^{2k}, i = 1, \ldots, p-1$$

$$\varepsilon_i^{2k+1} = \frac{n-i+1}{n-l+1}\varepsilon_l^{2k+1}, i = l+1, \ldots, n.$$

Proof. Because we use exact solves for the subproblems, the equations corresponding to the unknowns inside the subdomains are exactly verified and we have

$$A^{(1)} \begin{pmatrix} \varepsilon_1^{2k} \\ \vdots \\ \varepsilon_{p-1}^{2k} \end{pmatrix} = \begin{pmatrix} 0 \\ \vdots \\ 0 \\ \varepsilon_p^{2k-1} \end{pmatrix}.$$

The components 1 to $p-1$ of the error at iteration $2k$ are the components of the last column of the inverse of $A^{(1)}$ times ε_p^{2k-1}. Since the inverse is known explicitly, we obtain the result. The proof is the same for the other relation.
□

Note that $\varepsilon_p^{2k} = \varepsilon_p^{2k-1}$ and $\varepsilon_l^{2k+1} = \varepsilon_l^{2k}$. Then, it is clear how the error is reduced during the iterations. At the end of the first half step, the error is maximum for the node p and linear (being 0 at the ends of the interval). At the end of the second half step, the error is maximum for the node l and linear, see figure 10.3 where the errors for three half steps are shown. The error in the first half step is the solid line. The error in the second half step is the dashed line (partially hidden by the others). The error in the third half step is the dash–dotted line. Using the previous results, we can relate the maxima of the error at odd steps.

Theorem 10.4

At odd steps, the maximum of the (absolute value) of the error is obtained for node l and

$$\|\varepsilon^{2k+1}\|_\infty = \frac{l}{p}\frac{n-p+1}{n-l+1}\|\varepsilon^{2k-1}\|_\infty.$$

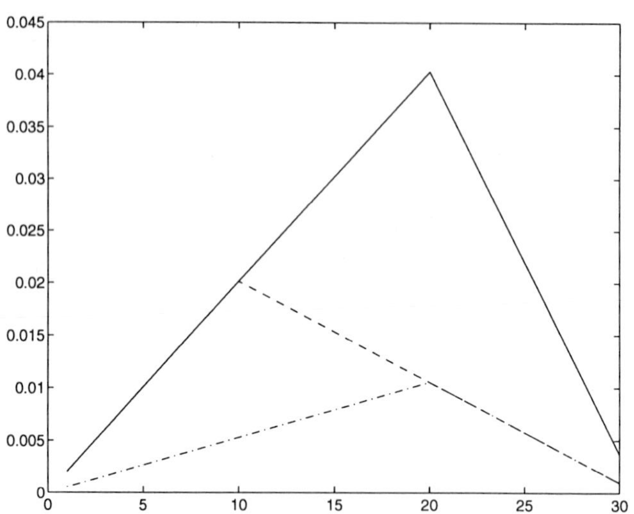

fig. 10.3: Errors for three half steps of the Schwarz method

Proof. The result is obvious from the previous discussion. We note that both factors are less than 1 since

$$\frac{n-p+1}{n-l+1} = 1 - \frac{p-l}{n-l+1}.$$

□

This result shows that the larger the overlap $(p-l)$ the faster the convergence. The result is also directly linked to the amount of diagonal dominance of the matrix. With a strictly diagonally dominant matrix (for instance having a diagonal with elements strictly larger than 2) the convergence rate would be better. The same analysis can be done on this problem for a larger number of subdomains as the error is still linear on each subdomain. Unfortunately, the rate of convergence is slower when we have a large number of subdomains as shown on figure 10.4 where we show the number of iterations for two and three subdomains as a function of the overlap (number of points in the overlapping region) for the one dimensional model problem. We see that when we increase the extent of overlapping we have a large decrease in the number of iterations at first and then they level off meaning that it is only marginally better to have a large overlap.

Being a `Gauss-Seidel`–like method the multiplicative Schwarz method cannot be directly accelerated by `CG`. However, it can be symmetrized in

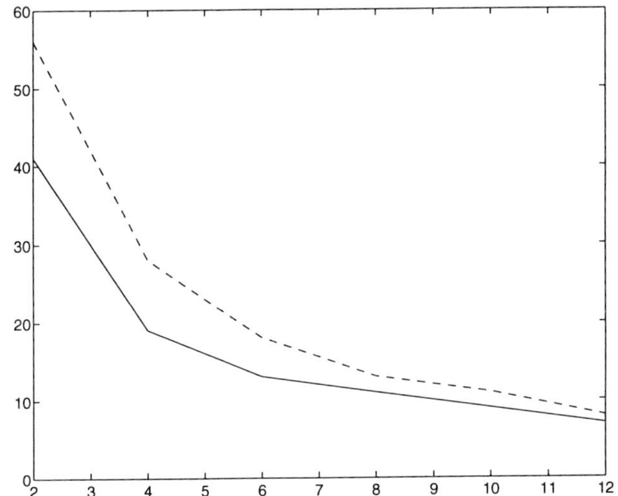

fig. 10.4: Number of iterations as a function of the overlap solid line: two subdomains, dashed: three subdomains

the same way as SSOR is derived from SOR. We simply have, at the end of a subdomain sweep, to add another sweep with the subproblems solved in the reverse order.

10.2.4. Other boundary conditions

A way to reduce the overlap while maintaining a good convergence rate is to use other inner boundary conditions than Dirichlet for the subproblems. This issue was considered by W.P. Tang [1235]. He proposed using inner mixed boundary conditions like continuity of

$$\omega u + (1-\omega)\frac{\partial u}{\partial n}.$$

Numerical results show that this can substantially improve the rate of convergence for small overlaps.

10.2.5. Parallelizing multiplicative Schwarz

Note that there is no parallelism in the Schwarz alternating method as we must solve for the subproblems sequentially. To get a parallel algorithm the same trick as for the SSOR preconditioner can be used, namely a coloring of the subdomains such that a subdomain of one color is only connected to subdomains of other colors. For a domain divided into strips (see figure 10.5) a red–black ordering is used, every other strip is black, and red strips alternate with black strips. This means red subdomains are only related to black subdomains, so we can compute simultaneously, in parallel, all red subdomains say, and then all black subdomains. Now the method is parallel, the degree of parallelism being half the number of subdomains.

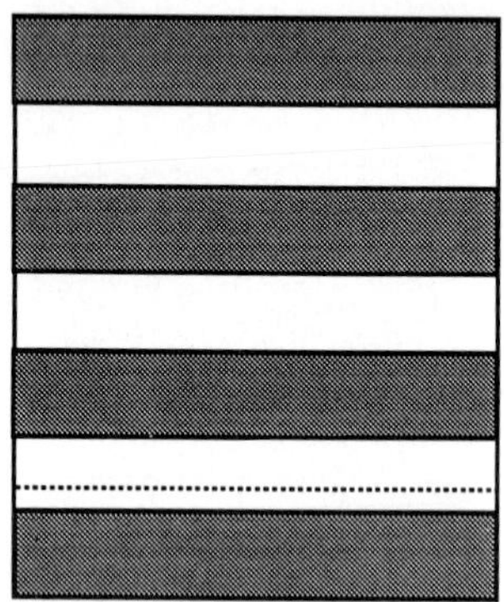

fig. 10.5: Red–Black partitioning of the rectangle

10.2.6. The additive Schwarz method

We have seen that the alternating Schwarz method can be considered as a kind of `Gauss-Seidel` algorithm. A way to get a parallel algorithm is to use instead a block `Jacobi`–like method. This is known as the Additive Schwarz method, see Dryja and Widlund [455]. Roughly speaking, we solve independently for each subdomain using the boundary conditions from the previous iteration, extend the results to the whole domain and sum them. When used as an iterative method, this is generally not convergent. However, this method is intended to be a preconditioner for `CG` which is defined as

$$M^{-1} = \sum_i R_i^T (R_i A R_i^T)^{-1} R_i,$$

where the summation is over the number of overlapping subdomains. More generally, one can replace the exact solves for each subdomain by approximations and define

$$M^{-1} = \sum_i R_i^T M_i^{-1} R_i.$$

10.2.7. Adding a coarse mesh correction

The rate of convergence of the multiplicative or additive Schwarz methods depends on the number of subdomains. An improvement on this is to add a coarse grid correction. The coarse grid corresponds to the interfaces in the partitioning. For instance, for additive Schwarz, we define

$$M^{-1} = \sum_i R_i^T (R_i A R_i^T)^{-1} R_i + R_0^T A_C^{-1} R_0,$$

the coarse grid operator may be chosen as a Galerkin approximation $A_C = R_0 A R_0^T$. Generally, if the extent of overlap is kept proportional to the "sizes" of the subdomains the number of iterations is independent of n and of the number of subdomains. Remark that the solution of the coarse grid problem is usually not parallel. This is a very general framework for defining preconditioners. It will also be used for DD methods without overlapping.

10.3. An additive Schwarz preconditioner for parabolic problems

In this section, we consider a method specially designed for parabolic PDEs like

$$\frac{\partial u}{\partial t} - \frac{\partial}{\partial x}(a(x,y)\frac{\partial u}{\partial x}) - \frac{\partial}{\partial y}(b(x,y)\frac{\partial u}{\partial y}) = f$$

in $\Omega \subset \mathbb{R}^2$,

$u\,|_{\partial\Omega} = 0, \quad u(x,0) = u_0(x).$

The model problem that is considered is the heat equation :

$$\frac{\partial u}{\partial t} - \Delta u = f,$$

with Dirichlet boundary conditions, Ω being the unit square. For stability and efficiency an implicit Crank–Nicolson scheme is used as described in the following framework. Consider the general parabolic partial differential equation,

$$\frac{\partial u}{\partial t} + Lu = f,$$

with L being a second order self–adjoint linear elliptic operator. The (space) operator L is discretized with a standard finite difference scheme with $m+1$ points on each side of the square domain Ω. The mesh size h is given by

$$h = \frac{1}{m+1}.$$

Then we obtain a sparse matrix A (of order $m^2 \times m^2$) from L,

$$L \Longrightarrow \frac{1}{h^2} A$$

where A is a block tridiagonal matrix (corresponding to the steady state problem $Lu = f$). Time is discretized with the usual centered scheme : with $t \in [0, T]$, $k = \Delta t$ being the time step, p referring to the values of unknowns at time pk, we have

$$\frac{u^{p+1} - u^p}{k} + \frac{1}{2h^2}(Au^{p+1} + Au^p) = \frac{1}{2}(f^{p+1} + f^p),$$

or

$$2\frac{h^2}{k}(u^{p+1} - u^p) + Au^{p+1} + Au^p = h^2(f^{p+1} + f^p).$$

At every time step this gives a linear system to solve:

$$(2\frac{h^2}{k}I + A)u^{p+1} = 2\frac{h^2}{k}u^p - Au^p + h^2(f^{p+1} + f^p).$$

The advantages of the implicit centered scheme are its second order accuracy (which means that the truncation error is $O(k^2) + O(h^2)$) and unconditional stability that allows larger time steps than more straightforward explicit schemes. In practice, one chooses to solve the problem in the form :

$$A_t x = h^2(f^{p+1} + f^p) - 2Au^p,$$

where $\theta = 2\frac{h^2}{k}$, and $A_t = \theta I + A$. Then $u^{p+1} = x + u^p$. We remark that for our problem, A_t is a symmetric strictly diagonally dominant M–matrix.

Although domain decomposition preconditioners for elliptic problems can also be used in our case, we shall use the peculiarities of the time dependent problems to derive a preconditioner that is both efficient and very easy to implement. This method was inspired by an algorithm of Y. Kuznetsov [873]. It belongs to the class of Additive Schwarz methods. However, contrary to

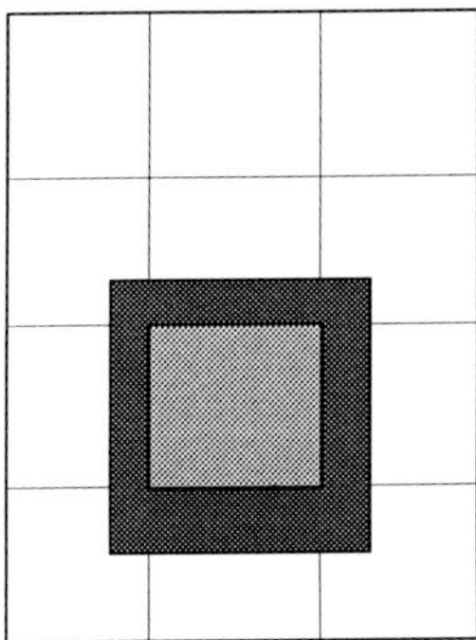

fig. 10.6: Example of one subdomain

Kuznetsov, we use CG to solve the linear system at each time step and domain decomposition, only to provide a preconditioner.

The domain Ω is divided into non–overlapping subdomains Ω_i, $i = 1, \ldots, l$. Each Ω_i is extended to a domain $\widehat{\Omega}_i$ that overlaps the neighbors of Ω_i (restricted of course to Ω), see figure 10.6, where Ω_i corresponds to the light gray area and $\widehat{\Omega}_i$ to the same plus the dark grey area.

To solve $Mz = r$, the following steps are performed : for each subdomain $\widehat{\Omega}_i$, $i = 1, \ldots, l$, let $\widehat{A_i}$ be the $n_i \times n_i$ matrix arising from the discretization of the problem on $\widehat{\Omega}_i$ with homogeneous Dirichlet boundary conditions. Let $\widehat{r_i}$ be the vector of length n_i that will be the right hand side on $\widehat{\Omega}_i$, whose entries are equal to those of r for components corresponding to mesh points in Ω_i and 0 elsewhere. Then let $\widehat{z_i}$ be defined by solving a problem on $\widehat{\Omega}_i$:

$$\widehat{A_i}\widehat{z_i} = \widehat{r_i}.$$

Then we extend $\widehat{z_i}$ to a vector z_i on the whole Ω with 0 components corresponding to mesh points outside $\widehat{\Omega}_i$. The solution z is simply defined as

$$z = \sum_{i=1}^{l} \widehat{z_i}.$$

The main problem is to know where to put the artificial boundaries, that is the definition of $\widehat{\Omega}_i$. Some hints and some bounds are given by the study of the decay of elements of the inverse A_t^{-1}, see Meurant [982]. However, usually a moderate amount of overlapping gives good results, see the numerical experiments in [979], [980]. This example shows that to develop efficient methods, one has to take into account the specific nature of the problem to be solved. In this case, the $1/\Delta t$ term on the diagonal of the matrix makes the matrix more diagonally dominant than for a stationary problem. Because of this we do not really need to use a coarse mesh to transport the information. Variants of this method can be designed using other inner boundary conditions than homogeneous Dirichlet. Preconditioners can be used instead of direct solvers for the subproblems, etc...

10.4. Algebraic domain decomposition methods without overlapping

In this section we introduce the main ingredients for domain decomposition methods without overlapping. For the sake of simplicity we shall first consider a square domain Ω decomposed into two subdomains. In the following sections, this will be generalized to more subdomains and to more general domains. Let us consider an elliptic second order PDE in a rectangle discretized by standard finite difference methods. Let Ω_1 and Ω_2 be the two subdomains and $\Gamma_{1,2}$ the interface which is a mesh line (see figure 10.7).

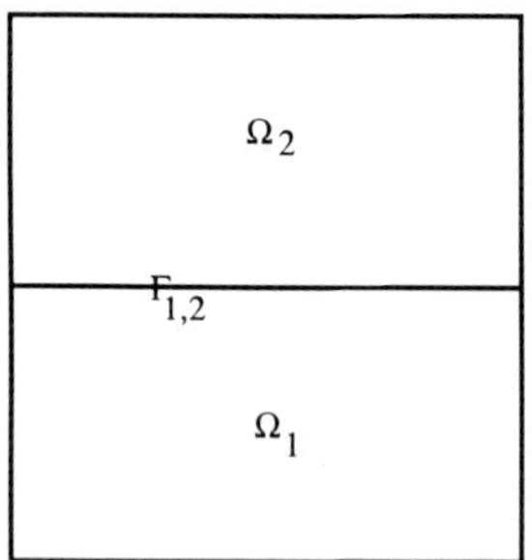

fig. 10.7: Partitioning of the rectangle with non–overlapping subdomains

We shall denote by m_1 (resp. m_2) the number of mesh lines in Ω_1 (resp. Ω_2), each mesh line having m mesh points ($m = m_1 + m_2 + 1$). To describe the domain decomposition method, we renumber the unknowns in Ω. Let x_1 (resp. x_2) be the vector of unknowns in Ω_1 (resp. in Ω_2) and $x_{1,2}$ be the vector of the unknowns on the interface. Within each subdomain we order the unknowns with the usual row–wise ordering. With this numbering of the unknowns, the linear system can be rewritten blockwise as

$$\begin{pmatrix} A_1 & 0 & E_1 \\ 0 & A_2 & E_2 \\ E_1^T & E_2^T & A_{12} \end{pmatrix} \begin{pmatrix} x_1 \\ x_2 \\ x_{1,2} \end{pmatrix} = \begin{pmatrix} b_1 \\ b_2 \\ b_{1,2} \end{pmatrix}. \tag{10.1}$$

It is important to understand the meaning of equation (10.1). A_1 (resp. A_2) is a block tridiagonal matrix that represents the coupling of the unknowns within Ω_1 (resp. Ω_2). A_{12} is a (point) tridiagonal matrix that represents the coupling of the unknowns on the interface. E_1 (resp. E_2) represents the coupling of Ω_1 (resp. Ω_2) with the interface. Note that when using the five point finite difference scheme, E_1 and E_2 are sparse matrices as the interface is only coupled with one mesh line above the interface in Ω_2 and one mesh line below in Ω_1. Therefore,

$$E_1 = (0\ 0\ \dots\ 0\ E_1^{m_1})^T, \qquad E_2 = (E_2^1\ 0\ \dots\ 0)^T,$$

where $E_1^{m_1}$ and E_2^1 are diagonal matrices.

Most algebraic DD methods are based on block Gaussian elimination (or approximate block Gaussian factorization) of the matrix in equation (10.1). The goal is to solve the system (10.1) by an iterative method. Basically, we have two possibilities depending on the fact that we can or cannot (or do not want to) solve linear systems corresponding to subproblems like

$$\begin{cases} A_1 y_1 = c_1 \\ A_2 y_2 = c_2 \end{cases} \tag{10.2}$$

"exactly" with a direct method (or with a fast solver). If we are not able to (or do not want to) solve exactly these systems then, we can use an iterative method or replace A_1 and A_2 by approximations. Let us consider successively these two possibilities.

10.4.1. Exact solvers for the subdomains

In this first class of methods, we eliminate the unknowns x_1 and x_2 in the subdomains. This gives a reduced system for the interface unknowns

$$S x_{1,2} = \overline{b_{1,2}}, \tag{10.3}$$

with

$$\overline{b_{1,2}} = b_{1,2} - E_1^T A_1^{-1} b_1 - E_2^T A_2^{-1} b_2$$

and

$$S = A_{12} - E_1^T A_1^{-1} E_1 - E_2^T A_2^{-1} E_2.$$

The matrix S is the Schur complement of A_{12} in A. Of course, A_1^{-1} and A_2^{-1} are dense matrices. Therefore, it is too costly to construct S. From the structure of E_1 and E_2, we see that if we consider A_1 and A_2 as block tridiagonal matrices (as we did before), a block corresponding to a mesh line, we only need to know the right lower block of the inverse of A_1 and the upper left block of the inverse of A_2. This can be computed by solving m linear systems with the matrix A_1 and m linear systems with the matrix A_2. These solves are usually done through the LU factorization of both matrices. As we shall see later on, since we only need the last block of unknowns in one case and the first one in the other case, it is more advantageous to use a block LU factorization for the former and a block UL factorization for the latter. This will lead to a number of operations of $O((m_1 + m_2)m^3)$. Of course, the two solves are independent from each other and can be performed in parallel.

Anyway, constructing and factoring S is costly (but it was done in the past in applications in structural mechanics where many domains are similar and the matrices need only be factored once; this method was called substructuring). A more economical solution is to solve the reduced system with matrix S on the interface with an iterative method. Then, we need to know the properties of the Schur complement to be able to choose a suitable iterative method. As we saw in Chapter 1, if A is a symmetric positive definite M–matrix, S is also a symmetric positive definite M–matrix. An efficient iterative method for solving such systems is `PCG`, the preconditioned Conjugate Gradient method. Solving the reduced Schur complement system (10.3) is interesting as we have the following result.

Theorem 10.5

For the Poisson model problem the condition number of the Schur complement is

$$\kappa(S) = O(\frac{1}{h}).$$

Proof. This result will be proved later when computing the eigenvalues of S. □

Therefore, the reduced system has two interesting properties: the dimension is $\sqrt{n}$ and the condition number is an order of magnitude better. One of the operations that we have to do for performing an iteration of PCG is the product of the matrix S by a given vector, say p. This may seem a costly operation and moreover, as we do not wish to compute the elements of S. However the product, Sp can be computed easily as

$$Sp = A_{1,2}p - E_1^T A_1^{-1} E_1 p - E_2^T A_2^{-1} E_2 p,$$

p being a vector defined on the interface. The vector $A_{1,2}p$ is easy and cheap to compute as $A_{1,2}$ is a symmetric tridiagonal matrix. The matrices E_1 and E_2 take p on the interface and extend it to Ω_1 and Ω_2, so that

$$E_1 p = (0 \ \dots 0 \ E_1^{m_1})^T p = (0 \ \dots 0 \ E_1^{m_1} p)^T,$$

$$E_2 p = (E_2^1 \ 0 \ \dots 0 \ )^T p = (E_2^1 p \ 0 \ \dots 0)^T.$$

Then $w^1 = A_1^{-1} E_1 p$ is computed by solving

$$A_1 w^1 = E_1 p, \tag{10.4}$$

This is solving a linear system corresponding to a problem in Ω_1. Note that only the last block of the right hand side is different from 0 and because we only need $E_1^T w^1$, the last block $w_{m_1}^1$ of the solution w^1 is what we must compute. Similarly, $w^2 = A_2^{-1} E_2 p$ is computed by solving

$$A_2 w^2 = E_2 p, \tag{10.5}$$

a problem in Ω_2 and we note that only the first block component of the right hand side is non–zero and that we only need the first block of the solution w_1^2. Finally, we have

$$Sp = A_{1,2}p - w_{m_1}^1 - w_1^2.$$

We have just shown that we do not need to explicitly construct S to be able to compute Sp as this can be done through subdomain solves. Regarding parallel computing, the important fact is that the linear systems (10.4) and (10.5) are independent and therefore can be solved in parallel. Of course, in this simple problem, the degree of parallelism is only two which is not really interesting, but the fact that the degree of parallelism is the number of subdomains will generalize when we consider more than two subdomains. This method can be seen as being semi–iterative (or semi–direct depending on one's preferences). Direct solvers can be used in parallel on the subdomains and an iterative method is used to patch these local solutions together to get the global solution of the problem. Moreover, this kind of method allows the use of fast solvers (like FFT) on domains where fast solvers cannot directly be

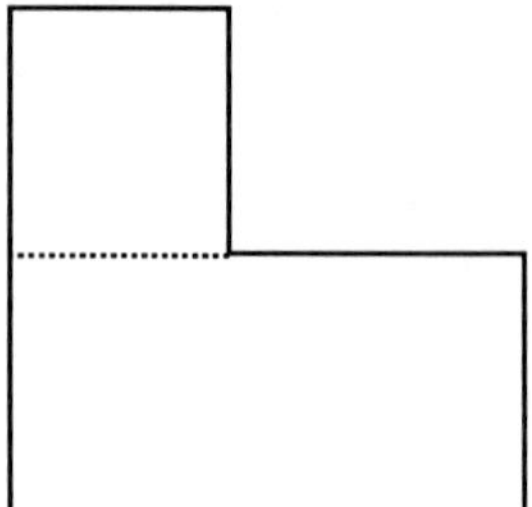

fig. 10.8: L–shaped domain

used. Consider, for instance, solving the Poisson equation in the L–shaped domain in figure 10.8 where the domain decomposition method may use fast solvers in each of the rectangular domains.

To improve the convergence rate of CG on the reduced system, a preconditioner M is needed. We have seen that the condition number of S is of the order of $1/h$, so we are looking for a preconditioner improving upon that. We would ideally like to find a preconditioner M such that

$$\kappa(M^{-1}S) = O(1).$$

This will imply that when h gets arbitrarily small the number of iterations stays approximately constant to reach a given precision. Then, the total number of operations will be proportional to the number of operations for one iteration. We must note too that to be able to use CG on a parallel computer, solving a linear system with M must be parallelizable. Therefore, our main problem with this approach is:

Find an approximation of the Schur complement S

10.4.2. Approximate solvers for the subdomains

In this case, we suppose that we are not able or do not want, because of the cost and storage, to use a direct method to "exactly" solve

$$\begin{cases} A_1 y_1 = c_1, \\ A_2 y_2 = c_2. \end{cases}$$

We suppose that these subproblems can only be solved approximately using an iterative method or that we would like to replace A_1 and A_2 by approximations (preconditioners). Let us consider the second possibility. To solve (10.2) we shall use PCG on all the unknowns that is on the whole of Ω. The problem is to find a global preconditioner M for the permuted form of A. This is the point where we shall use ideas of domain decomposition. The preconditioner M must be symmetric positive definite. Guided by the way block preconditioners were defined in Chapter 8 (see Concus, Golub & Meurant [342]), let us choose M in the form

$$M = L \begin{pmatrix} M_1^{-1} & & \\ & M_2^1 & \\ & & M_{1,2}^{-1} \end{pmatrix} L^T,$$

where M_1 (resp. M_2) is of the same order as A_1 (resp. A_2) and $M_{1,2}$ is of the same order as $A_{1,2}$. L is block lower triangular

$$L = \begin{pmatrix} M_1 & & \\ 0 & M_2 & \\ E_1^T & E_2^T & M_{1,2} \end{pmatrix}.$$

Then, we see how parallelism is introduced. At each PCG iteration, we must solve a linear system like

$$Mz = M \begin{pmatrix} z_1 \\ z_2 \\ z_{1,2} \end{pmatrix} = r = \begin{pmatrix} r_1 \\ r_2 \\ r_{1,2} \end{pmatrix}.$$

This is done by first solving $Ly = r$, where the first two steps are

$$M_1 y_1 = r_1, \quad M_2 y_2 = r_2.$$

This can be done in parallel. Then, we solve for the interface

$$M_{1,2} y_{1,2} = r_{1,2} - E_1^T y_1 - E_2^T y_2.$$

To obtain the solution, we have a backward solve step as

$$\begin{pmatrix} I & 0 & M_1^{-1}E_1 \\ & I & M_2^{-1}E_2 \\ & & I \end{pmatrix} \begin{pmatrix} z_1 \\ z_2 \\ z_{1,2} \end{pmatrix} = \begin{pmatrix} y_1 \\ y_2 \\ y_{1,2} \end{pmatrix}.$$

This implies that $z_{1,2} = y_{1,2}$ and

$$M_1 w_1 = E_1 z_{1,2}, \quad z_1 = y_1 - w_1,$$

$$M_2 w_2 = E_2 z_{1,2}, \quad z_2 = y_2 - w_2.$$

The last two stages can be done in parallel. Therefore, the behaviour of the algorithm is to perform two independent solves on the subdomains, a solve on the interface and two other independent solves on the subdomains. The problem we are facing is how to choose the approximations M_1, M_2 and $M_{1,2}$. If we multiply together the three matrices whose product defines M, we obtain

$$M = \begin{pmatrix} M_1 & 0 & E_1 \\ 0 & M_2 & E_2 \\ E_1^T & E_1^T & M_{1,2}^* \end{pmatrix},$$

where

$$M_{1,2}^* = M_{1,2} + E_1^T M_1^{-1} E_1 + E_2^T M_2^{-1} E_2.$$

Therefore, as we would like M to be an approximation of A, it makes sense to choose

$$M_1 \approx A_1, \quad M_2 \approx A_2,$$

and

$$M_{1,2}^* \approx A_{1,2} \Longrightarrow M_{1,2} \approx A_{12} - E_1^T M_1^{-1} E_1 - E_2^T M_2^{-1} E_2.$$

That is, if the inverse of M_1 (resp. M_2) is also a good approximation of the inverse of A_1 (resp. A_2), we are back to the same problem as before; that is to say, $M_{1,2}$ must be an approximation to the Schur complement S.

This shows that the two different classes of methods we distinguished before give rise to the same fundamental problem of constructing an approximate Schur complement. We shall now look at solving this problem. Of course, there are many ways of getting approximate Schur complements and we shall be looking only at a few of them.

10.5. Approximate Schur complements in the two subdomains case

In this section, we look at different possibilities of deriving approximations to Schur complements in the case of two subdomains. To be able to do so, we use some results on tridiagonal and block tridiagonal matrices that were established in Chapter 2. Then, we show how to compute the eigenvalues of the Schur complement in the case of separable equations and from there, we derive approximations that can be used in a more general setting. Next, we study other methods for algebraically constructing approximations for discontinuous coefficients problems arising from the discretization of general elliptic PDEs.

10.5.1. The Schur complement for block tridiagonal matrices

In this section we exploit the fact that A has a block tridiagonal structure,

$$A = \begin{pmatrix} D_1 & -B_2^T & & & \\ -B_2 & D_2 & -B_3^T & & \\ & \ddots & \ddots & \ddots & \\ & & -B_{m-1} & D_{m-1} & -B_m^T \\ & & & -B_m & D_m \end{pmatrix},$$

to derive other expressions for the Schur complement in the two subdomains case. If we use our domain decomposition ordering of the unknowns, we have

$$A_1 = \begin{pmatrix} D_1 & -B_2^T & & & \\ -B_2 & D_2 & -B_3^T & & \\ & \ddots & \ddots & \ddots & \\ & & -B_{m_1-1} & D_{m_1-1} & -B_{m_1}^T \\ & & & -B_{m_1} & D_{m_1} \end{pmatrix},$$

$$A_2 = \begin{pmatrix} D_{m_1+2} & -B_{m_1+3}^T & & & \\ -B_{m_1+3} & D_{m_1+3} & -B_{m_1+4}^T & & \\ & \ddots & \ddots & \ddots & \\ & & -B_{m-1} & D_{m-1} & -B_m^T \\ & & & -B_m & D_m \end{pmatrix},$$

and

$$A_{1,2} = D_{m_1+1}, \quad E_1^{m_1} = -B_{m_1+1}^T, \quad E_2^1 = -B_{m_1+2}.$$

To simplify the expression of the Schur complement S, we consider a block twisted factorization of A, the block in the center being $j = m_1 + 1$. For further reference, we introduce the block LU and UL factorizations. Let

$$A_1 = (\Delta + L_1)\Delta^{-1}(\Delta + L_1^T),$$

where Δ is a block diagonal matrix and L_1 is the block lower triangular part of A_1 which is of block order m_1. Similarly, we have

$$A_2 = (\Sigma + L_2^T)\Sigma^{-1}(\Sigma + L_2),$$

where Σ is a block tridiagonal matrix and L_2 is the block lower triangular part of A_2. Note that A_2 is of block order m_2. To simplify the notation, we denote the diagonal blocks of Σ by $\Sigma_{m_1+2}, \ldots, \Sigma_n$. By equating, we immediately obtain that

$$\begin{cases} \Delta_1 = D_1, \\ \Delta_i = D_i - B_i \Delta_{i-1}^{-1} B_i^T, \quad i = 2, \ldots, m_1 \end{cases} \tag{10.6}$$

and

$$\begin{cases} \Sigma_m = D_m, \\ \Sigma_i = D_i - B_{i+1}^T \Sigma_{i+1}^{-1} B_{i+1}, \quad i = m-1, \ldots, m_1 + 2 \end{cases} \tag{10.7}$$

Theorem 10.6

$$S = D_{m_1+1} - B_{m_1+1} \Delta_{m_1}^{-1} B_{m_1+1}^T - B_{m_1+2}^T \Sigma_{m_1+2}^{-1} B_{m_1+2}. \tag{10.8}$$

Proof. Straightforward. □

10.5.2. Eigenvalues of the Schur complement for separable problems

We specialize to the matrix arising from a separable problem (see Chapter 4). To simplify a bit more, we consider the matrix

$$A = \begin{pmatrix} T & -I & & & \\ -I & T & -I & & \\ & \ddots & \ddots & \ddots & \\ & & -I & T & -I \\ & & & -I & T \end{pmatrix}, \tag{10.9}$$

where we assume that we know the eigenvalues of the symmetric matrix T. Then

$$T = Q \Lambda Q^T,$$

Q being an orthogonal matrix ($QQ^T = I$) and Λ being a diagonal matrix whose diagonal elements are the eigenvalues of T. As usual we denote these eigenvalues by $\lambda_l, l = 1, \ldots, m$. The matrix in (10.9) is not the most general we can handle. Actually, we can replace the identity matrix I by another matrix J as long as J and T commute which implies that they share the same system of eigenvectors. Using (10.6) and (10.7), we can compute the eigenvalues of Δ_i and Σ_i.

Theorem 10.7

The spectral decompositions of matrices Δ_i and Σ_i are

$\Delta_i = Q\Lambda_i Q^T, \quad \Sigma_i = Q\Pi_i Q^T, \forall i$

where Λ_i and Π_i are diagonal matrices whose diagonal elements are given for $l = 1, \ldots, m$ by

$$\begin{cases} (\Lambda_1)_{l,l} = \Lambda_{l,l} = \lambda_l, \\ (\Lambda_i)_{l,l} = \lambda_l - \dfrac{1}{(\Lambda_{i-1})_{l,l}}, \quad i = 2, \ldots, m_1 \end{cases}$$

and

$$\begin{cases} (\Pi_m)_{l,l} = \lambda_l, \\ (\Pi_i)_{l,l} = \lambda_l - \dfrac{1}{(\Pi_{i+1})_{l,l}}, \quad i = m-1, \ldots, m_1+2 \end{cases}$$

Proof. As all the matrices in the block recurrences (10.6) and (10.7) share the same eigenvectors, we easily obtain the formulas for the eigenvalues. □

According to lemma 2.21, in the case of problem (10.9) we can solve analytically these recurrence relations.

Proposition 10.8

If $\lambda_l \neq 2$, then

$$(\Lambda_i)_{l,l} = \frac{(r_l)_+^{i+1} - (r_l)_-^{i+1}}{(r_l)_+^{i} - (r_l)_-^{i}}, \quad i = 1, \ldots, m_1$$

$$(\Pi_j)_{l,l} = \frac{(r_l)_+^{m-j+2} - (r_l)_-^{m-j+2}}{(r_l)_+^{m-j+1} - (r_l)_-^{m-j+1}}, \quad j = m, \ldots, m_1+2$$

where $(r_l)_\pm = \frac{\lambda_l \pm \sqrt{\lambda_l^2 - 4}}{2}$.

Proof. See lemma 2.21. □

From the previous result and using (10.8), we obtain the result for the Schur complement.

Theorem 10.9

The spectral decomposition of the Schur complement is

$$S = Q\Theta Q^T,$$

where Θ is a diagonal matrix whose diagonal elements θ_l are given by

$$\theta_l = \lambda_l - \frac{(r_l)_+^{m_1} - (r_l)_-^{m_1}}{(r_l)_+^{m_1+1} - (r_l)_-^{m_1+1}} - \frac{(r_l)_+^{m_2} - (r_l)_-^{m_2}}{(r_l)_+^{m_2+1} - (r_l)_-^{m_2+1}}, \quad l = 1, \ldots, m \tag{10.10}$$

where $(r_l)_\pm = \frac{\lambda_l \pm \sqrt{\lambda_l^2 - 4}}{2}$.

Proof. Since $D_{m_1+1} = T = Q\Lambda Q^T$, $B_{m_1+1} = B_{m_1+2} = I$ and

$$\Delta_{m_1}^{-1} = Q\Lambda_{m_1}^{-1}Q^T, \quad \Sigma_{m_1+2}^{-1} = Q\Pi_{m_1+2}^{-1}Q^T,$$

this proves that $S = Q\Theta Q^T$ with $\Theta = \Lambda - \Lambda_{m_1}^{-1} - \Pi_{m_1+2}^{-1}$. Therefore, S has the same eigenvectors as T and Θ is a diagonal matrix whose diagonal entries are the eigenvalues which are known from proposition 10.8. □

Note that we do not need to explicitly know the eigenvectors Q to compute the eigenvalues. Hence, this computation can be done for any separable problem, provided we are able to compute the eigenvalues of the block matrices in A. When we do not know explicitly the solution of the recurrence relations, they can be solved numerically. The analytical expression of the eigenvalues was also derived for the Poisson model problem by Chan and Resasco [292] using the explicit knowledge of the eigenvectors and in a more general setting by Meurant [969]. Let us give another expression for formula (10.10) that is more useful for the model problem.

Proposition 10.10

Let $\lambda_l = 2 + \sigma_l$ and $\gamma_l = \left(1 + \frac{\sigma_l}{2} - \sqrt{\sigma_l + \frac{\sigma_l^2}{4}}\right)^2$, then

$$\theta_l = \left(\frac{1 + \gamma_l^{m_1+1}}{1 - \gamma_l^{m_1+1}} + \frac{1 + \gamma_l^{m_2+1}}{1 - \gamma_l^{m_2+1}}\right)\sqrt{\sigma_l + \frac{\sigma_l^2}{4}}, \quad \forall l = 1, \ldots, m. \tag{10.11}$$

Proof. First, note that

$$\lambda_l = (r_l)_+ + (r_l)_-, \quad 2\sqrt{\sigma_l + \frac{\sigma_l^2}{4}} = (r_l)_+ - (r_l)_-.$$

Then, for $p = m_1$ or m_2,

$$\frac{\lambda_l}{2} - \frac{(r_l)_+^p - (r_l)_-^p}{(r_l)_+^{p+1} - (r_l)_-^{p+1}} = \frac{(r_l)_+^{p+1} + (r_l)_-^{p+1}}{(r_l)_+^{p+1} - (r_l)_-^{p+1}} \sqrt{\sigma_l + \frac{\sigma_l^2}{4}}.$$

This is because

$$\frac{((r_l)_+ + (r_l)_-)((r_l)_+^{p+1} - (r_l)_-^{p+1}) - 2((r_l)_+^p + (r_l)_-^p)}{2((r_l)_+^{p+1} - (r_l)_-^{p+1})}$$
$$= \frac{(r_l)_+^p((r_l)_+^2 - 1) - (r_l)_-^p((r_l)_-^2 - 1)}{2((r_l)_+^{p+1} - (r_l)_-^{p+1})}.$$

This gives the result since

$$(r_l)_+^2 - 1 = (r_l)_+((r_l)_+ - (r_l)_-), \quad (r_l)_-^2 - 1 = (r_l)_-((r_l)_- - (r_l)_+)$$

and we set

$$\gamma_l = \frac{(r_l)_-}{(r_l)_+} = (r_l)_-^2 = \left(1 + \frac{\sigma_l}{2} - \sqrt{\sigma_l + \frac{\sigma_l^2}{4}}\right)^2.$$

□

We note that if we assume $\lambda_l > 2, \forall l$, as $(r_l)_\pm > 0$ and $(r_l)_+ > (r_l)_-$, we have $0 < \gamma_l < 1$.

Remark. Consider the case of the model problem. We have

$$\lambda_l = 2 + \sigma_l = 4 - 2\cos(l\pi h), \quad l = 1 \ldots, m$$

where $h = \frac{1}{m+1}$. Then,

$$\sigma_{min} = 2 - 2\cos(\pi h) = 2\pi^2 h^2 + O(h^4),$$

$$\sigma_{max} = 2 - 2\cos(\pi \frac{m}{m+1}) = 4 - O(h^2).$$

This implies that, as a function of h,

$$\theta_{min} = C_1 h + O(h^2), \quad \theta_{max} = C_2 + O(h^2),$$

where C_1 and C_2 are two constants independent of h and

$$\kappa(S) = O\left(\frac{1}{h}\right).$$

This is a proof of theorem 10.5 for the model problem, but this result can be generalized.

Theorem 10.11

For the matrix in (10.9), if the eigenvalues of T are such that $\lambda_l > 2, \forall l$ and $\lambda_{min} = O(h^2)$, $\lambda_{max} = O(1)$, then

$$\kappa(S) = O\left(\frac{1}{h}\right).$$

Proof. With the hypothesis we made, $0 < \gamma_l < 1$. Moreover, γ_l is a decreasing function of σ_l. When h is small, γ_1 behaves like $1-2\sqrt{2}\pi h+O(h^2)$. This means that $\gamma_1^{p+1}, p = m_1$ or m_2 behaves like $1 - 2\sqrt{2}\pi(p+1)h$. But $(p+1)h$ is equal to a constant. This is the extent of Ω_1 or Ω_2 in the y direction. Therefore the term within parentheses in θ_l is bounded independently of h and θ_l can be written as

$$\theta_l = C(l)\sqrt{\frac{\lambda_l^2}{4} - 1},$$

where $C(l)$ is independent of h. Since the square root is an increasing function of λ_l, the result of the theorem follows. □

The hypothesis of theorem 10.11 gives only sufficient conditions and there are other situations where the same result is obtained. Let us now look at the eigenvalues of S when, for a fixed h, the domains Ω_1 and Ω_2 extend to infinity, that is $m_i \to \infty$.

Theorem 10.12

If $\lambda_l > 2$,

$$\theta_l \to 2\sqrt{\sigma_l + \frac{\sigma_l^2}{4}} \quad when \quad m_i \to \infty, \; i = 1, 2.$$

Proof. We have $0 < \gamma_l < 1$ and $\gamma_l^{m_i} \to 0$. □

Before defining the preconditioners for S, note that the knowledge of the eigenvalues of S allows us to construct a parallel fast Poisson solver (in the case of the Poisson model problem). We can compute the right hand side by using two fast solves in parallel, one for each subdomain. Then, since we know the eigenvalues and eigenvectors of S, we can solve the reduced problem. Note that for the model problem this is two fast Fourier transforms. Finally, knowing the solution on the interface, we can back solve in each domain by using two fast solvers in parallel.

10.5.3. Dryja's preconditioner

Let T_2 be the matrix corresponding to finite difference discretization of the one–dimensional Laplacian defined in Chapter 1. We have

$$T_2 = Q_2 \Sigma_2 Q_2^T,$$

where Σ_2 is the diagonal matrix of the eigenvalues whose diagonal elements are

$$\sigma_i = 2 - 2\cos(i\pi h), \quad i = 1, \ldots, m$$

Q_2 is the orthogonal matrix of the (normalized) eigenvectors whose components are

$$q_{i,j} = \sqrt{\frac{2}{m+1}} \sin(ij\pi h), \quad i, j = 1, \ldots, m.$$

Let $\sqrt{\Sigma_2}$ be the diagonal matrix whose diagonal elements are the square roots of the corresponding diagonal elements of Σ_2. Then, we define the Dryja's preconditioner M_D as

$$M_D = Q_2 \sqrt{\Sigma_2} Q_2^T.$$

Symbolically, we denote this by $\sqrt{-\Delta_{1D}}$. For the model problem, we can easily analyze the effect of this preconditioner. In that case S has the same eigenvectors as M_D (*i.e.*, $Q = Q_2$), therefore, the eigenvalues of $M_D^{-1}S$ are simply the ratios of the eigenvalues of S and M_D.

Proposition 10.13

For the model problem, the eigenvalues of $M_D^{-1}S$ are

$$\lambda_l(M_D^{-1}S) = \left(\frac{1+\gamma_l^{m_1+1}}{1-\gamma_l^{m_1+1}} + \frac{1+\gamma_l^{m_2+1}}{1-\gamma_l^{m_2+1}} \right) \sqrt{1+\frac{\sigma_l}{4}}, \quad l = 1, \ldots, m$$

where γ_l is defined as in proposition 10.10.

Proof. This statement is obvious as we know the eigenvalues of S from proposition 10.10 and because the eigenvalues of M_D are $\sqrt{\sigma_l}$. □

This implies that we have the following result which is what we were looking for.

Theorem 10.14

For the Poisson model problem $\kappa(M_D^{-1}S) = O(1)$.

Proof. Straightforward. □

Note that although the condition number does not depend on h, it depends on m_1 and m_2, but we shall look into this problem later. In a practical way, the action of M_D^{-1} on a vector can be implemented as two one dimensional FFTs and a division by the eigenvalues.

10.5.4. Golub and Mayers' preconditioner

The Golub and Mayers' preconditioner [663] is an improvement upon Dryja's preconditioner. We have seen that the exact eigenvalues in the case of the model problem are given by formula (10.11). The Golub and Mayers' preconditioner retains the term under the square root (which was replaced by $\sqrt{\sigma_l}$ in the Dryja's method). Therefore, it is defined as

$$M_{GM} = Q_2 \sqrt{\Sigma_2 + \frac{\Sigma_2^2}{4}} Q_2^T.$$

Symbolically, we denote this by $\sqrt{-\Delta_{1D} + \frac{\Delta_{1D}^2}{4}}$. Then, we have

Proposition 10.15

The eigenvalues of $M_{GM}^{-1}S$ are

$$\lambda_l(M_{GM}^{-1}S) = \left(\frac{1+\gamma_l^{m_1+1}}{1-\gamma_l^{m_1+1}} + \frac{1+\gamma_l^{m_2+1}}{1-\gamma_l^{m_2+1}}\right), \quad l = 1, \ldots, m$$

where γ_l is defined as in proposition 10.10.

Proof. Same as for proposition 10.13. □

Regarding the asymptotic analysis, we have a similar result as for the Dryja's preconditioner.

Theorem 10.16

For the Poisson model problem $\kappa(M_{GM}^{-1}S) = O(1)$.

Proof. Straightforward. □

10.5.5. The Neumann–Dirichlet preconditioner

This preconditioner was introduced by Bjørstad and Widlund [157]. If we remember that in the two subdomains case, A is written blockwise as

$$\begin{pmatrix} A_1 & 0 & E_1 \\ 0 & A_2 & E_2 \\ E_1^T & E_2^T & A_{1,2} \end{pmatrix} \begin{pmatrix} x_1 \\ x_2 \\ x_{1,2} \end{pmatrix} = \begin{pmatrix} b_1 \\ b_2 \\ b_{1,2} \end{pmatrix},$$

we can distinguish what in $A_{1,2}$ comes from subdomain Ω_1 and what comes from Ω_2. Let

$$A_{1,2} = A_{1,2}^{(1)} + A_{1,2}^{(2)}.$$

This is easy to do in the finite element setting. For finite difference methods, it is less natural. However, for instance, for the two dimensional model problem, we have

$$A_{1,2}^{(1)} = A_{1,2}^{(2)} = \begin{pmatrix} 2 & -\frac{1}{2} & & & \\ -\frac{1}{2} & 2 & -\frac{1}{2} & & \\ & \ddots & \ddots & \ddots & \\ & & -\frac{1}{2} & 2 & -\frac{1}{2} \\ & & & -\frac{1}{2} & 2 \end{pmatrix} = \frac{1}{2}A_{1,2}.$$

Since we know that

$$S = A_{1,2} - E_1^T A_1^{-1} E_1 - E_2^T A_2^{-1} E_2,$$

we can define

$$S^{(1)} = A_{1,2}^{(1)} - E_1^T A_1^{-1} E_1, \quad S^{(2)} = A_{1,2}^{(2)} - E_2^T A_2^{-1} E_2,$$

and $S = S^{(1)} + S^{(2)}$. The Neumann–Dirichlet preconditioner is defined as

$$M_{ND} = S^{(1)}. \tag{10.12}$$

Note, that we could also have chosen $S^{(2)}$ instead of $S^{(1)}$.

Proposition 10.17

The eigenvalues of M_{ND} are

$$\lambda_l(M_{ND}) = \left(\frac{1+\gamma_l^{m_1+1}}{1-\gamma_l^{m_1+1}}\right)\sqrt{\sigma_l + \frac{\sigma_l^2}{4}}, \quad l = 1, \dots, m$$

and the eigenvalues of $M_{ND}^{-1}S$ are

$$\lambda_l(M_{ND}^{-1}S) = 1 + \left(\frac{1-\gamma_l^{m_1+1}}{1+\gamma_l^{m_1+1}}\right)\left(\frac{1+\gamma_l^{m_2+1}}{1-\gamma_l^{m_2+1}}\right) \tag{10.13}$$

where γ_l is defined as in proposition 10.10.

Proof. The eigenvalues of $S^{(1)}$ are computed in the same way as those of S. We must consider only that which is coming from Ω_1. □

Theorem 10.18

For the Poisson model problem $\kappa(M_{ND}^{-1}S) = O(1)$.

Proof. Straightforward. □

Note that if $m_2 = m_1$, the preconditioner is exact, as it differs only from the exact result given in (10.11) by a constant factor 2. Therefore, if the conjugate gradient method is used on the interface, it must converge in one iteration. Of course, one does not use this method for the model problem as we know the exact solution and we can use fast solvers.

Let us explain why this preconditioner is called "Neumann–Dirichlet". This is because solving a linear system

$$S^{(1)}y_{12} = (A^{(1)}_{1,2} - E_1^T A_1^{-1} E_1)y_{1,2} = c_{1,2},$$

is equivalent to solving

$$\begin{pmatrix} A_1 & E_1 \\ E_1^T & A^{(1)}_{1,2} \end{pmatrix} \begin{pmatrix} y_1 \\ y_{1,2} \end{pmatrix} = \begin{pmatrix} 0 \\ c_{1,2} \end{pmatrix}.$$

For second order elliptic PDEs, it is easy to see that this is simply solving a problem in Ω_1 with given Neumann boundary conditions on the interface. When the solution is known on the interface, it is enough to solve a Dirichlet problem in Ω_2.

10.5.6. The Neumann–Neumann preconditioner

This preconditioner was introduced by Le Tallec [884]. With the same notation as before, it is defined as

$$M_{NN}^{-1} = \frac{1}{2}\left[(S^{(1)})^{-1} + (S^{(2)})^{-1}\right]. \tag{10.14}$$

Note that we directly define the inverse of the preconditioner as an average of inverses of "local" (to each subdomain) inverses of Schur complements.

Proposition 10.19

The eigenvalues of M_{NN}^{-1} are given by

$$\lambda_l(M_{NN}^{-1}) = \frac{1}{2}\left[\frac{1-\gamma_l^{m_1+1}}{1+\gamma_l^{m_1+1}} + \frac{1-\gamma_l^{m_2+1}}{1+\gamma_l^{m_2+1}}\right] \frac{1}{\sqrt{\sigma_l + \frac{\sigma_l^2}{4}}}$$

and the eigenvalues of $M_{NN}^{-1}S$ are

$$\begin{aligned}\lambda_l(M_{NN}^{-1}S) = 1 &+ \frac{1}{2}\left(\frac{1-\gamma_l^{m_1+1}}{1+\gamma_l^{m_1+1}}\right)\left(\frac{1+\gamma_l^{m_2+1}}{1-\gamma_l^{m_2+1}}\right) \\ &+ \frac{1}{2}\left(\frac{1-\gamma_l^{m_2+1}}{1+\gamma_l^{m_2+1}}\right)\left(\frac{1+\gamma_l^{m_1+1}}{1-\gamma_l^{m_1+1}}\right)\end{aligned}$$

Proof. We have seen in proposition 10.17 how to compute the eigenvalues of $S^{(1)}$. The computation for $S^{(2)}$ is similar and this gives the result. □

Theorem 10.20

For the Poisson model problem $\kappa(M_{NN}^{-1}S) = O(1)$.

Proof. Straightforward. □

We shall introduce some other preconditioners later on. For the moment, we shall study some of the properties of the ones we have just discussed.

10.5.7. Dependence on the aspect ratio

So far, we have more or less implicitly supposed that the domain Ω is evenly divided between Ω_1 and Ω_2. But this is not necessarily so. The situation might be as in figure 10.9 (a) where one subdomain is very thin. Of course, it seems silly to divide Ω in this way. But there are situations like figure 10.9 (b), where it is natural to have a thin subdomain.

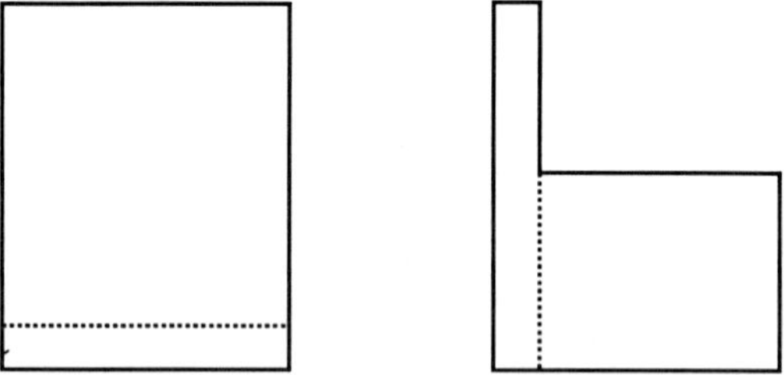

fig. 10.9: Examples of thin subdomains

Let $\alpha_i = (m_i + 1)h = \frac{m_i+1}{m+1}$, $i = 1, 2$ be the aspect ratios of the two subdomains. Since the mesh size h is supposed to be the same in both directions, α_i is small if m_i is and therefore the corresponding subdomain must be thin. In that case, even though the condition number of $M^{-1}S$ is independent of h, it can be large.

Theorem 10.21

Let $\alpha = \alpha_i$, $i = 1$ or 2 being small, then for the Dryja and Golub–Mayers preconditioners, we have for a fixed h,

$$\kappa(M^{-1}S) = O\left(\frac{1}{\alpha}\right).$$

Proof. This dependence arises from the analysis of the geometric factors that remain in both preconditioners. We introduce α in these factors and do an asymptotic expansion. □

For the Neumann–Dirichlet preconditioner, the result depends on which subdomain is thin and which one we choose in the definition of the preconditioner.

Theorem 10.22

Suppose α_1 is small and $\alpha_2 = O(1)$. Then,

1) If we choose Ω_1 in the definition, $\kappa(M_{ND}^{-1}S) = O(1)$,

2) otherwise, if we choose Ω_2, then $\kappa(M_{ND}^{-1}S) = O\left(\frac{1}{\alpha}\right)$.

Proof. This result comes from the asymptotic expansion of the geometric factor. □

For the Neumann–Neumann preconditioner, we get the following result.

Theorem 10.23

If one of the α_i, $i = 1, 2$ is small, then

$$\kappa(M_{NN}^{-1}S) = O\left(\frac{1}{\alpha}\right).$$

Proof. Suppose α_1 is small. Then, if we look at the eigenvalues of $M_{NN}^{-1}S$, there is a term $1 - \gamma_1^{m_1+1}$ in the denominator that gives the $1/\alpha$ behaviour.

Since there is such a term for both subdomains, we always get this result independently of which subdomain is small as long as there is at least one. □

From these results, it appears that if there is a small subdomain, the preconditioner of choice is Neumann–Dirichlet if we are careful in picking the right subdomain.

10.5.8. Dependence on the coefficients

Here, following Chan [264], we look into how the preconditioners depend on the coefficients of the PDE. Suppose, we want to solve the model problem, but in subdomain Ω_1 we multiply the coefficients by ρ. It means the problem we are looking at is

$$\begin{cases} -\rho\Delta u = f & \text{in } \Omega_1, \\ -\Delta u = f & \text{in } \Omega_2, \\ u|_{\partial\Omega} = 0. \end{cases}$$

When we discretize we can see that we have the following correspondence

$$\begin{aligned} A_1 &\to \rho A_1, \\ E_1 &\to \rho E_1, \\ A_{1,2}^{(1)} &\to \rho A_{1,2}^{(1)}. \end{aligned}$$

This implies that, if for the model problem ($\rho = 1$) we had

$$S = S^{(1)} + S^{(2)},$$

the Schur complement on the interface is now

$$S(\rho) = \rho S^{(1)} + S^{(2)}.$$

Suppose that we precondition $S(\rho)$ by S.

Theorem 10.24

$$\kappa(S^{-1}S(\rho)) \leq \max[\kappa(S^{-1}S^{(1)}), \kappa(S^{-1}S^{(2)})].$$

Proof. See Chan [351]. With the characterization of the eigenvalues, we have

$$\begin{aligned}\rho\lambda_{min}(S^{-1}S^{(1)}) &+ \lambda_{min}(S^{-1}S^{(2)}) \\ &\leq \frac{x^TS(\rho)x}{x^TSx} = \rho\frac{x^TS^{(1)}x}{x^TSx} + \frac{x^TS^{(2)}x}{x^TSx}, \\ &\leq \rho\lambda_{max}(S^{-1}S^{(1)}) + \lambda_{max}(S^{-1}S^{(2)}).\end{aligned}$$

Hence,

$$\kappa(S^{-1}S(\rho)) \leq \frac{\rho\lambda_{max}(S^{-1}S^{(1)}) + \lambda_{max}(S^{-1}S^{(2)})}{\rho\lambda_{min}(S^{-1}S^{(1)}) + \lambda_{min}(S^{-1}S^{(2)})}.$$

The right hand side is a monotone function of ρ. For $\rho = 0$, its value is $\kappa(S^{-1}S^{(2)})$. When $\rho \to \infty$, we obtain $\kappa(S^{-1}S^{(1)})$. This gives the result. □

Therefore, if we precondition with S, the condition number is bounded independently of ρ. Of course, as we have seen, we use an approximation of S.

10.5.9. Probing

There are other ways to construct preconditioners for the Schur complement S. Since we have

$$S = A_{1,2} - E_1^T A_1^{-1} E_1 - E_2^T A_2^{-1} E_2,$$

we note that we know how to compute Sv for a given vector v by solving two independent subproblems. Moreover, S depends only on the lower right block of A_1^{-1} and the upper left block of A_2^{-1}. For the kind of elliptic problems that we are considering, we know that the absolute values of the coefficients of these matrices decrease as we depart from the main diagonal on a given row. Therefore, the same property holds for S leading to the idea of approximating S by a banded matrix M (see figure 10.10).

We are going to compute the elements of M by requiring that

$$Mv^i = Sv^i, i = 1, \ldots, q$$

for a given set of vectors $v^i, i = 1, \ldots, q$. This idea (denoted as probing) was introduced in the preconditioning area by Chan [284] and Axelsson and Polman [82], [83]. Suppose, for instance that we would like to compute a

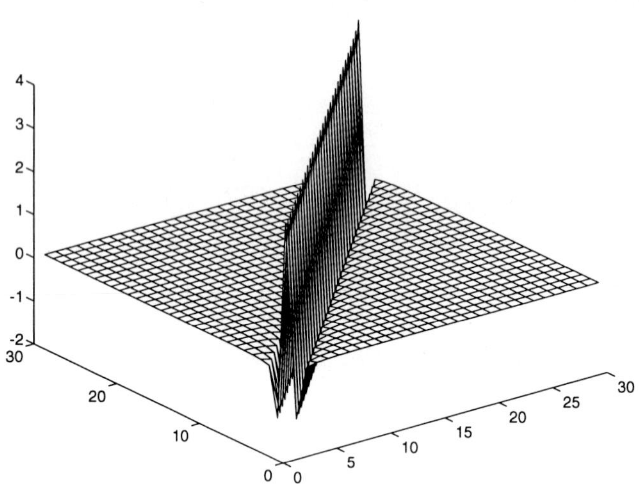

fig. 10.10: S for the model problem with two subdomains

tridiagonal approximation M of S. We set $q = 3$ and a possible choice of probing vectors is

$$v^1 = (1 \quad 0 \quad 0 \quad 1 \quad 0 \quad 0 \quad 1 \quad 0 \quad 0 \quad \ldots)^T,$$
$$v^2 = (0 \quad 1 \quad 0 \quad 0 \quad 1 \quad 0 \quad 0 \quad 1 \quad 0 \quad \ldots)^T,$$
$$v^3 = (0 \quad 0 \quad 1 \quad 0 \quad 0 \quad 1 \quad 0 \quad 0 \quad 1 \quad \ldots)^T.$$

If we denote $y^i = Mv^i, i = 1, 2, 3$, we have

$$y^1 = (m_{1,1} \quad m_{2,1} \quad m_{3,4} \quad m_{4,4} \quad m_{5,4} \quad \ldots)^T,$$
$$y^2 = (m_{1,2} \quad m_{2,2} \quad m_{3,2} \quad m_{4,5} \quad m_{5,5} \quad \ldots)^T,$$
$$y^3 = (0 \quad m_{2,3} \quad m_{3,3} \quad m_{4,3} \quad m_{5,6} \quad \ldots)^T.$$

We see that if we have computed Sv^1, Sv^2, Sv^3 by solving six subproblems, we obtain the non–zero coefficients of M right away. This can be generalized in a straightforward way to constructing a banded matrix M=Probe(S, d) with $2d + 1$ diagonals. However, this way of constructing the approximation leads to a non–symmetric M even when starting with a symmetric S as the following small example shows. Let

$$S = \begin{pmatrix} 10 & 0 & 0 & 5 \\ 0 & 10 & 0 & 2 \\ 0 & 0 & 10 & 0 \\ 5 & 2 & 0 & 10 \end{pmatrix},$$

then, we obtain

$$M = \text{Probe}(S,1) = \begin{pmatrix} 15 & 0 & 0 & 0 \\ 2 & 10 & 0 & 0 \\ 0 & 0 & 10 & 0 \\ 0 & 0 & 0 & 15 \end{pmatrix}.$$

Nevertheless this approximation possesses nice properties as T. Chan has shown that

$$\text{Probe}(\alpha S_1 + S_2, d) = \alpha\text{Probe}(S_1, d) + \text{Probe}(S_2, d)$$

and if a row of S is strictly diagonally dominant, then the corresponding row of $\text{Probe}(S, d)$ is also strictly diagonally dominant. A way to solve the symmetry problem is to use

$$M = \tfrac{1}{2}(\text{Probe}(S,d) + \text{Probe}(S,d)^T),$$

but then, the diagonal dominance property is not preserved. A better way to symmetrize is to define $\text{MProbe}(S, d)$ by computing the i, j entry from M=Probe(S, d) as $m_{i,j}$ if $|m_{i,j}| = \min(|m_{i,j}|, |m_{j,i}|)$ and $m_{j,i}$ otherwise. This preserves symmetry and strict diagonal dominance. For our previous small example we obtain a diagonal matrix.

By the way it is constructed, we cannot expect the tridiagonal probing approximation to give a condition number independent of h. Chan and Mathew [285] proved that, for the model problem

$$\kappa(M^{-1}S) = O\left(\frac{1}{\sqrt{h}}\right).$$

However, probing is an interesting idea that may be applied in many different problems.

10.5.10. INV and MINV approximations

For constructing a preconditioner of the Schur complement for the special case we are considering, one can use the same ideas as in Chapter 8 for the `INV` preconditioner. We have seen that

$$A_1 = (\Delta + L_1)\Delta^{-1}(\Delta + L_1^T),$$

where Δ is a block diagonal matrix and

$$A_2 = (\Sigma + L_2^T)\Sigma^{-1}(\Sigma + L_2),$$

where Σ is a block tridiagonal matrix and the diagonal blocks are given by

$$\begin{cases} \Delta_1 = D_1, \\ \Delta_i = D_i - B_i \Delta_{i-1}^{-1} B_i^T, \quad i = 2, \ldots, m_1 \end{cases}$$

and

$$\begin{cases} \Sigma_m = D_m, \\ \Sigma_i = D_i - B_{i+1}^T \Sigma_{i+1}^{-1} B_{i+1}, \quad i = m-1, \ldots, m_1 + 2. \end{cases}$$

The Schur complement is given by

$$S = D_{m_1+1} - B_{m_1+1} \Delta_{m_1}^{-1} B_{m_1+1}^T - B_{m_1+2}^T \Sigma_{m_1+2}^{-1} B_{m_1+2}.$$

To derive an approximation to S we do the same as in INV by taking tridiagonal approximations of the inverses of Δ_i and Σ_i. Let

$$\begin{cases} \bar{\Delta}_1 = D_1, \\ \bar{\Delta}_i = D_i - B_i trid(\bar{\Delta}_{i-1}^{-1}) B_i^T, \quad i = 2, \ldots, m_1 \end{cases}$$

$$\begin{cases} \bar{\Sigma}_m = D_m, \\ \bar{\Sigma}_i = D_i - B_{i+1}^T trid(\bar{\Sigma}_{i+1}^{-1}) B_{i+1}, \quad i = m-1, \ldots, m_1 + 2 \end{cases}$$

where *trid* gives the three main diagonals of a matrix. Then, we define

$$M = D_{m_1+1} - B_{m_1+1} trid(\bar{\Delta}_{m_1}^{-1}) B_{m_1+1}^T - B_{m_1+2}^T trid(\bar{\Sigma}_{m_1+2}^{-1}) B_{m_1+2}.$$

Note that the computation of this approximation is cheaper than probing as we do not need to solve subproblems. However, the drawback is that there is no theory supporting this choice. But, for the model problem the elements of the involved matrices converge when we increase the dimension of the problem. Therefore, we can analyze the limit problem where all matrices have constant diagonals. If we do so, it turns out that all the eigenvalues of the INV approximation of the Schur complement are bounded independently of h giving a condition number $\kappa(M^{-1}S) = O(1/h)$ which is of the same order as for S. Nevertheless, $\kappa(M^{-1}A)$ is much smaller than $\kappa(S)$. This can be improved by using a MINV approximation instead of INV. Then, we can obtain a smallest eigenvalue that depends on h. Figure 10.11 compares the eigenvalues of the Schur complement and its approximations by probing, INV and MINV for the model problem and $m = 30$. Figure 10.12 shows the condition numbers as a function of m. We can see that probing and MINV give almost the same results. We note that the MINV approximation is computed

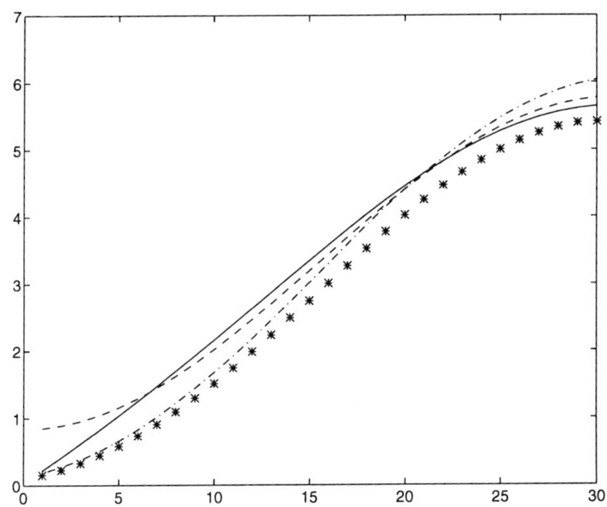

fig. 10.11: Eigenvalues of the Schur complement and its approximations
solid line: Schur, dash–dotted: probing, dashed: INV, stars: MINV

such that the row sums of M are the same as those of S. For the example, the same property is obtained by probing since the sum of the three probing vectors has all its components equal to 1, therefore the remainder of probing approximation has also zero row sums.

10.5.11. The Schur complement for more general problems

Let A_Δ be the matrix arising from the finite difference approximation of the Poisson model problem,

$$A_\Delta = \begin{pmatrix} B_1 & 0 & F_1 \\ 0 & B_2 & F_2 \\ F_1^T & F_2^T & B_{1,2} \end{pmatrix}$$

and A be another matrix

$$A = \begin{pmatrix} A_1 & 0 & E_1 \\ 0 & A_2 & E_2 \\ E_1^T & E_2^T & A_{1,2} \end{pmatrix}$$

which is spectrally equivalent to A_Δ *i.e.*, there exist constants C_0 and C_1 independent of h such that

$$C_0(A_\Delta x, x) \leq (Ax, x) \leq C_1(A_\Delta x, x), \ \forall x.$$

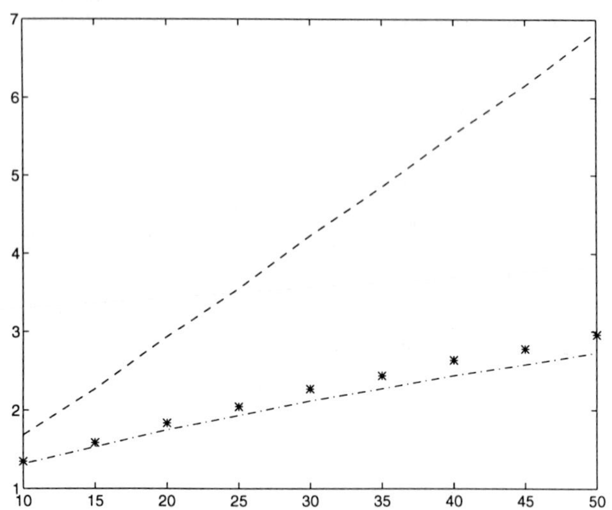

fig. 10.12: Condition numbers $\kappa(M^{-1}S)$ as a function of the problem dimension dash–dotted: probing, dashed: INV, stars: MINV

This implies that

$$\kappa(A_\Delta^{-1}A) \leq \frac{C_1}{C_0}.$$

Let $x = (x_1 \quad x_2 \quad x_{1,2})^T$ and $y = (-A_1^{-1}E_1x_{1,2} \quad -A_2^{-1}E_2x_{1,2} \quad x_{1,2})^T$, then

$$Ay = \begin{pmatrix} 0 \\ 0 \\ (A_{1,2} - E_1^TA_1^{-1}E_1 - E_2^TA_2^{-1}E_2)x_{1,2} \end{pmatrix} = \begin{pmatrix} 0 \\ 0 \\ Sx_{1,2} \end{pmatrix}.$$

The matrices $A_1^{-1}E_1$ and $A_2^{-1}E_2$ are the discrete analogues of what is called the harmonic extension. They take values on the inner boundaries and extend them to the subdomains. With this choice of y we have

$$(Ay, y) = (Sx_{1,2}, x_{1,2}),$$

and

$$A_\Delta y = \begin{pmatrix} (-B_1A_1^{-1}E_1 + F_1)x_{1,2} \\ (-B_2A_2^{-1}E_2 + F_2)x_{1,2} \\ (B_{1,2} - F_1^TA_1^{-1}E_1 - F_2^TA_2^{-1}E_2)x_{1,2} \end{pmatrix}.$$

Let us compute $(A_\Delta y, y)$:

$$\begin{aligned}(A_\Delta y, y) &= (y_1, (-B_1 A_1^{-1} E_1 + F_1) x_{1,2}) \\ &+ (y_2, (-B_2 A_2^{-1} E_2 + F_2) x_{1,2}) \\ &+ (x_{1,2}, (B_{1,2} - F_1^T A_1^{-1} E_1 - F_2^T A_2^{-1} E_2) x_{1,2}).\end{aligned}$$

Suppose that there exist non–singular matrices D_1 and D_2 such that

$$A_i = D_i B_i, \quad E_i = D_i F_i, \ i = 1, 2$$

then, the first and second terms in $(A_\Delta y, y)$ are zero and the third one is $(S_\Delta x_{1,2}, x_{1,2})$, S_Δ being the Schur complement of A_Δ. This shows that we have

$$C_0 (S_\Delta x_{1,2}, x_{1,2}) \le (S x_{1,2}, x_{1,2}) \le C_1 (S_\Delta x_{1,2}, x_{1,2});$$

so S is spectrally equivalent to S_Δ. So, we can use the same preconditioners for S as for the model problem. This happens, for instance, for the case we considered before if $A_i = \rho_i B_i, E_i = \rho_i F_i, i = 1, 2$.

10.6. Approximations of Schur complements with many subdomains

We would like to extend the results for two subdomains by considering now the domain Ω being divided into k strips as in figure 10.13.

fig. 10.13: Multi strips partitioning

We denote by x_i the unknowns in subdomain Ω_i and by $x_{i,i+1}$ the unknowns on the interface between Ω_i and Ω_{i+1}. With this notation the problem is written as $Ax = b$ with

$$A = \begin{pmatrix} A_1 & & & & & C_1 & & & \\ & A_2 & & & & E_2 & C_2 & & \\ & & A_3 & & & & E_3 & \ddots & \\ & & & \ddots & & & & \ddots & C_{k-1} \\ & & & & A_k & & & & E_k \\ C_1^T & E_2^T & & & & A_{1,2} & & & \\ & C_2^T & E_3^T & & & & B_{2,3} & & \\ & & \ddots & \ddots & & & & \ddots & \\ & & & C_{k-1}^T & E_k^T & & & & A_{k-1,k} \end{pmatrix},$$

$$x = \begin{pmatrix} x_1 \\ x_2 \\ x_3 \\ \vdots \\ x_k \\ x_{1,2} \\ x_{2,3} \\ \vdots \\ x_{k-1,k} \end{pmatrix}, \quad b = \begin{pmatrix} b_1 \\ b_2 \\ b_3 \\ \vdots \\ b_k \\ b_{1,2} \\ b_{2,3} \\ \vdots \\ b_{k-1,k} \end{pmatrix}.$$

The notation we used for the two subdomains case for the matrices corresponding to the subdomains become too cumbersome when generalizing to $k > 2$. Therefore, we denote

$$A_i = \begin{pmatrix} D_i^1 & (A_i^2)^T & & & \\ A_i^2 & D_i^2 & (A_i^3)^T & & \\ & \ddots & \ddots & \ddots & \\ & & A_i^{m_i} & D_i^{m_i-1} & (A_i^{m_i})^T \\ & & & A_i^{m_i} & D_i^{m_i} \end{pmatrix}, \quad i = 1, \ldots, k$$

We eliminate the inner unknowns $x_1, \ldots, x_k$ to obtain the Schur complement.

Proposition 10.25

The Schur complement matrix S is block tridiagonal and we denote

$$S = \begin{pmatrix} A'_{12} & F_2^T & & & \\ F_2 & A'_{23} & F_3^T & & \\ & \ddots & \ddots & \ddots & \\ & & F_{k-2} & A'_{k-2,k-1} & F_{k-1}^T \\ & & & F_{k-1} & A'_{k-1,k} \end{pmatrix}.$$

The blocks are defined as

$$A'_{i,i+1} = A_{i,i+1} - C_i^T A_i^{-1} C_i - E_{i+1}^T A_{i+1}^{-1} E_{i+1},$$

$$F_i = -C_i^T A_i^{-1} E_i.$$

Proof. It is obvious that when we eliminate the inner unknowns, an interface can only be coupled with the interface above and the interface below. The expressions for the blocks follow from straightforward algebra. □

As in the two subdomains case, we can slightly simplify the expressions for the blocks by using both LU and UL block factorizations for each subdomain. For subdomain number i we denote the diagonal blocks of these factorizations as Δ_i^j and Σ_i^j.

Theorem 10.26

The blocks of the Schur complement can be written as

$$A'_{i,i+1} = A_{i,i+1} - (C_i^{m_i})^T (\Delta_i^{m_i})^{-1} C_i^{m_i} - (E_{i+1}^1)^T (\Sigma_{i+1}^1)^{-1} E_{i+1}^1.$$

If $G_i^l = (\Sigma_i^1)^{-1} E_i^1$ *and*

$$G_i^l = -(\Sigma_i^l)^{-1} A_i^l G_i^{l-1}, \quad l = 2, \ldots, m_i$$

then

$$F_i = -(C_i^{m_i})^T G_i^{m_i}.$$

Proof. This is simply solving the linear systems using the block factorizations and noticing the block sparsity of the right hand sides. □

To compute the eigenvalues, we consider a separable problem and, more specifically the model problem. Then it is easy to show that $A'_{i,i+1}$ and F_i have the same eigenvectors. To simplify a little further, let us take identical subdomains with m mesh lines. All the matrices $A'_{i,i+1}$ are the same and the same is true for matrices F_i. Let us suppose that $A_{i,i+1}$ is of order m. We have already computed the eigenvalues θ_l of $A'_{i,i+1}$,

$$\theta_l = 2\left(\frac{1+\gamma_l^{m+1}}{1-\gamma_l^{m+1}}\right)\sqrt{\sigma_l + \frac{\sigma_l^2}{4}}, \quad \forall l = 1, \ldots, m.$$

Proposition 10.27

The eigenvalues of F_i *are*

$$\delta_l = -\frac{(r_l)_+ - (r_l)_-}{(r_l)_+^{m+1} - (r_l)_-^{m+1}},$$

where $(r_l)_\pm = \frac{\lambda_l \pm \sqrt{\lambda_l^2 - 4}}{2}$. *This can be written*

$$\delta_l = -\frac{2\gamma_l^{\frac{m+1}{2}}}{1-\gamma_l^{m+1}}\sqrt{\sigma_l + \frac{\sigma_l^2}{4}}.$$

Proof. Let $G_i^l = Q\Xi_i^l Q^T$ and $\Sigma_i^l = Q\Pi_i^l Q^T$. Then
$\Xi_i^1 = (\Pi_i^1)^{-1}, \quad \Xi_i^l = (\Pi_i^l)^{-1}\Xi_i^{l-1}, \; l = 2, \ldots, m.$
This shows that $F_i = Q\Xi_i^m Q^T$ and gives the expression for the eigenvalues.
□

If we do a permutation of the unknowns, we see that S is similar to a block diagonal matrix with m blocks, each block being a tridiagonal Toeplitz matrix of order $k-1$

$$\beta_l \begin{pmatrix} 1+\gamma_l^{m+1} & -\gamma_l^{\frac{m+1}{2}} & & & \\ -\gamma_l^{\frac{m+1}{2}} & 1+\gamma_l^{m+1} & -\gamma_l^{\frac{m+1}{2}} & & \\ & \ddots & \ddots & \ddots & \\ & & -\gamma_l^{\frac{m+1}{2}} & 1+\gamma_l^{m+1} & -\gamma_l^{\frac{m+1}{2}} \\ & & & -\gamma_l^{\frac{m+1}{2}} & 1+\gamma_l^{m+1} \end{pmatrix},$$

$$\beta_l = \frac{2}{1-\gamma_l^{m+1}}\sqrt{\sigma_l + \frac{\sigma_l^2}{4}}, \quad l = 1, \ldots, m$$

Theorem 10.28

The eigenvalues of S are

$$\omega_{l,j} = \frac{2}{1-\gamma_l^{m+1}}\sqrt{\sigma_l + \frac{\sigma_l^2}{4}}\left(1+\gamma_l^{m+1} - 2\gamma_l^{\frac{m+1}{2}}\cos\left(\frac{j\pi}{k}\right)\right),$$
$$l = 1,\ldots,m \quad j = 1,\ldots k-1$$

Proof. We verify by computing the eigenvalues of the Toeplitz tridiagonal matrices. □

As with two subdomains we can analyze the condition number of S. The maximum eigenvalue is bounded and the minimum is obtained for $l = 1, j = 1$. We have

$$\kappa(S) = O\left(\frac{k}{h}\right).$$

Therefore, if k is fixed and $h \to 0$, κ increases as $1/h$. If h is given and we increase the number of subdomains, κ increases as k. For a given h, the more subdomains we use, the smaller they are and then, the condition number of S (which is larger and larger) is proportional to k.

The problem now is to define preconditioners for S. A first idea is to use block diagonal preconditioners, the diagonal blocks being the preconditioners we define for the two subdomains case. If we do this with Dryja's preconditioner, remembering that the eigenvalues were $\sqrt{\sigma_l}$, we can compute the eigenvalues of $\lambda(M_D^{-1}S)$:

$$\frac{2}{1-\gamma_l^{m+1}}\sqrt{1 + \frac{\sigma_l}{4}}\left(1+\gamma_l^{m+1} - 2\gamma_l^{\frac{m+1}{2}}\cos\left(\frac{j\pi}{k}\right)\right).$$

This preconditioner removes the h dependency but not that on k and, in fact, we have

$$\kappa(M_D^{-1}S) = O(k^2).$$

The same result is obviously true for the Golub and Mayers' preconditioner although the condition number is a little smaller. It is more difficult to generalize the Neumann–Dirichlet preconditioner to many subdomains One obvious way is to use as preconditioner a block diagonal matrix whose blocks are the two subdomains Neumann–Dirichlet preconditioner. If we do this, we have $\kappa = O(k^2)$. Another way to derive a preconditioner is to apply the same principle as with two subdomains, that is to decompose the Schur complement into $S^{(1)} + S^{(2)}$. We describe the method on an example with $k = 5$. Then, we have 3 Neumann subdomains and 4 interfaces. Let $\bar{A}_{i,j}$ be the part of $A_{i,j}$ coming from the Neumann subdomains. Then, the preconditioner is defined as

$$\begin{pmatrix} \bar{A}_{12} - C_1^T A_1^{-1} C_1 & 0 & 0 & 0 \\ 0 & \bar{A}_{23} - E_3^T A_3^{-1} E_3 & -E_3^T A_3^{-1} C_3 & 0 \\ 0 & -C_3^T A_3^{-1} E_3 & \bar{A}_{34} - C_3^T A_3^{-1} C_3 & 0 \\ 0 & 0 & 0 & \bar{A}_{45} - E_5^T A_5^{-1} E_5 \end{pmatrix}$$

More generally, the preconditioning matrix has 2×2 blocks and the first and last blocks are either 2×2 or 1×1. As before, this preconditioner is equivalent to solving problems with Neumann boundary conditions on the interfaces on the "Neumann" subdomains. Dryja and Proskurowski [451] proved that the condition number is of order k^2. The generalization is easier if a red–black coloring of the subdomains can be done.

The Neumann–Neumann preconditioner can be easily extended to many subdomains, the inverses of partial Schur complements have to be weighted by the inverse of the number of subdomains which share a given node.

Another possibility is to use probing or INV to obtain a block incomplete factorization of S but, then the condition number cannot be h independent.

10.7. Inexact subdomain solvers

If we cannot solve exactly for the subproblems, we are not able to use an iterative method with S as we cannot compute the matrix×vector product Sv. Hence, we have to iterate on all the unknowns and we need a global parallel preconditioner. We define

$$M = L \begin{pmatrix} M_1^{-1} & & & & & & \\ & M_2^{-1} & & & & & \\ & & \ddots & & & & \\ & & & M_k^{-1} & & & \\ & & & & M_{1,2}^{-1} & & \\ & & & & & \ddots & \\ & & & & & & M_{k-1,k}^{-1} \end{pmatrix} L^T,$$

$$L = \begin{pmatrix} M_1 & & & & & & & & \\ & M_2 & & & & & & & \\ & & \ddots & & & & & & \\ & & & \ddots & & & & & \\ & & & & M_k & & & & \\ C_1^T & E_2^T & & & & M_{1,2} & & & \\ & C_2^T & E_3^T & & & H_2 & M_{2,3} & & \\ & & \ddots & & & & \ddots & \ddots & \\ & & & C_{k-1}^T & E_k^T & & & H_{k-1} & M_{k-1,k} \end{pmatrix}.$$

The matrices M_i can be chosen without problem as for the two subdomains case. For matrices $M_{i,i+1}$ and H_i, we have many possible choices as there are many different ways to construct an incomplete factorization of a block tridiagonal matrix. The simplest choice is to take an INV–like decomposition

$$M_{1,2} = A_{1,2} - (C_1^{m_1})^T \, trid((\Delta_1^{m_1})^{-1}) \, C_1^{m_1} - (E_2^1)^T \, trid((\Sigma_2^1)^{-1}) \, E_2^1.$$

$$M_{i,j} = A_{i,j} - (C_i^{m_i})^T \, trid((\Delta_i^{m_i})^{-1}) \, C_i^{m_i}$$
$$- (E_j^1)^T \, trid((\Sigma_j^1)^{-1}) \, E_j^1 - H_i \, trid(M_{i-1,j-1}^{-1}) \, H_i^T.$$

$$G_i^1 = diag((\Sigma_i^1)^{-1} \, (E_i^1)^T),$$

$$G_i^j = -diag((\Sigma_i^j)^{-1} \, A_i^j \, G_i^{j-1}),$$

$$H_i = -(C_i^{m_i})^T G_i^{m_i},$$

where *diag* defines a diagonal approximation. Then, H_i is diagonal. M_i is chosen as an INV block LU or UL approximation of A_i. We call this preconditioner INVDD.

There are two main problems with this preconditioner. First of all, it can be somewhat inefficient when, for a fixed mesh size, the number of subdomains is increased. Then, their aspect ratio decreases and the influence between the interfaces becomes stronger, which means that having a good approximation to the outer diagonals of the reduced system is more important. The simple diagonal approximation that we chose for simplicity would not be good enough to account for this strong coupling. A possible way to remedy to this problem is to keep more non–zero entries in H_i, for instance a tridiagonal approximation.

The second problem is related to parallel computation. For every conjugate gradient iteration, we must solve a linear system with matrix M. First, we solve a system $Ly = c$ and then a block upper triangular system whose matrix has the same structure as L^T. It is clearly seen from the structure we chose for M that, in both cases, we can solve independently for the subdomains, assigning for instance one subdomain to one processor. But, because we introduce a block factorization of the reduced system for the interfaces, there is, both in the forward and backward solves, a block recurrence. The interfaces cannot be solved in parallel. This is the sequential bottleneck of this method. Let us analyze the parallel complexity of this method and propose some more parallel variants.

We consider the analysis of the solve $Mz = r$ that is to be done every iteration. First, we count the number of floating point operations (Flops) needed to solve for one subdomain. Consider a subdomain Ω_i with p lines, each with m unknowns. The flops count depends on whether we use a vector method or not. The basic elimination method is recursive as we have point tridiagonal systems to solve. However, we saw in Chapter 8 how to vectorize `INV` block solves. We give both the scalar (`INV`) and vector (`INVV`) operation counts. For `INV` on one subdomain, we have $11pm - 7m$ multiplications and $6pm - 4m$ additions. For the vector algorithm `INVV`, we need $17pm - 10m$ multiplications and $14pm - 8m$ additions. Now, we must include these results in our DD algorithm. Suppose for the sake of simplicity that all k subdomains have p lines each, thus $kp + k - 1 = m$. Suppose also that we have q processors, k being a multiple of q. In the following, we are counting the "elapsed" number of flops, that is to say, the number of operations divided by the number of processors we can use in parallel. This will be proportional to the (elapsed) computing time on a dedicated parallel machine with no communication or memory contention problems.

The parallel part of the work for the subdomains is $2k$ solves, that is $\frac{2k(11pm-7m)}{q}$ multiplications and $\frac{2k(6pm-4m)}{q}$ additions for `INV`. For `INVV`, the corresponding count is $\frac{2k(17pm-10m)}{q}$ multiplications and $\frac{2k(14pm-8m)}{q}$ additions. Note that $\frac{k}{q}$ is an integer. For the interfaces the parallel work is $2(\lceil \frac{2(k-1)}{q} \rceil + 1)m$ multiplications and the same number for the additions. The balance of the work between the processors depends on the relative values of k and q and the details of the implementation. Unfortunately, as we said before, there is some sequential work associated with the reduced system : $11(k-1)m - 7m$ multiplications, $6(k-1)m - 4m$ additions if `INV` is used, $17(k-1)m - 10m$ multiplications, $14(k-1)m - 8m$ additions with `INVV`. The summary of these results is the following:

The total "elapsed" work for `INVDD` is

i) with the use of INV:

$$\frac{2k(11pm-7m)}{q}+2(\left\lceil\frac{2(k-1)}{q}\right\rceil+1)m+11(k-1)m-7m \text{ multiplications,}$$

$$\frac{2k(6pm-4m)}{q}+2(\left\lceil\frac{2(k-1)}{q}\right\rceil+1)m+6(k-1)m-4m \text{ additions,}$$

ii) with the use of INVV:

$$\frac{2k(17pm-10m)}{q}+2(\left\lceil\frac{2(k-1)}{q}\right\rceil+1)m+11(k-1)m-7m \text{ multiplications,}$$

$$\frac{2k(14pm-8m)}{q}+2(\left\lceil\frac{2(k-1)}{q}\right\rceil+1)m+6(k-1)m-4m \text{ additions.}$$

The problem is that, when m, the number of points in one direction, is fixed, as the number of subdomains increase (leading to more parallelism), p the number of lines per subdomain decreases (leading to less work to be done for each processor). The sequential work also increases as the number of interfaces grows. So there must be an optimum to be found. Suppose $q = k$, the number of processors is always the same as the number of subdomains then, for the number of multiplications, we should achieve a balance between the term $22pm$ which is decreasing and the term $11(k-1)m$ which is increasing. The approximate minimum of the sum is given by $2p = k - 1$. The same is true for the number of additions. Suppose that $m = 2^l - 1$, then $m = 2^{l-i} - 1$ and $k = 2^i, i = 1, \dots, l-1$; for large values of l the solution is $i = \frac{l+1}{2}$, for instance if $m = 127$, $l = 7$ and the "optimum" number of subdomains is 16. The following table gives the optimum number of subdomains as a function of the problem size :

n	*# of subdomains*
15	8
63	16
511	32
2047	64

How could we modify our INVDD preconditioner to introduce more parallelism?

The simplest idea is to set $H_i = 0, i = 2, \dots, k-1$, we call this method INVDDH. Doing so, we obtained a perfectly parallel method which only requires one synchronization point between the solve for the subdomains and the one for the interfaces. During the interface solve there are only $k-1$ processors active, so one processor is idle but the apparent number of multiplications is $34pm - 20m + 4m + m = 34pm - 9m$. Unfortunately, the numerical experiments show that, because we neglect the coupling between the interfaces, when the aspect ratio becomes small, the number of iterations increases much faster than for INVDD, see Meurant [970].

Another possibility is to compute H_i as in INVDD and then to set only a "few" H_i to 0. This leads to the solution of independent block systems for the interface solve. We call this method INVDDS. If, for instance, we zero out every other H_i, this allows us to use roughly $\frac{k-1}{2}$ processors in parallel, reducing the number of multiplications to approximately $34pm - 20m + 4m + 22m = 34pm + 6m$. The number of iterations is expected to be somewhere in between the result for INVDD and INVDDH. A third possibility for the preconditioner is to go back to the general form

$$M = \begin{pmatrix} M_S & 0 \\ M_{SI} & M_I \end{pmatrix} \begin{pmatrix} I & M_S^{-1} M_{SI}^T \\ 0 & I \end{pmatrix},$$

where M_I is the interface system. As soon as we have computed the approximate reduced system

$$R = \begin{pmatrix} \overline{M}_{1,2} & H_2^T & & & \\ H_2 & \overline{M}_{2,3} & H_3 & & \\ & \ddots & \ddots & \ddots & \\ & & H_{k-2} & \overline{M}_{k-2,k-1} & H_{k-1} \\ & & & H_{k-1} & \overline{M}_{k-1,k} \end{pmatrix},$$

we can choose for M_I whatever approximation we would like. The latter method was to use for M_I a usual block INV preconditioner of R. Another method is to take $M_I = R$ and to solve the system with matrix M_I applying the same DD method in a recursive way. This is very close to block cyclic reduction. Of course, the number of processors that we can use in parallel decreases at each reduction stage.

A choice which is more parallel for M_I is to use a polynomial preconditioner (see Chapter 8). The easiest choice is to take a least squares polynomial. The bounds for the eigenvalues are computed using the Gerschgorin disks. We call this method INVDPOL. The number of operations depends of course on the degree of the polynomial.

The last approach is to use the generalized incomplete twisted factorization, see Meurant [970]. In this combined method, we see domain decomposition as giving coarse grain parallelism and twisted factorization as providing finer grain parallelism for the interfaces. It is easier to explain this factorization with an example. Consider $k = 8$; we factor the reduced system as $\widetilde{L}\widetilde{D}\widetilde{L}^T$ with

$$\widetilde{L} = \begin{pmatrix} M_{12} & & & & & & \\ H_2 & M_{23} & H_3 & & & & \\ & & M_{34} & & & & \\ & & H_4 & M_{45} & H_5 & & \\ & & & & M_{56} & & \\ & & & & H_6 & M_{67} & H_7 \\ & & & & & & M_{78} \end{pmatrix},$$

with

$$M_{12} = \overline{M}_{12}, \quad M_{34} = \overline{M}_{34}, \quad M_{56} = \overline{M}_{56}, \quad M_{78} = \overline{M}_{78},$$

$$M_{23} = \overline{M}_{23} - H_2 trid(M_{12}^{-1})H_2 - H_3 trid(M_{34}^{-1})H_3,$$

$$M_{45} = \overline{M}_{45} - H_4 trid(M_{34}^{-1})H_4 - H_5 trid(M_{56}^{-1})H_5,$$

$$M_{67} = \overline{M}_{67} - H_6 trid(M_{56}^{-1})H_6 - H_7 trid(M_{78}^{-1})H_7.$$

In this small example, in the forward solve we can use first 4 processors in parallel to solve for M_{12}, M_{34}, M_{56} and M_{78} and then 3 processors to solve for M_{23}, M_{45} and M_{67}. In general the degree of parallelism is roughly $\frac{k-1}{2}$. We call this method `INVDTW`.

10.8. Domain decomposition with boxes

A domain decomposition with strips can be done for more general domains (than rectangles) by finding pseudo–peripheral nodes and constructing the level structure corresponding to one of these nodes, see Chapters 1 and 3. This is similar to one–way dissection. However, except for very large problems, when partitioning in this way, we cannot use many subdomains. A way to partition with many subdomains is to use so–called boxes, as in figure 10.14 for a rectangle and a general two dimensional domain.

This type of partitioning introduces new problems as the interfaces are now made of two kinds of sets: the edges and the points where the interfaces cross which are known as vertices or cross points. We shall use an index E_i for the edges, $E = \cup E_i$ and V for the vertices. We remark that with the most standard approximations the cross points are not linked to nodes inside the subdomains but only to nodes on the edges. Moreover, there is no connection between the cross points as we require at least one mesh point per edge. The

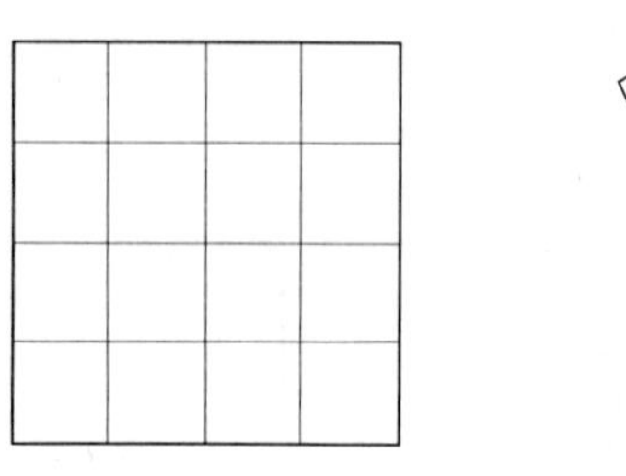
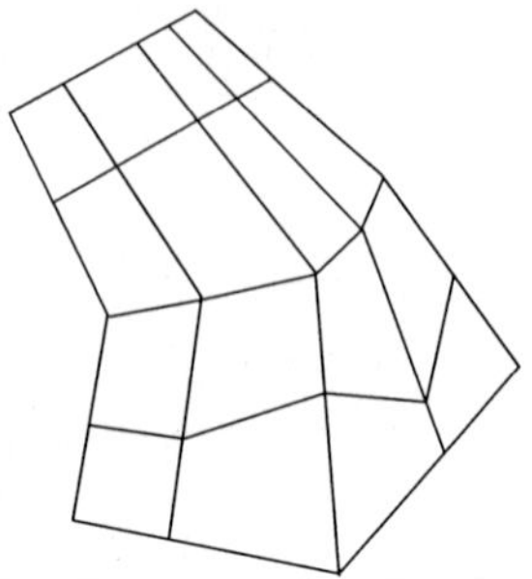

fig. 10.14: Partitioning with boxes

structure of the Schur complement S depends on the numbering of the nodes we choose. Let us consider an example for the model problem with 121 nodes and 9 subdomains arranged as 3×3, each subdomain having 3×3 nodes. Let us use the following ordering of the nodes:

61	62	63	114	70	71	72	117	79	80	81
58	59	60	113	67	68	69	116	76	77	78
55	56	57	112	64	65	66	115	73	74	75
103	104	105	120	106	107	108	121	109	110	111
34	35	36	99	43	44	45	102	52	53	54
31	32	33	98	40	41	42	101	49	50	51
28	29	30	97	37	38	39	100	46	47	48
88	89	90	118	91	92	93	119	94	95	96
7	8	9	84	16	17	18	87	25	26	27
4	5	6	83	13	14	15	86	22	23	24
1	2	3	82	10	11	12	85	19	20	21

With this ordering the structure of S is given in figure 10.15 with 3×3 blocks.

One block is only linked to a maximum of 5 blocks. Moreover, the elements of S outside a tridiagonal structure are quite small except for the last 4 rows and 4 columns corresponding to the links with the cross points, see figure 10.16. For the Poisson model problem, the magnitude of the elements depends only on the distance between nodes. However, this is not true for more general problems. We see that there is some coupling between the nodes on different (neighboring) edges and this must be taken into account when designing preconditioners.

10.8.1. The Bramble, Pasciak and Schatz preconditioner

In this section, we consider a preconditioner (known as BPS) that was proposed by Bramble, Pasciak and Schatz in [182] and the subsequent papers

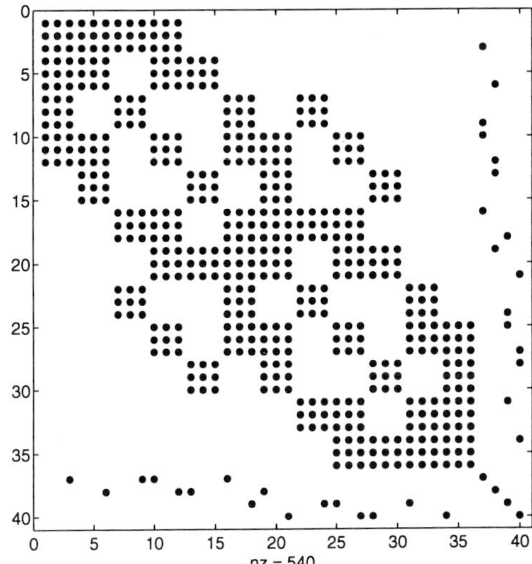

fig. 10.15: Structure of S with boxes

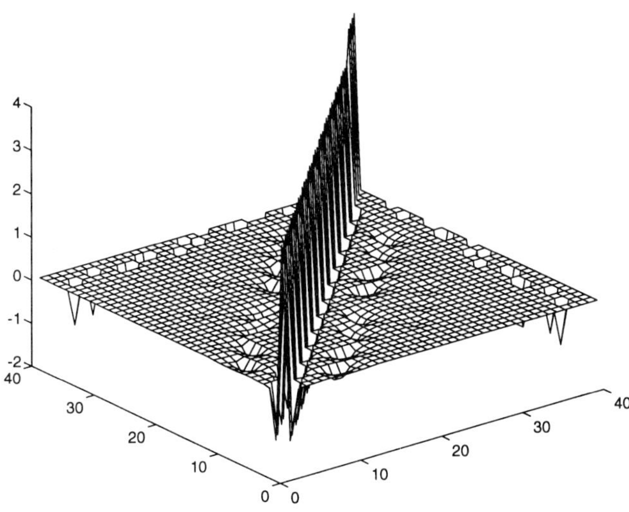

fig. 10.16: S with boxes

[183], [184], [185] for extensionsfor finite element problems in two dimensions. This is an algorithm that has had an important influence on the development of DD methods. The method relies on deriving a simpler finite element ap-

proximation of the problem. Suppose, we wish to solve

$$a(u, v) = (f, v), \ \forall v \in V$$

where a is a coercive bilinear form arising from a second order elliptic PDE, V being a Hilbert space, say $H_0^1(\Omega)$ for homogeneous Dirichlet boundary conditions. The aim of the algorithm is to construct another spectrally equivalent bilinear form $b(u, v)$ such that

$$\lambda_0 b(v, v) \leq a(v, v) \leq \lambda_1 b(v, v), \ \forall v \in V$$

and to use b as a preconditioner. Ω is divided into non–overlapping subdomains Ω_k, the edges between two subdomains being denoted as $\Gamma_{l,m}$. Another intermediate form is introduced to eventually allow for some averaging of the coefficients,

$$\tilde{a}(u, v) = \sum_k \sum_{i,j} \int_{\Omega_k} a_{i,j}^k \frac{\partial u}{\partial x_i} \frac{\partial v}{\partial x_j} \, dx = \sum_k \tilde{a}_k(u, v).$$

The method then separates interior, edges and vertices unknowns in the following way:

$$u = u_P + u_H,$$

where u_P is in $\sum_{\oplus} V^0(\Omega_k)$ where functions in $V^0(\Omega_k)$ have homogeneous Dirichlet boundary conditions and

$$u_P = 0 \text{ on } \Gamma_{l,m}.$$

u_P is defined by

$$\tilde{a}_k(u_P, \phi) = \tilde{a}_k(u, \phi), \ \forall \phi \in V^0(\Omega_k).$$

This takes care of the right hand side and u_H is defined by

$$\tilde{a}_k(u_H, \phi) = 0 \ \forall \phi \in V^0(\Omega_k).$$

The method goes one step further and decomposes u_H on the interfaces as

$$u_H = u_E + u_V,$$

where u_E stands for edge unknowns, u_V for vertices unknowns, $u_V(v_j) = u(v_j)$ and $u_V|_{\Gamma_{ij}}$ is linear, $u_E(v_j) = 0$. Bramble, Pasciak and Schatz defined an operator l_0 on the edges: $V^0(\Gamma_{i,j}) \rightarrow V^0(\Gamma_{i,j})$ by

$$\int_{\Gamma_{i,j}} c^{-1} l_0(w) \phi = \int_{\Gamma_{i,j}} c w' \phi', \ \forall \phi \in V^0(\Gamma_{i,j}),$$

where c is piecewise constant. More simply, this defines something which behaves like the one dimensional Laplace operator that we have already used with finite difference schemes. The bilinear form b is defined as

$$\begin{aligned} b(w,\phi) =& \tilde{a}(u_P,\phi_P) \\ &+ \sum_{\Gamma_{i,j}} \int_{\Gamma_{i,j}} \alpha_{i,j} c^{-1} l_0^{1/2}(u_E)\phi_E \\ &+ \sum_{\Gamma_{i,j}} (u_V(v_i) - u_V(v_j))(\phi_V(v_i) - \phi_V(v_j)). \end{aligned}$$

The basis functions that are used are the usual ones for the interior nodes, one dimensional hat functions for the edges (vanishing at the vertices) and functions which are linear on each edge, 1 at one vertex, 0 at the other ones for the vertices. This BPS preconditioner requires us to perform the following steps:

1) solve Dirichlet problems on each subdomain in parallel for $\longrightarrow u_P$,

2) solve one dimensional edge equations in parallel for $\longrightarrow u_E$

3) solve a coarse mesh system on vertices for $\longrightarrow u_V$. From u_E and u_V we obtain the boundary values of u_H. Therefore, the last step is

4) solve Dirichlet problems on each subdomain in parallel for $\longrightarrow u_H$.

The solution is $u_P + u_H$. For possible choices for the coefficients, see Bramble, Pasciak and Schatz [182].

Theorem 10.29

Under suitable hypotheses, the condition number for the preconditioned system in two dimensions is

$$\kappa \leq C\left(1 + \log^2\left(\frac{H}{h}\right)\right),$$

where H is the coarse mesh size.

Proof. See Bramble, Pasciak and Schatz [182]. □

Therefore, there is only a slight h dependency in the condition number and this gives a good rate of convergence with CG. This dependency comes from the fact that the vertices are not directly linked to the neighboring edge nodes. Variants of the BPS preconditioner can also be denoted as

$$M^{-1}v = \sum_{edges} R_{E_i}^T(\alpha_i M_i)^{-1} R_{E_i} v + R_H^T A_H^{-1} R_H v,$$

where R_{E_i} denotes the restriction to the edge E_i and R_H is a weighted restriction onto the coarse mesh, M_i being one of the preconditioners for two subdomain case: either Dryja or Golub–Mayers.

10.8.2. Vertex space preconditioners

One way to improve on BPS is to allow for some coupling between the vertices and the edge nodes. Some points are considered around each vertex on each of the edges. Let V_k be this set of points. Then the preconditioner is defined as

$$M^{-1}v = R_H^T A_H^{-1} R_H v + \sum_{edges} R_{E_i}^T(M_{E_i})^{-1} R_{E_i} v + \sum_{vertices} R_{V_k}^T(M_{V_k})^{-1} R_{V_k} v.$$

This includes some coupling between neighboring edges. The edge preconditioner can be chosen as a weighting of Dryja's or Golub–Mayers' preconditioners. Chan and Mathew [284] proposed the use of probing to define M_{V_k}. The restriction to the edges is tridiagonal and an edge is only linked to the crosspoint and to the two nodes adjacent to the crosspoint on the neighboring edges. Five probing vectors are chosen to construct this approximation. If enough points are used around each vertex, then the condition number is independent of h and of the number of subdomains. The vertex space algorithm was developed by B. Smith (see [1188], [1189], [1190], [1192]).

In his Ph.D. thesis L. Carvalho [261] considered some preconditioners whose spirit is quite close to the vertex space preconditioners. Because they involved some kind of overlapping between the edge and vertex parts, they are denoted as algebraic additive Schwarz (AAS). He studied several local block preconditioners for the subdomains and several coarse space preconditioners. For the local preconditioners, the main difference with the vertex space preconditioner is that the edge and the adjacent vertices are considered together, see figure 10.17. For anisotropic problems, it is sometimes enough to consider overlapping in one direction only.

Another proposal was to consider the complete boundary of one subdomain, to be able to retrieve all the couplings between the edge nodes and the vertices when the interior nodes are eliminated. In any case, the local preconditioners are obtained by probing. A careful implementation is described in

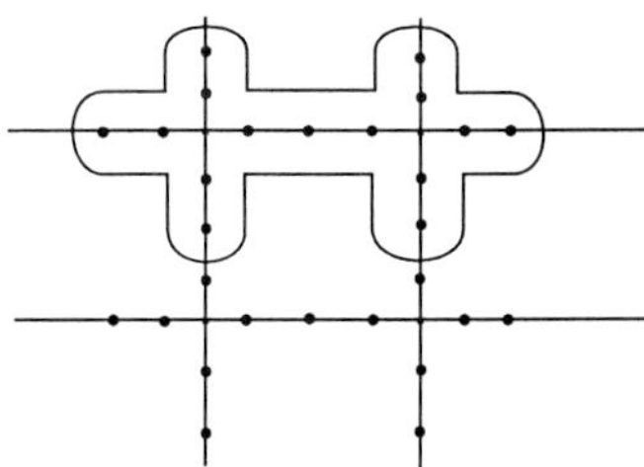

fig. 10.17: The points in the vertex space

[261]. A set of numerical experiments on elliptic problems shows that there is a $1/H^2$ dependency in the number of iterations. Therefore, it is necessary to add a coarse space component in the algorithm. A restriction operator R_0 is defined (depending on the choice of the coarse part of the preconditioner). The coarse component of the preconditioner is defined as $R_0^T A_0^{-1} R_0$ where A_0 is the Galerkin coarse space operator $A_0 = R_0 S R_0^T$. Several possibilities were considered:

i) a subdomain–based coarse space where all the boundary points of a subdomain are considered. The coarse space is spanned by vectors which have non–zero components for the points around a subdomain, for all subdomains.

ii) a vertex–based coarse space where the vertices and some few adjacent edge points are considered.

iii) an edge–based coarse space where the points of an edge and the adjacent vertices are considered.

When combining these coarse space preconditioners with the local parts, a preconditioner for which the condition number is insensitive to the mesh size or the number of subdomains is obtained except for very highly anisotropic problems. An efficient parallel implementation of these preconditioners is described in [261].

10.9. A block Red–Black DD preconditioner

The construction of the preconditioner is based on partitioning the domain as indicated in figure 10.18. This type of partitioning was introduced by Proskurowski et al [452]. We consider four types of unknowns:

(1) the nodes in the white boxes (W),

(2) the nodes in the black boxes (B),

(3) the nodes in the rectangular boxes (interfaces) called separators (S),

(4) the nodes in the small dark grey boxes called crosspoints (C).

fig. 10.18: The Red–Black decomposition of the domain

Classically we rewrite A as a 4×4 block matrix, the indices corresponding to the types previously defined. Thus, we have:

$$A = \begin{pmatrix} A_{WW} & 0 & A_{WS} & 0 \\ 0 & A_{BB} & A_{BS} & 0 \\ A_{WS}^T & A_{BS}^T & A_{SS} & A_{SC} \\ 0 & 0 & A_{SC}^T & A_{CC} \end{pmatrix}.$$

The preconditioner M is defined as

$$M = L\ D^{-1}\ L^T,$$

where D is block diagonal and has the same block diagonal as L and, with the same numbering as before,

$$L = \begin{pmatrix} M_{WW} & & & \\ 0 & M_{BB} & & \\ A_{WS}^T & A_{BS}^T & M_{SS} & \\ 0 & 0 & A_{SC}^T & M_{CC} \end{pmatrix}.$$

At each iteration of the PCG method, we have to solve a problem with matrix M. Now, the solution of the forward step

$$\begin{pmatrix} M_{WW} & & & \\ 0 & M_{BB} & & \\ A^T_{WS} & A^T_{BS} & M_{SS} & \\ 0 & 0 & A^T_{SC} & M_{CC} \end{pmatrix} \begin{pmatrix} y_W \\ y_B \\ y_S \\ y_C \end{pmatrix} = \begin{pmatrix} c_W \\ c_B \\ c_S \\ c_C \end{pmatrix}$$

is obtained by

1) Parallel solves on the white and the black boxes $M_{i,i}y_i = c_i$ $(i = W, B)$.

2) A solve on the interface $M_{SS}\; y_S = c_S - A^T_{WS}\; y_W - A^T_{BS}\; y_B$.

3) A solve on the cross points $M_{CC}y_C = c_C - A^T_{SC}y_S$.

Therefore, the total amount of work involved to solve a problem with matrix M is: two parallel solves on the boxes (matrices $M_{i,i}$, $i = W, S$), two solves on the interface (matrix M_{SS}) and one solve on the cross points (matrix M_{CC}). There are many choices for matrices $M_{i,i}$. We shall consider the following choices:

$M_{i,i} = A_{i,i}$, $i = W, S$; that is, a direct solver is used for the subproblems,

$M^1_{SS} = 2\{A^{(2)}_{SS} - A^T_{BS}\; M^{-1}_{BB} A_{BS}\}$, or $M^2_{SS} = B_{SS} - A^T_{BS}\; M^{-1}_{BB} A_{BS}$.

Here, $A^{(2)}_{SS}$ is the part of the assembly matrix computed on the black boxes, and B_{SS} is equal to A_{SS} except on the diagonal where

$(B_{SS})_{i,i} = (A_{SS})_{i,i} - |(A_{WS})_{j_i,i}|$, j_i such that $(A_{WS})_{j_i,i} \neq 0$.

This corresponds to a "Neumann" boundary condition on the black boxes. In both cases, the coupling with the white nodes is discarded. It is easy to see that if the unknowns in the black boxes are numbered in a "natural" way, then M_{SS} is a block diagonal matrix (each block corresponding to the nodes around a black box) and that the elements of these blocks can be computed cheaply. Indeed, the matrix M_{SS} corresponds to the elimination of the interior nodes of the black boxes. This gives a clique that connects all the nodes on the boundary of each box. Thus, the diagonal blocks of M_{SS} are dense matrices.

Finally, we choose $M_{CC} = A_{CC} - A^T_{SC}\; M^{-1}_{SS}\; A_{SC}$. This matrix is sparse and it is also easily seen that it has the structure given by a nine point stencil, two coefficients of which are zero. This particular point requires more explanation. The matrix M_{SS} is block diagonal, each block corresponding to a clique. A cross point is related to exactly two cliques by A_{SC} and its transpose. Thus the term $-A_{SC}^T\ M_{SS}^{-1}\ A_{SC}$ links the cross points as shown in figure 10.19. We solve this "coarse grid" system with a direct method or a diagonally Preconditioned Conjugate Gradient solver.

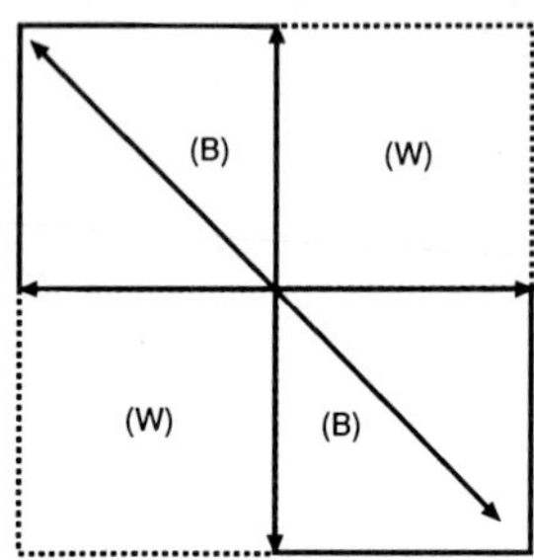

fig. 10.19: Coarse grid links

Further improvements of the method are to use this algorithm recursively in a multilevel way or to define an approximation of the inverse of the matrix M_{CC} for the cross points (a polynomial or a sparse inverse). This last method seems better suited for massively parallel architectures. Numerical experiments in Ciarlet and Meurant [351] show that the condition number is constant when the ratio of the number of points to the number of subdomains is kept constant.

10.10. Multilevel preconditioners

We have seen that it is useful to add a coarse space component to the additive Schwarz preconditioners. It is relatively easy to generalize these two level methods to a multilevel algorithm. There are also many multilevel preconditioners. In some sense, they are very close to the multigrid algorithms, specially to the algebraic multigrid methods. There are many different possible methods. For instance, it depends if we are looking at a problem where the mesh is given and for which there is a nested family of triangulations or if the coarse meshes are not nested with the finer ones. It depends also if the coarse meshes are given (as when the fine meshes are computed by refinement of coarse meshes) or if the coarse meshes have to be constructed from the fine ones by a process named "coarsening". One can also look at the problem purely algebraically trying to define "coarse" problems by looking only at the matrix coefficients. We shall describe only a few of all the possible methods. Let us start by straightforward generalizations of the additive Schwarz preconditioners.

10.10.1. Additive multilevel Schwarz preconditioners

Suppose we have L different levels, each level being decomposed into $N^{(l)}$ subdomains denoted as Ω_i^l. Then, the fully additive Schwarz preconditioner is defined as

$$M^{-1} = \sum_{l=0}^{L} \sum_{i=1}^{N^{(l)}} (R_i^l)^T (A_i^l)^{-1} R_i^l.$$

The index $l = 0$ corresponds to the coarsest grid (eventually one node). Note that the subdomains Ω_i^l overlap each other as in the one level case. A particularly simple case is the multilevel diagonal scaling preconditioner. Then, if the coarsest grid has only a single subdomain

$$M^{-1} = (R^0)^T (A^0)^{-1} R^0 + \sum_{l=1}^{L-1} (R^l)^T (D^l)^{-1} R^l + (D^L)^{-1},$$

where D^l is the diagonal of A^l. A closely related preconditioner was developed by Bramble, Pasciak and Xu [189] which is known as BPX. It is noted that in finite element methods with linear approximations, the diagonal elements of the matrix at level l must be of order $(h^l)^{d-2}$ where h is the mesh size and d is the dimension (1, 2 or 3). The BPX preconditioner is defined as

$$M^{-1} = (R^0)^T (A^0)^{-1} R^0 + \sum_{l=1}^{L-1} (h^l)^{2-d} (R^l)^T R^l + (h^L)^{2-d} I.$$

Note that the BPX preconditioner disregards all information about the coefficients of the problem except for the coarsest mesh, therefore it is not expected to be robust when there are large jumps in the coefficients. However, it has been proved that the BPX is theoretically optimal, the condition number being $O(1)$.

These additive Schwarz methods can be mixed with multiplicative methods in different ways. One can define as before fully additive methods which are additive among subdomains and between levels. Another possibility is to be multiplicative between subdomains on one level and additive between levels. A third kind of algorithm is being multiplicative between both subdomains and levels. This is very close to a V–cycle multigrid (without smoothing). For details, see Smith, Bjorstad and Gropp [1192].

10.10.2. Multilevel ILU preconditioners

The use of multilevel orderings for ILU preconditioners has been considered for a while (at least since multigrid has become popular). Research in this direction has been done by Axelsson and Vassilevski [85], [86], [87] and Axelsson and Eijkhout [73]. Here we refer to the work of Notay [1024], [1025]. There are basically two ways to derive optimal preconditioners using MILU: the first uses some sort of W–cycle by introducing polynomial acceleration at each level, the second uses a V–cycle like algorithm but relies also on smoothing as in multigrid algorithms.

Suppose that A is an M–matrix. Both methods start by looking at the two level algorithm, the matrix being partitioned as

$$A = \begin{pmatrix} A_{1,1} & A_{1,2} \\ A_{2,1} & A_{2,2} \end{pmatrix}.$$

The block $(1,1)$ refers to unknowns on the "fine" grid and $(2,2)$ to unknowns on the "coarse" grid. Then, the two level preconditioner is defined as

$$M = \begin{pmatrix} M_{1,1} & \\ A_{2,1} & S \end{pmatrix} \begin{pmatrix} I & M_{1,1}^{-1}A_{1,2} \\ & I \end{pmatrix},$$

where $M_{1,1}$ is an MILU incomplete factorization of $A_{1,1}$ (without fill–in) and $S = A_{2,2} - A_{2,1}KA_{1,2}$ is an approximation of the Schur complement of A with K being a diagonal matrix to be defined. By choosing this approximation, the reduced system has the same structure as the original problem. Since $M_{1,1}$ is an MILU preconditioner, we have $M_{1,1}e = A_{1,1}e$, where $e = (1, 1, \ldots, 1)^T$. Let D be a diagonal matrix such that $De = A_{1,1}e$. Then, $k_{i,i}$ is equal to $d_{i,i}^{-1}$ if the diagonal element is non–zero and zero otherwise. If A is symmetric, S is a symmetric M–matrix. Notay [1024] gave some bounds for the condition number of $M^{-1}A$. The multilevel method is obtained by a recursive application of the two level preconditioner. Let M_S be the preconditioner on the next coarsest level. A solve with S is exchanged for a multiplication by $P_2(M_S^{-1}S)S^{-1}$ where P_2 is a second order polynomial such that $P_2(0) = 0$. Usually, shifted Chebyshev polynomials are used. Using a second order polynomial leads to a kind of W–cycle algorithm. Notay [1024] proved that the condition number is bounded independently of h and that the complexity of one iteration step is proportional to n. Numerical results in [1024] show that this method seems quite robust. In [1025] Notay proposed using a V–cycle with the previous algorithm by replacing the use of the second order polynomial by some smoothing (dampened `Jacobi` or `RILU`). Note that there is no smoothing on the finest level. Results show that the number of iterations is independent of h.

R. Bank also proposed multilevel algorithms: the incomplete factorization multigraph algorithm, see Bank and Kent Smith [105] and the multilevel ILU decomposition, see Bank and Wagner [110] which are designed for general matrices. Unfortunately, these algorithms are quite complex particularly in defining the coarse levels. But, they give an h independent number of iterations.

Other proposals of this sort were made by Meurant, Chan and Ciarlet [985]. The idea was to use a multilevel ordering of the unknowns as this could allow for introducing a coupling between mesh points that are far away from each other and then to do an incomplete Cholesky factorization. Of course, to obtain the coupling, some fill–in has to be kept between different levels. A model problem in a square domain was considered. The multilevel ordering of 15×15 mesh is given in figure 10.20.

The coarsest level with just one node being denoted as 1, the next coarsest level with eight nodes being 2 and so on, the levels of the nodes are shown in figure 10.21.

Within each level the nodes are ordered with a lexicographic ordering. Let us denote this ordering as `ML`. More generally on an $m \times m$ mesh where $m = 2^L - 1$, `ML` is defined in the following way : the coarsest mesh G_1 (level 1) has only one node located at $(0.5, 0.5)$ in our example (the other 8 nodes are on the boundary and as such are not labeled). Level 2 corresponds to a mesh G_2 with mesh size of $1/2^2$ excluding the node already labeled. Level 3 (mesh G_3) has a mesh size of $1/2^3$ and so on up to the last level L corresponding to the finest mesh G_L. When the level of each node has been found, the nodes are labeled beginning with the finest level in descending order finishing with level 1. Within one level the nodes are numbered using a natural row ordering.

By itself ML is not a good ordering from the point of view of the number of fill–ins in the complete Cholesky decomposition. For our example, the structure of the lower triangular part of A has 645 non–zero entries, the `ROW` ordering gives 2744 fill–ins, the minimum degree ordering `MIND` gives 1203 fills and `ML` gives 7007 fills. When used with `IC(0)`, `ML` is also not a very good ordering. This could be explained by the theory we reviewed in Chapter 8.

176	175	174	173	172	171	170	169	168	167	166	165	164	163	162
161	216	160	215	159	214	158	213	157	212	156	211	155	210	154
153	152	151	150	149	148	147	146	145	144	143	142	141	140	139
138	209	137	224	136	208	135	223	134	207	133	222	132	206	131
130	129	128	127	126	125	124	123	122	121	120	119	118	117	116
115	205	114	204	113	203	112	202	111	201	110	200	109	199	108
107	106	105	104	103	102	101	100	99	98	97	96	95	94	93
92	198	91	221	90	197	89	225	88	196	87	220	86	195	85
84	83	82	81	80	79	78	77	76	75	74	73	72	71	70
69	194	68	193	67	192	66	191	65	190	64	189	63	188	62
61	60	59	58	57	56	55	54	53	52	51	50	49	48	47
46	187	45	219	44	186	43	218	42	185	41	217	40	184	39
38	37	36	35	34	33	32	31	30	29	28	27	26	25	24
23	183	22	182	21	181	20	180	19	179	18	178	17	177	16
15	14	13	12	11	10	9	8	7	6	5	4	3	2	1

fig. 10.20: The multilevel ordering

For the Poisson model problem, if the stopping criterion is defined as

$$\|r^k\|_2 \leq 10^{-6}\|r^0\|_2,$$

with x^0 random and $f \equiv 0$, we obtain 13 iterations with the `ROW` ordering and 17 iterations with `ML`. However, our purpose is to use `ML` not with `IC(0)` but to keep some fill–in during the incomplete factorization. Therefore we have to define the rules that will allow us to decide which fill–in we are going to keep. Before doing so, let us remark that we shall not be forced to store all the fill–in if we accept some extra work during the solution phase. For the sake of simplicity, let us suppose that we have only two levels. Then, the permuted matrix PAP^T can be written

$$A_P = PAP^T = \begin{pmatrix} A_2 & A_{21}^T \\ A_{21} & A_1 \end{pmatrix},$$

where A_2 represents the coupling of nodes belonging to level 2 (the fine mesh), A_1 represents the coupling of nodes of level 1 (the coarse mesh). In fact, in

4	4	4	4	4	4	4	4	4	4	4	4	4	4	4
4	3	4	3	4	3	4	3	4	3	4	3	4	3	4
4	4	4	4	4	4	4	4	4	4	4	4	4	4	4
4	3	4	2	4	3	4	2	4	3	4	2	4	3	4
4	4	4	4	4	4	4	4	4	4	4	4	4	4	4
4	3	4	3	4	3	4	3	4	3	4	3	4	3	4
4	4	4	4	4	4	4	4	4	4	4	4	4	4	4
4	3	4	2	4	3	4	1	4	3	4	2	4	3	4
4	4	4	4	4	4	4	4	4	4	4	4	4	4	4
4	3	4	3	4	3	4	3	4	3	4	3	4	3	4
4	4	4	4	4	4	4	4	4	4	4	4	4	4	4
4	3	4	2	4	3	4	2	4	3	4	2	4	3	4
4	4	4	4	4	4	4	4	4	4	4	4	4	4	4
4	3	4	3	4	3	4	3	4	3	4	3	4	3	4
4	4	4	4	4	4	4	4	4	4	4	4	4	4	4

fig. 10.21: The level of nodes for the multilevel ordering

our problem A_1 is diagonal, and A_{21} is the coupling between level 1 and level 2. Let LL^T (resp. $L_2L_2^T$, $L_SL_S^T$) be an incomplete decomposition of A_P (resp. A_2, $A_1 - A_{21}L_2^{-T}L_2^{-1}A_{21}^T$). Then we can write

$$LL^T = \begin{pmatrix} L_2 & 0 \\ A_{21}L_2^{-T} & L_S \end{pmatrix} \begin{pmatrix} L_2^T & L_2^{-1}A_{21}^T \\ 0 & L_S^T \end{pmatrix}.$$

Therefore, it is not necessary to store the fill–in that arises in the coupling between levels 1 and 2, i.e. the matrix $A_{21}L_2^{-T}$ since when we solve

$$L\begin{pmatrix} x_2 \\ x_1 \end{pmatrix} = \begin{pmatrix} y_2 \\ y_1 \end{pmatrix},$$

we can do $L_2x_2 = y_2$, $L_2^Tw_2 = x_2$, and $L_Sx_1 = y_1 - A_{21}w_2$. If we agree to solve two systems with matrix L_2 instead of one, we have to store only the diagonal blocks L_2 and L_S that is a part of the fill–in. This can be important in terms of storage. The previous technique can be applied recursively if the mesh is divided into more levels.

Let us now explain some of the strategies we tried to keep some fill–in. Let us first define $lev(i)$ as being the level of node i, $N_p(i)$ (p odd) as being the set of neighbors of node i on mesh G_L (the finest level) in a $p \times p$ square whose center is node i, $N_p^l(i)$ as being the same definition except that we consider the neighbors on mesh G_l. We shall denote the node we eliminate by i and the two nodes that are concerned with the fill–in by j and k (that is j and k are neighbors of i in the elimination graph and therefore the elimination of i may create a non–zero entry in position (j, k)). For each considered strategy, we give the resulting number of iterations nit, nel the total number of elements in $\mathcal{L}$, $nsto$ the number of elements that it is enough to store, nr the number of elements in the remainder R and the Frobenius norm of R, $frob_R$. The following results are for the model problem on the 15×15 mesh and the `ML` ordering.

1– we keep the fill–in if

$lev(j) = lev(k)$.

This gives

$nit = 16, \quad nel = 1919, \quad nsto = 1919, \quad nr = 1065, \quad frob_R = 48.7$

These results show that it is not enough to keep the fill–in within each level. Even though there is communication between nodes that are far away through the fill–in in the coarse levels, the number of iterations is almost as large as with no fill at all. Therefore, we must introduce in some way intra–level fill–in that is fill–in coupling different levels.

2– we keep the fill–in if

$lev(j) < lev(i) \quad \text{or} \quad lev(k) < lev(i)$.

$nit = 9, \quad nel = 4761, \quad nsto = 1453, \quad nr = 811, \quad frob_R = 7.96$

In this condition, at least one of the nodes is on a coarser level than the one being eliminated. This shows that adding some fill–in coupling the different levels helps a lot in reducing the number of iterations (See the reduction in the Frobenius norm). However, in this recipe we add quite a lot of fill and it will not be practical to use such a large amount of storage.

3– we keep the fill–in if

$j, k \in N_5^{lev(i)}(i)$.

$nit = 6, \quad nel = 2039, \quad nsto = 1181, \quad frob_R = 0.16$

This gives a good improvement in the number of iterations but the storage is quite large.

4– we keep the fill–in if

$\left\{j, k \in N_5^{lev(i)}(i) \quad \text{and} \quad [lev(k) \neq lev(j) \text{ or } lev(k) \neq L]\right\}$

or

$\{j, k \in N_3(i)\}$.

$nit = 6, \quad nel = 1531, \quad nsto = 865, \quad nr = 844, \quad frob_R = 0.69$

5– we keep the fill–in if

$j, k \in N_3(i) \text{ or } \left[lev(i) > lev(k) \text{ or } \left\{lev(i) > lev(j) \text{ and } j, k \in N_3^{lev(i)}(i)\right\}\right].$

$nit = 7, \quad nel = 1220, \quad nsto = 795, \quad frob_R = 1.17$

Unfortunately, the condition number $\kappa(M^{-1}A)$ is still h dependent as shown in the following results. We give the number of fill–ins and *nop* which is the number of multiplications needed to solve $Lw = c$ times the number of iterations. This is proportional to the work done for the preconditioner and gives an idea of the respective costs.

As a matter of comparison, we may look at results obtained with the `ROW` ordering when one uses a "level 1" fill and with the same ordering when the fill–in is kept or rejected according to a threshold parameter t.

We see that even with choosing the fill–in according to its magnitude, it is difficult to beat the `ROW` ordering with no fill. The only method that does better is $u = 0.2$ but the gain is quite small. When the threshold is decreased, there is too much additional fill–in even though the number of iterations is small. To try to see where the useful fill–ins are located with the `ML` ordering, we use the same method but with the `ML` ordering, i.e. we discard the fill–ins that are smaller than a given threshold.

condition	m	nit	nel	nop
`ML` $t = 10^{-4}$	15	3	4051	12153
`ML` $t = 10^{-3}$	15	4	2980	11920
`ML` $t = 5\ 10^{-3}$	15	6	2033	12198
`ML` $t = 10^{-2}$	15	6	1769	10614
`ML` $t = 2\ 10^{-2}$	15	7	1499	10493
`ML` $t = 5\ 10^{-2}$	15	11	1142	12562

condition	m	nit	nel	nop
1	15	16	1919	30704
	31	29	14747	427663
	63	51	114611	5845161
2	15	9	4761	42849
	31	13	77521	1007773
	63	?	?	?
3	15	6	2039	12234
	31	10	10332	103320
	63	16	46553	744848
4	15	6	1531	9186
	31	10	7808	78080
	63	17	35389	601613
5	15	7	1120	7840
	31	12	5841	70092
	63	20	25586	511720
ROW	15	13	645	8385
	31	23	2821	64883
	63	41	11781	483021
MIND	15	21	645	13545
	31	40	2821	112840
	63	70	11781	824670
ML	15	17	645	10965
	31	32	2821	90272
	63	57	11781	671517

In strategy number **3**, we eliminate some of the fills that are smaller than a threshold parameter. From these results, we see we can still save some storage and get the same number of iterations. Therefore, one way to improve this multilevel preconditioner is to use some threshold to discard some of the fill–ins which are too small to be interesting. Another way will be to use a modified incomplete factorization `MIC` (or a relaxed one). This was done by Van der Ploeg, Botta and Wubs [1273]. The main difference with what we have described is that a threshold variable with the level is used. The threshold decreases by a constant multiplicative factor smaller than 1 when

condition	m	nit	nel	nop
LEV1	15	11	840	9240
	31	18	3720	66960
$t = 10^{-4}$	15	4	3211	12844
	31	4	24838	99352
$t = 5\ 10^{-3}$	15	5	2156	10780
	31	6	11242	67452
$t = 10^{-2}$	15	5	1847	9235
	31	7	9207	64449
$t = 5\ 10^{-2}$	15	8	1022	8176
	31	12	4590	55080
$t = 0.2$	15	10	840	8400
	31	15	3720	49050
$t = 0.3$	15	14	645	9030
	31	24	2821	67704

condition	n	nit	nel	nop
$3\ t = 10^{-2}$	15	6	2039	12234
$3\ t = 2\ 10^{-2}$	15	6	1550	9300
$3\ t = 2\ 10^{-2}$	15	7	1499	10493
$3\ t = 3\ 10^{-2}$	15	8	1364	10912
$3\ t = 510^{-2}$	15	11	1142	12562

going to coarse levels. Therefore, more fill–in is kept in the coarse levels than in the fine ones. Good numerical results are reported in [1273].

10.11. Bibliographical comments

As we said at the beginning of this Chapter, even without going back to Schwarz, Domain Decomposition has been in use for quite a long time by engineers, mainly in structural mechanics, see Przemieniecki [1086]. This type of methods was also used by Russian applied mathematicians (in the former

Soviet Union) in the sixties and seventies because it has some advantages for solving large problems on computers with small memories.

Some work was done during the eighties by Glowinski, Périaux and their co–workers, see for instance Dinh, Glowinski and Périaux in 1980 [418]. One can also quote the papers of Widlund and his co–workers: Bjørstad, Dryja and Proskurowski, see for instance [157]. All this led to the organization of the first Domain Decomposition conference in Paris in 1987. Since then, annual conferences have taken place in different countries. Their proceedings are one of the best sources of information about domain decomposition.

A series of papers that have been really influential are those by Bramble, Pasciak and Schatz [181], [182], [183], [184], [185]. Many papers about DD were written in the nineties mainly coming from France, United States and Russia. Most methods and some theory is summarized in the book by Smith, Bjørstad and Gropp [1192] and the review paper by Chan and Mathew [351].

Domain Decomposition has been applied to more and more difficult problems starting from elliptic model problems in square domains in the eighties to complex three dimensional industrial problems at the end of the nineties. The search for preconditioners having a conditioner number independent of the mesh size leads to a merger of Domain Decomposition with multilevel ideas. There are now very efficient methods, although they are not always so easy to implement for general sparse matrices.

References

[1] AASEN,J.O., (1971), *On the reduction of a symmetric matrix to tridiagonal form,* BIT, vol 11, pages 233–242.

[2] ACHDOU,Y., KUZNETSOV,Y.A., (1994), *Algorithms for a non conforming domain decomposition method,* Report RI 296, Centre de Mathématiques Appliquées, Ecole Polytechnique.

[3] ADAMS,L.M., (1985), *M-step preconditioned conjugate gradient methods,* SIAM J. Sci. Comput., vol 6, no 2, pages 452–463.

[4] ADAMS,L.M., JORDAN,H.F., (1986), *Is SOR color blind?,* SIAM J. Sci. Comput., vol 7, pages 490–506.

[5] ADAMS,L.M., ONG,E.G., (1988), *Additive polynomial preconditioners for parallel computers,* Parallel Comp., vol 9, pages 333–345.

[6] ADVELAS,G., HADJIDIMOS, A., (1976), *On improving the convergence rates of extrapolated alternating direction implicit schemes,* BIT, vol 16, pages 363–373.

[7] ADVELAS,G., HADJIDIMOS, A., (1981), *Optimum accelerated overrelaxation method in a special case,* Math. Comp., vol 36, no 153, pages 183–187.

[8] AGAPOV,V.K., KUZNETSOV,Y.A., (1988), *On some versions of the domain decomposition method,* Sov. J. Numer. Anal. Math. Modelling, vol 3, no 4, pages 245–265.

[9] AHAC,A.A., BUONI,J.J., OLESKY,D.D., (1988), *Stable LU factorization of H–matrices,* Linear Algebra and its Appl., vol 99, pages 97–110.

[10] AHAC,A.A., OLESKY,D.D., (1986), *A stable method for the LU factorization of M–matrices,* SIAM J. Alg. Disc. Meth., vol 7, no 3, pages 368–378.

[11] ALAGHBAND,G., (1989), *Parallel pivoting combined with parallel reduction and fill–in control,* Parallel Comp., vol 11, pages 201–221.

[12] ALCOUFFE,R.E., BRANDT,A., DENDY,J.E., PAINTER,J.W., (1981), *The multi–grid method for the diffusion equation with strongly discontinuous coefficients,* SIAM J. Sci. Comput., vol 2, pages 430–454.

[13] ALEFELD,G., (1976), *Zur konvergenz des Peaceman–Rachford verfahrens,* Numer. Math., vol 26, pages 409–419.

[14] ALEFELD,G., VARGA,R.S., (1976), *Zur konvergenz des symmetrischen relaxationsverfahrens,* Numer. Math., vol 25, pages 291–295.

[15] ALEFELD,G., VOLKMANN,P., (1985), *Regular splittings and monotone iteration functions,* Numer. Math., vol 46, pages 213–228.

[16] ALOUGES,D., LOREAUX,P., (1997), *Massively parallel preconditioners,* Numer. Alg., vol 14, no 4, pages 361–375.

[17] ALVARADO,F.L., SCHREIBER,R., (1993), *Optimal parallel solution of sparse triangular systems,* SIAM J. Sci. Comput., vol 14, no 2, pages 446–460.

[18] AMESTOY,P., DAVIS,D.A., DUFF,I.S., (1996), *An approximate minimum degree ordering algorithm,* SIAM J. Matrix Anal. Appl., vol 17, no 4, pages 886–905.

[19] AMESTOY,P.R., (1990), *Factorisation de grandes matrices creuses non symétriques basée sur une methode multifrontale dans un environnement multiprocesseur,* Ph.D. Thesis, CERFACS report TH/PA/91/2.

[20] AMESTOY,P.R., DUFF,I.S., (1989), *Vectorization of a multiprocessor multifrontal code,* Int. J. Supercomputer Appl., vol 3, no 3, pages 41–59.

[21] AMODIO,P., BRUGNANO,L., POLITI,T., (1993), *Parallel factorizations for tridiagonal matrices,* SIAM J. Numer. Anal., vol 30, no 3, pages 813–823.

[22] ANAND,I.M., (1980), *Numerical stability of nested dissection orderings,* Math. Comp., vol 35, no 152, pages 1235–1249.

[23] ANDERSON,C.R., (1988), *Manipulating fast solvers, changing their boundary conditions and putting them on multiple processor computers,* Report CAM 88-37,Dept. of Mathematics, UCLA.

[24] ANDERSON,C.R., (1989), *Domain decomposition techniques and the solution of Poisson's equation in infinite domains,* in Second International Symposium on Domain Decomposition Methods for Partial Differential Equations, CHAN,T.F., GLOWINSKI,R., PERIAUX,J., and WIDLUND,O.B. Eds. SIAM, pages 129–139.

[25] ANDERSON,D.V., FRY,A.R., GRUBER,R., ROY,A., (1987), *Gigaflop speed algorithm for the direct solution of large block tridiagonal systems in 3D physics applications,* Report UCRL–96034, Lawrence Livermore Nat. Laboratory.

[26] ANDERSON,E., BAI,Z., BISCHOF,C., DEMMEL,J., DONGARRA,J., DUCROZ, J., GREENBAUM,A., HAMMARLING,S., McKENNEY,A., OSTROUCHOV,S. and SORENSEN,D., (1992), *LAPACK User's guide,* SIAM.

[27] ANDERSON,E., SAAD,Y., (1989), *Solving sparse triangular linear systems on parallel computers,* Int. J. High Speed Comp., vol 1, no 1, pages 73–95.

[28] ANDERSSEN,R.S., GOLUB,G.H., (1972), *Richardson's non–stationary matrix iterative procedure,* STAN-CS-72-304, Computer Science Dept., Stanford University.

[29] ANGELACCIO,M., CLAJANNI,M., (1994), *The row/column pivoting strategy on multicomputers,* Parallel Comp., vol 20, no 2, pages 197–214.

[30] ANGELACCIO,M., CLAJANNI,M., (1994), *Subcube matrix decomposition, a unifying view for LU factorization on multicomputers,* Parallel Comp., vol 20, no 2, pages 257–270.

[31] APELT,C.J., ISAACS,L.T., (1977), *On the estimation of the optimum accelerator for SOR applied to finite element methods,* Com. Meth. Appl. Mech. Eng., vol

12, pages 383–391.

[32] APPLEYARD,J.R., CHESHIRE,I.M., (1983), *Nested factorization,* Paper SPE 12264, Reservoir Simulation Symposium, SPE 1983.

[33] APPLEYARD,J.R., CHESHIRE,I.M., POLLARD,R.K., (1981), *Special techniques for fully implicit simulators,* Report CSS 112, AERE Harwell.

[34] ARIOLI,M., DEMMEL,J.W., DUFF,I.S., (1989), *Solving sparse linear systems with backward error,* SIAM J. Matrix Anal. Appl., vol 10, pages 165–190.

[35] ARIOLI,M., DUFF,I.S., NOAILLES,J., RUIZ, D., (1990), *A block projection method for sparse matrices,* RAL-90-093, Rutherford Appleton Laboratory.

[36] ARIOLI,M., DUFF,I.S., RUIZ,D., (1991), *Stopping criteria for iterative solvers,* SIAM J. Matrix Anal. Appl., vol 13, pages 138–144.

[37] ARIOLI,M., PTAK,V., STRAKOS,Z., (1997), *Krylov sequences of maximal length and convergence of GMRES,* Report ICS AS CR, Prague.

[38] ARIOLI,M., ROMANI,F., (1985), *Relations between condition numbers and the convergence of the Jacobi method for real positive definite matrices,* Numer. Math., vol 46, pages 31–42.

[39] ARIOLI,M., ROMANI,F., (1992), *Stability, convergence and conditioning of stationary iterative methods of the form $x(i+1)=Px(i)+q$ for the solution of linear systems,* IMA J. Num. Anal., vol 12, pages 21–30.

[40] ARNOLDI,W.E., (1951), *The principle of minimized iterations in the solution of the matrix eigenvalue problem,* Quart. Appl. Math., vol 9, pages 17–29.

[41] ASHBY,S.F., (1987), *Polynomial preconditioning for conjugate gradient methods,* Ph.D. Thesis, Report UIUCDCS-R-87-1355, Dept. of Computer Science, University of Illinois.

[42] ASHBY,S.F., (1991), *Minimax polynomial preconditioning for hermitian linear systems,* SIAM J. Matrix Anal. Appl., vol 12, no 4, pages 766–789.

[43] ASHBY,S.F., BROWN,P.N., DORR,M.R., HINDMARSH,A.C., (1990), *Preconditioned iterative methods for discretized transport equations,* UCRL JC–104901, Lawrence Livermore National Laboratory.

[44] ASHBY,S.F., GUTKNECHT,M.H., (1993), *A matrix analysis of conjugate gradient algorithms,* Report UCRL-JC-113560, Lawrence Livermore National Laboratory.

[45] ASHBY,S.F., HOLST,M.J., MANTEUFFEL,T.A., SAYLOR,P.E., (1992), *The role of the inner product in stopping criteria for conjugate gradient iterations,* Report UCRL-JC-112586, Lawrence Livermore National Laboratory.

[46] ASHBY,S.F., MANTEUFFEL,T.A., OTTO,J.S., (1991), *A comparison of adaptive Chebyshev and least squares polynomial preconditioning for Hermitian positive def-*

inite linear systems, SIAM J. Sci. Comput., vol 13, pages 1–29.

[47] ASHBY,S.F., MANTEUFFEL,T.A., SAYLOR,P.E., (1989), *Adaptive polynomial preconditioning for hermitian indefinite linear systems,* BIT, vol 29, pages 583–609.

[48] ASHBY,S.F., MANTEUFFEL,T.A., SAYLOR,P.E., (1990), *A taxonomy for conjugate gradient methods,* SIAM J. Numer. Anal., vol 27, no 6, pages 1542–1568.

[49] ASHCRAFT,C., (1991), *A taxonomy of distributed dense LU factorization methods,* Report ECA–TR–161, Boeing Computer Services.

[50] ASHCRAFT,C., (1995), *Compressed graphs and the minimum degree algorithm,* SIAM J. Sci. Comput., vol 16, pages 1404–1411.

[51] ASHCRAFT,C., EISENSTAT,S.C, LIU,J.W–H., (1990), *A fan–in algorithm for distributed sparse numerical factorization,* SIAM J. Sci. Comput., vol 11, no 3, pages 593–599.

[52] ASHCRAFT,C., GRIMES,R.G., (1989), *The influence of relaxed supernode partitions on the multifrontal method,* ACM Trans. Math. Soft., vol 15, pages 291–309.

[53] ASHCRAFT,C., GRIMES,R.G., LEWIS,J.G., PEYTON,B.W., SIMON,H.D., (1987), *Progress in sparse matrix methods for large linear systems on vector supercomputers,* Int. J. Supercomputer Appl., vol 1, no 4, pages 10–30.

[54] ASHCRAFT,C.C., GRIMES,R.G., (1988), *On vectorizing incomplete factorization and SSOR preconditioners,* SIAM J. Sci. Comput., vol 9, pages 122–151.

[55] ASHCRAFT,C.C., LIU,J.W–H., (1996), *Applications of the Dulmage–Mendelsohn decomposition and network flow to graph bisection improvement,* Report Department of Computer Science, York University.

[56] ASHCRAFT,C.C., LIU,J.W–H., (1996), *Robust ordering of sparse matrices using multisection,* Report Department of Computer Science, York University.

[57] ASPLUND,E., (1959), *Inverse of matrices $\{a_{ij}\}$ which satisfy $a_{ij} = 0$, $j > i + p$,* Math. Scand., vol 7, pages 57–60.

[58] AXELSSON,O., (1972), *A generalized SSOR method,* BIT, vol 13, pages 443–467.

[59] AXELSSON,O., (1974), *On preconditioning and convergence acceleration for sparse matrix problems,* Report 74-10, CERN.

[60] AXELSSON,O., (1976), *A class of iterative methods for finite element equations,* Com. Meth. Appl. Mech. Eng., vol 9, pages 123–137.

[61] AXELSSON,O., (1979), *Preconditioning of indefinite problems by regularization,* SIAM J. Numer. Anal., vol 16, no 1, pages 58–69.

[62] AXELSSON,O., (1980), *Conjugate gradient type methods for unsymmetric and*

inconsistent systems of linear equations, Linear Algebra and its Appl., vol 29, pages 1–16.

[63] AXELSSON,O., (1985), *A survey of preconditioned iterative methods for linear systems of algebraic equations,* BIT, vol 25, pages 166–187.

[64] AXELSSON,O., (1985), *Incomplete block matrix factorization preconditioning methods. The ultimate answer?,* J. Comp. Appl. Math., vol 12, pages 3–18.

[65] AXELSSON,O., (1986), *A general incomplete block matrix factorization method,* Linear Algebra and its Appl., vol 74, pages 179–190.

[66] AXELSSON,O., (1987), *A generalized conjugate gradient, least squares method,* Numer. Math., vol 51, pages 209–227.

[67] AXELSSON,O., (1994), *Iterative solution methods,* Cambridge University Press.

[68] AXELSSON,O., (1996), *Stabilization of algebraic multilevel iteration; additive methods,* in AMLI'96: Proceedings of the Conference on Algebraic Multilevel Iteration Methods with Applications, AXELSSON,O. and POLMAN,B. Eds. University of Nijmegen, pages 49–62.

[69] AXELSSON,O., BARKER,V.A., (1984), *Finite element solution of boundary value problems,* Academic Press.

[70] AXELSSON,O., BRINKKEMPER,S., IL'IN,V.P., (1984), *On some versions of incomplete block matrix factorization iterative methods,* Linear Algebra and its Appl., vol 58, pages 3–15.

[71] AXELSSON,O., EIJKHOUT,V., (1986), *A note on the vectorization of scalar recursions,* Parallel Comp., vol 3, pages 73–83.

[72] AXELSSON,O., EIJKHOUT,V., (1987), *Robust vectorizable preconditioners for three dimensional elliptic difference equations with anisotropy,* in Algorithms and applications on vector and parallel computers, RIELE,H.J.J., DEKKER,T.J. and VAN der VORST,H.A. Eds.. Elsevier, pages 279–306.

[73] AXELSSON,O., EIJKHOUT,V., (1991), *The nested recursive two level factorization method for nine point difference matrices,* SIAM J. Sci. Comput., vol 12, pages 1373–1400.

[74] AXELSSON,O., GUSTAFSSON,I., (1977), *Iterative methods for the solution of the Navier equations of elasticity,* Com. Meth. Appl. Mech. Eng., vol 15, pages 241–258.

[75] AXELSSON,O., GUSTAFSSON,I., (1979), *An iterative solver for a mixed variable variational formulation of the (first) biharmonic problem,* Com. Meth. Appl. Mech. Eng., vol 20, pages 9–16.

[76] AXELSSON,O., GUSTAFSSON,I., (1979), *A modified upwind scheme for convective transport equations and the use of a conjugate gradient method for the solution*

of nonsymmetric systems of equations, J. Inst. Maths. Applics., vol 23, pages 321–337.

[77] AXELSSON,O., GUSTAFSSON,I., (1980), *On the use of preconditioned conjugate gradient methods for Red/Black ordered five point difference schemes,* J. Comp. Physics, vol 35, pages 284–289.

[78] AXELSSON,O., LINDSKOG,G., (1986), *On the rate of convergence of the preconditioned conjugate gradient method,* Numer. Math., vol 48, pages 499–523.

[79] AXELSSON,O., LINDSKOG,G., (1986), *On the eigenvalue distribution of a class of preconditioning methods,* Numer. Math., vol 48, pages 479–498.

[80] AXELSSON,O., LU,H., (1994), *A survey of some estimates of eigenvalues and condition numbers for certain preconditioned matrices,* J. Comp. Appl. Math., vol 80, pages 241–264.

[81] AXELSSON,O., MUNKSGAARD,N., (1979), *A class of preconditioned conjugate gradient methods for the solution of a mixed finite element discretization of the biharmonic operator,* Int. J. Num. Meth. Eng., vol 14, pages 1001–1019.

[82] AXELSSON,O., POLMAN,B., (1986), *On approximate factorization methods for block matrices suitable for vector and parallel processors,* Linear Algebra and its Appl., vol 77, pages 3–26.

[83] AXELSSON,O., POLMAN,B., (1988), *Block preconditioning and domain decomposition methods, II,* J. Comp. Appl. Math., vol 24, pages 55–72.

[84] AXELSSON,O., STEIHAUG,T., (1978), *Some computational aspects in the numerical solution of parabolic equations,* J. Comp. Appl. Math., vol 4, no 2, pages 129–141.

[85] AXELSSON,O., VASSILEVSKI,P.S., (1989), *Algebraic multilevel preconditioning methods I,* Numer. Math., vol 56, pages 157–177.

[86] AXELSSON,O., VASSILEVSKI,P.S., (1989), *A survey of multilevel preconditioned iterative methods,* BIT, vol 29, pages 769–793.

[87] AXELSSON,O., VASSILEVSKI,P.S., (1990), *Algebraic multilevel preconditioning methods, II,* SIAM J. Numer. Anal., vol 27, no 6, pages 1569–1590.

[88] AXELSSON,O., VASSILEVSKI,P.S., (1991), *A black box generalized conjugate gradient solver with inner iterations and variable step preconditioning,* SIAM J. Matrix Anal. Appl., vol 12, pages 625–644.

[89] AXELSSON,O., VASSILEVSKI,P.S., (1991), *Algebraic multilevel preconditioning methods III,* in Fourth International Symposium on Domain Decomposition Methods for Partial Differential Equations, GLOWINSKI,R., KUZNETSOV,Y.A., MEURANT,G., PERIAUX,J. and WIDLUND,O.B. Eds. SIAM, pages 163–177.

[90] AYACHOUR,E., (1997), *Avoiding the look–ahead in the Lanczos method,* Rapport

Université de Lille, France.

[91] BABUSKA,I., (1958), *On the Schwarz algorithm in the theory of differential equations of mathematical physics,* Tchecosl. Math. J., vol 8, pages 328–342.

[92] BABUSKA,I., ELMAN,H.C., (1989), *Some aspects of parallel implementation of the finite element method on message passing architectures,* J. Comp. Appl. Math., vol 27, pages 157–187.

[93] BADEA,L., (1989), *A generalization of the Schwarz alternating method to an arbitrary number of subdomains,* Numer. Math., vol 55, pages 61–81.

[94] BAGLAMA,J., CALVETTI,D., GOLUB,G.H., REICHEL,L., (1996), *Adaptively preconditioned GMRES algorithms,* Report Computer Science Dept., Stanford University.

[95] BAI,Z., (1994), *Error analysis of the Lanczos algorithm for the nonsymmetric eigenvalue problem,* Math. Comp., vol 62, no 205, pages 209–226.

[96] BAI,Z.Z., (1996), *A class of hybrid algebraic multilevel preconditioning methods,* Appl. Numer. Math., vol 19, pages 389–399.

[97] BAI,Z., GOLUB,G.H., (1997), *Bounds for the trace of the inverse and the determinant of symmetric positive definite matrices,* Ann. Numer. Math., vol 4, pages 29–38.

[98] BAI,Z., HU,D., REICHEL,L., (1994), *A Newton basis GMRES implementation,* IMA J. Num. Anal., vol 14, pages 563–581.

[99] BANEGAS,A., (1978), *Fast Poisson solvers for problems with sparsity,* Math. Comp., vol 32, no 142, pages 441–446.

[100] BANK,R.E., (1981), *A comparison of two multi–level iterative methods for nonsymmetric and indefinite elliptic finite element equations,* SIAM J. Numer. Anal., vol 18, pages 724–743.

[101] BANK,R.E., (1996), *Hierarchical bases and the finite element method,* in Acta Numerica 5. Cambridge University Press, pages 1–43.

[102] BANK,R.E., CHAN,T.F., (1993), *An analysis of the composite step bi-conjugate gradient method,* Numer. Math., vol 66, pages 295–319.

[103] BANK,R.E., CHAN,T.F., (1994), *A composite step bi-conjugate gradient algorithm for nonsymmetric linear systems,* Numer. Alg., vol 7, pages 1–16.

[104] BANK,R.E., DUPONT,T.F., YSERENTANT,H., (1988), *The hierarchical basis multigrid method,* Numer. Math., vol 52, pages 427–458.

[105] BANK,R.E., KENT SMITH,R., (1997), *The incomplete factorization multigraph algorithm,* Report University of California, San Diego, Dept. of Mathematics.

[106] BANK,R.E., ROSE,D.J., (1975), *An $O(n^2)$ method for solving constant coefficient boundary value problems in two dimensions,* SIAM J. Numer. Anal., vol 12, no 4, pages 529–540.

[107] BANK,R.E., ROSE,D.J., (1977), *Marching algorithms for elliptic boundary value problems. I: the constant coefficient case,* SIAM J. Numer. Anal., vol 14, no 5, pages 792–829.

[108] BANK,R.E., ROSE,D.J., (1990), *On the complexity of sparse Gaussian elimination via bordering,* SIAM J. Sci. Comput., vol 11, no 1, pages 145–160.

[109] BANK,R.E., SMITH,R.K., (1987), *General sparse elimination requires no permanent integer storage,* SIAM J. Sci. Comput., vol 8, no 4, pages 574–584.

[110] BANK,R.E., WAGNER,C., (1997), *Multilevel ILU decomposition,* Report University of California, San Diego, Dept. of Mathematics.

[111] BANK,R.E., XU,J., (1994), *The hierarchical basis multigrid method and incomplete LU decomposition,* in Domain Decomposition Methods in Scientific and Engineering Computing: Proceedings of the Seventh International Conference on Domain Decomposition, KEYES,D.E. and XU,J. Eds. American Mathematical Society, pages 163–173.

[112] BANK,R.E., XU,J., (1996), *An algorithm for coarsening unstructured meshes,* Numer. Math., vol 73, pages 1–36.

[113] BARANGER,J., DUC-JACQUET,M., (1971), *Matrices tridiagonales symétriques et matrices factorisables,* RIRO, vol 3, pages 61–66.

[114] BARKAI,D., MORIARTY,K.J.M., REBBI,C., (1985), *A modified conjugate gradient solver for very large systems,* Comp. Phys. Comm., vol 36, pages 1–8.

[115] BARLOW,J.L., (1986), *A note on monitoring the stability of triangular decomposition of sparse matrices,* SIAM J. Sci. Comput., vol 7, no 1, pages 166–168.

[116] BARNARD,S.T., POTHEN,A., SIMON,H.D., (1995), *A spectral algorithm for envelope reduction of sparse matrices,* Numer. Lin. Alg. with Appl., pages 317–334.

[117] BARRETT,R., BERRY,M., DONGARRA,J., EIJKHOUT,V., ROMINE,C., (1994), *Algorithmic bombardment for the iterative solution of linear systems: a polyiterative approach,* Report Dept. of Computer Science, University of Tennessee.

[118] BARRETT,W.W., (1979), *A theorem on inverses of tridiagonal matrices,* Linear Algebra and its Appl., vol 27, pages 211–217.

[119] BARRETT,W.W., FEINSILVER,P.J., (1981), *Inverses of banded matrices,* Linear Algebra and its Appl., vol 41, pages 111–130.

[120] BARRETT,W.W., JOHNSON,C.R., (1984), *Determinantal formulae for matrices with sparse inverses,* Linear Algebra and its Appl., vol 56, pages 73–88.

[121] BARRETT,W.W., JOHNSON,C.R., OLESKY,D.D., VAN DEN DRIESSCHE,P., (1987), *Inherited matrix entries: principal submatrices of the inverse,* SIAM J. Alg. Disc. Meth., vol 8, no 3, pages 313–322.

[122] BARTH,T., MANTEUFFEL,T., (1996), *Variable metric conjugate gradient methods,* Report University of Colorado.

[123] BARWELL,V., GEORGE,A., (1976), *A comparison of algorithms for solving symmetric indefinite systems of linear equations,* ACM Trans. Math. Soft., vol 2, no 3, pages 242–251.

[124] BAUER,F.L., (1963), *Optimally scaled matrices,* Numer. Math., vol 5, pages 73–87.

[125] BAUER,F.L., (1974), *Computational graphs and rounding error,* SIAM J. Numer. Anal., vol 11, pages 87–96.

[126] BEAUWENS,R., (1984), *Upper eigenvalue bounds for pencils of matrices,* Linear Algebra and its Appl., vol 62, pages 87–104.

[127] BEAUWENS,R., (1985), *On Axelsson's perturbations,* Linear Algebra and its Appl., vol 68, pages 221–242.

[128] BEAUWENS,R., BEN BOUZID,M., (1988), *Existence and conditioning properties of sparse approximate block factorizations,* SIAM J. Numer. Anal., vol 25, no 4, pages 941–956.

[129] BEAUWENS,R., QUENON,L., (1976), *Existence criteria for partial matrix factorization in iterative methods,* SIAM J. Numer. Anal., vol 13, no 4, pages 615–643.

[130] BEAUWENS,R., WILMET,R., (1989), *Conditioning analysis of positive definite matrices by approximate factorizations,* J. Comp. Appl. Math., vol 26, pages 257–269.

[131] BEGUE,C., GLOWINSKI,R., PERIAUX,J., (1988), *Détermination d'un opérateur de préconditionnement pour la résolution itérative du problème de Stokes dans la formulation d'Helmholtz,* C.R. Acad. Sci. Paris, vol 306, pages 247–252.

[132] BEHIE,A., FORSYTH,P., (1983), *Comparison of fast iterative methods for symmetric systems,* IMA J. Num. Anal., vol 3, pages 41–63.

[133] BELLMAN,R., (1970), *Introduction to matrix analysis,* Mc Graw–Hill.

[134] BENNETT,J.M., (1965), *Triangular factors of modified matrices,* Numer. Math., vol 7, pages 217–221.

[135] BENSON,A., EVANS,D.J., (1972), *The successive peripheral block overrelaxation method,* J. Inst. Maths. Applics., vol 9, pages 68–79.

[136] BENZI,M., GOLUB,G.H., (1998), *Bounds for the entries of matrix functions with applications to preconditioning,* Report NA–98–??, Computer Science Dept., Stanford University.

[137] BENZI,M., MEYER,C.D., TUMA,M., (1996), *A sparse approximate inverse preconditioner for the conjugate gradient method,* SIAM J. Sci. Comput., vol 17, pages 1135–1149.

[138] BENZI,M., SZYLD,B., VAN DUIN,A., (1997), *Orderings for incomplete factorization preconditioning of nonsymmetric problems,* Report LA–UR–97–3525, Los Alamos National Laboratory, to appear in SIAM J. Sci. Comp..

[139] BENZI,M., TUMA,M., (1995), *A sparse approximate inverse preconditioner for nonsymmetric linear systems,* Report 653, Institute of Computer Science, Academy of Sciences of the Czech Republic.

[140] BENZI,M., TUMA,M., (1998), *A comparative study of sparse approximate inverse preconditioners,* Report LA–UR–98–0024, Los Alamos National Laboratory.

[141] BERMAN,A., NEUMANN,M., (1976), *Consistency and splittings,* SIAM J. Numer. Anal., vol 13, no 6, pages 877–888.

[142] BERMAN,A., PLEMMONS,R.J., (1979), *Nonnegative matrices in the mathematical sciences,* Academic Press.

[143] BERMAN,P., SCHNITGER,G., (1990), *On the performance of the minimum degree ordering for Gaussian elimination,* SIAM J. Matrix Anal. Appl., vol 11, no 1, pages 83–89.

[144] BERNARDI,C., MADAY,Y., (1992), *Approximations spectrales de problemes aux limites elliptiques,* Mathematiques & Applications. Springer.

[145] BERNHARDT,P.A., BRACKBILL,J.U., (1984), *Solution of elliptic equations using fast Poisson solvers,* J. Comp. Physics, vol 53, pages 382–394.

[146] BETTE,S., DIAZ,J.C., JINES,W.R., STEIHAUG,T., (1986), *A block preconditioned conjugate gradient type iterative solver for linear systems in thermal reservoir simulation,* J. Comp. Physics, vol 67, pages 37–58.

[147] BEVILACQUA,R., CODENOTTI,B., ROMANI,F., (1988), *Parallel solution of block tridiagonal linear systems,* Linear Algebra and its Appl., vol 104, pages 39–57.

[148] BICKLEY,W.G., McNAMEE,J., (1960), *Matrix and other direct methods for the solution of systems of linear difference equations,* Philos. Trans. Roy. Soc. London Ser. A, vol 252, pages 69–131.

[149] BIRKHOFF,G., VARGA,R.S., (1959), *Implicit alternating direction methods,* Trans. of AMS, vol 92, no 1, pages 13–24.

[150] BIRKHOFF,G., VARGA,R.S., YOUNG,D.M., (1962), *Alternating direction implicit methods,* in Advances in Computers, F. Alt and M. Rubinoff Eds., vol 3. Academic Press, pages 189–273.

[151] BISCHOF,C.H., (1990), *Incremental condition estimation,* SIAM J. Matrix Anal. Appl., vol 11, pages 644–659.

[152] BISCHOF,C.H., (1990), *Incremental condition estimation for sparse matrices,* SIAM J. Matrix Anal. Appl., vol 11, pages 312–322.

[153] BJORCK,A., (1994), *Numerics of Gram–Schmidt orthogonalization,* Linear Algebra and its Appl., vol 197, pages 297–316.

[154] BJORCK,A., (1996), *Numerical methods for least squares problems,* SIAM.

[155] BJORSTAD,P., (1983), *Fast numerical solution of the biharmonic Dirichlet problem on rectangles,* SIAM J. Numer. Anal., vol 20, no 1, pages 59–71.

[156] BJORSTAD,P., SKOGEN,M., (1992), *Domain decomposition algorithms of Schwarz type designed for massively parallel computers,* in Fifth International Symposium on Domain Decomposition Methods for Partial Differential Equations, KEYES,D.E., CHAN,T.F., MEURANT,G., SCROGGS,J.S. and VOIGT,R.G. Eds. SIAM.

[157] BJORSTAD,P., WIDLUND,O.B., (1986), *Iterative methods for the solution of elliptic problems on regions partitioned into substructures,* SIAM J. Numer. Anal., vol 23, no 6, pages 1097–1120.

[158] BJORSTAD,P., WIDLUND,O.B., (1989), *To overlap or not to overlap: A note on a domain decomposition method for elliptic problems,* SIAM J. Sci. Comput., vol 10, pages 1053–1061.

[159] BLAIR,J.R., PEYTON,B., (1993), *An introduction to chordal graphs and clique trees,* in Graph theory and sparse matrix computation, GEORGE,A., GILBERT,J. R. and LIU,J.W–H. Eds. Springer Verlag.

[160] BODEWIG,E., (1959), *Matrix calculus,* second edition. North–Holland.

[161] BOISVERT,R.F., (1991), *Algorithms for special tridiagonal systems,* SIAM J. Sci. Comput., vol 12, no 2, pages 423–442.

[162] BOLEY,D., BUZBEE,B.L., PARTER,S.V., (1978), *On block relaxation techniques,* MRC TR–1860, Mathematics Research Center, University of Wisconsin.

[163] BOLEY,D.L., ELHAY,S., GOLUB,G.H., GUTKNECHT,M.H., (1990), *Nonsymmetric Lanczos and finding orthogonal polynomials associated with indefinite weights,* Report NA-90-09, Computer Science Dept., Stanford University.

[164] BOLLEN,J.A.M., (1984), *Numerical stability of descent methods for solving linear equations,* Numer. Math., vol 43, pages 361–377.

[165] BOMAN,E.G., HENDRICKSON,B., (1996), *A multilevel algorithm for reducing the envelope of sparse matrices,* Report SCCM-96–14, Computer Science Dept., Stanford University.

[166] BONNET,M., MEURANT,G., (1980), *Resolution de systemes d'equations lineaires par la methode du gradient conjugue avec preconditionnement,* Rapport CEA/DAM 80069.

[167] BORGERS,C., (1989), *The Neumann–Dirichlet domain decomposition method with inexact solvers on the subdomains,* Numer. Math., vol 55, pages 123–136.

[168] BORGERS,C., (1990), *Domain imbedding methods for the Stokes equation,* Numer. Math., vol 57, pages 435–451.

[169] BORGERS,C., WIDLUND,O.B., (1990), *On finite element domain imbedding methods,* SIAM J. Numer. Anal., vol 27, pages 963–978.

[170] BOTA,E.F., VELDMAN,A.E., (1981), *On local relaxation methods and their application to convection–diffusion equations,* J. Comp. Physics, vol 48, pages 127–149.

[171] BOTHE,Z., (1975), *Bounds for rounding errors in the Gaussian elimination for band systems,* J. Inst. Maths. Applics., vol 16, pages 133–142.

[172] BOTTA,E.F.F., WUBS,F.W., (1993), *The convergence behaviour of iterative methods on severely stretched grids,* Int. J. Num. Meth. Eng., vol 36, pages 3333–3350.

[173] BRACHA-BARAK,A., SAYLOR,P.E., (1973), *A symmetric factorization procedure for the solution of elliptic boundary value problems,* SIAM J. Numer. Anal., vol 10, no 1, pages 190–206.

[174] BRAESS,D., (1981), *The contraction number of a multigrid method for solving the Poisson equation,* Numer. Math., vol 37, pages 387–404.

[175] BRAESS,D., (1995), *Towards algebraic multigrid for elliptic problems of second order,* Computing, vol 55, pages 379–393.

[176] BRAMBLE,J.H., (1993), *Multigrid methods,* Pitman Research Notes in Mathematical Sciences, 294. Longman Scientific and Technical.

[177] BRAMBLE,J.H., EWING,R.E., PASCIAK,J.E., SCHATZ,A.H., (1988), *A preconditioning technique for the efficient solution of problems with local grid refinement,* Com. Meth. Appl. Mech. Eng., vol 67, pages 149–159.

[178] BRAMBLE,J.H., PASCIAK,J.E., (1984), *Preconditioned iterative methods for nonselfadjoint or indefinite elliptic boundary value problems,* Technical Report YALEU/DCS/RR 309, Dept. of Computer Science, Yale University.

[179] BRAMBLE,J.H., PASCIAK,J.E., (1992), *The analysis of smoothers for multigrid algorithms,* Math. Comp., vol 58, pages 467–488.

[180] BRAMBLE,J.H., PASCIAK,J.E., SAMMON,P.M., TOMEE,V., (1988), *Incomplete iterations in multistep backward difference methods for parabolic problems with smooth and nonsmooth data,* Report BNL-41279, Brookhaven National Laboratory.

[181] BRAMBLE,J.H., PASCIAK,J.E., SCHATZ,A.H., (1986), *An iterative method for elliptic problems on regions partitioned into substructures,* Math. Comp., vol 46, no 174, pages 361–369.

[182] BRAMBLE,J.H., PASCIAK,J.E., SCHATZ,A.H., (1986), *The construction of pre-*

conditioners for elliptic problems by substructuring I, Math. Comp., vol 47, no 175, pages 103–134.

[183] BRAMBLE,J.H., PASCIAK,J.E., SCHATZ,A.H., (1987), *The construction of preconditioners for elliptic problems by substructuring II,* Math. Comp., vol 49, no 179, pages 1–16.

[184] BRAMBLE,J.H., PASCIAK,J.E., SCHATZ,A.H., (1988), *The construction of preconditioners for elliptic problems by substructuring III,* Math. Comp., vol 51, no 184, pages 415–430.

[185] BRAMBLE,J.H., PASCIAK,J.E., SCHATZ,A.H., (1989), *The construction of preconditioners for elliptic problems by substructuring IV,* Math. Comp., vol 53, no 187, pages 1–24.

[186] BRAMBLE,J.H., PASCIAK,J.E., WANG,J., XU,J., (1991), *Convergence estimates for product iterative methods with applications to domain decomposition,* Math. Comp., vol 57, no 195, pages 1–21.

[187] BRAMBLE,J.H., PASCIAK,J.E., WANG,J., XU,J., (1991), *Convergence estimates for multigrid algorithms without regularity assumptions,* Math. Comp., vol 57, no 195, pages 23–45.

[188] BRAMBLE,J.H., PASCIAK,J.E., XU,J., (1988), *The analysis of multigrid algorithms for nonsymmetric and indefinite elliptic problems,* Math. Comp., vol 51, pages 389–414.

[189] BRAMBLE,J.H., PASCIAK,J.E., XU,J., (1990), *Parallel multilevel preconditioners,* Math. Comp., vol 55, no 191, pages 1–22.

[190] BRAMBLE,J.H., PASCIAK,J.E., XU,J., (1991), *The analysis of multigrid algorithms with nonnested spaces or noninherited quadratic forms,* Math. Comp., vol 56, pages 1–34.

[191] BRAND,C.W., (1992), *An incomplete factorization preconditioning using repeated Red/Black ordering,* Numer. Math., vol 61, pages 433–454.

[192] BRANDT,A., (1973), *Multi-level adaptive technique (MLAT) for fast numerical solution to boundary value problems,* in Proceedings of the Third International Conference on Numerical Methods in Fluid Mechanics, CABANNES,H. and TEMAM,R. Eds, Lecture Notes in Physics 18. Springer, pages 82–89.

[193] BRANDT,A., (1977), *Multi-level adaptive solutions to boundary-value problems,* Math. Comp., vol 31, pages 333–390.

[194] BRANDT,A., (1982), *Guide to multigrid development,* in Multigrid methods, Proceedings of the Koln-Porz Conference, HACKBUSCH,W. and TROTTENBERG,U. Eds, Lecture Notes in Mathematics 960. Springer, pages 220–312.

[195] BRANDT,A., (1986), *Algebraic multigrid theory: The symmetric case,* Appl. Math. Comput, vol 19, pages 23–56.

[196] BRANDT,A., (1994), *Rigorous quantitative analysis of multigrid, I: Constant coefficients two-level cycle with L_2-norm,* SIAM J. Numer. Anal., vol 31, pages 1695–1730.

[197] BRANDT,A., DISKIN,B., (1994), *Multigrid solvers on decomposed domains,* in Proceedings of the Sixth International Conference on Domain Decomposition, QUARTERONI,A., PERIAUX,J., KUZNETSOV,Y.A. and WIDLUND,O.B. Eds. AMS, pages 135–155.

[198] BRAYTON,R.K., GUSTAVSON,F.G., WILLOUGHBY,R.A., (1970), *Some results on sparse matrices,* Math. Comp., vol 24, no 115, pages 937–954.

[199] BRENNER,S.C., (1996), *Preconditioning complicated finite elements by simple finite elements,* SIAM J. Sci. Comput., vol 17, pages 1269–1274.

[200] BREZINSKI,C., (1991), *Biorthogonality and conjugate gradient-type algorithms,* WSSIA, vol 2, pages 55–70.

[201] BREZINSKI,C., (1995), *Hybrid procedures and semi-iterative methods for linear systems,* Rapport LANO, Université de Lille.

[202] BREZINSKI,C., (1996), *Hybrid methods for solving systems of equations,* Rapport LANO, Université de Lille.

[203] BREZINSKI,C., REDIVO-ZAGLIA,M., (1994), *Treatment of near-breakdown in the CGS algorithm,* Numer. Alg., vol 7, pages 33–73.

[204] BREZINSKI,C., REDIVO-ZAGLIA,M., (1994), *Hybrid procedures for solving linear systems,* Numer. Math., vol 67, pages 1—19.

[205] BREZINSKI,C., REDIVO-ZAGLIA,M., (1995), *Look-ahead in BiCGSTAB and other methods for linear systems,* BIT, vol 35, pages 169–201.

[206] BREZINSKI,C., REDIVO-ZAGLIA,M., SADOK,H., (1991), *Avoiding breakdown and near-breakdown in Lanczos type algorithms,* Numer. Alg., vol 1, pages 261–284.

[207] BREZINSKI,C., REDIVO-ZAGLIA,M., SADOK,H., (1992), *A breakdown-free Lanczos type algorithm for solving linear systems,* Numer. Math., vol 63, pages 29–38.

[208] BREZINSKI,C., REDIVO-ZAGLIA,M., SADOK,H., (1992), *Addendum to "Avoiding breakdown and near breakdown in Lanczos type algorithms",* Numer. Alg., vol 2, pages 133–136.

[209] BREZINSKI,C., REDIVO-ZAGLIA,M., SADOK,H., (1996), *Problems of breakdown in Lanczos-based algorithms,* Rapport LANO, Université de Lille.

[210] BREZINSKI,C., SADOK,H., (1991), *Avoiding breakdown in the CGS algorithm,* Numer. Alg., vol 1, pages 199–206.

[211] BREZINSKI,C., SADOK,H., (1993), *Lanczos type algorithms for solving systems of linear equations,* Applied Numerical Math., vol 11, pages 443–473.

[212] BRIGGS,W.L., (1987), *A multigrid tutorial,* SIAM.

[213] BRIGGS,W.L., HENSON,V.E., (1990), *FFT as a multigrid algorithm,* SIAM Review, vol 32, pages 252–261.

[214] BRIGGS,W.L., HENSON,V.E., (1993), *Wavelets and multigrid,* SIAM J. Sci. Comput., vol 14, pages 506–510.

[215] BRINKKEMPER,S., (1984), *Two classes of preconditioning matrices,* Ph.D. Thesis, Dept. of Informatics, Catholic University of Nijmegen.

[216] BRINKKEMPER,S., (1985), *A class of incomplete block factorization methods,* Report 71, Faculteit der Wiskunde, Katholieke Universiteit Nijmegen.

[217] BRISTEAU,M.O., GLOWINSKI,R., PERIAUX,J., (1987), *Numerical methods for the Navier–Stokes equations,* Computer Physics Reports, vol 6, pages 73–187.

[218] BROWN,P.N., (1991), *A theoretical comparison of the Arnoldi and GMRES algorithms,* SIAM J. Sci. Comput., vol 12, pages 58–78.

[219] BROYDEN,C.G., (1973), *Some condition number bounds for the Gaussian elimination process,* J. Inst. Maths. Applics., vol 12, pages 273–286.

[220] BRUASET,A.M., TVEITO,A., WINTHER,R., (1990), *On the stability of relaxed incomplete LU factorizations,* Math. Comp., vol 54, no 190, pages 701–719.

[221] BRUASET,A.R., (1995), *A survey of preconditioned iterative methods,* Longman Scientific and Technical.

[222] BRUGNANO,L., MARRONE,M., (1990), *Vectorization of some block preconditioned conjugate gradient methods,* Parallel Comp., vol 14, pages 191-198.

[223] BRUSSINO,G., SONNAD,V., (1989), *A comparison of direct and preconditioned iterative techniques for sparse unsymmetric systems of linear equations,* Int. J. Num. Meth. Eng., vol 28, pages 801–815.

[224] BUKHBERGER,B., EMEL'YANENKO,G.A., (1973), *Methods of inverting tridiagonal matrices,* USSR Comp. Math. and Math. Phys., vol 13, pages 10–20.

[225] BULEEV,N.I., (1960), *A numerical method for the solution of two dimensional and three dimensional equations of diffusion,* Math. Sbornik, vol 51, pages 227–238.

[226] BUNCH,J.R., (1971), *Analysis of the diagonal pivoting method,* SIAM J. Numer. Anal., vol 8, no 4, pages 656–680.

[227] BUNCH,J.R., (1973), *Complexity of sparse elimination,* in Complexity of sequential and parallel numerical algorithms, TRAUB,J.F. Ed. Academic Press, pages 197–220.

[228] BUNCH,J.R., (1974), *Partial pivoting strategies for symmetric matrices,* SIAM J. Numer. Anal., vol 11, no 3, pages 521–528.

[229] BUNCH,J.R., (1974), *Analysis of sparse elimination,* SIAM J. Numer. Anal., vol 11, no 5, pages 847–873.

[230] BUNCH,J.R., (1987), *The weak and strong stability of algorithms in numerical linear algebra,* Linear Algebra and its Appl., vol 88, pages 49–66.

[231] BUNCH,J.R., DEMMEL,J.W., VAN LOAN,C., (1989), *The strong stability of algorithms for solving symmetric linear systems,* SIAM J. Matrix Anal. Appl., vol 10, no 4, pages 494–499.

[232] BUNCH,J.R., DONGARRA,J.J., MOLER,C., STEWART,G.W., (1979), *Linpack User's guide,* SIAM.

[233] BUNCH,J.R., HOPCROFT,J.,E., (1974), *Triangular factorization and inversion by fast matrix multiplication,* Math. Comp., vol 28, no 125, pages 231–236.

[234] BUNCH,J.R., KAUFMAN,L., (1977), *Some stable methods for calculating inertia and solving symmetric linear systems,* Math. Comp., vol 31, pages 162–179.

[235] BUNCH,J.R., PARLETT,B.N., (1971), *Direct methods for solving symmetric indefinite systems of linear equations,* SIAM J. Numer. Anal., vol 8, no 4, pages 639–655.

[236] BUNCH,J.R., ROSE,D.J., (1976), *Sparse matrices computations,* Academic Press.

[237] BUNEMAN,O., (1969), *A compact non-iterative Poisson solver,* Report 294, Stanford University Institute for Plasma Research, Stanford University.

[238] BUONI,J.J., (1990), *A stable method for the incomplete factorization of H–matrices,* Linear Algebra and its Appl., vol 129, pages 143–154.

[239] BUONI,J.J., VARGA,R.S., (1979), *A theorem of Stein–Rosenberg type,* in Advances in Computer Methods for Partial Differential Equations III, IMACS. pages 11–12.

[240] BURGESS,I.W., LAI,P.K.F., (1986), *A new node renumbering algorithm for bandwidth reduction,* Int. J. Num. Meth. Eng., vol 23, pages 1693–1704.

[241] BURRAGE,K., ERHEL,J., POHL,B., (1998), *A deflation technique for linear systems of equations,* SIAM J. Sci. Comput., vol 19, no 4, pages 1425–1460.

[242] BUZBEE,B.L., (1973), *A fast Poisson solver amenable to parallel computation,* IEEE Trans. on Comp., vol 22, no 8, pages 793–796.

[243] BUZBEE,B.L., DORR,F.W., (1974), *The direct solution of the biharmonic equation on rectangular regions and the Poisson equation on irregular regions,* SIAM J. Numer. Anal., vol 11, no 4, pages 753–763.

[244] BUZBEE,B.L., DORR,F.W., GEORGE,A., GOLUB,G.H., (1971), *The direct solu-*

tion of the discrete Poisson equation on irregular regions, SIAM J. Numer. Anal., vol 8, no 4, pages 722–736.

[245] BUZBEE,B.L., GOLUB,G.H., NIELSON,C.W., (1970), *On direct methods for solving Poisson's equations,* SIAM J. Numer. Anal., vol 7, no 4, pages 627–656.

[246] CAI,X-C., GROPP,W.D., KEYES,D.E., (1992), *A comparison of some domain decomposition algorithms for nonsymmetric elliptic problems,* in Fifth International Symposium on Domain Decomposition Methods for Partial Differential Equations, KEYES,D.E., CHAN,T.F., MEURANT,G., SCROGGS,J.S. and VOIGT,R.G. Eds. SIAM, pages 224–234.

[247] CAI,X-C., WIDLUND,O.B., (1992), *Domain decomposition algorithms for indefinite elliptic problems,* SIAM J. Sci. Comput., vol 13, no 1, pages 243–258.

[248] CAI,X-C., XU,J., (1992), *A preconditioned GMRES method for nonsymmetric or indefinite problems,* Math. Comp., vol 59, pages 311–319.

[249] CAI,Z., LAZAROV,R.D., MANTEUFFEL,T.A., McCORMICK,S.F., (1994), *First order system least squares for second-order partial differential equations: Part I,* SIAM J. Numer. Anal., vol 31, pages 1785–1802.

[250] CAI,Z., McCORMICK,S.F., (1989), *Computational complexity of the Schwarz alternating procedure,* Int. J. High Speed Comp., vol 1, pages 1–28.

[251] CALVETTI,D., GOLUB,G.H., REICHEL,L., (1994), *An adaptive Chebyshev iterative method for nonsymmetric linear systems based on modified moments,* Numer. Math., vol 67, pages 21–40.

[252] CALVETTI,D., REICHEL,L., SORENSEN,D.C., (1994), *An implicitly restarted Lanczos method for large symmetric eigenvalue problems,* ETNA, vol 2, pages 1–21.

[253] CAMION,P., HOFFMAN,A.J., (1966), *On the nonsingularity of complex matrices,* Pacific J. Math., vol 17, no 2, pages 211–214.

[254] CAMPBELL,S.L., IPSEN,I.C.F., KELLEY,C.T., MEYER,C.D., (1995), *GMRES and the minimal polynomial,* Report Dept. of Mathematics, North Carolina State University.

[255] CAMPBELL,Y.E., DAVIS,T.A., (1995), *Incomplete LU factorization: a multifrontal approach,* Report TR-95-024, Computer Science Dept., University of Florida.

[256] CANUTO,C., FUNARO,D., (1988), *The Schwarz algorithm for spectral methods,* SIAM J. Numer. Anal., vol 25, no 1, pages 24–40.

[257] CAO,F., GILBERT,J.R., TENG,S–H., (1996), *Partitioning meshes with lines and planes,* Report Xerox Parc, Palo Alto.

[258] CARLENZOLI,C., GERVASIO,P., (1991), *Effective numerical algorithms for the*

solution of algebraic systems arising in spectral methods, Report UMSI 91/137, Supercomputer Institute, University of Minnesota.

[259] CARNEVALI,P., RADICATI,G., ROBERT,Y., SGUAZZERO,P., (1987), *Efficient Fortran implementation of the Gaussian elimination and Householder reduction algorithms on the IBM 3090 vector multiprocessor,* Report ICE–0012, IBM European Center for Scientific and Engineering Computing.

[260] CARTER,R., (1991), *Y–MP floating point and Cholesky factorization,* Int. J. High Speed Comp., vol 3, no 3, pages 215–222.

[261] CARVALHO,L.M., (1997), *Preconditioned Schur complement methods in distributed memory environments,* Rapport TH/PA/97/41, Ph.D. Thesis, CERFACS, Toulouse, France.

[262] CHAITIN-CHATELIN,F., (1995), *Le calcul qualitatif. Comment donner un sens a des resultats faux ?,* Rapport CERFACS TR/PA/95/10.

[263] CHAN,T.F., (1984), *On the existence and computation of LU factorizations with small pivots,* Math. Comp., vol 42, no 168, pages 535–547.

[264] CHAN,T.F., (1987), *Analysis of preconditioners for domain decomposition,* SIAM J. Numer. Anal., vol 24, no 2, pages 382–390.

[265] CHAN,T.F., (1989), *Domain decomposition algorithms and computational fluid dynamics,* in Vector and Parallel Computing: issues in applied research and development, DONGARRA,J. J. , DUFF,I. S., GAFFNEY,P. and McKEE,S. Eds. Ellis Horwood, pages 65–82.

[266] CHAN,T.F., (1991), *Fourier analysis of relaxed incomplete factorization preconditioners,* SIAM J. Sci. Comput., vol 12, pages 668–680.

[267] CHAN,T.F., De PILLIS,L., VAN DER VORST,H.A., (1991), *A transpose free squared Lanczos algorithm and application to solving nonsymmetric linear systems,* CAM Report 91-17, Dept. of Mathematics, UCLA.

[268] CHAN,T.F., ELMAN,H.C., (1989), *Fourier analysis of iterative methods for elliptic problems,* SIAM Review, vol 31, no 1, pages 20–49.

[269] CHAN,T.F., FOULSER,D.E., (1988), *Effectively well conditioned linear systems,* SIAM J. Sci. Comput., vol 9, pages 963–969.

[270] CHAN,T.F., GALLOPOULOS,E., SIMONCINI,V., SZETO,T., TONG,C.H., (1993), *QMRCGSTAB: A quasi-minimal residual variant of the Bi-CGSTAB algorithm for nonsymmetric systems,* SIAM J. Sci. Comput., vol 15, pages 338–347.

[271] CHAN, T.F., GILBERT,J.R., TENG,S–H., (1995), *Geometric spectral partitioning,* Report Xerox Parc, Palo Alto.

[272] CHAN,T.F., GLOWINSKI,R., PERIAUX,J., WIDLUND,O.B., (1989), *Domain decomposition methods,* SIAM.

[273] CHAN,T.F., GLOWINSKI,R., PERIAUX,J., WIDLUND,O.B., (1990), *Third international symposium on domain decomposition methods for partial differential equations,* SIAM.

[274] CHAN,T.F., GO,S., ZIKATANOV,L., (1997), *Lecture notes on multilevel methods for elliptic problems on unstructured grids,* Report University of California, Los Angeles, Dept. of Mathematics.

[275] CHAN,T.F., GOOVAERTS,D., (1988), *Schwarz=Schur: overlapping versus non overlapping domain decomposition,* CAM report 88-21, Dept. of Mathematics, UCLA.

[276] CHAN,T.F., GOOVAERTS,D., (1990), *A note on the efficiency of domain decomposed incomplete factorizations,* SIAM J. Sci. Comput., vol 11, no 4, pages 794–803.

[277] CHAN,T.F., GOOVAERTS,D., (1991), *On the relationship between overlapping and nonoverlapping domain decomposition methods,* SIAM J. Matrix Anal. Appl., vol 13, pages 663–670.

[278] CHAN,T.F., HOU,T.Y., (1990), *Domain decomposition interface preconditioners for general second order elliptic problems,* CAM report 88-16, modified 1990, Dept. of Mathematics, UCLA.

[279] CHAN,T.F., HOU,T.Y., LIONS,P.L., (1991), *Geometry related convergence results for domain decomposition algorithms,* SIAM J. Numer. Anal., vol 28, no 2, pages 378–391.

[280] CHAN,T.F., JACKSON,K.R., ZHU,B., (1983), *Alternating direction incomplete factorizations,* SIAM J. Numer. Anal., vol 20, no 2, pages 239–257.

[281] CHAN,T.F., KEYES,D.E., (1990), *Interface preconditionings for domain decomposed convection-diffusion operators,* in Third International Symposium on Domain Decomposition Methods for Partial Differential Equations, CHAN,T.F., GLOWINSKI,R., PERIAUX,J. and WIDLUND,O.B. Eds. SIAM, pages 245–262.

[282] CHAN,T.F., KUO,C.C.J., TONG,C., (1989), *Parallel elliptic preconditioners: Fourier analysis and performance on the Connection Machine,* Comp. Phys. Comm., vol 53, pages 237–252.

[283] CHAN,T.F., KUO,C.C.J., TONG,C., (1990), *Multilevel filtering preconditioners,* SIAM J. Matrix Anal. Appl., vol 11, no 3, pages 403–429.

[284] CHAN,T.F., MATHEW,T., (1991), *An application of the probing technique to the vertex space method in domain decomposition,* in Fourth International Symposium on Domain Decomposition Methods for Partial Differential Equations, GLOWINSKI,R., KUZNETSOV,Y.A., MEURANT,G., PERIAUX,J., and WIDLUND,O.B. Eds. SIAM, pages 101–111.

[285] CHAN,T.F., MATHEW,T., (1992), *The interface probing in domain decomposition,* SIAM J. Matrix Anal. Appl., vol 13, pages 212–238.

[286] CHAN,T.F., MATHEW,T., (1994), *Domain Decomposition Algorithms,* in Acta Numerica 3. Cambridge University Press, pages 61–143.

[287] CHAN,T.F., MATHEW,T.P., SHAO,J-P., (1992), *Fourier and probe variants of the vertex space domain decomposition algorithm,* in Fifth International Symposium on Domain Decomposition Methods for Partial Differential Equations, KEYES,D.E., CHAN,T.F., MEURANT,G., SCROGGS,J.S. and VOIGT,R.G. Eds. SIAM, pages 236–249.

[288] CHAN,T.F., MEURANT,G., (1990), *Fourier analysis of block preconditioners,* Report CAM 90-04, Dept. of Mathematics, UCLA.

[289] CHAN,T.F., RESASCO,D.C., (1985), *A survey of preconditioners for domain decomposition,* Research Report YALEU/DCS/RR-414, Dept. of Computer Science, Yale University.

[290] CHAN,T.F., RESASCO,D.C., (1986), *Generalized deflated block elimination,* SIAM J. Numer. Anal., vol 23, no 5, pages 913–924.

[291] CHAN,T.F., RESASCO,D.C., (1987), *Analysis of domain decomposition preconditioners on irregular regions,* in Proceedings of the Sixth IMACS Int'l Symp. on Computer Methods for Partial Differential Equations, VICHNEVETSKY,R. and STEPLEMAN,R..

[292] CHAN,T.F., RESASCO,D.C., (1987), *A domain-decomposed fast Poisson solver on a rectangle,* SIAM J. Sci. Comput., vol 8, no 1,

[293] CHAN,T.F., RESASCO,D.C., (1988), *A framework for the analysis and construction of domain decomposition preconditioners,* in First International Symposium on Domain Decomposition Methods for Partial Differential Equations, GLOWINSKI,R., GOLUB,G.H., MEURANT,G. and PERIAUX,J. Eds. SIAM, pages 217–230.

[294] CHAN,T.F., SMITH,B.F., (1994), *Domain decomposition and multigrid algorithms for elliptic problems on unstructured meshes,* in Domain Decomposition Methods in Scientific and Engineering Computing: Proceedings of the Seventh International Conference on Domain Decomposition, KEYES,D.E. and XU,J. Eds. AMS, pages 175–189.

[295] CHAN,T.F., TANG,W.P., WAN,W.L., (1997), *Wavelet sparse approximate inverse preconditioners,* Report CAM 97–34, Mathematics Dept., UCLA.

[296] CHAN,T.F., TUMINARO,R.S., (1989), *Analysis of a parallel multigrid algorithm,* Report CAM 89–29, Mathematics Dept., UCLA.

[297] CHAN,T.F., VAN DER VORST,H.A., (1994), *Approximate and incomplete factorizations,* Report 871, Dept. of Mathematics, University of Utrecht.

[298] CHAN,T.F., VASSILEVSKI,P.S., (1993), *A framework for block ILU factorizations using block-size reduction,* SIAM J. Matrix Anal. Appl.,

[299] CHAN,T.F., WAN,W.L., (1994), *Analysis of projection methods for solving linear systems with multiple right-hand sides,* Report CAM 94–26, Dept. of Mathematics, University of California.

[300] CHAN,T.F., YE,Q., (1997), *A mixed product Krylov subspace method for solving nonsymmetric linear systems,* Asian J. Math., vol 1, no 3,

[301] CHAN,T.F., ZOU,J., (1994), *Additive Schwarz domain decomposition methods for elliptic problems on unstructured meshes,* Numer. Alg., vol 8, see also v. 9, no 2–4,

[302] CHAN,W.M., GEORGE,A., (1980), *A linear time implementation of the reverse Cuthill–McKee algorithm,* BIT, vol 20, pages 8–14.

[303] CHANDRA,R., (1978), *Conjugate gradient methods for partial differential equations,* Ph.D. Thesis, Report 129, Dept. of Computer Science, Yale University.

[304] CHANDRA,R., EISENSTAT,S.C., SCHULTZ,M.H., (1975), *Conjugate gradient methods for partial differential equations,* in Advances in Computer Methods for Partial Differential Equations, VICHNEVETSKY,R. Ed. AICA, pages 60–64.

[305] CHANDRASEKARAN,S., IPSEN,I., (1995), *On the sensitivity of solution components in linear systems of equations,* SIAM J. Matrix Anal. Appl., vol 16, no 1, pages 93–112.

[306] CHAPMAN,A., SAAD,Y., (1995), *Deflated and augmented Krylov subspace techniques,* Report UMSI 95-181, Supercomputer Institute, University of Minnesota.

[307] CHAPMAN,A., SAAD,Y., WIGTON,L., (1996), *High order ILU preconditioners for CFD problems,* Report UMSI 96-14, Supercomputer Institute, University of Minnesota.

[308] CHARRIER,P., ROMAN,J., (1989), *Algorithmique et calculs de complexité pour un solveur de type dissection emboîtée,* Numer. Math., vol 55, pages 463–476.

[309] CHARRIER,P., ROMAN,J., (1991), *Analysis of refined partitions for a parallel implementation of nested dissection,* Report LABRI 91–38, Universite de Bordeaux I.

[310] CHATELIN,F., FRAYSSE,V., (1996), *Lecture on finite precision computation,* SIAM.

[311] CHATELIN,F., MIRANKER,W.L., (1982), *Acceleration by aggregation of successive approximation methods,* Linear Algebra and its Appl., vol 43, pages 17–47.

[312] CHEN,S.C., KUCK,D.J., SAMEH,A.H., (1978), *Practical parallel band triangular system solvers,* ACM Trans. Math. Soft., vol 1, no 3, pages 270–277.

[313] CHEN,S.S., DONGARRA,J.J., HSIUNG,C.C., (1984), *Multiprocessing linear algebra algorithms on the CRAY X-MP-2: experiences with small granularity,* J. Par. Dist. Comp., vol 1, pages 22–31.

[314] CHEN,T., TEMAM,R., (1991), *Incremental unknowns for solving partial differential equations,* Numer. Math., vol 59, pages 255–271.

[315] CHERNESKY,M.P., ELMAN,H.C., (1993), *Costs of solving the discrete convection diffusion equation by relaxation methods,* CS-TR-3028, Computer Science Dept., University of Maryland.

[316] CHIN,R.C.Y., MANTEUFFEL,T.A., (1988), *An analysis of block successive over-relaxation for a class of matrices with complex spectra,* SIAM J. Numer. Anal., vol 25, no 3, pages 564–585.

[317] CHOI,J., DONGARRA,J., POZO,R., WALKER,D.W., (1994), *ScaLAPACK: a scalable linear algebra library for distributed memory concurrent computers,* in Proceedings of Fourth Symposium on the Frontiers of Massively Parallel Computation (McLean, Virginia). IEEE Computer Society Press.

[318] CHOW,E., SAAD,Y., (1994), *Approximate inverse preconditioners for general sparse matrices,* Report UMSI 94-101, Supercomputer Institute, University of Minnesota.

[319] CHOW,E., SAAD,Y., (1995), *Approximate inverse techniques for block partitioned matrices,* Report UMSI 95-13, Supercomputer Institute, University of Minnesota.

[320] CHRONOPOULOS,A.T., (1986), *A class of parallel iterative methods implemented on multiprocessors,* Ph.D. Thesis, report UIUCDCS-R-86-1267, Dept. of Computer Science, University of Illinois.

[321] CHRONOPOULOS,A.T., (1989), *S-step iterative methods for (non)symmetric (in)definite linear systems,* Report TR 89-35, Computer Science Dept., University of Minnesota.

[322] CHRONOPOULOS,A.T., (1990), *Krylov subspace iterative methods for nonsymmetric indefinite linear systems,* Report TR 90-21, Computer Science Dept., University of Minnesota.

[323] CHRONOPOULOS,A.T., GEAR,C.W., (1989), *On the efficient implementation of preconditioned s-step conjugate gradient methods on multiprocessors with memory hierarchy,* Parallel Comp., vol 11, pages 37–53.

[324] CHU,M.T., FUNDERLIC,R.E., GOLUB,G.H., (1995), *A rank one reduction formula and its applications to matrix factorizations,* SIAM Rev., vol 37, pages 512–530.

[325] CHU,E., GEORGE,A., (1987), *Gaussian elimination with partial pivoting and load balancing on a multiprocessor,* Parallel Comp., vol 5, pages 65–74.

[326] CIARLET Jr,P., (1992), *Etude de préconditionnements parallèles pour la résolution d'équations aux dérivées partielles elliptiques,* Ph.D. Thesis, Universite Paris 6.

[327] CIARLET Jr, P., (1994), *Implementation of a domain decomposition method well suited for (massively) parallel architectures,* Int. J. High Speed Comp., vol 6, pages

157–182.

[328] CIARLET Jr,P., (1994), *A comparison of three iterative algorithms based on domain decomposition methods,* in Domain Decomposition Methods in Scientific and Engineering Computing: Proceedings of the Seventh International Conference on Domain Decomposition, KEYES,D.E. and XU,J. Eds. AMS, pages 387–393.

[329] CIARLET Jr,P., (1994), *Repeated Red–Black ordering: a new approach,* Numer. Alg., vol 7, pages 295–324.

[330] CIARLET Jr,P., LAMOUR,F., (1994), *Spectral partitioning methods and greedy partitioning methods: a comparison on finite element graphs,* CAM Report 94-9, Dept. of Mathematics, UCLA.

[331] CIARLET Jr,P., LAMOUR,F., (1994), *An efficient low cost greedy graph partitioning heuristic,* CAM Report 94-1, Dept. of Mathematics, UCLA.

[332] CIARLET Jr,P., LAMOUR,F., SMITH,B.F., (1995), *On the influence of partitioning schemes on the efficiency of overlapping domain decomposition methods,* in Proceedings of the 5th Symposium on the Frontiers of Massively Parallel Computation 1995. IEEE, pages 375–383.

[333] CIARLET Jr.,P., LAMOUR,F., (1996), *On the validity of a front oriented approach to partitioning large sparse graphs with a connectivity constraint,* Numer. Alg., vol 12, no 1–2,

[334] CIARLET Jr.,P., LAMOUR,F., (1996), *Does contraction preserve triangular meshes?,* Numer. Alg., vol 13, no 3–4, pages 201–223.

[335] CIARLET Jr,P., MEURANT,G., (1994), *A class of domain decomposition preconditioners for massively parallel computers,* in Proceedings of the Sixth International Conference on Domain Decomposition, QUARTERONI,A., PERIAUX,J., KUZNETSOV,Y.A. and WIDLUND,O.B. Eds. AMS, pages 353–360.

[336] CLEARY,A.J., (1990), *Parallelism and fill–in in the Cholesky factorization of reordered banded matrices,* Report SAND90–2757, Sandia Nat. Lab., Albuquerque.

[337] CLINE,A.K., CONN,A.R., VAN LOAN,C.F., (1982), *Generalizing the LINPACK condition estimator,* in Lecture Notes in Mathematics 909. Springer, pages 73–83.

[338] CLINE,A.K., MOLER,C.B., STEWART,G.W., WILKINSON,J.H., (1979), *An estimate of the condition number of a matrix,* SIAM J. Numer. Anal., vol 16, pages 368–375.

[339] COHEN,A.M., (1974), *A note on pivot size in Gaussian elimination,* Linear Algebra and its Appl., vol 8, pages 361–368.

[340] CONCUS,P., GOLUB,G.H., (1973), *Use of fast direct methods for the efficient numerical solution of nonseparable elliptic equations,* SIAM J. Numer. Anal., vol 10, no 6, pages 1103–1120.

[341] CONCUS,P., GOLUB,G.H., (1976), *A generalized conjugate gradient method for non symmetric systems of linear equations,* Report STAN–CS–76–535, Computer Science Dept., Stanford University.

[342] CONCUS,P., GOLUB,G.H., MEURANT,G., (1985), *Block preconditioning for the conjugate gradient method,* SIAM J. Sci. Comput., vol 6, no 1, pages 220–252.

[343] CONCUS,P., GOLUB,G.H., O'LEARY,D.P., (1976), *A generalized conjugate gradient method for the numerical solution of elliptic partial differential equations,* Report STAN–CS–76–533, Comp. Science Dept., Stanford University.

[344] CONCUS,P., MEURANT,G., (1986), *On computing INV block preconditionings for the conjugate gradient method,* BIT, vol 26, pages 493–504.

[345] COOLEY,J.W., TUKEY,J.W., (1965), *An algorithm for the machine calculation of complex Fourier series,* Math. Comp., vol 19, pages 297–301.

[346] COSGROVE,J.D., DIAZ,J.C., GRIEWANK,A., (1991), *Approximate inverse preconditionings for sparse linear systems,* Report MCS-P268-1091, Math. and Computer Science Div., Argonne National Laboratory.

[347] COSNARD, M., ROBERT,Y., QUINTON,P., TCHUENTE,M., (1986), *Parallel algorithms and architectures,* North–Holland.

[348] COSNARD,M., ROBERT,Y., TRYSTRAM,D., (1986), *Resolution parallele de systemes lineaires denses par diagonalisation,* EDF, Bulletin DER, serie C no 2. pages 67–88.

[349] COTTLE,R.W., (1974), *Manifestations of the Schur complement,* Linear Algebra and its Appl., vol 8, pages 189–211.

[350] CROUZET,L., (1994), *Résolution des équations de Maxwell tridimensionnelles en régime fréquentiel par élements finis conformes, multiplicateurs de Lagrange et méthodes itératives,* Ph.D. Thesis, Université Paris 6.

[351] CROUZET,L., MEURANT,G., (1994), *Solving the 3D time harmonic Maxwell equations with finite elements, Lagrange multipliers and iterative methods,* in Proceedings of the meeting "50 years of the Courant element", Jyvaskyla, Finland.

[352] CRYER,C.W., (1968), *Pivot growth in Gaussian elimination,* Numer. Math., vol 12, pages 335-345.

[353] CSANKY,L., (1976), *Fast parallel matrix inversion algorithms,* SIAM J. Comp., vol 5, no 4, pages 618–623.

[354] CSORDAS,G., VARGA,R.S., (1984), *Comparisons of regular splittings of matrices,* Numer. Math., vol 44, pages 23–35.

[355] CULLUM,J., WILLOUGHBY,R.A., (1977), *The equivalence of the Lanczos and the conjugate gradient algorithms,* Report RC 6903, IBM Research Division.

[356] CULLUM,J., WILLOUGHBY,R.A., (1980), *The Lanczos phenomenon – An interpretation based upon conjugate gradient optimization,* Linear Algebra and its Appl., vol 29, pages 63–90.

[357] CULLUM,J., WILLOUGHBY,R.A., (1985), *Lanczos algorithms for large symmetric eigenvalue computations, Volume 1: theory,* Birkhauser.

[358] CULLUM,J., WILLOUGHBY,R.A., (1985), *Lanczos algorithms for large symmetric eigenvalue computations, Volume 2: programs,* Birkhauser.

[359] CURTIS,A.R., REID,J.K., (1971), *The solution of large sparse unsymmetric systems of linear equations,* J. Inst. Maths. Applics., vol 8, pages 344–353.

[360] CURTIS,A.R., REID,J.K., (1972), *On the automatic scaling of matrices for Gaussian elimination,* J. Inst. Maths. Applics., vol 10, pages 118–124.

[361] CUTHILL,E., (1972), *Several strategies for reducing the bandwidth of matrices,* in Sparse matrices and their applications, ROSE,D.J. and WILLOUGHBY,R.A. Eds. Plenum Press.

[362] CUTHILL,E., McKEE,J., (1969), *Reducing the bandwidth of sparse symmetric matrices,* in Proceedings of 24th Nat. Conf. Assoc. Comp. Mach.. ACM Publications.

[363] DAY,D., (1996), *Semi–duality in the two–sided Lanczos algorithm,* Report Sandia National Laboratories.

[364] D'AZEVEDO,E.F., EIJKHOUT,V., ROMINE,C., (1992), *Reducing communication costs in the conjugate gradient algorithm on distributed memory multiprocessors,* submitted to SIAM J. Sci. Comput..

[365] D'AZEVEDO,E.F., FORSYTH,P.A., TANG,W-P., (1990), *Towards a cost effective ILU preconditioner with high level fill,* Report CS-90-50,Dept. of Computer Science, University of Waterloo.

[366] D'AZEVEDO,E.F., FORSYTH,P.A., TANG,W-P., (1990), *An automatic ordering method for incomplete factorization iterative solvers,* Report CS-90-32, Computer Science Dept., University of Waterloo.

[367] D'AZEVEDO,E.F., FORSYTH,P.A., TANG,W-P., (1992), *Ordering methods for preconditioned conjugate gradient methods applied to unstructured grid problems,* SIAM J. Matrix Anal. Appl., vol 92, pages 944–961.

[368] D'SYLVA,E., MILES,G.A., (1963), *The SSOR iteration scheme for equations with σ_1 ordering,* Comput. J., vol 6, pages 366–367.

[369] DAHLQUIST,G., (1983), *On matrix majorants and minorants with applications to differential equations,* Linear Algebra and its Appl., vol 52, pages 199–216.

[370] DAHLQUIST,G., (1993), *A "multigrid" extension of the FFT for the numerical inversion of Fourier and Laplace transforms,* BIT, vol 33, pages 85–112.

[371] DAHLQUIST,G., BJÖRCK,A., (1974), *Numerical methods,* Prentice–Hall.

[372] DAHLQUIST,G., EISENSTAT,S.C., GOLUB,G.H., (1972), *Bounds for the error of linear systems of equations using the theory of moments,* J. Math. Anal. Appl., vol 37, no 1, pages 151–166.

[373] DAHMEN,W., KUNOTH,A., (1992), *Multilevel preconditioning,* Numer. Math., vol 63, pages 315–344.

[374] DAUBECHIES,I., (1992), *Ten Lectures on Wavelets,* in CBMS–NSF Regional Conference Series in Applied Mathematics, 61. SIAM.

[375] DAUTRAY,R., LIONS,J.L., (1985), *Analyse mathématique et calcul numérique pour les sciences et techniques,* Masson.

[376] DAVE,A.K., DUFF,I.S., (1987), *Sparse matrix calculations on the Cray 2,* Parallel Comp., vol 5, pages 55–64.

[377] DAVIS,P.J., RABINOWITZ,P., (1984), *Methods of Numerical Integration,* Academic Press.

[378] DAVIS,T.A., (1992), *Performance of an unsymmetric pattern multifrontal method for sparse LU factorization,* Report TR–92–014, University of Florida.

[379] DAVIS,T.A., (1993), *Users' guide for the unsymmetric pattern multifrontal package,* Report TR–93–020, University of Florida.

[380] DAVIS,T.A., (1994), *A combined unifrontal/multifrontal method for unsymmetric sparse matrices,* Report TR–94–005, University of Florida.

[381] DAVIS,T.A., DUFF,I.S., (1997), *An unsymmetric pattern multifrontal method for sparse LU factorization,* SIAM J. Matrix Anal. Appl., vol 18, no 1, pages 140–158.

[382] DAVIS,T.A., YEW,P.C., (1990), *A nondeterministic parallel algorithm for unsymmetric sparse LU factorization,* SIAM J. Matrix Anal. Appl., vol 11, pages 383–402.

[383] DAWSON,C., DU,Q., DUPONT,T.F., (1991), *A finite difference domain decomposition algorithm for numerical solution of the heat equation,* Math. Comp., vol 57, no 195, pages 63–71.

[384] DAWSON,C.N., DUPONT,T.F., (1990), *Explicit/implicit conservative Galerkin domain decomposition procedures for parabolic problems,* TR90-26, Dept. of Mathematical Sciences, Rice University.

[385] DAWSON,C.N., DUPONT,T.F., (1994), *Explicit/implicit, conservative domain decomposition procedures for parabolic problems based on block-centered finite differences,* SIAM J. Numer. Anal., vol 31, pages 1045–1061.

[386] DAYDE,M.J., DUFF,I.S., (1989), *Use of Level 3 BLAS in LU factorization on the Cray 2, the ETA–10P and the IBM 3090/VF,* Int. J. Supercomputer Appl., vol 3, no 2, pages 40–70.

[387] DAYDE,M.J., DUFF,I.S., PETITET,A., (1993), *A parallel block implementation of level 3 BLAS for MIMD vector processors,* Report RAL–93–037, Rutherford Appleton Laboratory.

[388] De BOOR,C., RICE,J.R., (1982), *Extremal polynomials with application to Richardson iterations for indefinite linear systems,* SIAM J. Sci. Comput., vol 3, no 1, pages 47–57.

[389] De PILLIS,J., (1978), *Faster convergence for iterative solutions to systems via three–part splittings,* SIAM J. Numer. Anal., vol 15, no 5, pages 888–911.

[390] De PILLIS,J., NEUMANN,M., (1981), *Iterative methods with k–part splittings,* IMA J. Num. Anal., vol 1, pages 65–79.

[391] De STURLER,E., (1993), *Nested Krylov methods based on GCR,* Report 93-50, Dept. of Mathematics and Informatics, Delft University of Technology.

[392] De STURLER,E., (1994), *A performance model for Krylov subspace methods on mesh-based parallel computers,* Report CSCS-TR-94-05, Swiss Scientific Computing Center , ETH.

[393] De STURLER,E., (1994), *IBLU preconditioners for massively parallel computers,* in Domain Decomposition Methods in Scientific and Engineering Computing: Proceedings of the Seventh International Conference on Domain Decomposition, KEYES,D.E. and XU,J. Eds. AMS, pages 395–400.

[394] De STURLER,E., FOKKEMA,D.R., (1994), *Nested Krylov methods and preserving the orthogonality,* Report, Faculty of Mathematics, Delft University of Technology.

[395] De STURLER,E., VAN DER VORST,H.A., (1993), *Reducing the effect of global communication in GMRES(m) and CG on parallel distributed memory computers,* Report 832, Dept. of Mathematics, Utrecht University.

[396] DEMKO,S., (1977), *Inverses of band matrices and local convergence of spline projections,* SIAM J. Numer. Anal., vol 14, pages 616–619.

[397] DEMKO,S., MOSS,W.F., SMITH,P.W., (1984), *Decay rates for inverses of band matrices,* Math. Comp., vol 43, no 168, pages 491–499.

[398] DEMMEL,J.W., (1982), *The condition number of equivalence transformations that block diagonalize matrix pencils,* in Matrix pencils, KAGSTROM,B. and RUHE,A. Eds, Lecture Notes in Mathematics 973. Springer.

[399] DEMMEL,J.W., (1984), *Underflow and the reliability of numerical software,* SIAM J. Sci. Comput., vol 5, pages 887–919.

[400] DEMMEL,J.W, (1987), *On condition numbers and the distance to the nearest ill–posed problem,* Numer. Math., vol 51, no 3, pages 251–289.

[401] DEMMEL,J.W., (1989), *On floating point errors in Cholesky,* Lapack Working note 14, University of Tennessee.

[402] DEMMEL,J.W., (1992), *Open problems in numerical linear algebra,* Report CS–92–164 Univ. of Tennessee.

[403] DEMMEL,J.W., (1992), *Trading off parallelism and numerical stability,* Report CS–92–179 Univ. of Tennessee.

[404] DEMMEL,J.W., (1992), *The componentwise distance to the nearest singular matrix,* SIAM J. Matrix Anal. Appl., vol 13, no 1, pages 10–19.

[405] DEMMEL,J.W., (1997), *Applied numerical linear algebra,* SIAM.

[406] DEMMEL,J.W., GILBERT,J.R., LI,X.S., (1997), *An asynchronous parallel supernodal algorithm for sparse Gaussian elimination,* Report Xerox Parc, Palo Alto.

[407] DEMMEL,J.W., HEATH,M.T., VAN DER VORST,H.A., (1993), *Parallel Numerical linear algebra,* in Acta Numerica. pages 111–197.

[408] DEMMEL,J.W., HIGHAM,N.J., SCHREIBER,R.S., (1995), *Stability of block LU factorization,* Numer. Lin. Alg. with Appl., vol 2, no 2, pages 173–190.

[409] DENDY,J.E., (1982), *Black box multigrid,* J. Comp. Physics, vol 48, pages 366–386.

[410] DENDY,J.E., HYMAN,J.M., (1980), *Multigrid and ICCG for problems with interfaces,* Report LA-UR-80-2127, Los Alamos Scientific Laboratory.

[411] DENDY,J.E., HYMAN,J.M., (1981), *Multi–grid and ICCG for problems with interfaces,* in Elliptic Problem Solvers, SCHULTZ,M.H. Ed. Academic Press, pages 247–253.

[412] DENDY,J.E., TAZARTES,C.C., (1995), *Grandchild of the frequency decomposition multigrid method,* SIAM J. Sci. Comput., vol 16, pages 307–319.

[413] DENNIS,J.E.,Jr., TURNER,K., (1987), *Generalized conjugate directions,* Linear Algebra and its Appl., vol 88, pages 187–209.

[414] DESHPANDE,V.R., GROTE,M.J., MESSMER,P., SAWYER,W.B., (1996), *Parallel sparse approximate inverse preconditioner,* Report TR-96-14, Swiss Center for Scientific Computing.

[415] DESPREZ,F., TOURANCHEAU,B., DONGARRA,J.J., (1994), *Performance complexity of LU factorization with efficient pipelining and overlap on a multiprocessor,* Report University of Tennessee 1994.

[416] DIAMOND,M.A., FERREIRA,D.L.V., (1976), *On a cyclic reduction method for the solution of Poisson's equation,* SIAM J. Numer. Anal., vol 13, no 1, pages 54–70.

[417] DINH,Q.V., GLOWINSKI,R., HE,J., KWOCK,V., PAN,T.W. , PERIAUX,J., (1992), *Lagrange multiplier approach to fictitious domain methods: application to fluid dynamics and electro–magnetics,* in Fifth International Symposium on Domain Decomposition Methods for Partial Differential Equations, KEYES,D.E.,

CHAN,T.F., MEURANT,G., SCROGGS,J.S. and VOIGT,R.G. Eds. SIAM, pages 151–194.

[418] DINH,Q.V., GLOWINSKI,R., PERIAUX,J., (1980), *Application of domain decomposition technique to the numerical solution of the Navier–Stokes equations,* Numer. Meth. for Engng., vol 1, pages 383–404.

[419] DINH,Q.V., GLOWINSKI,R., PERIAUX,J., TERRASSON,G., (1988), *On the coupling of viscous and inviscid models for incompressible fluid flows via domain decomposition,* in First International Symposium on Domain Decomposition Methods for Partial Differential Equations, GLOWINSKI,R., GOLUB,G.H., MEURANT,G. and PERIAUX,J. Eds. SIAM, pages 350–369.

[420] DOI,S., (1990), *Ordering effects for convergence of some preconditioned iterative methods,* SIAM J. Numer. Anal.,

[421] DOI,S., (1991), *On parallelism and convergence of incomplete LU factorizations,* Applied Num. Math., vol 7, pages 417–436.

[422] DOI,S., (1991), *A Gustafsson–type modification for parallel ordered incomplete LU factorizations,* in Advances in Numerical Methods for Large Sets of Linear Systems, NODERA,T. Ed.. Keio University

[423] DOI,S., HOSHI,A., (1991), *Large numbered multicolor MILU preconditioning on SX-3/14,* in Proc. Numerical Analysis Symposium. Information Processing Society of Japan.

[424] DOI,S., LICHNEWSKY,A., (1990), *Some parallel and vector implementations of preconditioned iterative methods on Cray-2,* Int. J. High Speed Comp., vol 2, pages 143–179.

[425] DOI,S., LICHNEWSKY,A., (1990), *An ordering theory for incomplete LU factorizations on regular grids,* submitted to SIAM J. Sci. Comput.,

[426] DONATO,J.M., (1991), *Iterative methods for scalar and coupled systems of elliptic equations,* Ph.D. Thesis, Dept. of Mathematics, UCLA.

[427] DONATO,J.M., CHAN,T.F., (1992), *Fourier analysis of incomplete factorization preconditioners for 3D anisotropic problems,* SIAM J. Sci. Comput., vol 13, no 1, pages 319–338.

[428] DONGARRA,J., DEMMEL,J.W., (1991), *LAPACK: a portable high–performance numerical library for linear algebra,* Supercomputer, vol 46, pages 33–38.

[429] DONGARRA,J., DU CROZ,J., HAMMARLING,S., DUFF,I.S., (1990), *A set of level 3 basic linear algebra subprograms,* ACM Trans. Math. Soft., vol 16, no 1, pages 1–17.

[430] DONGARRA,J., DU CROZ,J., HAMMARLING,S., HANSON,R., (1988), *An extended set of Fortran basic linear algebra subprograms,* ACM Trans. Math. Soft., vol 14, no 1, pages 1–17.

[431] DONGARRA,J., DUFF,I.S., GAFFNEY,P., McKEE,S., (1989), *Vector and parallel computing, issues in applied research and development,* Ellis Horwood.

[432] DONGARRA,J., DUFF,I.S., SORENSEN,D.C., VAN DER VORST,H.A., (1991), *Solving linear systems on vector and shared memory computers,* SIAM.

[433] DONGARRA,J., GUSTAVSON,F.G., KARP,A., (1984), *Implementing linear algebra algorithms for dense matrices on a vector pipeline machine,* SIAM Review, vol 26, pages 91–112.

[434] DONGARRA,J., HEWITT,T., (1986), *Implementing dense linear algebra algorithms using multitasking on the Cray X–MP–4 (or approaching the Gigaflop),* SIAM J. Sci. Comput., vol 7, no 1, pages 347–305.

[435] DONGARRA,J., SAMEH,A.H., (1984), *On some parallel banded system solvers,* Parallel Comp., vol 1, pages 223–235.

[436] DONGARRA,J., WALKER,D., (1993), *The design of linear algebra libraries for high performance computers,* Report University of Tennessee.

[437] DONGARRA,J.J., EISENSTAT,S.C., (1983), *Squeezing the most out of an algorithm in CRAY Fortran,* Report ANL/MCS-TM-9, Argonne National Laboratory.

[438] DORR,F.W., (1970), *The direct solution of the discrete Poisson equation on a rectangle,* SIAM Review, vol 12, no 2, pages 248–263.

[439] DORR,F.W., (1975), *The direct solution of the discrete Poisson equation in $O(n^2)$ operations,* SIAM Review, vol 17, no 3, pages 412–415.

[440] DOUGLAS,C.C., (1984), *Multi–grid algorithms with applications to elliptic boundary value problems,* SIAM J. Numer. Anal., vol 21, pages 236–254.

[441] DOUGLAS,C.C., (1990), *A variation of the Schwarz alternating method: the domain decomposition reduction method,* in Third International Symposium on Domain Decomposition Methods for Partial Differential Equations, CHAN,T.F., GLOWINSKI,R., PERIAUX,J. and WIDLUND,O.B. Eds. SIAM, pages 191–201.

[442] DOUGLAS,C.C., (1994), *Some remarks on completely vectorizing point Gauss–Seidel while using the natural ordering,* Advances Comput. Math., vol 2, pages 215–222.

[443] DOUGLAS,C.C., (1996), *A review of numerous parallel multigrid methods,* in Applications on Advanced Architecture Computers, ASTFALK,G. Ed. SIAM, pages 187–202.

[444] DOUGLAS,C.C., DOUGLAS,J., (1993), *A unified convergence theory for abstract multigrid or multilevel algorithms, serial and parallel,* SIAM J. Numer. Anal., vol 30, pages 136–158.

[445] DOUGLAS,C.C., DOUGLAS,J., FYFE,D.E., (1994), *A multigrid unified theory for non-nested grids and/or quadrature,* E. W. J. Numer. Math., vol 2, pages 285–294.

[446] DOUGLAS,C.C., MALHOTRA,M., SCHULTZ,M.H., (1997), *Parallel multigrid with ADI–like smoothers in two dimensions,* Report available in MGNET.

[447] DRAUX,A., (1983), *Polynômes orthogonaux formels – applications,* Lecture Notes in Mathematics, vol 974, Springer–Verlag.

[448] DRISCOLL,T.A., TOH,K-C., TREFETHEN,L.N., (1996), *Matrix iterations: the six gaps between potential theory and convergence,* Report Cornell University.

[449] DRMAC,Z., OMLADIC,M., VESELIC,K., (1994), *On the perturbation of the Cholesky factorization,* SIAM J. Matrix Anal. Appl., vol 15, no 4, pages 1319–1332.

[450] DRYJA,M., (1983), *A finite element capacitance matrix method for the elliptic problem,* SIAM J. Numer. Anal., vol 20, no 4, pages 671–680.

[451] DRYJA,M., PROSKUROWSKI,W., (1985), *Capacitance matrix method using strips with alternating Neumann and Dirichlet boundary conditions,* Appl. Numer. Math., vol 1, pages 285–298.

[452] DRYJA,M., PROSKUROWSKI,W., WIDLUND,O.B., (1986), *Numerical experiments and implementation of a domain decomposition method with cross points for the model problem,* CRI-86-37, Dept. of Computer Science, University of Southern California.

[453] DRYJA,M., PROSKUROWSKI,W., WIDLUND,O.B., (1986), *Method of domain decomposition with cross points for elliptic finite element problems,* in Proceedings of the International Symposium on Optimal Algorithms, Blaqoevgrad, Bulgaria. pages 97–111.

[454] DRYJA,M., SMITH,B.F., WIDLUND,O.B., (1994), *Schwarz analysis of iterative substructuring algorithms for elliptic problems in three dimensions,* SIAM J. Numer. Anal., vol 31, no 6, pages 1662–1694.

[455] DRYJA,M., WIDLUND,O.B., (1990), *Towards a unified theory of domain decomposition algorithms for elliptic problems,* in Third International Symposium on Domain Decomposition Methods for Partial Differential Equations, CHAN,T.F., GLOWINSKI,R., PERIAUX,J., and WIDLUND,O.B. Eds. SIAM, pages 3–21.

[456] DRYJA,M., WIDLUND,O.B., (1992), *Additive Schwarz methods for elliptic finite element problems in three dimensions,* in Fifth International Symposium on Domain Decomposition Methods for Partial Differential Equations, KEYES,D.E., CHAN,T.F., MEURANT,G., SCROGGS,J.S. and VOIGT,R.G. Eds. SIAM, pages 3–18.

[457] DRYJA,M., WIDLUND,O.B., (1994), *Domain decomposition algorithms with small overlap,* SIAM J. Sci. Comput., vol 15, pages 604–620.

[458] DU CROZ,J.J., HIGHAM,N.J., (1992), *Stability of methods for matrix inversion,* J. Inst. Maths. Applics., vol 12, pages 1–19.

[459] DUBOIS,P.F., GREENBAUM,A., RODRIGUE,G.H., (1979), *Approximating the inverse of a matrix for use in iterative algorithms on vector processors,* Computing, vol 22, pages 257–268.

[460] DUFF,I.S., (1974), *On the number of nonzeros added when Gaussian elimination is performed on sparse random matrices,* Math. Comp., vol 28, no 125, pages 219–230.

[461] DUFF,I.S., (1977), *A survey of sparse matrix research,* Proc. of IEEE, vol 65, no 4, pages 500–535.

[462] DUFF,I.S., (1981), *Sparse matrices and their use,* Academic Press.

[463] DUFF,I.S., (1981), *MA32, a package for solving sparse unsymmetric systems using the frontal method,* Report AERE R10079, Harwell Laboratory.

[464] DUFF,I.S., (1983), *The solution of sparse linear equations on the Cray 1,* Report CSS125, Harwell Laboratory.

[465] DUFF,I.S., (1983), *The solution of nearly symmetric sparse linear systems,* Report CSS 150, AERE Harwell.

[466] DUFF,I.S., (1984), *Design features of a frontal code for solving sparse unsymmetric linear systems out–of–core,* SIAM J. Sci. Comput., vol 5, pages 270–280.

[467] DUFF,I.S., (1984), *Direct methods for solving sparse systems of linear equations,* SIAM J. Sci. Comput., vol 5, no 3, pages 605–619.

[468] DUFF,I.S., (1984), *Data structures, algorithms and software for sparse matrices,* Report CSS 158, Harwell Laboratory.

[469] DUFF,I.S., (1986), *Parallel implementation of multifrontal schemes,* Parallel Comp., vol 3, pages 193–204.

[470] DUFF,I.S., (1986), *The use of vector and parallel computers in the solution of large sparse linear equations,* Report AERE R12393, Harwell Laboratory.

[471] DUFF,I.S., (1986), *The influence of vector and parallel processors on numerical analysis,* Report AERE R12329, Harwell Laboratory.

[472] DUFF,I.S., (1988), *Multiprocessing a sparse matrix code on the Alliant FX/8,* Report CSS 210, Harwell Laboratory.

[473] DUFF,I.S., (1993), *The solution of augmented systems,* Report RAL–93–084 Rutherford Appleton Laboratory.

[474] DUFF,I.S., ERISMAN,A.M., REID,J.K., (1976), *On George's nested dissection method,* SIAM J. Numer. Anal., vol 13, no 5, pages 686–695.

[475] DUFF,I.S., ERISMAN,A.M., REID,J.K., (1986), *Direct methods for sparse matrices,* Oxford University Press.

[476] DUFF,I.S., GOULD,N.I., LESCRENIER,M., REID,J.K., (1987), *The multifrontal method in a parallel environment,* Report CSS 211, Harwell Laboratory.

[477] DUFF,I.S., GOULD,N.I., REID,J.K., SCOTT,J.A., TURNER,K., (1990), *The factorization of sparse symmetric indefinite matrices,* Report RAL–90–084 Rutherford Appleton Laboratory.

[478] DUFF,I.S., JOHNSON,S.L., (1989), *Node ordering and concurrency in structurally symmetric sparse problems,* in Parallel supercomputing: methods, algorithms and applications, G. Carey Ed..

[479] DUFF,I.S., LAMINIE,J., LICHNEWSKY,A., THOMASSET,F., (1987), *An experiment with arithmetic precision in linear algebra computations,* Int. J. Num. Meth. in Fluids, vol 7, pages 1077–1092.

[480] DUFF,I.S., MEURANT,G., (1989), *The effect of ordering on preconditioned conjugate gradients,* BIT, vol 29, pages 635–657.

[481] DUFF,I.S., REID,J.K., (1974), *A comparison of sparsity orderings for obtaining a pivotal sequence in Gaussian elimination,* J. Inst. Maths. Applics., vol 14, pages 281–291.

[482] DUFF,I.S., REID,J.K., (1979), *Some design features of a sparse matrix code,* ACM Trans. Math. Soft., vol 5, no 1, pages 18–35.

[483] DUFF,I.S., REID,J.K., (1983), *The multifrontal solution of indefinite sparse symmetric linear systems,* ACM Trans. Math. Soft., vol 9, pages 302–325.

[484] DUFF,I.S., REID,J.K., (1984), *The multifrontal solution of unsymmetric sets of linear systems,* SIAM J. Sci. Comput., vol 5, no 3, pages 633–641.

[485] DUFF,I.S., REID,J.K., (1993), *MA48, a Fortran code for direct solution of sparse unsymmetric linear systems of equations,* Report RAL–93–072 Rutherford Appleton Laboratory.

[486] DUFF,I.S., REID,J.K., (1995), *MA47, a Fortran code for direct solution of indefinite sparse symmetric linear systems,* Report RAL–95–001 Rutherford Appleton Laboratory.

[487] DUFF,I.S., REID,J.K., MUNSKGAARD,N., NIELSEN,B., (1979), *Direct solution of sets of linear equations whose matrix is sparse, symmetric and indefinite,* J. Inst. Maths. Applics., vol 23, pages 235–250.

[488] DUFF,I.S., REID,J.K., SCOTT,J.A., (1989), *The use of profile reduction algorithms with a frontal code,* Int. J. Num. Meth. Eng., vol 28, pages 2555–2568.

[489] DUFF,I.S., SCOTT,J.A., (1993), *MA42– A new frontal code for solving sparse unsymmetric systems,* Report RAL–93–064, Rutherford Appleton Laboratory.

[490] DUFF,I.S., SCOTT,J.A., (1994), *The use of multiple fronts in Gaussian elimination,* Report RAL–94–040, Rutherford Appleton Laboratory.

[491] DUFF,I.S., STEWART,G.W., (1979), *Sparse matrix proceedings 1978,* SIAM.

[492] DUPONT,T., (1968), *A factorization procedure for the solution of elliptic difference equations,* SIAM J. Numer. Anal., vol 5, no 4, pages 753–782.

[493] DUPONT,T., STONE,H.L., RACHFORD,H.H., (1970), *Factorization techniques for elliptic difference equations,* in SIAM-AMS Proceedings, vol 2. American Mathematical Society, pages 168–174.

[494] DUPONT,T.F., KENDALL,R.P., RACHFORD,H.H., (1968), *An approximate factorization procedure for solving self–adjoint elliptic difference equations,* SIAM J. Numer. Anal., vol 5, pages 559–573.

[495] EDELMAN,A., (1993), *Large dense numerical linear algebra in 1993, the parallel computing influence,* Int. J. Supercomputer Appl., vol 7, no 2, pages 113–128.

[496] EDELMAN,A., McCORQUDALE,P., TOLEDO,S., (1997), *The future Fast Fourier Transform?,* Proceedings of the Eigth SIAM Conf. on Parallel Processing.

[497] EDELMAN,A., OHLROCH,M., (1991), *Editor's Note,* SIAM J. Matrix Anal. Appl., vol 12,

[498] EHRLICH,L.W., (1964), *The block symmetric successive overrelaxation method,* J. Soc. Indust. Appl. Math., vol 12, no 4, pages 807–826.

[499] EHRLICH,L.W., (1971), *Solving the biharmonic equations as coupled finite difference equations,* SIAM J. Numer. Anal., vol 8, no 2, pages 278–287.

[500] EHRLICH,L.W., (1972), *Coupled harmonic equations, SOR and Chebyshev acceleration,* Math. Comp., vol 26, no 118, pages 335–343.

[501] EHRLICH,L.W., (1979), *Solving the biharmonic equation on irregular regions,* ACM Trans. Math. Soft., vol 5, no 3, pages 251–258.

[502] EHRLICH,L.W., GUPTA,M.M., (1975), *Some difference schemes for the biharmonic equation,* SIAM J. Numer. Anal., vol 12, no 5, pages 773–790.

[503] EIERMANN,M., NIETHAMMER,W., (1983), *On the construction of semiiterative methods,* SIAM J. Numer. Anal., vol 20, no 6, pages 1153–1160.

[504] EIERMANN,M., NIETHAMMER,W., VARGA,R.S., (1985), *A study of semiiterative methods for nonsymmetric systems of linear equations,* Numer. Math., vol 47, pages 505-533.

[505] EIJKHOUT,V., (1988), *A general formulation for incomplete blockwise factorizations,* Comm. Appl. Num. Methods, vol 4, pages 161–164.

[506] EIJKHOUT,V., (1991), *Analysis of parallel incomplete point factorizations,* Linear Algebra and its Appl., vol 154, pages 723–740.

[507] EIJKHOUT,V., (1994), *Computational variants of the CGS and BiCGSTAB meth-*

ods, Report Dept. of Computer Science, University of Tennessee.

[508] EIJKHOUT,V., POLMAN,B., (1988), *Decay rates of inverses of banded M-matrices that are near to Toeplitz matrices,* Linear Algebra and its Appl., vol 109, pages 247–277.

[509] EIJKHOUT,V., VASSILEVSKI,P., (1986), *Positive definiteness aspects of vectorizable preconditioners,* Parallel Comp., vol 10, pages 93–100.

[510] EISENSTAT,S.C., (1981), *Efficient implementation of a class of preconditioned conjugate gradient methods,* SIAM J. Sci. Comput., vol 2, no 1, pages 1–4.

[511] EISENSTAT,S.C., (1983), *A note on the generalized conjugate gradient method,* SIAM J. Numer. Anal., vol 20, no 2, pages 358–361.

[512] EISENSTAT,S.C., ELMAN,H.C., SCHULTZ,M.H., (1983), *Variational iterative methods for nonsymmetric systems of linear equations,* SIAM J. Numer. Anal., vol 20, pages 345–357.

[513] EISENSTAT,S.C., ELMAN,H.C., SCHULTZ,M.H., (1985), *Block preconditioned conjugate gradient like methods for numerical reservoir simulation,* Paper SPE 13534, SPE Reservoir Simulation Symposium Proceedings. pages 397–405.

[514] EISENSTAT,S.C., HEATH,M.T., HENKEL,C.S., ROMINE,C.H., (1988), *Modified cyclic algorithms for solving triangular systems on distributed memory multiprocessors,* SIAM J. Sci. Comput., vol 9, no 3, pages 589–600.

[515] EISENSTAT,S.C., LEWIS,J.W., SCHULTZ,M.H., (1982), *Efficient implementation of sparse symmetric Gaussian elimination,* in Advances in Computer Methods for Partial Differential Equations, R. VICHNEVETSKY Ed.. pages 33–39.

[516] EISENSTAT,S.C., SCHULTZ,M.H., SHERMAN,A.H., (1975), *Optimal block diagonal scaling of block 2–cyclic matrices,* Linear Algebra and its Appl., vol 44, pages 181–186.

[517] EISENSTAT,S.C., SCHULTZ,M.H., SHERMAN,A.H., (1975), *Application of sparse matrix methods to partial differential equations,* in Advances in Computer Methods for Partial Differential Equations, R. Vichnevetsky Ed.. pages 40–45.

[518] EISENSTAT,S.C., SCHULTZ,M.H., SHERMAN,A.H., (1981), *Algorithms and data structures for sparse symmetric Gaussian elimination,* SIAM J. Sci. Comput., vol 2, no 2, pages 225–237.

[519] ELFVING,T., (1980), *Block–iterative methods for consistent and inconsistent linear equations,* Numer. Math., vol 35, pages 1–12.

[520] ELMAN,H.C., (1986), *A stability analysis of incomplete LU factorizations,* Math. Comp., vol 47, no 175, pages 191–217.

[521] ELMAN,H.C., (1989), *Relaxed and stabilized incomplete factorizations for non self adjoint linear systems,* Report CS-TR-2169, Computer Science Dept., University

of Maryland.

[522] ELMAN,H.C., (1989), *Approximate Schur complement preconditioners on serial and parallel computers,* SIAM J. Sci. Comput., vol 10, pages 581–605.

[523] ELMAN,H.C., (1994), *Multigrid and Krylov subspace methods for the discrete Stokes equations,* Report CS TR-3302, Computer Science Dept., University of Maryland.

[524] ELMAN,H.C., CHERNESKY,M.P., (1992), *Ordering effects on relaxation methods applied to the discrete one–dimensional convection–diffusion equation,* CS-TR-2875, Computer Science Dept., University of Maryland.

[525] ELMAN,H.C., GOLUB,G.H., (1990), *Iterative methods for cyclically reduced non–self–adjoint linear systems,* Math. Comp., vol 54, no 190, pages 671–700.

[526] ELMAN,H.C., GOLUB,G.H., (1990), *Line iterative methods for cyclically reduced discrete convection–diffusion problems,* CS-TR-2403, Computer Science Dept., University of Maryland.

[527] ELMAN,H.C., GOLUB,G.H., (1990), *Iterative methods for cyclically reduced non-self-adjoint linear systems II,* Math. Comp., vol 56, pages 215–242.

[528] ELMAN,H.C., GOLUB,G.H., (1993), *Inexact and preconditioned Uzawa algorithms for saddle point problems,* Report CS-TR-.

[529] ELMAN,H.C., GOLUB,G.H., STARKE,G., (1992), *On the convergence of line iterative methods for cyclically reduced non–symmetrizable linear systems,* CS-TR-2860, Computer Science Dept., University of Maryland.

[530] ELMAN,H.C., GUO,X., (1991), *Performance enhancements and parallel algorithms for two multilevel preconditioners,* Report CS-TR-91-143, Computer Science Dept., University of Maryland.

[531] ELMAN,H.C., SCHULTZ,M.H., (1986), *Preconditioning by fast direct methods for non self adjoint nonseparable elliptic equations,* SIAM J. Numer. Anal., vol 23, no 1, pages 44–57.

[532] ELMAN,H.C., SILVESTER,D., (1994), *Fast nonsymmetric iterations and preconditioning for Navier-Stokes equations,* Report CS TR-3283, Computer Science Sept., University of Maryland.

[533] ELSNER,L., (1984), *A note on optimal block–scaling of matrices,* Numer. Math., vol 44, pages 127–128.

[534] ERHEL,J., (1990), *Sparse matrix multiplication on vector computers,* Int. J. High Speed Comp., vol 2, no 2, pages 101–116.

[535] ERHEL,J., (1995), *A parallel GMRES version for general sparse matrices,* ETNA, vol 3, pages 160–176.

[536] ERHEL,J., BURRAGE,K., POHL,B., (1996), *Restarted GMRES preconditioned by deflation,* J. Comp. Appl. Math., vol 69, pages 303–318.

[537] ERHEL,J., GUYOMARC'H,F., (1997), *An augmented subspace conjugate gradient,* Rapport INRIA 1136, IRISA, France.

[538] ERISMAN,A.M., GRIMES,R.G., LEWIS,J.G., POOLE,W.G., SIMON,H.D., (19-87), *Evaluation of orderings for unsymmetric sparse matrices,* SIAM J. Sci. Comput., vol 8, no 4, pages 600–624.

[539] EVANS,D.J., (1967), *The use of preconditioning in iterative method for solving linear equations with symmetric positive definite matrices,* J. Inst. Maths. Applics., vol 4, pages 295–314.

[540] EVANS,D.J., (1985), *Sparsity and its applications,* Cambridge University Press.

[541] FABER,V., JOUBERT,W., KNILL,E., MANTEUFFEL,T., (1996), *Minimal residual method stronger than polynomial preconditioning,* SIAM J. Matrix Anal. Appl., vol 17, no 4, pages 707–729.

[542] FABER,V., MANTEUFFEL,T., (1984), *Necessary and sufficient condition for the existence of a conjugate gradient method,* SIAM J. Numer. Anal., vol 21, no 2, pages 352–362.

[543] FABER,V., MANTEUFFEL,T., (1987), *Orthogonal error methods,* SIAM J. Numer. Anal., vol 24, pages 170–187.

[544] FADDEEV,D.K., (1981), *Properties of the inverse of a Hessenberg matrix,* in Numerical methods and computational issues, Vol 5, ILIN,V.P. and KUBLANOVSKAYA,V.N. Eds.

[545] FADDEEV,D.K., FADDEEVA,V.N., (1963), *Computational methods of linear algebra,* W.H. Freeman and company.

[546] FARHAT,C., (1988), *A simple and efficient automatic FEM domain decomposer,* Computer and structures, vol 28, no 5, pages 579–602.

[547] FARHAT,C., (1990), *Redesigning the skyline solver for parallel/vector supercomputers,* Int. J. High Speed Comp., vol 2, no 3, pages 223–238.

[548] FARHAT,C., LANTERI,S., SIMON,H.D., (1995), *TOP/DOMDEC a software tool for mesh partitioning and parallel processing,* Comput. Syst. Eng., vol 6, pages 13–26.

[549] FEDORENKO,R.P., (1962), *A relaxation method for solving elliptic difference equations,* U.S.S.R. Comput. Math. and Math. Phys., vol 1, pages 1092–1096.

[550] FERGUSON, JR., W.E., (1986), *The rate of convergence of a class of block Jacobi schemes,* SIAM J. Numer. Anal., vol 23, no 2, pages 297–303.

[551] FIDUCCIA,C., MATTHEYSES,R., (1982), *A linear time heuristic for improving*

network partitions, in ACM-IEEE 19th Design Automation Conference. IEEE Press, pages 175–181.

[552] FIEDLER,M., (1975), *A property of eigenvectors of nonnegative symmetric matrices and its application to graph theory,* Czechoslovak Math. J., vol 25, pages 619–633.

[553] FIEDLER,M., (1986), *Special matrices and their applications in numerical mathematics,* Martinus Nijhoff.

[554] FIEDLER,M., PTAK,V., (1962), *On matrices with non-positive off-diagonal elements and positive principal minors,* Czechoslovak Math. J., vol 12, pages 123–128.

[555] FINOGENOV,S.A., KUZNETSOV,Y.A., (1988), *Two-stage fictitious components method for solving the Dirichlet boundary value problem,* Sov. J. Numer. Anal. Math. Modelling, vol 3, no 4, pages 301–323.

[556] FISCHER,B., (1996), *Polynomial based iteration methods for symmetric linear systems,* Wiley Teubner.

[557] FISCHER,B., FREUND,R.W., (1994), *An inner product-free conjugate gradient-like for Hermitian positive definite systems,* in Proceedings of the Cornelius Lanczos International Centenary Conference, BROWN,J., CHU,M., ELLISON,D. and PLEMMONS,R. Eds. SIAM, pages 288–290.

[558] FISCHER,B., GOLUB,G.H., (1991), *On generating polynomials which are orthogonal over several intervals,* Math. Comp., vol 56, no 194, pages 711–730.

[559] FISCHER,B., GOLUB,G.H., (1992), *On the error computation for polynomial based iteration methods,* Report NA 92–21,Computer Science Dept., Stanford University.

[560] FISCHER,B., REICHEL,L., (1988), *A stable Richardson iteration method for complex linear systems,* Numer. Math., vol 54, pages 225–242.

[561] FISCHER,C.F., USMANI,R.A., (1969), *Properties of some tridiagonal matrices and their application to boundary value problems,* SIAM J. Numer. Anal., vol 6, no 1, pages 127–142.

[562] FISCHER,D., GOLUB,G.H., HALD,O., LEIVA,C., WIDLUND,O., (1974), *On Fourier-Toeplitz methods for separable elliptic equations,* Math. Comp., vol 28, no 126, pages 349–368.

[563] FIX, G.J., LARSEN, K., (1971), *On the convergence of SOR iterations for finite element approximations to elliptic boundary value problems,* SIAM J. Numer. Anal., vol 8, no 3, pages 536–547.

[564] FLETCHER,R., (1975), *Conjugate gradient methods for indefinite systems,* in Proceedings of the Dundee Conference on Numerical Analysis, WATSON,G.A. Ed. Springer, pages 73–89.

[565] FLETCHER,R., (1976), *Factorizing symmetric indefinite matrices,* Linear Algebra and its Appl., vol 14, pages 257–272.

[566] FOKKEMA,D.R., SLEIJPEN,G.L., VAN DER VORST,H.A., (1996), *Generalized conjugate gradient squared,* J. Comp. Appl. Math., vol 71, pages 125–146.

[567] FONG,K., JORDAN,T.L., (1977), *Some linear algebraic algorithms and their performance on CRAY–1,* Report LA–6774, Los Alamos Scientific Laboratory.

[568] FORNBERG,B., (1981), *A vector implementation of the fast Fourier transform algorithm,* Math. Comp., vol 36, no 153, pages 189–191.

[569] FORSYTHE,G.E., (1953), *Solving linear systems can be interesting,* Bull. AMS, vol 59, pages 299–329.

[570] FORSYTHE,G.E., (1953), *Tentative classification of methods and bibliography on solving systems of linear equations,* Nat. Bur. Stand. Appl. Math., vol 29, pages 1–28.

[571] FORSYTHE,G.E., MALCOLM,M.A., MOLER,C.B., (1977), *Computer methods for mathematical computations,* Prentice Hall.

[572] FORSYTHE,G.E., MOLER,C.B., (1967), *Computer solution of linear algebraic systems,* Prentice–Hall.

[573] FORSYTHE,G.E., STRAUS,E.G., (1955), *On best conditioned matrices,* Proc. AMS, vol 6, pages 340–345.

[574] FORSYTHE,G.E., WASOW,W., (1960), *Finite difference methods for partial differential equations,* Wiley.

[575] FOSTER,L.V., (1994), *Gaussian elimination with partial pivoting can fail in practice,* SIAM J. Matrix Anal. Appl., vol 15, no 4, pages 1354–1362.

[576] FOX,L., (1964), *An introduction to numerical linear algebra,* Oxford University Press.

[577] FOX,L., HUSKEY,H.D., WILKINSON,J.H., (1948), *Notes on the solution of algebraic linear simultaneous equations,* Quart. J. Appl. Math., vol 1, pages 149–173.

[578] FRANCESCATTO,J., (1995), *Resolution de l'equation de Poisson sur des maillages etires par une methode multigrille,* Report 2712, INRIA, France.

[579] FRANCESCATTO,J., DERVIEUX,A., (1997), *A semi coarsening strategy for unstructured MG with agglomeration,* Report 2950, INRIA, France.

[580] FREDERICKSON,P.O., McBRYAN,O.A., (1988), *Parallel superconvergent multigrid,* in Multigrid methods, McCORMICK,S.F. ed, Marcel Dekker.

[581] FREUND,R.W., (1990), *On conjugate gradient type methods and polynomial preconditioners for a class of complex non Hermitian matrices,* Numer. Math., vol

57, pages 285–312.

[582] FREUND,R.W., (1991), *Krylov subspace methods for complex non Hermitian linear systems,* RIACS Tech Report 91.11, NASA Ames.

[583] FREUND,R.W., (1991), *Quasi kernel polynomials and their use in non Hermitian matrix iterations,* RIACS Tech Report 91.20, NASA Ames.

[584] FREUND,R.W., (1991), *A transpose free quasi minimal residual algorithm for non Hermitian linear systems,* RIACS Tech Report 91.18, NASA Ames.

[585] FREUND,R.W., (1992), *Conjugate gradient type methods for linear systems with complex symmetric coefficient matrices,* SIAM J. Sci. Comput., vol 13, pages 425–448.

[586] FREUND,R.W., GOLUB,G.H., NACHTIGAL,N.M., (1991), *Iterative solution of linear systems,* Acta Numerica, vol 1, pages 57–100.

[587] FREUND,R.W., GUTKNECHT,M.H., NACHTIGAL,N.M., (1993), *An implementation of the look-ahead Lanczos algorithm for non Hermitian matrices,* SIAM J. Sci. Comput., vol 14, pages 137–158.

[588] FREUND,R.W., HOCHBRUCK,M., (1991), *A biconjugate gradient type algorithm on massively parallel architectures,* RIACS Tech Report 91.07, NASA Ames.

[589] FREUND,R.W., MALHOTRA,M., (1996), *A block-QMR algorithm for non-hermitian linear systems with multiple right hand sides,* Report SCCM 96-01, Computer Science Dept., Stanford University.

[590] FREUND,R.W., NACHTIGAL,N.M., (1991), *QMR: a quasi-minimal residual method for non Hermitian linear systems,* Numer. Math., vol 60, pages 315–339.

[591] FREUND,R.W., SZETO,T., (1991), *A quasi minimal residual squared algorithm for non Hermitian linear systems,* RIACS Tech Report 91.26, NASA Ames.

[592] FRIGO,M., JOHNSON,S.G., (1997), *FFTW user's manual,* Report Massachusetts Institute of Technology.

[593] FUJINO,S., DOI,S., (1991), *Optimizing multicolor ICCG methods on some vector computers,* Proceedings IMACS Int. Symp. Iterative Methods in Linear Algebra, BEAUWENS,R. Ed.. North Holland.

[594] FUNDERLIC,R.E., NEUMANN,M., PLEMMONS,R.J., (1982), *LU decompositions of generalized diagonally dominant matrices,* Numer. Math., vol 40, pages 57–69.

[595] FUNDERLIC,R.E., PLEMMONS,R.J., (1981), *LU decomposition of M–matrices by elimination without pivoting,* Linear Algebra and its Appl., vol 41, pages 41–99.

[596] GALLIVAN,K.A., HEATH,M.T., NG,E., ORTEGA,J.M., PEYTON,B.W., PLEMMONS, R.J., ROMINE,C.H., SAMEH,A.H., VOIGT,R.C., (1990), *Parallel algo-*

rithms for matrix computations, SIAM.

[597] GALLIVAN,K.A., PLEMMONS,R.J., SAMEH,A.H., (1990), *Parallel algorithms for dense linear algebra computations,* SIAM Review, vol 32, no 1, pages 54–135.

[598] GALLOPOULOS,E., SAAD,Y., (1989), *A parallel block cyclic reduction algorithm for the fast solution of elliptic equations,* Parallel Comp., vol 10, pages 143–159.

[599] GANDER,W., GOLUB,G.H., GRUNTZ,D., (1989), *Solving linear equation by extrapolation,* NA-89-11, Computer Science Dept., Stanford University.

[600] GANNON,D., VAN ROSENDALE,J., (1986), *On the structure of parallelism in a highly concurrent PDE solver,* J. Par. Dist. Comp., vol 3, pages 106–135.

[601] GANTMACHER,F.R., (1959), *Matrix theory, vol I,* Chelsea.

[602] GARBEY,M., (1996), *A Schwarz alternating procedure for singular perturbation problems,* SIAM J. Sci. Comput., vol 17, no 5, pages 1175–1201.

[603] GARY,J., McCORMICK,S.F., SWEET,R.A., (1983), *Successive overrelaxation, multigrid, and preconditioned conjugate gradients algorithms for solving a diffusion problem on a vector computer,* Appl. Math. Comput., vol 13, pages 285–310.

[604] GAUTSCHI,W., (1968), *Construction of Gauss–Christoffel quadrature formulas,* Math. Comp., vol 23, pages 221–230.

[605] GAUTSCHI,W., (1985), *Orthogonal polynomials – constructive theory and applications,* J. Comp. Appl. Math., vol 12–13, pages 61–76.

[606] GEIST,G.A., (1985), *Efficient parallel LU factorization with pivoting on a hypercube multiprocessor,* Report ORNL–6211 Oak Ridge Nat. Laboratory.

[607] GEIST,G.A., NG,E., (1989), *Task scheduling for parallel sparse Cholesky factorization,* Int. J. Parallel Prog., vol 18, no 4, pages 291–314.

[608] GEIST,G.A., ROMINE,C.H., (1988), *LU factorization algorithms on distributed memory multiprocessor architectures,* SIAM J. Sci. Comput., vol 9, no 4, pages 639–649.

[609] GENTZSCH,W., (1987), *A fully vectorizable SOR variant,* Parallel Comp., vol 4, pages 349–353.

[610] GEORGE,A., (1973), *Nested dissection of a regular finite element mesh,* SIAM J. Numer. Anal., vol 10, no 2, pages 345–363.

[611] GEORGE,A., (1974), *On block elimination for sparse linear systems,* SIAM J. Numer. Anal., vol 11, no 3, pages 585–603.

[612] GEORGE,A., (1977), *Numerical experiments using dissection methods to solve n by n grid problems,* SIAM J. Numer. Anal., vol 14, no 2, pages 161–179.

[613] GEORGE,A., (1980), *An automatic one way dissection algorithm for irregular finite element problems,* SIAM J. Numer. Anal., vol 17, no 6, pages 740–751.

[614] GEORGE,A., GILBERT,J.R., LIU,J.W–H., (1993), *Graph theory and sparse matrix computation,* Springer Verlag.

[615] GEORGE,A., HEATH,M., LIU,J.W–H., (1986), *Parallel Cholesky factorization on a shared memory multiprocessor,* Linear Algebra and its Appl., vol 77, pages 165–187.

[616] GEORGE,A., HEATH,M.T., LIU,J.W–H., NG,E., (1988), *Sparse Cholesky factorization on a local memory multiprocessor,* SIAM J. Sci. Comput., vol 9, no 2, pages 327–340.

[617] GEORGE,A., LIU,J.W–H., (1975), *A note on fill for sparse matrices,* SIAM J. Numer. Anal., vol 12, no 3, pages 452–455 .

[618] GEORGE,A., LIU,J.W–H., (1978), *Algorithms for matrix partitioning and the numerical solution of finite element systems,* SIAM J. Numer. Anal., vol 15, no 2, pages 297–327.

[619] GEORGE,A., LIU,J.W–H., (1978), *An automatic nested dissection algorithm for irregular finite element problems,* SIAM J. Numer. Anal., vol 15, no 5, pages 1053–1069.

[620] GEORGE,A., LIU, J.W–H., (1979), *An implementation of a pseudoperipheral node finder,* ACM Trans. Math. Soft., vol 5, no 3, pages 284–295.

[621] GEORGE,A., LIU,J.W–H., (1979), *The design of a user interface for a sparse matrix package,* ACM Trans. Math. Soft., vol 5, no 2, pages 139–162.

[622] GEORGE,A., LIU,J.W–H., (1980), *A minimal storage implementation of the minimum degree algorithm,* SIAM J. Numer. Anal., vol 17, no 2, pages 282–299.

[623] GEORGE,A., LIU,J.W–H., (1980), *An optimal algorithm for symbolic factorization of symmetric matrices,* SIAM J. Comp., vol 9, no 3, pages 583–593.

[624] GEORGE,A., LIU,J.W–H., (1980), *A fast implementation of the minimum degree algorithm using quotient graphs,* ACM Trans. Math. Soft., vol 6, no 3, pages 337–358.

[625] GEORGE,A., LIU,J.W–H., (1981), *Computer solution of large sparse positive definite systems,* Prentice–Hall.

[626] GEORGE,A., LIU,J.W–H., (1989), *The evolution of the minimum degree ordering algorithm,* SIAM Review, vol 31, no 1, pages 1–19.

[627] GEORGE,A., LIU,J.W–H., NG,E., (1988), *A data structure for sparse QR and LU factorizations,* SIAM J. Sci. Comput., vol 9, no 1, pages 100–121.

[628] GEORGE,A., LIU,J.W–H., NG,E., (1989), *Communication results for parallel*

sparse Cholesky factorization on a hypercube, Parallel Comp., vol 10, no 3, pages 287–298.

[629] GEORGE,A., McINTYRE,D.R., (1978), *On the application of the minimum degree algorithm to finite element systems,* SIAM J. Numer. Anal., vol 15, no 1, pages 90–112.

[630] GEORGE,A., NG,E., (1985), *An implementation of Gaussian elimination with partial pivoting for sparse systems,* SIAM J. Sci. Comput., vol 6, no 2, pages 390–409.

[631] GEORGE,A., NG,E., (1987), *Symbolic factorization for sparse Gaussian elimination with partial pivoting,* SIAM J. Sci. Comput., vol 8, no 6, pages 877–898.

[632] GEORGE,A., NG,E., (1988), *On the complexity of sparse QR and LU factorization for finite element matrices,* SIAM J. Sci. Comput., vol 9, no 5, pages 849–861.

[633] GEORGE,A., POOLE,W.G., VOIGT,R.G., (1978), *Incomplete nested dissection for solving n by n grid problems,* SIAM J. Numer. Anal., vol 15, no 4, pages 662–673.

[634] GEORGE,A., POTHEN,A., (1995), *Analysis of the spectral approach to envelope reduction via a quadratic assignment formulation,* SIAM J. Matrix Anal. Appl.,

[635] GEORGE,A., RASHWAN,H., (1980), *On symbolic factorization of partitioned sparse symmetric matrices,* Linear Algebra and its Appl., vol 34, pages 145–157.

[636] GIBBS,N.E., POOLE,W.G., STOCKMEYER,P.K., (1976), *An algorithm for reducing the bandwidth and profile of a sparse matrix,* SIAM J. Numer. Anal., vol 13, no 2, pages 236–250.

[637] GIBBS,N.E., POOLE,W.G., STOCKMEYER,P.K., (1976), *A comparison of several bandwidth and profile reduction algorithms,* ACM Trans. Math. Soft., vol 2, pages 322–330.

[638] GILADI,E., GOLUB,G.H., KELLER,J.B., (1995), *Inner and outer iterations for the Chebyshev algorithm,* Report SCCM 95-12, Computer Science Dept., Stanford University.

[639] GILBERT,J.R., (1994), *Predicting structure in sparse matrix computations,* SIAM J. Matrix Anal. Appl., vol 15, no 1, pages 62–79.

[640] GILBERT,J.R., HAFSTEINSSON,H., (1990), *Parallel symbolic factorization of sparse linear systems,* Parallel Comp., vol 14, pages 151–162.

[641] GILBERT,J.R., LIU,J.W–H., (1993), *Elimination structures for unsymmetric sparse LU factors,* SIAM J. Matrix Anal. Appl., vol 14, no 2, pages 334–352.

[642] GILBERT,J.R., MILLER,G.L., TENG,S–H., (1994), *Geometric mesh partitioning: implementation and experiments,* Report Xerox Parx, Palo Alto.

[643] GILBERT,J.R., NG,E.G., PEYTON,B.W., (1994), *An efficient algorithm to compute row and column counts for sparse Cholesky factorization,* SIAM J. Matrix Anal. Appl., vol 15, no 4, pages 1075–1091.

[644] GILBERT,J.R., PEIERLS,T., (1988), *Sparse partial pivoting in time proportional to arithmetic operations,* SIAM J. Sci. Comput., vol 9, no 5, pages 862–874.

[645] GILBERT,J.R., SCHREIBER,R., (1992), *Highly parallel sparse Cholesky factorization,* SIAM J. Sci. Comput., vol 13, no 5, pages 1151–1172.

[646] GILBERT,J.R., ZMIJEWSKI,E., (1990), *A parallel graph partitioning algorithm for a message passing multiprocessor,* Int. J. Parallel Prog., vol 16, pages 427–449.

[647] GILL,P.E., GOLUB,G.H., MURRAY,W., SAUNDERS,M.A., (1974), *Methods for modifying matrix factorizations,* Math. Comp., vol 28, no 126, pages 505–535 .

[648] GILL,P.E., MURRAY,W., SAUNDERS,M.A., (1975), *Methods for computing and modifying the LDV factors of a matrix,* Math. Comp., vol 29, pages 1051–1077.

[649] GLOWINSKI,R., GOLUB,G.H., MEURANT,G.A., PERIAUX,J., (1988), *First international symposium on domain decomposition methods for partial differential equations,* SIAM.

[650] GLOWINSKI,R., KUZNETSOV,Y.A., MEURANT,G.A., PERIAUX,J., WIDLUND,O.B., (1991), *Fourth international symposium on domain decomposition methods for partial differential equations,* SIAM.

[651] GLOWINSKI,R., PERIAUX,J., TERRASSON,G., (1990), *On the coupling of viscous and inviscid models for compressible fluid flows via domain decomposition,* in Third International Symposium on Domain Decomposition Methods for Partial Differential Equations, CHAN,T.F., GLOWINSKI,R., PERIAUX,J., and WIDLUND,O.B. Eds. SIAM, pages 64–97.

[652] GOLBERG,D., (1991), *What every computer scientist should know about floating point arithmetic,* ACM Comp. surveys, vol 23, no 1, pages 5–48.

[653] GOLDFARB, D., (1972), *Modification methods for inverting matrices and solving systems of linear algebraic equations,* Math. Comp., vol 26, no 120, pages 829–852.

[654] GOLDSTINE,H.H., (1977), *A history of numerical analysis from the 16th through the 19th century,* Springer.

[655] GOLUB,G.H., (1962), *Bounds for the round–off errors in the Richardson second order method,* BIT, vol 2, pages 212–223.

[656] GOLUB,G.H., (1973), *Some modified matrix eigenvalue problems,* SIAM Review, vol 15, no 2, pages 318–334.

[657] GOLUB,G.H., (1974), *Bounds for matrix moments,* Rocky Mountain J. of Math., vol 4, no 2, pages 207–211.

[658] GOLUB,G.H., (1984), *Studies in numerical analysis,* The Mathematical Association of America.

[659] GOLUB,G.H., De PILLIS,J., (1989), *Toward an effective two–parameter SOR method,* NA-89-02, Computer Science Dept., Stanford University.

[660] GOLUB,G.H., GUTKNECHT,M.H., (1990), *Modified moments for indefinite weight functions,* Numer. Math., vol 57, pages 607–624.

[661] GOLUB,G.H., KANNAN,R., (1986), *Convergence of a two–stage Richardson process for non linear equations,* BIT, vol 26, pages 209–216.

[662] GOLUB,G.H., KAUTSKY,J., (1983), *Calculation of Gauss quadratures with multiple free and fixed knots,* Numer. Math., vol 41, pages 147–163.

[663] GOLUB,G.H., MAYERS,D., (1984), *The use of preconditioning over irregular regions,* in Computing methods in applied science and engineering VI, GLOWINSKI,R. and LIONS,J.L. Eds. North-Holland, pages 3–14.

[664] GOLUB,G.H., MEURANT,G., (1983), *Résolution numérique des grands systèmes linéaires,* vol 49, Eyrolles, (Paris).

[665] GOLUB,G.H., MEURANT,G., (1994), *Matrices, moments and quadrature,* in Numerical Analysis 1994, vol 303, Pitman research notes in mathematics series, GRIFFITHS,D.F. and WATSON,G.A. Eds. Longman Scientific and Technical, pages 105–156.

[666] GOLUB,G.H., MEURANT,G., (1997), *Matrices, moments and quadrature II or how to compute the norm of the error in iterative methods,* BIT, vol 37, no 3, pages 687–705.

[667] GOLUB,G.H., O'LEARY,D.P., (1989), *Some history of the conjugate gradient and Lanczos algorithms: 1948–1976,* SIAM Review, vol 31, no 1, pages 50–102.

[668] GOLUB,G.H., ORTEGA,J.M., (1992), *Scientific computing and differential equations, an introduction to numerical methods,* Academic Press.

[669] GOLUB,G.H., OVERTON,M., (1981), *Convergence of a two–stage Richardson iterative procedure for solving systems of linear equations,* NA-81-17, Computer Science Dept., Stanford University.

[670] GOLUB,G.H., OVERTON,M., (1988), *The convergence of inexact Chebyshev and Richardson methods for solving linear systems,* Numer. Math., vol 53, pages 571–593.

[671] GOLUB,G.H., VARAH,J.M., (1974), *On a characterization of the best l_2 scaling of a matrix,* SIAM J. Numer. Anal., vol 11, pages 472–479.

[672] GOLUB,G.H., STRAKOS,Z., (1994), *Estimates in quadratic formulas,* Numer. Alg., vol 8,

[673] GOLUB,G.H., UNDERWOOD,R., (1977), *The block Lanczos method for computing eigenvalues,* in Mathematical Software III, RICE,J. Ed. Academic Press, pages 364–377.

[674] GOLUB,G.H., VAN DER VORST,H.A., (1996), *Closer to the solution: iterative linear solvers,* Report Computer Science Dept., Stanford University, to appear in "The state of the art in numerical analysis", DUFF,I.S. and WATSON,G.A., Eds. Oxford University Press.

[675] GOLUB,G.H., VAN LOAN,C., (1979), *Unsymmetric positive definite linear systems,* Linear Algebra and its Appl., vol 28, pages 85–97.

[676] GOLUB,G.H., VAN LOAN,C., (1989), *Matrix computations, second edition,* Johns Hopkins University Press.

[677] GOLUB,G.H., VARGA,R.S., (1961), *Chebyshev semi–iterative methods, successive overrelaxation iterative methods and second order Richardson iterative methods, Part I,* Numer. Math., vol 3, pages 147–156.

[678] GOLUB,G.H., VARGA,R.S., (1961), *Chebyshev semi–iterative methods, successive overrelaxation iterative methods and second order Richardson iterative methods, Part II,* Numer. Math., vol 3, pages 157–168.

[679] GOLUB,G.H., WELSCH,J.H., (1969), *Calculation of Gauss quadrature rules,* Math. Comp., vol 23, pages 221–230.

[680] GOLUB,G.H., YE,Q., (1997), *Inexact preconditioned conjugate gradient method with inner–outer iteration,* Report NA–97–??, Computer Science Dept., Stanford University.

[681] GONZALEZ,R., (1986), *Domain decomposition for two dimensional elliptic operators on vector and parallel machines,* Ph.D. Thesis, TR 86-4, Dept. of Mathematical Sciences, Rice University.

[682] GOOD,I.J., (1958), *The interaction algorithm and practical Fourier analysis,* J. Roy. Statist. Soc. Ser. B, vol 20, pages 361–372.

[683] GOOD,I.J., (1971), *The relationship between two Fast Fourier Transforms,* IEEE Trans. Comput., vol 20, pages 310–317.

[684] GOOVAERTS,D., (1990), *Domain decomposition for elliptic partial differential equations,* Ph.D. Thesis, Katholieke Universiteit Leuven, Belgium.

[685] GOULD,N., (1991), *On growth in Gaussian elimination with complete pivoting,* SIAM J. Matrix Anal. Appl., vol 12, no 2, pages 354–361.

[686] GOULD,N.I., SCOTT,J.A., (1995), *On approximate inverse preconditioners,* Report RAL-TR-95-026, Rutherford Appleton Laboratory.

[687] GOURLAY,A.R., (1968), *The acceleration of the Peaceman–Rachford method by Chebyshev polynomials,* Comput. J., vol 10, pages 378–382.

[688] GOURLAY,A.R., (1970), *On Chebychev acceleration procedures for alternating direction iterative methods,* J. Inst. Maths. Applics., vol 6, pages 1–11.

[689] GRAGG,W.B., REICHEL,L., (1987), *On the application of orthogonal polynomials to the iterative solution of linear systems of equations with indefinite or non–hermitian matrices,* Linear Algebra and its Appl., vol 88, pages 349–371.

[690] GRAUSCHOPF,T., GRIEBEL,M., REGLER,H., (1996), *Additive multilevel preconditioners based on bilinear interpolation, matrix dependent geometric coarsening and algebraic multigrid coarsening for second order elliptic PDEs,* Report TUM–I9602, Institut fur Informatik, Technische Universitat Munchen.

[691] GRCAR,J.F., (1981), *Analysis of the Lanczos algorithm and of the approximation problem in Richardson's method,* Ph.D. Thesis, Report UIUCDCS–R–81–1074, University of Illinois.

[692] GRCAR,J.F., (1989), *Operator coefficient methods for linear equations,* Report SAND89-8691, Sandia National Laboratory.

[693] GRCAR,J.F., (1990), *Matrix stretching for linear equations,* SAND90–8723, Sandia National Labs..

[694] GREENBAUM,A., (1979), *Comparison of splittings used with the conjugate gradient algorithm,* Numer. Math., vol 33, pages 181–194.

[695] GREENBAUM,A., (1981), *Convergence properties of the conjugate gradient algorithm in exact and finite precision arithmetic,* Ph.D. Thesis, Dept of Maths., University of California, Berkeley.

[696] GREENBAUM,A., (1981), *Behavior of the conjugate gradient algorithm in finite precision arithmetic,* Report UCRL–85752, Lawrence Livermore Laboratory.

[697] GREENBAUM,A., (1984), *Analysis of a multigrid method as an iterative technique for solving linear systems,* SIAM J. Numer. Anal., vol 21, no 3, pages 473–485.

[698] GREENBAUM,A., (1989), *Behavior of slightly perturbed Lanczos and conjugate gradient recurrences,* Linear Algebra and its Appl., vol 113, pages 7–63.

[699] GREENBAUM,A., (1995), *Estimating the attainable accuracy of recursively computed residual methods,* Report Courant Institute of Mathematical Sciences.

[700] GREENBAUM,A., (1997), *Iterative methods for solving linear equations,* SIAM.

[701] GREENBAUM,A., LI,C., CHAO,H.Z., (1989), *Parallelizing preconditioned conjugate gradient algorithms,* Comp. Phys. Comm., vol 53, pages 295–309.

[702] GREENBAUM,A., PTAK,V., STRAKOS,Z., (1997), *Any nonincreasing curve is possible for GMRES,* SIAM J. Matrix Anal. Appl., vol 17, pages 465–469.

[703] GREENBAUM,A., RODRIGUE,G.H., (1989), *Optimal preconditioners of a given sparsity pattern,* BIT, vol 29, pages 610–634.

[704] GREENBAUM,A., STRAKOS,Z., (1992), *Predicting the behavior of finite precision Lanczos and conjugate gradient computations,* SIAM J. Matrix Anal. Appl., vol 13, no 1, pages 121–137.

[705] GREENBAUM,A., TREFETHEN,L.N., (1994), *GMRES/CR and Arnoldi /Lanczos as matrix approximation problems,* SIAM J. Sci. Comput., vol 15, no 2, pages 359–368.

[706] GRIEBEL,M., OSWALD,P., (1995), *On the abstract theory of additive and multiplicative Schwarz algorithms,* Numer. Math., vol 70, pages 163–180.

[707] GRIFFITHS,D.F., WATSON,G.A., (1989), *Numerical Analysis 1989, vol 228, Pitman research notes in mathematics series,* Longman Scientific and Technical.

[708] GRIFFITHS,D.F., WATSON,G.A., (1994), *Numerical Analysis 1994, vol 303, Pitman research notes in mathematics series,* Longman Scientific and Technical.

[709] GRIMES,R.G., LEWIS,J.G., (1981), *Condition number estimation for sparse matrices,* SIAM J. Sci. Comput., vol 2, pages 384–388.

[710] GRIMES,R.G., PIERCE,D.J., SIMON,H.D., (1990), *A new algorithm for finding a pseudoperipheral node in a graph,* SIAM J. Matrix Anal. Appl., vol 11, no 2, pages 323–334.

[711] GROPP,W.D., KEYES,D.E., (1988), *Complexity of parallel implementation of domain decomposition techniques for elliptic partial differential equations,* SIAM J. Sci. Comput., vol 9, pages 312–326.

[712] GROPP,W.D., KEYES,D.E., (1992), *Domain decomposition with local mesh refinement,* SIAM J. Sci. Comput., vol 13, pages 967–993.

[713] GUILINGER,Jr, W.H., (1965), *The Peaceman–Rachford method for small mesh increments,* J. Math. Anal. Appl., vol 11, pages 261–277.

[714] GROTE,M., SIMON,H.S., (1993), *Parallel preconditioning and approximate inverses on the Connection Machine,* Proceedings of the Sixth SIAM conference on parallel processing for scientific computing, SINCOVEC,R. Ed. SIAM, pages 519–523.

[715] GUILLARD,H., (1993), *Convergence analysis of a multilevel relaxation method,* Rapport 1884, INRIA, France.

[716] GUO,G-H., (1991), *Some results on sparse block factorization iterative methods,* Linear Algebra and its Appl., vol 145, pages 187–199.

[717] GUO,R., SKEEL,R.D., (1992), *An algebraic hierarchical basis preconditioner,* Appl. Numer. Math., vol 9, pages 21–32.

[718] GUPTA,A., KUMAR,V., SAMEH,A., (1992), *Performance and scalability of conjugate gradient methods on parallel computers,* Report TR 92-64, Dept of Computer Science, University of Minnesota.

[719] GUPTA,M.M., (1986), *A spectrum envelope technique for iterative solution of central difference approximations of convection–diffusion equations,* SIAM J. Alg. Disc. Meth., vol 7, no 4, pages 513–526.

[720] GUSTAFSSON,I., (1978), *A class of first order factorization methods,* BIT, vol 18, pages 142–156.

[721] GUSTAFSSON,I., (1979), *On modified incomplete Cholesky factorization methods for the solution of problems with mixed boundary conditions and problems with discontinuous material coefficients,* Int. J. Num. Meth. Eng., vol 14, pages 1127–1140.

[722] GUSTAFSSON,I., (1984), *A preconditioned iterative method for the solution of the biharmonic problem,* IMA J. Num. Anal., vol 4, pages 55–67.

[723] GUSTAFSSON,I., LINDSKOG,G., (1986), *A preconditioning technique based on element matrix factorization,* Com. Meth. Appl. Mech. Eng., vol 55, pages 201–220.

[724] GUTKNECHT,M.H., (1988), *Stationary and almost stationary iterative (k,l)-step methods for linear and nonlinear systems of equations,* IPS-88-01, ETH Zurich.

[725] GUTKNECHT,M.H., (1991), *On certain types of (k,l)–step methods for solving linear systems of equations,* IPS-91-01, ETH Zurich.

[726] GUTKNECHT,M.H., (1992), *A completed theory of the unsymmetric Lanczos process and related algorithms, Part I,* SIAM J. Matrix Anal. Appl., vol 13, pages 594–639.

[727] GUTKNECHT,M.H., (1993), *Changing the norm in conjugate gradient type algorithms,* SIAM J. Numer. Anal., vol 30, pages 40–56.

[728] GUTKNECHT,M.H., (1993), *Variants of BiCGSTAB for matrices with complex spectrum,* SIAM J. Sci. Comput., vol 14, pages 1020–1033.

[729] GUTKNECHT,M.H., (1994), *A completed theory of the unsymmetric Lanczos process and related algorithms, Part II,* SIAM J. Matrix Anal. Appl., vol 15, pages 15–58.

[730] GUTKNECHT,M.H., (1994), *The Lanczos process and Pade approximation,* IPS Research Report 94-06, ETH.

[731] GUTKNECHT,M.H., (1997), *Lanczos–type solvers for nonsymmetric linear systems of equations,* Report TR–97–04, Swiss Center for Scientific Computing.

[732] GUTKNECHT,M.H., STRAKOS,Z., (1997), *Accuracy of three–term and two–term recurrences for Krylov space solvers,* Report Swiss Center for Scientific Computing.

[733] HAASE,G., LANGER,U., MEYER,A., (1991), *Domain decomposition methods with inexact subdomain solvers,* Numer. Lin. Alg. with Appl., vol 1, pages 27–41.

[734] HACKBUSCH,W., (1978), *On the multigrid method applied to difference equations,* Computing, vol 20, pages 291–306.

[735] HACKBUSCH,W., (1980), *Convergence of multi–grid iterations applied to difference equations,* Math. Comp., vol 34, pages 425–440.

[736] HACKBUSCH,W., (1985), *Multigrid methods and applications,* Springer.

[737] HACKBUSCH,W., (1989), *The frequency decomposition multigrid method, Part I: Application to anisotropic equations,* Numer. Math., vol 56, pages 229–245.

[738] HACKBUSCH,W., (1992), *The frequency decomposition multigrid method, II. Convergence analysis based on the additive Schwarz method,* Numer. Math., vol 63, pages 433–453.

[739] HACKBUSCH,W., (1993), *Iterative solution of large sparse systems of equations,* Springer.

[740] HACKBUSCH,W., TROTTENBERG,U., (1982), *Multigrid Methods,* Lecture Notes in Mathematics 960. Springer.

[741] HACKBUSCH,W., WITTUM,G., (1993), *Incomplete Decomposition Algorithms, Theory and Applications,* Notes on Numerical Fluid Mechanics 41. Vieweg.

[742] HADFIELD,S.M., DAVIS,T.A., (1992), *Analysis of potential parallel implementations of the unsymmetric pattern multifrontal method for sparse LU factorization,* Report TR–92–017, University of Florida.

[743] HADFIELD,S.M., DAVIS,T.A., (1994), *A parallel unsymmetric pattern multifrontal method,* Report TR–94–??, University of Florida.

[744] HADFIELD,S.M., DAVIS,T.A., (1994), *Potential and achievable parallelism in the unsymmetric pattern LU factorization method for sparse matrices,* Report TR–94–006, University of Florida.

[745] HADFIELD,S.M., DAVIS,T.A., (1994), *Potential and achievable parallelism in the unsymmetric pattern LU factorization,* Report TR–94–027, University of Florida.

[746] HADJIDIMOS,A., (1978), *Accelerated Overrelaxation method,* Math. Comp., vol 32, no 141, pages 149–157.

[747] HAGEMAN,L.A., LUK,F.T., YOUNG,D.M., (1980), *On the equivalence of certain iterative acceleration methods,* SIAM J. Numer. Anal., vol 17, no 6, pages 852–873.

[748] HAGEMAN,L.A., YOUNG,D.M., (1981), *Applied iterative methods,* Academic Press.

[749] HAGER,W.W., (1984), *Condition estimates,* SIAM J. Sci. Comput., vol 5, no 2, pages 311–316.

[750] HAGHOO,M., PROSKUROWSKI,W., (1988), *Parallel implementation of a do-*

main decomposition method, Report CRI 88-06, University of Southern California.

[751] HALEY,S.B., (1980), *Solution of band matrix equations by projection-recurrence,* Linear Algebra and its Appl., vol 32, pages 33–48.

[752] HANKE,M., HOCHBRUCK,M., NIETHAMMER,W., (1993), *Experiments with Krylov subspace methods on a massively parallel computer,* Appl. Math., vol 38, pages 440–451.

[753] HARARY,F., (1971), *Sparse matrices and graph theory,* in Large sparse sets of linear equations, REID,J.K. Ed.. Academic Press, pages 139–150.

[754] HARROD,W.J., (1986), *LU decompositions of tridiagonal irreducible H–matrices,* SIAM J. Alg. Disc. Meth., vol 7, no 2, pages 180–187.

[755] HEATH,M.T., NG,E., PEYTON,B.W., (1991), *Parallel algorithms for sparse linear systems,* SIAM Review, vol 33, no 3, pages 420–460.

[756] HEATH,M.T., RAGHAVAN,P., (1993), *A cartesian parallel nested dissection algorithm,* SIAM J. Matrix Anal. Appl., vol 16, no 1, pages 235–253.

[757] HEATH,M.T., RAGHAVAN,P., (1993), *Distributed solution of sparse linear system,* Report UT CS-93-201, Dept. of Computer Science, University of Tennessee.

[758] HEATH,M.T., ROMINE,C.H., (1988), *Parallel solution of triangular systems on distributed memory multiprocessors,* SIAM J. Sci. Comput., vol 9, no 3, pages 558–588.

[759] HEGLAND,M., (1991), *On the parallel solution of tridiagonal systems by wrap-around partitioning and incomplete LU factorization,* Numer. Math., vol 59, pages 453–472.

[760] HEGLAND,M., SAYLOR, P.E., (1990), *Optimization exercise: block Jacobi for a model problem,* IPS CD-90-4, ETH Zurich.

[761] HELLER,D., (1976), *Some aspects of the cyclic reduction algorithm for block tridiagonal linear systems,* SIAM J. Numer. Anal., vol 13, pages 484–496.

[762] HELLER,D., (1977), *Direct and iterative methods for block tridiagonal linear systems,* Ph.D. Thesis, Carnegie–Mellon University.

[763] HELLER,D., (1978), *A survey of parallel algorithms in numerical linear algebra,* SIAM Review, vol 20, no 4, pages 740–777.

[764] HEMKER,P.W., (1980), *The incomplete LU–decomposition as a relaxation method in multi–grid algorithms,* in Boundary and Interior Layers—Computational and Asymptotic Methods, MILLER,J.J.H. Ed. Boole Press, pages 306–311.

[765] HENDERSON,H.V., SEARLE,S.R., (1981), *On deriving the inverse of a sum of matrices,* SIAM Review, vol 23, no 1, pages 53–60.

[766] HENDRICKSON,B., LELAND,R., (1995), *An improved spectral graph partitioning algorithm for mapping parallel computations,* SIAM J. Sci. Comput., vol 16, pages 452–469.

[767] HENDRICKSON,B.A., WOMBLE,D.E., (1994), *The torus–wrap mapping for dense matrix calculations on massively parallel computers,* SIAM J. Sci. Comput., vol 15, no 5, pages 1201–1226.

[768] HENRICI,P., (1964), *Elements of numerical analysis,* John Wiley.

[769] HENSON,V.E., (1989), *Parallel compact symmetric FFTs,* in Vector and Parallel Computing: issues in applied research and development, DONGARRA,J. J. , DUFF,I. S., GAFFNEY,P. and McKEE,S. Eds. Ellis Horwood, pages 153–164.

[770] HESTENES,M.R., (1954), *The conjugate gradient method for solving linear systems,* Report INA 54–11, National Bureau of Standards.

[771] HESTENES,M.R., (1980), *Conjugate direction methods in optimization,* Springer.

[772] HESTENES,M.R., STIEFEL,E., (1952), *Methods of conjugate gradients for solving linear systems,* J. Nat. Bur. Stand., vol 49, no 6, pages 409–436.

[773] HEYOUNI,M., SADOK,H., (1995), *On a variable smoothing procedure for conjugate gradient type methods,* Rapport LANO 348, Université de Lille.

[774] HIGHAM,N.J., (1986), *Efficient algorithms for computing the condition number of a tridiagonal matrix,* SIAM J. Sci. Comput., vol 7, no 1, pages 150–165.

[775] HIGHAM,N.J., (1987), *A survey of condition number estimation for triangular matrices,* SIAM Review, vol 29, no 4, pages 575–596.

[776] HIGHAM,N.J., (1989), *The accuracy of solutions to triangular systems,* SIAM J. Numer. Anal., vol 26, no 5, pages 252–1265.

[777] HIGHAM,N.J., (1989), *How accurate is Gaussian elimination?,* in Numerical Analysis 1989, vol 228, Pitman research notes in mathematics series, GRIFFITHS,D.F. and WATSON,G.A. Eds. Longman Scientific and Technical.

[778] HIGHAM,N.J., (1990), *Bounding the error in Gaussian elimination for tridiagonal systems,* SIAM J. Matrix Anal. Appl., vol 11, no 4, pages 521–530.

[779] HIGHAM,N.J., (1994), *A survey of componentwise perturbation theory in numerical linear algebra,* in Mathematics of computation 1943-1993, Proceedings of Symposia in Applied Mathematics. American Mathematical Society.

[780] HIGHAM,N.J., (1995), *The test matrix toolbox for Matlab,* Numerical analysis report 276, Dept. of Mathematics, University of Manchester.

[781] HIGHAM,N.J., (1996), *Accuracy and stability of numerical algorithms,* SIAM.

[782] HIGHAM,N.J., (1997), *Testing linear algebra software,* in The quality of numerical

software: assessment and enhancement, BOISVERT,R. Ed.. Chapman and Hall.

[783] HIGHAM,N.J., HIGHAM,D.J., (1989), *Large growth factors in Gaussian elimination with pivoting,* SIAM J. Matrix Anal. Appl., vol 10, no 2, pages 155–164.

[784] HIGHAM,N.J., HIGHAM,D.J., (1992), *Backward error and condition of structured linear systems,* SIAM J. Matrix Anal. Appl., vol 13, no 1, pages 162–175.

[785] HIGHAM,N.J., POTHEN,A., (1994), *Stability of the partitioned inverse method for parallel solution of sparse triangular systems,* SIAM J. Sci. Comput., vol 15, no 1, pages 139–148.

[786] HOCHBRUCK,M., (1996), *The Pade table and its relation to certain numerical algorithm,* Habilitationsschrift, Universitat Tubingen.

[787] HOCHBRUCK,M., LUBICH,C., (1996), *Error analysis of Krylov methods in a nutshell,* Report Mathematisches Institut, Universitat Tubingen.

[788] HOCKNEY,R.W., (1965), *A Fast direct solution of Poisson's equation using Fourier analysis,* J. ACM, vol 12, pages 95–113.

[789] HOCKNEY,R.W., (1970), *The potential calculation and some applications,* Methods in Comp. Phys., vol 9, pages 135–211.

[790] HOCKNEY,R.W., JESSHOPE,C.R., (1988), *Parallel computers 2,* Adam Hilger.

[791] HOFFMAN,A.J., MARTIN,M.S., ROSE.D.J., (1973), *Complexity bounds for regular finite difference and finite element grids,* SIAM J. Numer. Anal., vol 10, no 2, pages 364–369.

[792] HOOD,P., (1976), *Frontal solution program for unsymmetric matrices,* Int. J. Num. Meth. Eng., vol 10, pages 379–400.

[793] HOUSEHOLDER,A.S., (1953), *Principles of numerical analysis,* McGraw–Hill.

[794] HOUSEHOLDER,A.S., (1955), *Terminating and nonterminating iterations for solving linear systems,* J. Soc. Indust. Appl. Math., vol 3, no 2, pages 67–72.

[795] HOUSEHOLDER,A.S., (1958), *The approximate solution of matrix problems,* J. ACM, vol 5, pages 204–243.

[796] HOUSTIS,E.N., PAPATHEODOROU,T.S., (1979), *High order fast elliptic equation solver,* ACM Trans. Math. Soft., vol 5, no 4, pages 431–441.

[797] HUCKLE,T., (1996), *Efficient computation of sparse approximate inverses,* Report TUM-19608, Institut fur Informatik, Technische Universitat Munchen.

[798] HUCKLE,T., GROTE,M., (1994), *A new approach to parallel preconditioning with sparse approximate inverses,* Report SCCM-94-03, Computer Science Dept., Stanford University.

[799] HUCKLE,T., GROTE,M., (1997), *Parallel preconditioning with sparse approximate inverses,* SIAM J. Sci. Comput., vol 18, pages 838–853.

[800] HUGHES,T.J., LEVIT,I., WINGET,J., (1983), *An element by element solution algorithms for problems of structural and solid mechanics,* Com. Meth. Appl. Mech. Eng., vol 36, pages 241–254.

[801] HULBERT,L., ZMIJEWSKI,E., (1991), *Limiting communication in parallel sparse Cholesky factorization,* SIAM J. Sci. Comput., vol 12, no 5, pages 1184–1197.

[802] IFRAH,G., (1994), *Histoire universelle des chiffres,* Robert Laffont, Paris.

[803] IKEBE,Y., (1979), *On inverses of Hessenberg matrices,* Linear Algebra and its Appl., vol 24, pages 93–97.

[804] IKEUCHI,M., KOBAYASHI,H., SAWAMI,H., NIKI, H., (1979), *On the spectral radius of the Jacobi iteration matrix for a rectangular region with two different media,* J. Comp. Appl. Math., vol 5, no 4, pages 247–258.

[805] IL'IN,V.P., (1988), *Incomplete factorization methods,* Sov. J. Numer. Anal. Math. Modelling, vol 3, no 3, pages 179–198.

[806] ILIASH,J., ROSSI,T., TOIVANEN,J., (1993), *Two iterative methods for solving the Stokes problem,* Report 23/1993, Dept. of Mathematics, University of Jyvaskyla.

[807] IPSEN,I.C., SAAD,Y., SCHULTZ,M.H., (1986), *Complexity of dense linear system solution on a multiprocessor ring,* Linear Algebra and its Appl., vol 77, pages 205–239.

[808] IRONS,B.M., (1970), *A frontal solution program for finite element analysis,* Int. J. Num. Meth. Eng., vol 2, pages 5–32.

[809] ITO,K., (1984), *An iterative method for indefinite systems of linear equations,* Report 84-13, ICASE, NASA Langley.

[810] JACOBS,D.A.H., (1986), *A generalization of the conjugate gradient method to solve complex systems,* IMA J. Num. Anal., vol 6, pages 447–452.

[811] JALBY,W., MEIER,U., SAMEH,A., (1987), *The behavior of conjugate gradient based algorithms on a multi vector processor with a memory hierarchy,* Report CSRD 607, University of Illinois.

[812] JAMESON,A., (1993), *Computational algorithms for aerodynamic analysis and design,* Appl. Numer. Math., vol 13, pages 383–422.

[813] JANKOWSKI,M., WOZNIAKOWSKI,H., (1977), *Iterative refinement implies numerical stability,* BIT, vol 17, pages 303–311.

[814] JAY KUO,C.C., CHAN,T.F., (1990), *Two color Fourier analysis of iterative algorithms for elliptic problems with Red/BLack ordering,* SIAM J. Sci. Comput., vol 11, pages 767–793.

[815] JAY KUO,C.C., CHAN,T.F., TONG,C., (1990), *Multilevel filtering elliptic preconditioners,* SIAM J. Sci. Comput., vol 11, pages 403–429.

[816] JAY KUO,C.C., CHAN,T.F., TONG,C., (1992), *Multilevel filtering elliptic preconditioners: extensions to more general elliptic problems,* SIAM J. Sci. Comput., vol 13, pages 227–243.

[817] JEA,K.C., (1982), *Generalized conjugate gradient acceleration of iterative methods,* Ph.D. Thesis, Center for Numerical Analysis, The University of Texas at Austin.

[818] JEA,K.C., YOUNG,D.M., (1983), *On the simplification of generalized conjugate gradient methods for nonsymmetrizable linear systems,* Linear Algebra and its Appl., vol 52, pages 399–417.

[819] JENNINGS, A., (1971), *Accelerating the convergence of matrix iterative processes,* J. Inst. Maths. Applics., vol 8, pages 99–110.

[820] JENNINGS,A., (1977), *Matrix computation for engineers and scientists,* John Wiley.

[821] JENNINGS,A., (1977), *Influence of the eigenvalue spectrum on the convergence rate of the conjugate gradient method,* J. Inst. Maths. Applics., vol 20, pages 61–72.

[822] JENNINGS,A., MALIK,G.M., (1977), *Partial elimination,* J. Inst. Maths. Applics., vol 20, pages 307–316.

[823] JENNINGS,A., MALIK,G.M., (1978), *The solution of sparse linear equations by the conjugate gradient method,* Int. J. Num. Meth. Eng., vol 12, pages 141–158.

[824] JENNINGS,A., TUFF,A.D., (1971), *A direct method for the solution of large sparse symmetric simultaneous equations,* in Large sparse sets of linear equations, REID,J.K. Ed.. Academic Press, pages 97–104.

[825] JESPERSEN,D.C., (1984), *Multigrid methods for partial differential equations,* in Studies in Numerical Analysis, GOLUB,G.H. Ed. MAA, pages 270–318.

[826] JESS,J., KEES,H., (1982), *A data structure for parallel LU decomposition,* IEEE Trans. on Comp., vol C, no 31, pages 231–239.

[827] JOHAN,Z., (1992), *Data parallel finite element techniques for large-scale computational fluid dynamics,* Ph.D. Thesis, Dept. of Mechanical Engineering, Stanford University.

[828] JOHNSON,C., (1987), *Numerical Solution of Partial Differential Equations by the Finite Element Method,* Cambridge University Press.

[829] JOHNSON,C.R., (1982), *Inverse M-matrices,* Linear Algebra and its Appl., vol 47, pages 195–216.

[830] JOHNSON,O.G., MICHELLI,C.A., PAUL,G., (1983), *Polynomial preconditioners*

for conjugate gradient calculations, SIAM J. Numer. Anal., vol 20, no 2, pages 362–376.

[831] JOHNSSON,L., (1984), *Odd–even cyclic reduction on ensemble architectures and the solution of tridiagonal systems of equations,* Report CSD/RR–339 Yale University.

[832] JOHNSSON,L., (1987), *Solving tridiagonal systems on ensemble architectures,* SIAM J. Sci. Comput., vol 8, no 3, pages 354–392.

[833] JOHNSSON,L., MATHUR,K.K., (1990), *Data structures and algorithms for the finite element method on a data parallel supercomputer,* Int. J. Num. Meth. Eng., vol 29, pages 881–908.

[834] JOLY,P., (1991), *Résolution de systèmes linéaires avec plusieurs seconds membres par la méthode du gradient conjugué,* Rapport R91000, Lab. d'Analyse Numerique, Universite Pierre et Marie Curie.

[835] JOLY,P., EYMARD,R., (1990), *Preconditioned biconjugate gradient methods for numerical reservoir simulation,* J. Comp. Physics, vol 91, no 2, pages 298–309.

[836] JOLY,P., MEURANT,G., (1993), *Complex conjugate gradient methods,* Numer. Alg., vol 4, pages 379–406.

[837] JONES,M.T., PATRICK,M.L., (1994), *Factoring symmetric indefinite matrices on high–performance architectures,* SIAM J. Matrix Anal. Appl., vol 15, no 1, pages 273–283.

[838] JONES,M.T., PLASSMANN,P.E., (1992), *Solution of large sparse systems of linear equations in massively parallel applications,* in Proceedings of "Supercomputing 92". IEEE, pages 551–560.

[839] JONES,M.T., PLASSMANN,P.E., (1992), *The effect of many color orderings on the convergence of iterative methods,* Report MCS-P292-0292, Math. and Computer Science Div., Argonne National Laboratory.

[840] JONES,M.T., PLASSMANN,P.E., (1993), *The efficient parallel iterative solution of large sparse linear systems,* in Graph theory and sparse matrix computation, GEORGE,A., GILBERT,J.R. and LIU,J.W–H. Eds. Springer, pages 229–245.

[841] JONES,M.T., PLASSMANN,P.E., (1994), *Scalable iterative solution of sparse linear systems,* Parallel Comp., vol 20, pages 753–773.

[842] JORDAN,T.L., (1974), *Gaussian elimination for dense systems on STAR and a new parallel algorithm for diagonally dominant tridiagonal systems,* Report LA–5803, Los Alamos Scientific Laboratory.

[843] JOUBERT,W., (1994), *A robust GMRES-based adaptive polynomial preconditioning algorithm for nonsymmetric linear systems,* SIAM J. Sci. Comput., vol 15, no 2, pages 427–439.

[844] JOUBERT,W.D., YOUNG,D.M., (1987), *Necessary and sufficient condition for the simplification of generalized conjugate gradient algorithms,* Linear Algebra and its Appl., vol 88, pages 449–485.

[845] JUNG,M., (1995), *Parallelization of multigrid methods based on domain decomposition ideas,* Report SPC 95-27, Technische Universitat Chemnitz.

[846] KAASSCHIETER,E.F., (1986), *The solution of nonsymmetric linear systems by biconjugate gradients or conjugate gradients squared,* Report 86-21, Dept. of Mathematics and Informatics, Delft University of Technology.

[847] KAASSCHIETER,E.F., (1988), *A general finite element preconditioning for the conjugate gradient method,* Report PN 88-14, TNO Institute of Applied Geoscience, Delft.

[848] KAASSCHIETER,E.F., (1988), *A practical termination criterion for the conjugate gradient method,* BIT, vol 28, pages 308–322.

[849] KAASSCHIETER,E.F., (1990), *Preconditioned conjugate gradients and mixed–hybrid finite elements for the solution of potential flow problems,* Ph.D. Thesis, Technische Universiteit Delft.

[850] KAHAN,W., PARLETT,B.N., (1976), *How far should you go with the Lanczos process?,* in Sparse matrix computations, BUNCH,J. and ROSE,D. Eds. Academic Press, pages 131–144.

[851] KAHANER,D., MOLER,C.B., NASH,S., (1988), *Numerical methods and software,* Prentice–Hall.

[852] KAMATH,C., SAMEH,A., (1988), *A projection method for solving nonsymmetric linear systems on multiprocessors,* Parallel Comp., vol 9, pages 291–312.

[853] KARLSON,R., (1991), *A study of some roundoff effects in the GMRES method,* Report LiTH-MAT-R-1990-11, Institute of Technology, Linkoping University.

[854] KARYPIS,G., KUMAR,V., (1994), *A high performance sparse Cholesky factorization algorithm for scalable parallel computers,* Report 94–41, University of Minnesota.

[855] KELLEY,C.T., (1995), *Iterative methods for linear and nonlinear equations,* SIAM.

[856] KERNIGHAN,B.W., LIN,S., (1970), *An efficient heuristic procedure for partitioning graphs,* The Bell System Technical Journal, pages 291–307.

[857] KERSHAW,D.S., (1978), *The incomplete Cholesky conjugate gradient method for the iterative solution of systems of linear equations,* J. Comp. Physics, vol 26, pages 43–65.

[858] KERSHAW,D.S., (1980), *On the problem of unstable pivots in the incomplete LU conjugate gradient method,* J. Comp. Physics, vol 38, pages 114–123.

[859] KEYES,D., (1989), *Domain decomposition methods for the parallel computation of reacting flows,* Comp. Phys. Comm., vol 53, pages 181–200.

[860] KEYES,D., GROPP,W., (1987), *A comparison of domain decomposition techniques for elliptic partial differential equations,* SIAM J. Sci. Comput., vol 8, no 2, pages 166–202.

[861] KEYES,D.E., CHAN,T.F., MEURANT,G.A., SCROGGS,J.S., VOIGT,R.G., (1992), *Fifth international symposium on domain decomposition methods for partial differential equations,* SIAM.

[862] KEYES,D.E., XU,J., (1994), *Domain Decomposition Methods in Scientific and Engineering Computing: Proceedings of the Seventh International Conference on Domain Decomposition,* Contemporary Mathematics 180. AMS.

[863] KICKINGER,F., (1997), *Algebraic multigrid for discrete elliptic second order problems,* Report J. Kepler University, Linz.

[864] KINCAID,D.R., (1972), *Norms of successive overrelaxation,* Math. Comp., vol 26, no 118, pages 345–357.

[865] KINCAID,D.R., OPPE,T.C., YOUNG,D.M., (1986), *Vector computations for sparse linear systems,* SIAM J. Alg. Disc. Meth., vol 7, no 1, pages 99–112.

[866] KOLOTILINA,L., POLMAN,B., (1992), *On incomplete block factorization methods of generalized SSOR type for H-matrices,* Linear Algebra and its Appl., vol 177, pages 111–136.

[867] KORN,D.G., LAMBIOTTE Jr,J.J., (1979), *Computing the fast Fourier transform on a vector computer,* Math. Comp., vol 33, no 147, pages 977–992.

[868] KOWALIK,J.S., (1984), *High speed computation,* NATO ASI Series vol 7. Springer.

[869] KRATZER,D., PARTER,S.V., STEUERWALT,M., (1983), *Block splittings for the conjugate gradient method,* Comp. and fluids, vol 11, no 4, pages 255–279.

[870] KRISHNA,L.B., (1984), *On the convergence of the symmetric successive overrelaxation method,* Linear Algebra and its Appl., vol 56, pages 185–194.

[871] KULISCH,U., (1968), *Uber regulare zerlegungen von matrizen und einige anwendungen,* Numer. Math., vol 11, pages 444–449.

[872] KUMFERT,G., POTHEN,A., (1997), *Two improved algorithms for envelope and wavefront reduction,* BIT, vol 37, no 3, pages 1–32.

[873] KUZNETSOV,Y., (1988), *New algorithms for approximate realization of implicit difference schemes,* Sov. J. Numer. Anal. Math. Modelling, vol 3, no 2,

[874] KUZNETSOV,Y.A., (1989), *Algebraic multigrid domain decomposition methods,* Sov. J. Numer. Anal. Math. Modeling, vol 4, pages 361–392.

[875] KUZNETSOV,Y.A., (1989), *Multi–level domain decomposition methods,* Appl. Numer. Math., vol 6, pages 303–314.

[876] KUZNETSOV,Y.A., (1990), *Multigrid domain decomposition methods,* in Third International Symposium on Domain Decomposition Methods for Partial Differential Equations, CHAN,T.F., GLOWINSKI,R., PERIAUX,J. and WIDLUND,O.B. Eds. SIAM, pages 290–313.

[877] KUZNETSOV,Y.A., (1995), *Efficient iterative solvers for elliptic finite element problems on nonmatching grids,* Russ. J. Numer. Anal. Math. Modeling, vol 10, pages 187–211.

[878] KUZNETSOV,Y.A., MANNINEN,P., VASSILEVSKI,Y., (1993), *On numerical experiments with Neumann–Neumann and Neumann–Dirichlet domain decomposition preconditioners,* Report 3/1993, Dept. of Mathematics, University of Jyvaskyla.

[879] LANCZOS,C., (1952), *Solutions of systems of linear equations by minimized iterations,* J. Nat. Bur. Stand., vol 49, pages 33–53.

[880] LANGER,U., QUECK,W., (1986), *On the convergence factor of Uzawa's algorithm,* J. Comp. Appl. Math., vol 15, pages 191–202.

[881] LASCAUX,P., THEODOR,R., (1993), *Analyse numérique matricielle appliquée a l'art de l'ingénieur, tome 1, second edition,* Masson, Paris.

[882] LAWSON,C.L., HANSON,R.J., KINCAID,D.R., KROGH,F.T., (1979), *Basic linear algebra subprograms for Fortran usage,* ACM Trans. Math. Soft., vol 5, no 3, pages 308–323.

[883] LE BAIL,R.C., (1972), *Use of fast Fourier transforms for solving partial differential equations in physics,* J. Comp. Physics, vol 9, pages 440–465.

[884] LE TALLEC,P., (1992), *Neumann–Neumann domain decomposition algorithms for solving 2D elliptic problems with nonmatching grids,* Report 9235, CEREMADE, Universite Paris-Dauphine.

[885] LE TALLEC,P., (1994), *Domain decomposition methods in computational mechanics,* Comp. Mech. Adv., vol 2, pages 121–220.

[886] LE TALLEC,P., De ROECK,Y.H., VIDRASCU,M., (1991), *Domain decomposition methods for large linearly elliptic three–dimensional problems,* J. Comp. Appl. Math., vol 34, pages 93–117.

[887] LE TALLEC,P., De ROECK,Y.H., VIDRASCU,M., (1992), *A domain decomposed solver for nonlinear elasticity,* Com. Meth. Appl. Mech. Eng., vol 99, pages 187–207.

[888] LE TALLEC,P., SASSI,T., VIDRASCU,M., (1994), *Three-dimensional domain decomposition methods with nonmatching grids and unstructured coarse solvers,* in Domain Decomposition Methods in Scientific and Engineering Computing:

Proceedings of the Seventh International Conference on Domain Decomposition, KEYES,D.E. and XU,J. Eds. AMS, pages 61–74.

[889] LEE,H-C., WATHEN,A.J., (1989), *On element by element preconditioning for general elliptic problems,* Report NA-89-10, Computer Science Dept., Stanford University.

[890] LEHOUCQ,R.B., (1995), *Analysis and implementation of an implicitly restarted Arnoldi iteration,* Ph.D. Thesis, Rice University.

[891] LEHOUCQ,R.B., SORENSEN,D.C., (1996), *Deflation techniques for an implicitly re-started Arnoldi iteration,* SIAM J. Matrix Anal. Appl., vol 17, no 4, pages 789–821.

[892] LEUZE,M.R., (1989), *Independent set orderings for parallel matrix factorization by Gaussian elimination,* Parallel Comp., vol 10, pages 177–191.

[893] LEVEQUE,R.J., TREFETHEN,L.N., (1988), *Fourier analysis of the SOR iteration,* IMA J. Num. Anal., vol 8, pages 273–279.

[894] LEWIS,J.G., GRIMES,R.G., (1981), *Condition number estimation for sparse matrices,* SIAM J. Sci. Comput., vol 2, pages 384–388.

[895] LEWIS,J.G., PEYTON,B.W., POTHEN,A., (1989), *A fast algorithm for reordering sparse matrices for parallel factorization,* SIAM J. Sci. Comput., vol 10, no 6, pages 1146–1173.

[896] LEWIS,J.G., REHM,R.G., (1980), *The numerical solution of a non separable elliptic partial differential equation by preconditioned conjugate gradients,* J. Nat. Bur. Stand., vol 85, no 5, pages 367–390.

[897] LEWIS,J.G., SIMON,H.D., (1988), *The impact of hardware gather/scatter on sparse Gaussian elimination,* SIAM J. Sci. Comput., vol 9, no 2, pages 304–311.

[898] LI,G., COLEMAN,T.F., (1988), *A parallel triangular solver for a distributed-memory multiprocessor,* SIAM J. Sci. Comput., vol 9, pages 485–502.

[899] LI,G., COLEMAN,T.F., (1989), *A new method for solving triangular systems on distributed memory message passing multiprocessors,* SIAM J. Sci. Comput., vol 10, no 2, pages 382–396.

[900] LICHNEWSKY,A., (1984), *Some vector and parallel implementations for preconditioned conjugate gradient algorithms,* in NATO ASI Series, vol F7, High Speed Computation, KOWALIK,J.S. Ed.. Springer.

[901] LICHTENSTEIN,W., JOHNSSON,L., (1993), *Block–cyclic dense linear algebra,* SIAM J. Sci. Comput., vol 14, no 6, pages 1259–1288.

[902] LINDSKOG,G., (1982), *The continued fraction methods for the solution of systems of linear equations,* BIT, vol 22, pages 519–527.

[903] LIONS,P.L., (1988), *On the Schwarz alternating method I,* in Domain decomposition methods for partial differential equations, GLOWINSKI,R., GOLUB,G.H., MEURANT,G. and PERIAUX,J. Eds. SIAM, pages 1–42.

[904] LIONS,P.L., (1989), *On the Schwarz alternating method II: stochastic interpretation,* in Domain decomposition methods for partial differential equations, CHAN,T.F., GLOWINSKI,R., PERIAUX,J. and WIDLUND,O.B. Eds. SIAM, pages 47–70.

[905] LIONS,P.L., (1990), *On the Schwarz alternating method III: a variant for nonoverlapping subdomains,* in Third International Symposium on Domain Decomposition Methods for Partial Differential Equations, CHAN,T.F., GLOWINSKI,R., PERIAUX,J., and WIDLUND,O.B. Eds. SIAM, pages 202–223.

[906] LIPTON,R.J., ROSE,D.J., TARJAN,R.E., (1979), *Generalized nested dissection,* SIAM J. Numer. Anal., vol 16, no 2, pages 346–358.

[907] LIPTON,R.J., TARJAN,R.E., (1980), *Applications of a planar separator theorem,* SIAM J. Comp., vol 9, no 3, pages 615–627.

[908] LIU,J.W–H., (1985), *Modification of the minimum degree algorithm by multiple elimination,* ACM Trans. Math. Soft., vol 11, pages 141–153.

[909] LIU,J.W–H., (1987), *An adaptive general sparse out–of–core Cholesky factorization scheme,* SIAM J. Sci. Comput., vol 9, no 4, pages 585–599.

[910] LIU,J.W–H., (1987), *A note on sparse factorization in a paging environment,* SIAM J. Sci. Comput., vol 8, no 6, pages 1085–1088.

[911] LIU,J.W–H., (1988), *Equivalent sparse matrix reordering by elimination tree rotations,* SIAM J. Sci. Comput., vol 9, no 3, pages 424–444.

[912] LIU,J.W–H., (1989), *Reordering sparse matrices for parallel elimination,* Parallel Comp., vol 11, pages 73–91.

[913] LIU,J.W–H., (1989), *The minimum degree ordering with constraints,* SIAM J. Sci. Comput., vol 10, no 6, pages 1136–1145.

[914] LIU,J.W–H., (1990), *The role of elimination trees in sparse factorization,* SIAM J. Matrix Anal. Appl., vol 11, pages 134–172.

[915] LIU,J.W–H., (1992), *The multifrontal method for sparse matrix solution: theory and practice,* SIAM Review, vol 34, no 1, pages 82–109.

[916] LIU,J.W–H., MIRZAIAN,A., (1989), *A linear reordering algorithm for parallel pivoting of chordal graphs,* SIAM J. Alg. Disc. Meth., vol 2, pages 100–107.

[917] LIU,J.W–H., NG,E.G., PEYTON,B.W., (1993), *On finding supernodes for sparse matrix computations,* SIAM J. Matrix Anal. Appl., vol 14, no 1, pages 242–252.

[918] LIU,J.W–H., SHERMAN,A.H., (1976), *Comparative analysis of the Cuthill–McKee*

and the reverse Cuthill–McKee ordering algorithms for sparse matrices, SIAM J. Numer. Anal., vol 13, no 2, pages 198–213.

[919] LORD,R.E., KOWALIK,J.S, KUMAR,S.P., (1983), *Solving linear algebraic equations on an MIMD computer,* J. ACM, vol 30, no 1, pages 103–117.

[920] MADSEN,N.K., RODRIGUE,G.H., (1977), *Odd even reduction for pentadiagonal matrices,* in Parallel computers, parallel mathematics, FEILMEIER,M. Ed.. IAMCS, pages 103–105.

[921] MAITRE,J.F., MUSY,F., (1982), *The contraction number of a class of two–level methods; an exact evaluation for some finite element subspaces and model problems,* in Multigrid methods, Proceedings of the Koln-Porz Conference, HACKBUSCH,W. and TROTTENBERG,U. Eds, Lecture Notes in Mathematics 960. Springer, pages 535–544.

[922] MAITRE,J.F., MUSY,F., (1984), *Multigrid methods: convergence theory in a variational framework,* SIAM J. Numer. Anal., vol 21, pages 657–671.

[923] MANDEL,J., (1990), *On block diagonal and Schur complement preconditioning,* Numer. Math., vol 58, pages 79–93.

[924] MANDEL,J., BREZINA,M., VANEK,P., (1998), *Energy optimization of algebraic multigrid bases,* Report University of colorado, Denver.

[925] MANDEL,J., McCORMICK,S.F., RUGE,J.W., (1988), *An algebraic theory for multigrid methods for variational problems,* SIAM J. Numer. Anal., vol 25, no 1, pages 91–110.

[926] MANSFIELD,L., (1990), *On the conjugate gradient solution of the Schur complement system obtained from domain decomposition,* SIAM J. Numer. Anal., vol 27, no 6, pages 1612–1620.

[927] MANTEUFFEL,T.A., (1977), *The Tchebychev iteration for nonsymmetric linear systems,* Numer. Math., vol 28, pages 307–327.

[928] MANTEUFFEL,T.A., (1978), *Adaptive procedure for estimating parameters for the nonsymmetric Tchebychev iteration,* Numer. Math., vol 31, pages 183–208.

[929] MANTEUFFEL,T.A., (1979), *The shifted incomplete Cholesky factorization,* in Sparse matrix proceedings 1978, DUFF,I.S. and STEWART,G.W. Eds. SIAM.

[930] MANTEUFFEL,T.A., (1980), *An incomplete factorization technique for positive definite linear systems,* Math. Comp., vol 34, no 150, pages 473–497.

[931] MANTEUFFEL,T., OTTO,J., (1993), *On the roots of the orthogonal polynomials and residual polynomials associated with a conjugate gradient method,* J. Num. Lin. Alg.,

[932] MANTEUFFEL,T., OTTO,J., (1993), *Optimal equivalent preconditioners,* SIAM J. Numer. Anal., vol 30, no 3, pages 790–812.

[933] MANTEUFFEL,T.A., PARTER,S.V., (1990), *Preconditioning and boundary conditions,* SIAM J. Numer. Anal., vol 27, no 3, pages 656–694.

[934] MARCHUK,G.I., KUZNETSOV,Y.A., (1988), *Approximate algorithms for implicit difference schemes,* in Analyse Mathématique et Applications. Gauthiers-Villars, Paris, pages 357–371.

[935] MARCHUK,G.I., KUZNETSOV,Y.A., MATSOKIN,A.M., (1986), *Fictitious domain and domain decomposition methods,* Sov. J. Numer. Anal. Math. Modeling, vol 1, pages 3–35.

[936] MARKHAM,G., (1990), *Conjugate gradient type methods for indefinite, asymmetric and complex systems,* IMA J. Num. Anal., vol 10, pages 155–170.

[937] MARKOWITZ,H.M., (1957), *The elimination form of the inverse and its application to linear programming,* Manag. Sci. , vol 3, pages 255–269.

[938] MARRAKCHI,M., ROBERT,Y., (1989), *Optimal algorithms for Gaussian elimination on an MIMD computer,* Parallel Comp., vol 12, no 2, pages 183–194.

[939] MARTIN,R.S, WILKINSON,J.H., (1965), *Symmetric decomposition of positive definite band matrices,* Numer. Math., vol 7, pages 355–361.

[940] MARTIN,R.S., WILKINSON,J.H., (1967), *Solution of symmetric and unsymmetric band equations and the calculation of eigenvectors of band matrices,* Numer. Math., vol 9, pages 279–301.

[941] MARTINS,M.M., (1980), *On an accelerated overrelaxation iterative method for linear systems with strictly diagonally dominant matrix,* Math. Comp., vol 33, no 152, pages 1269–1273.

[942] MARTINS,M.M., (1981), *Note on irreducible diagonally dominant matrices and the convergence of the AOR iterative method,* Math. Comp., vol 37, no 155, pages 101–103.

[943] MATHEW,T.P., (1989), *Domain decomposition and iterative refinement methods for mixed finite element discretizations of elliptic problems,* Ph.D. Thesis, TR 463, Dept. of Computer Science, Courant Institute of Mathematical Sciences.

[944] MATHEW,T.P., (1993), *Schwarz alternating and iterative refinement methods for mixed formulations of elliptic problems, part I: algorithms and numerical results,* Numer. Math., vol ?, pages 445–468.

[945] MATHEW,T.P., (1993), *Schwarz alternating and iterative refinement methods for mixed formulations of elliptic problems, part II: theory,* Numer. Math., vol ?, pages 469–492.

[946] MATHEW,T.P., POLYAKOV,P.L., RUSSO,G., WANG,J., (1996), *Domain decomposition operator splittings for the solution of parabolic equations,* to appear in SIAM J. Sci. Comput., pages 469–492.

[947] McBRYAN,O.A., VAN DE VELDE,E.F., (1985), *Parallel algorithms for elliptic equations,* Comm. Pure Appl. Math, vol 38, pages 769–795.

[948] McCORMICK,S.F., (1977), *The methods of Kaczmarz and row orthogonalization for solving linear equations and least squares problems in Hilbert space,* Indiana Univ. Math. J., vol 26, no 6, pages 1137–1150.

[949] McCORMICK,S.F., (1982), *An algebraic interpretation of multigrid methods,* SIAM J. Numer. Anal., vol 19, pages 548–560.

[950] McCORMICK,S.F., (1985), *Multigrid methods for variational problems: general theory for the V–cycle,* SIAM J. Numer. Anal., vol 22, pages 634–643.

[951] McCORMICK,S.F., (1986), *The fast adaptive composite (FAC) method for elliptic equations,* Math. Comp., vol 46, pages 439–456.

[952] McCORMICK,S.F., (1987), *Multigrid methods,* SIAM.

[953] McCORMICK,S.F., (1988), *Multigrid methods: theory, applications, and supercomputing,* Lecture Notes in Pure and Applied Mathematics 110. Marcel Dekker.

[954] McCORMICK,S.F., (1989), *Multilevel adaptive methods for partial differential equations,* SIAM.

[955] McCORMICK,S.F., RUGE,J.W., (1983), *Unigrid for multigrid simulation,* Math. Comp., vol 41, no 163, pages 43–62.

[956] McDONALD,B.E., (1980), *The Chebychev method for solving nonself-adjoint elliptic equations on a vector computer,* J. Comp. Physics, vol 35, pages 147–168.

[957] McKENNEY,A., GREENGARD,L., MAYO,A., (1995), *A fast Poisson solver for complex geometries,* J. Comp. Physics, vol 118, pages 348–355.

[958] MEIER,U., (1985), *A parallel partition method for solving banded systems of linear equations,* Parallel Comp., vol 2, pages 33–43.

[959] MEIJERINK,J.A., VAN DER VORST,H.A., (1977), *An iterative solution method for linear systems of which the coefficient matrix is a symmetric M-matrices,* Math. Comp., vol 31, no 137, pages 148–162.

[960] MEIJERINK,J.A., VAN DER VORST,H.A., (1981), *Guidelines for the usage of incomplete decompositions in solving sets of linear equations as they occur in practical problems,* J. Comp. Physics, vol 44, pages 134–155.

[961] MELHEM,R.G., (1987), *Towards efficient implementation of PCG methods on vector supercomputers,* Int. J. Supercomputer Appl., vol 1, pages 70–98.

[962] MELHEM,R.G., (1988), *A modified frontal technique suitable for parallel systems,* SIAM J. Sci. Comput., vol 9, no 2, pages 289–303.

[963] MELHEM,R.G., (1988), *Parallel solution of linear systems with striped sparse ma-*

trices, Parallel Comp., vol 6, pages 165–184.

[964] MESINA,M., (1977), *Convergence acceleration for the iterative solution of the equations $x=Ax+f$,* Com. Meth. Appl. Mech. Eng., vol 10, pages 165–173.

[965] MEURANT,G., (1983), *The Fourier/tridiagonal method for the Poisson equation from the point of view of block Cholesky factorization,* Report LBID–764 , Lawrence Berkeley Laboratory, CA.

[966] MEURANT,G., (1984), *The block preconditioned conjugate gradient method on vector computers,* BIT, vol 24, pages 623–633.

[967] MEURANT,G., (1986), *The conjugate gradient method on supercomputers,* Supercomputer, vol 13, pages 9–17.

[968] MEURANT,G., (1987), *Multitasking the conjugate gradient method on the CRAY X–MP/48,* Parallel Comp., vol 5, pages 267–280.

[969] MEURANT,G., (1988), *Domain decomposition vs block preconditioning,* in Domain decomposition methods for partial differential equations, GLOWINSKI,R., GOLUB,G.H., MEURANT,G. and PERIAUX,J. Eds. SIAM. pages 231–249.

[970] MEURANT,G., (1988), *Domain decomposition preconditioners for the conjugate gradient method,* Calcolo, vol 25, no 1, pages 103–119.

[971] MEURANT,G., (1988), *Incomplete domain decomposition preconditioners for the conjugate gradient method,* in Proceedings of the third SIAM conference on parallel processing for scientific computing, Los Angeles 1987, RODRIGUE,G. Ed. SIAM, pages 100–106.

[972] MEURANT,G., (1989), *Incomplete domain decomposition preconditioners for nonsymmetric problems,* in Second International Symposium on Domain Decomposition Methods for Partial Differential Equations, CHAN,T.F., GLOWINSKI,R., PERIAUX,J., and WIDLUND,O.B. Eds. SIAM, pages 219–225.

[973] MEURANT,G., (1989), *Domain decomposition methods for partial differential equations on parallel computers,* Int. J. Supercomputer Appl., vol 2, no 4, pages 5–12.

[974] MEURANT,G., (1989), *Domain decomposition methods for partial differential equations on parallel computers,* in Vector and Parallel Computing: issues in applied research and development, DONGARRA,J. J. , DUFF,I. S., GAFFNEY,P. and McKEE,S. Eds. Ellis Horwood, pages 241–253.

[975] MEURANT,G., (1989), *Practical use of the conjugate gradient method on parallel supercomputers.,* Comp. Phys. Comm., vol 53, pages 467–477.

[976] MEURANT,G., (1989), *Iterative methods for multiprocessor vector computers,* Computer Physics Reports, vol 11, pages 51–80.

[977] MEURANT,G., (1990), *The conjugate gradient method on vector and parallel su-*

percomputers, in Computational techniques and applications, HOGARTH,W.L. and NOYE,B.J. Eds . Hemisphere Publishing Corp..

[978] MEURANT,G., (1991), *Parallel algorithms for supercomputers,* in Linear algebra and digital signal processing, GOLUB,G.H. and VAN DOOREN,P., Eds, NATO ASI Series. Springer.

[979] MEURANT,G., (1991), *Numerical experiments with domain decomposition methods for parabolic problems on parallel computers,* in Fourth International Symposium on Domain Decomposition Methods for Partial Differential Equations, GLOWINSKI,R., KUZNETSOV,Y.A., MEURANT,G., PERIAUX,J., and WIDLUND,O.B. Eds. SIAM, pages 394–408.

[980] MEURANT,G., (1991), *A domain decomposition method for parabolic problems,* Appl. Num. Math., vol 8, pages 427–441.

[981] MEURANT,G., (1991), *Domain decomposition methods for solving large sparse linear systems,* in Computer algorithms for solving linear algebraic equations, the state of the art, Proceedings of the NATO Advanced Study Institute on linear systems, Castelvecchio Pascoli (Italy), SPEDICATO,E. Ed., NATO ASI Series v 77. Springer.

[982] MEURANT,G., (1992), *A review on the inverse of symmetric tridiagonal and block tridiagonal matrices,* SIAM J. Matrix Anal. Appl., vol 13, no 3, pages 707–728.

[983] MEURANT,G., (1992), *Os computadores paralelos e o futuro da computacao cientifica,* Boletim da SBMAC, vol 3, no 1, pages 36–52.

[984] MEURANT,G., (1997), *The computation of bounds for the norm of the error in the conjugate gradient algorithm,* Numer. Alg., vol 16, pages 77–87.

[985] MEURANT,G., CHAN,T.F., CIARLET,P. Jr., (1990), *Multilevel incomplete Cholesky preconditioners,* Unpublished manuscript,

[986] MEYER,JR.,C.D., PLEMMONS,R.J., (1977), *Convergent powers of a matrix with applications to iterative methods for singular linear systems,* SIAM J. Numer. Anal., vol 14, no 4, pages 699–705.

[987] MICHIELSE,P.H., (1994), *Parallel multigrid using PVM,* Supercomputer, vol 10, pages 10–23.

[988] MICHIELSE,P.H., VAN DER VORST,H.A., (1988), *Data transport in Wang's partition method,* Parallel Comp., vol 7, pages 87–95.

[989] MIELLOU,J.-C., CORTEY-DUMONT,P., BOULBRACHENE,M., (1990), *Perturbation of fixed-point iterative methods,* in Advances in Parallel Computing, vol 1. pages 81–122.

[990] MIELLOU,J.-C., SPITERI,P., EVANS,D.J., (1990), *Le préconditionnement SSOR : une estimation du conditionnement obtenu; optimisation du paramètre de relaxation,* C.R. Acad. Sci. Paris, vol 311, pages 817–823.

[991] MILASZEWICZ,J.P., (1981), *On criticality and the Stein –Rosenberg theorem,* SIAM J. Numer. Anal., vol 18, no 3, pages 559–564.

[992] MILLER,K., (1965), *Numerical analogs to the Schwarz alternating procedure,* Numer. Math., vol 7, pages 91–103.

[993] MISSIRLIS,N.M., (1985), *A parallel iterative system solver,* Linear Algebra and its Appl., vol 65, pages 25–44.

[994] MITCHISON,G., DURBIN,R., (1986), *Optimal numberings of an N by N array,* SIAM J. Alg. Disc. Meth., vol 7, no 4, pages 571–582.

[995] MORE,J., (1971), *Global convergence of Newton Gauss Seidel methods,* SIAM J. Numer. Anal., vol 8, pages 325–336 .

[996] MOSKOVITZ,D., (1944), *The numerical solution of Laplace's and Poisson's equations,* Quart. Appl. Math., vol 2, pages 148–163.

[997] MUNKSGAARD,N., (1978), *Solution of general sparse symmetric sets of linear equations,* Report NI-78-02, Numerik Institut, Danmarks Tekniske Hojskole.

[998] MUNKSGAARD,N., (1979), *New factorization codes for sparse, symmetric and positive definite matrices,* BIT, vol 19, pages 43–52.

[999] MUNKSGAARD,N., (1980), *Solving sparse symmetric sets of linear equations by preconditioned conjugate gradients,* ACM Trans. Math. Soft., vol 6, no 2, pages 206–219.

[1000] MUNTHE-KAAS,H., (1987), *The convergence rate of inexact preconditioned steepest descent algorithm for solving linear systems,* Report NA–87–04, Computer Science Dept., Stanford University.

[1001] NABEN,R., (1997), *Decay rates of the inverse of nonsymmetric tridiagonal and band matrices,* submitted to SIAM J. Matrix Anal. Appl..

[1002] NACHTIGAL,N.M., (1991), *A look ahead variant of the Lanczos algorithm and its application to the quasi minimal residual method for non Hermitian linear systems,* Ph.D. Thesis MIT also RIACS Tech Report 91.19, NASA Ames.

[1003] NACHTIGAL,N.M., REDDY,S.C., TREFETHEN,L.N., (1992), *How fast are nonsymmetric matrix iterations?,* SIAM J. Matrix Anal. Appl., vol 13, pages 778–795.

[1004] NACHTIGAL,N.M., REICHEL,L., TREFETHEN,L.N., (1990), *A hybrid GMRES algorithm for nonsymmetric linear systems,* SIAM J. Matrix Anal. Appl.,

[1005] NAIMAN,A.E., BABUSKA,I.M., ELMAN,H.C., (1995), *A note on conjugate gradient convergence,* Report Computer Science Dept., University of Maryland.

[1006] NASH,S.G., (1990), *A history of scientific computing,* ACM Press.

[1007] NATAF,F., ROGIER,F., De STURLER,E., (1994), *Optimal interface conditions for domain decomposition methods,* Rapport RI 301, Centre de Mathématiques Appliquées, Ecole Polytechnique.

[1008] NEAL,L., POOLE,G., (1992), *A geometric analysis of Gaussian elimination,* Linear Algebra and its Appl., vol 173, pages 239–264.

[1009] NEUMAIER,A., VARGA,R.S., (1984), *Exact convergence and divergence domains for the symmetric successive overrelaxation iterative (SSOR) method applied to H–matrices,* Linear Algebra and its Appl., vol 58, pages 261–272.

[1010] NEUMANN,M., (1981), *The Kahan SOR convergence bound for nonsingular and irreducible M–matrices,* Linear Algebra and its Appl., vol 39, pages 205–222.

[1011] NEUMANN,M., PLEMMONS,R.J., (1987), *Convergence of parallel multisplitting iterative methods for M-matrices,* Linear Algebra and its Appl., vol 88, pages 559–573.

[1012] NEUMANN,M., VARGA,R.S., (1980), *On the sharpness of some upper bounds for the spectral radii of SOR iteration matrices,* Numer. Math., vol 35, pages 69–79.

[1013] NG,E., (1993), *Supernodal symbolic Cholesky factorization on a local memory multiprocessor,* Parallel Comp., vol 19, pages 153–162.

[1014] NG,E., PEYTON,B.W., (1993), *A supernodal Cholesky factorization algorithm for shared memory multiprocessors,* SIAM J. Sci. Comput., vol 14, no 4, pages 761–769.

[1015] NG,E., PEYTON,B.W., (1993), *Block sparse Cholesky algorithms on advanced uniprocessor computers,* SIAM J. Sci. Comput., vol 14, no 5, pages 1034–1056.

[1016] NICOLAIDES,R.A., (1975), *On multiple grid and related techniques for solving discrete elliptic systems,* J. Comp. Physics, vol 19, pages 418–431.

[1017] NICOLAIDES,R.A., (1979), *On some theoretical and practical aspects of multigrid methods,* Math. Comp., vol 33, no 147, pages 933–952.

[1018] NICOLAIDES,R.A., (1987), *Deflation of conjugate gradients with application to boundary value problems,* SIAM J. Numer. Anal., vol 24, no 2, pages 355–365.

[1019] NIETHAMMER,W., VARGA,R.S., (1983), *The analysis of k–step iterative methods for linear systems from summability theory,* Numer. Math., vol 41, pages 177–206.

[1020] NOTAY,Y., (1991), *Conditioning analysis of modified block incomplete factorizations,* Linear Algebra and its Appl., vol 154, pages 711–722.

[1021] NOTAY,Y., (1993), *On the convergence rate of the conjugate gradients in presence of rounding errors,* Numer. Math., vol 65, pages 301–317.

[1022] NOTAY,Y., (1993), *Ordering methods for approximate factorization preconditioning,* Report IT/IF/14–11, Universite libre de Bruxelles.

[1023] NOTAY,Y., (1994), *DRIC: a dynamic version of the RIC method,* Numer. Lin. Alg. with Appl., vol 1, pages 511–532.

[1024] NOTAY,Y., (1997), *Optimal order preconditioning of finite difference matrices,* Report GANMN 97–02, Université libre de Bruxelles.

[1025] NOTAY,Y., (1997), *Optimal V cycle algebraic multilvel preconditioning,* Report GANMN 97–05, Université libre de Bruxelles.

[1026] NOUR-OMID,B., PARLETT,B.N., (1982), *Element preconditioning,* Report PAM-103, University of California, Berkeley.

[1027] NOUR-OMID,B., PARLETT,B.N., (1985), *Element preconditioning using splitting techniques,* SIAM J. Sci. Comput., vol 6, pages 761–771.

[1028] O'CARROLL,M.J., (1973), *Inconsistencies and SOR convergence for the discrete Neumann problem,* J. Inst. Maths. Applics., vol 11, pages 343–350.

[1029] O'CARROLL,M.J., (1976), *Diagonal dominance and SOR performance with skew nets,* Int. J. Num. Meth. Eng., vol 10, pages 225–240.

[1030] O'LEARY,D., STEWART,G., (1986), *Assignment and scheduling in parallel matrix factorization,* Linear Algebra and its Appl., vol 77, pages 275–299.

[1031] O'LEARY,D.P., (1976), *Hybrid conjugate gradient algorithms,* Ph.D. Thesis, STAN–CS–76–548, Computer Science Dept., Stanford University.

[1032] O'LEARY,D.P., (1980), *Estimating matrix condition numbers,* SIAM J. Sci. Comput., vol 1, no 2, pages 205–209.

[1033] O'LEARY,D.P., (1980), *The block conjugate gradient algorithm and related methods,* Linear Algebra and its Appl., vol 29, pages 293–322.

[1034] O'LEARY,D.P., (1984), *Ordering schemes for parallel processing of certain mesh problems,* SIAM J. Sci. Comput., vol 5, pages 620–632.

[1035] O'LEARY,D.P., WHITE,R.E., (1985), *Multisplittings and parallel solution of linear systems,* SIAM J. Alg. Disc. Meth., vol 6, no 4, pages 630–640.

[1036] O'LEARY,D.P., WIDLUND,O.B., (1979), *Capacitance matrix methods for the Helmholtz equation on general three-dimensional regions,* Math. Comp., vol 33, no 147, pages 849–879.

[1037] OETTLI,W., PRAGER,W., (1964), *Compatibility of approximate solution of linear equations with given error bounds for coefficients and right hand sides,* Numer. Math., vol 6, pages 405–409.

[1038] OGIELSKI,A.T., AIELLO,W., (1993), *Sparse matrix computations on parallel processor arrays,* SIAM J. Sci. Comput., vol 14, no 3, pages 519–530.

[1039] OLDENBURGER,R., (1940), *Infinite powers of matrices and characteristic roots,*

Duke Math. J., vol 6, pages 357–361.

[1040] OLIGER,J., SKAMAROCK,W., TANG,W.P., (1986), *Convergence analysis and acceleration of the Schwarz alternating method,* Report CLaSSiC-86-12, Center for Large Scale Scientific Computation, Stanford University.

[1041] OLIPHANT,T.A., (1962), *An extrapolation procedure for solving linear systems,* SIAM J. Numer. Anal., vol 20, no 3, pages 257–265.

[1042] OLVER,F.W., WILKINSON,J.H., (1982), *A posteriori error bounds for Gaussian elimination,* J. Inst. Maths. Applics., vol 2, pages 377–406.

[1043] ONG,H.L., (1984), *Fast approximate solution of large scale sparse linear systems,* J. Comp. Appl. Math., vol 10, pages 45–54.

[1044] ONG,M.E., (1989), *Hierarchical basis preconditioner for second order elliptic problems in three dimensions,* Ph.D. Thesis, Dept. of Applied Mathematics, University of Washington.

[1045] ONG,M.E., (1991), *Linear hierarchical basis preconditioners in three dimensions,* Report CAM 91-15, Dept. of Mathematics, UCLA.

[1046] OPFER,G., SCHOBER,G., (1984), *Richardson's iteration for nonsymmetric matrices,* Linear Algebra and its Appl., vol 58, pages 343–361.

[1047] ORTEGA,J.M., (1988), *Introduction to parallel and vector solution of linear systems,* Plenum Press.

[1048] ORTEGA,J.M., (1988), *The ijk forms of factorization methods I. Vector computers,* Parallel Comp., vol 7, no 2, pages 135–148.

[1049] ORTEGA,J.M., PLEMMONS,R.J., (1979), *Extensions of the Ostrowski–Reich theorem for SOR iterations,* Linear Algebra and its Appl., vol 28, pages 177–191.

[1050] ORTEGA,J.M., ROMINE,C.H., (1988), *The ijk forms of factorization methods II. Parallel systems,* Parallel Comp., vol 7, no 2, pages 149–162.

[1051] ORTEGA,J.M., VOIGT,R.G., (1985), *Solution of partial differential equations on vector and parallel computers,* SIAM Review, vol 27, pages 149–240.

[1052] ORTEGA,J.M., VOIGT,R.G., ROMINE,C.H., (1988), *A bibliography on parallel and vector numerical algorithms,* ICASE Report NASA 181764.

[1053] OSTERBY,O., ZLATEV,Z., (1983), *Direct methods for sparse matrices,* Lecture Notes in Computer Science 157. Springer.

[1054] OSWALD,P., (1994), *On the convergence rate of SOR: a worst case estimate,* Computing, vol 52, pages 245–255.

[1055] PAIGE,C.C., (1974), *Bidiagonalization of matrices and solution of linear equations,* SIAM J. Numer. Anal., vol 11, no 1, pages 197–209.

[1056] PAIGE,C.C., (1976), *Error analysis of the Lanczos algorithm for tridiagonalizing a symmetric matrix,* J. Inst. Maths. Applics., vol 18, pages 341–349.

[1057] PAIGE,C.C., SAUNDERS,M.A., (1973), *Solution of sparse indefinite systems of equations and least squares problems,* Report STAN-CS-73-399, Computer Science Dept., Stanford University.

[1058] PAIGE,C.C., SAUNDERS,M.A., (1975), *Solution of sparse indefinite systems of linear equations,* SIAM J. Numer. Anal., vol 12, no 4, pages 617–629.

[1059] PAIGE,C.C., SAUNDERS,M.A., (1978), *A bidiagonalization algorithm for sparse linear equations and least squares problems,* Report SOL 78–19, Systems Optimization Lab., Stanford University.

[1060] PAPAMICHAEL,N., SMITH,G.D., (1975), *The determination of consistent orderings for the SOR iterative method,* J. Inst. Maths. Applics., vol 15, pages 239–248.

[1061] PARLETT,B.N., (1980), *A new look at the Lanczos algorithm for solving symmetric systems of linear equations,* Linear Algebra and its Appl., vol 29, pages 323–346.

[1062] PARLETT,B.N., (1980), *The symmetric eigenvalue problem,* Prentice–Hall.

[1063] PARLETT,B.N., REID,J.K., (1970), *On the solution of a system of linear equations whose matrix is symmetric but not definite,* BIT, vol 10, pages 386–397.

[1064] PARLETT,B.N., SCOTT,D.S., (1979), *The Lanczos algorithm with selective orthogonalization,* Math. Comp., vol 33, pages 217–238.

[1065] PARTER,S.V., (1961), *The use of linear graphs in Gauss elimination,* SIAM Review, vol 3, no 2, pages 119–130.

[1066] PARTER,S.V., STEUERWALT,M., (1980), *On k–line and k x k block iterative schemes for a problem arising in 3-D elliptic difference equations,* Report TR–374, Computer Sciences Dept, University of Wisconsin.

[1067] PARTER,S.V., STEUERWALT,M., (1985), *Block iterative methods for elliptic finite element equations,* SIAM J. Numer. Anal., vol 22, no 1, pages 146–179.

[1068] PAULINO,G.H., MENEZES,I.F., GATTASS,M., MUKHERJEE,S., (1994), *Node and element resequencing using the Laplacian of a finite element graph: part I-general concepts and algorithm,* Int. J. Num. Meth. Eng., vol 37, pages 1511–1530.

[1069] PEREYRA,V., (1967), *Accelerating the convergence of discretization algorithms,* SIAM J. Numer. Anal., vol 4, no 4, pages 508–533.

[1070] PEREYRA,V., PROSKUROWSKI,W., WIDLUND,O.B., (1977), *High order Laplace solvers for the Dirichlet problem on general regions,* Math. Comp., vol 31, no 137, pages 1–16.

[1071] PERLOT,O., (1995), *Préconditionnements de systèmes linéaires sur machines massivement parallèles CM-2 et CM-5,* Ph.D. Thesis, Universite Paris VI.

[1072] PETRAVIC,M., KUO-PETRAVIC,G., (1979), *An ILUCG algorithm which minimizes in the Euclidean norm,* J. Comp. Physics, vol 32, pages 263–269.

[1073] PEYTON,B., (1986), *Some applications of clique trees to the solution of sparse linear systems,* Ph.D. thesis Clemson University.

[1074] PISSANETZKY,S., (1984), *Sparse matrix technology,* Academic Press.

[1075] POOLE,E.L., ORTEGA,J.M., (1987), *Multicolor ICCG methods for vector computers,* SIAM J. Numer. Anal., vol 24, no 6, pages 1394–1418.

[1076] POOLE,G., NEAL,L., (1991), *A geometric analysis of Gaussian elimination, I,* Linear Algebra and its Appl., vol 149, pages 249–272.

[1077] POOLE,G., NEAL,L., (1992), *Gaussian elimination: when is scaling beneficial?,* Linear Algebra and its Appl., vol 162–164, pages 309–324.

[1078] POTHEN,A., (1996), *Graph partitioning algorithms with application to scientific computing,* in "Parallel numerical algorithms", KEYES,D.E., SAMEH,A.H., VENKATAKRISHNAN,V. Eds. Academic Press.

[1079] POTHEN,A., ALVARADO,F.L., (1992), *A fast reordering algorithm for parallel sparse triangular solution,* SIAM J. Sci. Comput., vol 13, no 2, pages 645–653.

[1080] POTHEN,A., SIMON,H.D., LIOU,K.P., (1990), *Partitioning sparse matrices with eigenvectors of graphs,* SIAM J. Matrix Anal. Appl., vol 11, no 3, pages 430–452.

[1081] POTHEN,A., SUN,C., (1993), *A mapping algorithm for parallel sparse Cholesky factorization,* SIAM J. Sci. Comput., vol 14, no 5, pages 1253–1257.

[1082] PROSKUROWSKI,W., (1979), *Numerical solution of Helmholtz's equation by implicit capacitance matrix methods,* ACM Trans. Math. Soft., vol 5, no 1, pages 36–49.

[1083] PROSKUROWSKI,W., SHA,S., (1990), *Performance of the Neumann-Dirichlet preconditioner for substructures with intersecting interfaces,* in Third International Symposium on Domain Decomposition Methods for Partial Differential Equations, CHAN,T.F., GLOWINSKI,R., PERIAUX,J., and WIDLUND,O.B. Eds. SIAM, pages 322–338.

[1084] PROSKUROWSKI,W., WIDLUND,O.B., (1976), *On the numerical solution of Helmholtz's equation by the capacitance matrix method,* Math. Comp., vol 30, no 135, pages 433–468.

[1085] PROSKUROWSKI,W., WIDLUND,O.B., (1980), *A finite element capacitance matrix method for the Neumann problem for Laplace's equation,* SIAM J. Sci. Comput., vol 1, no 4, pages 410–425.

[1086] PRZEMIENIECKI,J. S. , (1963), *Matrix structural analysis of substructures,* Am. Inst. Aero. Astro. J., vol 1, pages 138–147.

[1087] PRZEMIENIECKI,J. S. , (1968), *Theory of Matrix Structural Analysis,* McGraw-Hill.

[1088] QUARTERONI,A., (1990), *Domain decomposition method for the numerical solution of partial differential equations,* Report UMSI 90/246, Supercomputer Research Institute, University of Minnesota.

[1089] QUARTERONI,A., (1991), *Domain decomposition and parallel processing for the numerical solution of partial differential equations,* Surv. Math. Industry, vol 1, pages 75–118.

[1090] QUARTERONI,A., ZAMPIERI,E., (1991), *Finite element preconditioning for Legendre spectral collocation approximations to elliptic equations and systems,* Report UMSI 91/181, Supercomputer Institute, University of Minnesota.

[1091] RADICATI di BROZOLO,G., ROBERT,Y., (1989), *Parallel conjugate gradient like algorithms for solving sparse nonsymmetric linear systems on a vector multiprocessor,* Parallel Comp., vol 11, pages 223–239.

[1092] RAGHAVAN,P., (1995), *Distributed sparse Gaussian elimination and orthogonal factorization,* SIAM J. Sci. Comput., vol 16, pages 1462–1477.

[1093] REID,J.K., (1971), *Large sparse sets of linear equations,* Academic Press.

[1094] REID,J.K., (1971), *A note on the stability of Gaussian elimination,* J. Inst. Maths. Applics., vol 8, pages 374–375.

[1095] REID,J.K., (1971), *On the method of conjugate gradients for the solution of large sparse systems of linear equations,* in Large sparse sets of linear equations, REID,J.K. Ed. Academic Press, pages 231–254.

[1096] REID,J.K., (1972), *The use of conjugate gradients for systems of linear equations possessing "Property A",* SIAM J. Numer. Anal., vol 9, no 2, pages 325–332.

[1097] REID,J.K., (1986), *Sparse matrices,* Report CSS 201, Harwell Laboratory.

[1098] REITER,E., RODRIGUE,G., (1983), *An incomplete Cholesky factorization by a matrix partition algorithm,* Report UCRL-88989, Lawrence Livermore National Laboratory.

[1099] RESASCO,D., (1990), *Domain decomposition algorithms for elliptic partial differential equations,* Ph.D. Thesis, YALEU/DCS/RR-776, Dept. of Computer Science, Yale University.

[1100] REUSKEN,A., (1996), *A multigrid method based on incomplete Gaussian elimination,* Numer. Lin. Alg. with Appl., vol 3, no 5, pages 369–390.

[1101] RICHTMYER,R.D., MORTON,K.W., (1967), *Difference methods for initial value problems, second edition,* Wiley.

[1102] RIGAL,J.L., GACHES,J., (1967), *On the compatibility of a given solution with the data of a linear system,* J. ACM, vol 14, pages 543–548.

[1103] RIZVI,S.A., (1984), *Inverses of quasi tridiagonal matrices,* Linear Algebra and its Appl., vol 56, pages 177–184.

[1104] ROBERT,F., (1969), *Blocs H-matrices et convergence des methodes iteratives classiques par blocs,* Linear Algebra and its Appl., vol 2, pages 223–265.

[1105] ROBERT,Y., (1982), *Regular incomplete factorizations of real positive definite matrices,* Linear Algebra and its Appl., vol 48, pages 105–117.

[1106] ROBINSON,P.D., WATHEN,A.J., (1992), *Variational bounds on the entries of the inverse of a matrix,* IMA J. Num. Anal., vol 12, pages 463–486.

[1107] RODRIGUE,G., (1982), *Parallel Computations,* Academic Press.

[1108] RODRIGUE,G., (1985), *Inner/outer iterative methods and numerical Schwarz algorithms,* Parallel Comp., vol 2, pages 205–218.

[1109] RODRIGUE,G., (1989), *Parallel processing for scientific computing,* SIAM.

[1110] RODRIGUE,G., SAYLOR,P.E., (1985), *Inner/outer iterative methods and numerical Schwarz algorithms - II,* Proceedings of the IBM Conference on vector and parallel computations for scientific computing, Rome.

[1111] RODRIGUE,G., SIMON,J., (1984), *Jacobi splittings and the method of overlapping domains for solving elliptic PDE's,* Report UCRL-90656, Lawrence Livermore National Laboratory.

[1112] RODRIGUE,G., SIMON,J., (1984), *A generalization of the numerical Schwarz algorithm,* in Computing methods in applied sciences and engineering VI, GLOWINSKI,R. and LIONS,J.L. Eds. North-Holland, pages 273–283.

[1113] RODRIGUE,G., WOLITZER,D., (1984), *Preconditioning by incomplete block cyclic reduction,* Math. Comp., vol 42, no 166, pages 549–565.

[1114] RODRIGUE,G.H., MADSEN,N.K., KARUSH,J.I., (1979), *Odd–even reduction for banded linear equations,* J. ACM, vol 26, no 1, pages 72–81.

[1115] ROHN,J., (1990), *Nonsingularity under data rounding,* Linear Algebra and its Appl., vol 139, pages 171–174.

[1116] ROMAN,J., (1985), *Calcul de complexité relatifs à une méthode de dissection emboîtée,* Numer. Math., vol 47, pages 175–190.

[1117] ROMANI,F., (1986), *On the additive structure of the inverses of banded matrices,* Linear Algebra and its Appl., vol 80, pages 131–140.

[1118] ROMINE,C.H., ORTEGA,J.M., (1988), *Parallel solution of triangular systems of equations,* Parallel Comp., vol 6, pages 109–114.

[1119] ROSE,D.J., (1970), *Triangulated graphs and the elimination process,* J. Math. Anal. Appl., vol 32, pages 597–609.

[1120] ROSE,D.J., (1980), *Matrix identities of the fast Fourier transform,* Linear Algebra and its Appl., vol 29, pages 423–443.

[1121] ROSE,D.J., TARJAN,R.E., (1978), *Algorithmic aspects of vertex elimination on graphs,* SIAM J. Comp., vol 5, no 2, pages 266–283.

[1122] ROSE,D.J., TARJAN,R.E., LUEKER,G.S., (1976), *Algorithmic aspects of vertex elimination on directed graphs,* SIAM J. Appl. Math., vol 34, no 1, pages 176–197.

[1123] ROSE,D.J., WILLOUGHBY,R.A., (1972), *Sparse matrices and their applications,* Plenum Press.

[1124] ROTHBERG,E., (1993), *Exploiting the memory hierarchy in sequential and parallel sparse Cholesky factorization,* Ph.D. thesis, Stanford University.

[1125] ROTHBERG,E., GUPTA,A., (1994), *An efficient block–oriented approach to parallel sparse Cholesky factorization,* SIAM J. Sci. Comput., vol 15, no 6, pages 1413–1439.

[1126] ROUX,F.X., TROMEUR-DERVOUT,D., (1994), *Parallelization of a multigrid solver via a domain decomposition method,* in Domain Decomposition Methods in Scientific and Engineering Computing: Proceedings of the Seventh International Conference on Domain Decomposition, KEYES,D.E. and XU,J. Eds. AMS, pages 439–444.

[1127] RUGE,J.W., STUBEN,K., (1985), *Efficient solution of finite difference and finite element equations by algebraic multigrid (AMG),* in Multigrid methods for integral and differential equations, Paddon,D.J. and Holstein,H. eds., Clarendon Press, pages 169–212.

[1128] RUHE,A., (1983), *Numerical aspects of Gram-Schmidt orthogonalization of vectors,* Linear Algebra and its Appl., vol 52, pages 591–601.

[1129] RUIZ,D., (1992), *Solution of large sparse unsymmetric linear systems with a block iterative method in a multiprocessor environment,* Ph. D. Thesis, CERFACS Report, Toulouse, France.

[1130] SAAD,Y., (1981), *Krylov subspace methods for solving large unsymmetric linear systems,* Math. Comp., vol 37, no 155, pages 105–126.

[1131] SAAD,Y., (1982), *The Lanczos biorthogonalization algorithm and other oblique projection methods for solving large unsymmetric systems,* SIAM J. Numer. Anal., vol 19, no 3, pages 485–506.

[1132] SAAD,Y., (1983), *Iterative solution of indefinite symmetric linear systems by meth-*

ods using orthogonal polynomials over two disjoint intervals, SIAM J. Numer. Anal., vol 20, no 4, pages 784–811.

[1133] SAAD,Y., (1985), *Practical use of polynomial preconditionings for the conjugate gradient method,* SIAM J. Sci. Comput., vol 6, no 4, pages 865–881.

[1134] SAAD,Y., (1986), *Communication complexity of the Gaussian elimination algorithm on multiprocessors,* Linear Algebra and its Appl., vol 77, pages 315–340.

[1135] SAAD,Y., (1987), *On the Lanczos method for solving symmetric linear systems with several right hand sides,* Math. Comp., vol 48, pages 651–662.

[1136] SAAD,Y., (1987), *Least squares polynomials in the complex plane with applications to solving sparse nonsymmetric matrix problems,* SIAM J. Numer. Anal., vol 24, pages 155-169.

[1137] SAAD,Y., (1992), *Numerical methods for large eigenvalue problems,* Halsted Press.

[1138] SAAD,Y., (1993), *A flexible inner-outer preconditioned GMRES algorithm,* SIAM J. Sci. Comput., vol 14, no 2, pages 461–469.

[1139] SAAD,Y., (1996), *Iterative methods for sparse linear systems,* PWS Publishing Company.

[1140] SAAD,Y., SCHULTZ,M.H., (1985), *Conjugate gradient like algorithms for solving nonsymmetric linear systems,* Math. Comp., vol 44, no 170, pages 417–424.

[1141] SAAD,Y., SCHULTZ,M.H., (1986), *GMRES: a generalized minimal residual algorithm for solving nonsymmetric linear systems,* SIAM J. Sci. Comput., vol 7, pages 856–869.

[1142] SAAD,Y., SCHULTZ,M.H., (1987), *Parallel direct methods for solving banded linear systems,* Linear Algebra and its Appl., vol 88, pages 623–650.

[1143] SAAD,Y., SCHULTZ,M.H., (1989), *Data communication in parallel architectures,* Parallel Comp., vol 11, no 2, pages 131–150.

[1144] SAAD,Y., WU,K., (1995), *DQGMRES: a direct quasi-minimal residual algorithm based on incomplete orthogonalization,* Report UMSI 95-131, Supercomputer Institute, University of Minnesota.

[1145] SADOK,H., (1993), *New algorithms for solving linear systems by direct methods,* Rapport ANO, Université de Lille.

[1146] SADOK,H., (1993), *CMRH: a new method for solving nonsymmetric linear systems based on the Hessenberg reduction algorithm,* Rapport ANO, Université de Lille.

[1147] SADOK,H., (1993), *A unified approach to conjugate gradient algorithms for solving nonsymmetric linear systems,* Rapport ANO, Université de Lille.

[1148] SADOK,H., (1997), *Analysis of the convergence of the minimal and the orthogonal residual methods,* Rapport Université de Lille, France.

[1149] SAMEH,A.H., BRENT,R.P., (1977), *Solving triangular systems on a parallel computer,* SIAM J. Numer. Anal., vol 14, no 6, pages 1101–1113.

[1150] SAMEH,A.H., KUCK,D.J., (1978), *On stable parallel linear system solvers,* J. ACM, vol 25, no 1, pages 81–91.

[1151] SARIDAKIS,Y.G., (1986), *Generalized consistent orderings and the accelerated overrelaxation method,* BIT, vol 26, pages 369–376.

[1152] SARIDAKIS,Y.G., (1986), *On the analysis of the unsymmetric successive overrelaxation method when applied to p–cyclic matrices,* Numer. Math., vol 49, pages 461–473.

[1153] SAUNDERS,M.A., SIMON,H.D., YIP,E.L., (1988), *Two conjugate gradient type methods for unsymmetric linear equations,* SIAM J. Numer. Anal., vol 25, no 4, pages 927–940.

[1154] SAYLOR,P.E., (1988), *Leapfrog variants of iterative methods for linear algebraic equations,* J. Comp. Appl. Math., vol 24, pages 169–193.

[1155] SAYLOR,P.E., (1990), *Use of the singular value decomposition with the Manteuffel algorithm for nonsymmetric linear systems,* SIAM J. Sci. Comput., vol 1, no 2, pages 210–222.

[1156] SAYLOR,P.E., SKEEL,R.D., (1990), *Linear iterative solvers for implicit ODE methods,* ICASE Report 90-51, NASA Langley Research Center.

[1157] SAYLOR,P.E., SMOLARSKI,D.C., (1990), *Implementation of an adaptive algorithm for Richardson's method,* TR–139, Dept. Informatik, ETH Zurich.

[1158] SCHONAUER,W., WEISS,R., (1995), *An engineering approach to generalized conjugate gradient methods and beyond,* Appl. Num. Meth., vol 19, no 3, pages 175–206.

[1159] SCHREIBER,R.S., (1982), *A new implementation of sparse Gaussian elimination,* ACM Trans. Math. Soft., vol 8, pages 256–276.

[1160] SCHREIBER,R.S., (1984), *Generalized iterative methods for semidefinite linear systems,* Linear Algebra and its Appl., vol 62, pages 219–230.

[1161] SCHUMANN,U., SWEET,R.A., (1976), *A direct method for the solution of Poisson's equation with Neumann boundary conditions on a staggered grid of arbitrary size,* J. Comp. Physics, vol 20, no 2, pages 171–182.

[1162] SCHUMANN,U., SWEET,R.A., (1988), *Fast Fourier transforms for direct solution of Poisson's equation with staggered boundary conditions,* J. Comp. Physics, vol 75, pages 123–137.

[1163] SCHWARZ,H.A., (1869), *Uber einige abbildungsaufgaben,* Ges. Math. Abh., vol

11, pages 65–83.

[1164] SCOTT,D.S., (1979), *How to make the Lanczos algorithm converge slowly,* Math. Comp., vol 33, pages 239–247.

[1165] SHAHID,J.N., TUMINARO,R.S., (1992), *Sparse iterative algorithm software for large scale MIMD machines: an initial discussion and implementation,* Concurrency, Pract. Exp., vol 4, pages 481–497.

[1166] SHAPIRA,Y., (1997), *A multilevel method for sparse linear systems,* Report Los Alamos National Laboratory.

[1167] SHAPIRA,Y., (1997), *Multigrid for locally refined meshes,* Report Los Alamos National Laboratory.

[1168] SHAPIRO A., (1985), *Optimal block diagonal l_2–scaling of matrices,* SIAM J. Numer. Anal., vol 22, no 1, pages 81–94.

[1169] SHERMAN,A.H., (1978), *Algorithms for sparse Gaussian elimination with partial pivoting,* ACM Trans. Math. Soft., vol 4, no 4, pages 330–338.

[1170] SHEWCHUK,J.R., (1994), *An introduction to the conjugate gradient method without the agonizing pain,* Report CMU–CS–94–125, Carnegie Mellon University.

[1171] SHIEH,A.S.L., (1978), *On the convergence of the conjugate gradient method for singular capacitance matrix equations from the Neumann problem of the Poisson equation,* Numer. Math., vol 29, pages 307–327.

[1172] SIMON,H.D., (1982), *The Lanczos algorithm for solving symmetric linear systems,* Ph.D. Thesis, University of California, Berkeley.

[1173] SIMON,H.D., (1984), *The Lanczos algorithm with partial reorthogonalization,* Math. Comp., vol 42, no 165, pages 115–142.

[1174] SIMON,H.D., (1984), *Analysis of the symmetric Lanczos algorithm with reorthogonalization methods,* Linear Algebra and its Appl., vol 61, pages 101–131.

[1175] SIMON,H.D., (1985), *Incomplete LU preconditioners for conjugate gradient type iterative methods,* Paper SPE 13533, SPE 1985 Reservoir Simulation Symposium .

[1176] SIMON,H.D., (1991), *Partitioning of unstructured problems for parallel processing,* Comput. Syst. Eng., vol 2, pages 135–148.

[1177] SKEEL,R.D., (1979), *Scaling for numerical stability in Gaussian elimination,* J. ACM, vol 26, no 3, pages 494–526.

[1178] SKEEL,R.D., (1980), *Iterative refinement implies numerical stability for Gaussian elimination,* Math. Comp., vol 35, no 151, pages 817–832.

[1179] SKEEL,R.D., (1981), *Effect of equilibration on residual size for partial pivoting,* SIAM J. Numer. Anal., vol 18, pages 449–455.

[1180] SLEIJPEN,G.L., FOKKEMA,D.R., (1993), *BiCGSTAB(l) for linear equations involving unsymmetric matrices with complex spectrum,* ETNA, vol 1, pages 11–32.

[1181] SLEIJPEN,G.L., VAN DER VORST,H.A., (1993), *Krylov subspace methods for large linear systems of equations,* Report 803, Dept. of Mathematics, University of Utrecht.

[1182] SLEIJPEN,G.L., VAN DER VORST,H.A., (1995), *Maintaining convergence properties of BiCGSTAB methods in finite precision arithmetic,* Numer. Alg., vol 10, pages 203–223.

[1183] SLEIJPEN,G.L., VAN DER VORST,H.A., (1995), *An overview of approaches for the stable computation of hybrid BiCG methods,* Appl. Numer. Math., vol 19, pages 235–254.

[1184] SLEIJPEN,G.L., VAN DER VORST,H.A., (1996), *Reliable updated residuals in hybrid BiCG methods,* Computing, vol 56, pages 141–163.

[1185] SLEIJPEN,G., VAN DER VORST,H., FOKKEMA,D., (1994), *biCGSTAB(l) and other hybrid BiCG methods,* Numer. Alg., vol 7, pages 75–108.

[1186] SLEIJPEN,G., VAN DER VORST,H., MODERSITSKI,J., (1997), *The main effects of rounding errors in Krylov solvers for symmetric linear systems,* Report 1006, University of Utrecht.

[1187] SLOAN,S.W., (1986), *An algorithm for profile and wavefront reduction of sparse matrices,* Int. J. Num. Meth. Eng., vol 23, pages 239–251.

[1188] SMITH,B.F., (1991), *A domain decomposition algorithm for elliptic problems in three dimensions,* Numer. Math., vol 60, pages 219–234.

[1189] SMITH,B.F., (1992), *An iterative substructuring algorithm for problems in three dimensions,* in Fifth International Symposium on Domain Decomposition Methods for Partial Differential Equations, KEYES,D.E., CHAN,T.F., MEURANT,G., SCROGGS,J.S. and VOIGT,R.G. Eds. SIAM, pages 91–98.

[1190] SMITH,B.F., (1992), *An optimal domain decomposition preconditioner for the finite element solution of linear elasticity problems,* SIAM J. Sci. Comput., vol 13, pages 364–378.

[1191] SMITH,B.F., (1993), *A parallel implementation of an iterative substructuring algorithm for problems in three dimensions,* SIAM J. Sci. Comput., vol 14, pages 406–423.

[1192] SMITH,B.F., BJORSTAD,P.E., GROPP,W.D., (1996), *Domain decomposition: parallel multilevel methods for elliptic partial differential equations,* Cambridge University Press.

[1193] SMITH,B.F., WIDLUND,O.B., (1990), *A domain decomposition algorithm using a hierarchical basis,* SIAM J. Sci. Comput., vol 11, pages 1212–1226.

[1194] SMOLARSKI,D.C., SAYLOR,P.E., (1988), *An optimum iterative method for solving any linear system with a square matrix,* BIT, vol 28, pages 163–178.

[1195] SONNEVELD,P., (1989), *CGS: a fast Lanczos–type solver for nonsymmetric linear systems,* SIAM J. Sci. Comput., vol 10, pages 36–52.

[1196] SORENSEN,D.C., (1992), *Implicit application of polynomial filters in a k-step Arnoldi method,* SIAM J. Matrix Anal. Appl., vol 13, no 1, pages 357–385.

[1197] SOUTHWELL,R.V., (1940), *Relaxation Methods in Engineering Science,* Oxford University Press.

[1198] STARK,S., BERIS,A.N., (1992), *LU decomposition optimized for a parallel computer with a hierarchical memory,* Parallel Comp., vol 18, no 9, pages 959–972.

[1199] STEINKE,R.G., (1979), *An approximate inversion method for five point difference matrix equations,* Report LA-UR-79-162, Los Alamos Scientific Laboratory.

[1200] STEWART,G.W., (1973), *Introduction to matrix computations,* Academic Press.

[1201] STEWART,G.W., (1973), *Conjugate direction methods for solving systems of linear equations,* Numer. Math., vol 21, pages 285–297.

[1202] STEWART,G.W., (1975), *The convergence of the method of conjugate gradients at isolated extreme points in the spectrum,* Numer. Math., vol 24, pages 85–93.

[1203] STEWART,G.W., (1990), *Communication and matrix computations on large message passing systems,* Parallel Comp., vol 16, no 1, pages 27–40.

[1204] STEWART,G.W., (1991), *Maybe we should call it "Lagrangian elimination",* Nanet message of Friday 21 June 91.

[1205] STEWART,G.W., (1991), *Lanczos and linear systems,* Report CS–TR 2641, Dept. of Computer Science, University of Maryland.

[1206] STEWART,G.W., (1993), *On the perturbation of LU, Cholesky and QR factorizations,* SIAM J. Matrix Anal. Appl., vol 14, no 4, pages 1141–1145.

[1207] STEWART,G.W., (1994), *Gauss, statistics and Gaussian elimination,* Report TR-94-78, Dept. of Computer Science, University of Maryland.

[1208] STEWART,G.W., (1995), *On the perturbation of LU and Cholesky factors,* Report TR–95–93, University of Maryland.

[1209] STEWART,G.W., (1995), *The triangular matrices of Gaussian elimination and related decompositions,* Report TR-3533, Dept. of Computer Science, University of Maryland.

[1210] STEWART,G.W., SUN,J–G., (1990), *Matrix perturbation theory,* Academic Press.

[1211] STOER,J., BULIRSCH,R., (1980), *Introduction to numerical analysis,* Springer.

[1212] STONE,H.S., (1968), *Iterative solution of implicit approximations of multidimensional partial differential equations,* SIAM J. Numer. Anal., vol 5, no 3, pages 530–573.

[1213] STONE,H.S., (1975), *Parallel tridiagonal equation solvers,* ACM Trans. Math. Soft., vol 1, no 4, pages 289–307.

[1214] STRAKOS,Z, (1991), *On the real convergence rate of the conjugate gradient method,* Linear Algebra and its Appl., vol 154, pages 535–549.

[1215] STRAKOS,Z., ROZLOZNIK,M., DRKOSOVA,J., (1994), *Numerical stability of GMRES,* Technical report 579, Institute of Computer Science, Academy of Sciences of the Czech Republic.

[1216] STRANG,G., (1976), *Linear algebra and its applications,* Academic Press.

[1217] STRASSEN,V., (1969), *Gaussian elimination is not optimal,* Numer. Math., vol 13, pages 354–356.

[1218] STRIKWERDA,J., (1980), *Iterative methods for the numerical solution of second order elliptic equations with large first order terms,* SIAM J. Sci. Comput., vol 1, no 1, pages 119–130.

[1219] STUBEN,K., (1983), *Algebraic multigrid (AMG): experiences and comparisons,* Appl. Math. Comput., vol 13, pages 419–452.

[1220] STUBEN,K., TROTTENBERG,U., (1982), *Multigrid methods: Fundamental algorithms, model problem analysis and applications,* in Multigrid methods, Proceedings of the Koln-Porz Conference, HACKBUSCH,W. and TROTTENBERG,U. Eds, Lecture Notes in Mathematics 960. Springer, pages 1–176.

[1221] SUNDHOLM,D., (1988), *A block preconditioned conjugate gradient method for solving high order finite element matrix equations,* Comp. Phys. Comm., vol 49, pages 409–415.

[1222] SWARZTRAUBER,P.N., (1972), *On the numerical solution of the Dirichlet problem for a region of general shape,* SIAM J. Numer. Anal., vol 9, no 2, pages 300–306.

[1223] SWARZTRAUBER,P.N., (1973), *The direct solution of the discrete Poisson equation on the surface of a sphere,* J. Comp. Physics, vol 15, no 1, pages 46–54.

[1224] SWARZTRAUBER,P.N., (1974), *A direct method for the discrete solution of separable elliptic equations,* SIAM J. Numer. Anal., vol 11, no 6, pages 1136–1150.

[1225] SWARZTRAUBER,P.N., (1977), *The methods of cyclic reduction, Fourier analysis and the FACR algorithm for the discrete solution of Poisson's equation on a rectangle,* SIAM Review, vol 19, no 3, pages 490–501.

[1226] SWARZTRAUBER,P.N., (1982), *Vectorizing the FFTs,* in Parallel Computations, RODRIGUE,G ed, Academic Press, pages 51–84.

[1227] SWARZTRAUBER,P.N., (1984), *FFT algorithms for vector computers,* Par. Comp., vol 1, pages 45–63.

[1228] SWARZTRAUBER,P.N., SWEET,R.A., (1973), *The direct solution of the discrete Poisson equation on a disk,* SIAM J. Numer. Anal., vol 10, no 5, pages 900–907

[1229] SWARZTRAUBER,P.N., SWEET,R.A., (1979), *Algorithm 541, efficient Fortran subprograms for the solution of separable elliptic partial differential equations,* ACM Trans. Math. Soft., vol 5, no 3, pages 352–364.

[1230] SWARZTRAUBER,P.N., SWEET,R.A., (1989), *Vector and parallel methods for the direct solution of Poisson's equation,* J. Comp. Appl. Math., vol 27, pages 241–263.

[1231] SWEET,R.A., (1973), *Direct methods for the solution of Poisson's equation on a staggered grid,* J. Comp. Physics, vol 12, no 3, pages 422–428.

[1232] SWEET,R.A., (1974), *A generalized cyclic reduction algorithm,* SIAM J. Numer. Anal., vol 11, no 3, pages 506–520.

[1233] SWEET,R.A., (1977), *A cyclic reduction algorithm for solving block tridiagonal systems of arbitrary dimension,* SIAM J. Numer. Anal., vol 14, no 4, pages 706–720.

[1234] TANG,W.P., (1987), *Schwarz splitting and template operators,* Ph.D. Thesis, Computer Science Dept., Stanford University.

[1235] TANG,W.P., (1992), *Generalized Schwarz splittings,* SIAM J. Sci. Comput., vol 13, pages 573–595.

[1236] TARTAR,L., (1971), *Une nouvelle caractérisation des M-matrices,* RIRO, vol 3, pages 127–128.

[1237] TEMPERTON,C., (1979), *Direct methods for the solution of the discrete Poisson equation: some comparisons,* J. Comp. Physics, vol 31, pages 1–20.

[1238] TEMPERTON,C., (1980), *On the FACR(l) algorithm for the discrete Poisson equation,* J. Comp. Physics, vol 34, pages 314–329.

[1239] TEMPERTON,C., (1983), *Self-sorting mixed-radix Fast Fourier Transforms,* J. Comp. Physics, vol 52, pages 1–23.

[1240] TEMPERTON,C., (1983), *A note on prime factor FFT algorithms,* J. Comp. Physics, vol 52, pages 198–204.

[1241] TEMPERTON,C., (1985), *Implementation of a self sorting in place prime factor FFT algorithm,* J. Comp. Physics, vol 58, pages 283–299.

[1242] TEMPERTON,C., (1988), *A self-sorting in-place prime factor real/half-complex FFT algorithm,* J. Comp. Physics, vol 75, pages 199–216.

[1243] TEMPERTON,C., (1991), *Self sorting in place Fast Fourier Transforms,* SIAM J. Sci. Comput., vol 12, pages 808–823.

[1244] TEMPERTON,C., (1992), *A generalized prime factor FFT algorithm for any* $n = 2^p 3^q 5^r$, SIAM J. Sci. Comput., vol 13, no 3, pages 676–686.

[1245] TEWARSON,R.P., (1967), *The product form of the inverse of sparse matrices and graph theory,* SIAM Review, vol 9, no 1, pages 91–99.

[1246] TEWARSON,R.P., (1970), *Computations with sparse matrices,* SIAM Review, vol 12, no 4, pages 527–543.

[1247] TEWARSON,R.P., (1973), *Sparse matrices,* Academic Press.

[1248] TEZDUYAR,T.E., LIOU,J., (1989), *Grouped element by element iteration schemes for incompressible flow computations,* Comp. Phys. Comm., vol 53, pages 441–453.

[1249] TINNEY,W.F., MEYER,W.S., (1973), *Solution of large sparse systems by ordered triangular factorization,* IEEE Trans. Aut. Control, vol 18, no 4, pages 333–346.

[1250] TINNEY,W.F., WALKER,J.W., (1967), *Direct solutions of sparse network equations by optimally ordered triangular factorization,* Proc. of IEEE, vol 55, pages 1801–1809.

[1251] TISMENETSKY,M., (1986), *A direct method for solving linear systems,* Technical report 88.179, IBM Israel.

[1252] TISMENETSKY,M., (1989), *A nested block preconditioning linear solver,* Report 88.262, IBM Israel.

[1253] TISMENETSKY,M., (1989), *A robust preconditioning technique for solving large sparse linear systems,* Report 88.264, IBM Israel.

[1254] TISMENETSKY,M., (1990), *A note on a block preconditioner,* Report 90.007, IBM Germany, Heidelberg Scientific Center.

[1255] TISMENETSKY,M., EFRAT,I., (1985), *An efficient preconditioning algorithm and its analysis,* Report 88.172, IBM Israel.

[1256] TISMENETSKY,M., EFRAT,I., (1987), *A modified generalized minimal residual algorithm for nonsymmetric linear systems,* Technical Report 88.219, IBM Israel.

[1257] TONG,C.H., (1990), *Parallel preconditioned conjugate gradient methods for elliptic partial differential equations,* Ph.D. Thesis.

[1258] TONG,C.H., (1992), *A comparative study of preconditioned Lanczos methods for non symmetric linear systems,* Report SAND91-8240, Sandia National Laboratory.

[1259] TONG,C.H., CHAN,T.F., JAY KUO,C.C., (1991), *A domain decomposition preconditioner based on a change to a multilevel nodal basis,* SIAM J. Sci. Comput., vol 12, no 6,

[1260] TONG,C.H., YE,Q., (1995), *Analysis of the finite precision biconjugate gradient algorithm for nonsymmetric linear systems,* Report SCCM 95-11, Computer Science Dept., Stanford University.

[1261] TONG,C.H., YE,Q., (1996), *A linear system solver based on a modified Krylov subspace method for breakdown recovery,* Numer. Alg., vol 12, pages 233–252.

[1262] TRAUB,J.F., (1973), *Complexity of sequential and parallel numerical algorithms,* Academic Press,

[1263] TREFETHEN,L.N., (1985), *Three mysteries of Gaussian elimination,* SIGNUM Newsletter, vol 20, no 4, pages 2–5.

[1264] TREFETHEN,L.N., (1997), *Pseudospectra of linear operators,* SIAM Rev., vol 39,

[1265] TREFETHEN,L.N., BAU,III,D., (1997), *Numerical linear algebra,* SIAM.

[1266] TREFETHEN,L.N., SCHREIBER,R.S., (1990), *Average–case stability of Gaussian elimination,* SIAM J. Matrix Anal. Appl., vol 11, no 3, pages 335–360.

[1267] TUFF,A.D., JENNINGS,A., (1973), *An iterative method for large systems of linear structural equations,* Int. J. Num. Meth. Eng., vol 7, pages 175–183.

[1268] TUMINARO,R.S., (1989), *Multigrid algorithms on parallel processing systems,* Ph.D. Thesis, STAN-CS-90-1299, Dept. of Computer Science, Stanford University.

[1269] TUMINARO,R.S., (1992), *A highly parallel multigrid like method for the solution of the Euler equations,* SIAM J. Sci. Comput., vol 13, pages 88–100.

[1270] TUMINARO,R.S., WOMBLE,D.E., (1993), *Analysis of the multigrid FMV cycle on large scale parallel machines,* SIAM J. Sci. Comput., vol 14, pages 1159–1173.

[1271] TURING,A.M., (1948), *Rounding–off errors in matrix processes,* Quart. J. Mech. Appl. Math., vol 1, pages 287–308.

[1272] UNDERWOOD,R., (1976), *An approximate factorization procedure based on the block Cholesky decomposition and its use with the conjugate gradient method,* Report NEDE–11386, General Electric.

[1273] VAN DER PLOEG, A., BOTTA,E.F.F., WUBS,F.W., (1996), *Nested grids ILU decomposition (NGILU),* J. Comp. Appl. Math., vol 66, pages 515–526.

[1274] VAN DER SLUIS,A., (1969), *Condition numbers and equilibration of matrices,* Numer. Math., vol 14, pages 14–23.

[1275] VAN DER SLUIS,A., VAN DER VORST,H.A., (1986), *The rate of convergence of conjugate gradients,* Numer. Math., vol 48, pages 543–560.

[1276] VAN DER SLUIS,A., VAN DER VORST,H.A., (1986), *The convergence behaviour of Ritz values in the presence of close eigenvalues,* Report 86–08, Dept of Maths.,

Delft University.

[1277] VAN DER VORST,H.A., (1982), *A vectorizable variant of some ICCG methods,* SIAM J. Sci. Comput., vol 3, no 3, pages 350–356.

[1278] VAN DER VORST,H.A., (1986), *Analysis of a parallel solution method for tridiagonal systems,* Parallel Comp., vol 5, pages 303–311.

[1279] VAN DER VORST,H.A., (1986), *The performance of Fortran implementations for preconditioned conjugate gradients on vector computers,* Parallel Comp., vol 3, pages 49–58.

[1280] VAN DER VORST,H.A., (1986), *(M)ICCG for 2D problems on vector computers,* Report A-17, Data Processing Center, Kyoto University.

[1281] VAN DER VORST,H.A., (1987), *ICCG and related methods for 3D problems on vector supercomputers,* Report A-18, Data Processing Center, Kyoto University.

[1282] VAN DER VORST,H.A., (1987), *Large tridiagonal and block tridiagonal linear systems on vector and parallel computers,* Parallel Comp., vol 5, pages 45–54.

[1283] VAN DER VORST,H.A., (1988), *Practical aspects of parallel scientific computing,* Fut. Gen. Comp. Sys., vol 4, pages 285–291.

[1284] VAN DER VORST,H., (1992), *BiCGSTAB: a fast and smoothly converging variant of BiCG for the solution of nonsymmetric linear systems,* SIAM J. Sci. Comput., vol 13, pages 631–644.

[1285] VAN DER VORST,H.A., (1993), *Lecture notes on iterative methods,* Report Mathematical Institute, University of Utrecht.

[1286] VAN DER VORST,H.A., (1995), *Parallel iterative solution methods for linear systems arising from discretized PDE's,* AGARD-FDP-VKI Workshop Lecture Notes.

[1287] VAN DER VORST,H.A., CHAN,T.F., (1994), *Linear system solvers: sparse iterative methods,* Report 869, Mathematical Institute, University of Utrecht.

[1288] VAN DER VORST,H.A., VUIK,C., (1993), *The superlinear convergence behaviour of GMRES,* J. Comp. Appl. Math., vol 48, pages 327–341.

[1289] VAN DER VORST,H.A., VUIK,C., (1994), *GMRESR: a family of nested GMRES methods,* Numer. Lin. Alg. with Appl., vol 1, no 4, pages 369–386.

[1290] VANEK,P., KRIZKOVA,J., (1995), *Two level method on unstructured meshes with convergence rate independent of the coarse space size,* Report University of colorado, Denver.

[1291] VANEK,P., MANDEL,J., BREZINA,M., (1994), *Algebraic multigrid on unstructured meshes,* Report University of colorado, Denver.

[1292] VANEK,P., MANDEL,J., BREZINA,M., (1994), *Algebraic multigrid by smoothed aggregation for second and fourth order elliptic problems,* Report University of colorado, Denver.

[1293] VAN LOAN,C.F., (1992), *Computational frameworks for the Fast Fourier Transform,* SIAM.

[1294] VAN NOOYEN,R.R.P., VUIK,C., WESSELING,P., (1996), *Some parallelizable preconditioners for GMRES,* Report 96-50, Delft University of Technology.

[1295] VAN ROSENDALE,J., (1983), *Minimizing inner product data dependencies in conjugate gradient iteration,* Report 172178, ICASE, NASA Langley.

[1296] VANDEWALLE,S., HORTON,G., (1995), *Fourier mode analysis of the multigrid waveform relaxation and time parallel multigrid methods,* Computing, vol 54, pages 317–330.

[1297] VARGA,R.S., (1957), *A comparison of the successive overrelaxation method and semi–iterative methods using Chebyshev polynomials,* J. Soc. Indust. Appl. Math., vol 5, no 2, pages 39–46.

[1298] VARGA,R.S., (1959), *Orderings of the successive overrelaxation scheme,* Pacific J. Math., vol 9, no 3, pages 925–939.

[1299] VARGA,R.S., (1960), *Factorization and normalized iterative methods,* in Boundary problems in differential equations, Langer,R. Ed.. The University of Wisconsin Press.

[1300] VARGA,R.S., (1962), *Matrix iterative analysis,* Prentice Hall.

[1301] VARGA,R.S., CAI,D–Y., (1981), *On the LU factorization of M–matrices,* Numer. Math., vol 38, pages 179–192.

[1302] VARGA,R.S., NIETHAMMER,W., (1984), *P-cyclic matrices and the symmetric successive overrelaxation method,* Linear Algebra and its Appl., vol 58, pages 425–439.

[1303] VARGA,R.S., NIETHAMMER,W., CAI,D.–Y., (1984), *P–cyclic matrices and the symmetric successive overrelaxation method,* Linear Algebra and its Appl., vol 58, pages 425–439.

[1304] VARGA,R.S., SAFF,E.B., MEHRMANN,V., (1980), *Incomplete factorizations of matrices and connections with H-matrices,* SIAM J. Numer. Anal., vol 17, no 6, pages 787–793.

[1305] VASSILEVSKI,P.S., (1992), *Hybrid V-cycle algebraic multilevel preconditioners,* Math. Comp., vol 58, pages 489–512.

[1306] VASSILEVSKI,P.S., LAZAROV,R., (1992), *Preconditioning saddle point problems arising from mixed finite element discretization of elliptic equations,* Report CAM 92-46, Dept. of Mathematics, UCLA.

[1307] VATSYA,S.R., (1988), *Convergence of conjugate residual-like methods to solve linear equations,* SIAM J. Numer. Anal., vol 25, no 4, pages 957–964.

[1308] VINSOME,P.K.W., (1976), *Orthomin, an iterative method for solving sparse sets of simultaneous linear equations,* in Proceedings of the fourth symposium on reservoir simulation, SPE, pages 149–159.

[1309] VOEVODIN,V.V., (1980), *On methods of conjugate direction,* USSR Comput. Math. Phys., vol 19, pages 228–233.

[1310] VON NEUMANN,J., GOLDSTINE,H.H., (1947), *Numerical inverting of matrices of high order,* Bull. AMS, vol 53, no 11, pages 1021–1099.

[1311] VUIK,C., (1992), *Further experiences with GMRESR,* Report 92-12, Faculty of Mathematics, Delft University of Technology.

[1312] VUIK,C., (1993), *New insights in GMRES-like methods with variable preconditioners,* Report 93-10, Faculty of Mathematics, Delft University of Technology.

[1313] WACHSPRESS,E.L., (1963), *Extended application of alternating direction implicit iteration model problem theory,* J. Soc. Indust. Appl. Math., vol 11, no 4,

[1314] WALKER,H.F., (1988), *Implementation of the GMRES method using Householder transformations,* SIAM J. Sci. Comput., vol 9, pages 152–163.

[1315] WALLIS,J.R., KENDALL,R.P., LITTLE,T.E., (1985), *Constrained residual acceleration of conjugate residual methods,* Paper SPE 1536 in Proceedings of the SPE 1985 Reservoir Simulation Symposium. pages 415–428.

[1316] WAN,W.L., CHAN,T.F., SMITH,B., (1997), *An energy minimizing interpolation for robust multigrid,* Report CAM, University of California, Los Angeles.

[1317] WANG,H.H., (1981), *A parallel method for tridiagonal equations,* ACM Trans. Math. Soft., vol 7, no 2, pages 170–183.

[1318] WATANABE,H., DOI,S., (1991), *A comparison of BiCGSTAB and CGS methods for convection-diffusion equations,* in Advances in Numerical Methods for Large Sparse Sets of Linear Systems, NODERA,T. Ed no 7. Keio University.

[1319] WATHEN,A., SILVESTER,D., (1993), *Fast iterative solution of stabilised Stokes systems, part I: using simple diagonal preconditioners,* SIAM J. Numer. Anal., vol 30, no 3, pages 630–649.

[1320] WEINBERGER,H.F., (1966), *A posteriori error bounds in iterative matrix inversion,* in Numerical solution of partial differential equations, J.H. Bramble Ed. Academic Press, pages 153–163.

[1321] WEISS,R., (1995), *A theoretical overview of Krylov subspace methods,* Appl. Num. Meth., vol 19, no 3, pages 33–56.

[1322] WEISS,R., (1996), *Parameter-free iterative linear solvers,* Akademie Verlag.

[1323] WESSELING,P., (1982), *Theoretical and practical aspects of a multigrid method,* SIAM J. Sci. Comput., vol 3, pages 387–407.

[1324] WESSELING,P., (1982), *A robust and efficient multigrid method,* in Multigrid methods, Proceedings of the Koln-Porz Conference, HACKBUSCH,W. and TROTTENBERG,U. Eds, Lecture Notes in Mathematics 960. Springer, pages 614–630.

[1325] WESSELING,P., (1991), *A survey of Fourier smoothing analysis results,* in Multigrid methods III, HACKBUSCH,W. and TROTTENBERG,U. Eds, International Series of Numerical Mathematics 98. Birkhauser, pages 105–127.

[1326] WESSELING,P., (1992), *An Introduction to Multigrid Methods,* John Wiley.

[1327] WIDLUND,O.B., (1965), *On the rate of convergence of an alternating direction implicit method in a non commutative case,* Math. Comp., vol 20, pages 500–515.

[1328] WIDLUND,O.B., (1971), *On the effects of scaling of the Peaceman–Rachford method,* Math. Comp., vol 25, no 113, pages 33–41.

[1329] WIDLUND,O.B., (1978), *A Lanczos method for a class of nonsymmetric systems of linear equations,* SIAM J. Numer. Anal., vol 15, no 4, pages 801–812.

[1330] WILHELMSON,R.B., ERICKSEN,J.H., (1977), *Direct solutions for Poisson's equation in three dimensions,* J. Comp. Physics, vol 25, pages 319–331.

[1331] WILF,H.S., (1962), *Mathematics for the physical sciences,* Wiley.

[1332] WILKINSON,J.H., (1961), *Error analysis of direct methods of matrix inversion,* J. ACM, vol 10, pages 281–330.

[1333] WILKINSON,J.H., (1965), *The algebraic eigenvalue problem,* Oxford University Press.

[1334] WILKINSON,J.H., (1971), *Modern error analysis,* SIAM Review, vol 13, pages 548–568.

[1335] WINGET,J.M., HUGHES,T.J., (1985), *Solution algorithms for nonlinear transient heat conduction analysis employing element by element iterative strategies,* Com. Meth. Appl. Mech. Eng., vol 52, pages 711–815.

[1336] WONG,Y.S., (1980), *Preconditioned conjugate gradient methods for large sparse matrix problems,* in Numerical methods in thermal problems, Lewis,R.W. and Morgan,K. Eds.. Pineridge Press, pages 967–979.

[1337] WOZNIAKOWSKI,H., (1980), *Roundoff error analysis of a new class of conjugate gradient algorithms,* Linear Algebra and its Appl., vol 29, pages 507–529.

[1338] WRIGHT,S.J., (1991), *Parallel algorithms for banded linear systems,* SIAM J. Sci. Comput., vol 12, no 4, pages 824–842.

[1339] WRIGHT,S.J., (1993), *A collection of problems for which Gaussian elimination with partial pivoting is unstable,* SIAM J. Sci. Comput., vol 14, no 4, pages 231–238.

[1340] XU,J., (1989), *Theory of multilevel methods,* Ph.D. Thesis, report AM 48, Dept. of Mathematics, Penn State University.

[1341] XU,J., (1992), *A new class of iterative methods for nonselfadjoint or indefinite problems,* SIAM J. Numer. Anal., vol 29, pages 303–319.

[1342] XU,J., (1992), *Iterative methods by SPD and small subspace solvers for nonsymmetric or indefinite problems,* in Fifth International Symposium on Domain Decomposition Methods for Partial Differential Equations, KEYES,D.E., CHAN,T.F., MEURANT,G., SCROGGS,J.S. and VOIGT,R.G. Eds. SIAM, pages 106–118.

[1343] XU,J., (1992), *Iterative methods by space decomposition and subspace correction: a unifying approach,* SIAM Review, vol 34, pages 581–613.

[1344] XU,J., (1996), *The auxiliary space method and optimal multigrid preconditioning techniques for unstructured meshes,* Computing, vol 56, pages 215–235.

[1345] XU,J., CAI,X-C., (1990), *A preconditioned GMRES method for nonsymmetric or indefinite problems,* Report AM 79, Dept. of Mathematics, Penn State University.

[1346] XU,J., QIN,J., (1994), *Some remarks on a multigrid preconditioner,* SIAM J. Sci. Comput., vol 15, pages 172–184.

[1347] YAMAMOTO,T., IKEBE,Y., (1979), *Inversion of band matrices,* Linear Algebra and its Appl., vol 24, pages 105–111.

[1348] YANG,W.H., (1977), *A method for updating Cholesky factorization of a band matrix,* Com. Meth. Appl. Mech. Eng., vol 12, pages 281–288.

[1349] YANNAKAKIS,M., (1981), *Computing the minimum fill–in is NP–complete,* SIAM J. Alg. Disc. Meth., vol 2, no 1, pages 77–79.

[1350] YE,Q., (1994), *A breakdown-free variation of the nonsymmetric Lanczos algorithms,* Math. Comp., vol 62, no 205, pages 179–207.

[1351] YOUNG,D.M., (1971), *Iterative solution of large linear systems,* Academic Press.

[1352] YOUNG,D.M., (1977), *On the accelerated SSOR method for solving large linear systems,* Advances in Maths., vol 23, pages 215–271.

[1353] YOUNG,D.M., FRANK,T.G., (1963), *A survey of computer methods for solving elliptic and parabolic partial differential equations,* in ICC Bulletin, vol 2 no 1, Rome. pages 3–61.

[1354] YOUNG,D.M., JEA,K.C., (1980), *Generalized conjugate gradient acceleration of nonsymmetrizable iterative methods,* Linear Algebra and its Appl., vol 34, pages 159–194.

[1355] YSERENTANT,H., (1986), *On the multilevel splitting of finite element spaces,* Numer. Math., vol 49, pages 379–412.

[1356] YSERENTANT,H., (1986), *On the multi–level splitting of finite element spaces,* Numer. Math., vol 49, pages 379–412.

[1357] YSERENTANT,H., (1986), *On the multi–level splitting of finite element spaces for indefinite elliptic boundary value problems,* SIAM J. Numer. Anal., vol 23, pages 581–595.

[1358] YSERENTANT,H., (1986), *The convergence of multi–level methods for solving finite–element equations in the presence of singularities,* Math. Comp., vol 47, pages 399–409.

[1359] YSERENTANT,H., (1986), *Hierarchical bases give conjugate gradient type methods a multigrid speed of convergence,* Appl. Math. Comp., vol 19, pages 347–358.

[1360] YSERENTANT,H., (1988), *Preconditioning indefinite discretization matrices,* Report SC 87-6, Konrad Zuse Zentrum fur Informationstechnik.

[1361] YSERENTANT,H., (1990), *Two preconditioners based on the multilevel splitting of finite element spaces,* Numer. Math., vol 58, pages 163–184.

[1362] ZHANG,G., ELMAN,H.C., (1992), *Parallel sparse Cholesky factorization on a shared memory multiprocessor,* Parallel Comp., vol 18, no 9, pages 1009–1022.

[1363] ZHANG,X., (1992), *Multilevel Schwarz methods,* Numer. Math., vol 63, pages 521–539.

[1364] ZHOU,L., WALKER,H., (1994), *Residual smoothing techniques for iterative methods,* SIAM J. Sci. Comput., vol 15, pages 297–312.

[1365] ZIENKIEWICZ,O.C., LOHNER,R., (1985), *Accelerated relaxation or direct solution? Future prospects for FEM,* Int. J. Num. Meth. Eng., vol 21, pages 1–11.

[1366] ZLATEV,Z., (1980), *On some pivotal strategies in Gaussian elimination by sparse technique,* SIAM J. Numer. Anal., vol 17, no 1, pages 18–30.

[1367] ZMIJEWSKI,E., GILBERT,J.R., (1988), *A parallel algorithm for sparse symbolic Cholesky factorization on a multiprocessor,* Parallel Comp., vol 7, pages 199–210.

[1368] ZYVOLOSKI,G., (1986), *Incomplete factorization for finite element methods,* Int. J. Num. Meth. Eng., vol 23, pages 1101–1109.

Index